# Water in Crisis

# Water in Crisis

## A Guide to the World's Fresh Water Resources

*Edited by*

Peter H. Gleick

*Foreword by*

*Gilbert F. White*

Pacific Institute for Studies in Development, Environment, and Security
Stockholm Environment Institute

*New York*        *Oxford*
OXFORD UNIVERSITY PRESS
1993

*Oxford University Press*

*Oxford   New York   Toronto*
*Delhi   Bombay   Calcutta   Madras   Karachi*
*Kuala Lumpur   Singapore   Hong Kong   Tokyo*
*Nairobi   Dar es Salaam   Cape Town*
*Melbourne   Auckland   Madrid*

*and associated companies in*
*Berlin   Ibadan*

*Published by Oxford University Press, Inc.,*
*200 Madison Avenue, New York, New York 10016*

*Oxford is a registered trademark of Oxford University Press*

*Library of Congress Cataloging-in-Publication Data*

*Water in crisis : a guide to the world's fresh water resources / edited by*
*Peter H. Gleick, Pacific Institute for Studies in Development, Environment,*
*and Security, Stockholm Environment Institute.*
*p.   cm.      Includes bibliographical references.*
*ISBN 0-19-507627-3. — ISBN 0-19-507628-1 (pbk.)*
*1. Fresh water.   2. Water quality.   3. Water-supply.*
*I. Gleick, Peter H.    II. Pacific Institute for Studies in Development,*
*Environment, and Security.   III. Stockholm Environment Institute.*
*TD345.W264 1993*
*333.91—dc20          92-30061*

*9 8 7 6 5 4 3 2 1*

*Printed in the United States of America*
*on acid-free paper*

For Nicki Norman and our boys,

Daniel Donen and Jeremy Robert.

*May your water always be pure and plentiful.*

# Foreword

This volume is a guide to the world's fresh water resources in the best sense of the term. It provides any serious explorer of fresh water problems on either a global or a regional scale with two sets of information basic to that quest – background information on critical water issues and extensive, carefully chosen water data.

In a world that is increasingly aware of the limitations and fragility of the earth's resources, water is distinguished from resources of soil and vegetation in at least one significant respect. The total quantity of water molecules cannot be destroyed. Water can be transformed from atmospheric vapor to flowing water to the frozen state, it can be transported over great distances, aquifers can be exhausted, and it can be changed in quality, but the basic resource remains undiminished in the biosphere.

At the same time, opportunities to alter the distribution and quality of water are immense. These continue to be the product of human decisions – individual and public – that shape water use and manipulation for good or ill.

This guide, carefully compiled and edited by Peter Gleick, provides a solid, discriminating array of data on the spatial and temporal distribution of the world's fresh water resources. The tables selected for this collection encompass more completely than any other effort to date the basic statistical information available. More important, each table is accompanied by a thoughtful statement of the limitations of the data in its validity, accuracy, and mode or presentation. These are welcome cautions that should help readers avoid misunderstandings, misinterpretations, and misstatements. No other water book offers this valuable background.

In reviewing past and emerging water policy issues, this guide helps the reader understand the range and complexities of the options that lie ahead. Again, the limitations of such analysis are recognized and discussed. Social values, political organizations, and economic factors all play a role in determining the future of our use of fresh water.

Only with sane realistic recognition of the physical and social characteristics of water can sound water policy be expected to emerge on a global scale. Gleick's guide offers vitally important information to help us reach this goal.

Boulder, Colorado                                                                    Gilbert F. White

# Acknowledgments

A book of this magnitude and scope is the work of many people. Several of them deserve special recognition, for without their contributions this book would have taken much longer and looked much different.

Special thanks go to my research assistant, Sharon Brooks, who arrived at the Pacific Institute without knowing what she was getting into. She brought with her a persistence, a desire to learn, and a set of skills that proved invaluable to the project, including a working knowledge of the intricacies of the University of California library system, an interest and skill in a range of computer languages and graphics, and the ability to seek out and find obscure references. Her efforts continued even after her entire life's belongings, except for her computer, went up in flames in the tragic East Bay Hills fire in 1991. She also found herself in charge of an eager set of interns and managed to keep them all interested and active. I hope when she looks at the final product, she is proud of her accomplishments.

A group of persistent and reliable research interns helped to collect and enter data, to proofread tables, and to do innumerable tedious but necessary tasks getting the book ready for publication. These included Cynthia Chiang, Virginia Esperanza, David Fifer, Debbie Goldenberg, Brent Grossman, Todd Lappin, Jeremy Levin, Yolanda Todd, Laura Wallace, and Devron Weiss. David Fifer managed to coordinate the complicated process of getting permissions for the many tables of the book, in addition to many other jobs he readily took on and completed. The entire staff of the Pacific Institute provided invaluable support throughout the project, particularly Pat Brenner and Nancy Levin, who put up with chaos in the office for many months.

The book has benefited from the comments, additions, and reviews of many individuals, including Eric Barron, Charles Bourne, Gunther Craun, Carel de Rooy, Kurt Fausch, Earl Foote, John Hobbie, Henry Jacoby, Andrew Jones, Andrew Keller, Jim Kirchner, Lars Kristoferson, Owen Lammars, Curtis Lomax, Bruce Macler, Juliet Major, Deborah Moore, Thomas Naff, Philip Micklin, Howard Perlman, Khaled Rahman, Paul Raskin, David Rib, Eric Rodenburg, Peter Rogers, Arno Rosemarin, Lisa Ross, S.J. Siméant, Wayne Solley, Nick Sundt, Ted W. Mermel, Celinda Verano, and Zongping Zhu.

Many other people made valuable contributions to the book. Kent Anderson compiled the comprehensive section on units, data conversions, and constants, and proofread portions of the text. Valentina Yanuta in St. Petersburg managed to keep open Konstantin Vinnikov's line of electronic communication between Russia and the United States, even during the most tumultuous periods there, permitting the regular and efficient exchange of information, reviews, and comments on Professor Shiklomanov's chapter. Larry Cannon's professional translation of that chapter made the readable Russian into more readable English. Caron Cooper provided hard-to-find data from the Soviet Union and initial translations of these data. Jim Williams tried his best to standardize the English transliterations for major Chinese rivers and lakes. Steve McElroy's copyediting greatly improved the flow and language of the Falkenmark and Lindh chapter. Linda Nash did far more than write a fine chapter on water quality and health; she reviewed other papers, helped compile and edit tables, and provided numerous insights into the quality of water data. Joyce Berry at Oxford University Press helped navigate the book through the many shoals of the publishing world. The librarians at the University of California, Berkeley, particularly Gerald Gieffer and Susan Greer in the invaluable Water Resources Center Archives, provided guidance and data that simply cannot be found elsewhere. I hope we've returned all our overdue library books.

Finally, the financial support of the Stockholm Environment Institute and the institutional support of the Pacific Institute for Studies in Development, Environment, and Security have been critical to this project. Gordon Goodman and Lars Kristoferson of Stockholm encouraged me to pursue this project and without the aid of these organizations and individuals this project would never have been begun.

# About the authors

**Alan P. Covich**, long-time Professor of Zoology at the University of Oklahoma, and now at the Department of Fishery and Wildlife Biology at Colorado State University, is a limnologist with wide research experience in temperate and tropical fresh water ecosystems. Dr. Covich received his PhD from Yale University and has been a visiting scientist at the Center for Energy and Environmental Research at the University of Puerto Rico. Dr. Covich is a specialist in foodweb dynamics and recently co-edited a book on *Ecology and Classification of North American Freshwater Invertebrates* (Academic Press, New York).

**Malin Falkenmark** is Professor of International Hydrology at the Natural Science Research Council in Stockholm. Dr. Falkenmark served as Rapporteur General at the United Nations Water Conference in 1977 in Mar del Plata and is active in the International Union for the Conservation of Nature (IUCN), the World Bank and the International Water Resources Association, where she presently chairs the committee on Water Strategies for the 21st Century. She is a Global 500 Laureate and scientific advisor to the World Bank, United Nations Environment Programme, and United Nations Development Programme Global Environmental Facility. She has worked extensively on water issues in developing countries, particularly Africa, and is widely published.

**Peter H. Gleick** is one of the founders of the Pacific Institute for Studies in Development, Environment, and Security in Oakland, California, where he directs the Global Environment Program. Dr. Gleick has served on the Climate and Water Panel of the American Association for the Advancement of Science (AAAS) and co-chaired Working Group 2 of the Advisory Group on Greenhouse Gases (AGGG), sponsored by the United Nations Environment Programme (UNEP) and the World Meteorological Organization (WMO). Dr. Gleick received a MacArthur Foundation Research and Writing Fellowship to explore the implications of global environmental changes for water and international security and currently works on a wide range of water resources problems and the links between global environmental issues and international security.

**Gunnar Lindh**, Professor Emeritus at Lund University, Sweden, is an expert on water use in developing countries, including the question of definitions of sustainable development and water. Dr. Lindh is a member of both the Royal Swedish Academy of Engineering Sciences and the Swedish Committee for Hydrology. His work on water and urban areas has received wide international attention, and he is the author of *Water and the City*, now published in six languages. Recently he has contributed to the United Nations Environment Programme study of the Mediterranean region and he is working on the implications of sea-level rise for water resources in coastal urban areas.

**Stephen McCaffrey** is Professor of Law at McGeorge University in California. Professor McCaffrey has been on the McGeorge Law School faculty since 1977 and served two terms as a member of the United Nations International Law Commission, from 1982 to 1991. From 1985 to 1991, he was Special Rapporteur for the United Nations International Law Commission, negotiating "The Law of the Non-Navigational Uses of International Watercourses." He has been a consultant or adviser to various governmental agencies, including the Office of the Legal Adviser, US Department of State, the International Union for the Conservation of Nature and Natural Resources, the Ministry of External Affairs, India, and the United Nations Environment Programme.

**Linda Nash** is a Senior Associate of the Pacific Institute. She has extensive experience in water resources management and environmental issues. Her recent research has included a study of the ecological impacts of the California drought, an analysis of the impacts of climate change on water resources in the Colorado River Basin, and an assessment of the impacts of contaminated sediments on water quality. Ms. Nash has worked as an environmental scientist specializing in water management for the California Water Resources Control Board and the US Environmental Protection Agency. She holds an MS degree from the Energy and Resources Group at the University of California, Berkeley, and a BS in Civil Engineering from Stanford University.

**Sandra Postel** is Vice President for Research at the Worldwatch Institute in Washington, DC, a private non-profit research organization devoted to analyzing global environmental trends. Ms. Postel is a leading water resources analyst focusing on water and agriculture in the United States and internationally. She has written and lectured extensively on water issues and is a contributing author to all nine of the Worldwatch Institute's annual *State of the World* reports. She serves on the Board of Advisors to the Seventh Generation company, the Environmental Media Association, and the International Water Resources Association. She has just completed *Last Oasis: Facing Water Scarcity* (Norton, New York).

**Igor A. Shiklomanov**, Doctor of Geographical Sciences, is a Corresponding Member of the Russian Academy of Natural Sciences and Director of the State Hydrological Institute in St. Petersburg. Dr. Shiklomanov has played an active role in assessing water resources and river runoff as well as the impacts of human activities on water. He is the author of more than 140 publications, including 6 monographs containing hydrological analysis for the territory of the former Soviet Union and for the globe. Dr. Shiklomanov actively participates in the international hydrologic programs of the United Nations Educational, Scientific, and Cultural Organization (Unesco) and the World Meteorological Organization (WMO), and he is a coordinator of several international projects.

# Contents

**Contents**

# List of Part I tables

# List of figures

# Part I
# Essays on fresh water issues

# Chapter 1
# An introduction to global fresh water issues

**Peter H. Gleick**

*Water is the best of all things.*
*Pindar, Greek poet*

Fresh water is a fundamental resource, integral to all environmental and societal processes. Water is a critical component of ecological cycles. Aquatic ecosystems harbor diverse species and offer many valuable services. Human beings require water to run industries, to provide energy, and to grow food. In order to mobilize water for human needs, we build huge reservoirs to store water for dry periods and to hold back flood waters; we build aqueducts to transport water thousands of kilometers from water-rich to water-poor regions; we burn oil to generate electricity to desalinate salt water in arid regions; and we dream of towing icebergs from the polar regions and of reversing the flow of massive rivers.

Harsh realities intrude on these dreams. As we approach the 21st century we must now acknowledge that many of our efforts to harness water have been inadequate or misdirected. We remain ignorant of the functioning of basic hydrologic processes. Rivers, lakes, and ground water aquifers are increasingly contaminated with biological and chemical wastes. Vast numbers of people lack clean drinking water and rudimentary sanitation services. Millions of people die every year from water-related diseases such as malaria, typhoid, and cholera. Massive water developments have destroyed many of the world's most productive wetlands and other aquatic habitats. The economic and hydrologic resources for major new irrigation projects cannot be found. And expected changes in global climatic conditions will alter future water supply, demand, and quality.

Scientists, planners, health officials, and the public are slowly beginning to perceive the intimate connections between the availability and quality of fresh water resources and other resource and environmental problems. This book highlights these links, offers an overview of critical fresh water issues, and provides detailed data on the water resources of the earth and how we use and abuse them. Part I consists of essays written by eight leading water experts, addressing important fresh water problems, including the global water balance, water quality and health, natural ecosystems, agriculture, energy, economic development, and politics and international law. Part II is a compilation of over 200 tables and figures on all aspects of fresh water resources. Each table is accompanied by a description and a discussion of definitions, data limitations, and sources.

Problems with the supply and quality of fresh water should be of concern to everyone. My hope is that the discussions and data offered here provide a starting point in the debates to come.

## The world's fresh water

Hindu and Buddhist traditions place mythical Mount Meru – the dwelling place of the gods – at the center of the universe. Here originate the rivers of the earth, including the Indus, the Ganges, and the Brahmaputra. In early Christian tradition, the waters of the earth originate in the fountains of the Garden of Eden, which divide into the world's great streams: the Nile, the Tigris, the Euphrates, the Indus, and the Ganges.[1] These myths arise in part from the sacred role of water in sustaining life, and in part from our early ignorance of the functioning of the global water cycle.

By the end of the 20th century, our understanding of the stocks, flows, and condition of global water resources is still distressingly imperfect. The International Hydrological Decade, coordinated by the United Nations Educational, Scientific, and Cultural Organization (UNESCO), beginning in the mid-1960s, was successful in elucidating many unknown aspects of the global hydrological cycle and some of the best references on global water resources still date from that period.[2] Unfortunately, with the exception of the United Nations International Drinking Water Supply and Sanitation Decade from 1981 to 1990, there has been little follow-up on understanding and solving critical water issues, and much of the momentum of that earlier period has been lost.

Samuel Taylor Coleridge (1772–1834), in his classic poem "The Rime of the Ancient Mariner," effectively described the principal characteristic of the earth's water resources when he wrote, "Water, water, everywhere, nor any drop to drink." Ninety-seven per cent of all the water on earth is salt water – unsuitable for drinking or growing crops. The remaining 3% is fresh water, comprising a total volume of about 35 million $km^3$. If this water were spread out evenly over the surface of the earth it would make a layer 70 m thick. Yet almost all of this fresh water is effectively locked away in the ice caps of Antarctica and Greenland and in deep underground aquifers, which remain technologically or economically beyond our reach. Less than 100,000 $km^3$ – just 0.3% of total fresh water reserves on earth – is found in the rivers and lakes that constitute the bulk of our usable supply. Details of the global water balance are described by Igor Shiklomanov in Chapter 2, and a wide variety of water balance data can be found in the tables of Part II, Sections A and B.

Fresh water is a renewable resource, made continuously available by the constant flow to the earth of solar energy, which evaporates water from the oceans and land and redistributes it around the globe. More water evaporates from the oceans than falls on them as precipitation; thus there is a continuous transfer of fresh water from the oceans to the continents. This water runs off in the rivers and streams that sustain our natural ecosystems and societies and recharge our aquifers. On an annual basis, about 45,000 $km^3$ per year are returned to the oceans as river and ground water runoff (see Figure 1.1). If that runoff were evenly distributed, each person on earth would receive more than 8,000 $m^3$ per year.

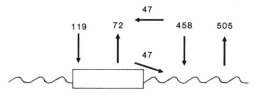

Fig. 1–1. Flows of the hydrologic cycle. This simplified diagram of the hydrologic cycles shows (from left to right): precipitation to land; evaporation from land; runoff from land to oceans (lower arrow); transfers of atmospheric water to land (upper arrow); precipitation to ocean; and evaporation from ocean. All flows are in $km^3 \times 10^3$ per year.

In fact, these fresh water resources are *not* evenly distributed. Rainfall and runoff are apportioned in both space and time in a grossly irregular manner. We often don't get water when we need it or where we need it. Some places receive enormous quantities; others receive almost none. The Atacama Desert in South America is one of the world's driest areas. The rain gauge at Arica, Chile routinely records zero annual precipitation. Pictures from space reveal cloudless skies over the great Sahara desert of Africa, which extends from 15°W to 30°E longitude and from 10° to 35°N. At the other extreme, Mount Waialeale, on the island of Kauai, Hawaii has recorded more than 11.5 m of rainfall in a single year.[3]

Other examples of these severe distribution problems abound. Twenty per cent of global average runoff comes from the Amazon River alone. All of Europe accounts for only 7% of global runoff; all of Australia produces only 1%. Thirty per cent of the total runoff in Africa comes from a single river basin, the Congo/Zaire. Many regions get nearly 100% of

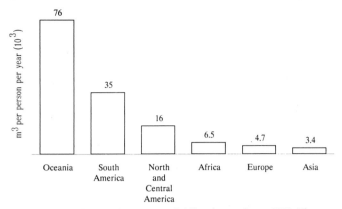

Fig. 1–2. Annual per capita water availability, by continent, 1990. The availability of fresh water varies significantly, as shown by this comparison of continental per capita fresh water availability. Oceania has over twenty times the water, per person, of Asia because of both rich water resources and low populations. These inequalities in distribution persist at every spatial and temporal scale. Data from Table A.10.

their precipitation during a brief, intense rainy season. In Cheerapunji, India, over 10.5 m of rain per year may fall during the brief period of the monsoon. Much of California gets no rainfall at all from May through September, which is precisely the period of greatest water demand. Figure 1.2 shows the annual per capita availability of fresh water by continent. Oceania, because of its low population and large water resources, offers more than twenty times as much fresh water per person as does Asia.

This tremendous variability is a natural feature of our climate and helps define many of the problems facing us today. Shiklomanov argues that solving our water problems requires that we expand our study of these natural characteristics, improve our ability to assess both static and dynamic hydrologic processes, and incorporate the effects of human economic activities and growing populations into our understanding of the global water cycle. These actions will require far more effective international cooperation on water issues than we have seen before.

## Water quality and health

A problem long facing civilization has been how to provide the water we need to drink and to remove our wastes, particularly in large cities far from adequate and reliable sources of water. The Greeks and the Romans were among the first to take water management seriously and to introduce long-distance water pipelines. The ancient cities of Jericho, Ur, Memphis, Babylon, Athens, Carthage, and Alexandria relied to a great extent on artificial water systems to provide drinking water.[4]

The ancient Romans built elaborate aqueducts and water distribution systems hundreds of kilometers long that survive to this day, and they developed the first major closed sewage system over 2,500 years ago. Yet today, over a thousand million people throughout the world are served by sewage treatment systems not up to the standards of ancient Rome, or by no systems at all. Even in industrialized nations, adequate waste treatment is a relatively recent development.

In her chapter on water quality and health, Linda Nash describes how we have continued to improve our understanding of the links between water quality, water-related diseases, and the hydrologic cycle, but have failed to reduce the suffering these water quality problems cause. There are two major water quality concerns: microbiologic contamination, responsible for many of the world's most persistent and widespread diseases; and chemical contamination, which poses risks for human beings and aquatic ecosystems. On a global scale, the risks of toxic chemicals, metals, and radionuclides in water are still dwarfed by the health effects of viral and bacterial contamination, but the increasing prevalence and magnitude of chemical contamination is leading to the concern that human health risks from these pollutants will rise in the future.

Water-related diseases have their roots in bacteria, viruses, or insects that spend part or all of their lives in water. These diseases can be con-

trolled by improving water quality and sanitation services, and in the early parts of the 19th century better water supply led to dramatic improvements in public health in the industrialized nations of North America and Europe. In September 1854, Dr. John Snow of London traced an outbreak of cholera to one particular water well in London that had caused over 500 deaths in a single neighborhood. He ended the outbreak by having the handle of the contaminated pump removed, proving the connection between contaminated water supply and cholera.[5]

These improvements in understanding have not led to the protection of health in much of the world. Nearly 140 years later, new epidemics continue to rage unchecked in many parts of the world because of inadequate sanitation facilities, lack of access to safe drinking water, and continued ignorance about the links between water and disease. Worldwide, more than 250 million new cases of water-related diseases are now reported each year, resulting in approximately 10 million deaths annually. The vast majority of these deaths occur in the tropical countries of the developing world. In Latin America, the unprecedented outbreak of cholera in 1991 shows how vulnerable poorer regions are to these diseases and highlights our failure to control the contamination of drinking water and to educate people about sanitation, water, and health. In Peru, for example, the women who dress the victims of the current cholera epidemic for burial also prepare the food and drink shared at the services for the dead. Figures 1.3 and 1.4 show the number of cases and deaths from cholera reported worldwide from 1979 to 1991. Data on water-related diseases, and the serious gaps in the official reporting of these diseases, are shown in the tables of Part II, Section C.

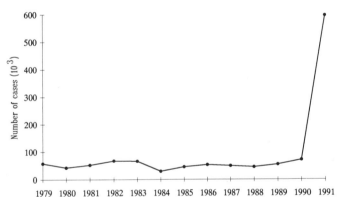

Fig. 1–3. Cases of cholera worldwide, 1979–1991. This figure plots the number of cholera cases reported to the World Health Organization. In 1991, severe cholera epidemics broke out in Latin America, which had been free of cholera for a century. Cholera also increased dramatically in parts of Asia in 1991. See Tables C.22 and C.23 for data.

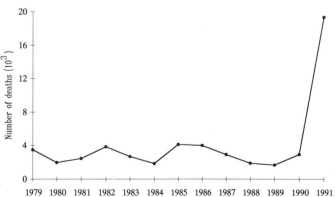

Fig. 1–4. Cholera deaths worldwide, 1979–1991. This figure plots the number of cholera deaths reported to the World Health Organization. In 1991, severe cholera epidemics broke out in Latin America, which had been free of cholera for a century. Cholera also increased dramatically in parts of Asia in 1991. See Tables C.22 and C.23 for data.

Other water quality problems abound. Nitrate contamination comes from the runoff of agricultural fertilizers and industrial wastes. Heavy metal contamination results from mine drainage, industrial processing of ores, and leaching from solid wastes. The contamination of water by organic materials results from many manufacturing processes and from the production of pesticides. For example, many tributaries of the Amazon River have been severely contaminated with mercury from gold mining operations, and concentrations of methyl mercury in many Amazon fish species exceed the World Health Organization (WHO) criteria for human consumption. In 1984, a study of the Cree Indian inhabitants of Chiasabi, Canada found that 64% of all the residents had mercury concentrations exceeding WHO standards. The contamination resulted from Quebec's James Bay hydroelectric project, which caused mercury levels to rise dramatically in the fish that are the principal staple of the Cree Indians. DDT and other pesticides continue to be sold throughout the developing world, often by corporations from industrialized nations that have themselves banned the use of these chemicals because of their devastating ecological effects.

Public concern over the connections between these pollutants and human and ecological health is growing rapidly. Yet our knowledge about these links remains poor. Most water pollutants have been inadequately studied for their health effects, though the rising incidence of cancer in the developed world and growing number of "cancer clusters," miscarriages, birth defects, and sterility in regions with contaminated water continue to raise unanswered questions about the adequate control of water pollutants.

The primary obstacles to effective water quality management, according to Nash, are cost and the lack of information. The provision of low-cost technologies proven effective in preventing disease and improving public health is urgently needed, as are new and better data on even the most common water quality problems such as fecal contamination, microbiologic disease, and the types and extent of chemical contamination. But the lack of information must not stop us from acting to prevent further contamination of our fresh waters on an international level.

## Water and ecosystems

The lakes, ponds, rivers, streams, wetlands, and ground water of the world are interconnected parts of the global fresh water ecosystem. The diversity and scarcity of many of these systems are described by Alan Covich in Chapter 4, together with the persistent threats they face from human intervention in the water cycle. Only 1% of the earth's surface is covered by inland waters; 70% is covered by the oceans. Yet the oceans contain only 7% of the animal species alive today, while inland waters contain 12%. Some estimates say that the total faunal diversity of rivers is 65 times greater than that of the sea. Unfortunately, many fresh water aquatic habitats and their biota are being destroyed faster than they can be studied or protected, because of ignorance, competing demands for limited water, and industrial or commercial developments.[6]

Covich argues that water quality, species distributions, and ecosystem processes will be altered in the future at rates faster than ever before. These accelerated effects will be caused by increases in human populations, decreases in water quality, invasions of non-native species, dam construction, eutrophication, acid precipitation, and global climatic changes.

When we grow cotton or alfalfa in the desert, or take water from a distant river for use in our cities, we use scarce water resources that maintain delicate ecosystems. Our release of sulfur and nitrogen into the atmosphere leads to acid rain that alters the chemistry of natural waters. Global climate change will alter the temperature and water quality of balanced ecosystems. And habitat destruction and wetlands loss is leading to the impoverishment of biotic diversity.

We know distressingly little about the impacts of our activities on natural aquatic ecosystems. One-tenth of total world fish yield comes from inland waters. Yet some yields are declining as a result of overfishing, the introduction of non-native species of aquatic organisms, and the regulation and diversion of rivers and fresh water flows. A well-documented example is the unintentional fisheries mismanagement of the Laurentian Great Lakes of North America that occurred following the

construction of the Erie Canal in the north-eastern United States. There were unexpected and undesired impacts of this project, mostly to natural ecosystems. The canal brought sea lampreys – an ocean parasite – that killed off most of the Great Lakes salmonids and the associated fishing industry. Additional overfishing of Great Lakes sturgeon and destruction of spawning habitat further decimated native fish populations, and unintentional ecological modifications continue to this day. Most recently, the accidental introduction of exotic invertebrates, such as the zebra mussel and the spiny water flea, is leading to disruptions of natural ecological communities of mussels and zooplankton, and the blocking of cooling water intakes and heat exchangers in power plants and industries.

Covich describes how the introduction of new predatory fish species into the deep oligotrophic Lake Atitlan in Guatemala wiped out the smaller fishes that were used by the indigenous population. The introduction of the peacock bass from the Amazon River into Gatun Lake in the Panama Canal altered the pelagic food web and eliminated the smaller native fishes, leading to increased numbers of mosquitos, important vectors of malaria in the region. The introduction of the Nile perch into Lake Victoria around 1959 has led to the extinction of many endemic species of fish and a great reduction in the catch of native species. Many of the extinct cichlid species have never been scientifically studied and named.

In addition to species introductions, modification of free-flowing rivers for energy or water supply has many effects on aquatic ecosystems, including losses in species diversity and floodplain fertility. New dams replace free-flowing rivers with standing pools of water, with concomitant ecological effects. In the western United States, where large volumes of water are diverted from natural ecosystems into enormous water supply and hydroelectric systems, fisheries populations have been severely depleted. These drops are attributed to decreases in the flows of fresh water into natural ecosystems, toxic contamination of rivers, and a natural drought made more severe by diversions of limited water, in particular the export of water out of the Sacramento delta to supply irrigation and municipalities in southern California.[7]

When several salmon populations in northern California, Oregon, and Washington collapsed in the late 1980s and early 1990s, proposed causes included overfishing, changes in river flow regimes by the myriad of dams on the rivers, destruction of spawning habitat by dam construction and operation, entrainment of young fish trying to reach the open ocean by enormous pumps transferring water from one river basin to another, and extreme temperature fluctuations caused by other human activities.[8] In fact, each of these stresses has played a role in the fisheries decline.[9]

The diversion of water from one basin to another can have enormously destructive ecological effects. The transfer of water out of the Mono Basin in eastern California by the city of Los Angeles is leading to the destruction of a unique closed-basin lake vitally important for many endemic species. On a much larger scale, the diversion of almost 100% of the water from the Amu Dar'ya and Syr Dar'ya rivers in the former Soviet Union to grow cotton and other crops has led to the desiccation of the Aral Sea, the destruction of the fisheries there, local health problems, and the economic collapse of the region. Between 1926 and 1990, the surface area of the Aral Sea dropped 40% and the volume decreased 65%, and salinity has more than tripled (see Figures 1.5 and 1.6). All 24 native fish species have been killed. Reviving the Aral Sea may not be possible at all; it certainly cannot be done without a complete change in the style and form of water use and management in the region. Mukhammed Salikh, an Uzbek poet, said, "You cannot fill the Aral with tears."[10]

Finally, Covich argues that distinguishing and resolving the many sources of natural and anthropogenic disturbances of fresh water ecosystems will require far more effort than has been expended to date. Technological advances, such as remote sensing and geographic information systems modeling will help, but many important changes are occurring rapidly, and many problems require institutional, not technological, solutions. If our sustainable biological resources are, in fact, to be sustained, we must ensure that adequate water is provided for natural habitats, alterations of natural ecosystem processes and losses of overall biodiversity are minimized, remaining natural fresh water habitats with high levels of biodiversity and endemic species are preserved, and better multidisciplinary research projects on regionally representative biotas and hydrologic regimes are undertaken.

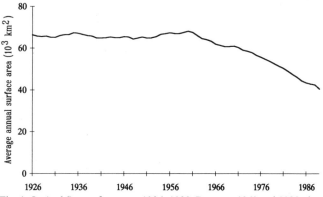

Fig. 1–5.  Aral Sea surface area, 1926–1989. Between 1960 and 1989, the surface area of the Aral Sea decreased by more than 40%, from 68,000 km² to 40,000 km². This situation, brought about by agricultural consumption of water that used to flow into the sea, has destroyed the natural ecosystem of the sea and the fishing communities along its margins. See Table F.20 for data.

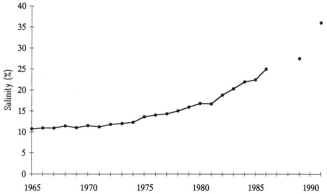

Fig. 1–6.  Aral Sea salinity, 1965–1991. Reductions in the volume of incoming fresh water have caused Aral Sea salinity to increase dramatically in the last two decades. From 1965–1991, salinity in the sea has more than tripled, and the once-substantial fish catch has dropped to zero. All 24 endemic species of fish are now thought to have been killed. See Table F.20 for data.

## Agriculture

Plants combine solar energy from the sun, carbon dioxide from the atmosphere, and water to form carbohydrates – the basic food supply of all life. Without dependable water supplies, growing food for the earth's population is at best a marginal affair. The earliest forms of civilization developed in the arid lands of the Middle East, Asia, and the Americas along major rivers and streams such as the Tigris, Euphrates, Nile, Indus, and Colorado, and used these rivers to provide reliable irrigation water. In Egypt, irrigation was being practiced along the Nile in 3,400 BC. The civilizations that arose in the foothills of the Himalayan mountains used the waters of the Indus basin for irrigation as much as 5,000 years ago, constructing networks of canals for this purpose.[11] In the New World, the Incas in Peru were using irrigation in 1,000 BC,[12] and there is archeological evidence that irrigation canals were built and maintained by the ancient Hohokam tribe in the Salt River valley of the lower Colorado River basin 2,000 years ago.[13]

The Green Revolution of the 20th century was accomplished in large part by greatly expanding the availability and use of irrigation technology and water. Between 1950 and the end of the 1980s, global irrigated area more than doubled, from 94 million hectares to over 230 million hectares. One-third of the total global harvest of food comes from the 17% of the world's cropland that is irrigated. Figure 1.7 shows the percentage of cropland irrigated for each continent. Today, two-thirds of the fresh water used worldwide goes to agriculture, and more will be required in the future to meet the growing food needs of the earth's swelling population.

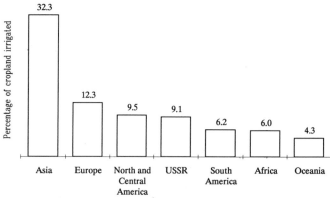

Fig. 1–7.  Fraction of cropland that is irrigated, by continent, 1989. Globally about 16% of total cropland is irrigated, but there are wide disparities among the major regions. Asia irrigates nearly one-third of all cropland; Africa and Oceania irrigate about one-twentieth of their cropland. See Table E.2 for data.

Unfortunately, we are not keeping up with growing population. Sandra Postel in Chapter 5 describes how world irrigated area per capita increased throughout the 20th century up until the late 1970s, when it peaked at 48 hectares for every 1,000 people. Since then, the amount of irrigated area per person has fallen, as population growth outstripped additions to irrigated area.

Postel argues that significant increases in the rate of additions to irrigated area in the future are unlikely, in part because of rising economic and environmental costs. She notes that the cost of developing new irrigation capacity in Africa, where such capacity is especially needed, is often exceptionally high because of lack of roads and other infrastructure, the small parcels of irrigable land, and the seasonal nature of river flows.

Solutions exist, but it remains to be seen how effective they will be in solving agricultural water problems and providing food for our growing population. In particular, there is enormous potential for increasing the efficiency of existing systems, so that crops can be grown with less water. In many irrigation networks, less than half the water actually benefits crops, and the rest is lost through seepage from unlined canals, evaporation and runoff from poorly applied water, and poor management that fails to deliver water to crops at the right time and in suitable quantities.

All of these factors mean that the world's existing irrigation system yields far less than its potential. While the good news is that this gives us room to meet our food needs by improving irrigation efficiency, the bad news is that dramatic improvements will be needed to meet the rapid population growth now occurring.

Irrigation also has a profound impact on the quality of water resources and on the cropland that is watered. Among the costs of massive irrigation projects are waterlogged and salinized lands, declining and contaminated ground water supplies, shrinking lakes and inland seas, and the destruction of aquatic habitats. These environmental problems contribute to concern over new irrigation projects.

These are not new problems. Soil salinization and waterlogging related to irrigation development are thought to have caused the collapse 2,500 years ago of the Mesopotamian civilization. Massive soil erosion is also thought to have contributed to the decline of Greek civilization. Pliny, the Greek philosopher, natural scientist, and historian wrote:

> ...like a body of whom because of a wasting disease only the bones are left, the fertile and soft soil is everywhere eroded and only the sterile skeleton is left. But in those [old] times, when the land was still undamaged, its mountains were high and covered with earth and likewise its plains, which are now called stony fields, consisted of fertile soil.[14]

At the other extreme, overpumping of ground water in many regions is beginning to make continued irrigation too costly, forcing land out of irrigation. Postel estimates that roughly one-fifth of the irrigated area of the United States is watered by pumping ground water at rates faster than the aquifers are being recharged. The situation is similar in many other regions of the world, including parts of China, India, and the Middle East.

Ultimately, overpumping ground water for irrigation is unsustainable and must fail. In some regions, such as the Great Plains of the United States and fields in Saudi Arabia, farms have already been abandoned because of declining well yields and rising pumping costs. Irrigated area in the Texas High Plains region over the Ogallala aquifer has decreased by 34% between 1974 and 1989 because the cost of pumping water from continuously declining aquifers now exceeds the value of the food produced.

Further pressures on irrigation water are coming from the growing demands for water from other sectors, particularly rapidly growing urban populations, also discussed by Falkenmark and Lindh in Chapter 7. Shifting water from farms to cities is already being done in the western United States, in China to meet the urgent needs of Beijing and other cities, in Mexico to supply the needs of Mexico City, and elsewhere. In regions with water ''markets,'' some farmers can earn more money by selling water to nearby cities than by using it to irrigate crops. The extent to which such water transfers reduce irrigated area depends on the severity of the needs of other sectors, the institutional and legal barriers to such transfers, and the political and social implications. Because of the enormous volume of water used by the agricultural sector, however, it seems likely that pressure to transfer irrigation water to other sectors will grow.

Slowing expansion of irrigated area, growing environmental damage from irrigation projects, new pressure to transfer water away from the agricultural sector to the cities, and future threats such as global climatic change all complicate the rational management of agriculture and will affect our ability to grow food sustainably for the world's population. Postel argues that these trends point to the need for creative and diverse approaches, including raising water use efficiency, integrating irrigation more fully with basic development goals, and improving the productivity of rainfed farming, which still produces two-thirds of our food, through a better understanding of successful traditional agricultural practices. A secure water future requires us to recognize natural limits to water availability and the need to bring human numbers and demands into line with them.

## Energy

We use water to help us produce the energy we need to run our societies. We use energy to help us clean, pump, and transport the water we need. And we are running up against physical and environmental constraints in our use of both resources. Limitations on the availability of fresh water in some regions of the world may restrict the type and extent of energy development, and limitations on the availability of energy will constrain our ability to provide adequate clean water and sanitation services to the thousands of millions of people who lack those basic services. Chapter 6 explores the connections between our demand for and use of energy and water, and suggests that there are strong parallels between both the problems and the solutions to the growing water crisis and conflicts over energy resources. Part II, Section G presents data on many aspects of water and energy.

The production and use of energy resources often requires a significant commitment of water resources. The amount of water needed to produce energy varies greatly with the type of facility and the characteristics of the fuel cycle. Fossil fuel, nuclear, and geothermal power plants require enormous amounts of water for cooling. Some of this water may be lost to evaporation or contamination; much of it may be returned to a watershed for use by other sectors of society. Solar photovoltaic power systems, wind turbines, and other renewable energy sources often require minimal amounts of water. Chapter 6 provides estimates of the water consumed per unit energy produced for a wide range of energy facilities and fuels used.

In water-poor regions, the use of large quantities of water for energy production can lead to changes in natural hydrological and ecological systems and increase the pressure for inter-basin transfers of water. In some places, absolute constraints on water availability will limit choices of sites and types of energy facilities.

Several examples presented in Chapter 6 highlight the vulnerability of our energy systems to fluctuations in water supply. The severe drought in California between 1987 and 1991 greatly reduced hydroelectric produc-

tion and forced electric utilities there to burn more fossil fuels than normal, at an added cost of approximately $3,000 million to electricity consumers and a 25% increase in carbon dioxide emissions from the state's electric utility sector. Similarly, the decade-long drought in the 1980s in north-eastern Africa caused a reduction in hydroelectric generation from the Aswan Dam in Egypt, which supplies nearly half of Egypt's electricity demand.

As the amount of water available for energy production decreases in the future due to growing demands from other sectors of society or growing populations, we are likely to face further regional constraints on our ability to produce and use energy. At the same time, growing energy demands will place additional pressures on water supplies in regions where supply is constrained by either nature or humanity. Other unknowns, such as global climate changes that alter regional water availability, further complicate the picture.

There are some bright spots. Many of the renewable energy sources, particularly photovoltaics, wind generation, and solar thermal systems, require far less water per unit energy produced than do conventional systems. For example, the drought in California that curtailed hydroelectric production had no effect on the continued reliable operation of the world's largest wind electric and solar thermal electric systems. In water-short regions, sources of energy with low water requirements may increasingly be the systems of choice.

In addition to using water to produce energy, we use energy to produce, move, and clean water. Energy permits us to make use of water that was previously considered undrinkable or unobtainable. We can now remove salts and other contaminants using desalination and wastewater treatment techniques, and we pump water from deep underground aquifers or distant sources. The availability and price of energy sets limits on the extent to which unusual sources of water can be tapped. As a result, understanding the links between water supply and quality and energy will help us evaluate constraints on meeting future water needs.

Society's first answer to the problem of the grossly uneven distribution of fresh water resources was to build large aqueducts and dams to even out variations in river runoff over time and to move water from regions of surplus to regions of deficit. While significant amounts of water were often provided by early irrigation and water supply systems, they were ultimately limited in the amount of water that could be supplied, and where that water could go, by the force of gravity. Water could be transferred from one place to another only as long as the source was uphill of the demand.

The availability of cheap fossil fuels and modern technology has changed all of that. Civilization has greatly increased its ability to transfer water from farther and farther away by using energy to pump water over hills and mountains. Throughout the 20th century, large-scale water transfer projects have permitted continued growth in arid and semi-arid regions that would otherwise have been constrained by natural limits long ago. And larger and larger projects are constantly being proposed and evaluated as populations and industrial water requirements increase.

Cheap energy has also made it possible to exploit deep underground water stocks never before available. Some of these aquifers consist of ground water that has accumulated over hundreds to thousands of years, or longer, and we are now using them faster than they can be replenished. The limits to how much water can be extracted from a finite ground water aquifer are economic and environmental. When water is pumped out faster than it is recharged by natural processes, the water level in an aquifer drops and the distance water must be raised to the surface increases. Eventually, either the energy costs of pumping that water rise to a point that exceeds the value of the water, or the water quality in the aquifer falls below acceptable levels. At this point, pumping must cease.

Ground water overdrafting is occurring in many parts of the United States, Mexico, India, China, the former Soviet Union, and the Middle East. This unsustainable practice reflects both the urgent need for water in many regions and the failure of traditional economics to consider long-term, multi-generational interests when valuing certain non-renewable resources. Ultimately, these resources will be depleted, and future generations will be forced to make the difficult and expensive choices being avoided today.

Another water supply alternative with significant energy requirements

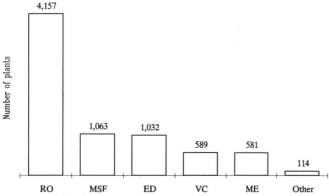

Fig. 1–8. Number of desalination plants worldwide, by process. There were 7,536 desalination plants worldwide at the beginning of 1990 (excluding shipboard units). The most common choice for desalination plants today is reverse osmosis, though multiple-stage flash distillation is still more common in large facilities. More than half of all reverse osmosis plants are in the United States, Saudi Arabia, and Japan. See Table H.32 for data on desalination plants. RO, reverse osmosis; MSF, multiple-stage flash distillation; ED, electrodialysis; VC, vapor-compression; ME, multiple-effect evaporation; "Other" includes freezing, hybrid processes, ultrafiltration, and other unspecified processes.

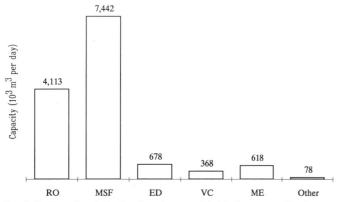

Fig. 1–9. Capacity of desalination plants worldwide, by process. By the beginning of 1990, worldwide desalination capacity exceeded 13 million m$^3$ of fresh water per day for plants capable of producing more than 100 m$^3$ per day. Nearly two-thirds of this capacity uses the multiple-stage flash distillation process; much of the remaining desalinated water is produced using reverse osmosis techniques. Electrodialysis, vapor-compression, and multiple-effect evaporation each produce less than 1 million m$^3$ per day. See Table H.32. RO, reverse osmosis; MSF, multiple-stage flash distillation; ED, electrodialysis; VC, vapor-compression; ME, multiple-effect evaporation; "Other" includes freezing, hybrid processes, ultrafiltration, and other unspecified processes.

is the desalination of salt and brackish water. This process is technologically mature, but energy intensive and expensive. Total global desalting capacity by the end of 1989 exceeded 13.2 million m$^3$ of fresh water per day produced from over 7,500 facilities, as shown in Figures 1.8 and 1.9. More than 60% of all capacity was in the oil-rich Middle Eastern countries of Saudi Arabia, Kuwait, the United Arab Emirates, Libya, Iraq, Qatar, Bahrain, and Oman. Fifteen per cent was in the United States and Japan.[15] A series of tables in Part II, Section H describes the current state of desalination technology and development. Overall, desalination provides just one-thousandth of present world water use.[16]

The economic attractiveness of desalination is directly tied to the cost of energy. Today, desalinated water costs between $1 and $8 per cubic meter, depending on the technology used, compared to between $0.01 and $0.05 per cubic meter paid by farmers in the western United States, and about $0.30 paid by urban users. Thus desalination is a realistic alternative only in water-poor and energy-rich regions, such as the Middle East.

Rational energy and water policies cannot be formulated independently because of the many and intricate connections between energy and water resources. Where water is scarce, energy choices will be limited; where energy is expensive, opportunities for improving and expanding water supplies will be limited.

## Water for development

Nearly a quarter century has passed since the United Nations Conference on the Environment was held in Stockholm in 1970. Since that gathering, concern over a wide range of environmental problems has grown tremendously. Many people now understand that the condition of our environment and the condition of human society cannot be separated. The wrong kind of economic development worsens our environmental problems, and environmental degradation often complicates economic and human development.

In Chapter 7, Malin Falkenmark and Gunnar Lindh explore the role of clean, adequate water supplies for human health, welfare, and a decent quality of life. They argue that the world is approaching a breaking point in terms of socioeconomic development and its relation to water resources, and that new water management policies and approaches are necessary. Rapid population growth in regions where the climate is dry is leading to increasing water scarcity, and within a few decades, unrestrained water demand could outstrip the amount that can be sustainably provided in many areas.

The presence or absence of water can mean life or death, prosperity or poverty. Water is a necessary commodity in household and municipal activities and a critical factor in agricultural and industrial production. In today's world of 5,400 million people, providing this water is already a serious challenge, straining our water management systems and institutions. Indeed, according to Falkenmark and Lindh, some eighty countries, with 40% of the world's population, already suffer from serious water shortages in some regions or at some times during the year.

In Africa, shortages are related to both underdevelopment of potentially available water resources and to their uneven distribution. In the Americas and the republics of the former Soviet Union, water resources are abundant in relation to demand, although wide regional disparities exist. Europe, endowed with a generally temperate climate and many small rivers with reliable stream flows, uses a relatively high proportion of runoff to meet its needs for water. Figure 1.10 shows the annual withdrawal of water per capita by continent, graphically demonstrating the great disparities in water use worldwide. In all regions, we find water bodies exposed to a variety of pollution loads that render the water unusable.

Access to water must not be considered an end in itself, but rather a means to other ends: providing decent health, industrial and agricultural production, and maintaining our natural ecosystems. In many countries, easily accessible water resources have been developed and the costs of using remaining sources will be increasingly expensive. Population-

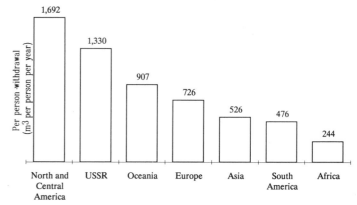

Fig. 1–10. Annual fresh water withdrawal per capita, by region. These columns show the great disparity in water withdrawn in the regions of the world. There is virtually no correspondence between this figure, and that showing total available renewable resources.

driven increases in drinking water demand will force governments to find more efficient ways of allocating the water at their disposal.

Like Postel in Chapter 5, Falkenmark and Lindh stress the need for a more comprehensive approach to water use for irrigation. More efficient use of water in agriculture would permit an increase in the availability of water for other domestic and municipal uses, which may, in turn, do far more to improve a nation's standard of living than the production of crops that can be produced more efficiently elsewhere.

The more efficient use of available water resources is not possible without: (i) the technical competence to compensate for variations in water availability over time and space, (ii) the presence of geological and topographical features for stored water to be economically and safely retrieved, and (iii) the ability to manage shared water resources successfully, including a way of including the true cost of water in the price people pay for it. In the real world, the water a nation can mobilize usually falls far short of water potential: geology may not permit the easy retrieval of stored water, technical know-how may be missing, the cost of new water developments may be high, and the cooperation and goodwill essential to sharing resources may be entirely lacking. These problems are most severe in developing nations with struggling economies, growing populations, and other development problems.

Today, the realization that renewable fresh water availability is finite, while population growth will continuously increase water demand, is starting to cause serious concern in developing nations. Concern is particularly high in arid and semi-arid Arab countries striving to achieve food security through irrigated agricultural production, which is extremely water intensive. By the year 2000, the Middle East and North Africa will be facing overall water deficits if maintaining food security remains an important political goal. By 2025, the available water will be far less than that needed, forcing the region to import food anyway to feed its populations.

Conventional wisdom typically focuses on evaluating water needs for different purposes. It asks: how much water do we need and how should that amount be made accessible? Instead, Falkenmark and Lindh take a non-traditional approach to evaluating water needs and ask: how much water is there and what future demand levels could realistically be met given regional constraints on availability?

In their analysis, a country's water predicament is related to its population, total water availability, and the fraction of gross water availability that can be made accessible, which depends on technology, economy, topography, and climate. As populations continue to grow, the principal options are demand reduction through rationing, improved distribution of available supplies, increased efficiency of water use, restructuring of societal activities, such as cultivating low-water use crops, and avoiding water pollution as human activities per unit of water intensify.

Another critical water and development issue is the enormous growth in urban populations. By the year 2000, half the world's population will be living in urban areas. This urban growth constitutes what Falkenmark and Lindh term an "urban revolution" – equally dramatic in its implications for the history of civilization as have been its predecessors, the agricultural and industrial revolutions. This unprecedented growth in urban areas is especially evident in developing countries. Between 1950 and 1980, cities in Latin America, such as Bogota, Mexico City, São Paulo, and Managua have tripled or quadrupled in population, and cities like Nairobi, Dar es Salaam, Lagos, and Kinshasa have increased over sevenfold.

Some of the largest cities are increasingly unable to provide an adequate quality of life for their citizens, including providing clean drinking water and sanitation services. A good example of this predicament is Mexico City, which has grown from 1 million inhabitants in 1920 to 15 million in 1980, and is expected to have 25 million people by the year 2000. The difficulty of supporting the city with water results in part from its geography. In 1982, Mexico City had to pump water from a distance of 100 km and from 1,000 m below the city. A decade later, rapid population growth now requires the city to withdraw additional water from 200 km away and 2,000 m lower. And unfortunately, to some extent the many leaks and breaks in the water system defeat the enormous effort expended. The amount of water estimated to be lost through leaks in the system corresponds to the gross water needs of Rome.

Sustainable development of our water resources will require careful management of the interactions among water, natural ecosystems, and society. The scale of the problems makes it impossible to rely solely on the top-down solutions with which most water experts in developed nations are generally familiar, and local initiatives must also be encouraged and facilitated.

## Water, politics, and international law

The growth in global population and the level of industrial development and water demand during the second half of the 20th century has led to an increase in the risks of political conflict over the Earth's finite fresh water resources. A substantial portion of fresh water resources is contained in international drainage basins, basins shared by two or more nations. These basins make up nearly 50% of the Earth's land area, and some 60% of the area of both Africa and Latin America. While there are numerous treaties regulating the use of shared water resources, international agreements are often either inadequate or lacking entirely in some parts of the world where water is in greatest demand. Some observers believe that the political tension over shared rivers, lakes, and ground water aquifers may escalate to the point of war, even before we move into the 21st century.

In Chapter 8, Stephen McCaffrey surveys some actual and potential disputes over fresh water, provides an overview of international water law, and assesses the adequacy of international law to regulate water use and to avoid and resolve controversies concerning international fresh water resources.

No region of the world with shared international waters is exempt from water-related controversies, though the most serious problems occur in water-scarce regions. Wallace Stegner, one of the most eloquent writers to look at water and politics, wrote: "Water is the true wealth in a dry land; without it, land is worthless or nearly so. And if you control water, you control the land that depends on it."[17] McCaffrey presents several case studies, including quarrels over the Ganges, Indus, Paraná, Rio Grande, Colorado, and several rivers in the Middle East, where a combination of water scarcity and political frictions produces what is perhaps the most volatile situation in the world.

Competition among the nations in the Middle East for scarce water resources has already led to hostilities on a number of occasions. For example, in 1964, the Arab states announced plans to divert and use the headwaters of the Jordan, partly in response to Israel's construction of a major water transfer system, the National Water Carrier. Israel considered the diversion of the Jordan headwaters to be a violation of its sovereign rights and launched a series of military strikes against the diversion works. These water-related hostilities are considered a major factor, though certainly not the only one, leading to the 1967 June War.

One consequence of that war was the occupation by Israel of the West Bank of the Jordan River. A large ground water aquifer in this area now provides at least one-fourth of Israel's water. Israel is now so dependent on the waters of the disputed territories that control of the West Bank aquifers has become a major political issue in the ongoing peace talks.

Some water disputes, like the conflicts in the Jordan River basin, can lead to violent conflict; others are resolved peaceably through existing or new institutions and agreements; many are still ongoing and show no sign of an imminent resolution. In some cases, existing political tensions unrelated to water may result in recourse to violence over questions that would have been resolved peacefully between other countries. In all of the cases reviewed by McCaffrey, whether or not actual violent conflict occurred, the water disputes were matters of high national interest – high politics, in the language of international security specialists – engaging the attention of government officials, diplomats, and scholars alike. And several of the rivers will probably receive increasing international attention as expanding populations, climate change, and upstream development efforts combine to place further demands on already scarce water resources.

In addition to describing several important water-related disputes, McCaffrey presents an overview of the general principles of international law governing the use and protection of international water resources. Even the most serious and complex resource controversies can be resolved in an equitable manner, provided the states concerned have the political will to do so. The 1960 agreement between India and Pakistan on the Indus River, which successfully diffused a major source of tension between the two countries, is an example of such a resolution.

The history of controversies concerning international water resources suggests that law will frequently be invoked by at least one, and probably all, of the parties to any dispute over these resources. Most rules of international water law derive from one of two categories of sources: treaties or international custom. Treaty-based rules are relatively easy to ascertain, although there is always the possibility of differing interpretations of individual provisions. Norms of customary international law are more difficult to establish, but recent efforts at codifying these rules have greatly assisted the process.

The sheer number of international water treaties is large. An index prepared by the Food and Agriculture Organization (FAO) of the United Nations contains more than 2,000 examples, some dating back over one thousand years.[18] Most agreements concerning shared water resources are bilateral, and relate to specific rivers that form or cross boundaries, or lakes that straddle them. McCaffrey describes several major principles regularly enunciated by these water treaties, including two he considers vitally important: the principle of equitable utilization, which states that the apportionment of the uses and benefits of a shared watercourse should be in an *equitable* manner, and the requirement that a state, through its actions affecting an international watercourse, may not significantly harm other states.

The role of international law in resolving conflicts over water resources differs from case to case, and it must be observed that major international water controversies have generally not been resolved solely, or even principally, on the basis of applicable legal principles. On the other hand, McCaffrey points out that states rarely defy generally accepted principles of international law and, indeed, usually rely on those principles in their diplomatic exchanges. The more concrete and generally accepted the applicable legal principles become, the more likely it is that they will play a major role in the resolution of international water controversies. In this connection, the work of organizations such as the Institut de Droit International, the International Law Association, and the International Law Commission contributes to the resolution of existing disputes and the avoidance of future controversies.

The key to peaceful solutions of disputes over shared water resources is continued communication between the states concerned, preferably on the technical level, over everything from hydrologic and meteorological data to basin-wide development plans. Experience has shown that such communication can most effectively occur through some form of joint mechanism, such as a commission composed of experts from all basin states. Unfortunately, political frictions often prevent the formation of such bodies precisely in the cases where they are needed most. The end of the Cold War now offers an opportunity for a new era of cooperation, even where tensions have been highest, by permitting renewed attention to the resolution of water controversies in many regions of the world.

## Water in the 21st century

Since the beginning of civilizations, humans have harnessed fresh water resources for food production, for transportation, and for energy. In arid and semi-arid regions, the earliest irrigation systems permitted the establishment of civilizations that were stable for centuries. The ready availability of energy from falling water helped to power the industrial revolutions of the 1700s and 1800s. The Green Revolution of the 20th century, which permitted us to feed our growing population in the 1960s, 1970s, and 1980s, depended in large part on the availability of fresh water for the dramatic expansions of irrigated agriculture.

As we now look to the 21st century, several challenges face us. Foremost among them is how to satisfy the food, drinking water, sanitation, and health needs of ten or twelve or fifteen billion people, when we have failed to do so in a world of five billion. In 1990, 1,230 million people lacked access to clean drinking water, and 1,740 million people lacked access to adequate sanitation services. The United Nations estimates that by the turn of the century, population growth alone will increase these numbers by nearly 900 million people, and we have no clear plan for how

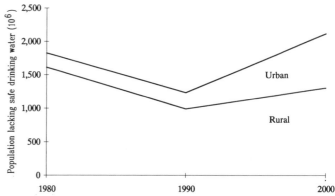

Fig. 1–11. Population lacking access to safe drinking water in developing countries, 1980–2000. This graph shows the estimated number of people in developing countries without access to fresh water for 1980, 1990, and 2000. In most cases, the total population with access to safe water increased between 1980 and 1990, but population growth, particularly in urban areas, erased any substantial gain. Between 1990 and 2000, an additional 900 million people are projected to be born in regions without safe drinking water. See Tables C.1–C.4 for more data.

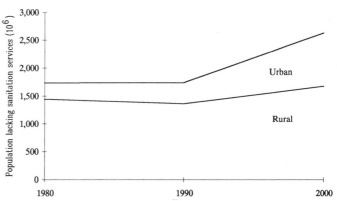

Fig. 1–12. Population lacking access to sanitation services in developing countries, 1980–2000. This graph shows the estimated number of people in developing countries without access to basic sanitation services for 1980, 1990, and 2000. In most cases, the total population with access to these services increased between 1980 and 1990, but population growth, particularly in urban areas, erased any substantial gain. Between 1990 and 2000, an additional 900 million people are projected to be born in regions without the basic services. See Tables C.1–C.4 for more data.

to satisfy these additional water needs. Figures 1.11 and 1.12 show United Nations estimates of the populations in developing countries without adequate sanitation services or access to clean drinking water in 1980, 1990, and 2000.

Additional water will also be required for new industrial development, new energy projects, and expanded agricultural efforts. Yet the total amount of irrigated land per person has already begun to fall, and is poised to fall even farther and faster as population growth continues to outpace efforts to bring new land – while maintaining the quality of existing land – under irrigation. Despite all of our efforts, we are falling farther and farther behind. Meeting these human needs without conflicts over resources will take considerable ingenuity and commitment. Baba Amte wrote in 1989, regarding the divisive Narmada water project in India, "The government has completely lost track of what must be regarded as its basic objective: finding the best possible way of providing water to the people – a path, moreover, not laden with blood and tears, sorrow and suffering."[19]

Adding to our difficulty, however, is a problem of uncertain but potentially enormous magnitude: the alteration of the earth's atmosphere, including both the destruction of the ozone layer and global climatic change. Human beings have become such a large force on earth that we are now able to manipulate, albeit unintentionally, the very atmosphere that permits life on earth and that shapes our lives, our civilizations, and our environment. The ozone layer in the upper reaches of the earth's atmosphere screens out the devastating ultraviolet radiation streaming out from the sun. Scientists now believe that one reason life has not been found on the surface of Mars has to do with the absence of such a screen in the Martian atmosphere. Any earth-like microbe would die instantly if exposed to the ultraviolet radiation that reaches all the way to the surface of Mars. Yet we are now destroying our own protective ozone shield.

Equally important, yet even more complex, is our climate system, which provides our fresh water, permits us to grow crops, defines where and how we live, and supports the natural ecosystems upon which we depend for clean air, food, medicines, and other goods and services. The earth's climate is its most complex geophysical characteristic, transferring vast quantities of heat and water from one region to another, driving the enormous ocean currents and atmospheric winds, and determining the extent and severity of the tropical storms, hurricanes, and monsoons, which bring both death and life to many regions. Of all natural disasters, floods and tropical cyclones are the two largest killers. Ninety-five per cent of these deaths occur in developing countries. The Huanghe (Yellow River) in China carries more silt than any other river in the world except the Brahmaputra, which builds up banks and worsens the risks of major flooding. Floods along the Huanghe have killed more people than any other single feature of the earth. In 1887 between 900,000 and 2 million people died from flooding or subsequent starvation; in 1931, between 1 million and 3.7 million people died.[20] The intensification of the hydrologic cycle, expected due to global climatic changes, may worsen some of these problems.

We now know that we are on the verge of changing our climate in an uncontrolled manner through the emissions of trace gases from our burning of fossil fuels and the tropical forests. The scientific community believes that global climatic changes are in the offing unless we quickly begin to reduce our emissions of these gases. Indeed, many scientists believe that even quick actions to reduce emissions will not be sufficient to avoid at least some climatic changes.[21]

Among the expected changes are higher global and regional temperatures, increases in global average precipitation, changes in the regional patterns of rainfall, snowfall, and snowmelt, changes in the intensity, severity, and timing of major storms, and a wide range of other geophysical effects. These changes will have many secondary impacts on fresh water resources, altering both the demand and supply of water, and changing its quality.[22]

Many uncertainties remain about the timing, direction, and extent of these climatic changes, as well as about their societal implications. These uncertainties greatly complicate rational water resource planning for the future, and have contributed to the intense political debate over how to respond to the problem of climate change.

Until recently, fresh water has been considered subsidiary to today's other major environmental problems, such as climate change, food production, forest destruction, and energy development. The Brandt Commission report, published in 1980, has only one page devoted to water resources.[23] The work of the World Commission on Environment and Development (the Brundtland Commission), which has played such an important role in bringing international environment and development issues onto the political agenda, included separate sections on population, food, species and ecosystems, energy, industry, the urban challenge, peace and security, and even Antarctica, but no section on water resources.[24] Part of the problem is that no single international agency is responsible for water resources. Even at the national level, rarely is one agency or ministry responsible for water; rather water issues are often distributed widely among resource and mining agencies, agricultural agencies, or environmental protection ministries. But these problems also stem from our failure to see the connections among environmental problems. We treat individual waterborne diseases, like cholera or malaria, rather than addressing questions of widespread poverty, adequate water supply, illiteracy, and community health. We try to preserve selected species of endangered fish rather than the entire ecosystem that is being modified or destroyed. We build massive hydroelectric facilities without

proper attention to the social, cultural, and ecological ramifications that such projects inevitably bring.

The water problems facing us as the new millennium begins can be solved if we can muster the foresight to deal with long-term environmental problems, the political will to take actions today to reduce the risk of future catastrophes, and the willingness to invest in our future. Whether we will do so remains to be seen.

## Notes

1. For an historical overview of the role of water in early religious myths, see D. Boorstin, 1983, *The Discoverers*, Random House, Inc., New York.
2. See, for example, UNESCO, 1978, World Water Balances and Water Resources of the Earth, USSR Committee for the International Hydrological Decade (English translation of a 1974 publication), Paris; M.I. L'vovich, 1979, *World Water Resources and their Future* (English translation by R. Nace of a 1974 USSR publication, Mysl' Publishing, Moscow), American Geophysical Union, Washington DC; A. Baumgartner and E. Reichel, 1975, *The World Water Balance*, Elsevier Scientific Publishing Co., Munich; and R. Nace, 1967, *Are we running out of water?*, U.S. Geological Survey Circular no. 536, Washington, DC.
3. U.S. Department of Commerce, 1977 and updates, *Climates of the World*, Environmental Science Service Administration, Washington, DC.
4. Described by R. Clarke, 1991, *Water: The International Crisis*, Earthscan Publications Ltd, London.
5. This incident, and the mapping method Snow used to identify the source of the outbreak, are neatly described in E.R. Tufte, 1983, *The Visual Display of Quantitative Information*, Graphics Press, Cheshire, CT.
6. P.B. Moyle and J.J. Cech, Jr., 1982, *Fishes: An Introduction to Ichthyology*, Prentice-Hall, Englewood Cliffs, NJ, pp. 1–593; D.H. Stansbery, 1973, Why preserve rivers?, *The Explorer* **15**(3), 14–16; R.J. Wootton, 1990, *Ecology of Teleost Fishes*, Chapman and Hall, New York, pp. 1–404.
7. B. Herbold, A.D. Jassby and P.B. Moyle, 1992, *Status and Trends Report on Aquatic Resources in the San Francisco Estuary*, San Francisco Estuary Project, San Francisco.
8. P.H. Gleick and L. Nash, 1991, *The Environmental and Societal Costs of the Continuing California Drought: A Report to Congress*, Pacific Institute for Studies in Development, Environment, and Security, Berkeley, California.
9. B. Herbold, A.D. Jassby, and P.B. Moyle, 1992, *Status and Trends Report on Aquatic Resources in the San Francisco Estuary*, San Francisco Estuary Project, San Francisco; P.H. Gleick and L. Nash, 1991, *The Environmental and Societal Costs of the Continuing California Drought: A Report to Congress*, Pacific Institute for Studies in Development, Environment, and Security, Berkeley, California.
10. M. Salikh, Uzbek poet, quoted in W.S. Ellis, 1990, The Aral: A Soviet sea lies dying, *National Geographic*, **177**(2), 73–92, The National Geographic Society, Washington, DC.
11. See G.C. Taylor Jr., 1965, Water, history, and the Indus plain, *Natural History*, **74**, p. 40; H. Rouse and S. Ince, 1957, *History of Hydraulics*, Iowa Institute of Hydraulic Research, State University of Iowa.
12. G.V. Skogerboe, 1985, Problems of water reallocation in the Colorado – America's most fully utilized river, in J. Lundqvist, U. Lohm and M. Falkenmark (eds.) *Strategies for River Basin Management*, D. Reidel Publishing Co., Dordrecht, pp. 219–227.
13. W. Stegner, 1954, *Beyond the Hundredth Meridian: John Wesley Powell and the Second Opening of the West*, University of Nebraska Press, Lincoln, NB.
14. This quote is from Plato's *Critias*, and is cited in R. Clarke, 1991, *Water: The International Crisis*, Earthscan Publications Ltd, London. Pliny was one of the earliest writers to describe the natural environment, and he was killed during an eruption of Mount Vesuvius, while trying to learn more about volcanoes. Unfortunately, few of his writings have survived.
15. K. Wangnick, 1990, *1990 IDA Worldwide Desalting Plants Inventory Report No. 11*, Wangnick Consulting, Gnarrenburg, Germany.
16. Total global water withdrawals are estimated to be 3,240 km³ per year. The total annual production of desalinated water in 1989 was approximately 4.8 km³ (see Part II, Section H).
17. W. Stegner, 1954, *Beyond the Hundredth Meridian: John Wesley Powell and the Second Opening of the West*, University of Nebraska Press, Lincoln, NB.
18. See, for example, the index of the FAO, 1978 (as updated), *Systematic Index of International Water Resources Treaties, Declarations, Acts and Cases by Basin*, Legislative Study no. 15, FAO, Rome. This index has been

updated several times by the United Nations and, among many other treaties, describes a grant of freedom of navigation to a monastery in the year
805 and a bilateral treaty concerning the Weser River, which flows through
present-day Germany, of 2 October 1221.

19.   B. Amte, 1989, *Cry, the Beloved Narmada*, Maharogi Sewa Samiti,
      Chandrapur, Maharashtra, India.

20.   A. Wijkman and L. Timberlake, 1984, *Natural Disasters: Acts of God or
      Acts of Man?*, Earthscan Publications Ltd, London; Adrian T. McDonald
      and David Kay, *1988, Water Resources: Issues and Strategies*, Longman
      Scientific and Technical Publishers Ltd, London.

21.   The scientific consensus around these issues, as well as a discussion of the
      uncertainties, can be found in the extensive report of the Intergovernmental
      Panel on Climate Change, 1990, *Climate Change: The IPCC Scientific
      Assessment*, Cambridge University Press, Cambridge. See also I.M.
      Mintzer (ed.), 1992, *Confronting Climate Change: Risks, Implications and
      Responses*, Cambridge University Press, Cambridge.

22.   Some of these effects are described in detail in P.H. Gleick, 1989, Climate
      change, hydrology, and water resources, *Reviews of Geophysics*, **27**(3),
      329–344; P.H. Gleick, 1992, The effects of climate change on shared fresh
      water resources, in I.M. Mintzer (ed.) *Confronting Climate Change: Risks,
      Implications and Responses*, Cambridge University Press, Cambridge, pp.
      127–140.

23.   Independent Commission on International Development Issues (The
      Brandt Commission), 1980, *North–South: A Programme for Survival: The
      Report of the Independent Commission on International Development Is-
      sues*, Pan Books, London.

24.   The World Commission on Environment and Development, 1987, *Our
      Common Future*, Oxford University Press, Oxford.

# Chapter 2
# World fresh water resources

Igor A. Shiklomanov

## Introduction

The rational use and protection of water resources and supplying human-kind with adequate clean fresh water, or, to be concise, "water problems," are among today's most acute and complex scientific and technical problems. They increasingly reach beyond national and regional borders and are becoming global in nature. Shortages of fresh water and the increasing pollution of water bodies are becoming limiting factors in the economic and social development of many countries throughout the world, even countries not located in arid zones. Under these conditions the reliable assessment of water resources is extremely important, particularly the quantitative estimation and calculation of annual stream flows and their fluctuations in time and space. This is due to the fact that the mean annual river runoff and annually renewable ground water resources almost everywhere support the bulk of water consumption and determine the available water supply for a given region.

Scientists from around the world have prepared assessments of global water resources and their use. Information on water resources and the water balance, water use, and the impact of economic activity on water resources is available from several publications prepared over the last 20–25 years.[1] Additional information can be found in the tables in Part II, and in the references described there. More recently, three major monographs on global problems of water resources and the water balance have been published.[2] The conclusions presented in this chapter are based primarily on the results of the assessments given in the cited works, which remain pertinent today.

## World water stocks

Reliable estimates of water resources stored in various water bodies and in different physical states are critical to a clear understanding of the natural water cycle and the effect that human activities might have. Data on global water resources collected by Soviet scientists are presented in Table 2.1.

It should be noted that the data on the amount of water on earth (as the authors of the cited monograph themselves note) should not be considered very accurate; they are only approximations of the actual values. The data on water stored in underground ice in permafrost regions, on the amount of soil moisture, and on water stored in bogs and marshes, which were obtained computationally under fairly crude assumptions, are especially rough. At the same time, much more reliable estimates are now available of water stored in the oceans, in lakes and reservoirs, in polar ice, and in mountain glaciers, and of stocks of fresh and saline ground water.

According to the data in Table 2.1, the total volume of (long-term) fresh water stocks is 35 million km$^3$, or just 2.5% of the total stock of water in the hydrosphere. A large fraction of the fresh water (24 million km$^3$, or 68.7%) is in the form of ice and permanent snow cover in the Antarctic and Arctic regions. Fresh water lakes and rivers, which are the main sources for human water consumption, contain on average about 90,000 km$^3$ of water, or just 0.26% of total global fresh water reserves.

The total area of all fresh water lakes in the world is about 1.5 million km$^2$; basic morphometric data on the 28 largest fresh water lakes in the

**TABLE 2.1** Water reserves on the earth

| | Distribution area (10$^3$ km$^2$) | Volume (10$^3$ km$^3$) | Layer (m) | Percentage of global reserves Of total water | Of fresh water |
|---|---|---|---|---|---|
| World ocean | 361,300 | 1,338,000 | 3,700 | 96.5 | – |
| Ground water | 134,800 | 23,400 | 174 | 1.7 | – |
| Fresh water | | 10,530 | 78 | 0.76 | 30.1 |
| Soil moisture | | 16.5 | 0.2 | 0.001 | 0.05 |
| Glaciers and permanent snow cover | 16,227 | 24,064 | 1,463 | 1.74 | 68.7 |
| Antarctic | 13,980 | 21,600 | 1,546 | 1.56 | 61.7 |
| Greenland | 1,802 | 2,340 | 1,298 | 0.17 | 6.68 |
| Arctic islands | 226 | 83.5 | 369 | 0.006 | 0.24 |
| Mountainous regions | 224 | 40.6 | 181 | 0.003 | 0.12 |
| Ground ice/permafrost | 21,000 | 300 | 14 | 0.022 | 0.86 |
| Water reserves in lakes | 2,058.7 | 176.4 | 85.7 | 0.013 | – |
| Fresh | 1,236.4 | 91 | 73.6 | 0.007 | 0.26 |
| Saline | 822.3 | 85.4 | 103.8 | 0.006 | – |
| Swamp water | 2,682.6 | 11.47 | 4.28 | 0.0008 | 0.03 |
| River flows | 148,800 | 2.12 | 0.014 | 0.0002 | 0.006 |
| Biological water | 510,000 | 1.12 | 0.002 | 0.0001 | 0.003 |
| Atmospheric water | 510,000 | 12.9 | 0.025 | 0.001 | 0.04 |
| Total water reserves | 510,000 | 1,385,984 | 2,718 | 100 | – |
| Total fresh water reserves | 148,800 | 35,029 | 235 | 2.53 | 100 |

**TABLE 2.2** Large fresh lakes of the world (with surface area greater than 5,000 km²)

| Lake | Area (km²) | Volume (km³) | Maximum depth (m) | Continent |
|---|---|---|---|---|
| Superior | 82,680 | 11,600 | 406 | North America |
| Victoria | 69,000 | 2,700 | 92 | Africa |
| Huron | 59,800 | 3,580 | 299 | North America |
| Michigan | 58,100 | 4,680 | 281 | North America |
| Tanganyika | 32,900 | 18,900 | 1,435 | Africa |
| Baikal | 31,500 | 23,000 | 1,741 | Asia |
| Nyasa | 30,900 | 7,725 | 706 | Africa |
| Great Bear | 30,200 | 1,010 | 137 | North America |
| Great Slave | 27,200 | 1,070 | 156 | North America |
| Erie | 25,700 | 545 | 64 | North America |
| Winnipeg | 24,600 | 127 | 19 | North America |
| Ontario | 19,000 | 1,710 | 236 | North America |
| Ladoga | 17,700 | 908 | 230 | Europe |
| Chad | 16,600 | 44.4 | 12 | Africa |
| Maracaibo | 13,300 | – | 35 | South America |
| Tonlé Sap | 10,000 | 40 | 12 | Asia |
| Onega | 9,630 | 295 | 127 | Europe |
| Rudolf | 8,660 | – | 73 | Africa |
| Nicaragua | 8,430 | 108 | 70 | North America |
| Titicaca | 8,110 | 710 | 230 | South America |
| Athabasca | 7,900 | 110 | 60 | North America |
| Reindeer | 6,300 | – | – | North America |
| Tung Ting | 6,000 | – | 10 | Asia |
| Vänern | 5,550 | 180 | 100 | Europe |
| Zaisan | 5,510 | 53 | 8.5 | Asia |
| Winnipegosis | 5,470 | 16 | 12 | North America |
| Albert | 5,300 | 64 | 57 | Africa |
| Mweru | 5,100 | 32 | 15 | Africa |

world, each with an area of more than 5,000 km², are presented in Table 2.2. Additional data on lakes can be found in Table B.10. These lakes together account for approximately 85% of the volume and 40% of the water surface area of all fresh water lakes on earth.

The largest accumulations of fresh water lakes are concentrated in regions of ancient and recent glaciations and in regions of large tectonic fractures in the earth's crust. The Great Lakes in North America form the largest lake complex. Lake Baikal, in the former Soviet Union, which contains approximately 25% of global lacustrine fresh water, is the largest in volume.

The creation of artificial lakes and reservoirs is of great importance to the use of water resources and the control of river flow. The earliest water reservoirs were built on rivers thousands of years ago, in the heyday of ancient civilizations, but they have become global-scale objects only within the last few decades. The total volume of reservoirs in the world increased nearly tenfold from 1951 to 1980; at present it exceeds 5,000 km³. The total surface area of these reservoirs is more than 400,000 km².[3] The largest volume of water in artificial reservoirs is found in three countries: the United States, the former Soviet Union, and Canada, where approximately half the total volume of reservoirs in the world is concentrated.

The largest reservoirs in the world by volume are Owen Falls off Lake Victoria in Uganda (205 km³), the Kariba reservoir on the Zambesi river (182 km³), and the Bratsk reservoir on the Angara river (169 km³). Among the largest reservoirs by surface area are the Volta in Ghana (8,500 km²) and the Kuybyshev reservoir in the former Soviet Union (6,500 km²). According to long-range plans that exist in many countries, the total volume of the world's reservoirs may reach 7,000–7,500 km³ by the end of this century.

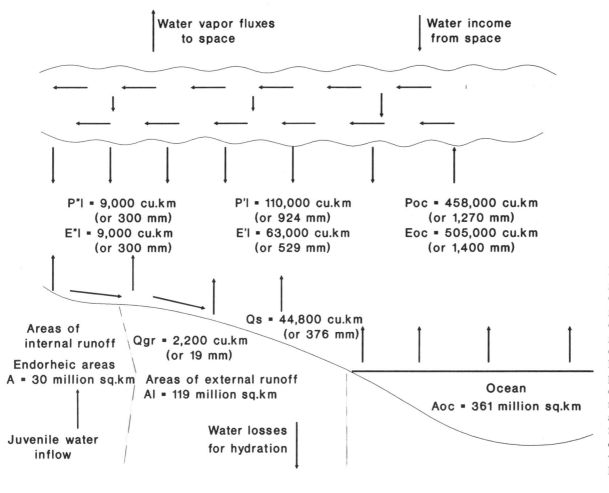

Fig. 2–1. The global water circulation. A generalized description of the global hydrologic cycle is shown here, including precipitation (P), evaporation (E), runoff (Q), and areas (A). Subscripts: oc, oceans; gr, ground water; s, surface; l, land.

**TABLE 2.3** Water balance of the land

| Continent | Precipitation (mm) | Precipitation (km³) | Evaporation (mm) | Evaporation (km³) | Runoff (mm) | Runoff (km³) |
|---|---|---|---|---|---|---|
| Europe | 790 | 8,290 | 507 | 5,320 | 283 | 2,970 |
| Asia | 740 | 32,200 | 416 | 18,100 | 324 | 14,100 |
| Africa | 740 | 22,300 | 587 | 17,700 | 153 | 4,600 |
| North America | 756 | 18,300 | 418 | 10,100 | 339 | 8,180 |
| South America | 1,600 | 28,400 | 910 | 16,200 | 685 | 12,200 |
| Australia and Oceania | 791 | 7,080 | 511 | 4,570 | 280 | 2,510 |
| Antarctica | 165 | 2,310 | 0 | 0 | 165 | 2,310 |
| Land as a whole | 800 | 119,000 | 485 | 72,000 | 315 | 47,000 |
| Areas of external runoff | 924 | 110,000 | 529 | 63,000 | 395 | 47,000[a] |
| Areas of internal runoff | 300 | 9,000 | 300 | 9,000 | 34 | 1,000[b] |

[a] Including underground water not drained by rivers.
[b] Lost in the region through evaporation.

## The hydrologic cycle and the water balance

All types of water on earth interact closely as water passes from one form to another and moves from the ocean to land and back under the influence of solar energy and gravity.

All types of natural waters are renewed annually, but the rates of renewal differ sharply. Water present in rivers is completely renewed every 16 days on average and water in the atmosphere is renewed every 8 days, but the renewal periods of glaciers, ground water, ocean water, and the largest lakes run to hundreds or thousands of years. When slowly renewed resources are used by humans at a rapid rate, they effectively become non-renewable resources with subsequent disruptions of the natural cycle.

The main link in the water cycle in nature is exchange between the oceans and land, which includes not only quantitative renewal but – what is especially important – qualitative restoration as well. An overall diagram of the global hydrologic cycle with its main quantitative indices is presented in Figure 2.1.[4]

An enormous amount of water, equal to about 505,000 km³, or a layer 1,400 mm thick according to current estimates, evaporates annually from the ocean surface. Of this amount, nearly 90% (458,000 km³ per year) returns to the ocean in the form of atmospheric precipitation that falls on the ocean surface, and 10% (50,500 km³ per year) falls on to dry land. Together with atmospheric precipitation of local origin (68,500 km³ per year), the total precipitation falling on dry land and supplying all types of land water is 119,000 km³ per year, or on average about 1,000 mm per year. Of this amount, 47,000 km³ per year (about 35%) is returned to the oceans in the form of river, ground, and glacial runoff. On the whole, an average of 577,000 km³ of precipitation, or 1,130 mm, falls annually on the earth's surface (this is 177,000 km³ per year more than the value used in the hydrometeorological literature prior to 1965). Approximately the same amount of water evaporates annually from the ocean surface and dry land, i.e., the world water balance may be considered closed. At present, there are no data whatever to indicate a significant one-way outflow of water vapor from the earth's atmosphere into outer space. Similarly, an inflow of juvenile water from the earth's interior as a result of outgassing of the mantle probably does not play a significant part in the present world water balance as a whole. Although these processes do take place in some geologically active zones, there is every reason to believe that the influx of juvenile water is compensated by corresponding water consumption for hydration. Rounded data on the water balance of continents and for dry land as a whole are presented in Table 2.3.[5]

It should be noted that in addition to the quantitative characteristics of the components of the world water balance mentioned above, which are based on research conducted by Soviet scientists, other data that differ substantially can also be found in the hydrometeorological literature. Other estimates are presented in the Tables of Part II, Section A.[6]

## Natural river flow and water supply

The main source of fresh water is surface runoff, which is used extensively to satisfy widely varying human needs. The most important feature of runoff compared with other natural resources is its annual renewal in the course of the hydrologic cycle in nature. Runoff is a dynamic part of long-term water reserves, and serves as a characteristic of potential renewable water resources, not only of rivers proper but also of lakes, reservoirs, and glaciers. The stable portion of runoff (about 25% of the total) is at the same time an index of potential renewable ground water resources (excluding deep subsurface waters that are located below the active zone and that directly enter the ocean and interior closed water bodies).

The total global runoff averages (excluding the ice flow of Antarctica) 44,500 km³ per year (Table 2.4). Of this volume, 43,500 km³ per year enters the oceans directly. Runoff to interior hydrologically-closed regions (such as the basins of the Caspian Sea and the Aral Sea, the lakes of Chad, the Great Salt Lake, and so on), where the stream flow goes entirely to evaporation, account for about 1,000 km³ per year, or approximately 2.4%. It is important to emphasize that from the standpoint of practical use, the stream flows of exterior and interior (closed) regions differ significantly. The use of the flow from exterior regions has little effect on the oceans, entails essentially no pronounced changes in their water–salt balance, and can affect to one degree or another only the regime of marginal and interior seas (the Baltic Sea, Black Sea, etc.). The (unrecoverable) use of the runoff of interior regions can lead to a fundamental change in the water–salt balance of water bodies into which the streams flow (the Caspian Sea, Aral Sea, etc.).

It is well known that global runoff is distributed extremely unevenly – a fact apparent from a simple comparison of runoff by continent (Table 2.4). More than half the global runoff occurs in Asia and South America (31% and 25%, respectively), while Europe accounts for just 7% and Australia for only 1%. Most of the runoff (over 80%) is concentrated in the northern and equatorial zones, which have fairly small populations.

**TABLE 2.4** River runoff resources in the world

| Territory | Annual river runoff (mm) | Annual river runoff (km³) | Portion of total runoff (%) | Area (10³ km²) | Specific discharge (l/s/km²) |
|---|---|---|---|---|---|
| Europe | 306 | 3,210 | 7 | 10,500 | 9.7 |
| Asia[a] | 332 | 14,410 | 31 | 43,475 | 10.5 |
| Africa[b] | 151 | 4,570 | 10 | 30,120 | 4.8 |
| North America[c] | 339 | 8,200 | 17 | 24,200 | 10.7 |
| South America | 661 | 11,760 | 25 | 17,800 | 21.0 |
| Australia[d] | 45 | 348 | 1 | 7,683 | 1.44 |
| Oceania | 1,610 | 2,040 | 4 | 1,267 | 51.1 |
| Antarctica | 160 | 2,230 | 5 | 13,977 | 5.1 |
| Total land area | 314 | 46,770 | 100 | 149,000 | 10.0 |

[a] Asia includes Japan, the Philippines, and Indonesia.
[b] Africa includes Madagascar.
[c] North and Central America.
[d] Australia includes Tasmania.

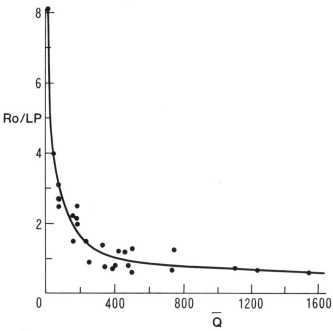

Fig. 2–2. Long-term runoff by region. The dependence of mean long-term river runoff ($Q$) is plotted for physiographic and economic regions of the world as a function of the aridity index ($R_0/(LP)$).

**TABLE 2.5** Large rivers of the world (with mean annual runoff greater than 200 km³)

| River | Average runoff (km³/ yr) | Area of basin (10³ km²) | Length (km) | Continent |
|---|---|---|---|---|
| Amazon | 6,930 | 6,915 | 6,280 | South America |
| Congo | 1,460 | 3,820 | 4,370 | Africa |
| Ganges (with Brahmaputra | 1,400 | 1,730 | 3,000 | Asia |
| Yangzijiang | 995 | 1,800 | 5,520 | Asia |
| Orinoco | 914 | 1,000 | 2,740 | South America |
| Paraná | 725 | 2,970 | 4,700 | South America |
| Yenisei | 610 | 2,580 | 3,490 | Asia |
| Mississippi | 580 | 3,220 | 5,985 | North America |
| Lena | 532 | 2,490 | 4,400 | Asia |
| Mekong | 510 | 810 | 4,500 | Asia |
| Irrawaddy | 486 | 410 | 2,300 | Asia |
| St. Lawrence | 439 | 1,290 | 3,060 | North America |
| Ob | 395 | 2,990 | 3,650 | Asia |
| Chutsyan | 363 | 437 | 2,130 | Asia |
| Amur | 355 | 1,855 | 2,820 | Asia |
| Mackenzie | 350 | 1,800 | 4,240 | North America |
| Niger | 320 | 2,090 | 4,160 | Africa |
| Columbia | 267 | 669 | 1,950 | North America |
| Magdalena | 260 | 260 | 1,530 | South America |
| Volga | 254 | 1,360 | 3,350 | Europe |
| Indus | 220 | 960 | 3,180 | Asia |
| Danube | 214 | 817 | 2,860 | Europe |
| Salween | 211 | 325 | 2,820 | Asia |
| Yukon | 207 | 852 | 3,000 | North America |
| Nile | 202 | 2,870 | 6,670 | Africa |

In the temperate zones (the southern forest zone, the forest–steppe zone, the steppe zone) that are most suitable for human life and activities (especially for agricultural production), where most of the earth's population lives, flowing water resources are extremely limited.

On the continents and in large regions of the earth, the distribution of water resources is still more uneven. An analysis of the shortfall and excess of water resources in the world shows that within the confines of each continent there are vast regions with arid climate and regions with limited water resources; these regions occupy 33% of Europe, 60% of Asia, a large fraction of Africa, the south-western regions of North America, about 30% of South America, and the overwhelming majority of Australia. On the other hand, on all the continents listed there are wet regions with abundant water resources.[7]

With respect to the major natural and economic regions of the earth, it is not hard to identify quantitative relationships between total runoff and climatic factors. For instance, such a relationship is presented in Figure 2.2. This relationship shows the close dependence between the total perennial stream flow layer $Q_{mm}$ determined for major natural and economic regions of the earth, on a complex climatic parameter expressed in the form of the "dryness index":

$$R_0/(LP)$$

where $R_0$ is the radiation balance of the wet surface, $L$ is the specific heat of evaporation, and $P$ is precipitation. This index has been determined on an approximate basis for each region by using very detailed global charts of the radiation balance and precipitation.[8] The names of the regions and the values of the runoff layer and dryness index for each are presented in Table 2.9.

According to Figure 2.2, the smallest value of a region's natural water resources corresponds to the largest dryness index; conversely, the largest flow layer is associated with regions with minimum values of the dryness index.

The development of the stream network and the water capacity of rivers depend on climatic factors, ruggedness of terrain, and geologic structure. The main characteristics of the world's largest rivers with an average flow of more than 200 km³ per year are presented in Table 2.5. The Amazon, the largest river in the world, carries more than 15% of the annual global runoff, and the total annual flow of all rivers presented in Table 2.5 is approximately 40% of global runoff.

In most regions on earth the values of natural annual flow are not a realistic index of water supply, since runoff is distributed quite unevenly throughout the year. Most of it (60%–70%) occurs in the flood (high water) period. Naturally, for the continents as a whole total stream flow varies significantly throughout the year; these variations are shown by month in Figure 2.3. Most of the runoff (48%) in Europe occurs in April through July, in Asia in May through October (80%), in Africa in January through June (74%), in North America in May through August (54%), in South America in March through September (70%), and in Australia in January through March (68%). For all dry land as a whole, the highest water period extends from May to October, when the rivers transport 63% of the annual flow.

Naturally, for regions that are small in area, the non-uniformity of the stream flow increases sharply in both time (from year to year and within the year) and space. The values of the variability generally are largely dependent on total moisture and water resources. The higher the dryness index of a region, the greater the variability of water resources from year to year and by season and the more unevenly they are distributed in space.

To estimate the quantitative characteristics of the water resources of a given country or region, relative or specific indices of potential water supply are usually used in addition to absolute indices (such as km³ per year or in a season). Relative or specific indices represent the volume of annual (or seasonal) runoff per unit area (for example km³ per year per km²) or per population (m³ per year per capita).

Among the countries of the world, first place in absolute volume of runoff belongs to Brazil, whose water resources exceed 20% of all global renewable water resources (Table 2.6). This is nearly twice as much as the water resources of the former Soviet Union, which is second among the other countries (10.6%). China (5.7%) and Canada (5.6%) hold third and fourth place.

In terms of specific water supply, Norway is first both per unit area (1,250,000 m³ per year [per km²]) and per capita (98,800 m³ per year per capita). For comparison, the potential specific water supply per unit area is 213,000 m³ per year per km² in the former Soviet Union, and 207,000 m³ per year per km² in the United States, or approximately one-sixth as much as in Norway. Some smaller countries, such as Iceland, exceed even

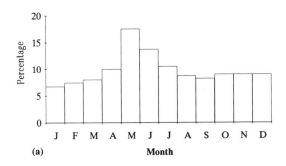

(a)   **Month**

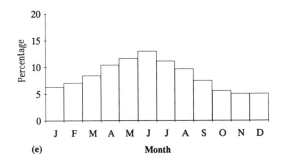

(e)   **Month**

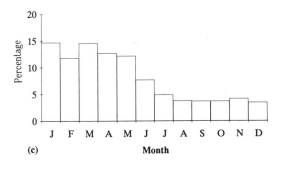

(b)   **Month**

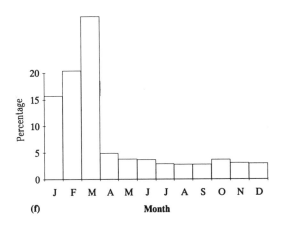

(f)   **Month**

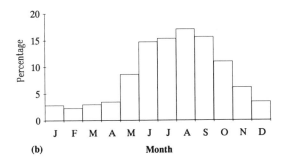

(c)   **Month**

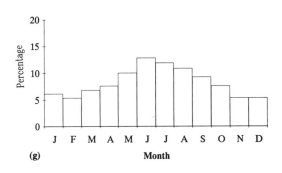

(g)   **Month**

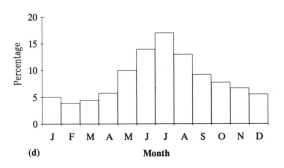

(d)   **Month**

Fig. 2–3. Average monthly runoff as a fraction of annual runoff, by continent. Variations of total river runoff of the continents during a year: (a) Europe, (b) Asia, (c) Africa, (d) North America, (e) South America, (f) Australia, (g) total land area.

**TABLE 2.6**   Water availability in some countries, late 1980s

| Country | Area ($10^3$ km$^2$) | Population ($10^6$) | Runoff total (km$^3$) | Runoff per unit area ($10^3$ m$^3$ per km$^2$) | Runoff per capita ($10^3$ m$^3$ per person) | Percentage of global runoff |
|---|---|---|---|---|---|---|
| Brazil | 8,512 | 129.9 | 9,230 | 1,084 | 71.1 | 20.7 |
| USSR | 22,274 | 275 | 4,740 | 213 | 17.2 | 10.6 |
| China | 9,561 | 1,024 | 2,550 | 267 | 2.49 | 5.7 |
| Canada | 9,976 | 24.9 | 2,470 | 248 | 99.2 | 5.6 |
| India | 3,288 | 718 | 1,680 | 511 | 2.34 | 3.8 |
| United States | 9,363 | 234.2 | 1,940 | 207 | 8.28 | 4.4 |
| Norway | 324 | 4.1 | 405 | 1,250 | 98.8 | 0.9 |
| Yugoslavia | 256 | 22.8 | 256 | 1,000 | 11.2 | 0.6 |
| France | 544 | 54.6 | 183 | 336 | 3.35 | 0.4 |
| Finland | 337 | 4.9 | 110 | 326 | 22.4 | 0.2 |
| World total[a] | 134,800 | 4,665 | 44,500 | 330 | 9.54 | 100 |

The header "Long-term mean annual river runoff" spans the Runoff total, Runoff per unit area, Runoff per capita, and Percentage of global runoff columns.

[a] World total does not include Antarctica.

the values of Norway, while many countries have far fewer water resources per capita or per unit area.

And these numbers are not fixed. As we know, the earth's population is growing rapidly. Whereas it was 1,170 million in 1850, today it already exceeds 5,400 million, and according to demographic forecasts it will exceed 6,000 million by the turn of the century. Population has been growing especially quickly since the 1950s. Population growth is accompanied by a progressive reduction in specific per capita water supply in the world, which has decreased from 33,300 m$^3$ per year per capita in 1850 to 8,500 m$^3$ per year per capita today. A change in specific water supply due to human impact on water resources and on the quality of natural water is occurring to no less a degree in many parts of the world.

## Impact of economic activities on water resources

The problem of studying water resources includes not only an assessment of their natural state, territorial distribution, and fluctuations in time, but also of changes due to human economic activities. Despite the ability of stream flow to renew and self-purify, in recent decades the intensive development of industry and agriculture throughout the world, population growth, the opening of new territories, the associated sharp increase in water withdrawals on all continents (except most recently in the United States and parts of Europe), and the transformation of the earth's natural cover have begun to exert a significant impact on the natural fluctuations of the stream flow and the state of fresh water resources.

In the regions that have been most developed to date, no major river systems remain with a regime that has not been disturbed by human activities to one degree or another.

Within major river drainage basins and vast territories located in regions that have been most developed economically, runoff is usually affected by a host of anthropogenic factors that have various effects on the characteristics of the water regime, total annual flow, and water quality. According to the nature of the effect on hydrologic processes (quantitative characteristics of the regime and the quality of natural waters), factors of economic activities may be combined into the following groups:[9]

1. Factors that principally affect flow due to direct diversions of water from water sources (the stream network, lakes, reservoirs, aquifers), the use of these stocks and flows, and the discharge of water back into the river system (water intakes for irrigation, industrial and municipal water use, agricultural water supply, and runoff diversions).
2. Factors that affect the hydrologic cycle and water resources as a result of direct transformations of the stream network (construction of reservoirs and ponds, damming and straightening of channels, excavation of earth from river channels, etc.).
3. Factors that alter the conditions of formation of flow and other

components of the water balance by affecting the surface of drainage basins (agrotechnical measures, drainage of swamps and marshlands, cutting and planting of trees, urbanization, etc.).
4. Factors of economic activities that affect the flow, water balance, and hydrologic cycle through alterations of overall climatic characteristics on global and regional scales as a result of anthropogenic modifications of atmospheric gas composition and air pollution, as well as changes in characteristics of the hydrologic cycle due to incremental evaporation resulting from the development of large-scale water management measures.

In examining anthropogenic changes in the characteristics of runoff on the global scale and within continents and major natural and economic regions, it is practically impossible to give a quantitative assessment of the importance of all the factors of economic activity listed above, and indeed there is scarcely any need to do so. We believe that one may full well ignore the effect of factors that act on drainage basins (the third group). These factors have their primary impact on the flow of small and medium-sized drainage basins, and usually not on annual flow but on the distribution within the year, the extreme characteristics of the flow, and water quality. Here, depending on specific geophysical conditions, the anthropogenic factors indicated above usually affect the flow in different directions, i.e., under certain conditions they may even promote some increase in the perennial flow of small and medium-sized rivers by reducing total evaporation.

Of the anthropogenic factors acting in the stream network (the second group) as applicable to the discharge of large basins and regions, we may well confine ourselves to an assessment of the role of large reservoirs; other factors are of local importance, and their impact on the quantitative characteristics of water resources is limited.

Thus, to assess the impact of economic activities on the state of global water resources, we must first take account of anthropogenic factors related to direct water consumption, as well as flow regulation by large reservoirs. These factors, which cause a one-way decrease in surface and subsurface flows, are ubiquitous, develop most intensively, and are capable of exerting an especially large impact on the state of water resources in large regions.

In the past 20 years many attempts have been made in various countries to estimate the size of current and future water withdrawals and consumption throughout the world for various economic needs.

Reliable and detailed data were obtained in 1974 at the State Hydrological Institute of the former Soviet Union and in 1980 by the US Geological Survey.[10] These estimates were made independently, and for this reason comparing them with each other is of particular interest (Table 2.7).

In analyzing the data in Table 2.7, let us note first that the data on total global water withdrawals match very well (to within 5%), and in general, water diversions throughout the world are in quite good agreement for

**TABLE 2.7** Water withdrawals by continents and individual countries, 1975–1977

| Continent and country | Source | Industrial and energy water withdrawals (km³ per year) | Irrigation Area (10⁶ hectares) | Irrigation Water withdrawals (km³ per year) | Commercial-residential water withdrawals (km³ per year) | Total water withdrawals (km³ per year) |
|---|---|---|---|---|---|---|
| Africa | SHI[a] | 6 | 10 | 120 | 6 | 170 |
| | USGS[b] | 15.4 | 6.4 | 60.8 | 12 | 88 |
| Asia | SHI | 80 | 187 | 1,500 | 50 | 1,700 |
| | USGS | 98.7 | 147.4 | 1,400 | 98 | 1,597 |
| Australia and Oceania | SHI | 10 | 1.8 | 14 | 1.5 | 30 |
| | USGS | 13.6 | 1.4 | 13 | 2 | 29 |
| Europe | SHI | 195 | 24 | 150 | 36 | 380 |
| | USGS | 350.6 | 12.2 | 116 | 40 | 516 |
| USSR | SHI | 83 | 14 | 181 | 14 | 290 |
| | USGS | 182.8 | 9.9 | 94 | 18 | 295 |
| North America | SHI | 340 | 27 | 230 | 46 | 640 |
| | USGS | 308.6 | 21.6 | 205 | 38 | 551 |
| United States | SHI | 305 | 21 | 181 | 42 | 540 |
| | USGS | 285.4 | 16.9 | 160 | 32 | 477 |
| South America | SHI | 12 | 8 | 60 | 7 | 90 |
| | USGS | 10.8 | 3.7 | 35 | 11 | 57 |
| World total | SHI | 630 | 260 | 2,100 | 150 | 3,020 |
| | USGS | 887 | 192.7 | 1,830 | 201 | 2,838 |

[a] State Hydrological Institute, forecasts published in 1974.
[b] United States Geological Survey, estimates published in 1980.

individual water consumers (the differences are 13%–25%). The values of total water withdrawals by continent are in fair agreement, with the greatest uncertainties in the data for Africa and South America. There are significant differences (up to a factor of 2) in the assessment of the importance of individual consumers for countries and continents. Unfortunately, comparison of the data presented in these two works is possible only for a few indices – the work of most US hydrologists does not contain any data on consumptive water use or on water losses to incremental evaporation from reservoirs, nor does it give an analysis of the dynamics and trends of global water withdrawals in past periods or for the long term.

The situation is considerably worse for the continents, although some data (including long-term forecasts) are available.[11] Nevertheless, having been calculated 15 years ago, those data undoubtedly need to be refined and updated by using more complete data gathered in recent years. Data on the dynamics of water withdrawals and consumption for individual natural and economic regions of the world were, until recently, almost entirely unavailable. In 1985 new data on these variables were published in a monograph.[12]

The values of water use in various large regions of the earth are determined by three main factors: the level of economic development, population, and the geophysical (especially climatic) peculiarities of the territory in question.

To analyze the temporal and spatial variability of global water withdrawals within each continent, we identified major natural and economic regions characterized by more or less uniform geophysical conditions and a more or less uniform level of development of economic activity; in all, 26 such regions were identified, each continent having between 3 and 8.

For each region estimates were obtained of total water withdrawals and consumptive water use for needs of the urban and rural population, industry (including the heat and power industry), and irrigation, as well as water losses to incremental evaporation from reservoirs. All estimates were made for different years from 1900 to 2000. This made it possible to investigate the dynamics of global water use in space and time throughout the present century, with some extrapolation beyond the year 2000.

In the absence of data on water withdrawals for each large country or for a region as a whole, the calculations made use of indirect methods of estimation, using data on countries with similar geophysical conditions that are close in their level and features of economic development.

Water use by the public was determined separately for cities and inhab-

ited rural points by using actual data available for each country on the dynamics and predicted total size of the urban and rural population, as well as specific water withdrawals per capita, as obtained from similar countries. For extrapolation into the future, allowance was made for trends in the change in per capita water use by the urban and rural populace and for consumptive use of water as a percentage of total water withdrawals.

The estimate of water used for irrigation was based on data on irrigation areas and specific water withdrawals for irrigation that had been obtained for many countries and averaged for certain regions. Irrigation is the primary consumer of water on earth. Therefore, the accuracy of the determination of global water use overall, and especially for continents such as Asia, Africa, and South America, where irrigation determines 70%–90% of total water withdrawals, largely hinges on the accuracy with which irrigated areas are taken into account.

In calculations of irrigated areas for continents and natural and economic regions throughout the world for the period 1990–2000, the author used long-term forecasts and data on irrigation-development programs, which are available for many countries. To estimate the required volumes of water for irrigation needs, we took into account decreases in per capita water consumption because of measures that are being taken to improve the efficiency of production processes, equipment, and methods of irrigation. According to available detailed analyses, recoverable water from irrigation was assumed to be 20%–50% of water diversion (see Chapter 5, this volume).

Industrial water use was calculated on the basis of the dynamics of industrial production in various regions of the earth. Here, available data on the dynamics of this type of water use in the countries listed above, including countries with different levels of economic development that have the most disparate geophysical conditions, were adopted as counterparts. The calculations were done separately for the heat and power industry and for all other branches of industry, which have significantly different trends and rates of development and consumptive use, and then were summed for each region.

Consumptive use of water in the electric utility industry was assumed to be 1%–4%, and in other branches of industry from 10% to 40%, depending on the level of industrial development, the presence of circulating water supply systems, and climatic conditions. In the long term, industrial production (and accordingly water consumption) will develop at a much faster pace in developing countries than in developed ones.

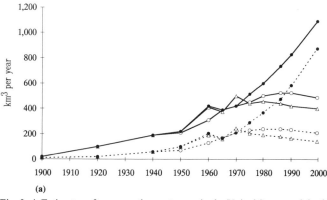
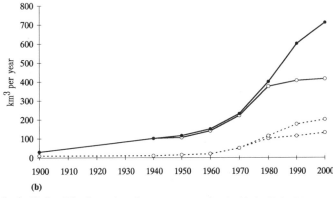

Fig. 2–4. Estimates of consumptive water use in the United States and the former Soviet Union. The dynamics of water consumption in (a) the United States, and (b) the area of the former Soviet Union, computed and predicted in different years in km$^3$ per year. Computations and forecasts: (•), 1965–1970; (Δ), 1977–1981; (O), 1983–1985. Total water consumption is indicated by solid lines; industrial water consumption (including power generation) by broken lines.

Incremental losses to evaporation from reservoirs were calculated for all of the world's largest reservoirs with a volume of more than 5 km$^3$ from the difference in mean evaporation from the water surface and from dry land; here allowance was made for a factor that gives the ratio of the incremental surface area of a reservoir to its total area. The total volume of evaporative loss was computed for each region by adding the data for each large reservoir (larger than 5 km$^3$) and increasing the result by 20%, since reservoirs larger than 5 km$^3$ in volume account for approximately 80% of the total reservoir volume and surface area. For the long term, losses to evaporation from reservoirs in each region were computed, with consideration for the rates of and plans for construction of large reservoirs in different countries and regions and for their geophysical peculiarities.

Before turning to an analysis of water use by region and continent, it would be of interest to consider the dynamics of water use in different countries throughout the world, especially the United States and the former Soviet Union, for which detailed analyses are available (Figure 2.4).

In the United States a detailed estimate and long-range forecasts (for the period 2000–2020) of water required for various economic needs were first carried out in the 1960s.[13] According to predictions from that period, between 1970 and 2000 there would be an increase of 100%–150% in annual fresh water use in the United States (Figure 2.4a) to a total of 850–1,100 km$^3$ per year. Most of the increase was expected to be for water supply for industry and power plant cooling. In the United States, the period since 1975 has, in fact, been characterized by fundamental changes

in the approach to the use of water resources, with a great deal of attention being given to problems of conservation and reuse of water resources, and a transition from extensive to intensive and comprehensive use taking place. All these factors have led to the stabilization of the volume of water withdrawals and have been the basis for a fundamental review of predictions of water needs for the future. Data from actual accounting of water withdrawals attest to this stabilization, beginning in the 1980s, of the amount of fresh water withdrawals in the United States and even to a slight decrease in withdrawals, mainly through a reduction of water use in agriculture, industry, and the electric utility sector (Figure 2.4a).[14]

Analogous trends also hold in the former Soviet Union: whereas in the 1960s and 1970s an increase in water withdrawals to 600–700 km$^3$ per year by the turn of the century was planned for the Soviet Union, present forecasts now postulate a very slight increase to 400–450 km$^3$ per year (Figure 2.4b). It should be noted that in addition to the United States, beginning in the 1970s and 1980s, the volume of total water withdrawals has stabilized in a number of countries in northern and western Europe (Sweden, Great Britain, and the Netherlands, for example), and will even decrease somewhat by the turn of the century.

Despite the progressive trends toward stabilization of water needs that have clearly taken shape in a number of countries, for the world as a whole water requirements are growing and will continue to grow through the turn of the century in all types of economic activity (Table 2.8). Present (as of 1990) gross water withdrawals in the world are 4,100 km$^3$ per year,

**TABLE 2.8** Dynamics of water use in the world by human activity

| Water users[a] | 1900 (km$^3$ per year) | 1940 (km$^3$ per year) | 1950 (km$^3$ per year) | 1960 (km$^3$ per year) | 1970 (km$^3$ per year) | 1975 (km$^3$ per year) | 1980 (km$^3$ per year) | 1980 (%) | 1990[b] (km$^3$ per year) | 1990[b] (%) | 2000[b] (km$^3$ per year) | 2000[b] (%) |
|---|---|---|---|---|---|---|---|---|---|---|---|---|
| Agriculture | | | | | | | | | | | | |
| Withdrawal | 525 | 893 | 1,130 | 1,550 | 1,850 | 2,050 | 2,290 | 69.0 | 2,680 | 64.9 | 3,250 | 62.6 |
| Consumption | 409 | 679 | 859 | 1,180 | 1,400 | 1,570 | 1,730 | 88.7 | 2,050 | 86.9 | 2,500 | 86.2 |
| Industry | | | | | | | | | | | | |
| Withdrawal | 37.2 | 124 | 178 | 330 | 540 | 612 | 710 | 21.4 | 973 | 23.6 | 1,280 | 24.7 |
| Consumption | 3.5 | 9.7 | 14.5 | 24.9 | 38.0 | 47.2 | 61.9 | 3.2 | 88.5 | 3.8 | 117 | 4.0 |
| Municipal supply | | | | | | | | | | | | |
| Withdrawal | 16.1 | 36.3 | 52.0 | 82.0 | 130 | 161 | 200 | 6.0 | 300 | 7.3 | 441 | 8.5 |
| Consumption | 4.0 | 9.0 | 14 | 20.3 | 29.2 | 34.3 | 41.1 | 2.1 | 52.4 | 2.2 | 64.5 | 2.2 |
| Reservoirs | | | | | | | | | | | | |
| Withdrawal | 0.3 | 3.7 | 6.5 | 23.0 | 66.0 | 103 | 120 | 3.6 | 170 | 4.1 | 220 | 4.2 |
| Consumption | 0.3 | 3.7 | 6.5 | 23.0 | 66.0 | 103 | 120 | 6.2 | 170 | 7.2 | 220 | 7.6 |
| Total (rounded off) | | | | | | | | | | | | |
| Withdrawal | 579 | 1,060 | 1,360 | 1,990 | 2,590 | 2,930 | 3,320 | 100 | 4,130 | 100 | 5,190 | 100 |
| Consumption | 417 | 701 | 894 | 1,250 | 1,540 | 1,760 | 1,950 | 100 | 2,360 | 100 | 2,900 | 100 |

[a] Total water withdrawal is shown in the first line of each category; consumptive use (irretrievable water loss) is shown in the second line.
[b] Estimated.

of which 2,300 km$^3$ per year (56% of total water withdrawals) are unrecoverable (consumed). By the turn of the century we should expect an increase in total water withdrawals to 5,200 km$^3$ per year, and an increase in consumptive use of about 30%, to 2,900 km$^3$ per year.

Agriculture currently accounts for approximately 69% of total water use and 89% of consumptive water use in the world. In the long term, the fraction accounted for by agriculture will decrease somewhat, principally as a result of an increase in the fraction used by industry. Incremental evaporation from reservoirs plays a prominent part in total unrecoverable water losses throughout the world: it exceeds consumptive use by industry and municipal services combined.

Absolute amounts of water use by region vary quite significantly. For example, in 1980, withdrawals ranged from 530–670 km$^3$ in the United States and South Asia to 2.4–2.8 km$^3$ in Central Africa and Oceania (Table 2.9).

In considering the dynamics of water use throughout the world, we should note that a continual increase in water withdrawals during this century has been characteristic of all regions, the largest growth occurring in the 1950s and 1960s.[15] A significant increase in water requirements over 1980 levels is also expected through the turn of the century, with the largest increases expected to occur in South America and Africa (95% and 70%). Decreases are possible in many major industrialized countries.

The volume of water needed in each region depends on population, climatic factors, and the level of economic and social development; here climatic characteristics are especially important. The graphs presented in Figure 2.5 confirm this relationship. In these graphs we see the direct relationships between the volumes of consumptive water use per capita and the dryness index – the higher the dryness index, the greater the consumptive use in a region.

Analysis of the relationships presented in Figures 2.2 and 2.5 shows convincingly that under the conditions of a dry, hot climate, where water resources are minimal, all other conditions being the same, water consumption for economic needs increases sharply, creating a shortage of water resources and an exceptionally low actual level of water supply. The reverse picture is observed in moist regions, where there is a surfeit of water resources under natural conditions. Here the dryness index has its minimum value and consumptive water use is small.

Thus, under the conditions of intensive economic activity the impact of climatic factors on water resources is not diminished but rather is significantly enhanced. In arid regions, climatic factors determine not only the natural stream flow but also, to a significant extent, the degree of reduction of natural runoff as a result of human activities.

It is of interest to compare the amounts of water withdrawals and consumption throughout the world with stream flow resources; this is not hard to do by using the data in Table 2.9.

For the entire earth, total water withdrawn for use in 1990 was 9.3% of total surface runoff and unrecoverable consumptive use was 5.2%; by the year 2000 these values will be 11.6% and 6.5%, respectively. At the same

**TABLE 2.9**  Annual runoff and water consumption by continents and by physiographic and economic regions of the world

| | Mean annual runoff | | Aridity index (R/LP) | Water consumption (km$^3$ per year) | | | | | |
| | | | | 1980 | | 1990 | | 2000 | |
| Continent and region | (mm) | (km$^3$ per year) | | Total | Irretrievable | Total | Irretrievable | Total | Irretrievable |
|---|---|---|---|---|---|---|---|---|---|
| *Europe* | 310 | 3,210 | | 435 | 127 | 555 | 178 | 673 | 222 |
| North | 480 | 737 | 0.6 | 9.9 | 1.6 | 12 | 2.0 | 13 | 2.3 |
| Central | 380 | 705 | 0.7 | 141 | 22 | 176 | 28 | 205 | 33 |
| South | 320 | 564 | 1.4 | 132 | 51 | 184 | 64 | 226 | 73 |
| European USSR (North) | 330 | 601 | 0.7 | 18 | 2.1 | 24 | 3.4 | 29 | 5.2 |
| European USSR (South) | 150 | 525 | 1.5 | 134 | 50 | 159 | 81 | 200 | 108 |
| *North America* | 340 | 8,200 | | 663 | 224 | 724 | 255 | 796 | 302 |
| Canada and Alaska | 390 | 5,300 | 0.8 | 41 | 8 | 57 | 11 | 97 | 15 |
| United States | 220 | 1,700 | 1.5 | 527 | 155 | 546 | 171 | 531 | 194 |
| Central America | 450 | 1,200 | 1.2 | 95 | 61 | 120 | 73 | 168 | 93 |
| *Africa* | 150 | 4,570 | | 168 | 129 | 232 | 165 | 317 | 211 |
| North | 17 | 154 | 8.1 | 100 | 79 | 125 | 97 | 150 | 112 |
| South | 68 | 349 | 2.5 | 23 | 16 | 36 | 20 | 63 | 34 |
| East | 160 | 809 | 2.2 | 23 | 18 | 32 | 23 | 45 | 28 |
| West | 190 | 1,350 | 2.5 | 19 | 14 | 33 | 23 | 51 | 34 |
| Central | 470 | 1,909 | 0.8 | 2.8 | 1.3 | 4.8 | 2.1 | 8.4 | 3.4 |
| *Asia* | 330 | 14,410 | | 1,910 | 1,380 | 2,440 | 1,660 | 3,140 | 2,020 |
| North China and Mongolia | 160 | 1,470 | 2.2 | 395 | 270 | 527 | 314 | 677 | 360 |
| South | 490 | 2,200 | 1.3 | 668 | 518 | 857 | 638 | 1,200 | 865 |
| West | 72 | 490 | 2.7 | 192 | 147 | 220 | 165 | 262 | 190 |
| South-east | 1,090 | 6,650 | 0.7 | 461 | 337 | 609 | 399 | 741 | 435 |
| Central Asia and Kazakhstan | 70 | 170 | 3.1 | 135 | 87 | 157 | 109 | 174 | 128 |
| Siberia and Far East | 230 | 3,350 | 0.9 | 34 | 11 | 40 | 17 | 49 | 25 |
| Trans-Caucasus | 410 | 77 | 1.2 | 24 | 14 | 26 | 18 | 33 | 21 |
| *South America* | 660 | 11,760 | | 111 | 71 | 150 | 86 | 216 | 116 |
| Northern area | 1,230 | 3,126 | 0.6 | 15 | 11 | 23 | 16 | 33 | 20 |
| Brazil | 720 | 6,148 | 0.7 | 23 | 10 | 33 | 14 | 48 | 21 |
| West | 740 | 1,714 | 1.3 | 40 | 30 | 45 | 32 | 64 | 44 |
| Central | 170 | 812 | 2.0 | 33 | 20 | 48 | 24 | 70 | 31 |
| *Australia and Oceania* | 270 | 2,390 | | 29 | 15 | 38 | 17 | 47 | 22 |
| Australia | 39 | 301 | 4.0 | 27 | 13 | 34 | 16 | 42 | 20 |
| Oceania | 1,560 | 2,090 | 0.6 | 2.4 | 1.5 | 3.3 | 1.8 | 4.5 | 2.3 |
| Land area (rounded off) | | 44,500 | | 3,320 | 1,450 | 4,130 | 2,360 | 5,190 | 2,900 |

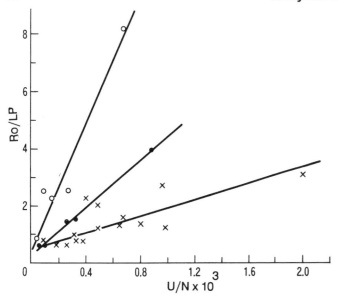

Fig. 2–5. Consumptive water use by region as a function of climatic conditions. Estimates of the consumptive use of water (''specific irretrievable water losses'') are shown for different physiographic and economic regions of the world (1980) as a function of the aridity index ($R_0/LP$). (○), Africa; (•) Europe and Australia; (×) North and South America, and Asia.

time, in many major regions of the world total water withdrawals are already 20%–65% of annual runoff (North Africa, Central Asia and Kazakhstan, West and South Asia, the Trans-Caucasus, the United States, southern and central Europe, the southern part of the European part of the former Soviet Union), and in the long term will reach 40%–100% by the turn of the century, i.e., in some regions total demand will equal the entire stream flow (Table 2.9).

Thus, in the long term the change in the evaporation regime as a result of economic activities may lead to some transformation of the ratios between elements of the water balance in different parts of continents and large regions. Quantitative estimates of these changes are of great scientific and practical importance to long-term planning of large-scale measures to ensure the rational use of water resources. These problems should become a subject of research based on international collaboration between hydrologists and climatologists.

Population dynamics as well as climatic factors and economic activities contribute to the extremely uneven distribution of water supply in various regions throughout the earth. The unevenness of the distribution of water resources and the fact that they are ill matched to the disposition of the population and economy can be vividly illustrated by comparing the actual or residual water supply per capita of individual regions for the same periods of time. For each period the specific (per capita) actual water supply of regions was determined by dividing the total runoff of a region, less the volume of unrecoverable water consumption, by the number of inhabitants.[16]

The values obtained for the actual water supply (in $10^3$ m³ per year per capita) are presented in Table 2.10 for all regions and continents at the levels for the years 1950, 1960, 1970, 1980, and 2000. To analyze them, it is convenient to group them on the following scale ($10^3$ m³ per year per

**TABLE 2.10** Dynamics of actual water availability in different regions of the world

| Continent and region | Area ($10^6$ km²) | Actual water availability ($10^3$ m³ per year per capita) | | | | |
|---|---|---|---|---|---|---|
| | | 1950 | 1960 | 1970 | 1980 | 2000 |
| *Europe* | 10.28 | 5.9 | 5.4 | 4.9 | 4.6 | 4.1 |
| North | 1.32 | 39.2 | 36.5 | 33.9 | 32.7 | 30.9 |
| Central | 1.86 | 3.0 | 2.8 | 2.6 | 2.4 | 2.3 |
| South | 1.76 | 3.8 | 3.5 | 3.1 | 2.8 | 2.5 |
| European USSR (North) | 1.82 | 33.8 | 29.2 | 26.3 | 24.1 | 20.9 |
| European USSR (South) | 3.52 | 4.4 | 4 | 3.6 | 3.2 | 2.4 |
| *North America* | 24.16 | 37.2 | 30.2 | 25.2 | 21.3 | 17.5 |
| Canada and Alaska | 13.67 | 384 | 294 | 246 | 219 | 189 |
| United States | 7.83 | 10.6 | 8.8 | 7.6 | 6.8 | 5.6 |
| Central America | 2.67 | 22.7 | 17.2 | 12.5 | 9.4 | 7.1 |
| *Africa* | 30.10 | 20.6 | 16.5 | 12.7 | 9.4 | 5.1 |
| North | 8.78 | 2.3 | 1.6 | 1.1 | 0.69 | 0.21 |
| South | 5.11 | 12.2 | 10.3 | 7.6 | 5.7 | 3.0 |
| East | 5.17 | 15.0 | 12 | 9.2 | 6.9 | 3.7 |
| West | 6.96 | 20.5 | 16.2 | 12.4 | 9.2 | 4.9 |
| Central | 4.08 | 92.7 | 79.5 | 59.1 | 46.0 | 25.4 |
| *Asia* | 44.56 | 9.6 | 7.9 | 6.1 | 5.1 | 3.3 |
| North China and Mongolia | 9.14 | 3.8 | 3.0 | 2.3 | 1.9 | 1.2 |
| South | 4.49 | 4.1 | 3.4 | 2.5 | 2.1 | 1.1 |
| West | 6.82 | 6.3 | 4.2 | 3.3 | 2.3 | 1.3 |
| South-east | 7.17 | 13.2 | 11.1 | 8.6 | 7.1 | 4.9 |
| Central Asia and Kazakhstan | 2.43 | 7.5 | 5.5 | 3.3 | 2.0 | 0.7 |
| Siberia and Far East | 14.32 | 124 | 112 | 102 | 96.2 | 95.3 |
| Trans-Caucasus | 0.19 | 8.8 | 6.9 | 5.4 | 4.5 | 3.0 |
| *South America* | 17.85 | 105 | 80.2 | 61.7 | 48.8 | 28.3 |
| North | 2.55 | 179 | 128 | 94.8 | 72.9 | 37.4 |
| Brazil | 8.51 | 115 | 86 | 64.5 | 50.3 | 32.2 |
| West | 2.33 | 97.9 | 77.1 | 58.6 | 45.8 | 25.7 |
| Central | 4.46 | 34 | 27 | 23.9 | 20.5 | 10.4 |
| *Australia and Oceania* | 8.59 | 112 | 91.3 | 74.6 | 64.0 | 50.0 |
| Australia | 7.62 | 35.7 | 28.4 | 23 | 19.8 | 15.0 |
| Oceania | 1.34 | 161 | 132 | 108 | 92.4 | 73.5 |

capita): ≤1 extremely low; 1.1–2.0 very low; 2.1–5.0 low; 5.1–10.0 average; 10.1–20.0 above average; 20.1–50 high; >50 very high.

In 1950 (Table 2.10) the level of per capita water supply throughout most of the earth was average or higher, and the level was low (from 2.1 to $5.0 \times 10^3$ m³ per year per capita) only in North Africa, central and southern Europe, China, and South Asia. In no region on earth was the level of water supply very low or extremely low.

Thirty years later, in 1980, the actual level of per capita water supply had decreased sharply in many regions throughout the world due to population increases. It had become extremely low in North Africa, very low in North China and Mongolia, Central Asia, Asia, and Kazakhstan, and low in six other regions (Table 2.10). By the turn of the century, a low actual level of per capita availability is anticipated in three regions (North Africa, Central Asia, and Kazakhstan), very low in three (North China and Mongolia, South Asia, and West Asia), and low in seven (central and southern Europe, the southern European part of the former Soviet Union, South-east Asia, and West, East, and South Africa). At the same time, in all periods in question there is a high or very high level of water supply in North Europe, the northern European part of the former Soviet Union, Canada, and Alaska, nearly all of South America, Central Africa, Siberia, the Far East, and Oceania. It is important to note here that the dynamics of the actual level of water supply is such that the rate of decrease in it is especially significant in regions with a low absolute level of water supply, i.e., where there is a shortage of water resources. For example, in regions with the lowest water supply (Central Asia, Kazakhstan, and North Africa) per capita availability decreases by a factor of 11 in the period 1950–2000, but in regions with a larger supply (Siberia, the Far East, North Europe, Canada, Alaska, Central Africa) it decreases by a factor of just 1.5–5 over the same period. Thus, the very high natural non-uniformity in the distribution of water supply throughout the earth is increasing still more with time as a result of human economic activities and population change, and at an extremely rapid rate. Because of this, the urgency of the problem of the territorial redistribution of water resources on the global scale will increase significantly with time.

In conclusion, it should be noted that the values presented above for water resources and for volumes of water withdrawals were calculated for the long term on the basis of the assumption of steady climatic fluctuations, and are characteristic of the mean climatic conditions of each region, i.e., it was assumed that the possible anthropogenic global-scale climatic changes through the year 2000 are not included (see Chapter 9). Allowing for the impact of possible anthropogenic changes in global climate on world water resources and water consumption is a problem for the next decade, and, the author believes, the role of effective international cooperation in resolving it would be hard to overestimate.

## Conclusions and tasks for further research

Considerable advances have been made in the study of the global water balance and water resources, but as more complete observational data are collected and the requirements for water use and environmental protection grow, the imperfection and inadequacy of our knowledge of water resources become increasingly apparent.

In the development of research on water resources and regional and world water balances, a transition is needed to a qualitatively new stage: expansion of theoretical and experimental research on the mechanism of hydrologic phenomena and processes, all components of the water, heat, and salt balances, and the changes therein as a result of anthropogenic factors. A transition is needed from the study of water resources and the water balance under static conditions based on perennial data to the pursuit of an immeasurably more complex undertaking: the consideration of hydrologic processes under dynamic conditions, over shorter time intervals – the year, season, or month, on a global scale, and for the most important natural and economic regions of the world. To date no patterns have been identified in perennial fluctuations in water resources, and no trends in or causes of the appearance of protracted periods of abundant precipitation and drought that often have catastrophic consequences to the public and the economy have been elucidated. The role of the oceanic and atmospheric component of the hydrologic cycle in the formation of water resources of large regions is not entirely clear; no summarization

of the dynamics of the world water balance has been performed with consideration for changes in water reserves in glaciers, ground waters and soils, or in the lake and stream network; and no reliable assessments or predictions are available for anthropogenic changes in the water resources of many regions and river basins as a result of the use of fresh water, the conversion of the surface of drainage basins, and pollution of water bodies.

Allowance for the impact of economic activities on the hydrologic cycle is of very great importance in the problem of studying water resources and their fluctuations in time and space, and of evaluating the dynamics of the level of water supply in various regions.

Human economic activities and population growth already have led, in the most oversettled regions of the world, to a sharp decrease in per capita water availability, some actual decreases in surface runoff, a decrease in the levels of interior closed water bodies, and contamination of both surface and ground waters. The impact of anthropogenic factors on surface runoff depends not only on their scale and rates of development, but also largely on natural fluctuations of climatic characteristics. These circumstances largely govern the attitude toward large water management projects.

In arid and semi-arid regions and during hot, dry periods the impact of economic activities has an especially pernicious effect on water resources and the level of water supply, greatly aggravating the water management situation and providing an incentive for the planning and development of measures intended to provide a fundamental solution to water supply problems.

In cooler regions and during cold, moist periods lasting many years, the impact of economic activities manifests itself to a much lesser degree in a decrease in the water content of rivers, the water management situation improves sharply, the development of water management measures often grinds to a halt, and projects that have been drawn up are subjected to bitter criticism, with attention focused on their negative aspects.

As economic activity increases, the dependence of water resources and the level of water supply on climatic characteristics increases significantly, especially in zones of variable moisture and arid regions. Here climatic conditions determine not only natural runoff, but also largely the degree of the reduction of flow as a result of all anthropogenic factors. The great natural non-uniformity in the distribution of water supply on earth is increasing still further as a result of population growth and the intensification of human economic activities, and at an extremely rapid pace. Plans for new large-scale flow diversions are now extremely unpopular in many developed countries because of their substantial ecological and economic costs. However, there may be objective grounds for further development, particularly in poorer countries. They offer one way of reducing both the shortage of water resources in particular countries and regions and the impacts of severe floods. For this reason, comprehensive research should be continued and developed in this area, including research within the framework of international cooperation, especially to provide a reliable assessment of the impact of flow diversions on the environment under the conditions of anthropogenic changes in global climate.

The ever-growing dependence of water resources and water supply on climatic factors accounts for the very close linkage between modern problems of supplying fresh water to humankind and problems of natural and anthropogenic changes in climate.

From the standpoint of future water resources, changes in global climate due to an increase in the carbon dioxide concentration in the atmosphere as a result of human activities are of primary importance. The anthropogenic changes in climate predicted by climatologists in the next 20–30 years are so significant in scale, especially for the temperate and high latitudes, that scientists already are confronted with major scientific problems: above all, to estimate water resources and the water balance in the future and changes in them as a result of economic activities; water supply for the public, industry, and agriculture in the long term; and the territorial redistribution of water resources and regulation of runoff.

The resolution of all these problems requires effective international cooperation in conducting comprehensive research, with participation by hydrologists, climatologists, and specialists in water management and environmental protection.

## Notes

1.   See, for example, C.A. Doxiadis, 1967, Water for human environment, in *Proceedings of the Conference "Water for Peace," vol. I, Washington, DC, pp. 33–60; G.P. Kalinin, 1968, Global Hydrology Problems*, Gidrometeoizdat, Leningrad (in Russian); M.I. L'vovich, 1969, *Water Resources in the Future*, Prosveshcheniye, USSR (in Russian); M.I. L'vovich, 1974, *World Water Resources and their Future*, Mysl' Publishing, Moscow (in Russian), 1979 US translation available from the American Geophysical Union, Washington, DC; R. Nace, 1968, *Water of the World*, US Department of the Interior, US Geological Survey, Washington, DC (July); IAHS/UNESCO/WMO, 1972, *World Water Balance, Proceedings of the International Symposium on World Water Balance*, Reading, July 1970, vols. 1–3; and M. Holy, 1974, *Water and the Environment*, Food and Agricultural Organization, United Nations, Rome.

2.   These include a book by a team of Soviet scientists dealing with Earth's water balance and water resources: *World Water Balance and the Water Resources of the Earth*, 1974, Gidrometeoizdat, Leningrad (in Russian), later translated into English by UNESCO, 1978, *World Water Balance and Water Resources of the Earth*, R. Nace (ed.), Paris; a review of the world water balance, published by West German scientists: A. Baumgartner and E. Reichel, 1975, *The World Water Balance*, R. Oldenburg Verlag, Munich, and Elsevier Scientific Publishing Co., New York; and a monograph on the use of the earth's water resources, prepared under the supervision of the author of the present chapter: I.A. Shiklomanov and O.A. Markova, 1987, *Specific Water Availability and River Runoff Transfers in the World*, Gidrometeoizdat, Leningrad (in Russian).

3.   I.A. Shiklomanov, 1989, *Man's Impact on River Runoff*, Gidrometeoizdat, Leningrad (in Russian).

4.   These were drawn from *World Water Balance and Water Resources of the Earth*, 1974, Gidrometeoizdat, Leningrad (in Russian), and I.A. Shiklomanov, 1989, *Man's Impact on River Runoff*, Gidrometeoizdat, Leningrad (in Russian).

5.   I.A. Shiklomanov and A.A. Sokolov, 1983, Methodological basis of world water balance investigation and computation, in *Proceedings of the Hamburg Workshop*, August 1981, IAHS Publication no. 148, pp. 77–91.

6.   For example, a monograph by Baumgartner and Riechel presents data on the water balance of the continents, which for some components differ from the data presented above by up to 20%–30% (A. Baumgartner and E. Reichel, 1975, *The World Water Balance*, R. Oldenburg Verlag, Munich, and Elsevier Scientific Publishing Co., New York). These differences are explained by different raw data and methodological approaches. With respect to the evaluation of water resources of regions, the author believes that the data presented in *World Water Balance and Water Resources of the Earth*, 1974, Gidrometeoizdat, Leningrad, (in Russian), which were obtained directly by summarizing observational data from the world hydrological network, are more reliable. These issues are also discussed in I.A. Shiklomanov and A.A. Sokolov, 1983, Methodological basis of world water balance investigation and computation, In *Proceedings of the Hamburg Workshop*, August 1981, IAHS Publication no. 148, pp. 77–91.

7.   I.A. Shiklomanov, 1985, Large-scale water transfer, in J. Rodda (ed.) *Facets of Hydrology*, vol. II, pp. 345–387, John Wiley & Sons, London.

8.   According to data of the *World Water Balance and Water Resources of the Earth*, 1974, Gidrometeoizdat, Leningrad (in Russian) and M.I. Budyko, 1956, *The Heat Balance of the Earth's Surface*, Gidrometeoizdat, Leningrad (in Russian).

9.   I.A. Shiklomanov, 1988, *Studying Land and Water Resources: Results, Problems, and Outlook*, Gidrometeoizdat, Leningrad (in Russian).

10.  Data on total withdrawals and consumptive use for all continents and water consumers from 1900 to 2000 are available from *World Water Balance and Water Resources of the Earth*, Gidrometeoizdat, Leningrad (in Russian). Data on total water withdrawals for all water consumers and continents and selected countries as of 1977 are available from *The Global 2000: Entering the 21st Century, 1980*, Report to the President of the USA, vols. 2–9, US Government Printing Office, Washington, DC, pp. 137–159.

11.  *World Water Balance and Water Resources of the Earth*, 1974, Gidrometeoizdat, Leningrad (in Russian).

12.  I.A. Shiklomanov and O.A. Markova, 1987, *Specific Water Availability and River Runoff Transfers in the World*, Gidrometeoizdat, Leningrad (in Russian).

13.  H.H. Landsberg, L.L. Fishman and J.L. Fisher, 1963 *Resources in America's Future: Patterns of Requirements and Availabilities 1960–2000*, Johns Hopkins, Baltimore.

14.  US Geological Survey, 1984, *National Water Summary 1983: Hydrologic Events and Issues*, US Geological Survey Water Supply Paper 2250, Washington, DC.

15.  I.A. Shiklomanov and O.A. Markova, 1987, *Specific Water Availability and River Runoff Transfers in the World*, Gidrometeoizdat, Leningrad (in Russian).

16.  This last quantity was determined from data of the United Nations Food and Agricultural Organization (FAO) and from forecasts in I.A. Shiklomanov and O.A. Markova, 1987, *Specific Water Availability and River Runoff Transfers in the World*, Gidrometeoizdat, Leningrad (in Russian).

# Chapter 3
# Water quality and health

Linda Nash

## Introduction

During the last decade, it has become commonplace to speak of the links between water quantity and water quality. As population has continued to rise throughout most of the world, it has become increasingly difficult to provide an adequate supply of water for all uses. While in a few highly arid regions, such as the Sahel and the Arabian peninsula, this is the result of naturally scarce supplies, in most countries the crux of the problem lies with water quality rather than water quantity.

In the industrialized countries, industrial effluents and agricultural chemicals have contaminated surface waters, posing a threat to aquatic life and necessitating costly treatment before fresh water can be tapped as a source for municipal and industrial supply. Similarly, ground water resources, once considered better protected from anthropogenic contamination, have become widely polluted with chemicals, many of which are known or suspected carcinogens.

In the developing world, chemicals are being introduced at an ever-increasing rate; however, traditional water-related disease, particularly the enteric and diarrheal pathogens, remain the primary health concern. Diarrheal disease is the leading cause of morbidity in the world, responsible for the deaths of three million young children each year.[1] In Latin America, a recent outbreak of cholera, last seen in the western hemisphere over a century ago, suggests that in some regions progress in controlling the microbiologic contamination of drinking water has been completely inadequate. Yet while still grappling with the most basic water quality issues, the less-developed countries are at the same time having to face the water quality problems posed by industrialization and chemically reliant agriculture.

The water quality challenges that the world currently faces are enormous. This chapter aims to provide an overview of global water quality in relation to human health by presenting an inventory and some examples of specific water quality problems. Subsequently, emerging issues and trends are described, and in the last section approaches to water quality management in both the industrialized and developing worlds are presented and discussed.

## Water quality problems

**Background.** The quality of natural waters varies both temporally and spatially, making it difficult to compare water quality across broad geographic regions or time-scales. For instance, dissolved oxygen concentration, an important parameter for aquatic life, varies inversely with temperature and so changes continuously as the temperature of the atmosphere changes. Water quality in rivers may change substantially on a seasonal basis in response to contaminant inputs that are carried by runoff, such as metals and petroleum products associated with urban stormwater runoff, and chemicals leached from agricultural fields. Over longer time periods, water quality will vary in response to natural events, both catastrophic changes such as volcanic explosions and gradual changes such as the maturation of lakes. There are also vast differences among natural waterbodies depending on their location, their origin, and the surrounding climate. For instance, some water sources have naturally high concentrations of radionuclides, which make them unsuitable for most human uses.

To some extent, water quality concerns are specific to the type of waterbody, i.e., river, lake, or underground aquifer. Eutrophication and sedimentation are problems that affect only surface waters. Lakes and reservoirs are more prone to long-term acidification than are rivers, while the latter are more susceptible to sudden pulses of high acidity due to their lower buffering capacity. In some regions, ground water is better protected from sewage pollution yet more susceptible to contamination from fertilizers and agricultural chemicals.

The adequacy of water quality for human use, however, depends not only upon the absolute concentrations of given parameters but also upon how the water will be used. The highest quality water is required for drinking. Beyond that, water of progressively lesser quality is required for swimming and recreational use, industrial use, and irrigation. The water quality needs of aquatic life vary; however, aquatic species are frequently more sensitive to organic and chemical pollution than are human beings.

Worldwide, natural processes still dominate the chemistry of natural waters. In only a few cases has anthropogenic pollution been truly global in scale; these include increases in lead, DDT, and atmospheric $CO_2$. On a local, and increasingly on a regional scale, however, human influences have begun to dominate the quality of natural waters.[2] For centuries surface waters have been used as receptacles for human waste because the natural processes of waterbodies – sedimentation, aeration, mixing, bacterial processing – help to break down and to return wastes to the natural environment. But the ever-increasing human population and its growing wasteload have clearly overtaxed the natural recycling capabilities of many waterways. In areas of eastern Europe, for example, rapid and uncontrolled industrialization has resulted in grossly polluted waterways. According to a recent report released by the government of Czechoslovakia, 70% of all surface waters in the country are heavily polluted and 30% are not capable of sustaining fish.[3]

Anthropogenic pollution may be categorized as emanating from either municipal, industrial, or agricultural sources. Municipal waste is composed primarily of human excreta and generally contains few or no chemical contaminants but may carry numerous pathogenic microorganisms. Industrial wastes vary tremendously depending upon the type of industry or processing activity, and may contain a wide variety of both organic and inorganic chemicals. Agricultural pollution is comprised of the excess phosphorus and nitrogen present in fertilizers, which pollute ground water and accelerate the eutrophication of surface waters, as well as numerous different organic pesticides.

A further distinction is often made between "point" and "non-point" sources of pollution. Both municipal and most industrial effluents are labeled point sources because they are emitted from one specific and identifiable place (e.g., a sewage or industrial outfall). Agricultural pollution, on the other hand, enters waterways in a diffuse manner as chemicals percolate into ground water or are washed from the soil into nearby surface waters. Other examples of non-point sources include runoff from mining operations, uncollected sewage (particularly in rural areas), and urban stormwater runoff. The distinction between point and non-point sources is particularly relevant from a management perspective because it is much more difficult both to identify and to measure the impact of diffuse sources. Finally, pollutants may be categorized not by their source but by their chemical, physical, and biological characteristics – such as bacteria, metals, and organic chemicals. The next several sections of this chapter describe the major classes of water quality pollutants, the threats they pose to human health, their sources, and what is known about their extent.

**Microbiologic contamination and water-related disease.** For more than half a century, water-related disease has been of minor consequence in the industrialized world, yet globally microbiologic contamination remains the most pressing water quality concern. Estimates of the global occurrence of waterborne disease are highly uncertain, but on the order of 250 million new cases of waterborne disease are reported each year, resulting in roughly 10 million deaths. Seventy-five per cent of these cases occur in tropical countries, where both climatic conditions and the poor state of water supply and sanitation infrastructure combine to spread disease.[4]

Microbiologic diseases associated with water are not necessarily waterborne, but are more properly classified among four different categories (see Table 3.1). The term "waterborne" includes those diseases in which water is the passive agent for transmission, and generally refers to those pathogens that are passed from excreta to water and then subsequently ingested. Most important in this category are the waterborne enteric and diarrheal diseases. The diarrheal diseases are the result of so many different etiologies that the pathogenic agents are often not identified, but can be roughly grouped into those diseases caused by bacteria (e.g., *Vibrio cholera*), those caused by parasites (e.g., *Giardia lamblia*), and those caused by viruses (e.g., Norwalk, rotaviruses).

Water-washed, or water-hygiene, diseases result from unsanitary conditions and can be avoided through the use of additional water for washing. This category includes the eye and louse-borne diseases (e.g.,

trachoma, typhus) as well as many of the diarrheal diseases which may be passed directly from person to person without first passing into water supplies.

Water-based diseases are passed through means other than ingestion by hosts that either live in water or require water for part of their life-cycle. Among the most common diseases in this category are schistosomiasis, which is contracted from dermal contact with infected snails, and guinea worm disease (dracunculiasis). Finally, water-vector diseases are spread by insects that either breed in or bite near water.

The principal pathway for the transmission of waterborne disease is through the contamination of drinking water supplies with pathogen-carrying animal or human excreta, and the subsequent ingestion of pathogens by uninfected people (fecal–oral transmission). Typhoid fever and cholera were the first diseases identified as waterborne, and cholera remains one of the more important diseases in this class. The first cholera pandemic spread all the way to the Americas from South Asia in 1817, and up until the mid-19th century, cholera outbreaks occurred repeatedly in the U.S. In 1970 the disease leaped across a vast expanse of unaffected territory to reach western Africa, where it rapidly invaded several countries. Today, cholera remains endemic in Asia and Africa, and in January of 1991 cholera arrived in Peru, making its reappearance in the western hemisphere for the first time in this century. Since then the disease has traveled to several other Latin American countries, claiming nearly 4,000 lives within a year. Water transmission has played a key role in the epi-

**TABLE 3.1** Classifications of water-related infections

| Category | Infection | Pathogenic agent |
|---|---|---|
| 1. Fecal–oral (waterborne or water-washed) | Diarrheas and dysentries | |
| | Amoebiasis | P |
| | *Campylobacter* interitis | B |
| | Cholera | B |
| | *E. coli* diarrhea | B |
| | Giardiasis | P |
| | Rotavirus diarrhea | V |
| | Salmonellosis | B |
| | Shigellosis (bacillary dysentry) | B |
| | Enteric fevers | |
| | Typhoid | B |
| | Paratyphoid | B |
| | Poliomyelitis | V |
| | Ascariasis (giant roundworm) | H |
| | Trichuriasis (whipworm) | H |
| | Strongyloidiasis | H |
| | Taenia solium taeniasis (pork tapeworm) | H |
| 2. Water-washed: | | |
| Skin and eye infections | Infectious skin diseases | M |
| | Infectious eye diseases | M |
| Other | Louse-borne typhus | R |
| | Louse-borne relapsing fever | S |
| 3. Water-based: | | |
| Penetrating skin | Schistosomiasis | H |
| Ingested | Dracunculiasis (guinea worm) | H |
| | Clonorchiasis | H |
| | Others | H |
| 4. Water-related insect vector: | Trypanosomiasis (sleeping sickness) | P |
| Biting near water | Filariasis | H |
| Breeding in water | Malaria | P |
| | Onchocerciasis (river blindness) | H |
| | Mosquito-borne viruses | |
| | Yellow fever | V |
| | Dengue | V |
| | Others | V |

*Source*: Adapted from R.G. Feachem, 1984, Infections related to water and excreta: The health dimension of the decade, in P.G. Bourne (ed.) *Water and Sanitation*, Academic Press, Inc., Orlando, pp. 21–47.

B, bacterium; P, protozoan; S, spirochete; M, miscellaneous; H, helminth; R, rickettsia; V, virus.

demic, especially in the slums of Lima and other Peruvian cities where most of the deaths have occurred.[5]

Although over 30 species of parasites infect the human intestine, only seven have global distributions or serious pathologies: amoebiasis, giardiasis, *Taenia solium* taeniasis, ascariasis, hookworm, trichuriasis and strongyloidiasis.[6] Globally, *Giardia* is the most common animal parasite of human beings and the most common cause of waterborne disease; its distribution is increasing because of its resistance to chlorine.[7] Global morbidity due to whipworms (trichuriasis) is estimated at roughly 100,000 annually, with 500 million people infected. In some regions, including Cameroon, the Caribbean, and Malaysia, whipworm infection rates are reported to be more than 90%. The parasite responsible for amoebiasis, *Entamoeba histolytica*, is estimated to infect as many as 500 million people worldwide, and it remains a major health problem in China, Mexico, eastern South America, western and southern Africa, and all of south-east Asia. *Ascaris* is estimated to infect 1 billion people worldwide.[8]

One of the more unusual waterborne diseases is Legionnaire's disease, first diagnosed in July of 1976 when an outbreak of acute respiratory illness occurred during an American Legion convention in Philadelphia, causing 221 cases of illness and 34 deaths. Infection is believed to have occurred through the inhalation of bacteria in aerosolized, contaminated water. Unlike most pathogens, *Legionella* does not have an animal host reservoir but is a common inhabitant of natural waters. More important from a health standpoint is its resistance to chlorine, which allows it to survive and multiply in water treated to meet U.S. and European drinking standards.[9]

Waterborne diseases can be controlled directly through improved water quality. In the early part of the 19th century, Europe and North America made dramatic improvements in public health through the protection and treatment of water supplies, ultimately bringing both cholera and typhoid under control. Currently, in the developed countries the microbial pathogens of greatest concern are those that demonstrate resistance to chlorine, including hepatitis A, *Giardia lamblia*, and *Cryptosporidium*. During the current century, the incidence of waterborne disease in the U.S. population has declined from roughly eight cases per 100,000 person-years during the period 1920–1940 to four cases in the period 1971–1980.[10] Although outbreaks of waterborne disease still occur with some frequency, such outbreaks are now limited to small water systems and consequently affect many fewer people. In 1989–1990, waterborne outbreaks in the U.S. caused 4,288 cases of illness and four deaths. Among these outbreaks giardiasis was the most common.[11]

Of the water-based diseases, schistosomiasis and dracunculiasis are the most widespread. Schistosomiasis currently infects 200 million people and puts another 400 million at risk in more than 70 countries. In many regions, the disease is spreading and intensifying due to new irrigation projects, which create a favorable environment for the host snails. In the Nile delta, the parasite quickly spread throughout the river region following the construction of the high dam at Aswan and related irrigation projects, resulting in infection rates of up to 100%. In Sudan the construction of the Sennâr dam in 1924 has had similar results. In the newly irrigated region of Gezira, the prevalence in children of two carrying parasites (*S. haematobium* and *S. mansoni*) increased from less than 5% to more than 45% and 77%, respectively. In West Africa, the construction of the Akosombo dam and Lake Volta resulted in infection rates in children rising from less than 10% to more than 90% within a year of the reservoir's filling.[12]

Dracunculiasis, or guinea worm disease, is caused by the ingestion of water containing a species of crustacean that is infected with guinea worm larvae (*Dracunculus medinensis*). Upon maturation, adult worms emerge through the skin and release new larvae into water. An extremely debilitating disease, guinea worm remains a major public health concern in Africa and South Asia, despite remarkable progress in controlling the vector. The total number of cases worldwide in 1990 was estimated at less than 3 million, compared to 5–10 million in the mid-1980s. Twelve countries in Africa and South Asia now account for more than 98% of the cases worldwide.[13]

Guinea worm disease is one of the few diseases that can be controlled completely through the provision of safe drinking water, and most researchers acknowledge that it could be eradicated from the globe within the next decade. Control measures include protecting sources of water from guinea worm larvae, filtration of drinking water to remove both larvae and infected crustaceans, and boiling or chlorination of drinking water to destroy larvae. In Africa, the most dramatic evidence of success comes from the Côte D'Ivoire, which has implemented an aggressive rural water supply and education program. In 1966, the country reported more than 67,000 cases of guinea worm disease. By 1976, this had fallen to 4,971 cases, and by 1990 to 1,360 cases, despite the fact that reporting and surveillance had become much more thorough. The Republic of Gambia has had similar success and may actually have succeeded in eliminating the disease. Overall, there remain ten ''core'' African countries with high levels of infection.[14]

Water-vector diseases form a slightly different class because they are not directly attributable to water quality, although in many cases it is the large-scale development of water resources for urban and agricultural use that has created favorable conditions for host insects. Of the diseases transmitted by water-related insect vectors, malaria is unquestionably the most important. More than 2 billion people, or 40% of the world's population, remain exposed to malaria risk in approximately 100 countries. Malaria is one of the most important killers of children under 5 years of age, accounting for 20%–30% of childhood deaths. Nine per cent of those exposed live in areas with intense transmission and no fully implemented control programs, while 32% live in regions where malaria was once controlled but has now re-emerged. The global incidence of malaria is estimated to be on the order of 120 million clinical cases each year, with roughly 300 million people carrying the parasite. Half of all cases recorded come from India and Brazil, and about one-fourth of reported cases originate from Thailand, Sri Lanka, Afghanistan, Vietnam, China, and Myanmar. Summarizing the malaria situation globally, however, tends to mask the great variations that exist between and within countries. In India, for example, 2 million cases were reported in 1989, but 55% of these cases came from only three states: Gujarat, Orissa, and Madhya Pradesh. In the Americas, where malaria incidence increased dramatically from 270,000 cases in 1974 to 1.1 million in 1989, 52% of the cases are from Brazil. Within Brazil, the Amazon region recorded 97% of all cases, the majority coming from only three states; even within these states cases are concentrated in particular villages.[15] Aside from malaria, mosquitos may transmit several other diseases, such as yellow fever, dengue fever, filariasis, and dozens of lesser-known maladies.

Several water-related vector diseases are transmitted by flies. African trypanosomiasis, or sleeping sickness, is a very deadly disease that is transmitted by the African tsetse fly. In 36 affected countries, 50 million people are at risk and 25,000 new cases are reported each year. During the last decade, severe outbreaks have occurred in Sudan and Uganda, but the extent of the disease, like many others, is poorly documented. American trypanosomiasis (Chagas disease) affects 18 million people in Latin America, and transmission has recently intensified as a result of the proliferation of slums in urban areas.[16] Onchocerciasis, or river blindness, affects seven West African countries, causing approximately 70,000 cases of blindness or impaired vision. After 15 years of vector control, however, the disease is no longer considered a major public health problem.[17]

The actual degree of disease and suffering caused by water-related disease in the developing world is unknown. Most illnesses go undiagnosed and unreported, while epidemiological and microbiological studies have been conducted for only a handful of these diseases. The wide variety of disease-carrying bacteria, parasites, and insects that exist in tropical regions, combined with the relatively poor state of health care and water supply, suggest that these long-standing water quality problems will not be easily overcome. The continuing costs in human life and suffering clearly make these diseases one of the foremost global problems of any kind.

**Nitrates.** Nitrate is one of the many forms of nitrogen that occur in the environment. Although nitrogen is an essential element for all forms of life, increasing concentrations of nitrate in drinking waters pose two human health concerns. First, nitrates may cause infant methemoglobinemia, or blue-baby syndrome, in which the oxygen-carrying capacity of hemoglobin is blocked, causing suffocation.[18] The onset of methemoglobinemia requires relatively high levels of nitrate, at least 10 mg/l but usually on the order of 100 mg/l.[19] Since 1945 about 3,000 cases of methemoglobinemia have been reported worldwide, some fatal. More than 1,300 cases occurred in Hungary between 1976 and 1982, all of which were associated with private wells with very high concentrations of nitrate, greater than 22 mg/l $NO_3$-N.[20] The other health concern posed by nitrates is their potential role in the formation of cancers of the digestive tract attributable to N-nitrogenated compounds (nitrosamines). Nitrosamines are among the most potent of the known carcinogens in laboratory animals; but the contribution of ingested nitrates to their formation is unclear, and the epidemiological evidence is considered inconclusive. While nitrates currently do not pose a major threat to public health, the rapid and widespread increase in nitrate pollution raises serious concerns about the future quality of many waterbodies.

Nitrogen is always present in aquatic systems, most abundantly as a gas, but overall concentrations and input rates are intimately connected to the surrounding watershed and land use practices. Nitrate ions, the most common form of combined inorganic nitrogen in water, move easily through soils and are lost rapidly from land even in unperturbed watersheds. Stormwater runoff dissolves soil nitrate while erosion transfers particulate nitrate. Even moderate environmental disturbances, such as sensible farming or careful logging, release significant quantities of nitrates to surrounding waters.

The principal natural sources of nitrate in water include soil nitrogen, nitrogen-rich geologic deposits, and atmospheric deposition. The principal anthropogenic sources include fertilizers, septic tank drainage, feedlots, dairy and poultry farming, land disposal of municipal and industrial wastes, disturbance of mineralized soils by cultivation or logging, and the leaching of nitrates through irrigation. However, the amount of nitrate that actually enters aquatic systems is controlled by a complex set of hydrologic, chemical, and biological processes, referred to collectively as the nitrogen cycle. In aquatic systems, nitrogen fixation (the biologic conversion of ammonia to nitrite and subsequently to nitrate) and denitrification (the bacterial reduction of nitrate to nitrogen gas) are the most important transformations affecting nitrate concentrations.

Even though concentrations vary substantially under natural conditions, measurements indicate that the nitrate content of fresh water has been rising steadily in many countries over the last three decades. The major causes of this rise are increased industrial and urban waste, and the increased use of nitrogen fertilizers and manure in agriculture. Meybeck has estimated that about one-third of the total dissolved nitrogen (both organic and inorganic) in river water is the result of pollution.[21] The average nitrate concentration for pristine rivers was estimated to be 0.1 mg/l $NO_3$-N. Based on this estimate, less than 10% of the rivers in Europe can be classified as pristine.[22] The median level of $NO_3$-N in rivers outside of Europe monitored as part of the World Health Organization's (WHO) Global Environmental Monitoring System (GEMS) is 0.25 mg/l. For Europe, the median is 4.5 mg/l.[23]

Atmospheric loading has caused substantial increases in nitrate concentrations in Lake Superior. Estimates of springtime nitrate concentration were approximately 75 μg/l in 1906, rising to 311 μg/l in 1976. Moreover, these exponential increases will continue over the next several decades because the lake has not nearly attained equilibrium with existing nitrogen levels in the atmosphere.[24] While concentrations are still well below drinking water standards, this example nonetheless indicates the dramatic impact of human activities on water quality.

Levels of nitrate are generally much higher in ground waters than in surface waters. In the United States, a sampling network established by the Geological Survey showed that in more than one-third of the counties analyzed, 25% of the wells had nitrate concentrations that exceeded background levels, which were liberally estimated to be 3 mg/l $NO_3$-N. In 5% of the counties, more than 25% of the wells exceeded the federal drinking water criteria of 10 mg/l $NO_3$-N. The highest measured concentrations exceeded 100 mg/l $NO_3$-N.[25] Similarly, nitrate levels in Danish ground waters have nearly trebled since the 1940s. Mean concentrations were estimated to be approximately 1 mg/l $NO_3$-N in the 1940s and 1950s but had risen to 3 mg/l by 1980. Moreover, 8% of Danish waterworks tested exceeded the maximum admissible limit for drinking water of 11 mg/l.[26]

**Metals.** The earliest records of metal pollution in natural waters date from the 19th century and describe the effects of mine drainage on surface waters in Great Britain:

All these streams are turbid, whitened by the waste of the lead mines in their course; and flood waters in which, spreading over the adjoining flats, wither befoul grazing on the dirtied herbage, or, by killing the plants whose roots have held the land together, render the shores more liable to abrasion and destruction on the next occasion of high water.[27]

Like nitrates, metals are natural constituents of soil and water, but in recent decades the worldwide production and use of metals has expanded dramatically, and so has the associated problem of aquatic pollution. The dispersion of metals into the biosphere as a consequence of industrial and agricultural activity appears now to rival, and sometimes to exceed, natural mobilizations for certain elements. This is particularly true for lead. Present day levels of lead in industrialized countries have been estimated to be two to three orders of magnitude higher than those of the pre-technological age.[28] While no other metal has the same prevalence as lead, several are widespread in aquatic environments. The ocean influx of cadmium has increased by as much as 60% over natural levels.[29] Aluminum has been mobilized on a regional scale in soils and water affected by acid precipitation. Mercury contamination has also become a problem of regional significance.

It was not until the 1950s that several mass poisoning episodes brought the toxicity of heavy metals to the attention of the public and the scientific community. The first such episode occurred between 1947 and 1965 when the Jintsu River in Japan became heavily contaminated with cadmium from a mine located upstream. Water from the river flooded lowlying rice fields, and over a 20-year period, 100 deaths due to cadmium poisoning (Itai-itai disease) were reported in the region.[30] In 1953, the first mercury-related deaths were reported around Minamata Bay, also in Japan. Fish and foodstuffs in the villages surrounding the bay were contaminated with methyl mercury, which originated in the effluent of an upstream plastics manufacturing company.[31] Similar episodes were recorded in other parts of Japan, Sweden, and Canada, ultimately leading to several bans on fishing in mercury-contaminated waters.

These incidents are examples of the indirect poisoning of human beings that resulted from contaminated water affecting foodstuffs. The acute poisoning of human beings by metal-contaminated drinking water is rare because of the large doses required. The primary health concern posed by metals is either indirect poisoning, particularly through the formation of organic metal complexes in foodstuffs, or long-term chronic effects. In addition, however, certain metals are highly toxic to aquatic life at relatively low concentrations.

From a human health perspective, the metals of greatest concern are lead, mercury, arsenic, and cadmium. Lead is one of the most toxic metals found in aquatic systems; it is particularly toxic to young children, and its hazards include kidney damage, metabolic interference, central and peripheral nervous system toxicity, and depressed biosynthesis of protein, nerve, and red-blood cell formation.[32] Mercury is also highly toxic and unique among the metals because it is consistently biomagnified within the aquatic food chain. In its inorganic form, mercury may cause kidney damage and ulceration. In surface waters, however, mercury is frequently converted to its more toxic organic form, methyl mercury. Ingestion of methyl mercury affects the central nervous system and can cause death even at relatively low doses. Arsenic exhibits several chronic effects when it is ingested; the most characteristic are hyperkeratosis of the palms and soles of feet and hyperpigmentation. Arsenic also affects the gastrointestinal tract and liver, and may induce skin tumors and skin cancer. The major health effect associated with long-term ingestion of cadmium is renal disease. Other metals which pose chronic health threats include aluminum, which may be a confounding factor in Alzheimer's disease; chromium, which is associated with dermatitis, pulmonary congestion, and nephritis; and organotin complexes, which are neurotoxic.[33]

Many metals that are not particularly toxic to human beings are, nonetheless, highly toxic to aquatic life. Among these are copper, silver, selenium, zinc, and chromium. The toxicity of metals to aquatic plants and animals is controlled by their oxidation state and their ability to form particular complexes in the environment, which are largely determined by prevailing water chemistry (e.g., pH, salinity, pE, and temperature). For instance, the toxicity of copper is moderated by the formation of water-soluble ligands that bind the metal and reduce its bioavailability; the formation of these complexes is favored by circumneutral pH.[34]

Sources of metal pollution include geologic weathering, the industrial processing of ores and minerals, the use of metals and metal components, the leaching of metals from garbage and solid waste dumps, and animal and human excretions that contain metals. Domestic effluents constitute the largest single source of elevated metal concentrations in rivers and lakes.[35] More recently, atmospheric deposition has been recognized as a source of metal pollution in rural and alpine waters. A study of lake sediments in Denmark, for example, found that atmospheric deposition since 1945 had increased the heavy metal content of sediments in oligotrophic Lake Hampen to levels 180 times greater than background.[36] Nriagu and Pacyna have estimated the global anthropogenic inputs of trace metals into aquatic ecosystems (including oceans), concluding that the most important sources (in descending order) were domestic wastewater effluents (As, Cr, Cu, Mn, and Ni), coal-burning power plants (As, Hg, and Se), non-ferrous metal smelters (Cd, Ni, Pb, and Se), iron and steel plants (Cr, Mo, Sb, and Zn), and the dumping of sewage sludge (As, Mn, and Pb). On a weight basis, anthropogenic inputs are greatest for manganese, zinc, chromium, and lead.[37]

An inventory of Lake Erie conducted in the late 1970s found that among the various sources of metal inputs into the lake, atmospheric inputs accounted for 20% of the copper influx, 35% of the zinc influx, and 50% of the lead influx. Sewage effluents were responsible for 45%, 30%, and 20% of the inputs of copper, zinc, and lead, respectively.[38] A similar inventory of the Ruhr River, which flows through a highly industrialized region of Germany, found that 55% of the heavy metal loading came from wastewater treatment plants (industrial and sewage) and 45% from geochemical loading. Overall, 90% of the chromium entering the Ruhr was discharged by industries, predominantly from metal refining processes such as galvanizing. Industry was primarily responsible for nickel and cadmium pollution as well, but accounted for only 50% of the copper and zinc inputs.[39]

In urban areas, stormwater runoff is a major source of metal pollution in surface waters. The U.S. Environmental Protection Agency's (EPA) Nationwide Urban Runoff study concluded that heavy metals, especially copper, lead, and zinc, were by far the most prevalent pollutants found in urban runoff. End-of-pipe concentrations exceeded both EPA ambient water quality criteria and drinking water standards in many instances. All 13 metals on the EPA's priority pollutant list were detected in urban runoff samples, with a frequency that generally exceeded 10%.[40]

Mining is another important source of metal inputs into surface and ground waters, particularly in Latin America and parts of south-east Asia, where mining activities are both widespread and poorly controlled. Since the late 1970s, many rivers and waterways in the Amazon have been exploited for gold, resulting in extensive mercury pollution of a relatively pristine system. Mercury is used in the mining process as amalgamate to separate the fine gold particles from other mineral constituents in sediments and gravel. Official figures imply that for every kilogram of gold extracted, at least 1.32 kg of mercury is lost to the environment. The Brazilian state of Rondônia has estimated that from 1979 to 1985 approximately 100 tonnes of mercury were discharged into the Madeira River basin, of which about 45% will reach the river. Very high concentrations of methyl mercury have been found in fish in downstream reaches (up to 2.7 µg/g wet weight), exceeding the WHO's criterion for human consumption (0.5 µg/g).[41]

The rate at which metals are mobilized from surrounding soils has been increasing in many regions as a result of acid precipitation. Acid conditions are strongly associated with the mobilization of aluminum, a metal which is generally insoluble in neutral and alkaline water. Until recently, aluminum was regarded as being non-toxic; consequently, there are relatively few data on aluminum exposure and toxicity. Levels of aluminum in drinking water vary. For neutral water, concentrations are generally in the range 0.01–0.1 mg/l. In the southernmost parts of Norway the mean aluminum concentration in waterworks is about 0.4 mg/l, but levels as high as 1.3 mg/l have been found in some well water. Globally, the WHO estimates that aluminum intake via drinking water amounts to only about

3% of total dietary intake, but in regions with highly acidic water this may rise to 30% or 40%.[42]

A strong relationship also exists between lake water pH and mercury levels in fish. Mercury appears to be mobilized at a greater rate by low pH, and both its availability for methylation and its solubility also increase. A correlation between the pH of lake water and mercury body burdens in fish has been observed in North America and Scandinavia in regions where no local anthropogenic sources of mercury exist. In Sweden, more than 9,400 lakes contain fish with methyl mercury concentrations higher than 1 µg/g.[43] The mobility of cadmium and lead in soils is also highly dependent upon the pH of soil moisture.

A nationwide investigation of well water in Finland revealed that out of 100 wells surveyed, the average pH had dropped from 6.7 in 1958 to 6.3 in 1989. More than half of these wells had pH levels below the minimum recommended by the Finnish health authorities for drinking water (6.0). The most extreme declines in pH were observed in areas subject to the highest sulfur fallout. Similarly, distinct geographical variations in drinking water quality have been found in Norway and Sweden, with metal concentrations showing a north–south spatial gradient.[44] Of particular concern is the increased aggressiveness of acid water on pipes, cisterns, and tanks, which may leach lead and other metals into drinking water supplies, posing a direct risk to consumers.

An assessment of the global and regional extent of metals contamination is hampered by both a lack of data and inadequate sampling and analytical techniques. For example, the WHO's GEMS project monitored for dissolved metals in only 35% of the network rivers, and the significance of these limited measurements is still questionable, due to poor quality control procedures.[45] In general, however, metal concentrations in fresh waters vary greatly due to the localization of sources and the tendency of metals to concentrate in sediments and slowly partition into overlying waters as environmental conditions change.

A temporal and spatial analysis of metal concentrations conducted for the Rhine River shows that cadmium concentrations in sediments rose from less than 2 mg/kg before 1920 to about 40 mg/kg in the early 1970s. Following the 1970s, concentrations decline slightly, which may reflect increased controls on cadmium discharges but more likely reflects improved analytical techniques. Overall, dissolved cadmium concentrations increase dramatically from the source (12 ng/l) to the mouth of the river (400 ng/l), suggesting the predominance of anthropogenic sources.[46]

Table 3.2 shows concentrations of cadmium for various rivers, with values for dissolved cadmium ranging from 2 to 400 ng/l. Those rivers with the highest concentrations are not surprisingly those from the most industrialized areas, including the Hudson and the Rhine. Unquestionably, metal concentrations in fresh water have been rising rapidly over the

**TABLE 3.2** Concentration of cadmium in rivers

| River | Suspended particulate matter (mg/l) | Particulate cadmium (ng/l) | Particulate cadmium (mg/kg) | Dissolved cadmium (ng/l) |
|---|---|---|---|---|
| Elbe[a] | 75 | 150 | 2 | 100–200 |
|  |  | 85 |  | 35 |
| Weser | 50 | 100 | 2 | 100 |
|  |  |  |  | 70 |
| Varde Å | 40 | 80 | 2 | 70 |
| Rhine | 30 | 1800 | 60 | 400 |
| Amazon |  |  |  | 8 |
| Mississippi |  |  | 0.6 | 15 |
| St. Lawrence | 10 | 17 | 1.7 | 110 |
| Hudson | 10–20 | 70–80 | 4–8 | 200–300 |
| Orinoco |  |  |  | 2 |
| Yangzijiang |  |  |  | 2 |
| Gota | 25–100 | 14–64 | 0.3–0.6 | 9–25 |

*Source:* P.A. Yeats and J.M. Bewers, 1987, Evidence for anthropogenic modification of global transport of cadmium, in J.O. Nriagu and J.B. Sprague (eds.) *Cadmium in the Aquatic Environment*, John Wiley, New York, p. 22. (Copyright © 1987, Reprinted by permission of John Wiley and Sons, Inc.)

[a] Different values from the same river were measured by different investigators.

**TABLE 3.3**  Examples of organic pollutants, by class and typical use

| Class and pollutant | Typical use |
|---|---|
| Halogenated aliphatic (chain) hydrocarbons | |
| Dibromochloromethane (THM) | Trihalomethanes (THMs) are formed during the disinfection of drinking water |
| Trichloromethane (THM) | |
| Dichloromethane | Solvent used to decaffeinate coffee |
| Carbon tetrachloride | Used in the manufacture of fluorocarbon propellents, general purpose cleaner and solvent |
| Tetrachloroethylene (PCE) | Most widely used drycleaning chemical in the US; found in spot removers, rug cleaners, paint strippers, etc. |
| 1,1,1–Trichloroethane (TCA) | Very common solvent; used to clean electrical equipment; found in drain cleaners, shoe polish, spot removers, etc. |
| Trichloroethylene (TCE) | Solvent; formerly used to decaffeinate coffee; metal degreaser; found in dyes, ink spot removers, etc. |
| Aromatic (ring) hydrocarbons | |
| Benzene | Derived from petrolem and coal; common feedstock; found in paints, oils, adhesives, aspirin, asphalt, etc. |
| Toluene | Petroleum by-product and gasoline additive |
| Naphthalene | Found in moth balls, carpet cleaners, typewriter correction fluid, adhesives, asphalt, etc. |
| Anthracene (PAH) | Polyaromatic hydrocarbons (PAHs) are formed from incomplete fossil fuel combustion; |
| Benzo(a)pyrene (PAH) | benzo(a)pyrene is present at relatively high concentrations in cigarette smoke |
| Chloro– and nitro–aromatic hydrocarbons | |
| Hexachlorobenzene | Fungicide |
| Trinitrotoluene (TNT) | Explosive |
| Phthalates | |
| Dimethyl–phthalate | Phthalates are added to plastics to make them flexible; found in rainwear, footwear, |
| Bis(2–ethylhexyl)–phthalate | shower curtains, children's toys |
| Halogenated ethers | |
| Bis(2–chloroisopropyl)ether | Halogenated ethers are used in the production of plastics and resins and in research laboratories |
| 2–Chloroethylvinylether | |
| Phenols | |
| Pentachlorophenal | Fungicide; wood preservative |
| 2–Chlorophenol | Chloro–, dichloro–, and trichloro–phenols are by-products in the production of pentachlorophenol |
| 2,4–Dichlorophenol | |
| Organochlorines | |
| DDT | DDT, lindane, aldrin and chlordane are examples of the extremely persistent organochlorine |
| Lindane | pesticides widely used in the 1950s and 1960s |
| Aldrin/dieldrin | Dieldrin is a degradation product of the pesticide aldrin |
| Chlordane | |
| Alpha–hexachlorocyclohexane (HCH) | Common pesticide |
| 2,4–D | Common herbicide |
| Polychlorinated biphenyls (PCBs) | Used to insulate electrical equipment, hydraulic fluids and lubricants |
| 2,3,7,8–Tetrachlorodibenzon–p–dioxin | By-product in the production of certain herbicides; common contaminant of the defoliant Agent Orange |

last few decades. Although human health concerns are limited to a few key metals, the prevalence of metal pollution poses widespread risks to aquatic life and ecosystems.

**Synthetic and industrial organic pollutants.** Over the last three decades, the large-scale manufacture and release of organic chemicals have resulted in widespread environmental contamination. The most well-known examples remain the extremely persistent organochlorine compounds, including the pesticide DDT and its derivatives and the polychlorinated biphenyls (PCBs). DDT has been detected almost everywhere in the world, including in water from melted Antarctic snow.[47] Yet many other classes of organic contaminants now affect ground and surface waters. There are approximately 100,000 synthetic compounds currently in use, and more are introduced every year, often without a full understanding of the risk they pose to the environment and to human health.[48] It is impossible to generalize about the potential toxicity of organic micropollutants because there are so many, all with different derivatives, different solubilities, and different persistence. Yet an important aspect of organic chemicals is that many are thought to be hazardous to aquatic life and human health at concentrations much lower than those that can be reliably measured by common analytical methods. The uncertainty generated by the presence of organic chemicals and their potential health effects is perhaps their most distinctive characteristic. The more important classes of organic micropollutants from a drinking water perspective are given in Table 3.3.

Most organic pollutants originate in industrial activities such as petroleum refining, coal mining, the manufacture of synthetic chemicals and products, iron and steel production, textile production, and wood pulp processing. In addition, the domestic use of petroleum and heating oils, atomizers, pesticides, and fertilizers further diffuses synthetic pollutants into the environment. A study conducted in Los Angeles identified 101 organic substances in urban wastes, 36 of which were on the U.S. EPA list of priority pollutants.[49] Non-point sources also make a significant contribution to organic chemical pollution. For example, the EPA urban runoff study identified 63 organic pollutants in runoff from 19 cities. The most frequently identified were a phthalate plasticizer, the pesticide - hexachlorocyclohexane, and several polycyclic aromatic hydrocarbons (PAHs).[50] Agriculture, however, is the principal non-point source of organic micropollutants in drinking water, and the pollution of water with agricultural pesticides and herbicides has become a global concern.[51]

The term ''pesticide'' is a general one, and in the past 50 years several different classes of synthetic organic pesticides have been developed and used extensively throughout the world. Among these are the ''early'' chlorinated hydrocarbon insecticides, such as DDT, lindane, aldrin, dieldrin, etc. – most of which have been either banned or severely restricted in the industrialized world. Other classes include the organophosphorus and carbamate pesticides, chlorophenoxy acids, triazines, and bipyridilium herbicides, and the more recent pyrethroid insecticides. Although plants and soil are the recipients of most pesticides applied, water provides the principal transport pathway through both the erosion of contaminated soil and the dissolution of water-soluble compounds.

The environmental behavior of different pesticides is controlled by two principal characteristics: solubility in water and persistence in the environment. Chemicals with low solubility and high persistence, such as many of the older organochlorine insecticides, are generally found in association with particulate bed materials or suspended sediment and may not degrade for several years. These chemicals also have a high lipid solubility and tend to accumulate in aquatic organisms and their predators. Over time, they can reach harmful concentrations in organisms even though concentrations in water and sediment may remain low. Conversely, most newer generation pesticides are much less persistent and have a much lower tendency to bioaccumulate. They may last only days or weeks in the environment before breaking down. Even though these chemicals frequently degrade to a benign form and do not accumulate in organisms, most are more acutely toxic than the organochlorine insecticides. For many of these newer compounds, volatilization is the primary pathway for human exposure, and relatively high concentrations have been detected in rain and dew.[52] The highest concentrations, however, are found in surface water runoff from the edges of fields. A study of the

lower Mississippi River found that several triazine and chloroacetanilide herbicides and their degradation products were present in the river system. Daily loadings were found to be as high as thousands of kilograms.[53]

In the U.S., concentrations of organochlorine insecticides, including dieldrin, chlordane, and DDT, appear to have decreased erratically but significantly in both water and sediment since the mid-1970s, when their use was greatly curtailed.[54] The use of herbicides, however, has rapidly increased during the past 20 years, with atrazine and 2,4-D accounting for much of the rise. Other prevalent herbicides include alachlor, butylate, and metachlor, which account for over one-half of all herbicides used in the U.S.[55] These herbicides are both less adsorbent and more soluble in water and therefore have a greater potential to move out of the root zone and to contaminate ground water than their precursors.

In addition to pesticides and herbicides, there are several other important classes of organic micropollutants:

- PCBs, which were used extensively in the electrical industry as dielectrics in large transformers and capacitors. The term PCB is general, and is applicable to a wide range of related compounds. Those with a higher degree of chlorination pose the greatest environmental and health concerns.
- PAHs, which result from the incomplete combustion of organic materials, primarily fossil fuels. These are hydrophobic compounds that are associated with sediments and are relatively immobile in aquatic environments; however, they may be transported long distances in the atmosphere.[56] The toxicity varies among individual compounds, but some, such as benzo(a)pyrene, are known human carcinogens.
- Organic solvents, which are highly volatile, chain carbon compounds frequently found in ground water as the result of both industrial processes (e.g., plating and electronics manufacturing) and leaching from landfills. Many solvents, such as trichloroethane, are common household products (cleaners, paint thinners, etc.). Particular solvents have been associated with cancer, birth defects, and cardiovascular disease, although in most cases the epidemiological data are inadequate to establish causality.
- Phthalates, a large group of chemicals used as plasticizers, especially in the production of polyvinyl chloride resins. Two compounds are of particular concern for human health: butylbenzyl phthalate (BBP) and di(ethylhexyl)phthalate (DEHP).
- Disinfection by-products (DBPs), chemicals that are formed upon the addition of chlorine or a similar disinfectant (ozone, chloramines, chlorine dioxide) to water. For instance, chlorine reacts with naturally occurring humic substances to produce trihalomethanes (THMs), including the carcinogen chloroform. Most of the DBPs of concern are organic, including THMs, haloacetic acids, and chlorinated hydroxyfuranones. Several DBPs are known or suspected human carcinogens.

Since the mid-1970s, public concern has grown steadily over the link between cancer and organic chemicals in drinking water. While several organic chemicals are either known or suspected human carcinogens, many more have not been adequately studied. Moreover, while the risks posed by individual chemicals in drinking water are relatively low, the cumulative effects of long-term exposure to a variety of chemicals may be much larger but cannot be well quantified. However, the rising incidence of cancer in the industrialized world and the identification of both ''cancer clusters'' and high rates of miscarriages, birth defects, and sterility in areas with contaminated drinking water and high levels of chemical exposure have continued to raise questions about the safety of organic chemicals and the adequacy of their regulation.[57]

Several epidemiological studies have suggested a link between cancer or birth defects and the presence of organic chemicals in drinking water. For instance, a study conducted in 1983 in Santa Clara County, California confirmed a statistically significant increase in adverse pregnancy outcomes in a community exposed to solvent-contaminated ground water, but could not establish a causal connection.[58] Similarly, a correlation has been found between chlorinated drinking water and colon and bladder cancer. While no direct evidence points to DBPs as the cause, recent estimates of the attributable risk (i.e., the proportion of all cancer cases in both exposed and unexposed individuals that is presumed to be caused by the exposure of interest) for colon and bladder cancers are approximately

38% and 21%, respectively.[59] To date, the link between cancer and chlorination remains weak, and it is widely acknowledged that the benefits of chlorination far outweigh the risks. Nonetheless, considerable effort in the industrialized countries is being put towards identifying alternative means of disinfection and methods that reduce chlorine contact time and overall DBP concentrations.

Most studies that have attempted to define the extent of organic chemicals in water have been local in scale, and thus it is not possible to assess the status of fresh waters globally. The expense of laboratory analyses, the variable frequency of detection, regional patterns of use, and the constantly changing array of available pesticides and synthetic compounds make national or regional scale monitoring a very difficult undertaking. The GEMS project, for example, monitored only a few organochlorine pesticides and PCBs at only 25% of the stations. Noticeable was the absence of any measurements in Africa, in the Middle East, and in South America (except for one station in Colombia since 1982).[60] While concentrations of the chlorinated organics appear to be decreasing in many regions of the world, most of the other identified contaminants have not even been tested for. Certainly the increasing usage of organic chemicals in a variety of activities and industries suggests that their prevalence in fresh water is increasing as well. To date, the human health impacts of organic chemicals in water have been relatively minor or extremely uncertain. Nonetheless, the relatively high toxicity of these chemicals, the existence of so many synthetic compounds, and the chronic and cumulative nature of the threat all suggest that in the future the presence of organic chemicals in fresh water will emerge as a major public health concern unless the global trend toward increasing chemical production is curbed.

**Radionuclides.** Compared to the background exposure of human beings to radiation from natural causes, the radiation risks associated with drinking water are often negligible. Yet, in some areas, radioactive contamination of drinking water from either accidents, such as the Chernobyl disaster, unsafe disposal practices, or high levels of naturally occurring radioactivity, may cause significant health concerns at a local or regional scale. Natural sources of radioactivity and the disposal of nuclear waste may affect ground water and springs, while surface waters are subject to contamination from atmospheric testing, power plant accidents, the mining of radioactive materials, manufacturing and transportation accidents, and the disposal of radioactive wastes. The health effects of radiation are well established: developmental abnormalities, cancer, and death. The most abundant radionuclides in water are given in Table 3.4.

From a drinking water standpoint, naturally occurring radionuclides pose the greatest threat, although anthropogenically produced particles may also be a health concern in specific localities. The level of radioactivity in natural waters generally falls between 1 and 1,000 pCi/l; however, in some naturally radioactive springs, concentrations as high as 100,000 pCi/l have been detected.[61] The WHO guidelines for radioactivity in drinking water, established in 1984, are 0.1 Bq/l (0.003 pCi/l) for alpha radiation and 1 Bq/l (0.03 pCi/l) for beta radiation.[62]

The individual radionuclides of greatest concern are various isotopes

of naturally occurring uranium, radon, radium, thorium, lead, and polonium. Of these, radon, an alpha emitter, is the most prevalent and poses the greatest health risk. A water soluble gas, radon is produced by the decay of radium isotopes. The predominant form is $^{222}$Rn. The primary source of radon is soil, from which it emanates directly, while lesser sources include ground water, oceans, phosphate residues, uranium tailings, coal residues, natural gas combustion, coal combustion, and human exhalation. Ground water is the most significant of the secondary sources.[63]

The U.S. EPA estimates that indoor radon causes between 8,000 and 40,000 lung cancer deaths each year. In homes supplied by ground water, approximately 5% of these deaths are the result of radon released from water during showering, bathing, flushing toilets, cooking, and washing clothes and dishes.[64] Radon present in water accounts for about 10 times more morbidity than $^{226}$Ra or $^{228}$Ra, and 20 times more than natural uranium.[65] In fact, the calculated incidence of fatal cancers in the U.S. population from radon in water alone may be larger than the sum of all other carcinogens known to be present in existing water supplies, making radon in drinking water a significant health threat in some localities.[66]

Radon concentrations in drinking water range from less than 10 pCi/l in typical surface waters to a reported high of 2,000,000 pCi/l in ground water from a private well. A study undertaken by the American Water Works Association estimated a population-weighted average concentration of 106 pCi/l in U.S. drinking water, and a corresponding lifetime risk of death from cancer of $10^{-4}$. The average population-weighted concentration among ground water systems was considerably higher at 273 pCi/l.[67]

In local and regional areas, poor waste disposal practices have led to contamination of surface and ground waters with anthropogenic radionuclides. For example, tritium has been detected in ground water in Burke County, Georgia and Barnwell, South Carolina, and has been attributed to a nearby weapons manufacturing plant. And in December of 1991, a spill of tritium from the same plant entered the Savannah River and forced the temporary shut down of food processing plants that draw water from the river.[68]

## Current and emerging trends affecting water quality

**Population growth, urbanization, and poverty.** In the coming decades, the greatest challenges to water quality will come from continued population growth and the trend towards urbanization in the developing countries. In 1975, 38.5% of the world population lived in urban areas; in 1990 this rose to 42.7%, and it is expected to reach 46.7% by the year 2000 and 60.5% by 2025.[69] While the overall level of urbanization in the developing countries is not yet as high as in the industrialized nations, the trend toward urbanization in the former is even more pronounced. For instance, in Latin America and the Caribbean, which is the most urbanized region in the developing world, there are 215 cities that have populations of more than 100,000. Overall, the region's urban population comprises 69% of the total, and one-third of the urban population is located in 15 "mega-cities" with over 2 million inhabitants.[70]

Among the consequences of this urban influx are the overloading of water and sanitation infrastructure and a rapid deterioration in urban living conditions. In many overcrowded cities often only one-fourth to one-third of the population is served by garbage collection.[71] According to WHO, only 41% of the urban population in Latin America and the Caribbean has access to sewer systems, and over 90% of the wastewater that is collected is discharged directly into receiving waters without any treatment. Another 38% of urban residents is served only by on-site sanitation systems; thus, 59 million urban dwellers do not have access to adequate sanitation. Moreover, higher population densities make many low-cost sanitation options, such as pit latrines and cesspools, less effective in the prevention of disease. If projections of urban growth in the region prove accurate, an additional 141 million people in Latin America will need sanitation services by the year 2000.[72] Similar gaps in service are projected in Asia and Africa.

Improvements in health will be confounded by the pervasive link between urban poverty and water quality. The number of urban poor will have doubled in the period between 1975 and 2000, from 35% to 77% of

**TABLE 3.4**  Radionuclides most abundant in drinking water

| Low LET[a] | High LET |
|---|---|
| Potassium–40 | Radium–226 |
| Tritium | Daughters of Radium–228 |
| Carbon–14 | Polonium–210 |
| Rubidium–87 | Uranium |
|  | Thorium |
|  | Radon–220 |
|  | Radon–222 |

*Source:*  J. De Zuane, 1990, *Handbook of Drinking Water Quality: Standards and Controls*, Van Nostrand Reinhold, New York, p. 340.

[a]  Linear energy transfer = Average amount of energy lost by an ionizing particle per unit length of track in matter.

families, with even higher percentages for South Asia and Africa. Several studies have established the greater prevalence of diarrhea and helminthic infections in environments that have poorer housing, water, and sanitation facilities. For example, in São Paulo, the incidence of diarrhea in the lowest socioeconomic stratum was 13.1 episodes per 100 children-months compared with 9.6 episodes in the next stratum and 3.6 episodes in the upper stratum. With respect to helminths, studies in Durban, Singapore, Guatemala and Seoul have all found a higher prevalence of *Ascaris* and *Trichuris* in poorer parts of the city as compared to more wealthy parts.[73] Similarly, the rapid spread of Chagas disease in Latin America is attributed to the proliferation of urban slums where poor housing, overcrowding, and inadequate health services contribute to transmission.

The United Nations and WHO designated the 1980s as the "International Drinking Water Supply and Sanitation Decade" (the Water Decade) in an attempt to draw both attention and resources to the problem of drinking water supply in the developing world. During this decade, significant improvements were made, especially in rural water supply, where the number of persons served increased by 240% globally, while the number of persons with access to rural sanitation increased by 150%. The number of urban dwellers with access to water supply rose by 150% as well, as did the number with access to sanitation; however, this still translated into an increase in the number of urban dwellers without access to services. Put into absolute numbers, the additional people with water supply at the end of the decade totaled 1.3 billion, while those with sanitation increased by 750 million. Yet on the other side of the equation, this left 1.2 billion people without a safe water supply and 1.7 billion without adequate sanitation. In Latin America and Asia, the percentage of the urban population with access to sanitation remained essentially unchanged, which reflects the rapid growth of cities. Globally the percentage of urban persons with sanitation services increased only marginally, from 69% to 72%.[74]

The failure of the decade to meet its ambitious goal reflects the continuing problem posed by rapid population growth and urbanization in the developing world. Substantial investments in sanitation and water supply are now required merely to keep pace. While population continues to grow rapidly in many developing countries, little progress has been made in alleviating poverty or improving living conditions. Environmental conditions are deteriorating rapidly, particularly in overcrowded urban areas. More and more people are born into poverty and will eventually suffer from and spread waterborne disease. Yet in addition to the problems faced by the Third World, the declining economic situations of eastern Europe and the former Soviet Union are likely to lead to more serious water quality problems. During 1991, the incidence of certain infectious diseases in the Russian Republic increased, reversing previous trends. National incidence rates for bacterial dysentery and other enterically transmitted diseases increased substantially. In the Tom River basin in Siberia, inadequate maintenance of water-purification systems and organic contamination of drinking water supplies were associated with increased incidence rates of gastritis, hepatitis A, and bacterial dysentery. Prevalence rates for these diseases were 82%, 47%, and 22% higher, respectively, than the national incidence rates.[75]

In general, water quality deteriorates along with economic conditions. And to date, the wealthier countries of the world have failed to address the economic situation of poorer regions with the urgency required. Clearly, if current trends continue, the problems posed by deteriorating water quality could prove overwhelming.

**Industrial expansion.** Along with population growth, rapid industrialization puts pressure on water resources and creates two principal threats to water quality: the increase in industrial effluents, particularly in countries that lack adequate environmental controls, and the increased risk of chemical spills and accidents.

The major urban centers in the developing world have increasingly become centers of industrial concentration, and severe industrial pollution problems have arisen in most large cities. While this situation is similar to that faced by the countries of Europe and North America 50 years ago, the critical difference is the much greater pressure on water and other natural resources in many parts of the less developed world. It is

**TABLE 3.5** The production of selected classes of chemicals in India, 1950 to 1980 ($10^3$ tonnes)

| Class | 1950 | 1960 | 1970 | 1980 |
|---|---|---|---|---|
| Pesticides | – | 1.46 | 3.00 | 40.68 |
| Dyes/pigments | – | 1.15 | 13.55 | 30.85 |
| Pharmaceuticals | 0.25 | 1.23 | 1.79 | 5.07 |
| Organic chemicals | 200 | 580 | 17,100 | 24,100 |
| Fertilizers | 18 | 153 | 1,059 | 3,005 |
| Caustic soda | 11 | 101 | 304 | 457 |

*Source:* B.B. Sundaransan *et al.*, 1983, An overview of toxic and hazardous waste in India, *Industrial Hazardous Waste Management: Industry and Environment*, special issue no. 4, United Nations Environmental Programme, Paris, p. 70, Environmental Pollution Control in Relation to Development, cited in: World Health Organization, 1985, Technical Report Series no. 718, WHO, Geneva.

estimated that industrial effluents constitute 90% of non-agricultural water pollution in Mexico. The paper and steel industries, which rank among the most important industrial sources of water pollution in Latin America, have been growing twice as fast as the economy of these countries as a whole. Companies are finding it increasingly profitable to move manufacturing processes into regions with lower labor costs and less stringent environmental regulations. This is particularly true for very hazardous industries, such as those that manufacture asbestos products, certain dyes, and pesticides, as well as mineral processing. Table 3.5 shows the overall increase in the manufacture of certain classes of chemicals in India. And Latin America, which had only 5.3% of the world total in value-added industry, had proportionately higher shares in pollution-intensive industries such as petroleum refining (17.7%) and chemical production (14.7%).[76] In developing countries, advanced pollution control technologies are rarely adopted in newer, chemical-intensive industries such as high technology. But even in traditional industries such as auto manufacturing and food processing, there is now a greater reliance on chemicals.[77]

Food processing is an example of an industry widespread in the developing world in which large amounts of wastes are discharged into surface water with little regulation and no treatment. In Latin America, more than 17.5 million tonnes are processed annually and an estimated 10 million tonnes of waste are discarded, with little or no treatment.[78] Similarly, in Kenya coffee is the leading cash crop; over 120,000 tonnes are processed annually. There are over 1,200 coffee processing factories in the country, and all are located near watercourses, which become severely polluted during the coffee season. A survey of several rivers and streams between Nairobi and Thika during the processing season found gross pollution. Every river and stream surveyed was anaerobic, with all measurements of 5-day biological oxygen demand (BOD-5) exceeding 10 mg/l and some rising as high as 100 mg/l.[79] For the most part coffee wastes are treatable; however, economic constraints have slowed the adoption of environmental control technologies.

The expansion of industrial activities throughout the world also brings with it an increased risk of large-scale industrial accidents, such as the 1986 chemical fire which polluted the Rhine River with pesticides, solvents, dyes, and other hazardous chemicals. The fire burned a storehouse that contained at least 90 different chemicals, including a few highly toxic mercury-based pesticides. An estimated 1%–3% of the 1,300 tonnes of stored materials reached the river. Following the accident, biota in the river were heavily damaged for several hundred kilometers downstream. Benthic organisms were among the most severely affected, along with eels, which were eradicated over a distance of 400 km. Downstream, ground water was also contaminated, with many wells in the Alsace region showing the presence of organophosphate compounds and mercury.[80] Although the Rhine is a source of drinking water for approximately 12 million people, no direct health threats occurred because the water is extensively treated and well-monitored before use. Yet had the accident occurred in a region any less prepared, widespread human health impacts would have accompanied the destruction of the ecosystem. During November 1990, a tank-filling accident at a plastics plant in Novopolotsk, Byelorussia spilled several thousands gallons of acetocyanohydrin, a toxic chemical. Mishandling of what should have been a minor incident allowed toxic amounts of cyanide to flow into tributaries of the Daugava River, Latvia's chief waterway. The Daugava supplies drinking water to about 1 million people in the cities of Daugave and Riga.[81]

While these are only isolated examples of the threat posed by industrial accidents, they are indicative of both the magnitude and the variability of accidental pollution. In many regions, the technology for adequate monitoring of industrial pollution is not available, much less the regulatory infrastructure for responding to massive spills. Poorer countries are also unable to mitigate the effects of environmental catastrophes. As polluting industries shift into those regions that are less prepared to deal, both technically and financially, with environmental concerns, the result will be the increasing pollution of soil, air, and water.

**Agricultural pollution.** Traditional water quality problems associated with agriculture will continue, including sedimentation and eutrophication, but from the perspective of human health, the major concern in the future will be the increase in nitrate and pesticide contamination of ground and surface water as greater quantities of chemicals are introduced into agriculture. As discussed above, nitrate concentrations in water have been rising rapidly since the 1940s, corresponding with the increased use of fertilizers and the expansion of agriculture. The prospect of a similar rise in pesticide concentrations is particularly disturbing.

In the U.S., the realization that agricultural pesticides threatened ground water resources came in 1979 with the discovery of aldicarb in 96 wells on Long Island and 1,2-dibromon-3-chlorpropane (DBCP) in more than 2,000 wells in California.[82] To date, nearly 100 different pesticides have been detected in ground water. According to an assessment made by the U.S. EPA, 74 different pesticides were detected in ground water in 38 states. Of these, 32 pesticides in 12 states were thought to be related to point sources and 46 pesticides detected in 26 states were attributed to agricultural non-point sources.[83]

Globally, there are two disturbing trends in pesticide usage: the continued use of extremely hazardous and highly persistent organochlorine compounds in the developing countries, and the propagation of a new generation of herbicides in the industrialized countries for which current water quality monitoring is inadequate. The less developed countries have continued to use the organochlorine chemicals in agriculture and for the control of disease-carrying insect vectors. For example, India still uses DDT in cotton production and in malaria control. In Latin America, governments restrict only 20%–25% of the agricultural chemicals that appear on the United Nations list of chemicals that have been banned or restricted by other national governments (Table 3.6). Brazil used both the extremely toxic hexachlorocyclohexane (HCH) and DDT extensively from 1970 to 1980; currently Brazil bans these chemicals in agriculture but continues to allow their use in public health applications. In addition, overall fertilizer and pesticide application rates are rising throughout the world. While application rates are highest in the industrialized countries, the rate of increase is most rapid in the developing countries. The total volume of consumption is not known; however, estimates of global pesticide usage in 1985 yield a figure of approximately 3 million tonnes. The fastest

growing market is Africa, which had an increase in pesticide sales of 182% between 1980 and 1984. Latin America is also experiencing rapid growth in usage, with pesticide imports increasing by almost half between 1973 and 1985.[84] Total pesticide usage also continues to rise in the industrialized countries, despite the restriction of several of the most toxic compounds. In the U.S., pesticide use grew by roughly 170% between 1964 and 1985.[85]

The severity and extent of ground water contamination cannot be adequately assessed for several reasons. First, even in the industrialized world, most of the data that exist are for shallower drinking water wells, and little information is available for deeper wells. Second, contaminant concentrations vary over time, making it difficult to assess ground water quality without undertaking long-term, and costly, monitoring programs. Third, once the temporal and spatial variability of water quality are adequately described, it is difficult to relate contamination to selected agricultural practices, in part because ground water velocities are relatively low. Agricultural practices undertaken today may not be reflected in the hydrologic regime for several months or possibly years.[86]

Finally, almost all the data that exist come from the more developed countries. The rapid introduction of pesticides into less developed countries has not been accompanied by monitoring. Today's practices may be affecting vital water resources over large areas without our knowledge. This is unlikely to change in the near term because ground water monitoring is an expensive undertaking that may not seem economically justifiable in many poorer regions of the world. At this time, we may be only beginning to detect contamination resulting from agricultural practices of 30–40 years ago; the impact of the last 15 years of chemically dependent agriculture remains to be seen.

**Water supply development.** Ironically, the rapid increase in the number of large reservoir projects in recent years has contributed to declines in water quality. The most common water quality problem associated with reservoirs is accelerated eutrophication, which is attributable to the accumulation of nutrients, particularly nitrogen and phosphorus (see Chapter 4). While not usually harmful to human health in themselves, the presence of these nutrients in large quantities can lead to excessive plant growth and impaired water quality. Eutrophication poses a severe threat to fisheries and aquatic life. In addition, from a drinking water and human health perspective, eutrophication creates taste and odor problems, pipe corrosion, water treatment difficulties, and health hazards to bathers. And in some cases, toxin-producing algae such as cyanobacteria may cause diarrheal disease when ingested.

Reservoirs may also exacerbate both water-vector and water-habitat diseases, such as malaria and schistosomiasis. And more recently, concern has developed over the relationship between surface impoundments and methyl mercury concentrations. Numerous studies have demonstrated that mercury methylation is enhanced by the increased availability of organic carbon, and the increased decomposition of organic matter is a major cause of increased methylation in newly flooded reservoirs. A study in Manitoba, Canada found that mercury levels in northern pike and walleye in natural lakes increased rapidly from levels of 0.2–0.3 μg/g to 0.5–1.0 μg/g following impoundment.[87] In 1984, a study of the Chiasabi,

**TABLE 3.6**   Pesticides used in or sold to Latin American countries during the 1980s whose use has been banned or restricted elsewhere

| Product | Country[a] |
|---|---|
| Aldrin | Argentina, Ecuador, El Salvador, Guatemala, Guyana, Mexico, Suriname, Uruguay |
| Arsenicals | Uruguay |
| BHC | Argentina, El Salvador, Mexico, Suriname |
| DDT | Argentina, Ecuador, El Salvador, Guatemala, Mexico, Suriname |
| Lindane | Agentina, Guatemala, Honduras, Mexico, Uruguay |
| Parathion | Argentina, Ecuador, El Salvador, Guatemala, Honduras, Mexico, Uruguay |
| Toxaphene | El Salvador, Mexico |
| 2,4–D | Argentina, Ecuador, Honduras, Mexico, Suriname, Uruguay |
| 2,4,5–T | Argentina, El Salvador, Guatemala, Mexico, Suriname |

*Source:*   Adapted from UN Economic Commission for Latin America and the Caribbean (CEPAL), 1990, The Water Resources of Latin America and the Caribbean: Planning, Hazards, and Pollution, Santiago, Chile, pp. 157–158.

[a] Listing of countries is not necessarily comprehensive but reflects only those countries for which data are available.

a Cree Indian village located in the vicinity of Quebec's James Bay project, found that 64% of the residents, all of whom were heavy fish consumers, had mercury concentrations that exceeded WHO standards.[88]

Finally the development and utilization of water supplies for the purposes of irrigation may bring not only the economic benefits of agriculture to a region, but also associated water quality problems, such as:

- A reduction in the assimilative capacity of regulated streams where waste disposal competes seasonally with irrigation and energy uses.
- The increased use of fertilizers and pesticides as a result of expanded agriculture and the resulting pollution of surface and ground waters. For example, in Egypt the total amount of nitrate and phosphate fertilizers introduced increased from 0.88 million tonnes in 1951 before the construction of the Aswan High Dam to 6.19 million tonnes in 1987–1988.[89]
- The salinization of arid and semi-arid lands and adjacent waters due to intensive irrigation projects (see Chapter 5).

**Changes in land use.** Globally, many of the trends in water quality can be attributed to large-scale changes in land use, including deforestation, the conversion of grasslands and savannahs, and the loss of wetlands. Even though individual land uses may not significantly degrade water quality or aquatic habitat, the combined effects of several activities may be devastating. Widespread logging has several direct impacts on water quality. The most obvious is increased erosion and sediment loading, which increases turbidity, decreases dissolved oxygen concentrations, and destroys many of the more sensitive aquatic organisms. In many areas, logging has been correlated with declines in fish catch and long-term changes in species composition.[90] In addition, deforestation releases large quantities of nutrients from soils, which may change patterns of primary productivity in surface waters and cause accelerated eutrophication in lakes and reservoirs. By some estimates, deforestation is now proceeding at a rate of up to 20 million hectares per year, with the most rapid rates currently in tropical evergreen forests.[91] More generally, the conversion of wildlands to agricultural uses implies the introduction of fertilizers and pesticides, which eventually make their way into ground and surface waters.

The conversion of wetlands presents a different problem: the loss of natural pollutant sinks. As water floods into wetlands from rivers and streams, the loss in velocity causes sediments and their sorbed pollutants to settle out before they can enter waterbodies. In the U.S., artificial wetlands have been proposed as a means of controlling pollution from non-point sources. Brillion Marsh in Wisconsin, a cattail marsh, has received domestic sewage since 1923. After passing through the marsh, the biological oxygen demand of wastewater decreased by 80%, coliform bacteria decreased by 86%, nitrates by 50%, chemical oxygen demand by 40%, and phosphorus by 13%.[92] The dramatic decline in wetlands globally thus suggests not only loss of habitat but decreases in water quality (see Chapter 4). In sum, surface and ground waters are intimately connected to the land that surrounds them, and global trends in land use, such as deforestation and wetland filling, will inevitably be reflected in the quality of natural waters.

## Water quality management

Given the rapid spread of water pollution and the growing concern over resource availability, the links between water supply and water quality, while always present, have become more apparent. One obvious linkage is the potential impact of large-scale water resources development on water quality, described above. While some of these problems can be avoided through careful site selection and construction techniques, in other cases they are the unavoidable effects of large water projects.

Another link between water supply and quality materializes when water availability diminishes, making it increasingly inefficient to use surface waters for excessive waste disposal. When surface water is plentiful and wasteloads do not exceed the capacity of waterbodies to absorb them, discharging wastes to surface waters provides adequate treatment. Yet as population and waste volumes increase, waterbodies become more polluted, rendering them unfit for certain uses, such as drinking. Ulti-

mately, water quality may become the primary determinant of regional water availability.

The management of water quality is comprised of two principal components: the protection of fresh water sources from contamination and the treatment of drinking water prior to use. The protection of water sources requires the regulation of waste discharges. In the early phases of water pollution control in the industrialized countries, specific treatment techniques were required.

While the implementation of these "technology-based" standards has led to substantial improvements in water quality in some areas, it has also allowed the continued degradation of more pristine waterbodies. Thus, newer standards in the U.S. and in other industrialized countries are tailored to specific waterbodies. Regulations must take into account multiple pollution sources, the sensitivity of local aquatic species, and regional hydrologic conditions. Effluent limitations and treatment technologies are then specified so as to ensure that ambient water quality standards are not exceeded. Obviously, the development of such standards is both a lengthy and a costly process, and consequently it has not spread beyond the world's wealthiest countries.

Non-point sources of pollution, including agricultural and stormwater runoff, pose a more difficult problem. Because non-point sources are more difficult to monitor, the regulation of land use, or "watershed management", has emerged as the principal tool for protecting water quality. In the U.S., the regulation of non-point sources occurs primarily on the local level, and relies heavily on the identification of "best management practices", such as erosion control measures in agricultural areas, which are intended to minimize pollutant loadings. In essence, this is a new set of technology-based standards that specifies the practices that must be adopted in a region rather than setting absolute limits on the discharge of particular pollutants. The effectiveness of this approach in controlling non-point sources of pollution varies widely from region to region and depends upon the scope of the problem, the rigor of the regulations, and the level of enforcement. Yet the difficulty of adopting contaminant-specific standards for non-point sources suggests that best management practices and land use regulation will remain the primary components of non-point pollution control. However, a policy debate has nonetheless ensued from resistance to the mandatory regulation of land use on one hand, and doubt about the efficacy of purely voluntary regulation on the other.

The treatment of drinking water, the second principal aspect of water quality management, became common in the U.S. and Europe at the end of the 19th century when filtration was introduced in order to control the transmission of typhoid and cholera, as well as other waterborne diseases. Filtration removes many microbiologic contaminants and quickly proved effective in controlling the spread of these two major diseases. Chlorination for public water supplies was first introduced in the U.S. in 1908 and was rapidly adopted in most large towns and cities.

Filtration and chlorination remain the essential components of drinking water treatment in the industrialized world. However, because these treatments do not remove chemical contaminants, including dissolved pesticides and metals, governments have begun to impose monitoring requirements and standards for many different contaminants that occur in water, particularly those organic micropollutants that are known or suspected carcinogens. The emphasis on drinking water standards is the most notable trend in water quality management in the industrialized world. Until the last decade, standards existed only for fecal coliforms and a few other contaminants known to cause taste and odor problems. But by 1991, the U.S. EPA had adopted federal standards for 60 different contaminants, with many more currently proposed or under review. The adoption and enforcement of standards, however, presuppose regular water quality monitoring and treatment, which are costly and technology-intensive processes, making them infeasible for many poorer regions and countries. For example, in January of 1991, the U.S. EPA adopted new standards for 38 contaminants, and estimated that the total cost to water suppliers would be $88 million per year, which included $64 million annually for treatment and another $24 million for monitoring.[93] A recent assessment of EPA's proposed standard for radon estimates that the cost of compliance for this single contaminant could be as high as $2.5 billion per year;

the EPA's cost estimate, $307 million annually, is considerably less, but still substantial.[94]

One result of the proliferation of new standards has been the development and adoption of new technologies. Many water treatment plants currently employ advanced methods to remove metals, organic chemicals, and radionuclides. While new technologies are being rapidly developed, older technologies are being re-evaluated for their potential to address new problems. For instance, one alternative for the control of disinfection by-products is the optimization of conventional treatment. Up to 50% of the DBP precursors can be removed by pH adjustment, changing coagulant and coagulant dosage, and improved mixing conditions.[95] Similarly, ion-exchange processes, which have traditionally been employed to remove cations responsible for water hardness, may also be used to remove radium, barium, iron, manganese, cadmium, lead, chromium III (cation exchange), as well as nitrates, uranium, chromium VI, selenium, and arsenic V (anion exchange).

Of the newer technologies, granular activated carbon (GAC) is one of the most common. GAC is capable of removing many synthetic organic chemicals as well as mercury and radon. GAC readily adsorbs aromatic compounds, chlorinated aliphatics, pesticides, herbicides, and trihalomethanes, but is less effective at removing lower molecular weight compounds (simple aliphatics, ketones, acids, aldehydes). A closely related technology, powdered activated carbon, is effective at removing heavy metals. Air stripping, in which contaminants are transferred from a liquid to a gas phase, is widely used to remove volatile organics in ground water.

Membrane technology is a newer process that has not been widely used in water treatment, with the exception of reverse osmosis, which has been used in desalting plants. In addition to reverse osmosis, other pressure-driven membrane processes include membrane filtration and ultrafiltration. Membrane processes that utilize driving forces other than pressure include electrodialysis (electric potential), pervaporation (concentration), and membrane distillation (temperature gradient). Membrane technology may have important applications in the future. Reverse osmosis, for example, can be used to remove nitrate and other salts, organic chemicals, and microorganisms. Electrodialysis is very effective at removing nitrates. To date, the primary technical limitations of membrane technology have been the problem of frequent membrane blocking and the large volume of rejected water and consequent low efficiency. Possible solutions to these problems include the development of dynamic membranes, which are easily replaced, and the use of other pretreatment technologies in conjunction with membranes to reduce blocking and water rejection.[96]

Bioorganisms may also be used either to transform or to concentrate water pollutants. For instance, both aerobic and anaerobic bacteria have been used for many years to break down human wastes. In recent years, however, bacteria have been applied to remove petroleum products, PAHs, phenol compounds, nitrates, and certain chlorinated organics. A potentially new class of remediating bioorganism is that of "biosorbents", naturally occurring biological materials that are able to adsorb and concentrate pollutants, particularly metals, from aqueous solutions.[97] The major advantage of biosorbents over conventional treatment processes, such as chemical precipitation or ion exchange, is their ability to accumulate metals present at very dilute concentrations.

Despite the promise of new technologies, in almost every case they are limited by their costs. Membrane processes are particularly technology-intensive, with the estimated capital cost of a 0.10 mg reverse osmosis plant equal to $275,000 and operating costs equal to $35,000 per year.[98] This puts reverse osmosis beyond the reach of most smaller water supply systems and out of the question for most regions in the developing world. Moreover, the development of new technologies is complicated by the wide variety of pollutants now present in water and their very different chemical and physical characteristics. Membrane processes have the greatest potential to address multiple contaminants. Thus far, however, water sources frequently require multiple treatments, raising the cost of ensuring water quality still higher.

Because of the expense involved in monitoring for and removing so many chemicals and the relatively low levels of risk involved, a debate has emerged over the appropriate level of water quality. Acceptable risk levels for environmental pollution vary considerably in the U.S.; however, the permissible lifetime risk for carcinogens in drinking water is on the order of $10^{-4}$ to $10^{-6}$, which implies one additional death for every 10,000–1,000,000 persons exposed over a lifetime.[99] Yet compared to the enormous risk posed by microbiologic disease in the less developed countries, even the cumulative risk posed by multiple chemical contaminants in drinking water is trivial.

Much more attention is also being placed on the water quality needs of aquatic life in the industrialized countries. Most of the major lakes and waterways in the industrialized world are highly managed for human needs and chronically polluted. In many cases, water quality needs for aquatic life are greater than those for human usage, both because some species are more sensitive and because most surface waters are treated before being tapped as drinking water sources. The needs of aquatic life raise different management issues. For instance, ecosystem protection requires information on the impact of long-term exposure to contaminated water and sediments, the behavior of compounds under different physical and chemical conditions, and the relative toxicities of effluents and ambient water to numerous different species. These concerns are changing the focus of regulation from developing simple ambient water quality criteria for individual contaminants to assessing the cumulative impacts of aquatic pollution and ecosystem disturbance. Water quality management is becoming increasingly sophisticated, complex, and costly.

In the developing world, the most basic components of water quality management are not in place for large segments of the population, and while ecosystem concerns are acknowledged they are rarely addressed. Drinking water standards are almost non-existent except in major urban areas, where basic tests for fecal contamination may be carried out. While significant efforts have been made to provide well-head protection in many localities, water sources are still frequently unprotected. Municipal and industrial discharges are rarely treated or monitored. Surface waters are used directly for human waste disposal, industrial discharges, washing, recreation, drinking, and irrigation. Drinking water treatment prior to delivery may consist of filtration in larger towns and cities that have centralized supplies; chlorination, however, is rare.

The primary obstacle to more comprehensive water quality management is cost. Levels of investment in the water supply and sanitation sector in 1990 were estimated to be approximately $10 billion per year; yet, to achieve complete coverage using conventional piped supplies and centralized wastewater treatment, practitioners estimate that an investment at least five times greater will be required.[100] Thus, there is a compelling need to focus on low-cost technologies that can be easily maintained, so that coverage can be extended to more people, more quickly. Practitioners and funding agencies are at last moving away from the conventionally held belief that full-scale piped water and sewerage systems should be the standard in all areas. In fact, in rural areas the provision of simpler technologies such as pour-flush latrines and hand pumps has proven effective in preventing disease and improving community health.[101] These "appropriate" technologies are much more financially feasible. More to the point, however, is the need to prevent water quality deterioration because the costs of future remediation are likely to be prohibitive.

The problems of sanitation and water supply are further complicated by social and cultural factors. The provision of safe water supply and sanitation systems alone may not necessarily lead to improved health. In Egypt, the first developing country to extend potable water supplies to all its rural population, the infant mortality rate remains as high as that of countries with much less infrastructure, and diarrheal disease remains a major cause of death, estimated at 88 deaths per 1,000 in 1986.[102] The reasons for this failure lie in both behavioral patterns and a popular lack of knowledge about the cause and transmission of disease, pointing out that infrastructure improvements alone are not sufficient to improve water quality. Community education and involvement are critical components of any water quality improvement effort that have been too frequently overlooked in the past.

## Conclusions

Water quality problems fall into two broad categories: microbiologic contamination responsible for outbreaks of acute disease, and more recent chemical contamination, which poses cumulative and chronic health risks to human beings and aquatic life. On a global scale, the risks of toxic chemicals, metals, and radionuclides in water are clearly not comparable to those posed by viral and bacterial contamination. Yet although the human health risks of chemical pollutants in drinking water are still relatively small, the increasing prevalence and magnitude of chemical contamination suggest that these risks will rise considerably in the future. More importantly, chemicals have led to the contamination of water on a regional, and, in a few cases, on a global, scale. Atmospheric transport has become a key pathway, allowing chemicals to move long distances and to contaminate areas quite distant from their source.

Several regional trends are also affecting global water quality. In the developing world, rapid population growth and urbanization are straining infrastructure and eating up recent gains made in water supply and sanitation coverage, posing an enormous challenge that the world is unlikely to meet without a fundamental readjustment of priorities. Concurrently, the rapid proliferation and diffusion of chemicals are forcing the less developed countries to shoulder two distinct threats to water quality at one time. They will not have the luxury of addressing water quality problems sequentially and implementing control measures gradually, as Europe and North America did, without risking the unmanageable degradation and depletion of their water resources.

Globally, we are still in the assessment stage. In much of the developing world, little data exist on even the most common water quality problems, such as fecal contamination and microbiologic disease. The extent of chemical contamination is almost totally unknown, and much remains to be learned about water quality, such as how to assess the chronic health risks posed by organic chemicals, how to predict pollutant fate in ground water, and how multiple pollutants affect the long-term health of aquatic ecosystems. Thus, one of the major emphases in the coming decade will be the collection and interpretation of regional-level data. Yet at the same time, the rising costs of treatment and cleanup, combined with the increasing value of fresh water, point to a critical need to prevent further contamination using the information that we already have. The protection of water resources from further degradation must become an immediate focus of international efforts. For instance, better agricultural practices, a decreasing emphasis on pesticides, the recycling and reuse of wastewater, and more careful water development must all become central to water and resource planning. To date, the more developed countries have failed to address the question of source reduction and comprehensive watershed management, and have instead relied on much more costly monitoring and treatment strategies that will not be available to most of the less developed countries in the foreseeable future.

Water quality remains a key indicator of a country's ability to invest in the health of its population and its environment; it is but one of an integrated set of environment and development issues that require immediate attention. In the coming decades, governments face formidable challenges. Population growth continues apace. Infrastructure in many cities is stretched beyond capacity. The introduction and uncontrolled use of chemicals is putting ever greater stress on the natural environment, and the depletion and degradation of water resources are causing the costs of new water supplies to rise. Without fundamentally new approaches to both development and environmental protection, the widespread degradation of water quality that we currently face could become an unmanageable crisis.

## Notes

1. World Health Organization, 1992, *Annual Statistics 1991*, WHO, Geneva.
2. W. Stumm, 1986, Water, an endangered ecosystem, *Ambio*, **15**, 201–207.
3. B. Moldan, and J.L. Schnoor, 1991, Czechoslovakia: Examining a critically ill environment, *Environmental Science and Technology* **26**(1), 14–21.
4. T.C. Hazen and G.A. Toranzos, 1990, Tropical source water, in G.A. McFeters (ed.) *Drinking Water Microbiology*, Springer-Verlag, New York, pp. 32–54; J.D. Snyder and M.H. Merson, 1982, The magnitude of the global problem of acute diarrheal disease: A review of active surveillance data, *Bulletin of the World Health Organization*, **60**, 605–613.

5. Cholera in the Americas: An update, *Epidemiological Bulletin of the Pan American Health Organization*, **12**(4), 11–13 (December, 1991).
6. World Health Organization, Parasitic Diseases Programme, 1986, Major parasitic infections: A global review, *World Health Statistical Quarterly*, **39**, 145–160.
7. C.P. Hibler and C.M. Hancock, 1990, Waterborne giardiasis, in G.A. McFeters (ed.) *Drinking Water Microbiology*, Springer-Verlag, New York.
8. World Health Organization, Parasitic Diseases Programme, 1986, Major parasitic infections: A global review, *World Health Statistical Quarterly*, **39**, 145–160; World Health Organization, 1992, *Annual Statistics 1991*, WHO, Geneva, p. 15.
9. S.J. States, R.M. Wadowsky, J.M. Kuchta, R.S. Wolford, L.F. Conley and R.B. Yee, 1990, *Legionella* in drinking water, in G.A. McFeters (ed.) *Drinking Water Microbiology*, Springer-Verlag, New York, pp. 340-367.
10. G.F. Craun, 1988, Surface water supplies and health, *Journal of the American Water Works Association*, **80**(2), 40–52.
11. B.L. Herwaldt, G.F. Craun, S.L. Stokes and D.D. Juranek, 1991, Waterborne-disease outbreaks, 1989–90, *CDC Surveillance Summaries, MMWR*, **40**(SS-3), 1-13. In addition to outbreaks of infectious diseases, evidence exists for the endemic occurrence of waterborne disease in developed countries with conventional wastewater treatment. Payment *et al.* found that a water system meeting existing microbial drinking water standards could have endemic waterborne illness rates of 35%, although the severity of illness remained quite low. See P. Payment, L. Richardson, J. Siemiatycki, R. Dewar, M. Edwardes and E. Franco, 1991, A randomized trial to evaluate the risk of gastrointestinal disease due to consumption of drinking water meeting current microbiological standards, *American Journal of Public Health*, **81**(6), 703–707.
12. D.B. Warner, and L. Laugari, 1991, Health for all: Legacy of the water decade, *Water International*, **16**(3), 135–141; World Health Organization, Parasitic Diseases Programme, 1986, Major parasitic infections: A global review, *World Health Statistical Quarterly*, **39**, 145–160.
13. 1991, Dracunculiasis: Global surveillance summary, 1990, *Weekly Epidemiological Record*, **66**, 225–232.
14. D.R. Hopkins, 1987, Dracunculiasis eradication: A mid-decade status report, *American Journal of Tropical Medicine*, **37**(1), 115–118; 1991, Dracunculiasis: Global surveillance summary, 1990, *Weekly Epidemiological Record*, **66**, 225–232.
15. World malaria situation in 1989–Part 1, *Weekly Epidemiological Record*, **66**(22), 157–163 (31 May 1991); World malaria situation in 1990–Part 1, *Weekly Epidemiological Record*, **67**(22), 161–168 (29 May 1992).
16. World Health Organization, Parasitic Diseases Programme, 1986, Major parasitic infections: A global review, *World Health Statistical Quarterly*, **39**, 145–160; World Health Organization, 1992, *Annual Statistics 1991*, WHO, Geneva, p. 16.
17. World Health Organization, 1992, *Annual Statistics 1991*, WHO, Geneva, p. 16.
18. Methemoglobinemia is actually caused by nitrites, which result from the conversion of nitrates in the stomach. Infants are particularly susceptible because the relatively low acidity of their stomachs causes almost all ingested nitrates to be converted to nitrite. In adults, only about 5% of the ingested nitrates are converted. See J. Harte, C. Holdren, R. Schneider and C. Shirley, 1991, *Toxics A to Z: A Guide to Everyday Pollution Hazards*, University of California Press, Berkeley, p. 361.
19. G.F. Craun, D.G. Greathouse and D.H. Zunderson, 1981, Methemoglobin levels in young children consuming high nitrate well water in the United States, *International Journal of Epidemiology*, **10**, 309–317.
20. O.C. Bøckman and D.D. Bryson, 1989, Well-water methaemoglobinaemia and the bacteriological factor, in D. Wheeler, M.L. Richardson and J. Bridges (eds.) *Watershed 89: The Future for Water Quality in Europe, Proceedings of the IAWPRC Conference*, Guildford, UK, 17–20 April 1989, Pergamon Press, Oxford.
21. M. Meybeck, 1982, Carbon, nitrogen and phosphorus transport by world rivers, *American Journal of Science*, **282**, 401–450.
22. M. Meybeck, 1982, Carbon, nitrogen and phosphorus transport by world rivers, *American Journal of Science*, **282**, 401–450.
23. M. Meybeck, D.V. Chapman and R. Helmer, 1989, *Global Freshwater Quality: A First Assessment*, World Health Organization and United Nations Environment Program, Blackwell Reference, Oxford.
24. E.B. Bennett, 1986, The nitrifying of Lake Superior, *Ambio*, **15**, 272–275.
25. J.A. Goodrich, B.W. Lykins, Jr. and R.M. Clark, 1991, Drinking water from agriculturally contaminated groundwater, *Journal of Environmental Quality*, **20**(4), 707–717.
26. J. Forslund, 1986, Groundwater quality today and tomorrow, *World Health Statistical Quarterly*, **39**, 81–92.
27. From the Fifth Report of the British River Pollution Commission (1874),

cited in U. Förstner and G.T.W. Wittman, 1981, *Metal Pollution in the Aquatic Environment*, Springer-Verlag, New York.

28. U. Förstner, 1989, Metal releases from toxic wastes, in D. Wheeler, M.L. Richardson and J. Bridges (eds.) *Watershed 89: The Future for Water Quality in Europe, Proceedings of the IAWPRC Conference*, Guildford, UK, 17–20 April 1989, Pergamon Press, Oxford, pp. 87-106.

29. P.A. Yeats and J.M. Bewers, 1987, Evidence for anthropogenic modification of global transport of cadmium, in J.O. Nriagu and J.B. Sprague, *Cadmium in the Aquatic Environment*, John Wiley and Sons, New York, pp. 19–34.

30. K. Tsuchiya (ed.), 1978, *Cadmium Studies in Japan: A Review*, Elsevier, Amsterdam.

31. T. Tsubake, K. Hirota, K. Shirakaw, K. Kondo and T. Sato, 1978, Clinical, epidemiological, and toxicological studies on methylmercury poisoning, in G.L. Plaa and W.A.M. Duncan (eds.) *Proceedings of the First International Congress on Toxicology*, Academic Press, New York, pp. 339–357.

32. M.A. Bernarde, 1987, Health effects of acid rain: Are there any?, *Journal of the Royal Society of Health*, **4**, 139–145.

33. World Health Organization, 1984, *Guidelines for Drinking Water Quality*, WHO, Geneva; M.A. Bernarde, 1987, Health effects of acid rain: Are there any?, *Journal of the Royal Society of Health*, **4**, 139–145; J.W. Moore, 1991, *Inorganic Contaminants of Surface Water: Research and Monitoring Priorities*, Springer-Verlag, New York.

34. J.W. Moore, 1991, *Inorganic Contaminants of Surface Water: Research and Monitoring Priorities*, Springer-Verlag, New York.

35. U. Förstner and G.T.W. Wittman, 1981, *Metal Pollution in the Aquatic Environment*, Springer-Verlag, New York.

36. V.J. Larsen, 1983, The significance of atmospheric deposition of heavy metals in four Danish lakes, *Science of the Total Environment*, **30**, 111–127.

37. J.O. Nriagu and J.M. Pacyna, 1988, Quantitative assessment of worldwide contamination of air, water and soils by trace metals, *Nature*, **333**, 134–139; J.O. Nriagu, 1988, A silent epidemic of environmental metal poisoning?, *Environmental Pollution*, **50**(1–2), 137–161.

38. J.O. Nriagu, A.L.W. Kemp, H.K.T. Wong and N. Harper, 1979, Sedimentary record of heavy metal pollution in Lake Erie, *Geochim. Cosmochim. Acta*, **43**, 247–258; M. Meybeck, D.V. Chapman and R. Helmer, 1989, *Global Freshwater Quality: A First Assessment*, World Health Organization and United Nations Environment Program, Blackwell Reference, Oxford.

39. K.R. Imhoff, P. Koppe and R. Dietz, 1980, Heavy metals in the Ruhr River and their budget in the catchment area, *Prog. Water Technol.*, **12**, 735–749.

40. U.S. EPA, 1983, *Results of the Nationwide Urban Runoff Program, Final Report*, Vol. 1, U.S. GPO, Washington, DC (December).

41. O. Maim, W.C. Pfeiffer, C.M.M. Souza and R. Reuther, 1990, Mercury pollution due to gold mining in the Madeira river basin, Brazil, *Ambio*, **19**(1), 11–15.

42. World Health Organization Working Group, 1986, Health impacts of acidic deposition, *Science of the Total Environment*, **52**, 157–187; J.W. Moore, 1991, *Inorganic Contaminants of Surface Water: Research and Monitoring Priorities*, Springer-Verlag, New York.

43. Y. Lee and H. Hultberg, 1990, Methylmercury in some Swedish surface waters, *Environmental Toxicology and Chemistry*, **9**, 833–841; E. Steinnes, 1990, Lead, cadmium and other metals in Scandinavian surface waters, with emphasis on acidification and atmospheric deposition, *Environmental Toxicology and Chemistry*, **9**, 825–831; M.R. Winfrey and J.W.M. Rudd, 1990, Environmental factors affecting the formation of methylmercury in low pH lakes, *Environmental Toxicology and Chemistry*, **9**, 853–869.

44. E. Steinnes, 1990, Lead, cadmium and other metals in Scandinavian surface waters, with emphasis on acidification and atmospheric deposition, *Environmental Toxicology and Chemistry*, **9**, 825–831.

45. M. Meybeck, D.V. Chapman and R. Helmer, 1989, *Global Freshwater Quality: A First Assessment*, World Health Organization and United Nations Environment Program, Blackwell Reference, Oxford.

46. P.A. Yeats and J.M. Bewers, 1987, Evidence for anthropogenic modification of global transport of cadmium, in J.O. Nriagu and J.B. Sprague, *Cadmium in the Aquatic Environment*, John Wiley and Sons, New York, pp. 19–34; W. Salomons and U. Förstner, 1984, *Metals in the Hydrocycle*, Springer-Verlag, New York.

47. World Health Organization, 1979, *DDT and its Derivatives, WHO Environmental Health Criteria*, **9**, WHO, Geneva; R.R. Weber and R.C. Montone, 1990, Distribution of organochlorines in the atmosphere of the South Atlantic and Antarctic oceans, in D.A. Kurtz (ed.) *Long Range Transport of Pesticides*, Lewis Publishers, Chelsea, Michigan, pp. 185–198.

48. World Health Organization, 1991, *Water Quality*, prepared in cooperation with the United Nations Environment Program for the 1992 Earth Summit, WHO, Geneva.

49. R.W. Gosset, D.A. Brown and D.R. Young, 1983, Predicting the bioaccumulation of organic compounds in marine organisms using the octanol-water partition coefficient, *Marine Pollution Bulletin*, **14**(10), 387–392.

50. U.S. EPA, 1983, *Results of the Nationwide Urban Runoff Program, Final Report*, Vol. 1, U.S. GPO, Washington, DC (December).

51. U.S. Geological Survey, 1988, *National Water Quality Summary 1986*, U.S. GPO, Washington, DC, p. 139.

52. D.B. Baker and R.P. Richards, 1990, Transport of soluble pesticides through drainage networks in large agricultural river basins, in D.A. Kurtz (ed.) *Long Range Transport of Pesticides*, Lewis Publishers, Chelsea, Michigan, pp. 241–270.

53. W.E. Pereira and C.E. Rostad, 1990, Occurrence, distributions, and transport of herbicides and their degradation products in the lower Mississippi river and its tributaries, *Environmental Science and Technology*, **24**(9), 1400–1406.

54. U.S. Geological Survey, *National Water Summary 1984 – Water Quality Issues*, p. 87.

55. J.A. Goodrich, B.W. Lykins, Jr. and R.M. Clark, 1991, Drinking water from agriculturally contaminated groundwater, *Journal of Environmental Quality*, **20**(4), 707–717.

56. J.H. Carey, E.D. Ongley and E. Nagy, 1990, Hydrocarbon transport in the Mackenzie River, Canada, *Science of the Total Environment*, **97/98**, 69–88; F. van Hoof and T. van Rompuy, 1991, Polycyclic aromatic hydrocarbons in the River Meuse basin, in G. Angeletti and A. Bjørseth (eds.) *Organic Micropollutants in the Aquatic Environment, Proceedings of the 6th European Symposium*, Lisbon, 22–24 May 1990, Kluwer, Dordrecht.

57. J. Harte, C. Holdren, R. Schneider and C. Shirley, 1991, *Toxics A to Z: A Guide to Everyday Pollution Hazards*, University of California Press, Berkeley; G.F. Craun, 1990, Drinking water disinfection: Assessing health risks, *Health and Environment Digest*, **4**(3), 1–3.

58. M. Wrensch, S. Swan, J. Lipscomb, D. Epstein, L. Fenster, K. Claxton, P.J. Murphy, D. Shusterman and R. Neutra, 1990, Pregnancy outcomes in women potentially exposed to solvent-contaminated drinking water in San Jose, California, *American Journal of Epidemiology*, **131**(2), 283–300.

59. These estimates come from a statistical analysis conducted by R.D. Morris *et al.* reported in *New York Times*, 1 July 1992 and translate into 6,500 cases of rectal cancer and 4,200 cases of bladder cancer in the U.S. each year. Similar estimates by Craun, 1988 put the attributable risk of bladder cancer at 28%. (G.F. Craun, 1988, Surface water supplies and health, *Journal of the American Water Works Association*, **80**(2), 40–52.)

60. M. Meybeck, D.V. Chapman and R. Helmer, 1989, *Global Freshwater Quality: A First Assessment*, World Health Organization and United Nations Environment Program, Blackwell Reference, Oxford.

61. J. de Zuane, 1990, *Handbook of Drinking Water Quality: Standards and Controls*, Van Nostrand Reinhold, New York, p. 337.

62. World Health Organization, 1984, *Guidelines for Drinking Water Quality*, WHO, Geneva.

63. C.R. Cothern, 1987, Estimating the health risks of radon in drinking water, *Journal of the American Water Works Association*, **79**(4), 153–158.

64. U.S. EPA, 1990, *Radionuclides in Drinking Water Fact Sheet*.

65. H.M. Prichard, 1987, The transfer of radon from domestic water to indoor air, *Journal of the American Water Works Association*, **65**(4), 159–161.

66. D.J. Crawford-Brown, 1992, Cancer risk from radon, *Journal of the American Water Works Association*, **84**, 77–81.

67. The risk of death from cancer calculated by D.J. Crawford-Brown (Cancer risk from radon, *Journal of the American Water Works Association*, **84**, 77–81, 1992) is slightly lower than the risk of $2 \times 10^{-4}$ calculated by the U.S. EPA.

68. *U.S. Water News*, **8**(9), 9 (March 1992).

69. T. Harpham and C. Stephens, 1991, Urbanization and health in developing countries, *World Health Statistical Quarterly*, **44**, 62–69.

70. C.R. Bartone, 1990, Water quality and urbanization in Latin America, *Water International*, **15**(1), 2–14.

71. S. Cointreau, 1982, *Environmental Management of Urban Solid Wastes in Developing Countries*, International Bank for Reconstruction and Development, Washington, DC

72. C.R. Bartone, 1990, Water quality and urbanization in Latin America, *Water International*, **15**(1), 2–14.

73. T. Harpham and C. Stephens, 1991, Urbanization and health in developing countries, *World Health Statistical Quarterly*, **44**, 62–69.

74. D.B. Warner, and L. Laugari, 1991, Health for all: Legacy of the water decade, *Water International*, **16**(3), 135–141; J. Christmas and C. de Rooy,

1991, The decade and beyond: At a glance, *Water International*, **16**(3), 127–134.

75. *CDC Surveillance Summaries MMWR*, **14**(6), 1–2 (14 February 1992).

76. UN Economic Commission for Latin America and the Caribbean (CEPAL), 1990, *The Water Resources of Latin America and the Caribbean: Planning, Hazards, and Pollution*, Santiago, Chile.

77. World Health Organization, 1985, *Environmental Pollution Control in Relation to Development*, Technical Report Series, No. 718; L. Brown *et al.*, 1992, *State of the World*, Norton, New York, p. 167.

78. UN Economic Commission for Latin America and the Caribbean (CEPAL), 1990, *The Water Resources of Latin America and the Caribbean: Planning, Hazards, and Pollution*, Santiago, Chile.

79. B. Gathuo, P. Rantala and R. Määttä, 1991, Coffee industry wastes, in M. Viitasaari and J. Hukka (eds.) *Proceedings of the First IAWPRC E. African Regional Conference on Industrial Wastewaters*, Nairobi, Kenya, 25–28 October 1989, *Water Science and Technology*, **24**(1), 53–60.

80. P.D. Capel, W. Giger, P. Reichert and O. Wanner, 1988, Accidental input of pesticides into the Rhine River, *Environmental Science and Technology*, **22**, 992–997; Commission du Rhin, 1988, Rapport d'Activité 1986 de la Commission Internationale pour la Protection du Rhin Contre la Pollution, cited in M. Meybeck, D.V. Chapman and R. Helmer, 1989, *Global Freshwater Quality: A First Assessment*, World Health Organization and United Nations Environment Program, Blackwell Reference, Oxford.

81. Freshwater Foundation, *Facets of Freshwater*, **21**(8), 2 (February 1992).

82. J.A. Goodrich, B.W. Lykins, Jr. and R.M. Clark, 1991, Drinking water from agriculturally contaminated groundwater, *Journal of Environmental Quality*, **20**(4), 707–717.

83. W.M. Williams, P.W. Holden, D.W. Parsons and M.N. Lorber, 1988, *Pesticides in Ground Water Data Base: 1988 Interim Report*, U.S. EPA, Office of Pesticide Programs (December); J.P.G. Loch, A. van Dijk-Looyaard and B.C.J. Zoeteman, 1989, Organics in groundwater, in D. Wheeler, M.L. Richardson and J. Bridges (eds.) *Watershed 89: The Future for Water Quality in Europe, Proceedings of the IAWPRC Conference*, Guildford, UK, 17–20 April 1989, Pergamon Press, Oxford, pp. 39–53.

84. UN Economic Commission for Latin America and the Caribbean (CEPAL), 1990, *The Water Resources of Latin America and the Caribbean: Planning, Hazards, and Pollution*, Santiago, Chile; World Health Organization, 1990, *Public Health Impacts of Pesticides Used in Agriculture*, WHO, Geneva, pp. 26–29.

85. D.B. Baker and R.P. Richards, 1990, Transport of soluble pesticides through drainage networks in large agricultural river basins, in D.A. Kurtz (ed.) *Long Range Transport of Pesticides*, Lewis Publishers, Chelsea, Michigan, pp. 241–270.

86. J.A. Goodrich, B.W. Lykins, Jr. and R.M. Clark, 1991, Drinking water from agriculturally contaminated groundwater, *Journal of Environmental Quality*, **20**(4), 707–717.

87. R.A. Bodaly, R.E. Heday and R.J.P. Fudge, 1984, Increases in fish mercury levels in lakes flooded by the Churchill River diversion, Northern Manitoba, *Canadian Journal of Fisheries and Aquatic Sciences*, **41**, 682–691.

88. S.H. Verhovek, 1992, Power struggle, *New York Times Magazine*, 12 January 1992, pp. 16–21.

89. World Health Organization, 1991, *Annual Statistics 1990*, WHO, Geneva.

90. S.V. Gregory, G.A. Lamberti, D.C. Erman, K.V. Koski, M.L. Murphy and J.R. Sedell, 1986, Influence of forest practices on aquatic production, in *Streamside Management: Forestry and Fishery Interactions*, University of Washington, Seattle.

91. World Resources Institute, 1992, *World Resources 1992–93*, Oxford University Press, New York.

92. R.W. Tiner, Jr., 1984, *Wetlands of the United States: Current Status and Trends*, Department of the Interior, U.S. GPO, Washington, DC (March).

93. U.S. EPA, 1990, *Radionuclides in Drinking Water Fact Sheet*.

94. R.S. Raucher and J.A. Drago, 1992, Estimating the cost of compliance with the drinking water standard for radon, *Journal of the American Water Works Association*, **84**(3), 51–65.

95. D.W. Ferguson, J.T. Gramith and M.J. McGuire, 1991, Applying ozone for organics control and disinfection: A utility perspective. *Journal of the American Water Works Association*, **83**(5), 32–39.

96. B.T. Croll, 1992, Membrane technology: The way forward?, *Journal of the Institution of Water and Environmental Management*, **6**(2), 121–129.

97. M. Brown, 1992, Biosorbents for water and effluent treatment, *Water and Wastewater Treatment*, January 1992, p. 36.

98. G.S Logsdon, T.J. Sorg and R.M. Clark, 1990, Capability and cost of treatment technologies for small systems, *Journal of the American Water Works Association*, **82**(6), 60–66.

99. D.C. Kocher and F.O. Hoffman, 1991, Regulating environmental carcinogens: Where do we draw the line?, *Environmental Science and Technology*, **25**, 1987–1989. The proposed standard for radon would result in a lifetime risk of roughly $10^{-4}$; see D.J. Crawford-Brown, 1992, Cancer risk from radon, *Journal of the American Water Works Association*, **84**, 77–81.

100. J. Christmas and C. de Rooy, 1991, The decade and beyond: At a glance, *Water International*, **16**(3), 127–134.

101. C. Carnemark, 1989, The decade and after: Learning from the 80s for the 90s and beyond, in Institute of Civil Engineers, *World Water 89: Managing the Future, Learning from the Past*, Proceedings of a conference held in London, 14–16 November 1989, Thomas Telford, London, pp. 1–6.

102. S. El Katsha and A.U. White, 1989, Women, water, and sanitation: Household behavioral patterns in two Egyptian villages, *Water International*, **14**(3), 103–111.

# Chapter 4
# Water and ecosystems

Alan P. Covich

## Introduction

By looking at earth from space the importance of large lakes and rivers is evident. The great rivers and lakes of the world stand out as reflective surfaces whose shapes mirror both the natural and human processes that control their boundaries. Smaller lakes, ponds, streams, springs, wetlands, and ground waters are intrinsically interconnected parts of these larger ecosystems. Regional differences in topographic relief, rainfall, and runoff set the spatial limits for aquatic habitats. In addition to being bounded by geographical features, surface waters interact with their basins. These ecosystems transform the landscape through processes of nutrient cycling, bank erosion, sediment deposition, and accumulation. If air photos or satellite images of drainage areas are compared over time for nearly any location on earth, it is obvious that fresh water ecosystems are being rapidly altered; some are being lost while others are being created by natural or cultural events. Definition of causative relationships that alter biotic communities in large drainage basins is an ongoing challenge that will be especially important in the next century.

Here I will first review several abiotic and biotic relationships of inland waters to emphasize their dependence on terrestrial and atmospheric ecosystem processes. The integrity of land, air, and water becomes obvious as ecological questions emerge. I stress that the diversity and scarcity of fresh water ecosystems must be considered in conserving some habitats while managing others for a wide range of values. Many localized, unique species are irreversibly lost when certain habitats are altered. In other habitats the dominant species may be widely distributed. In all cases the total community diversity is high because multi-species and age–class interactions are typically complex. Species do not simply replace each other's functions when some are lost to extinction. Impoverishment of biotic diversity can lead to less productive, less stable habitats. Moreover, species have intrinsic and aesthetic values that transcend their ecological and economic productivity.[1]

In the second half of this chapter I examine six examples of persistent, interconnected, biological problems ranging from acid deposition to impacts of global climate change. I contend that solutions to these problems hinge on appropriate interpretations of abiotic, biotic, social, and political interactions. These six problems also illustrate how future sustainability of inland water resources may be enhanced if conflicting management objectives can be reconciled with global environmental changes.

Some ecologists liken the solution of complex environmental water resource problems to assembling an incomplete jigsaw puzzle where only a few pieces are precisely cut and snugly fit into place.[2] It is difficult to envision the whole picture under these conditions even if a few corner elements can be defined. The puzzle analogy is limited because we cannot force all the pieces to fit into a convenient, if incomplete, perspective when the pieces may change their shapes. The kinetics of fresh water habitats and biotic communities are dynamic; habitats move and shift as each component species responds to another's changes. Communities can have alternative states of assembly and reassembly following disturbances that are dependent, in part, on routes and rates of colonization, competition, predation, parasitism, and mutualism.[3]

We need to focus on alternative ecosystem responses to environmental changes in well studied persistent ecosystems if we are to learn how best to live within natural constraints. Response variables typically have been either changes in water quality (such as concentrations of dissolved oxygen) or shifts in numbers of top food web consumers. As discussed below, these measures are very important, especially when measured over the long term. However, monitoring a wider scope of variables, especially biotic community diversity, nutrient cycling, and decomposition rates, provides more detailed understanding of food web connections.[4]

Both flowing and standing fresh waters have been affected by a large number of different problems in the past, but future water quality, species distributions, and ecosystem processes will probably be altered in even more complex ways and possibly at faster rates than ever before. These accelerated effects will be caused by increased growth of human populations in areas where water resources are scarce or polluted and where biotic communities are already being altered by extinctions of native species, transfers of non-native species, increased channelization, dam construction, eutrophication, and acidification, as well as potential climatic changes, such as more frequent and extreme droughts and floods. Although the total global volume of fresh water is extremely small compared to oceanic waters (Table 4.1), our immediate and long-term dependence on the limited geographical distribution of these inland resources makes them politically, economically, and ecologically important.

## Ecosystem integrity

Aquatic ecologists are beginning to study how regional and global ecosystem dynamics are interconnected. The quality and quantity of runoff from drainage areas, ground waters, rivers, lakes, and wetlands are linked to many terrestrial and atmospheric processes. Local elevational differences along a series of interconnected tributaries, rivers, and lakes must be evaluated within an entire drainage basin if all the important connections are to be included. These connections include upstream and downstream processes as well as lateral and vertical interactions with floodplains, ground waters, and in-channel, subsurface waters (the hyporheic zone).[5]

Management of the highly coupled terrestrial, atmospheric, and aquatic ecosystems requires interdisciplinary and international cooperation because fresh water ecosystems often span large drainage areas and cross national boundaries. Nearly 90% of the world's inland surface water supply is contained in 253 large lakes that are widely distributed in 64 countries, although nearly half of these large lakes occur in North America. The Laurentian Great Lakes of North America contain about 20% of the world's surface fresh water. These waters are often subject to conflicting demands because of their political and economic importance. Globally, 214 river and lake basins are bordered by two or more countries. These aquatic borders can lead to complex legal disputes over fishing rights, water transfers, hydroelectric development, and other uses of international resources. Large sources of fresh water such as the Rio Grande, Colorado, Euphrates, Indus, Jordan, and La Plata rivers will continue to be major foci of development and conflict. For example, the Ganges basin is expected to support some 500 million people by the year 2000.[6]

On a global basis the average water balance is such that during any given year there are simultaneously both scarcities of water supplies in some regions and excesses of supply over demand in other regions. These disparities in regional and temporal distributions of fresh waters continue to intensify for many people and for many fresh water species. Floods and droughts cause habitats to change rapidly. Human responses to these hydrologic shifts cause even more intense disturbances. Some aquatic habitats have already been drained and destroyed because water was diverted to irrigate farms, produce power, cool machinery, and supply urban water needs. Other fresh waters are being quickly transformed by salinization, acidification, and alkalinization. Construction of dams, roads, and bridges, channelizing rivers, and draining wetlands can all cause major habitat alterations both upstream and downstream from the projects. Cumulative effects of these disturbances can limit dispersal routes for many types of fishes and invertebrates.

Our current understanding of the ecology and evolution of life in fresh waters is based mainly on short-term, small-scale studies in temperate zone drainages and wetlands. Information on the basic life-cycles and ecology of many species from both temperate and tropical regions is characterized by significant gaps. There are relatively few studies of tropical reservoirs, tributary streams, floodplain lakes, and whole catchments. The structure and function of some large tropical river and lake ecosystems are well documented but are being rapidly altered by a wide range of activities.[7]

Some of the least studied tropical fresh water ecosystems are also the largest. For example, the Gran Pantanal, a wetland that extends over 80,000 km[2] at the headwaters of the Rio Paraguay in South America, is highly productive and biologically diverse, with more than 600 species of fishes, 650 species of birds, and 80 species of mammals. It covers lowlands in western Brazil, eastern Bolivia, and north-eastern Paraguay. Rivers and lakes fill to overflowing each year after torrential rains and form a vast, continuous habitat of mixed types of savannah, and riverine, swamp forests. This alluvial plain is inundated from November to June each year until the rains stop. Evaporation slowly lowers the water level until isolated pools are distributed across an enormous grassland. Its immense capacity as a natural, seasonally temporary storage basin also functions as a flood-control and water quality-enhancing ecosystem as well as an extensive grazing area for cattle. Many of the major tropical rivers are characterized by seasonal flooding that extends considerable distances inland from the river's channel. The Zambezi River, for instance, floods up to 16 km from both of its banks and covers a large area during the wet season. Attempts to regulate these flood events have created numerous environmental problems and reduced biotic diversity.[8]

## Biotic diversity

To understand the repercussions of changing environments on fresh water ecosystems it is helpful to consider how aquatic species have come to live in the many distinct types of fresh water habitats around the world. A disproportionate richness of species lives in inland waters. In contrast with vast oceanic areas that comprise 70% of the earth's surface (to an average depth of 3,800 m), only 1% is covered with inland waters, an area of about 2.5 million km[2] (down to a maximal depth of 1,600 m and up to altitudes above 4,000 m). Yet the oceans contain only 7% of the animal species alive today; 12% of all animal species live in inland waters. Fresh waters form only 0.0093% of the total volume of water on earth (Table 4.1) but contain exceptional concentrations of the world's fauna. For example, about 40% of the 20,000 recognized fish species are found in fresh waters and about 1% migrate between fresh and salt waters. Rivers are rich in many types of organisms, especially insects, crustaceans, and mollusks. Stansbery estimates that on a per unit area basis the total faunal diversity of rivers is 65 times greater than that of the seas.[9]

Latitudinal and elevational patterns of taxonomic richness for invertebrate groups are regionally complex. Terrestrial and marine invertebrates typically are much more diverse in low-latitude, tropical habitats. This pattern is less commonly observed for fresh water species. For example, the fresh water bivalve fauna of temperate North America is the world's most diverse, with 260 native (and 6 introduced) species; there are about 500 species of gastropods, and 319 species and subspecies of North American crayfishes. The total fresh water crustacean diversity is about 4,000 species. Microscopic, zooplanktonic rotifers are especially diverse, with nearly 2,000 species distributed throughout the world's fresh waters; only about 50 species are marine.[10]

Distinct assemblages are characteristic of highly variable habitats such as seasonally intermittent springs, streams, and ponds. Other assemblages occupy temporary pool habitats that have rapidly changing environmental conditions over time-scales of days or weeks. In these temporary habitats, large populations of competing producer and consumer species often do not persist long enough to interact intensely. These species cope with fluctuating waters primarily by dispersing to other locations or by waiting for better conditions while in an inactive resting stage. Some specialized species, however, grow rapidly and do form highly interactive communities even in temporary habitats.[11]

Certain species comprise distinct assemblages that are especially adapted to ground waters, wetlands, or floodplain pools with variable water levels. Others occur only in temporary pools of grasslands or forests or in highly variable salinities found in estuaries and salt lakes. Regional faunal patterns often reflect distinctly different evolutionary pathways so that adaptations to various climatic factors can influence diversity. For example, the Australian inland fauna appears to be well adapted to variable water levels and intermediate ranges of salinities. Animals confined to temporary waters can burrow into the sediment and remain inactive for many months, while others produce resistant propagules such as encased eggs that remain viable even when dry, frozen, or held at warm temperatures. Resting stages of many invertebrates, algae, and seeds of aquatic plants are also adaptations for dispersal to other fresh water habitats.[12]

Many other species, however, require a continuous, relatively constant habitat and are acclimated to a narrow range of environmental conditions. They are not well adapted for dispersal and are dependent on the physical and chemical characteristics of water that minimize their exposure to large fluctuations in temperature, pH, or salinity. As discussed below, these assemblages are often characterized by intense biotic interactions that partially regulate which species coexist.

The more numerous, shallow habitats are highly clustered in areas with high rainfall and recent glaciations. These regions contain "lake districts", extensive wetlands, and numerous networks of river drainages. In these regions both aridity and the geographical orientation of river drainages can greatly influence dispersal, colonization, and persistence of species. For example, species diversity of fishes in the North American Great Plains decreases from east to west as aridity increases. Large North American rivers, east of the Rocky Mountains, are oriented primarily north and south. These drainages are characterized by diverse biotas. As the Pleistocene glacial ice sheets extended southward and contracted northward, several times over millions of years, many riverine species also moved back and forth into available river and lake habitats. West of the Rockies, the less biologically diverse rivers run east and west and had

**TABLE 4.1** Distribution of water in the biosphere

| Location | Volume (10³ km³) | Percentage of total | Renewal time |
|---|---|---|---|
| Oceans | 1,370,000 | 97.61 | 3,100 years |
| Polar ice, glaciers | 29,000 | 2.08 | 16,000 years |
| Ground water (active exchange) | 4,000 | 0.29 | 300 years |
| Fresh water lakes | 125 | 0.009 | 1–100 years |
| Saline lakes | 104 | 0.008 | 10–1,000 years |
| Soil and subsoil moisture | 67 | 0.005 | 280 days |
| Rivers | 1.2 | 0.00009 | 12–20 days |
| Atmospheric water vapor | 14 | 0.0009 | 9 days |

*Source:* R.G. Wetzel, 1983, *Limnology*, second edition, Saunders College Publishing, New York, p. 2.

few corridors for north–south migration during glacial and interglacial periods. Apparently, many of these western species became extinct because they lacked refugia and migratory corridors during Pleistocene climatic changes.

This same east–west pattern of river drainage orientation is characteristic of western and central Europe. These European rivers had similar glacial histories to rivers west of the Rockies and also have relatively low diversity. As discussed later in this chapter, north–south orientation of river drainages in complex north-temperate terrains may also become important migratory corridors for aquatic species facing global warming of rivers and lakes. Lowland and montane tropical areas have several other distinct patterns of drainages characterized by complex floodplain lakes, and a wide variety of wetlands and stream habitats that have also influenced their historical biogeography.[13]

Many of the relatively smaller, shallower lakes and streams have species that are well adapted for dispersal. However, certain species of aquatic plants and animals have very restricted geographic distributions. For instance, one family of flowering plants, the Podostemaceae, contains 206 species in 46 genera restricted to specific rapids and waterfalls. Some present-day, shallow habitats are remnants of what were once much larger lakes or rivers. The flora and fauna of these locations are not typically well adapted for dispersal or migration; many species in these remnant habitats may have already perished. For example, among the North American fishes 103 taxa are considered endangered, another 114 are threatened, and 147 others are of special concern because of patterns of previous extinctions. Many restricted distributions of endangered species are found in arid regions where increased demands on water resources and rising salinities are eliminating their habitats.[14]

Species richness is highly concentrated in some riverine groups such as the naiad mollusks. These pearly fresh water mussels are found in extremely diverse assemblages in several river drainages. Stansbery notes that of the 108 species of mussels known from the Ohio river basin, 42 (39%) are either extinct or endangered. At least 70 species were initially found at Mussel Shoals 70 years ago, but Stansbery notes that less than half of this unique fauna remains. He concluded that "we are gradually destroying nearly a thousand endemic species of fresh water mollusks. This fauna was millions of years in coming into being and is in the process of being eliminated in only a century or two."[15]

In general, the relatively old, deep, fresh water habitats have the highest species diversity of fishes and invertebrates. The oldest and deepest fresh water habitats are widely dispersed across the earth's surface. From ancient Lake Baikal in Siberia, to Lake Tanganyika and the Nile River in the African rift valley, to the Amazon River in South America, there are major evolutionary differences among the species that are adapted to these deep, long-lasting ecosystems (Table 4.2). Only 45 lakes have depths greater than 100 m and these relatively constant environments have distinct, stable species assemblages. These few, very deep lakes often have high concentrations of unique species that are restricted in their distribution to a single, ancient-lake ecosystem. For example, Lake Tanganyika has 247 species of fishes, significantly more than the total of 170 species of fishes in all the Laurentian Great Lakes. Because the Laurent-

ian Great Lakes were formed during the last ice age, they are relatively young in comparison with some of the deepest and most species-rich lakes such as Baikal, Tanganyika, Malawi, Titicaca, Biwa, and Ohrid. The richest zones for fish diversity are in south-east Asia, central Africa, and South and Central America.[16]

Large river ecosystems are generally more persistent over geologic and evolutionary time-scales than are lake basins. Although lacking the same degree of geographical isolation of most ancient lakes, the large rivers often have large numbers of unique species that are adapted to specific microhabitats. For example, the Amazon River flows some 6,500 km, has a maximal depth of more than 100 m, and transports about one-fifth of the fresh water that annually discharges into the oceans from all the world's rivers. This amount is 10 times that discharged by the Mississippi River and 3,500 times that of the Thames. The great morphological diversity of the river and its tributaries is associated with distinct optical properties resulting from suspended sediments and dissolved organic humic acids. The Amazon drainage contains many unusual species of fishes such as piranhas, neons, electric knife-fish, catfishes, and many others that make it the world's most diverse ichthyological region. Goulding estimates there are approximately 2,500 to 3,000 species of fishes in the Amazon. He suggests that this high diversity is about ten times as many fish species as is found in all of Europe. Goulding concluded that the neotropical characin fishes are one of the most extreme examples of adaptive radiations ever observed among living vertebrates. Many of these unique species, especially among the siluroid catfishes and characins, are restricted to limited biogeographic ranges.[17]

Impacts of large water control projects can be very detrimental to highly adapted, specialized species. Recently, a committee of conservationists from ten environmental groups in the United States proposed that the national wild and scenic river systems be greatly expanded by the year 2000. Criteria are needed for determining which unique and "representative" aquatic habitats will be preserved. Preservation of both unique and more generally typical habitats is important to ecologists and conservation biologists as well as to managers.[18]

To sustain ecosystem productivity we need to understand how life can adapt and persist in different, sometimes isolated or extreme, environments as well as in other, more interconnected, widely tolerable habitats. Unfortunately, numerous habitats and their associated biotas are being destroyed faster than they can be studied. Without some multiple baselines for studies of unmanaged, naturally variable ecosystems it is impossible to evaluate effectively the success or failure of any new management alternatives. Some "no-management controls" must be available for comparative analyses because important differences in natural variation can occur among smaller fresh water ecosystems within a region. Effects of man's impacts on highly variable ecosystems are difficult to evaluate without regional monitoring of both managed and unmanaged drainage areas.

We have often ignored the high species richness associated with inland waters and have allowed many fresh water habitats to be dammed, channelized, drained, eroded, and polluted with nutrients, salts, silt, and chemicals. Biodiversity and ecosystem integrity are declining in a wide range of locations throughout the world because we lack coordinated efforts to preserve and manage entire drainage areas. The listing of federally recognized, endangered, fresh water species is one important means of tracking total integrity. Global listings are available from international agencies such as the United Nations Environment Program (UNEP) and conservation groups such as the World Wide Fund for Nature (formerly the World Wildlife Fund). Calls to restore and preserve biotic integrity of fresh water ecosystems continue to point out the long-term costs of delaying implementation of adequate safeguards and remedial management.[19]

### Environmental adaptations

Aquatic organisms that live in stable, persistent habitats can adapt their optimal growth potential to compete and to coexist in particular environments. Generally, most animals are either well adapted for living in flowing waters of rivers and streams or in standing waters of deep lakes and ponds; relatively few species live in both flowing and standing waters. These differences in adaptations are important when rivers are dammed

**TABLE 4.2** Data on species-rich, deep lakes of the world

| Lake | Maximum depth (m) | Mean depth (m) | Area (km$^2$) | Volume (km$^3$) |
|---|---|---|---|---|
| Baikal | 1,620 | 740 | 31,500 | 23,000 |
| Tanganyika | 1,470 | 572 | 33,000 | 19,000 |
| Malawi | 704 | 273 | 371,000 | 67,500 |
| Crater | 608 | 364 | 55 | 20 |
| Superior | 307 | 144 | 83,300 | 12,000 |
| Tahoe | 501 | 313 | 499 | 156 |
| Titicaca | 281 | 107 | 8,100 | 866 |
| Victoria | 79 | 40 | 68,800 | 2,700 |

*Source:* Data from C.R. Goldman and A.J. Horne, 1983, *Limnology*, McGraw-Hill, New York, p. 14; G.E. Hutchinson, 1967, *A Treatise on Limnology*, Vol. 1., John Wiley and Sons, New York, pp. 168–169 .

and large, permanent, man-made lakes are formed in regions where permanent, standing-water habitats may not have occurred, or where small, shallow, temporary floodplain pools served as spawning areas. The lack of a stable littoral zone filled with submerged and emergent aquatic plants limits the growth and reproduction of many invertebrates and fishes in reservoirs.

Seasonal fluctuations of water levels in small, shallow ponds, saline lakes, and wetlands are often associated with rapid growth and reproduction, since life-cycles are relatively short. In contrast, life-cycles of long-lived species in stable, persistent habitats are characterized by relatively slow growth. Long-lived species require many years to reach reproductive maturity and are more vulnerable to physical and chemical disturbances. Rare and endangered species often require many years to mature and are highly adapted to live in specific habitats. Biotic interactions, such as competition and predation, are likely to influence community composition and ecosystem processes in these stable physical and chemical environments.[20]

Life-history traits include different adaptive modes for dealing with physiological stresses associated with dispersal. The biota in relatively shallow fresh water habitats is generally well adapted for dispersal because these habitats are transitory over evolutionary and ecological time-scales. The potential for widespread dispersal by many species has persisted since the earliest invasions of fresh water by marine-derived and land-derived organisms. Many fresh water species originated in coastal habitats so that important components of some inland ecosystems still have marine linkages while others require annual, long-distance migrations along rivers or over land.[21]

Because of these long evolutionary pathways fresh water ecological relationships and primary productivity can be spatially and temporally complex (Table 4.3). For example, in eastern North America the synchronously pulsed upstream migrations of salmon to spawn results in large seasonal shifts in community dynamics. First, predation pressure increases on the prey populations consumed by salmon. Later, when the adult salmon die, they provide a surplus of food for scavengers. In contrast, when adult Pacific salmon (*Oncorhynchus spp.*) migrate upstream to spawn in the relatively low nutrient rivers of western North America they stop feeding, their digestive tract atrophies, and all their energy is allocated to reproduction. Their spawning migration functions as an upstream nutrient "pump". If insufficient numbers of fish migrate upstream (because dams block their movement) the nursery needed for larval fish in the headwater ecosystem may not function adequately to sustain the species. As discussed below, the "natural" upstream movements of alewives and sea lampreys into coastal rivers and lakes were extended unintentionally far inland (3,800 km) to the Laurentian Great Lakes once canals were constructed for inland transportation. Construction of dams and canals can greatly alter aquatic communities and destabilize food webs that evolved over millions of years.

**Physical variables.** The unique physical nature of water causes its specific heat capacity to be among the highest of any element, ten times higher than that of iron. Because of this high heat capacity, life in fresh

**TABLE 4.3** Net primary productivity for regional fresh water ecosystems.

| Ecosystem | mg C/m² per day | g C/m² per year[a] |
|---|---|---|
| Tropical lakes | 100–7,600 | 30–2,500 |
| Temperate lakes | 5–3,600 | 2– 950[b] |
| Arctic lakes | 1– 170 | < 1– 35 |
| Antarctic lakes | 1– 35 | 1– 10 |
| Alpine lakes | 1– 450 | < 1– 100 |
| Temperate rivers | < 1–3,000 | < 1– 650 |
| Tropical rivers | < 1– ? | 1–1,000 ? |

*Source:* G.E. Likens, 1975, Primary production of inland aquatic ecosystems, in: H. Lieth and R.H. Whittaker (eds.) *Primary Productivity of the Biosphere*, Springer-Verlag, New York, pp. 185–202.

[a] In most cases, averaged over estimated "growing season."
[b] Naturally eutrophic lakes may reach a maximum of 450.

water occupies a much more stable thermal environment than that on land.

*Temperature.* Physiological limits for organisms are determined by a combination of genetic adaptations and acclimation. The lower thermal limit for nearly all aquatic species is the freezing point of water or 0°C in pure water and slightly below zero for saline waters. Only a few species of fishes have evolved natural physiological antifreezes. The upper lethal limits of most species of invertebrates and fish are rarely above 40°C. For example, a cichlid fish, *Sarotherodon grahami*, that is adapted for living in a soda lake (Lake Magadi, Kenya), occurs in waters nearly 40°C and at pH 11. The mosquito fish (*Gambusia spp.*) is distributed widely in habitats from ice-covered lakes and ponds to hot springs. They can tolerate water in the 42–44°C range. A desert-dwelling pupfish (*Cyprinodon macularius*) can survive in habitats up to nearly 50°C. Upper temperature limits are known for many species from a long series of ecological studies on thermal tolerances in inland waters. Remarkably, some aquatic invertebrates live at very high temperatures in geothermal hot springs and crater lakes; the highest thermal limit for metazoans in hot springs is 54°C for a small ostracod crustacean species. Some bluegreen algae and thermophilic fungi can tolerate water up to 60°C and various bacteria are known from hot springs with temperatures above 80°C.[22]

Temperatures and total numbers of degree days (sum of the mean daily water temperatures) directly influence rates of respiration and growth, while indirectly influencing metabolism by altering concentrations of dissolved oxygen and other gases. Large, deep lakes provide an especially stable thermal environment because of water's capacity to store heat. The wind-driven currents can only slowly alter the temperatures of the deepest waters in density-stratified lakes. Some running-water habitats, such as spring-fed streams and cave streams, are extremely constant. The same temperature (equal to mean annual air temperature) is maintained throughout the year and from one year to another. Such uniformly constant habitats provide site-specific thermal regimes for regional and altitudinal comparative studies. In general, lowland tropical streams also have relatively little daily and annual temperature fluctuations, although montane tropical streams and rivers have biologically significant temperature gradients along steep elevational profiles.[23]

In drainage-fed streams water temperatures generally decline with altitude and annual variations increase downstream. Short-term changes in temperatures are associated with flooding and seasonal events that influence floodplains and marginal wetlands. Dams that release colder, bottom waters during the summer for hydroelectric production or flood control can greatly alter the composition of riverine communities downstream of reservoirs and cause these lower sites to resemble colder, upstream locations.[24] Species vary greatly in their physiological ability to regulate their metabolic rates as temperatures increase. The rate of temperature increase is of great importance in determining how species respond. Many species can respond immediately, behaviorally, to warmer temperatures by seeking cooler microhabitats such as deeper waters of summer-stratified lakes or shady, deep pools in streams. Although behavioral responses are the most rapid these reactions are limited to areas where organisms can find thermal refugia. If environmental temperatures change slowly, then long-term evolutionary responses may occur through shifts in gene frequencies and selection for new combinations of genes that regulate biochemical pathways under varied thermal conditions. The physiological ecology of these relationships for numerous fresh water organisms is well documented. Optimal growth and reproduction typically peak at intermediate values of temperature tolerances that characterize different species. Diverse groups of species have distinct thermal preferenda.[25]

*Flow.* Stream flow variability is viewed by ecologists as a primary control on the relative roles of biotic and abiotic processes that determine the composition and dynamics of stream ecosystems. Infrequent and intense events such as ten-year floods and low or no flows during prolonged droughts act drastically to alter natural communities.[26] As discussed previously some organisms are well adapted to survive in intermittent waters such as streams and pools that are seasonally dry. A relatively small number of species (e.g., brine shrimp and fairy shrimp) can survive many years of complete dryness and the associated wind erosion during prolonged droughts. Because aquatic organisms are primarily composed of water, few species can live very long without it. Droughts typically cause severe reductions in stream flow, increased salinity, and increased water temperatures.

Despite numerous, well-funded attempts to control many rivers there are frequent and extensive floods. In the United States about 6% of the total land area (about 543,000 km$^2$) is within the 100-year flood zone and still routinely floods.[27] Floods can dislodge stream-dwelling species and flush out lake species that occupy open water habitats. Although many riverine communities are highly resistant to floods (and can often recolonize areas rapidly if they are displaced), unpredictable high-flow events can influence composition and persistence of biotic communities. Poff and Ward classified only three stream types ("winter rain", "snow melt" and "snow melt and rain") as characterized by highly predictable floods.[28]

As discussed below in more detail, human-created disruptions can override natural variability in stream flow. For instance, regulated flow regimes in many regions have demonstrated that some fishes and invertebrates are intolerant of major changes in flow variability. Modification of channels to facilitate flood control and navigation by dredging, removal of woody debris and snags, construction of locks and dams, and elimination of riparian habitats have caused many species to decline or become extinct. Deforestation and subsequent flooding may also have negative impacts on fish community diversity. In some habitats two types of alterations can cancel out each other's effects. For example, communities can be disrupted by both floods and transfers of non-native predators or competitors. Introduced predatory fish can displace native species in some food webs during periods of low flow (from prolonged droughts or flow regulation by dams) when prey are abundant and no predators are present to keep the non-native consumers in check. However, with increased frequencies of natural flooding the native species may persist because it is better adapted to such events than is the exotic predator.[29]

**Chemical variables.** Many chemical characteristics are biologically important in inland waters although only a few parameters (e.g., salinity, total hardness, acidity, dissolved nutrients, and dissolved oxygen) are widely used in distinguishing major types of biotic assemblages. All of these variables can be altered rapidly by competing uses of inland waters and their catchments. Listed below are brief descriptions of some general relationships.

Dissolved oxygen. Three main processes use dissolved oxygen in aquatic habitats: microbial decomposition of dead plants; respiration of living plants, animals, and microbes; and the chemical oxidation of sediments. Biological oxygen demand (BOD) is widely used to measure aquatic ecosystem responses to organic matter decomposition (by bacteria and fungi); similarly, the chemical oxygen demand (COD) is used to measure the non-biotic oxidation of the sediments. Throughout the growing season, any unconsumed biomass dies and sinks so that decay of accumulated debris can deplete the dissolved oxygen of bottom waters in thermally stratified lakes. Large numbers of fishes that require colder waters in the bottom of the lake cannot live without sufficient dissolved oxygen. This "summer kill" of cold-water fishes is typical of lakes that have received excessive nutrients from mismanagement of the lakes' drainage areas (as discussed below). If the dead fish are not consumed by littoral scavengers (such as crayfish or crabs) they are washed up on beaches or sink to the bottom waters and lead to further deoxygenation.[30]

Salinity. Different types of fishes and invertebrates are adapted to live in extremely different habitats, ranging from very dilute to highly saline waters. Nearly half of the volume of inland waters is naturally saline (Table 4.1). The sum of all ionic concentrations is the basis for salinity measurements; total dissolved solids and ionic conductivity of waters are the generally used measurements. "Saltern", or "thalassohaline" (more marine-like) waters are high in sodium chloride while "athalassohaline" lake waters are high in sodium carbonate and bicarbonates (also termed "soda" lakes) or sulfates. Waters with very low salinities and alkalinities are generally referred to as "soft" and are associated with dense, igneous rocks composed of minerals that dissolve slowly. "Hard" or alkaline

waters have higher concentrations of alkaline-earth minerals associated with relatively fast rates of weathering of sedimentary rocks such as calcareous limestones. The lower boundary for "saline" waters from a biotic perspective is between 3 and 5 o/oo (ppt, parts per thousand, or grams of total dissolved solids per liter).[31]

Extremely saline waters are dominated by a relatively low number of different species but the total productivity is often very high. Saline lakes typically occur in arid, closed drainage basins where evaporative processes increase salinity and nutrients. Many saline lakes occur in Africa, including the world's most saline habitat, the Dead Sea (Israel and Jordan). The extreme salinities (350 o/oo) of the Dead Sea were once thought to exclude all life, but several microbial species were found to dominate this habitat. However, no zooplankton or fish can live in these extreme, hypersaline waters.[32]

The consumer species adapted to live in saline waters include many types of fishes, insects, crustaceans, ciliate protozoans, foraminiferans, gastropods and oligochaetes. Among the organisms typical of inland waters with fluctuating salinities are various endemic species of pupfishes, brine flies, brine shrimp, mollusks, macrophytes, bluegreen algae, some green algae, and highly adapted bacteria. In some isolated localities there are unique species, sometimes near extinction, that have evolved in these unusual habitats. For example, in the Cuatro Cienegas Basin of Coahuila, Mexico, there are at least 12 endemic species of fishes, 12 endemic species of gastropods, and other groups that have adapted to live only in these spring-fed habitats. There are approximately 15 ecosystems in desert areas of North America that have been identified to have 83 endangered species of fishes; some 164 species of fishes and associated biota could be eliminated through habitat disruptions. On the other hand, other species are adapted to variable salinities and occur in many regions. Brine shrimp, for instance, are widely distributed and physiologically adapted for a wide range of salinities, from 10 o/oo (lower than sea water at 35 o/oo) to 220 o/oo.[33]

Highly variable salinities are typical of coastal habitats where the seasonal and interannual variations in rainfall and oceanic storm surges cause short-term fluctuations in coastal rivers, lagoons and wetlands. Salinities of smaller volumes of water in coastal ponds and streams are also seasonally influenced by rapid evaporation and evapotranspiration during prolonged dry periods. Slower rates of change in salinities characterize large, closed lake basins. These shifts in salinity are used to study long-term climatic changes in regional rainfall and evaporation. For example, the Great Salt Lake of Utah is the largest saline lake in North America. Its highest historic level was in 1873 when it covered 6,200 km$^2$ and had a maximal depth of 12.5 m; by 1963 it had shrunk to 2,400 km$^2$ and maximal depth of 8.5 m. Recently, increased rainfall and the resulting decreased salinity over several years in the Great Salt Lake caused major changes in distributions of planktonic, open water invertebrates, and fish. The decreased salinity of the southern basin in 1986 allowed a breeding population of killifish (*Lucania parva*) to occur in the lake for the first time in recorded history. The drought during 1987 increased salinity of the southern basin again and the ability of these fish to survive is in question. Mono Lake and Pyramid Lake are other well studied examples of western North American ecosystems where changing lake levels (generally lowered inflows and resulting higher salinities) have greatly altered biotic communities. Saline lakes are found in drier regions of nearly all continents, especially in south-west and north-west United States, western Canada, east Africa, Australia, Russia, the northern dry side of the Himalayas, and Antarctica. Volcanic crater lakes and geothermal hot springs often have high salinities and extremes in chemical compositions that limit species distributions.[34]

Nutrients. At least 20 elements are known to be required for optimal growth by most species of aquatic plants. Nutrient concentrations and specific nutrient ratios influence growth rates of different species. For instance, the nitrogen to phosphorus ratio is known to regulate the quantity and composition of some phytoplanktonic algae. Very low concentrations of several major nutrients can restrict growth of many types of aquatic plants and thereby limit food supplies for grazing animals and detritivores (consumers of dead organic matter). Extremely dilute lakes typically have small, sparsely vegetated drainage areas composed of highly resistant, slow-weathering bedrock such as granites. Their waters are derived mainly from precipitation falling directly onto the lake's surface. Lake Notasha in the Cascade mountain range of southern Oregon is considered to be the world's most dilute lake and derives much of its inflow from snow melt. Such extremely dilute, nutrient-poor lakes are excellent sites for long-term monitoring of eutrophication and atmospheric contamination.[35]

At the other extreme, high concentrations of some macronutrients and micronutrients can be toxic to many plant species or lead to excessive growth of the few plants that tolerate waters with high nutrient concentrations. This excessive growth is typically followed by rapid decay and oxygen depletion, as discussed previously. Micronutrients, including trace elements such as boron and copper, minerals such as iron and silica, and vitamins, are all required in very low concentrations by many types of algae and some littoral vegetation.

Short-term nutrient cycling (over minutes and hours) is regulated by several biotic interactions. These include: nutrient uptake by phytoplankton and bacteria; cellular breakdown by microbes (bacteria and fungi); and grazing and excretion by open-water, invertebrate zooplankton as well as by some bottom-dwelling, filter-feeding invertebrates (such as bivalves, bryozoans, and sponges), and by open-water and bottom-feeding fishes. Abiotic interactions such as release of nutrients from sediments are also important but only during periods when bottom waters are deoxygenated.

Long-term nutrient cycling (over months and years) is regulated by biotic uptake, biomass storage by long-living species, sediment mixing by burrowing organisms, and by abiotic factors such as sedimentation, resuspension, and mixing of stratified waters. In density-stratified lakes the ratio of silica to phosphorus partially controls the seasonal timing of dense "blooms" (pulses of high population densities) of diatoms. These algae are single-celled, free-floating, microscopic plants that have silicate cell walls. They are very widely distributed in lakes and streams at all latitudes. In very deep, nutrient-poor lakes these algae extend to depths of 200 m if light penetration is adequate (typically 1% of the incident surface radiation). Some species are restricted to nutrient-poor waters while others thrive under nutrient-rich conditions.

Nutrients accumulate seasonally in the bottom waters of stratified lakes. During brief periods of complete mixing of the lake these bottom stored nutrients (especially phosphorus, nitrogen, iron and silica) are transported by wind-driven currents up to the well-lit "photic zone" where light intensity is sufficient for plant growth. Lakes mix completely ("turnover") once, twice, or many times a year, depending on the latitude and altitude of the basin. The basin's location affects timing of strong winds, annual temperature changes, and relative densities of top and bottom waters. Some deep or saline lakes never mix completely.

Nutrients also accumulate over time in some lake basins where inflowing waters have long residence times and/or where nutrients are concentrated by evaporation. At late stages of eutrophication, restoration of a lake's community is possible only if the nutrients entering the lake from the drainage are reduced and if the flow through of water in the lake is relatively rapid and well mixed. The longer water remains in a lake, the more likely it is that nutrients will move from the sedimentary deposits into the overlying waters and be mixed back into the brightly illuminated surface waters.[36]

An important debate in nutrient studies focused on the role of dissolved phosphorus as the limiting factor for algal growth. This debate grew out of observations made in many locations around the world. Bluegreen algae begin to dominate phytoplankton communities when phosphorus concentrations are high because (unlike green algae and diatoms) bluegreen algae can fix (i.e., use) atmospheric nitrogen to maintain high nitrogen:phosphorus ratios. Thus, because bluegreens continue to grow rapidly, through a steady supply of nitrogen from the atmosphere and phosphorus from waters flowing into the lake, these bluegreen algae cause large algal "blooms" composed of only a few species. They build up large populations because some bluegreens are relatively less palatable to grazers and others produce toxins that affect consumer species.[37]

The mix of macronutrients and micronutrients that characterize eutrophication is regionally and temporally complex. Dissolved inorganic nitrogen (a macronutrient) can be the limiting factor in some lakes. Although less frequently studied, trace elements (micronutrients such as

iron, manganese, molybdenum, selenium, and cobalt) may be limiting in nitrogen- and phosphorus-poor lakes. Only in extremely nutrient-rich waters has carbon been observed to limit algal growth. Management of nutrient levels and ratios is a very important issue because bluegreen algae can outcompete other types of phytoplankton that provide a stable food supply to zooplankton grazers. Some species produce taste and odor problems in reservoirs and lakes used for drinking water supplies.[38]

Resolution of this debate became politically and economically important when laws were being passed to ban the use of phosphate-based detergents in some countries but not in others that shared common waters such as the Laurentian Great Lakes. Today, the detergents used in this drainage area contain no or relatively little phosphorus (<2%). Phosphate-based detergents used in other countries contain 5%–12% phosphorus by weight and are known to cause problems with bluegreen algal blooms.[39]

*Acidity.* Generally, the effects of acidification are opposite to those of nutrient accumulation or eutrophication. Naturally acidic waters tend to have low nutrient concentrations and very low biomass. Aquatic habitats have a wide range of pH, from acidic peat bogs and black-water streams, to highly buffered habitats in karst (limestone) regions or mineral-rich hot springs, and saline lakes. Sphagnum mosses grow well at low pH and dominate bogs. Organic, humic acids are produced by decomposition of terrestrial vegetation in sandy soils and these acids enter ground waters. Tea-colored, black-water habitats are relatively widespread in sandy, coastal-plain rivers and lakes, and in some wetlands. In contrast, karst habitats are weathered by weak carbonic acids produced in rainwater. These calcium carbonate-rich waters are well buffered, have relatively immobile, insoluble calcium phosphates, and high pH values. Some plants growing in karst aquatic habitats are typically limited by low concentrations of dissolved phosphorus and at times by low levels of dissolved carbon dioxide (unless they can use bicarbonate ions as a source of carbon during photosynthesis).

Successive acidification of poorly buffered, low-calcium waters has very deleterious effects on most species of fishes and aquatic invertebrates. Generally, when pH drops below 5.5 to 6.5 in fresh waters there are negative effects on ecosystems because nutrient cycling is altered and many species cannot tolerate the increased acidity. Many cold-water, salmonid species are generally more sensitive than others (such as brook trout or bass). Many fishes undergo physiological stress associated with low pH including lowered oxygen transport (by circulation of blood) and diminished regulation of internal salt balances. However, some fishes have adapted to living in the naturally acidic, black-water rivers of the Amazon that range from pH 3.8 to pH 4.9. In temperate zone bogs and other natural acidic waters various species of mosses, midges, flies, beetles, rotifers, and protozoa can survive at very low pH. Certain microbial species can live in extremely acidic habitats. However, the lower limit for bluegreen algae is pH 4.0, a limit similar to that of some warm water fishes.[40]

Most fresh water crustaceans and molluscans are highly intolerant of acidic waters because they require calcium carbonate to strengthen their outer skeletons or shells and calcium is generally not available in acidic habitats. Other invertebrate species such as some aquatic insects (e.g., whirligig beetles and water boatmen) and protozoans are tolerant of low pH. Most phytoplanktonic algae decline at pH below 5.8. In recently acidified Swedish lakes some zooplanktonic microcrustaceans (water fleas or *Daphnia*) are sensitive to low pH but other species (*Bythotrephes*) are found in European lakes with value of pH 5.4 and lower. Thus, in acidic lakes this latter crustacean can readily outcompete less tolerant species of zooplankton. As discussed later, this zooplankton species has recently invaded the Laurentian Great Lakes and is being closely monitored.[41]

## Present and future problems

*Acidification.* Although some concerns about acid deposition occurred as early as 1852, it has only been during the last 30 years that biological impacts have been well documented in the northern hemisphere, especially in eastern North America and in Europe. Aquatic ecosystems often show the detrimental effects of acid deposition sooner than their surrounding drainages. It is now clear that aquatic ecosystems reflect the changes in their drainages through a complex series of chemical alterations of rocks, soils, and vegetation. Recent acidification of streams and lakes from atmospheric pollution by sulfur dioxide and nitrous oxides has proven to be a major problem of both industrialized and downwind drainages. Acid deposition has altered surface waters in parts of Europe and North America while mining wastes and industrial effluents have polluted fresh waters over much of the biosphere.[42]

Until recently there was intense debate about how specific chemical interactions caused death of fishes, zooplankton, crayfishes, insects, and algae in acidified habitats. Interpretations were initially confounded by observations that some other species of algae and grazers were tolerant of low pH and expanded their populations. The timing of acidic inputs has turned out to be very important in measuring and interpreting acidic effects. Fish were found to be particularly at risk during the spring snow melt, which produced a pulse of highly acidic water that flowed quickly into streams and lakes. Accumulation of acid deposition during the winter is released over a period of weeks during the spring melting and coincides with the annual hatch of fish fry that are especially sensitive to low pH. Even in relatively well buffered waters the rapidly pulsed input of acidic water can reduce pH values to toxic levels.[43]

Experimental manipulations of lake and stream ecosystems together with wide-scale regional surveys have identified cause and effect relationships between declines in species abundances and acidity. The toxicity of inorganic, labile monomeric aluminum is now well established and associated with direct toxicity from low pH. Toxic metals such as mercury, manganese, zinc, and lead can also be leached from drainages and directly cause fish kills or indirectly reduce fish populations by killing fish-food resources such as insects, crustaceans, mollusks, and plants.[44]

**Sustainable fisheries yields.** One-tenth of the total recorded world fish yield is caught in inland waters, and this estimate from the UN's Food and Agriculture Organization does not include subsistence fishing. Much of this subsistence and commercial yield is dependent on floodplain rivers. Lateral lakes on floodplains of tropical rivers continue to yield extremely high fish production if natural flood frequencies are maintained. Large rivers produce 75% of the fish yield in the Amazon and 50% of the inland fisheries for Africa. Sustained annual yields of fishes range between 40 kg and 60 kg per hectare in some large African rivers. Fish are a significant source of high quality protein for poorer families in many developing countries and constitute a major export commodity in both coastal and inland regions.

Coastal river runoff also influences production of near-shore marine fisheries through controls over water-transported nutrients and salinity so that major diversions of river water for irrigation can influence both offshore and inland fisheries production. Attempts to develop management

strategies for high yields of fisheries and other associated resources must be based on ecological principles of long-term sustainability.[45]

Different types of inland waters are exploited for fisheries and yields range widely (Figure 4.1) in natural lakes, rivers, and streams as well as in managed reservoirs and aquacultural ponds. The main factors that regulate fish growth rates are: food availability; seasonal competition for food and spawning areas within and between species; density and types of predators; and dissolved oxygen, salinity, and temperatures of bottom and surface waters. Deep lakes and reservoirs have distinct vertical zonation based on nonlinear differences in light, temperature, dissolved oxygen, pH, and nutrients along depth profiles. The interannual and seasonal differences in the depths to which dissolved oxygen is mixed are extremely important in regulating fish production. In all inland and coastal ecosystems the management of sustained fish production requires a great deal of basic biological and ecological information on reproduction and growth in order to regulate size limits and timing of fishing effort.[46]

Lakes with relatively small drainage areas, low rainfall, and deep, steeply sloped basins generally have very low nutrients in their illuminated surface waters and thus have low productivity (i.e., the annual rate of forming organic carbon) of suspended algae (phytoplankton) and their grazers (zooplankton). These lakes typically also have very low yields of fishes. In contrast, relatively large drainage areas with gently sloped, shallow lake basins often have higher levels of dissolved nutrients and much higher productivity. Algae and invertebrates provide a sustainable supply of food for fishes. A positive correlation between fish yields and total production of algae is reported for some lakes while other predictive relationships occur between fish production and the amounts of bottom-dwelling invertebrates.[47]

In warm waters the growth rates of many species are relatively faster and life spans are shorter than in cooler, temperate waters. Because maximal algal and invertebrate productivity occurs in warm, shallow waters with high levels of nutrients, fish yields are also generally high in tropical ecosystems. Maximal fish production occurs in those tropical areas where seasonal rainfall is intense and rivers transport elevated amounts of dissolved and suspended nutrients. The high number of fish species that coexist in tropical habitats may also increase productivity because available foods can be used more effectively by a wide range of different foraging modes.[48]

The mean depth of lakes is often a good predictor of both algal and fish production as it relates to light penetration, thermal stratification, mixing of dissolved oxygen, and nutrient cycling. When mean depth, total dissolved solids, and (in some cases) drainage area measurements are combined to create various indices, there is improved predictability of productivity.[49] However, different types of fishes have different ecological requirements and can feed on a wide variety of foods, such as dead organic detritus, bottom-living (benthic) invertebrates, algae, and smaller

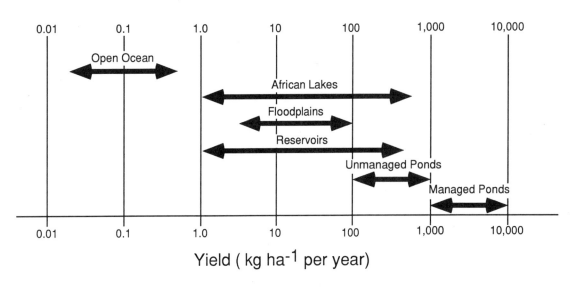

Fig. 4–1. Comparisons of fish yields from fresh water and marine ecosystems. Fresh water yields (harvestable weight per unit area per unit time) are much higher than open-ocean values, especially in the tropics. Shallow fresh water ponds, lakes, and reservoirs are generally more productive than deep lakes and deep reservoirs. (Redrawn from R.H. Lowe-McConnell, 1987, *Ecological Studies in Tropical Fish Communities*, Cambridge University Press, Cambridge, pp. 1–382).

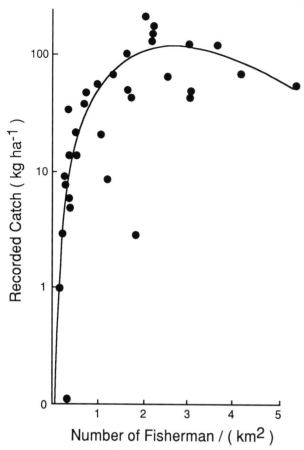

Fig. 4–2. The effect of increased numbers of fishermen on yields in open-access fisheries. (Data are compiled from the Food and Agriculture Organization of the United Nations for African lakes, redrawn from A.I. Payne, 1986, *The Ecology of Tropical Lakes and Rivers*, John Wiley and Sons, New York, pp. 1–301).

fishes of their own or other species. Moreover, endemic fishes of the ancient, deep lakes have unique adaptations for unusual modes of feeding and are highly subject to disruption from invasions by non-native species. These biotic considerations must be included when attempting to predict general productivity relationships.[50]

Human social factors also combine to alter fisheries production. The well-known "tragedy of the commons" occurs in lakes when individual fishing effort is increased after considering only individual costs and benefits. As fishing effort increases, the catch per unit effort declines (e.g., Figure 4.2). These intense levels of harvesting may rapidly deplete the "renewable" resources to sufficiently low levels that the fishes either go biologically extinct (through disruption of schooling and/or mating behavior), or become economically extinct (through reaching such a low level of production they cannot effectively be caught using standard net deployment and fishing methods). Early fisheries models for predicting maximal sustained yield from data based on catch per unit effort were too simplistic to consider the actual biological and economic dynamics in these "open" ecosystems and several important fisheries collapsed from overexploitation. Assumptions regarding fish growth rates, sources of natural mortality, and interannual variability in reproduction were not supported. More complex models are currently used but long-term data to calibrate these methods are lacking.[51]

**Non-native species transfers.** Historically, fisheries managers have focused on game fishes and have sometimes attempted to introduce north-temperate zone species into tropical and southern hemisphere lakes and streams. Fisheries management followed the same conventional methods of agriculture without fully considering the impacts on natural communities. The focus on yields of game fishes for sportsmen and their

impacts on commercial fishes for the market place often ignored the potential effects on non-game, native fishes and their intrinsic values. As discussed below these native fishes also play extremely important ecological and, in some cases, cultural roles. In his analyses of 1,354 transfers of 237 species into 140 countries, Welcomme concluded that introductions of non-native species into natural communities can be a valuable management option but that the risks are high. He recommended restricting transfers only to very specific, well-studied habitats. Some transfers have been economically beneficial when properly done in controlled aquacultural settings where the risks of escape and naturalization can be reduced through control of reproduction.[52]

Inappropriate non-native fish releases have been used widely for short-term increases in sports fisheries in many parts of the world without regard for impacts on native species or the long-term sustainability of the entire food web. The movement of fishes, invertebrates, algae, and aquatic macrophytes from their native habitats to other natural locations often causes a series of ecological problems and collapse of natural food webs and fisheries. The costs of reversing these effects can be extremely high and, in some cases, food web restoration may be impossible.[53]

Introduced species can increase rapidly where natural checks on their abundances are absent. For example, accidental transplants of aquatic plants (such as the tropical water weed *Hydrilla*) into habitats that lack their natural grazers or diseases often undergo excessive growth and out-compete native species for high-quality habitat. The explosive increase of filter-feeding zebra mussels (from Europe to the Laurentian Great Lakes, as discussed below) is a recent example of quick colonization. Non-native predators often increase beyond the habitat's sustainable carrying capacity and go extinct once they have eliminated the available native prey populations.

Beneficial effects such as biological control are rarely achieved when predators are transplanted into isolated ecosystems. The best studied cases are where various species of fishes (for example, mosquito fishes, *Gambusia affinis* and *Gambusia holbrooki*, are now found on every continent except Antarctica) have caused detrimental impacts on native fishes, sometimes without any control achieved on the target pest. It is not unusual for some governmental health agencies to propagate and periodically restock mosquito fish to control mosquitos in shallow-water habitats where other wildlife and conservation agencies (working for the same government) are actively attempting to eliminate the introduced mosquito fish. The problem occurs because once these aggressive predators have reduced the abundance of their target prey species (the mosquito), they shift to consuming non-target, native fish fry or other important species within the food web that previously consumed mosquitos. The spread of non-native fishes in arid regions has contributed to some of the highest extinction rates of native fresh water fishes on record, especially in areas where springs are also being drained for irrigation.[54]

A well documented example of fisheries mismanagement is now part of the history of the Laurentian Great Lakes of North America. These lakes contain over 100 native species including paddlefish, lake sturgeon, and bowfin. The major species of commercial fishes were salmonids such as lake trout, and whitefishes. Rainbow smelt were unintentionally introduced into the Great Lakes from Crystal Lake, Michigan, in the 1920s and 1930s with great commercial success. Overfishing of the Great Lakes sturgeon and destruction of their stream-spawning habitats caused greatly reduced yields. Furthermore, combined effects of environmental disruption and invasion of the Great Lakes by predatory sea lamprey and the alewife led to a series of major declines in salmonids from 1945 to 1955.[55]

The history of the sea lamprey and alewife invasions is complex. These marine species annually migrated upstream to spawn in many large coastal streams and lakes (as do many other anadromous fishes). Construction of the Erie Canal linked the Hudson River with the Great Lakes and provided the first invasion route for migratory marine species. The sea lamprey first migrated into Lake Ontario but was unable to swim past the Niagara Falls to enter Lakes Erie, Huron, Michigan, and Superior. After completion of the Welland Canal around the falls, sea lamprey invaded Erie, Huron and Michigan in the 1930s and then Superior in the 1940s. Canal and port construction continued and allowed large ocean-going ships to travel 3,800 km inland to Lake Superior. Sea lamprey and alewife disrupted the natural food web; lamprey attacked the larger salmonids while the alewife competed with smaller fishes for zooplankton prey. The cumulative effects of commercial overfishing and pollution together with the loss of larger, reproductive adult salmonids ended this fishery.

Although the lamprey had coexisted with the native fishes in Lake Ontario for years, their later impact on salmonids in some of the other Great Lakes was very detrimental. These marked dissimilarities in the impacts of sea lamprey in Lake Ontario and the other Great Lakes are not surprising. It is difficult to predict how any non-native species will respond to a large number of variables that differ in other lakes. Major differences among lakes in light penetration, depths of thermal stratification, and complex connections with different food webs make it very risky to generalize from short-term, localized sampling.

In 1985 some additional exotic invertebrates such as two species of zebra mussel and a spinose, crustacean zooplankton, the spiny water flea, *Bythotrephes cederstroemi*, invaded the region, apparently by ships emptying ballast waters from European ports. In Swedish lakes another species, *Bythotrephes longimanus*, has been found to be highly tolerant of acidic waters and can outcompete other crustaceans. *Bythotrephes* is very spiny and fish predators avoid consuming it. Simultaneously, *Bythotrephes* preys on *Daphnia*, an important zooplanktonic food item of fishes. The zebra mussels are expanding their populations rapidly as their planktonic larvae are moved from lake to lake. The mussels further disrupted the biotic communities of the Great Lakes and the upper Mississippi by competing with native species of mussels and zooplankton, and altering predator–prey relationships among coexisting species. The larvae attach to any hard surface (including intake pipes for cooling waters in electric power generators) and grow into filter-feeding adults that live up to four or more years. As they accumulate and block the cooling water intakes and heat exchangers, power plants and other industries must periodically shut down to clean out the mussels. Today the Great Lakes continue to recover from a series of accidental and intentional introductions of organisms as well as from a wide range of environmental mismanagement on a large scale.[56]

Although some transfers have been ''successful'' in providing game fish over the short run, they have generally had negative impacts on native fishes without any associated analyses of the costs and benefits over the long run. Following earlier practices, such as transfer of European carp first from the Danube River to Italy and finally to North America, economic development projects in the late 1940s and throughout the 1960s typically included attempts to establish non-native game fishes in lakes and reservoirs. Introductions of predatory species such as largemouth bass, black crappie, and sunfishes into the deep, oligotrophic Lake Atitlan, in the mountains of Guatemala, wiped out the smaller fishes that were used by the indigenous people living around this large lake. Similarly, the introduction of the peacock bass from the Amazon River into Gatun Lake, in the Panama Canal, greatly changed the pelagic food web. This large predator eliminated the smaller, native fishes and resulted in a cascade of detrimental effects. The most unexpected change may have been increased numbers of mosquitos, vectors of malaria in this region, because small minnows that previously consumed mosquito larvae in the shallow margins of the lake had been reduced in numbers by predation from the peacock bass. With the decline in small species of fishes many fish-eating birds also declined or moved to other lakes.[57]

Fifty years ago rainbow trout, brown trout, and lake trout were introduced into Lake Titicaca, a deep lake in the Andes of Peru and Bolivia. Of the 30 native species in Titicaca, 23 are endemic species of one genus of small killifish, *Orestias*. The lake is sufficiently large that most of the native species have persisted even though the rainbows are now quite widely distributed (and some are raised now in aquacultural cages); brown trout are confined to one tributary and the lake trout are very rare or absent. No native fishes have apparently gone extinct through competition or predation as a result of the fish introductions. However, many million have died from attacks by parasites brought into the lake with exotic fishes.[58] In temperate-zone springs, streams, and lakes both rainbow and brown trout seem to have been transplanted at one time or another to nearly every available habitat. The potential for disruptions of native fishes in many of these habitats is now very apparent but new

consequences may continue to appear for many years after an introduction.[59]

Long time delays often occur before any negative consequences of introductions are observed because it may take decades for generations of non-native species to reach high population densities and have a measurable effect on the native species they compete with and/or consume. An example of this lag is illustrated by the invasion of the Nile perch into Lake Victoria, one of the largest lakes in Africa with over 200 species of fishes.[60] Although relatively large, Victoria is not deep in comparison with rift valley lakes such as Tanganyika and Malawi (Table 4.2). Apparently, the Nile perch was introduced to Lake Victoria from Lake Albert about 1959. Its increased numbers and predatory pressure on smaller native fishes first led some fisheries managers to predict that a reduction in the numerous small, competing prey fishes would increase the growth rates of surviving prey and eventually lead to harvest of more profitable, larger native cichlid fishes. They did not anticipate the potential extinction of native species in this large, fluctuating lake that periodically floods the vegetated margins that surround it. For many years the fish productivity was high throughout the year. In the early 1980s, however, the catch of native species was greatly reduced and many endemic species are thought to have gone extinct. Some of these now extinct cichlid species may not even have been scientifically studied and named. Lake Victoria had been characterized by hundreds of endemic species of cichlid fishes that were restricted in their occurrence to this one lake which was isolated from Nile fishes by high waterfalls. There were another 38 non-cichlid species in Lake Victoria, including 16 endemics. Today up to a third of the total annual catch in Victoria is composed of the Nile perch; it may be following a similar trend to its expansion in nearby Lake Kyoga.[61]

Some of the most disruptive introductions have been of aquatic plants that float at the surface and shade out the phytoplankton needed by zooplanktonic grazers. Other plant species grow rapidly in lakes or streams because their structure or internal chemistry is such that native herbivores avoid eating them. The invasion of canals, reservoirs, and lakes by the water hyacinth, which floats on the surface and grows rapidly by vegetative reproduction, is well documented in many regions. In New Zealand the rapid spread of exotic plants such as the water net (*Hydrodictyon reticulatum*) is partly determined by people accidentally transporting the plant from lake to lake. The spread of floating ferns, small-leaved plants (such as duckweeds), and large-leaved plants (such as water chestnut) are a major problem in tropical lakes and reservoirs. In Papua New Guinea, for example, a few floating ferns (*Salvinia molesta*) were accidentally introduced into the Sepik River floodplain in 1972 and in eight years grew into floating mats that covered 250 km$^2$ and disrupted the fishing and transportation for 80,000 indigenous people.[62]

**Regulated rivers and reservoirs.** More than 20 years ago Wolman noted that the regulation of stream flow by impoundments has probably exerted a more profound, and possibly irreversible, effect on riverine ecosystems than pollutants. Ortmann provided clear evidence that species were disappearing as a result of dam construction at the turn of the century. He noted that ''dams prevent free migration, for instance of fishes, and thus they must be an obstacle to the natural restocking of the rivers.''[63] Although the majority of large rivers have already been dammed to produce hydroelectricity, there is interest in building more dams on both large and small rivers for low-flow augmentation and irrigation. Interest in dam building is accelerated by lending agencies that generally have had a better understanding of economics and construction costs than of ecological costs.[64]

In some catchments the construction of additional dams can be avoided through appropriate land use. For instance, the objectives of flood control could be attained by avoiding channelization of streams, protecting floodplains and vegetation in the drainage area, establishing riparian buffer strips, and by zoning flood-prone areas for parks and green belts. Additional dam construction for water storage could also be avoided in some regions if water use efficiency were increased. Improved planning of non-construction alternatives, appropriate maintenance of infrastructural components to reduce leakage, and establishing markets to transfer water at realistic values are some ways to increase efficiency of water usage.[65]

The continents differ greatly in their annual water balances and their capacities to store surface waters. Because of geological and hydrological differences, Africa has more than 20 times the volume of lakes and reservoirs found in South America even though the large rivers of South America drain an area that has more than twice the rainfall of Africa. These physical facts can lead to major disruptions of natural ecosystems and human populations in several ways. For example, to create additional water storage for drought relief and hydropower, many more large dams and reservoirs are rapidly being constructed in South America, while the high variability in rainfall and frequent droughts in many parts of Africa cause salinities to fluctuate and also create a need for additional water storage development. Biotic and cultural changes are occurring on a very large scale as a result of these engineering designs to regulate and redistribute water spatially and temporally.

Rivers have been modified in many parts of the world for centuries. However, during the last 50 years regulated rivers began to dominate many landscapes because of the accelerated development of hydropower. According to the World Register of Dams more than 12,000 large dams (greater than 15 m high) were built or under construction on major rivers by 1968 and this number continues to increase rapidly in many locations.[66] (See Tables in Part II, Section G.) Petts notes that the peak of dam construction occurred between 1950 and 1980. During this period more than 200 large dams per year were completed in North America alone. By 1980 China had completed 80,000 reservoirs, with their combined storage accounting for 16% of the annual runoff. Petts estimates that large dams are currently being completed at a rate of about 500 per year throughout the world. He predicts that by the year 2000, more than 60% of the world's total stream flow will be regulated.[67] Especially in mountainous and desert regions there are continued pressures for damming more streams to provide power, flood control, irrigation waters, and urban water supplies; relatively few free-flowing rivers remain.[68]

Plans to control the annual distribution of water through construction of storage reservoirs run into many types of environmental problems. Large projects such as the Aswan High Dam in Egypt have created many new ecological and economic problems.[69] Flood control projects may achieve limited goals of reducing loss of property but can cause large losses in floodplain fertility and species diversity. Many species of fishes use floodplain habitats for reproduction or for refuge from predators during key periods in their life cycle. Natural, seasonal flooding and drying are cues used by some river-dwelling animals to begin reproduction or migration. Some level of variable flooding is typically required in natural stream ecosystems to sustain functional floodplains and high productivity.[70]

The cascade of biotic effects from dam construction has been well documented (Figure 4.3) in several regions but transfer of information from one geographical region to another has proven difficult. Immediately after completion of dams there is often a decline in diversity of native fishes because the man-made lake environment is different from the original river ecosystem. Dams interrupt the seasonal upstream migrations of many species and alter the flow, dissolved oxygen, and temperatures that affect reproduction and survival.[71]

Once an assemblage of reservoir-adapted fishes becomes established, the fisheries harvests can match those of natural lakes. Yields can increase significantly after completion of large dams but rarely remain stable. With proper selection of fishes and management practices it may be possible to sustain relatively high productivity for several years or decades unless nutrient depletion, overfishing, or droughts become limiting. In some cases yields in tropical lakes may stabilize relatively quickly. Lake Kariba, the first large man-made tropical lake, stabilized in about 10 years after dam closure on the Zambezi River. In other cases 100 years could be necessary for any "stabilization" to occur. Unfortunately, the design of large reservoirs rarely includes sufficient planning for sustainable fisheries. The time-scales for ecological impacts of dam construction are thought to follow a general pattern (Figure 4.3 ).[72]

Reservoirs differ significantly from most natural lakes because their flooded river valleys usually have highly irregular shorelines and complex hydrologies. Dams alter the upstream/downstream hydrology and dynamics of riverine ecosystems through changes in nutrient retention, seasonal dynamics of water level fluctuations, and a wide variety of other

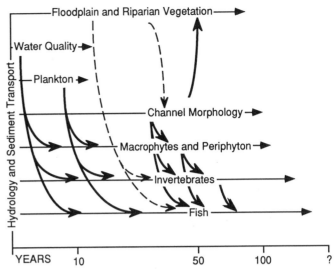

Fig. 4–3. Approximate time-scales for ecological relationships following dam construction. Immediate, direct impacts affect water quality. Intermediate, direct impacts alter aquatic plant producers and animal consumers. Long-term impacts alter river channel morphology and riparian plant communities. Solid arrows indicate major, direct relationships, and dashed arrows illustrate some feedback relationships within a general hierarchy of connections (modified and redrawn from G.E. Petts, 1989, Perspectives for ecological management of regulated rivers, in J.A. Gore and G.E. Petts (eds.) *Alternatives in Regulated River Management*, CRC Press, Boca Raton, Florida, pp. 3–24).

impacts. Because reservoir releases are managed for multiple needs (e.g., flood control, power production, water supply, recreation) there are often conflicts among competing objectives. Those reservoirs with only one primary management objective such as drinking water can include fisheries production easily. Some reservoir-management plans accommodate needs for stable water levels during spawning seasons and other natural requirements for stabilizing biotic communities.[73]

Large interannual lake-level fluctuations during droughts or non-seasonal drawdowns in flood control reservoirs can greatly alter littoral habitats where fish feed and spawn.[74] Nutrient cycling within the littoral zone is also greatly influenced by large water-level changes.[75] Development of a stable zone of submerged, near-shore vegetation is important because many reservoir-adapted fishes forage for their prey by moving in and out of the littoral zone. Many fishes migrate daily from the pelagic, open waters of deep lakes and reservoirs to feed in the near-shore vegetation. Poorly regulated water-level fluctuations in reservoirs and wetlands are deleterious to seasonal fish spawning, as well as to shoreline birds and mammals (such as beavers, muskrats, moose, and raccoons). Proper timing of water level drawdowns, however, can be used as a fisheries management tool to enhance spawning of certain species.[76]

Many existing dams presently have no capacity to produce electricity but could be retrofitted with generators and thereby decrease the need to build additional dams. For example, Cahn notes that "the vast majority of the 60,000 dams in the United States produce no power."[77] The relative locations of large and small reservoirs in a catchment can influence environmental impacts. If a single, large reservoir is used for generating hydroelectric power to meet daily consumer demands, large daily drawdowns cause enormous differences between day and night releases rather than natural, seasonally based fluctuations in downstream flows. Alterations of substrata and disruptions of reproduction among fishes and other biota can be avoided by modifying water releases from large reservoirs or by designing a sequence of larger reservoirs. In some dual hydropower reservoirs with pumped storage, the design can provide more controlled, variable downstream flows than a single, large reservoir and also increase mixing of waters within the reservoir. These operations include small storage reservoirs that are managed to hold effluent water that is pumped back into the main reservoir during the night or at other periods of off-peak demand.[78] If properly managed this mode of hydroelectric generation can minimize daily fluctuations for downstream aquatic communities. As multiple objectives for reservoirs are included in the management plans, such as flood control or water supply for irrigation, it becomes much more difficult to include needs for fisheries and wildlife. Future sustainable development of water resources will require management that can effectively incorporate spatial and temporal ecosystem dynamics while avoiding overexploitation during extreme conditions. Some rivers are especially diverse in their biotic communities and should be preserved rather than developed for hydropower.

Major concerns of aquatic ecologists, fisheries biologists, and public health workers regarding water regulation include: declines in reproduction of native fishes because of barriers to long-distance, upstream-spawning migrations;[79] spread of vectors of several human parasitic diseases such as schistosomiasis[80] and malaria; reduction in transport of suspended silts; increased bank erosion; and highly variable water temperatures downstream from the dam site. These large construction projects are often designed to provide water supply for irrigation so that increased densities of human populations are attracted to ecologically sensitive riverine corridors while often the lower density, more dispersed, indigenous populations may be flooded out of their ancestral lands. As discussed previously, there are "soft" alternatives to further dam construction (such as improved land use to control flooding and reduce the need for irrigation waters, or conservation of water that reduces the need for water storage) that can eliminate detrimental environmental impacts. Much more information on the biology of natural and reservoir-associated species is needed before sustainable management of these man-made habitats can be widely achieved.

**Deforestation and catchment land use.** Land use practices are known to strongly influence how nutrients, silt, and water move through any catchment. Changes in land use can rapidly alter nutrient inputs to tributary drainages and associated downstream rivers, lakes, and wetlands. Field experiments from temperate zone catchments demonstrate that after small storms the peak flows, associated erosion and nutrient losses, and flooding of tributaries are increased after deforestation.[81] Reforestation can increase water storage and slow down rates of runoff. However, reforestation of some large, temperate zone drainage basins apparently has resulted in relatively little control over the effects of very large, catastrophic storm events.[82]

In response to long-term, large-scale deforestation in many tropical areas there are projects attempting to restore forest cover and thereby increase water supplies for irrigation and urban use, especially in the driest months of the year, and to reduce flooding during the wettest months.[83] Deforestation is often associated with increased frequency and intensity of floods in tropical catchments, although interpretations of existing data sometimes vary greatly.[84]

Soil erosion and sediment transport by rivers resulting from intense rains cause many problems for aquatic organisms. Erosion and siltation eliminate hard, stable, surface substrata required by many bottom-dwelling species. Silt fills in crevices that serve as refugia from predators and eliminates interstitial spaces in gravel beds that are used as nesting sites for some fishes. Rapid rates of bank erosion, subsurface flows through "macro-pores" (burrows, crevices, root-generated cavities), and overland flows can move large quantities of sediment and modify downstream habitats. Even in undisturbed, forested catchments major storm events can cause landslides and stream bank erosion in steeply sloped regions. For example, in a study of southern Kenyan forested catchments, 20–30 metric tons/km$^2$/year of sediment were lost. This loss was relatively low in comparison with other streams. Approximately 600 metric tons/km$^2$/year were lost from a small, forested catchment on Barro Colorado Island, Panama, while several thousand metric tons/km$^2$/year were lost from some steeply sloped agricultural basins in Kenya.[85] As managers seek to reduce flooding and soil erosion, to provide more water for irrigation and drinking, and to restore biotic diversity of forests, wetlands, and streams, they face difficult decisions regarding land use, especially in steeply sloped catchment areas. In many regions long-term data and field testing at appropriate scales are inadequate. Research on effects of disturbance on aquatic habitats has focused on multiple[86] and cumula-

tive[87] impacts of soil erosion, pesticide contamination, and nutrient runoff resulting from extensive alterations of drainage areas.

**Global climate change.** Biotic responses to more extreme floods, droughts, and warm water temperatures may cause unexpected cumulative effects to result from relatively gradual changes in land use and climate.[88] Currently, only a few sites exist where hydrologic regimes, water quality, and associated biotic communities have been studied for more than 25 years. Even these records are much too brief a time span and too small a spatial coverage for understanding ecosystem responses to global change. Changing patterns of land use, species introductions, rainfall, and temperature are having major effects on the geographic distributions and abundances of many types of fresh water plants and animals. How biotic variables will be influenced by climatic change is now being studied in many locations throughout the world. Regional hydrological responses to climatic changes are relatively well known from many areas. How individuals, populations, communities, and ecosystems respond to these regional changes is not yet well documented. However, the relative importance of many factors can be identified from widespread examples of habitat abuse around the world and from a large number of detailed studies dealing with interrelated sources of disturbance.

Although much is known about particular, small-scale connections within catchments, there are many large-scale spatial and temporal relationships still undefined or in need of further study.[89] For example, we know that increased temperature will increase rates of many individual physiological responses and key ecosystem properties such as decomposition and nutrient cycling but we do not have much information about how species can respond behaviorally or genetically over long periods of time or large geographic areas. Recent work in biogeography, population genetics, and conservation biology is emphasizing a landscape level of analysis.[90]

Global circulation models of the atmosphere suggest that increased carbon dioxide and other greenhouse gases may cause regional effects such as warmer air temperatures at high latitudes and high altitudes, while more precipitation is expected from 30° south to 30° north latitude because of higher evapotranspiration rates.[91] More intense storms are also predicted for some regions such as hurricane zones. Other models of global change such as the El Niño Southern Oscillation (ENSO) suggest possible alterations in rainfall and stream flow. Sites extending across the southern hemisphere and into lower parts of the northern hemisphere may be affected as hurricanes and other storm patterns also respond to increased sea surface temperatures. Channel and basin infilling as well as fluctuations in water temperatures can be increased following changes in the seasonal and annual distributions of precipitation. Thermal and hydrologic changes in montane streams at various latitudes may become useful integrators of information on global warming. If global warming and/or ENSO cycles of sea surface temperature shifts continue to follow recent trends there may be predictive patterns of hydrologic change (such as higher evapotranspiration during more intense droughts) that require altered management of drainage areas.[92]

Numerous types of streams can be identified that are likely to be sensitive to drought, flooding, and thermal disruptions. For example, cool headwater streams may become warmer with longer growing seasons in the future. Predictions of rates of temperature change from global circulation models currently range from 1°C to 4°C over 10 to 100 years. Most slow-growing, long-maturing fresh water species, such as fishes and bivalves, may not have sufficient time to develop adaptive traits. Many species are already living near their maximal thermal limits of survival during present-day, long, hot droughts.[93]

High elevational and high latitudinal streams and lakes are also likely to show effects of global warming through alterations in growth rates and species compositions. For example, the geographical limits of introduced brown trout on the North Island of New Zealand might be more restricted than rainbow trout in some lakes and shallow, unshaded rivers because brown trout egg mortality is 100% in 15°C waters but rainbows show only about 25% mortality at this temperature.[94]

Knowledge of thermal characteristics of stratified lakes provides a basis for predicting where to find fish at certain depths. For example, the

22 main resident species of fish in Lake St. Clair (Ontario, Canada) can be grouped into two distinct assemblages that differ in their responses to maximal water temperatures during late summer. Changes in species abundances for these two assemblages can be expected if global warming conditions result in warmer summertime surface waters for this region.[95]

Prairie wetlands in the interior of North America currently have seasonal and cyclic fluctuations in water levels. Fluctuations in volume may become more severe and diminish these important waterfowl habitats.[96] Arid regions may show even more pronounced variability of intermittent flows and other types of biotic changes as salinities fluctuate.

Indirect effects of warmer temperatures and longer seasons on some prey and host populations could occur through increased exposure to more numerous predators and parasites. Some prey populations are held in dynamic balance with their predator populations through the small prey species' ability to grow faster and reproduce more quickly than their larger, slower-growing predators. Expansion of geographical range limits could also bring new predators and parasites into contact with populations that were previously spatially isolated. Pathogens may infect more animal hosts if increased temperatures result in higher transmission rates. Previous studies of an artificially heated reservoir in South Carolina demonstrate that elevated temperatures can influence infectivity of hosts by parasites.[97]

If global warming is accompanied by continued eutrophication of lakes it is likely that biotic communities will be altered. As discussed previously, the decay of excessive plant biomass can lead to suffocation of many types of bottom-dwelling animals such as bivalves, aquatic insects, and fishes.[98] Most cold-water fishes require relatively large volumes of water with at least 5 mg/l dissolved oxygen. Because oxygen is less soluble in warm water, the potential for deoxygenation is high in tropical waters or during the summer in shallow ponds and intermittent streams. Warm temperatures also increase rates of decomposition of organic matter and further deplete oxygen at fast rates. If oxygen concentrations drop below 3 mg/l then many fishes adapted for warm waters will not survive. Thus, any climatic shift toward more extreme warming can influence oxygen concentrations in both flowing and standing waters, especially in shallow lakes. In deep, low-nutrient, northern lakes global warming may only extend the growing season and lead to more optimal growth rather than to lethal deoxygenation.[99]

## Summary and future studies

As more data become available from a wider array of sites, it is clear that spatial scales of ecological interactions can extend from local and regional to global dimensions. Distinguishing the multiple, simultaneous sources of natural and man-induced disturbances of fresh water ecosystems will require study at several spatial and temporal scales that include sufficient heterogeneity across landscapes. Technological advances in remote sensing, radar, radiotelemetry, acoustical sounding, and geographic information systems modeling will continue to improve both spatial resolution and multi-scale analyses of inland aquatic ecosystems. Many dynamics of fresh water ecosystems, such as food web responses to non-native species introductions, succession, and eutrophication, can be quite long relative to human timescales and occur over large areas. Other local changes and responses occur rapidly. Some events can be missed entirely if studies are based on infrequent sampling. Event-driven, automated sampling methodologies, improved computer networks of data transmission, and sharing of large databases will improve ecological modeling of catchment dynamics.

As Tolba has concluded, "Fresh water lakes are a vital part of our natural resource heritage. Properly cared for, lakes will help support humankind indefinitely ... If lakes and other natural resources are not receiving enough attention in the development decisions made by governments, then that is a reflection of a failure by environmentalists to make their existence and value known. Greater understanding must be achieved, locally, nationally, and globally."[100] A great deal of scientific study has demonstrated important interconnections between inland waters and their catchments. Unfortunately, relatively few generalizations have emerged because of the many regional differences that alter sustainable productiv-

ity in inland waters. This regional, biotic, and physical diversity creates a major, but fascinating, challenge for managers and scientists in the years ahead. New solutions may appear once appropriately scaled studies are completed and more communication among specialists focuses on parameters that previously had been overlooked.

Large-scale water resource management is essential if sustainable biological resources are to be protected. Major issues in future management scenarios will be to: (i) provide adequate quantity and high quality water for human use and natural habitats through control of catchment land use; (ii) minimize alterations of natural ecosystem processes and losses of overall biodiversity and integrity through greater control over introduced, non-native species and improved site selection for hydroelectric projects; (iii) preserve remaining natural fresh water habitats that have high biodiversity and endemic species; and (iv) increase cooperative, multidisciplinary, international research projects on well selected sites that include regionally representative biotas and hydrologic regimes.

## Notes

1. For example, J.B. Callicott, 1989, The philosophical value of wildlife, in D.J. Decker and G.R. Goff (eds.) *Valuing Wildlife*, Westview Press, Boulder, Colorado, pp. 214–221; H. Rolston, III, 1985, Valuing wildlands, *Environmental Ethics*, 7(1), 23–48.
2. H.A. Regier and J.D. Meisner, 1990, Anticipated effects of climate change on freshwater fishes and their habitat, *Fisheries*, 15(6), 10–15.
3. J.A. Drake, 1990, Communities as assembled structures: Do rules govern pattern?, *Trends in Ecology and Evolution*, 5(5), 159–163.
4. See generally K.E. Cummins, 1988, The study of stream ecosystems: A functional view, in L.R. Pomeroy and J.J. Alberts (eds.) *Concepts of Ecosystems Ecology. A Comparative View*, Springer-Verlag, New York, pp. 247–262; K.D. Fausch, J. Lyons, J.R. Karr and P.L. Angermier, 1990, Fish communities as indicators of environmental degradation, in *Biological Indicators of Stress in Fish, American Fisheries Society Symposium*, 8, Bethesda, Maryland, pp. 123–144; J.R. Karr, 1991, Biological integrity: A long-neglected aspect of water resource management, *Ecological Applications* 1(1), 66–84.
5. For review, see S.J. Cohen, Great Lakes levels and climate change: Impacts, responses, and futures, in M.H. Glantz (ed.) *Societal Responses to Regional Climatic Change. Forecasting by Analogy*, Westview Press, Boulder, Colorado, pp. 143–167; M.H. Glantz, 1990, Does history have a future? Forecasting climate change effects on fisheries by analogy, *Fisheries*, 15(6), 39–45; M.C. Molles, Jr. and C.N. Dahm, 1990, A perspective on El Niño and La Niña: Global implications for stream ecology, *Journal of the North American Benthological Society*, 9(1), 69–76; J.V. Ward, 1989, The four-dimensional nature of lotic ecosystems, *Journal of the North American Benthological Society*, 8(1), 2–8.
6. See generally C.E. Herdendorf, 1990, Distribution of the world's large lakes, in M.M. Tilzer and C. Serruya (eds.) *Large Lakes. Ecological Structure and Function*, Springer-Verlag, New York, pp. 3–38; M.K. Tolba, 1987, *Sustainable Development. Constraints and Opportunities*, Butterworths, London, pp. 1–221.
7. For example, A.P. Covich, 1988, Geographical and historical comparisons of neotropical streams: Biotic diversity and detrital processing in highly variable habitats, *Journal of the North American Benthological Society*, 7(4), 361–386; C.H. Fernando, 1984, Reservoirs and lakes of Southeast Asia (Oriental Region), in F.B. Taub (ed.) *Lakes and Reservoirs. Ecosystems of the World*, vol. 23, Elsevier, New York, pp. 411–446; M. Goulding, 1990, *Amazon: The Flooded Forest*, Sterling Publishing, New York, pp. 11–53; R.H. Lowe-McConnell, 1987, *Ecological Studies in Tropical Freshwaters*, Cambridge University Press, Cambridge, pp. 1–382; R.L. Welcomme, 1989, Floodplain fisheries management, in J.A. Gore and G.E. Petts (eds.) *Alternatives in Regulated River Management*, CRC Press, Boca Raton, Florida, pp. 209–234.
8. See generally V. Banks, 1991, *The Pantanal, Brazil's Forgotten Wilderness*, Sierra Club Books, San Francisco, pp. 196–245; A. Bonetto, 1975, Hydrologic regime of the Paraná River and its influence on ecosystems, in A.D. Hasler (ed.) *Coupling of Land and Water Systems*, Springer-Verlag, New York, pp. 175–197; G.E. Petts, 1989, Perspectives for ecological management of regulated rivers, in J.A. Gore and G.E. Petts (eds.) *Alternatives in Regulated River Management*, CRC Press, Boca Raton, Florida, pp. 3–24; C. Serruya and U. Pollingher, 1983, *Lakes in the Warm Belt*, Cambridge University Press, Cambridge, pp. 1–569.
9. P.B. Moyle and J.J. Cech, Jr., 1982, *Fishes: An Introduction to Ichthyology*,

**Essays on fresh water issues**

Prentice-Hall, Englewood Cliffs, New Jersey, pp.1–593; D. H. Stansbery, 1973, Why preserve rivers?, *The Explorer*, **15**(3), 14–16; R.J. Wootton, 1990, *Ecology of Teleost Fishes*, Chapman and Hall, New York, pp. 1–404.

10. See chapters in J.H. Thorp and A.P. Covich (eds.) *Ecology and Classification of North American Freshwater Invertebrates*, Academic Press, San Diego, pp. 1–911; A.P. Covich, 1988, Geographical and historical comparisons of neotropical streams: Biotic diversity and detrital processing in highly variable habitats, *Journal of the North American Benthological Society*, **7**(4), 361–386; R.J. Shiel and W.D. Williams, 1990, Species richness in tropical freshwaters of Australia, *Hydrobiologia*, **202**, 175–183; P. Banaresu, 1990. Zoogeography of Fresh Waters, **1**, *General Distribution and Dispersal of Freshwater Animals*, AULA Verlag, Wiesbaden, pp. 1–511.

11. For recent reviews, see J.D. Thomas, 1990, Mutualistic interactions in freshwater modular systems with molluscan components, *Advances in Ecological Research*, **20**, 125–178; D.D. Williams, 1987, *The Ecology of Temporary Waters*, Croom Helm, London, pp. 1–193; J.V. Ward, 1992, "Aquatic Insect Ecology." 1. *Biology and Habitat*, Wiley, New York, pp. 1–438.

12. D.D. Williams, 1987, *The Ecology of Temporary Waters*, Croom Helm, London, pp. 1–193; W.D. Williams, A.J. Boulton and R.G. Taaffe, 1990, Salinity as a determinant of salt lake fauna in fresh water, *Hydrobiologia*, **197**, 257–266.

13. For example, F.B. Cross and R.E. Moss, 1987, Historic changes in fish communities and aquatic habitats in plains streams of Kansas, in W.J. Matthews and C.D. Heins (eds.) *Community and Evolutionary Ecology of North American Stream Fishes*, University of Oklahoma Press, Norman, Oklahoma, pp. 155–165; M. Goulding, 1981, Man and fisheries on an Amazon frontier, *Developments in Hydrobiology*, **4**, 1–137; C.H. Hocutt and E.O. Wiley, 1986, *The Zoogeography of North American Freshwater Fishes*, John Wiley and Sons, New York, pp. 1–866; G.E. Hutchinson, 1957, *A Treatise on Limnology. Vol. 1: Geography, Physics, and Chemistry*, John Wiley and Sons, New York, pp. 1–1015; A.I. Payne, 1986, *The Ecology of Tropical Lakes and Rivers*, John Wiley and Sons, New York, pp. 1–301.

14. For example, D.N. Riemer, 1984, *Introduction to Freshwater Vegetation*, AVI Publishers, Westport, Connecticut, pp. 1–207; R.R. Miller, J.D. Williams and J.E. Williams, 1989, Extinctions of North American fishes during the past century, *Fisheries*, **14**(6), 22–38; P.B. Moyle and J.E. Williams, 1990, Biodiversity loss in the temperate zone: Decline of the native fish fauna of California, *Conservation Biology*, **4**(3), 275–284; E.P. Pister, 1990, Desert fishes: An interdisciplinary approach to endangered species conservation in North America, *Journal of Fish Biology*, **37**, suppl. A, 183–187; J.E. Williams and R.R. Miller, 1990, Conservation status of the North American fish fauna in fresh water, *Journal of Fish Biology*, **37**, suppl. A, 79–85.

15. D. H. Stansbery, 1970, Eastern freshwater mollusks (I). The Mississippi and St. Lawrence River systems, *Malacologia*, **10**(1), 9–21.

16. See generally, M.J. Burgis and P. Morris, 1987, *The Natural History of Lakes*, Cambridge University Press, Cambridge, pp. 1–218; P.H. Greenwood, 1984, African cichlids and evolutionary theories, in A.A. Echelle and I. Kornfield (eds.) *Evolution of Fish Species Flocks*, University of Maine Press, Orono, pp. 141–154; G.E. Hutchinson, 1967, *A Treatise on Limnology. Vol. 2: Introduction to Lake Biology and Limnoplankton*, John Wiley and Sons, New York, pp. 1–1115; R.H. Lowe-McConnell, 1987, *Ecological Studies in Tropical Freshwaters*, Cambridge University Press, Cambridge, pp. 1–382; R.J. Wootton, 1990, *Ecology of Teleost Fishes*, Chapman and Hall, New York, pp. 1–404.

17. M. Goulding, 1985, Forest fishes of the Amazon, in G.T. Prance and T.E. Lovejoy (eds.) *Key Environments: Amazonia*, Pergamon Press, Oxford, pp. 267–276; M. Goulding, 1990, *Amazon: The Flooded Forest*, Sterling Publishing, New York, pp. 11–53.

18. For review, see A.C. Benke, 1990, A perspective on America's vanishing streams, *Journal of the North American Benthological Society*, **9**(1), 77–88; W. K. Olson, 1989, Introduction, in J.D. Echeverria, P. Barrow, and R. Roos-Collins (eds.) *Rivers at Risk*, Island Press, Washington, DC, pp. 1–13; R.I. Cahn (ed.), 1985, *An Environmental Agenda for the Future*, Island Press, Washington, D.C., pp. 1–155.

19. See generally, J.R. Karr, 1991, Biological integrity: A long-neglected aspect of water resource management, *Ecological Applications*, **1**(1), 66–84; J. Cairns (ed.), 1992. *Restoration of Aquatic Ecosystems*, National Academy Press, Washington, D.C., pp. 1–552.

20. For review, see B.L. Peckarsky, 1983, Biotic interactions or abiotic limitations? A model of lotic community structure, in T.D. Fontaine, III and S.M. Bartell (eds.) *Dynamics of Lotic Ecosystems*, Ann Arbor Press, Ann Arbor, Michigan, pp.303–323.

21. G.E. Hutchinson, 1967, *A Treatise on Limnology. Vol. 2: Introduction to Lake Biology and Limnoplankton*, John Wiley and Sons, New York, pp. 1–1115; for example, H.W. Li, C.B. Schreck, C.E. Bond, and E. Rexstad, 1987, Factors influencing changes in fish assemblages of Pacific Northwest streams, in W.J. Matthews and D.C. Heins (eds.) *Community and Evolutionary Ecology of North American Stream Fishes*, University of Oklahoma Press, Norman, pp. 193–202; W.S. Devick, J.M. Fitzsimmons, and R.T. Nishimoto, 1992, *Conservation of Hawaiian Freshwater Fishes*, Division of Aquatic Resources, Department of Land and Natural Resources, Honolulu, pp. 1–26.

22. For example, T.D. Brock, 1985, Life at high temperatures, *Science*, **230**, 132–138; G.A. Cole, 1968, Desert limnology, in G.W. Brown (ed.) *Desert Biology*, vol. 1, Academic Press, New York, pp. 423–486; C.C. Coutant, 1990, Temperature-oxygen habitat for freshwater and coastal striped bass in a changing climate, *Transactions of the American Fisheries Society*, **119**, 240–253; A.P. Covich and J.H. Thorp, 1991, Crustacea: Introduction and peracarida, in J.H. Thorp and A.P. Covich (eds.) *Ecology and Classification of North American Freshwater Invertebrates*, Academic Press, San Diego, pp. 665–689; J.C. Gottschal and R.A. Prins, 1991, Thermophiles: A life at elevated temperatures, *Trends in Ecology and Evolution*, **6**(5), 57–162; G.K. Meffe and F.F. Snelson, Jr., 1989, An ecological overview of poeciild fishes, in G.K. Meffe and F.F. Snelson, Jr. (eds.) *Ecology and Evolution of Live-Bearing Fishes (Poeciliidae)*, Prentice-Hall, Englewood Cliffs, New Jersey, pp. 13–31; N.D. Mundahl, 1990, Heat death of fish in shrinking stream pools, *American Midland Naturalist*, **123**, 40–46.

23. For example, A.P. Covich, 1988, Geographical and historical comparisons of neotropical streams: Biotic diversity and detrital processing in highly variable habitats, *Journal of the North American Benthological Society*, **7**(4), 361–386; J.V. Ward, 1985, Thermal characteristics of running waters, *Hydrobiologia*, **125**, 31–46.

24. For example, J.A. Stanford, F.R. Hauer and J.V. Ward, 1988, Serial discontinuity in a large river system, *Verhandlungen Internationale Vereinigung für Theoretische und Angewandte Limnologie*, **23**, 114–118; B. Statzner, 1987, Characteristics of lotic ecosystems and consequences for future research directions, in E.-D. Schulze and H. Zwolfer (eds.) *Potentials and Limitations of Ecosystem Analysis*, Springer-Verlag, New York, pp. 365–390; J.V. Ward, 1986, Altitudinal zonation in a Rocky Mountain stream, *Archiv für Hydrobiologie Supplement*, **74**, 133–199; J.V. Ward and J.A. Stanford, 1983, The intermediate-disturbance hypothesis: An explanation for biotic diversity patterns in lotic ecosystems, in T.D. Fontaine, III and S.M. Bartell (eds.) *Dynamics of Lotic Ecosystems*, Ann Arbor Press, Ann Arbor, Michigan, pp. 347–356.

25. See generally R.G. Danzmann, D.S. MacLennan, D.G. Hector, P.D. Hebert and J. Kolasa, 1991, Acute and final temperature preferenda as predictors of Lake St. Clair fish catchability, *Canadian Journal of Fisheries and Aquatic Sciences*, **48**, 1408–1418; W.J. Matthews and E.G. Zimmerman, 1990, Potential effects of global warming on native fishes of the southern great plains and the southwest, *Fisheries*, **15**(6), 26–32.

26. See generally N.L. Poff and J.V. Ward, 1990, Physical habitat template of lotic systems: Recovery in the context of spatiotemporal heterogeneity, *Environmental Management*, **14**, 629–645; A.H. Resh, A.V. Brown, A.P. Covich, M.E. Gurtz, H.W. Li, G.W. Minshall, S. R. Reice, A.L. Sheldon, J.B. Wallace and R. Wissmar, 1988, The role of disturbance in stream ecology, *Journal of the North American Benthological Society*, **7**(4), 433–455.

27. R.L. Welcomme, 1979, *Fisheries Ecology of Floodplain Rivers*, Longman, London, pp. 1–317.

28. N.L. Poff and J.V. Ward, 1989, Implications of streamflow variability and predictability for lotic community structure: A regional analysis of streamflow patterns, *Canadian Journal of Fisheries and Aquatic Sciences*, **46**, 1805–1818.

29. For example, G.K. Meffe, 1984, Effects of abiotic disturbance on coexistence of predatory fish species, *Ecology*, **65**(5), 1525–1534; W.L. Pflieger and T.B. Grace, 1987, Changes in the fish fauna of the lower Mississippi River, 1940–1983, in W.J. Matthews and D.C. Heins (eds.) *Community and Evolutionary Ecology of North American Stream Fishes*, University of Oklahoma Press, Norman, Oklahoma, pp. 166–177.

30. For review, see B. Henderson-Sellers and H.R. Markland, 1987, *Decaying Lakes. The Origins and Control of Cultural Eutrophication*, John Wiley and Sons, New York, pp. 1–254.

31. For discussion, see R.J. Naiman, 1981, An ecosystem overview: Desert fishes and their habitats, in R.J. Naiman and D.L. Soltz (eds.) *Fishes in North American Deserts*, John Wiley and Sons, New York, pp. 493–531; E. Vareschi, 1987, Saline lake ecosystems, in E.-D. Schulz and H. Zwolfer (eds.) *Potential and Limitations of Ecosystem Analysis*, Springer-Verlag, New York, pp. 347–364; W.D. Williams, A.J. Boulton and R. G. Taaffe,

1990, Salinity as a determinant of salt lake fauna: A question of scale, *Hydrobiologia*, **197**, 257–266.

32. See generally H.P. Hugster and L.A. Hardie, 1978, Saline lakes, in A. Lerman (ed.) *Lakes: Chemistry, Geology, Physics*, Springer-Verlag, New York, pp. 237–293; D.A. Livingstone and J.M. Melack, 1984, Some lakes of sub-Saharan Africa, in F.B. Taub (ed.) *Lakes and Reservoirs. Ecosystems of the World*, vol. 23, Elsevier, New York, pp. 467–497.

33. For example, G.A. Cole, 1968, Desert limnology, in G.W. Brown (ed.) *Desert Biology*, vol. 1, Academic Press, New York, pp. 423–486; J.E. Williams, D.B. Bowman, J.E. Brooks, A.A. Echelle, R.J. Edwards, D.A. Hendrickson and J.J. Landye, 1985, Endangered aquatic ecosystems in North American deserts with a list of vanishing fishes of the region, *Journal of Arizona-Nevada Academy of Science*, **20**(1), 1–61.

34. For example, G.L. Dana and P.H. Lenz, 1986, Effects of increasing salinity on an *Artemia* population from Mono Lake, California, *Oecologia*, **68**, 428–436; D.L. Galat, M. Coleman and R. Robinson, 1988, Experimental effects of elevated salinity on three benthic invertebrates in Pyramid Lake, Nevada, *Hydrobiologia*, **158**, 133–144; T.R. Karl and P.J. Young, 1986, Recent heavy precipitation in the vicinity of the Great Salt Lake: Just how unusual?, *Climate and Applied Meteorology*, **25**, 353–363; D.W. Stephens, 1990, Changes in lake levels, salinity, and the biological community of Great Salt Lake (Utah, U.S.A.), 1847–1987, *Hydrobiologia*, **197**, 139–146; W.A. Wurtsbaugh, 1992, Food-web modification by an invertebrate predator in the Great Salt Lake (U.S.A.), *Oecologia*, **89**(2), 168–175; W.A. Wurtsbaugh and T.S. Berry, 1990, Cascading effects of decreased salinity on the plankton, chemistry, and physics of the Great Salt Lake (Utah), *Canadian Journal of Fisheries and Aquatic Sciences*, **47**, 100–109.

35. For example, J.M. Eilers, T.J. Sullivan and K.C. Hurley, 1990, The most dilute lake in the world?, *Hydrobiologia*, **199**(1), 1–6; V.H. Smith, 1983, Low nitrogen to phosphorus ratios favor dominance by blue-green algae in lake phytoplankton, *Science*, **221**, 669–671.

36. See generally S.-O. Ryding and W. Rast (eds.), 1989, *The Control of Eutrophication of Lakes and Reservoirs. Man and the Biosphere Series*, vol. 1, UNESCO and Parthenon Publishing, Park Ridge, New Jersey, pp. 1–314.

37. For review, see W.T. Edmondson, 1991, *The Uses of Ecology*, University of Washington Press, Seattle, pp. 1–329. J.T. Lehman, 1986, Control of eutrophication in Lake Washington, in *Ecological Knowledge and Environmental Problem-Solving: Concepts and Case Studies*, National Research Council, National Academy of Sciences Press, Washington, D.C., pp. 302–312.

38. For review, see W.W. Carmichael, 1980, *The Water Environment: Algal Toxins and Health*, Plenum Press, New York, pp. 1–491; C.R. Goldman and A.J. Horne, 1983, *Limnology*, McGraw-Hill, New York, pp. 1–464.

39. B. Henderson-Sellers and H.R. Markland, 1987, *Decaying Lakes. The Origins and Control of Cultural Eutrophication*, John Wiley and Sons, New York, pp. 1–254.

40. For example, D.F. Brakke, D.H. Landers and J.M. Eilers, 1988, Chemical and physical characteristics of lakes in the northeastern United States, *Environmental Science and Technology*, **22**, 155–163; pH of the Amazon's Rio Negro is as low as 3.2 according to M. Goulding, *The Fishes and the Forest, Explorations in Amazonian Natural History*, University of California Press, Berkeley, California, p. 16.

41. B. Almer, W. Dickson, C. Ekstrom and E. Hornstrom, 1974, Effects of acidification in Swedish lakes, *Ambio*, **3**(1), 30–36.

42. See generally L.A. Baker, A.T. Herlihy, P.R. Kaufmann and J.M. Eilers, 1991, Acidic lakes and streams in the United States: The role of acidic deposition, *Science*, **252**, 1151–1154; I.P. Muniz, 1990, Freshwater acidification– its effects on species and communities of freshwater microbes, plants, and animals, *Proceedings of the Royal Society of Edinburgh, Section B, Biological Sciences*, **97**, 227–254.

43. See generally J.P. Baker and S.W. Christensen, 1991, Effects of acidification on biological communities in aquatic ecosystems, in D.F. Charles (ed.) *Acid Deposition and Aquatic Ecosystems*, Springer–Verlag, New York, pp. 83–106.

44. For example, P.L. Brezonik, L.A. Baker, J.R. Eaton, T.M. Frost, P. Garrison, T.K. Kratz, J.J. Magnuson, J.A. Perry, W.J. Rose, B.K. Shepard, W.A. Swenson, C.J. Watras and K.W. Webster, 1987, Artificial acidification of Little Rock Lake, Wisconsin, *Water, Air, and Soil Pollution*, **31**, 115–122; R.J. Hall, G.E. Likens, S.B. Fiance and G.R. Hendrey, 1980, Experimental acidification of a stream in the Hubbard Brook Experimental Forest, New Hampshire, *Ecology*, **61**(5), 976–989; D.W. Schindler, K.H. Mills, D.F. Malley, D.L. Findlay, J.A. Shearer, I.J. Davies, M.A. Turner, G.A. Linsey and D.R. Cruikshank, 1985, Long-term ecosystem stress: The effects of years of experimental acidification on a small lake, *Science*, **228**, 1395–1401.

45. For example, P.B. Bayley, 1988, Accounting for effort when comparing tropical fisheries in lakes, river-floodplains, and lagoons, *Limnology and Oceanography*, **33**(4), 963–972; A.I. Payne, 1990, The value of fish, in A. Speedy (ed.) *Developing World Agriculture*, Grosvenor Press International, London, pp. 244–253; R.L. Welcomme, 1976, Some general and theoretical considerations on the fish yields of African rivers, *Journal of Fish Biology*, **8**, 351–364.

46. For example, P.B. Bayley, 1981, Fish yields from Amazon in Brazil: Comparison with African river yields and management possibilities, *Transactions of the American Fisheries Society*, **110**, 351–359; S.P. Rissotto and R.E. Turner, 1985, Annual fluctuation in abundance of the commercial fisheries of the Mississippi River tributaries, *North American Journal of Fisheries Management*, **5**, 557–574.

47. For example, J.M. Hanson and W.C. Leggert, 1982, Empirical prediction of fish biomass and yield, *Canadian Journal of Fisheries and Aquatic Sciences*, **39**, 257–263; R.E. Hecky, E.J. Fee, H.J. Kling and J.W.M. Rudd, 1981, Relationship between primary production and fish production in Lake Tanganyika, *Transactions of the American Fisheries Society*, **110**, 64–71.

48. P.B. Bayley, 1981, Fish yields from Amazon in Brazil: Comparison with African river yields and management possibilities, *Transactions of the American Fisheries Society*, **110**, 351–359; R.H. Lowe-McConnell, 1987, *Ecological Studies in Tropical Freshwaters*, Cambridge University Press, Cambridge, pp. 1–382; A.I. Payne, 1986, *The Ecology of Tropical Lakes and Rivers*, John Wiley and Sons, New York, pp. 1–301.

49. For example, S.R. Kerr and R.A. Ryder, 1988, The applicability of fish yield indices in freshwater and marine ecosystems, *Limnology and Oceanography*, **33**(4), 973–981.

50. P.H. Greenwood, 1984, African cichlids and evolutionary theories, in A.A. Echelle and I. Kornfield (eds.) *Evolution of Fish Species Flocks*, University of Maine Press, Orono, pp. 141–154.

51. P.B. Bayley, 1988, Accounting for effort when comparing tropical fisheries in lakes, river-floodplains, and lagoons, *Limnology and Oceanography*, **33**(4), 963–972.

52. See generally W.L. Shelton, 1986, Strategies for reducing risks from introduction of aquatic organisms: An aquaculture perspective, *Fisheries*, **11**(2), 16–19; W.L. Shelton, 1986, Reproductive control of exotic fishes – A primary requisite for utilization in management, in R.H. Stroud (ed.) *Fish Culture in Fisheries Management*, American Fisheries Society, Bethesda, Maryland, pp. 427–434; R.L. Welcomme, 1988, *International Introductions of Inland Aquatic Species*, Fisheries Technical Paper 294, United Nations Food and Agricultural Organization, Rome, pp. 1–318.

53. For example, P.B. Moyle, 1986, Fish introductions into North America: Patterns and ecological impact, in H.A. Mooney and J.A. Drake (eds.) *Ecology of Biological Invasions of North America and Hawaii*, Springer-Verlag, New York, pp. 27–57; T.M. Zaret and R.T. Paine, 1973, Species introduction in a tropical lake, *Science*, **182**, 449–455.

54. See generally W.R. Courtnay, Jr. and G.K. Meffe, 1989, Small fishes in strange places: A review of introduced Poeciliids, in G.K. Meffe and F.F. Snelson, Jr. (eds.) *Ecology and Evolution of Livebearing Fishes (Poeciliidae)*, Prentice-Hall, Englewood Cliffs, New Jersey, pp. 319–331; W.L. Minckley and J.E. Deacon (eds.), 1991, *Battle Against Extinction: Native Fish Management in the American West*, University of Arizona Press, Tucson, pp. 1–517.

55. See generally W.J. Christie, 1974, Changes in the fish species composition of the Great Lakes, *Journal of the Fisheries Research Board of Canada*, **31**, 827–854; W.J. Christie, K.A. Scott, P.G. Sly and R.H. Strus, 1987, Recent changes in the aquatic food web of eastern Lake Ontario, *Canadian Journal of Fisheries Research and Aquatic Sciences*, **44**, 37–52; E.J. Crossman, 1991, Introduced freshwater fishes: A review of the North American perspective with emphasis on Canada, *Canadian Journal of Fisheries and Aquatic Sciences*, **48**, (Supplement 1), pp. 46–57.

56. For example, J.T. Lehman, 1987, Palearctic predator invades North American Great Lakes, *Oecologia*, **74**, 478–480; N.E. Mandrake, 1989, Potential invasion of the Great Lakes by fish species associated with climatic warming, *Journal of Great Lakes Research*, **15**, 306–316; R.W. Griffiths, D.W. Schloesser, J.H. Leach and W.P. Kovalak, 1991, Distribution and dispersal of the zebra mussel (*Dreissena polymorpha*) in the Great Lakes Region, *Canadian Journal of Fisheries and Aquatic Sciences*, **48**, 1381–1388; W.G. Sprules, H.P. Riessen and E.H. Jin, 1990, Dynamics of the *Bythotrephes* invasion of the St. Lawrence Great Lakes, *Journal of Great Lakes Research*, **16**, 346–351; D.L. Strayer, 1991, Projected distribution of the zebra mussel, *Dreissena polymorpha*, in North America, *Canadian Journal of Fisheries and Aquatic Sciences*, **48**, 1389–1395.

57. P.B. Moyle, 1986, Fish introductions into North America: Patterns and ecological impact, in H.A. Mooney and J.A. Drake (eds.) *Ecology of Bio-*

*logical Invasions of North America and Hawaii*, Springer-Verlag, New York, pp. 27–57; C.H. Fernando, 1991, Impacts of fish introductions in tropical Asia and America, *Canadian Journal of Fisheries and Aquatic Sciences* **48**, (Supplement 1), pp. 24–32, and related references, note 53.

58. For example, L.R. Parenti, 1984, Biogeography of the Andean killifish genus *Orestias* with comment on the species flock concept, in A.A. Echelle and I. Kornfield (eds.) *Evolution of Fish Species Flocks*, University of Maine Press, Orono, pp. 85–92; P. Vaux, W. Wurtsbaugh, H. Trevino, L. Marino, E. Bustamante, J. Torres, P. Richerson and R. Alfaro, 1988, Ecology of the pelagic fishes of Lake Titicaca, Peru-Bolivia, *Biotropica*, **20**, 220–229.

59. For example, R.M. McDowall, 1990, *New Zealand Freshwater Fishes*, Heinemann Redd MAF Publishing Group, Auckland, pp. 1–553; C.R. Townsend and T.A. Crowl, 1991, Fragmented population structure in a native New Zealand fish–An effect of introduced brown trout, *Oikos*, **61**, 347–354; C.C. Krueger and B. May, 1991, Ecological and genetic effects of salmonid introductions in North America, *Canadian Journal of Fisheries and Aquatic Sciences*, **48**, (Supplement 1), pp. 66–77; F.W. Allendorf, 1991, Ecological and genetic effects of fish introductions: Synthesis and recommendations, *Canadian Journal of Fisheries and Aquatic Sciences*, **48**. (Supplement 1), pp. 178–181.

60. See generally E.K. Balon and M.N. Burton, 1986, Introduction of alien species or why scientific advice is not heeded, *Environmental Biology of Fishes*, **16**, 225–230; P.H. Greenwood, 1974, The cichlid fishes in L. Victoria, East Africa: The biology and evolution of a species flock, *Bulletin of the British Museum, London, Zoological Supplement (Natural History)*, **6**, 1–134.

61. For example, J.C. Avise, 1990, Flocks of African fishes, *Nature*, **347**, 512–513.

62. For example, I. Hawes, C. Howard-Williams, R. Wells and J. Clayton, 1991, Invasion of water net, *Hydrodictyon reticulatum* – The surprising success of an aquatic plant new to our flora, *New Zealand Journal of Marine and Freshwater Research*, **25**, 227–230; R.M. Newman, 1991, Herbivory and detritivory on freshwater macrophytes by invertebrates: A review, *Journal of the North American Benthological Society*, **10**, 89–114.

63. See generally J.A. Stanford and J.V. Ward, 1979, Stream regulation in North America, in J.V. Ward and J.A. Stanford (eds.) *The Ecology of Regulated Streams*, Plenum Press, New York. pp. 215–236.

64. A.C. Benke, 1990, A perspective on America's vanishing streams, *Journal of the North American Benthological Society*, **9**(1), 77–88; C. Caufield, 1982, Brazil, energy and the Amazon, *New Scientist*, **96**, 240–243; M. Fineman, 1990, India's gamble on the Holy Narmada, *Smithsonian*, **21**(8), 118–133; P.D. Vaux and C.R. Goldman, 1990, "Dams and development in the tropics: the role of applied ecology," in R. Goodland (ed.) *Race to Save the Tropics: Ecology and Economics for a Sustainable Future*, Island Press, Washington, D.C., pp. 101–123.

65. For example, A. Brookes, 1989, Alternatives for channel modification, in J.A. Gore and G.E. Petts (eds.) *Alternatives in Regulated River Management*, CRC Press, Boca Raton, Florida, pp. 139–162.

66. E. Fels and R. Keller, 1973, World register on man-made lakes, in W.C. Ackerman, G.F. White and E.B. Worthington (eds.) *Man-Made Lakes: Their Problems and Environmental Effects*, American Geophysical Union, Geophysical Monograph 17, Washington, D.C.

67. G.E. Petts, 1989, Perspectives for ecological management of regulated rivers, in J.A. Gore and G.E. Petts (eds.) *Alternatives in Regulated River Management*, CRC Press, Boca Raton, Florida, pp. 3–24.

68. A.C. Benke, 1990, A perspective on America's vanishing streams, *Journal of the North American Benthological Society*, **9**(1), 77–88.

69. For example, S. Postel, 1991, Emerging water scarcities, in L.R. Brown (ed.) *The World Watch Reader on Global Environmental Issues*, W.W. Norton and Company, New York, pp. 127–143; and see generally A.F.A. Latif, 1984, Lake Nasser – The new man-made lake in Egypt, in F.B. Taub (ed.) *Lakes and Reservoirs. Ecosystems of the World*, vol. 23, Elsevier, New York, pp. 385–410.

70. For example, R.L. Welcomme and D. Hagborg, 1977, Towards a model of a floodplain fish population and its fishery, *Environmental Biology of Fishes*, **2**(1), 7–24; see generally R.J. Naiman (ed.), 1992, Watershed Management: Balancing Sustainability and Environmental Change, Springer-Verlag, New York, pp. 1–542.

71. For example, U. Fuchs and B. Statzner, 1990, Time scales for the recovery potential of river communities after restoration: Lessons to be learned from smaller streams, *Regulated Rivers: Research and Management*, **5**(1), 77–87; E.K. Balon, 1992, How dams on the River Danube might have caused hybridization and influenced the appearances of a new cyprinid taxon, *Environmental Biology of Fishes*, **33**(1–2), 167–180; M.R. Winston, C.M. Taylor, and J. Pigg, 1991, Upstream extirpation of four minnow species due to damming of a prairie stream, *Transactions of the American Fisher-*

*ies Society*, **120**, 98–195; D.W. Blinn and G.A. Cole, 1991, Algal and invertebrate biota in the Colorado River: Comparisons or pre- and post-dam conditions, in *Colorado River Ecology and Dam Management*, National Academy Press, Washington, D.C., pp. 102–123.

72. For example, E.K. Balon, 1974, Fish production of a tropical ecosystem, in E.K. Balon and A.G. Coche (eds.) *Lake Kariba, A Man-Made Tropical Ecosystem in Central Africa*, W. Junk, The Hague, pp. 253–676.

73. See generally O. Gaschignard and A. Berly, 1987, Impacts of large discharge fluctuations on the macroinvertebrate populations downstream of a dam, in J.F. Craig and J.B. Kemper (eds.) *Regulated Streams. Advances in Ecology*, Plenum Press, New York, pp. 145–161.

74. For example, A. Gasith and S. Gafny, 1990, Effects of water level fluctuation and the structure and function of the littoral zone, in M.M. Tilzer and C. Serruya (eds.) *Large Lakes. Ecological Structure and Function*, Springer-Verlag, New York, pp. 156–171.

75. For example, C. Howard-Williams and G.M. Lenton, 1975, The role of the littoral zone in the function of a shallow tropical lake ecosystem, *Freshwater Biology*, **5**, 445–459.

76. For example, D.H. Bennett, O.E. Maughn and D.B. Jester, Jr., 1985, Generalized model for predicting spawning success of fish in reservoirs with fluctuating water levels, *North American Journal of Fisheries Management*, **5**, 12–20.

77. R.I. Cahn (ed.), 1985, *An Environmental Agenda for the Future*, Island Press, Washington, D.C, pp. 1–155.

78. See generally D.E. Ford, 1990, Reservoir transport processes, in K.W. Thornton, B.L. Kimmel and F.E. Payne (eds.) *Reservoir Limnology: Ecological Perspectives*, John Wiley ..nd Sons, New York, pp. 15–41.

79. See generally K.W. Thornton, B.L. Kimmel and F.E. Payne (eds.), 1990, *Reservoir Limnology: Ecological Perspectives*, John Wiley and Sons, New York, pp. 1–246.

80. For example, J.D. Thomas, 1987, An evaluation of the interactions between freshwater pulmonate snail hosts of human schistosomes and macrophytes, *Philosophical Transactions of the Royal Society of London*, **B315**, 75–125.

81. See generally W.T. Swank and D.A. Crossley, Jr. (eds.), 1988, *Forest Hydrology and Ecology at Coweeta*, Springer-Verlag, New York, pp. 1–469.

82. See generally J.M. Bosch and J.D. Hewlett, 1982, A review of catchment experiments to determine the effect of vegetation changes on water yield and evapotranspiration, *Journal of Hydrology*, **55**, 3–23.

83. See generally L.A. Bruijneel, 1990, *Hydrology of Moist Tropical Forests and Effects of Conversion: A State of Knowledge Review*, UNESCO International Hydrological Program, Paris, pp. 1–224.

84. For examples of these debates, see W.J.H. Ramsay, 1987, *Deforestation and Erosion in the Nepalese Himalaya: Is the Link Myth or Reality?*, International Association of Hydrological Sciences Publication 167, pp. 239–250; J.E. Richey, C. Nobre and C. Deser, 1989, Amazon River discharge and climate variability: 1903–1985, *Science*, **246**, 101.

85. T. Dunne and W. Dietrich, 1982, Sediment sources in tropical drainage basins, in *Soil Erosion and Conservation in the Tropics*, American Society of Agronomy and Soil Science Society of America, Madison, Wisconsin, pp. 41–55.

86. For example, J.B. Wallace, 1990, Recovery of lotic macroinvertebrate communities from disturbance, *Environmental Management*, **14**, 605–620; R.C. Wissmar and F.J. Swanson, 1990, Landscape disturbances and lotic ecotones, in R.J. Naiman and H. Decamps (eds.) *The Ecology and Management of Aquatic–Terrestrial Ecotones. Man and the Biosphere Series*, vol. 4, UNESCO and Parthenon Publishing Group, Paris, pp. 65–89; M.A. Palmer, A.E. Bely and K.E. Berg, 1992, Response of invertebrates to lotic disturbance: A test of the hyporheic refuge hypothesis, *Oecologia*, **89**(2), 182–194; see related references, notes 24 and 26.

87. See generally B.L. Bedford and E.M. Preston, 1988, Developing the scientific basis for assessing cumulative effects of wetland loss and degradation on landscape functions: status, perspectives, and prospects, *Environmental Management*, **12**, 751–771; D.L. Childers and J.G. Gosselink, 1990, Assessment of cumulative impacts to water quality in a forested wetland landscape, *Journal of Environmental Quality*, **19**, 455–463.

88. See generally J.T. Hayes, 1991, Global climate change and water resources, in R.L. Wyman (ed.) *Global Climate Change and Life on Earth*, Routledge, Chapman and Hall, New York, pp. 18–42.

89. For example, J.V. Ward, 1989, The four-dimensional nature of lotic ecosystems, *Journal of the North American Benthological Society*, **8**(1), 2–8; J. Harte, M. Torn, and D. Jensen, 1992, The nature and consequences of indirect linkages between climate change and biological diversity, in R.L. Peters and T.E. Lovejoy (eds.) *Global Warming and Biological Diversity*, Yale University Press, New Haven, pp. 325–342.

90.   For example, R.T. Dillion, 1984, Geographic distance, environmental difference, and divergence between isolated populations, *Systematic Zoology*, **33**(1), 69–82; G.L. Gooch, 1990, Spatial genetic patterns in relation to regional history and structure; *Gammarus minus* (Amphipoda) in Appalachian watersheds, *American Midland Naturalist*, **124**, 93–104; R.L. France, 1991, Empirical methodology for predicting changes in species range extension and richness associated with climate warming, *International Journal of Biometeorology*, **34**, pp.211–216.

91.   See generally P.H. Gleick, 1990, Vulnerabilities of water systems, in P.E. Waggoner (ed.) *Climate Change and U.S. Water Resources*, John Wiley and Sons, New York, pp. 223–240.

92.   W.M. Gray, 1990, Strong association between West African rainfall and U.S. landfall of intense hurricanes, *Science*, **249**, 1251–1256; M. C. Molles, Jr. and C.N. Dahm, 1990, A perspective on El Niño and La Niña: Global implications for stream ecology, *Journal of the North American Benthological Society*, **9**(1), 69–76; C.N. Dahm and M.C. Moles, Jr., 1992, Streams in semiarid regions as sensitive indicators of global climate change, in P. Firth and S.G. Fisher (eds.) *Global Climate Change and Freshwater Ecosystems*, Springer-Verlag, New York, pp. 250–260; S. Hastenrath, 1990, Predictability of anomalous river discharge in Guyana, *Nature (London)*, **345**, 53–53; J.M. Melack, 1992, Reciprocal interactions among lakes, large rivers, and climate, in P. Firth and S.G. Fisher (eds.) *Global Climate Change and Freshwater Ecosystems*, Springer-Verlag, New York, pp. 68–87; S.R. Carpenter, S.G Fisher, N.B. Grimm and J.F. Kitchell, 1992. Global change and fresh water ecosystems, *Annual Review of Ecology and Systematics*, **23** 119–139.

93.   See review by R.F. McMahon, 1991, Mollusca: Bivalvia, in J.H. Thorp and A.P. Covich (eds.) *Ecology and Classification of North American Freshwater Invertebrates*, Academic Press, San Diego, pp. 315–399; W.J. Matthews and E.G. Zimmerman, 1990, Potential effects of global warming on native fishes of the southern great plains and the southwest, *Fisheries*, **15**(6), 26–32.

94.   D. Rowe and D. Scott, 1989, Effects of climate warming on trout fisheries in northern New Zealand, *Freshwater Catch*, **41**, 3–4.

95.   R.G. Danzmann, D.S. MacLennan, D.G. Hector, P.D. Hebert and J. Kolasa, 1991, Acute and final temperature prefenda as predictors of Lake St. Clair fish catchability, *Canadian Journal of Fisheries and Aquatic Sciences*, **48**, 1408–1418.

96.   For example, K.A. Poiani and W.C. Johnson, 1991, Global warming and prairie wetlands, *Bioscience*, **41**, 611–618.

97.   For example, A. Dobson and R. Carper, 1992, Global warming and potential changes in host–parasite and disease–vector relationships, in R.L. Peters and T.E. Lovejoy (eds.) *Global Warming and Biological Diversity*, Yale University Press, New Haven, Connecticut, pp. 201–217.

98.   See generally J.M. Elliot, 1981, Some aspects of thermal stress on freshwater teleosts, in A.D. Pickering (ed.) *Stress and Fish*, Academic Press, New York, pp. 209–245; and related references, note 93.

99.   For example, J.J. Magnuson, J.D. Meisner and D.K. Hill, 1990, Potential changes in the thermal habitat of great lakes fish after global warming, *Transactions of the American Fisheries Society*, **119**, 254–264; H.A. Regier and J.D. Meisner, 1990, Anticipated effects of climate change on freshwater fishes and their habitat, *Fisheries*, **15**(6), 10–15.

100.  See generally M.K. Tolba, 1987, *Sustainable Development. Constraints and Opportunities*, Butterworths, London, pp. 1–221.

# Chapter 5
# Water and agriculture

## Sandra Postel[*]

## Introduction

All one needs in order to grasp water's vital importance to life and growth on the planet are the few pages from any basic biology text describing the process of photosynthesis. Using the sun's energy, plants combine water and carbon dioxide to form carbohydrates. This conversion of solar energy into biochemical energy creates the earth's basic food supply, supporting all life.

Sufficient water is thus at the heart of agriculture, as African farmers in the drought-prone Sahelian belt know all too well. Whether relying solely on rainfall or on pumps and canals to deliver supplies, agriculturalists depend on enough moisture being available to nurture their crops. With human numbers headed toward 6 billion by the end of this decade, and food needs rising apace, global food security hinges in large part on this critical link between water and crops.

Today, agriculture accounts for about two-thirds of global water use. By enabling farmers to apply water when and where needed, irrigation has turned many of the earth's sunniest, warmest, and most fertile lands into important crop-producing regions. Egypt could grow virtually no food without water drawn from the Nile or from underground aquifers. California's Central Valley and the Aral Sea basin – the fruit and vegetable baskets of the United States and the former Soviet Union – could barely be cultivated without supplemental water supplies. And without irrigation, yields in the critical grain-growing areas of northern China, north-west India, and the western U.S. Great Plains would drop by one-third to one-half.

Though it grew steadily over the centuries, irrigation entered a heyday about 1950. The growth of oil on the world energy scene enhanced the flexibility and power of pumping technologies, and ground water wells proliferated. Relatively rapid global economic growth from mid-century through the early 1970s favored large capital investments. All but 7 of the 100 largest dams in the world were completed after World War II. Between 1950 and the late 1970s, irrigated area more than doubled.[1]

In recent years, however, a confluence of forces has begun to slow irrigation's expansion and to raise questions about agriculture's heavy claim on the world's rivers, streams, and underground aquifers. Rising real costs for new water supplies and a host of environmental problems are making new projects harder to justify. In several parts of the world, water demands are fast approaching the limits of the resource. Many areas could enter a period of chronic shortages during the 1990s, including northern China, virtually all of northern Africa, pockets of India, Mexico, much of the Middle East, and parts of the western United States. Where scarcities loom, competition is already brewing between cities and farms; when supplies tighten, typically farms lose water.

Moreover, the longer term threat of climate change from the buildup of atmospheric greenhouse gases casts ominous shadows over future water budgets. As temperatures rise and rainfall patterns shift with the onset of climatic change, water supplies will increase in some areas and decrease in others (see Chapter 9). Whatever the outcome for particular regions, though, adjusting the global irrigation base to the changes in water availability will be costly. It seems inevitable that for several decades, irrigation systems will be poorly matched to altered rainfall regimes and redistributed water supplies, jeopardizing agriculture's ability to provide enough food for the world's growing population.

Moving rapidly from profligacy to greater efficiency and equity in agriculture's use of water is the surest way to avert shortages and lessen irrigation's ecological toll. Most of the vast quantity of water diverted by and for farmers never benefits a crop: worldwide, the efficiency of irrigation systems averages less than 40%. The technologies and know-how exist to boost that figure substantially; what is needed are policies and incentives that foster efficiency instead of discouraging it. At the same time, greater attention to small-scale irrigation, rainfed farming, and better integration of water projects with other development goals is essential to give food production and farmers' livelihoods a more secure foundation in the Third World.[2]

## Irrigation trends

It was not until about two centuries ago that the science of irrigation, grounded in the principles of hydraulics, took root. In 1800, an estimated 8 million hectares, an area about the size of Austria, were equipped with watering facilities. A number of large projects constructed during the 19th century, especially in India and what is now Pakistan, brought the total area irrigated in 1900 to about 48 million hectares.[3]

As world population grew from 1.6 billion to more than 5 billion over

**TABLE 5.1** Net irrigated area, top 20 countries and world, 1989

| Country | Net irrigated area[a] (thousand hectares) | Share of cropland that is irrigated (%) |
|---|---|---|
| China | 45,349 | 47 |
| India | 45,039 | 25 |
| Soviet Union | 21,064 | 9 |
| United States | 20,162 | 11 |
| Pakistan | 16,220 | 78 |
| Indonesia | 7,550 | 36 |
| Iran | 5,750 | 39 |
| Mexico | 5,150 | 21 |
| Thailand | 4,230 | 19 |
| Romania | 3,450 | 33 |
| Spain | 3,360 | 17 |
| Italy | 3,100 | 26 |
| Japan | 2,868 | 62 |
| Bangladesh | 2,738 | 29 |
| Brazil | 2,700 | 3 |
| Afghanistan | 2,660 | 33 |
| Egypt | 2,585 | 100 |
| Iraq | 2,550 | 47 |
| Turkey | 2,220 | 8 |
| Sudan | 1,890 | 15 |
| Other | 36,664 | 7 |
| World | 235,299 | 16 |

*Source:* S. Postel, 1992, *Last Oasis: Facing Water Scarcity*, W.W. Norton & Company, New York.

[a] Area actually irrigated; does not take into account double cropping.

[*]This chapter is updated and adapted from S. Postel, 1989, Water for Agriculture: Facing the Limits, Worldwatch Paper 93, Worldwatch Institute, Washington, D.C. Further treatment of these issues can be found in S. Postel, 1992, *Last Oasis: Facing Water Scarcity*, W.W. Norton & Company, New York.

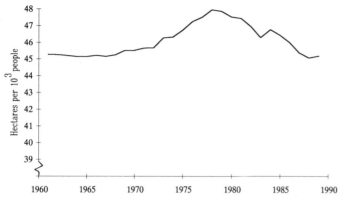

Fig. 5–1. World irrigated area per capita, 1960–1990. Increases in irrigated area are now failing to keep up with increases in population. After increasing through the 1970s, per capita irrigated area worldwide is now falling. If this trend continues, feeding the world's growing population will require even greater improvements in yields per hectare or greater food production from non-irrigated lands. Such improvements are likely to be increasingly difficult.

*Source:* S. Postel, 1992, *Last Oasis: Facing Water Scarcity*, W.W. Norton & Company, New York.

the last 90 years, irrigation became a cornerstone of global food security. The higher yields farmers could get with a controllable water supply proved vital to feeding the millions added to our numbers each year, especially as opportunities to cultivate new land dwindled.

Irrigated area worldwide nearly doubled during the first half of this century, rising to 94 million hectares in 1950. An even faster surge since 1950 has brought the total to more than 235 million hectares. Today, one-third of the global harvest comes from the 16% of the world's cropland that is irrigated (see Table 5.1.) Many countries – such as China, Egypt, India, Indonesia, Israel, Japan, North and South Korea, Pakistan, and Peru – rely on such land for more than half their domestic food production.[4]

Since the late 1970s, though, expansion has slowed markedly. Whereas net irrigated area grew by 2%–4% per year during the 1960s and 1970s, the annual rate since 1979 has been on average about 1% – far less than world population. (Net irrigated area is the land estimated to have actually been irrigated that particular year; it is less than the gross area, that estimated to be equipped with irrigation facilities.) Per capita irrigated area peaked at 48 hectares per thousand people in 1978, and then fell to 45 hectares in 1989, a drop of more than 6% (see Figure 5.1). Estimates compiled by the United Nations Food and Agriculture Organization (FAO) show that only 23.6 million hectares were added to the global irrigation base between 1980 and 1989, an average of 2.6 million hectares per year, roughly three-fifths the rate during the 1970s.[5]

This slowdown was brought about largely by low commodity prices, comparatively high energy costs (following the oil price hikes of the 1970s), and economic conditions during much of the 1980s that discouraged agricultural investments. The logical presumption is that when these conditions change, irrigation will pick up speed again. But, for several reasons, this now seems unlikely.

Lending for irrigation by the major international donors has declined sharply over the last decade. Total lending to 23 countries in Asia, North Africa, and the Middle East by the World Bank, the Asian Development Bank, the U.S. Agency for International Development, and the Japanese Overseas Economic Cooperation Fund fell more than 60% in real terms between 1977 and 1987. Given that large projects often take a decade or more to complete, this funding drop suggests that in a good deal of the Third World, growth in irrigated area will not quicken much during this decade.[6]

A related cause of the slowdown is that the costs of adding irrigation capacity – building new dams, reservoirs, canals, and distribution channels – have been rising in many countries, making it harder to justify such investments on economic grounds. Engineers naturally exploited the best sites for water development first. Recent projects have been technically

more complex, leading to greater expenditures. In India, for example, the cost in real terms of large canal schemes more than doubled between 1950 and 1980. Today, capital costs for new irrigation capacity run at about $1,500 per hectare in China, and between $1,500 and $4,000 per hectare for large projects in India, Indonesia, Pakistan, the Philippines, and Thailand. They approach $6,000 per hectare for public projects in Brazil, and $10,000 per hectare in Mexico.[7]

Developing new capacity costs the most in parts of Africa. Lack of roads and other infrastructure, the relatively small parcels of irrigable land, and the seasonal nature of river flows have driven per-hectare costs in many projects to between $10,000 and $20,000, or even higher. Not even double-cropping of higher-valued crops can make irrigation systems at the top end of this spectrum economical. Consequently, irrigated area in sub-Saharan Africa has expanded minimally since 1980, even though less than 5% of the region's cropland is irrigated.[8]

The debt burden hanging over many developing countries makes the situation even bleaker. Irrigated area in Mexico actually declined during the late 1980s, in large part because its faltering economy led to cuts in capital improvements, including expanding and rehabilitating these systems. Officials in Brazil view accelerated irrigation as critical to their goal of achieving greater food self-sufficiency. But their ambitious targets for the early 1990s seem unlikely to be met, given the required investments.[9]

As costs rise and public investments decline, a larger share of what new irrigation does come on-line seems likely to support production of higher-valued cash crops rather than rice, wheat, and other food staples, especially if governments cannot afford to subsidize this farming heavily. Should grain prices rise because of diminished global stocks – a possibility during the 1990s – investments will almost certainly pick up again. But the poor are unlikely to be able to afford crops produced with expensive water. Neither scenario is comforting if irrigation's ultimate objective is to enhance food security.[10]

For the foreseeable future, then, irrigation's contribution will have to come more from improving existing systems than from expanding them to new land. This requires a quite fundamental shift in thinking: engineers are trained to design and build physical infrastructure, not to operate it effectively from the farmers' point of view. As Indian water analyst Ashok K. Mitra points out, "Irrigation engineers tend to view the output of the canal system as 'water' and not the additional crop which the water should allow." And many government irrigation agencies, intent on maintaining their large budgets, consider their ultimate mission to be irrigation expansion, rather than to increase food output and raise rural incomes.[11]

Behind the raw numbers lie vast discrepancies in irrigation's performance and quality. In many large surface water networks, less than half the water diverted from the reservoir actually benefits crops. Typical efficiencies for large projects in Asia run at about 30%. Much water seeps through unlined canals before it gets to fields. An additional amount runs off the land or percolates unused through the soil because farmers apply water unevenly, excessively, or at the wrong times. Fortunately, not all this water is wasted: much of it returns to a nearby stream or joins underground supplies, and thus can be used again. But as it picks up salts, fertilizers, and pesticides along the way, it is often degraded in quality and contaminates water sources downstream.[12]

Lack of maintenance has caused many systems to fall into disrepair, further inhibiting performance. Over time, distribution channels fill with silt, increasing the likelihood of breaching, outlets break or are bypassed, and salts build up in the soil. In China, for example, more than 930,000 hectares of irrigated farmland have come out of production since 1980, an average loss of some 116,000 hectares per year. Similarly, in the former Soviet Union – which was the world's third largest irrigator – poor performance caused irrigation to cease on 2.9 million hectares between 1971 and 1985, equal to a quarter of the new area brought under irrigation. Worldwide, an estimated 150 million hectares, more than 60% of the world's total irrigated area, need some form of upgrading to remain in good working order. An increasing share of available capital will thus go to rehabilitating existing systems. While this is wise and necessary, it will further diminish funding for new projects.[13]

Poor management and inequities in the distribution of water are also to blame for irrigation's disappointing performance. Many systems deliver water on a fixed schedule that does not match crops' needs. In other cases, the supply is so unpredictable and unreliable that farmers cannot risk investing in other yield-raising inputs and activities, such as better seeds, fertilizers, or a leveling of their land. As a result, output remains far below its potential. A study in Pakistan, for example, found that the weighted average yield from irrigated fields of wheat, rice, and sugarcane was 60%–70% below the upper range of yields some farmers achieved. This ''yield gap'' can largely be explained by ineffective management of the water supply and its influence on farmers' other management decisions.[14]

The plight of ''tailenders'' – farmers last in line to receive supplies from large canals – underscores the consequences of inefficient and unfair distribution. Typically, farmers at the head of a project area begin irrigating before the canals and distribution channels reach the end. By planting water-intensive crops, such as sugarcane or rice, they stake a claim to more water than was planned for their plots. Often wealthier and wielding considerable political influence, they then lobby effectively to maintain their heavy withdrawals even when the project is completed. Poorer farmers at the end of the line end up without enough water for their crops.

One project in the Indian state of Maharashtra, for example, had sugarcane occupying 12% of the irrigated area when the design called for only 4%. Since the water needed for a hectare of sugarcane could irrigate about 10 hectares of wheat, this practice causes much less area actually to be irrigated than was initially planned. Food production falls short of expectations, and the gap between wealthy and poor farmers widens.[15]

All these factors add up to much of the world's existing irrigation yielding far below its potential, and, in some cases, failing to enhance food and income security for those who most need it – the rural poor. To be sure, irrigation has greatly expanded world food output, alleviating hunger and improving diets in many areas. But as future gains get harder to come by, an important question becomes: irrigation for whom and for what?

India, with more than a fifth of the world's irrigated area, perhaps best illustrates the need for a rethinking of ends and means. Most analysts credit irrigation for much of the progress Indian agriculture has made over the last three decades. To extend those benefits, the nation seems to be in an all-out race to harness its ultimate irrigation potential, estimated at 113 million hectares, or roughly double the area currently equipped with irrigation facilities. But much of this existing area is not in use or is performing poorly. Yields of irrigated wheat and rice are only half what Indian agronomists say they could be. Boosting yields closer to their potential on existing irrigated lands through better management could thus produce as much additional food as doubling the irrigated area, at less cost economically and environmentally.[16]

The late prime minister, Rajiv Gandhi, himself bluntly criticized the country's track record in a speech to state irrigation ministers in 1986. Of the 246 large surface irrigation projects started since 1951, only 65 had been completed, he said, and little benefit had come from projects started after 1970: ''For 16 years we have poured money out. The people have got nothing back, no irrigation, no water, no increase in production, no help in their daily life.''[17]

As the era of easy water development comes to an end, so do the rapid crop production gains from global irrigation expansion. It now seems likely that net irrigated area will increase by only about 2 million hectares per year, or 0.9%, for the near future, with most of the growth occurring in Bangladesh, Brazil, India, Nigeria, Pakistan, Turkey, and a few other countries. Given that world population is growing at 1.7% per year, a dramatic boost in yields on both existing irrigated and rainfed lands will be needed for food production to keep pace with human numbers.[18]

## Irrigation's environmental price

Each year, some 2,700 km$^3$ of water – about five times the annual flow of the Mississippi – are removed from the earth's rivers, streams, and underground aquifers to water crops. Practiced on such a scale, irrigation has had a profound impact on global water bodies and on the cropland that is watered. Waterlogged and salted lands, declining and contaminated aquifers, shrinking lakes and inland seas, and the destruction of aquatic habitats combine to hang on irrigation a high environmental price. Mounting concern about this damage is making large new water projects increas-

ingly unacceptable. And ironically, the degradation of irrigated land from poor water management is forcing some land to be retired completely, offsetting a portion of the gains costly new projects are intended to yield.[19]

By far the most pervasive damage stems from waterlogging and soil salinization brought about by poor water management. Without adequate drainage, seepage from unlined canals and overwatering of fields causes the ground water table to rise. Eventually, the root zone becomes waterlogged, starving plants of oxygen and inhibiting their growth. In dry climates, evaporation of water near the soil surface leaves behind a layer of salt that also reduces crop yields and eventually, if the buildup becomes excessive, kills the crops.

Salts also get added to the soil from the irrigation water itself. Even the best supplies typically have concentrations of 200–500 parts per million (ppm). (For comparison, ocean water has a salinity of about 35,000 ppm; water with less than 1,000 ppm is considered fresh; and the recommended limit for drinking water in the United States is 500 ppm.) Applying 10,000 m$^3$ of water to a hectare per year, a fairly typical irrigation rate, thus adds between 2 and 5 tons of salt to the soil annually. If it is not flushed out, enormous quantities can build up in a couple of decades, greatly damaging the land. Aerial views of abandoned irrigated areas in the world's dry regions reveal vast expanses of glistening white salt, land so destroyed it is essentially useless.[20]

No one knows for sure how large an area suffers from salinization. International irrigation consultant W. Robert Rangeley estimates that 15 million hectares in developing countries, primarily India, China, Pakistan, Iran, and Iraq, are experiencing serious reductions in crop yields because of salinization. Studies by the World Bank have found that in Egypt and Pakistan waterlogging and salinity are slashing yields of major crops by 30%. And in Mexico, salinization is estimated to be reducing crop output by the equivalent of 1 million tons of grain per year – enough to feed 5 million people, roughly a quarter of Mexico City.[21]

In the United States, salinity expert James Rhoades estimates that salt buildup lowers crop yields on 25%–30% of the nation's irrigated land, or over 5 million hectares. Some 2.5 million hectares are salinized in the former Soviet Union, most of them in the irrigated deserts of central Asia. All told, some 25 million hectares, more than 10% of world irrigated area, appear to suffer from yield-suppressing salt buildup, and this figure is expanding each year.[22]

In the western United States, excessive irrigation and poor drainage have spawned another set of problems. There, scientists have linked alarming discoveries of death, grotesque deformities, and reproductive failure in fish, birds, and other wildlife to agricultural drainage water laced with toxic elements. Since 1985, intensive investigations throughout the region have found lethal or potentially hazardous selenium concentrations at 22 different wildlife sites, including Kesterson National Wildlife Refuge, which has earned an unenviable reputation as the ''Three Mile Island of irrigated agriculture.''[23]

Selenium, a natural element, is needed by humans and some other animals in very small amounts, but is highly poisonous at greater concentrations. Irrigation has washed more selenium and other dangerous chemicals out of the soil in several decades than natural rainfall would have done in centuries. Potential solutions to the drainage dilemma, which include detoxification by soil microbes, mechanical filtration, or chemical treatment, range in cost from $30 million to more than $100 million per year for some indefinite period in California alone. With such price tags and the high ecological stakes of continued irrigation, it seems likely that some of this land will come out of production over the next decade. How much, and in which states, remains to be seen.[24]

Contamination of land and water by salts and toxic chemicals is only one indicator of irrigation's unsustainability. In much of the world, falling water tables signal that ground water withdrawals exceed the rate of replenishment. Such overpumping can eventually make irrigation too costly to continue, and can even drain some aquifers dry. In either case, land is forced out of irrigation, temporarily or permanently.

In the United States, more than 4 million hectares – roughly a fifth of the nation's irrigated area – are watered by pumping in excess of recharge. By the early 1980s, the depletion was already particularly severe in Texas, California, Kansas, and Nebraska, four important food producing states. Current year-to-year fluctuations in U.S. irrigated area reflect crop prices and government farm policies more than water availability and cost. But overpumping cannot continue indefinitely. The 4 million hectares currently watered unsustainably will eventually come out of irrigated production unless farmers reduce pumping to no more than the rate of recharge.[25]

No other country has systematically assessed the extent of excessive ground water pumping. But the situation is serious elsewhere as well, including in China and India, two of the three other major food producers. Ground water levels are falling up to a meter per year in parts of northern China, and heavy pumping in portions of the southern Indian state of Tamil Nadu reportedly dropped water levels by as much as 25–30 m in a decade. In the western state of Gujarat, overpumping by irrigators in the coastal districts has caused saltwater to invade the aquifer, contaminating village drinking supplies.[26]

Irrigation schemes in some of the world's driest regions rely on water drawn from so-called "fossil" aquifers – water supplies, often deep underground, that get very little recharge. Withdrawals from such sources constitute water mining, one-time extractions from a depletable reserve. While it is reasonable to put this water to productive use, much of it is being depleted rapidly and wastefully. Moreover, crops produced with mined water contribute to a false sense of food security, since the harvest cannot be sustained over the long term.[27]

Saudi Arabia represents one of the worst cases of water mining. Today, three-fourths of that nation's water supply comes from non-renewable ground water sources, and this share is expected to rise. During the latter half of the 1980s, ground water pumping exceeded estimated recharge more than five-fold. Projected water demands suggest that by 2010 these deep aquifers will hold 42% less water than they did in 1985. Some farms have already been abandoned because of declining well yields and high pumping costs.[28]

In its strivings toward greater food self-sufficiency, the north African nation of Libya plans to irrigate some 200,000–240,000 hectares with water drawn from fossil aquifers beneath its deserts. Dubbed "The Great Man-Made River Project," this scheme aims to pump water from deep wells and transfer it north through underground pipes to the Mediterranean coast. The total cost of the project, the first phase of which was completed in 1991, is estimated at $25 billion.[29]

In the United States, the Ogallala aquifer, which stretches from southern South Dakota to north-west Texas, has been heavily depleted in its southern portions. Overpumping in the Texas High Plains has diminished supplies there by 24%, from 550 million acre-feet prior to large-scale irrigation development to 417 million acre-feet in 1990. Partly as a result of declining well yields and increased pumping costs, irrigated area in the Texas High Plains decreased by 34% between 1974, the peak year, and 1989.[30]

Signs of irrigation overstepping ecological limits are evident among the world's rivers, lakes, and streams, as well. By far the most dramatic is the shrinking Aral Sea in the central Asian republics of the former Soviet Union. Fully 95% of the Soviet cotton harvest was grown in this region, as well as a third of the country's fruits, a quarter of its vegetables, and 40% of its rice.[31]

Owing to the dry climate, 90% of the crops in these former Soviet republics are irrigated. As production expanded during the 1960s and 1970s, increasing amounts of water were diverted from the region's two major rivers, the Amu Dar'ya and Syr Dar'ya, the only sources of replenishment for the Aral Sea other than the meager rainfall. By 1980, flows in these rivers' lower stretches had been reduced to a trickle, and the sea had contracted markedly.[32]

The Aral's surface area has shrunk by more than 40% since 1960, volume has dropped by two-thirds, and salinity levels have risen three-fold. Virtually all native fish species have disappeared, decimating the region's fishing industry. Winds pick up salt from the dry seabed and annually dump 43 million tons of it on surrounding cropland, damaging harvests. V.M. Kotlyakov, director of the Institute of Geography at the Soviet Academy of Sciences in Moscow, warns that without drastic measures, the Aral, once the world's fourth largest inland fresh water body, "is destined to become a small brine lake of between 4,000 and 5,000 square kilometers."[33]

Saving the Aral, albeit in some diminished form, is now a high priority in the former Soviet Union. In an early 1991 resolution, the Supreme Soviet placed most of the responsibility for dealing with the crisis with the republics. Officials and scientists who have studied the problem say that restoring the ecosystem will take major reductions in agricultural water use. Much could be achieved by improving irrigation efficiency and switching to less water-intensive crops. But it also may be necessary to take some of Central Asia's prized farmland out of irrigation.[34]

All of this visible damage from large-scale irrigation has spawned strong opposition to new dams and diversion projects in industrial and developing countries alike. In February 1989, 10,000 people demonstrated against the Sardar Sarovar Dam, a centerpiece of India's Narmada Valley Development Project, which encompasses 30 large dams, 135 medium-sized ones, and 3,000 small ones. Sardar Sarovar alone would flood nearly 40,000 hectares and displace up to 100,000 people. Government officials want to push ahead, even though an independent review commissioned by the World Bank, which has loaned $450 million for the project, found that the dam's ecological impact has not been adequately investigated nor the resettlement issues adequately resolved. Nor, apparently, have studies been completed on how to manage the watershed to prevent erosion from rapidly silting up the planned reservoir and canals.[35]

Sardar Sarovar has become a rallying point for a rapidly growing environmental movement in India. Another demonstration at Harsud, a village of 15,000 that would be submerged if the dam is completed, drew tens of thousands of people in late September 1989, some of whom walked for days to attend. Opponents criticize not only the dam's massive social and ecological ramifications, but also its failure to address the needs of the poorest villagers.[36]

In Gujarat, the main beneficiary state, 30 out of 52 *talukas* (subdistricts) classified as drought-prone or arid will receive no irrigation benefits from the project; most benefits will go to areas that are already relatively well-endowed with water. Moreover, as Baba Amte writes in *Cry, the Beloved Narmada*, Sardar Sarovar "will draw money away from various other schemes which could provide water to these areas [in need] .... The government has completely lost track of what must be regarded as its basic objective: finding the best possible way of providing water to the people – a path, moreover, not laden with blood and tears, sorrow and suffering."[37]

In Africa, similarly, planners often overestimate the benefits of large projects and underestimate social and environmental costs. Development anthropologist Thayer Scudder points out that large irrigation schemes have tended to benefit a small minority while often destroying local production systems vital to the poor, such as floodwater farming and fisheries. Following completion of the Kainji Dam in northern Nigeria, fish catches and dry-season harvests from traditional floodplain agriculture dropped by more than half. A similar pattern was noted when the Bakolori Dam came on-line.[38]

In the United States, where the era of dam building has largely ended and where water in many river basins is already overallocated, concerns center around the loss of free-flowing rivers, the destruction of fisheries from stream flow depletion, and damage to riverine and wildlife habitat. The American Fisheries Society lists 364 species or subspecies of fresh water fish as endangered, threatened, or of special concern. The vast majority are imperiled because of habitat loss. Likewise, pollution and habitat destruction from dam building are among the key factors behind documented declines in mollusk populations. Efforts to protect aquatic animal life and their natural ecosystems could well involve shifting water away from agriculture, by far the biggest consumer in the West.[39]

Heightened awareness about irrigation's environmental effects could also upset the renewal of federal irrigation contracts, under which farmers receive irrigation water from government projects at highly subsidized prices. Hundreds of these contracts come due over the next couple of decades. New legislation, signed into law in the fall of 1992, calls for 15 to 25% of the yield from the federal Central Valley Project in California to be reserved for protection and restoration of fisheries and wildlife.[40]

## Scarcity and competition

Regional shortages cropping up around the world form another growing threat to agriculture's claim on water resources. Scarcities loom in a number of naturally water-short areas where irrigation is critical to farming, and often accounts for 80% or more of consumption. Many of these regions are also experiencing rapid population growth and urbanization, setting the stage for heated competition between cities and farms, and imposing constraints on agricultural development.

In much of northern and eastern Africa, expanding human numbers are on a collision course with scarce water resources. Swedish hydrologist Malin Falkenmark has found that societies typically experience water stress when annual renewable supplies approach 2,000 m$^3$ per person. By the end of the 1990s, six out of seven East African countries (Burundi, Ethiopia, Kenya, Rwanda, Somalia, and Tanzania) and all five north African countries bordering the Mediterranean Sea (Algeria, Egypt, Libya, Morocco, and Tunisia) will be in this category. Six of these countries will have less than 1,000 m$^3$ to tap per person. All told, some 240 million Africans live in countries where per capita water availability is fast approaching, or is already below, a reasonable human support level. The prospect of these populations doubling within a generation undoubtedly means rising food imports, declining living standards, and more hunger and malnutrition.[41]

Egypt, where rain is sparse, represents the extreme of the region's dilemma. Its 55 million people depend almost entirely on the waters of the Nile, none of which originates within the desert nation's borders. Virtually all of its cropland requires irrigation. Because of its burgeoning population, drinking water and food needs are rising rapidly: water demands will likely exceed reliable supplies within a decade or so. Moreover, Ethiopia, which controls the headwaters of 80% of the Nile's flow, has plans to divert more upper basin water for itself. Supplies downstream could thus diminish, turning Egypt's situation dire.[42]

Just across the Suez Canal, the situation looks equally grim. Israel and Jordan get most of their water from the Jordan River basin. Israel is already running a water deficit, with annual usage exceeding renewable supplies by 15%. Demands in Jordan, too, recently surpassed the sustainable supply. That nation's water deficit is projected to mount rapidly under the pressures of a population growing at 4.1% per year (which implies a doubling of population in 17 years). According to Thomas Naff of the University of Pennsylvania, a specialist in Middle East water issues, it will take a fundamental restructuring of these nations' economies away from irrigated agriculture, coupled with efficiency improvements, for there even to be hope of achieving a long-term water balance. Indeed, Israel's water projections show a 15% reduction in water allocated to agriculture by the year 2000, but, ultimately, larger shifts will likely be necessary.[43]

Competition over water is heightening in China, as well, where dozens of cities already face acute shortages. The predicament is especially severe in and around Beijing, in the important industrial city of Tianjin, and in other portions of the North China Plain, a vast expanse of flat, fertile farmland that yields a quarter of the nation's grain. Water tables beneath the capital have been dropping 1–2 meters per year; already, a third of its wells have reportedly gone dry.[44]

Most supplies on the North China Plain have already been tapped, yet demands continue to grow rapidly. At current use rates, planners project Beijing's water needs to increase by 50% over the next decade. Demands in Tianjin, where water tables have also been falling precipitously, are expected to more than double. A long-awaited diversion of the Yangsijang from central China to the north plain would ease the crunch, but only slightly: even this grandiose scheme would meet only a fraction of Beijing's anticipated needs in 2000, still leaving a gap. Moreover, even if the project continues to be supported, it is probably a decade or more away from completion, and so will do nothing to alleviate shortages during the 1990s.[45]

Shifting water from farm to city use may be the only way to balance the region's water budget. A management study for Beijing suggests that farmers in the vicinity could lose 30%–40% of their current supply within the next 10 years. In dry years, some already lose out: when the levels of Beijing's two major reservoirs fell to record lows in 1985, supplies to farmers not growing vegetables were cut off.[46]

In pockets of the western United States, a region plagued by numerous water deficits, supplies are being siphoned away from agriculture by thirsty cities willing to pay a premium to ensure water for their future

growth. Where systems of water law and allocation establish clear property rights to water, markets can operate to transfer supplies between willing buyers and sellers for an agreed price. Marketing in the American West was quite common around the turn of the century, but the pace of transactions slowed when the government began building and subsidizing large water projects. As that era draws rapidly to a close, many western cities are finding that buying water from another user is an attractive alternative to developing costly new supplies.

In 1990, 121 major water transactions were reported, 73 of them for immediate or long-term municipal use. If a farmer can earn more by selling water to a nearby city than by using it to irrigate cotton, alfalfa, or wheat, shifting that water from farm to city use is economically beneficial. If it prevents the city from damming another river to increase supplies, the transfer can also benefit the environment. Sale prices have varied greatly, with perpetual water rights having been sold for less than $200 per acre-foot in the Salt Lake City area, but for as much as $3,000–$6,000 per acre-foot in the rapidly urbanizing Colorado Front Range. (An acre-foot equals 325,850 gallons or 1,233 m$^3$, enough to supply a typical four-person U.S. household for about two years.)[47]

In Arizona, Tucson, Phoenix, and other burgeoning cities have taken to ''water ranching.'' State law makes it difficult to buy rights to water independent of the land. Thus, to secure water for their future growth, cities are purchasing farmland for the water that comes with it. Passage of a new state law dealing with water transfers makes it difficult to project how much cropland will eventually be retired, but irrigated agriculture will certainly shrink as urban and industrial demands increase.[48]

The extent to which water marketing leads to reductions in irrigated area will depend on how desperate for water cities become, how quickly institutional and legal barriers to marketing are removed, and on how farmers respond. But whether marketing is encouraged or not, agriculture will likely lose supplies where demands are nearing water's natural limits. Almost everywhere, water's value in crop production is far less than in other activities. Planners in China calculate, for example, that a given amount of water used in industry generates more than 60 times the value of that same amount used in agriculture.[49]

## Improving irrigation efficiency

In the drive to expand irrigation, comparatively little attention has been paid to the efficiency with which irrigation systems operate. Much water is lost as it is conveyed from reservoirs to farmlands, distributed among farmers, and applied to fields. Worldwide, the efficiency of irrigation systems is estimated to average only 37%. Some of this ''lost'' water returns to a stream or aquifer where it can be tapped again, provided the necessary infrastructure is available. But much is rendered unproductive or becomes severely degraded in quality as it picks up salts, pesticides, and toxic elements from the land.[50]

Besides accounting for about two-thirds of water withdrawals worldwide, agriculture consumes the largest share of most nations' water budgets. Whereas 90% or more of the water supplied to industries and homes is available for reuse, return flows from agriculture are often only half the initial withdrawal. The rest is consumed through evaporation and transpiration, which depletes the local water supply. Though water can be saved only by reducing consumption, reducing withdrawals, whether they are consumed or not, can make a given reservoir or aquifer supply last longer or serve a larger area. Raising irrigation efficiency by 10% in the Indus region of Pakistan, for example, could provide enough water to irrigate an additional 2 million hectares.[51]

Most farmers still irrigate the way their predecessors did 5,000 years ago, by flooding or channeling water through parallel furrows. Water flows by gravity across a gently sloping field, seeping into the soil along the way. These gravity systems (also called surface systems) are typically the least expensive to install and by far the most common method in use today. Unfortunately, most fail to distribute water evenly. Farmers must often apply an excessive amount of water to ensure that enough reaches plants situated on higher ground or on the far side of the field. Some areas receive more water than the crops can use, and the excess percolates out of the root zone or simply runs off the field.[52]

Because of these problems, many gravity systems are less than 50%

efficient: only half the water applied to the field actually benefits the crops. Yet a number of practices can greatly improve their performance. Probably the most universally applicable is leveling the land so that water gets distributed more evenly. To water sufficiently crops sitting just 3 cm higher than the surrounding surface, farmers may have to apply as much as 40% more water to the entire field. Precise leveling can thus greatly reduce water needs, besides alleviating waterlogging, curbing erosion, and raising crop yields. It can be done with traditional equipment – a tractor or draft animals pulling a soil scraper and land plane – but most farmers will require training and assistance in carrying out the initial field surveys and leveling operations. In recent years, farmers in the United States and elsewhere have begun to use lasers to guide the leveling process, which can raise the efficiency of surface systems to as high as 90%.[53]

Farmers can also reduce water losses by capturing and recycling water that would otherwise run off the field. This typically requires constructing a pond to collect and store the runoff, and installing pumps to return the water to the head of the field. Some U.S. states now require these tailwater recycling systems, and for many farmers they pay for themselves in reduced energy costs. Especially where irrigation water is drawn from great depths, the energy needed to recycle the runoff is usually less than that needed to pump new supplies from the aquifer.[54]

Researchers have devised a technique called "surge" irrigation that can greatly improve traditional furrow systems. Under the surge method, instead of water being released in a continuous stream down the furrow, irrigation alternates between two rows at specific time intervals. The initial wetting somewhat seals the soil, allowing the next application to advance more quickly down the field. This surging effect reduces percolation losses at the head of the field and distributes water more uniformly. Though this principle could likely be applied in simple farming systems, surge units developed for the U.S. market include a valve and timer that automatically release water at established intervals. In field tests, surge irrigation has reduced water and energy use by 10%–40%. For farmers in the Texas plains, where savings have averaged 25%, the initial investment (about $30 per hectare) is typically recouped within a year.[55]

Over the last two decades, much new land has been brought under irrigation with a variety of high-pressure sprinkler designs. In some areas, farmers have used them to irrigate hilly and marginal lands unsuitable for gravity methods. Among the most common designs is the center pivot, in which a horizontal sprinkler arm circles around a fixed point. Each covering about 50 hectares, center pivots now irrigate much of the U.S. High Plains with water drawn from the Ogallala aquifer. Saudi Arabia has also adopted the technology in its drive for self-sufficiency in grain production. More than 12,000 center pivots have been installed in this desert nation over the last decade or so.[56]

Sprinklers tend to irrigate more uniformly than gravity systems, with efficiencies typically averaging 60%–70%. In windy, dry areas, much water can be lost to evaporation. A method known as low-energy precision application (LEPA) offers substantial improvements over conventional designs. Rather than spraying water high into the air, the LEPA method delivers water closer to the crops by means of drop tubes extending vertically from the sprinkler arm. When used in conjunction with water conserving land preparation methods, LEPA irrigation can achieve efficiencies as high as 95%. Since the system operates at low pressure, energy requirements may drop by 20%–50%. With LEPA, crop yields have tended to average 10%–20% higher than those from fields watered by conventional furrow or sprinkler systems. Retrofitting to a LEPA system costs Texas farmers in the range of $60–$160 per hectare, and the water, energy and yield gains typically make it a cost-effective investment.[57]

For fruits, vegetables, and orchard crops, a group of thrifty irrigation techniques collectively known as microirrigation has rapidly expanded over the last decade. The most familiar micro-method is drip (also known as trickle) irrigation, in which a network of porous or perforated piping, installed on or below the soil surface, delivers water directly to the crops' roots. This keeps evaporation and seepage losses extremely low. To sufficiently water the same crop, drip systems may apply 20%–25% less water to the field than conventional sprinklers and 40%–60% less than simple gravity systems. An important feature, especially for arid lands, is that drip is often better suited for irrigating with brackish water. The fairly

**TABLE 5.2** Use of microirrigation, leading countries and world 1991[a]

| Country | Area under microirrigation (hectares) | Share of total irrigated area under microirrigation[b] (%) |
|---|---|---|
| United States | 606,000 | 3.0 |
| Spain | 160,000 | 4.8 |
| Australia | 147,000 | 7.8 |
| Israel[c] | 104,302 | 48.7 |
| South Africa | 102,250 | 9.0 |
| Egypt | 68,450 | 2.6 |
| Mexico | 60,600 | 1.2 |
| France | 50,953 | 4.8 |
| Thailand | 41,150 | 1.0 |
| Colombia | 29,500 | 5.7 |
| Cyprus | 25,000 | 71.4 |
| Portugal | 23,565 | 3.7 |
| Italy | 21,700 | 0.7 |
| Brazil | 20,150 | 0.7 |
| China | 19,000 | <0.1 |
| India | 17,000 | <0.1 |
| Jordan | 12,000 | 21.1 |
| Taiwan | 10,005 | 2.4 |
| Morocco | 9,766 | 0.8 |
| Chile | 8,830 | 0.7 |
| Other | 39,397 | – |
| World[d] | 1,576,618 | 0.7 |

*Source:* S. Postel, 1992, *Last Oasis: Facing Water Scarcity*, W.W. Norton & Company, New York.

[a] Microirrigation includes primarily drip (surface and subsurface) methods and micro-sprinklers.
[b] Irrigated areas are for 1989, the latest available.
[c] Israel's drip and total irrigated area are down 18 and 15 percent, respectively, from 1986, reflecting water allocation cutbacks due to drought.
[d] 13,280 hectares (11,200 of them in the Soviet Union) were reported in 1981 by countries that did not report at all in 1991; world total does not include this area.

constant level of moisture maintained in the root zone helps prevent salt concentrations from rising to yield-reducing levels.[58]

Although the principles behind drip irrigation date back more than a century, the emergence of inexpensive plastic following World War II spurred the technology's commercial development. By the mid-1970s, six countries (Australia, Israel, Mexico, New Zealand, South Africa, and the United States) were irrigating substantial areas by drip methods, and drip area worldwide totaled about 56,600 hectares. Since then, its use has grown more than twenty-eight-fold, with nearly 1.6 million hectares watered by drip and micro-sprinklers in 1991 (see Table 5.2.) While this represents just 0.7% of world irrigated area, some countries have moved rapidly toward these thrifty irrigation methods in recent years. Israel, for example, waters more than 100,000 hectares by microirrigation, 49% of its total irrigated area.[59]

In many developing countries, improving the performance of canal systems is critical to conserving water and boosting irrigated crop yields closer to their potential. Better management alone could reduce water withdrawals for most surface canal systems by at least 10%–15%, allowing new land to be brought under irrigation for a much lower cost than developing new supplies.

What constitutes "better management" varies from project to project. Typical problems include the large water losses resulting from canal seepage as water is conveyed from reservoirs to fields, poor mechanisms for distributing water among the farmers served by a particular project, and, at the farm level, lack of control over the timing and amount of water applied to fields. Consequently, often less land is irrigated than was originally planned in a project design, contributing to the low rate of return from many irrigation investments. Some farmers get too much water, while others get too little, and few apply water to their crops in optimal amounts. These shortcomings diminish not only the productivity of the water supply, but food production and farmers' livelihoods.

Redesigning projects so that farmers get water when they want it rather than according to some fixed schedule is thus a key to increasing irrigation's productivity. In many cases, this requires bridging an unfortunate gap between irrigation ministries, who often view their mission as merely supplying water, and farmers, who can make optimum use of water only if they have some control over it. A reliable and flexible supply enables farmers to invest in fertilizer and other yield-raising agricultural inputs and to diversify cropping patterns, helping boost production and income. Especially in government-run projects, some form of farmers' organization is necessary to make farmers' concerns and needs known to decision-makers in the irrigation bureaucracy. Such an organization also provides a mechanism for collecting fees to cover operation and maintenance costs, without which expensive irrigation works fall into disrepair.[60]

Whatever the type of system used – flood or furrow, sprinklers, or drip methods – farmers can greatly increase their water efficiency on the farm by scheduling their irrigations to coincide more closely with their crops' water needs. This requires periodically monitoring soil moisture and irrigating just before crops would become stressed by lack of water. Farmers with limited financial resources may do fairly well by extracting a soil sample from the appropriate depth and estimating moisture content by its consistency. If data on evapotranspiration and rainfall are available, growers can keep a water budget, irrigating when their calculations show that their crops will soon need more water.[61]

Many devices are available to measure soil moisture, of which gypsum blocks are among the least costly and simplest to use. When buried in the root zone, the blocks acquire a moisture content roughly equal to that of the surrounding soil. Electrodes embedded inside them are connected to a meter that measures electrical resistance: the wetter the soil, the wetter the gypsum block, and the less it will impede an electrical current. When interpreted for the appropriate soil type, the meter reading tells the farmer how moist or dry the soil is. On test plots of alfalfa and corn, irrigation scheduling using gypsum blocks led to, respectively, a 14% and 27% reduction in water applications compared to neighboring control plots. One tomato grower estimated that the method could cut the number of irrigations needed during the growing season from his usual five or seven to three, with a probable 20% reduction in water use.[62]

Which of these myriad technologies and practices proves practical and economical will vary from place to place. But, if given sufficient incentives, most farmers could cut their water withdrawals by 10%–50% without reducing crop production. Experience shows that an investment in irrigation efficiency is usually also an investment in the productivity of crops and soils. With better water management, yields often increase, erosion is reduced, and soils are less likely to become waterlogged or sapped of nutrients. Encouraging more widespread adoption of water-saving methods would stretch scarce water supplies, lessen ecological damage to overtaxed rivers and streams, and help the growing number of farmers faced with rising water costs to stay in business.

## Strategies for the 1990s and beyond

Securing water to meet the world's growing food needs sustainably will not be easy. The slowdown in irrigation growth, mounting environmental damage from irrigation and from water projects generally, worsening regional scarcities, and the prospects of climate change all combine to severely constrain water use for crop production. Together these trends point to the need for a more creative and diverse approach to watering crops, one that focuses more on raising water efficiency, on integrating irrigation more fully with basic development goals, and on improving water's productivity in rainfed farming.

Much of the pervasive overuse and mismanagement of water in agriculture stems from the near-universal failure to price it properly. Irrigation systems are often built, operated, and maintained by public agencies that charge next to nothing for these services. Farmers' fees in Pakistan, for example, cover only about 13% of the government's costs. Indeed, in most of the Third World, government revenues from irrigation average no more than 10%–20% of the full cost of delivering water. Such undercharging deprives agencies of the funds needed to maintain canals and other irrigation works adequately.[63]

Variously called "irrigation associations" or "water user associations," organized groups of farmers who share a common water source can be instrumental in their effective involvement in projects and in collecting fees. The Philippines National Irrigation Agency now receives no federal funding for operating and maintaining irrigation systems: it depends on revenues from the irrigation associations. This has made the agency more responsive to farmers' needs and more apt to solicit their views. Similar approaches are followed in China and South Korea, where canal systems tend to perform quite well.[64]

Pricing reforms are equally needed in the United States. The federal Bureau of Reclamation supplies water to a quarter of the West's irrigated land – more than 4 million hectares – under long-term contracts (typically 40 years) at greatly subsidized prices. Farmers benefiting from the huge Central Valley Project in California, for example, have repaid only 5% of the project's costs over the last 40 years: $50 million out of $931 million. Largely because of this underpricing, one-third of the Bureau's water irrigates hay, pasture, and other low-value forage crops that are fed to livestock, even though so many other higher-valued activities need additional water.[65]

This free ride in large part explains why so few western farmers invest in efficiency improvements. Hundreds of federal irrigation contracts will be coming up for renewal during the 1990s, as noted earlier. If the Bureau seizes this opportunity to establish contract terms that foster efficiency rather than discourage it, water stresses in the American West could ease measurably. A number of important steps will be taken in this direction by the new Central Valley Project Improvement Act, part of a broader water bill signed into law in late 1992.

One example of what appropriate incentives can accomplish involves a water trade agreement in southern California. The Metropolitan Water District of Southern California (MWD), water wholesaler for roughly half the state's 30 million residents, has agreed to finance conservation projects in the neighboring Imperial Irrigation District in exchange for the estimated 100,000 acre-feet of water per year the investments will save. The annual cost per acre-foot conserved is estimated at $128, less than MWD's best new-supply option. Enough water is being traded this way to meet the needs of 800,000 Californians, yet no cropland is being taken out of production and no irrigation water rights are actually being sold.[66]

Greater attention to using ground water in association with large surface water systems can also improve water productivity. In many areas, seepage from canals and watering of fields creates a large reservoir underground. If not drawn out or drained away, this water can lead to waterlogging and salinization, as has happened in so many large irrigation schemes. Encouraging private development of wells to tap this underground source, however, can help manage this problem while at the same time augmenting farmers' dependable supplies. Especially if the ground water is drawn during the dry season, when surface waters are short, this joint management can make water available year-round, allowing farmers to plant two or three crops a year. In Punjab, Andhra Pradesh, and other parts of India, for example, irrigators with ground water wells produce 5–6 tons of grain per hectare, compared with 2.5–3 tons with canal-delivered surface water.[67]

Investing in smaller scale irrigation projects can help boost food output while reducing the need for expensive large dams with their attendant social and environmental costs. Smaller projects are decentralized, and focus on improving the productivity of water where it falls, rather than transporting it great distances. In contrast to the top-down character of large projects, they can be developed at the grassroots, inspiring local self-reliance. As local initiatives, they also tend to be more responsive to farmers' needs as they perceive them.[68]

An analysis by the Environmental Defense Fund in Washington DC shows that a variety of proven small-scale techniques collectively constitute a viable alternative to the irrigation component of the huge Sardar Sarovar Dam, discussed earlier. Even the most expensive small-scale methods, which include small reservoirs to store rainfall, percolation tanks to replenish ground water, and check dams and microcatchments, cost less than half as much per hectare as irrigation with water from Sardar Sarovar. Equally important, these alternatives would allow water development benefits to be distributed more equitably, reduce construction time, afford farmers more control over their supply, and promote local employment. They clearly deserve a careful and objective examination by Indian officials and the World Bank, which is helping to fund Sardar Sarovar.[69]

Though improvements in irrigation offer the greatest possibilities for saving water, better management of the 84% of cropland watered only by rainfall is of vital importance to increasing food production at the village level and raising incomes among the poorest farmers. Simple techniques aimed at increasing soil moisture in the root zones of crops can markedly boost yields and make production more reliable.

For instance, earthen or stone walls built along the contours of fields can check soil erosion and help store moisture. A survey of farmers in Yatenga Province in northern Burkina Faso showed a 37% increase in yields on plots applying this technique, along with the digging of small pits around each plant, compared with those left untreated. Vetiver grass, known in its native India as *khus*, also shows promise for increasing sustainable cropping on sloping land. When planted along the contours of a hillside, vetiver grass creates a vegetative barrier that slows runoff, allowing rainwater to seep into the soil. Sediment trapped behind it forms a terrace that further conserves soil and moisture for crops. Yields have increased by half over those in similar areas not using this technique.[70]

Although such down-to-earth practices rarely get the visibility and fanfare afforded a large, new dam, they nonetheless can produce the needed results. In Karnataka, India, for example, a watershed management effort involving some 600,000 hectares – nearly a third of the cultivated area – is based on low-cost techniques that farmers can use to increase soil moisture in their fields. Cropping intensity reportedly doubled in the project areas, to 220%.[71]

Besides directly funding more small-scale projects, development organizations can help foster private initiatives by giving small landholders access to credit and by bolstering local industries that supply needed tools and equipment. In Bangladesh, the progressive Grameen Bank is acquiring control of 100 tubewells and then offering loans to groups of five farmers who wish to buy one. The International Fund for Agricultural Development, a 12-year-old UN agency with an impressive track record, is among the global organizations now giving greater emphasis to small, farmer-managed systems. Its experience has been that they are ''more economical, easier to manage, and better targeted at the poor.''[72]

Tackling head-on the question of ''irrigation for whom and for what?'' is particularly critical in Africa. For much of the continent, the irrigation miracle will never work wonders the way it did in Asia. Water and irrigable land are too scarce, and the costs of developing water too high. Thus far, irrigation has tended to expand production of higher-valued crops grown for export or urban consumption rather than basic food crops.[73]

Projects that supplement rather than replace traditional agricultural practices may be the most promising ways to boost water productivity in Africa over the next couple of decades. African farmers' cultivation strategies tend to aim more toward minimizing risks of crop failure in bad years than maximizing yields in good ones. For many, complete dependence on irrigation would increase risk, since spare parts and power to run the systems may not always be available or reliable. Moreover, outside of Egypt and the Sudan, large-scale irrigation is foreign to most African cultures and has the potential to alienate farmers rather than benefit them.[74]

An important test of integrating modern water management with traditional practices is now under way in the Senegal River valley. Farmers there practice ''flood-recession agriculture,'' an indigenous cultivation method used by millions that involves planting crops on river floodplains after seasonal floodwaters recede. By engineering controlled floods, the reliability and productivity of this method can be improved – indeed, some say yields can be doubled or tripled – without incurring much of the ecological damage, exorbitant cost, and disruption of local traditions that full-scale irrigation often brings. At least until irrigation facilities are in place, the Manantali Dam is being operated for optimum recession cropping of sorghum. The first controlled flood was released in September 1988, and plans are to continue them through 1993. If successful and sufficiently high yielding, this modified version of traditional floodplain agriculture could become a permanent part of the dam's operations.[75]

Restoring deforested watersheds urgently needs greater international assistance. Especially in tropical regions, a large share of potentially useful water runs off in damaging floods, causing far more harm than good. Reforesting mountainous catchments will help slow runoff, enhance the infiltration of rainwater into the soil, and thereby increase downstream ground water supplies for the dry season. Revegetating watersheds is especially important if, as some climate models predict, tropical monsoons begin striking with even greater ferocity. An important step forward would be for the Asian Development Bank, the World Bank, or other international body to work with the countries sharing the Himalayan watershed on a restoration plan for this badly degraded region.[76]

As fresh water becomes increasingly scarce, and as cities bid more supplies away from farmers, the use of treated urban wastewater for irrigation is likely to become commonplace. This returns valuable nutrients to the land, where they belong, and helps keep them out of rivers and streams, where they become troublesome pollutants. With proper treatment, and with care in how and where reclaimed wastewater is applied, the practice can be very beneficial.

In California's Monterey County, a decade-long study concluded that irrigating vegetable crops with wastewater that had received advanced (tertiary) treatment proved just as safe as using fresh drinking quality supplies, and cost only a fifth as much as a new fresh water source. Israel, with virtually no new fresh sources to tap, is now reusing 66% of its municipal and industrial wastewater, more than half of it for crop irrigation. Some 19,000 hectares, many of them cultivating cotton, are now irrigated with the reclaimed water. Use of reclaimed effluent is projected to increase by nearly 80% by the year 2000.[77]

A relatively inexpensive way governments can buy insurance against future water constraints, especially in light of climatic change, is to increase funding of international agricultural research centers working to develop new strains of crops. Plants that are more salt-tolerant, drought-resistant, and water-efficient could play a vital part in securing adequate food supplies. Research suggests that wheat, for one, is a good candidate for breeding in greater tolerance to salt, which could allow this important grain to remain productive on salinized land where other common cultivars would not thrive.[78]

Salicornia, a succulent, can be irrigated with seawater and shows promise as a substitute for thirsty fodder crops in water-short regions. Its yield of oil seed compares well with soybeans, and it can contribute up to 10% of a fodder mix for cattle, sheep, and other livestock. Developing new strains that are sufficiently high yielding and profitable to grow, however, takes time. Greater support for such efforts now could pay back handsomely in the decades ahead.[79]

No quick fix is going to solve agriculture's water problems any time soon. Transforming crop production into a water-thrifty but still highly productive enterprise is a monumental task. Added up, the varied ways of using irrigation water more efficiently and of increasing water productivity on rainfed lands can go a long way toward preventing water scarcity from undermining food security. But these diverse technologies and measures will spread widely only if pricing policies, markets, regulations, and international development agencies promote them.

Crucial as they are, though, these measures are but stopgaps. Any hope for balancing water budgets for the long term hinges upon a slowing of population growth and more fundamental adaptation to water constraints. In a country such as Egypt, with a population leaping by 1 million every eight months, modernizing irrigation systems is simply not enough. Depleting scarce water to grow thirsty cotton crops in the desert, as in the Central Asian republics, or to irrigate hay for cattle to eat, as in the water-short American West, may simply not be possible for much longer. The struggle for a secure water future will not end until societies recognize water's natural limits, and begin to bring human numbers and demands into line with them.

## Notes

1.    Dam figure from F. van der Leeden, 1975, *Water Resources of the World*, Water Information Center, Inc., Port Washington, New York; P. Williams,

1983, Damming the world, *Not Man Apart*, October. Water use estimate from S. Postel, 1984, *Water: Rethinking Management in an Age of Scarcity*, Worldwatch Institute, Washington, DC, December.

2. S. Postel, 1984, *Water: Rethinking Management in an Age of Scarcity*, Worldwatch Institute, Washington, DC, December; W.R. Rangeley, 1987, Irrigation and drainage in the world, in W.R. Jordan (ed.) *Water and Water Policy in World Food Supplies*, Texas A & M University, College Station, Texas.

3. K.K. Framji and I.K. Mahajan, 1969, *Irrigation and Drainage in the World: A Global Review*, Caxton Press Private Limited, New Delhi.

4. Irrigated area today from United Nations Food and Agriculture Organization (FAO), 1991, *1990 Production Yearbook*, Rome adjusted for the United States and Taiwan with data from U.S. Department of Agriculture, Economic Research Service, 1990, *Agricultural Resources, Cropland, Water, and Conservation*, Washington D.C., and Sophie Hung, 1991, USDA, ERS, personal communication; W.R. Rangeley, 1987, Irrigation and drainage in the world, in W.R. Jordan (ed.) *Water and Water Policy in World Food Supplies*, Texas A & M University, College Station, Texas.

5. United Nations FAO, 1991, *1990 FAO Production Yearbook*, Rome; United Nations, 1991, Department of International Economical and Social Affairs, 1989, *World Population Prospects 1990*, New York.

6. G. Levine, R. Barker, M. Rosegrant and M. Svendsen, 1988, *Irrigation in Asia and the Near East in the 1990s: Problems and Prospects*, prepared for the Irrigation Support Project for Asia and the Near East at the request of the Asia/Neareast Bureau, U.S. Agency for International Development, Washington, DC, August; data for figure provided by M. Rosegrant, International Food Policy Research Institute, Washington, DC.

7. India figures from M. Svendsen, 1988, Sources of future growth in Indian irrigated agriculture, paper presented to the *IFPRI Planning Workshop on Policy Related Issues in Indian Irrigation*, Ootacamund, Tamil Nadu, India, 26–28 April 1988; China estimate from D. Gunaratnum, private communication, 1989, China Agriculture Operations Division, World Bank, Washington, DC; supporting figures and Mexico estimate from R. Repetto, 1986, *Skimming the Water: Rent-Seeking and the Performance of Public Irrigation Systems*, World Resources Institute, Washington, DC; Brazil figure from J.-L. Ginnsz, private communication, 1989, Brazil Agriculture Operations Division, World Bank; see also W.R. Rangeley, 1987, Irrigation and drainage in the world, in W.R. Jordan (ed.) *Water and Water Policy in World Food Supplies*, Texas A & M University, College Station, Texas; M. Yudelman, 1989, Sustainable and equitable development in irrigated environments, in H. J. Leonard (ed.) *Environment and the Poor: Development Strategies for a Common Agenda*, Transaction Books, New Brunswick, New Jersey.

8. W.R. Rangeley, 1987, Irrigation and drainage in the world, in W.R. Jordan (ed.) *Water and Water Policy in World Food Supplies*, Texas A & M University, College Station, Texas; T. Scudder, 1989, Conservation vs. development: River basin projects in Africa, *Environment*, **31**(2), 4–9, 27–31; United Nations FAO, 1987, *Consultation on Irrigation in Africa*, Rome; W.P. Field, 1990, World Irrigation, *Irrigation and Drainage Systems*, **4**, 15–23.

9. H. von Pogrell, private communication, 1989, Latin American Agriculture Operations Division, World Bank, Washington, DC; J.-L. Ginnsz, private communication, 1989, Brazil Agriculture Operations Division, World Bank; M. Yudelman, 1989, Sustainable and equitable development in irrigated environments, in H. J. Leonard (ed.) *Environment and the Poor: Development Strategies for a Common Agenda*, Transaction Books, New Brunswick, New Jersey.

10. Possibility of rising prices and dwindling stocks from L.R. Brown, 1988, *The Changing World Food Prospect: The Nineties and Beyond*, Worldwatch Paper 85, Worldwatch Institute, Washington, DC.

11. A.K. Mitra, 1986, Underutilisation revisited: Surface irrigation in drought prone areas of Western Maharashtra, *Economic and Political Weekly*, 26 April, Bombay.

12. W.R. Rangeley, 1987, Irrigation and drainage in the world, in W.R. Jordan (ed.) *Water and Water Policy in World Food Supplies*, Texas A & M University, College Station, Texas.

13. L. Zhuoyan and G. Jin'an, 1989, Neglect of water projects hurts farmland, *China Daily*, 10 March; U.S. Department of Agriculture, 1989, *USSR: Agriculture and Trade Report*, Economic Research Service (ERS), Washington, DC; 150 million figure from M.E. Jensen, W.R. Rangeley and P.J. Dieleman, 1990, Irrigation trends in world agriculture, in B.A. Stewart and D.R. Nielsen (eds.) *Irrigation of Agricultural Crops*, American Society of Agronomy, Madison, Wisconsin.

14. Manzur Ahmad, 1987, Water as a constraint to world food supplies, in W.R. Jordan (ed.) *Water and Water Policy in World Food Supplies*, Texas A & M University, College Station, Texas.

15. A.K. Mitra, 1986, Underutilisation revisited: Surface irrigation in drought prone areas of Western Maharashtra, *Economic and Political Weekly*, 26 April, Bombay; see also R. Chambers, 1987, Food and water as if poor people mattered: A professional revolution, in W.R. Jordan (ed.) *Water and Water Policy in World Food Supplies*, Texas A & M University, College Station, Texas.

16. M. Svendsen, 1988, Sources of future growth in Indian irrigated agriculture, paper presented to the *IFPRI Planning Workshop on Policy Related Issues in Indian Irrigation*, Ootacamund, Tamil Nadu, India, 26–28 April 1988; irrigation potential from S.C.G. Desai, 1988, Planning targets for irrigation development, paper presented to the *IFPRI Planning Workshop on Policy Related Issues in Indian Irrigation*, Ootacamund, Tamil Nadu, 26–28 April; potential yield increase from B.D. Dhawan, 1988, Indian irrigation: An assessment, *Economic and Political Weekly*, 7 May.

17. Quoted in O. Sattaur, 1989, India's troubled waters, *New Scientist*, **122**(1666), 46–51.

18. Population estimate from Population Reference Bureau, 1991, *1991 World Population Data Sheet*, Washington, DC.

19. Agriculture's use of water from I.A. Shiklomanov, 1990, Global Water Resources, *Nature and Resources*, Vol. 26, No. 3. See also Chapter 2, this book.

20. V.A. Kovda, 1983, Loss of productive land due to salinization, *Ambio*, **12**(2), 91–93.

21. W.R. Rangeley, private communication, 30 January 1989; reference to World Bank study from S. Barghouti and G. Le Moigne, 1991, Irrigation and the environmental challenge, *Finance and Development*, June; Mexico figure from M. Yudelman, 1989, Sustainable and equitable development in irrigated environments, in H. J. Leonard (ed.) *Environment and the Poor: Development Strategies for a Common Agenda*, Transaction Books, New Brunswick, New Jersey.

22. J. Rhoades, private communication, 1989, U.S. Salinity Laboratory, Riverside, California; Soviet figure from P.P. Micklin, private communication, 1989, Western Michigan University, Kalamazoo, Michigan.

23. T. Harris, 1989, A valley filled with selenium, *Sacramento Bee*, 16 July; The Wilderness Society, 1988, *Ten Most Endangered National Wildlife Refuges*, Washington, DC; see also E. Marshall, 1986, High selenium levels confirmed in six states, *Science*, **231**, 111; and C. Peterson, 1989, Toxic time bomb ticks in San Joaquin Valley, *Washington Post*, 19 March.

24. T. Harris, 1989, A valley filled with selenium, *Sacramento Bee*, 16 July; San Joaquin Valley Drainage Program, 1989, *Preliminary Planning Alternatives for Solving Agricultural Drainage and Drainage-Related Problems in the San Joaquin Valley*, Sacramento, California; National Research Council, 1989, *Irrigation-Induced Water Quality Problems: What Can be Learned from the San Joaquin Valley Experience?*, National Academy Press, Washington, DC.

25. C. Dickason, 1988, Improved estimates of groundwater mining acreage, *Journal of Soil and Water Conservation*, **43**(3), 239–240; C. Dickason, private communication, 1989, U.S. Department of Agriculture, Economic Research Service; Agricultural Outlook, May 1989, Advance census reports show irrigation rebound.

26. J.E. Nickum and J. Dixon, 1989, Environmental problems and economic modernization, in C.E. Morrison and R.F. Dernberger (eds.) *Focus: China in the Reform Era, Asia–Pacific Report*, East–West Center, Honolulu; reference to Tamil Nadu in C. Widstrand (ed.), 1980, *Water Conflicts and Research Priorities*, Pergamon Press, New York; R. Chengappa, 1986, India's water crisis, *India Today*, 31 May, excerpted in *World Press Review*, August 1986.

27. J. Margat and K.F. Saad, 1984, Deep-lying aquifers: Water mines under the desert?, *Nature and Resources*, 21, April/June.

28. A. Ali-Ibrahim, 1991, Excessive use of groundwater resources in Saudi Arabia: Impacts and policy options, *Ambio*, **20**(1), 34–37.

29. G.F. Seib, 1985, Libya launches $25 billion project to quench Sahara nation's thirst, *Wall Street Journal*, 3 October; 1986, Massive Libyan groundwater project unaffected so far, *The Groundwater Newsletter*, 17 March; 1990, Egypt moves to tap Nubian desert water, *World Water*, April.

30. E.D. Gutentag, F.J. Heimes, N.C. Krothe, R.R. Luckey and J.B. Weeks, 1984, *Geohydrology of the High Plains Aquifer in Parts of Colorado, Kansas, Nebraska, New Mexico, Oklahoma, South Dakota, Texas, and Wyoming*, U.S. Geological Survey Paper 1400-B, U.S. Government Printing Office, Washington, DC; W. Wyatt, 1991, Water Management – Southern High Plains of Texas, unpublished paper, High Plains Underground Water Conservation District No. 1, Lubbock, Texas, May; drop in irrigated area from Texas Water Development Board, 1991, *Surveys of Irrigation in Texas – 1958, 1964, 1969, 1974, 1979, 1984, and 1989*, Report 329, Austin, Texas.

31. P.P. Micklin, 1989, *The Water Management Crisis in Soviet Central Asia*,

final report to the National Council for Soviet and East European Research, Washington, DC, February.

32. P.P. Micklin, 1989, *The Water Management Crisis in Soviet Central Asia*, final report to the National Council for Soviet and East European Research, Washington, DC, February; Concerning the changes in the structure of the use of river flow in the irrigated zone of the Amu Dar'ya and Syr Dar'ya basins, *Vodnyye Resursy*, **3**; 1987 estimate from P.P. Micklin, private communication, 1989.

33. P.P. Micklin, 1989, *The Water Management Crisis in Soviet Central Asia*, final report to the National Council for Soviet and East European Research, Washington, DC, February; P.P. Micklin, 1988, Desiccation of the Aral Sea: A water management disaster in the Soviet Union, *Science*, **241**, 1170–1176; P.P. Micklin, private communication, 1989; V.M. Kotlyakov, 1991, The Aral Sea basin: A critical environmental zone, *Environment*, **33**(1), 4–9, 36–38.

34. V. Rich, 1991, A new life for the sea that died, *New Scientist*, Vol. 130, No 1764, p. 15, 13 April; V.M. Kotlyakov, 1991, The Aral Sea basin: A critical environmental zone, *Environment*, **33**(1), 4–9, 36–38; A.V. Yablokov, private communication, 1991, Deputy Chairman, Committee of Ecology, The USSR Supreme Soviet, Moscow.

35. L. Udall, 1989, *The Environmental and Social Impacts of the World Bank Financed Sardar Sarovar Dam in India*, Testimony before the Subcommittee on Natural Resources, Agricultural Research and Environment, Committee on Science, Space and Technology, U.S. House of Representatives, 24 October; O. Sattaur, 1989, India's troubled waters, *New Scientist*, **122**(1666), 46–51. B. Morse and T. Berger, 1992, *Sardar Sarovar*, Report of the Independent Review, Resources Futures International Inc, Ottowa, Ontario.

36. B. Crossette, 1989, Water, water everywhere? Many now say "No!", *The New York Times*, 7 October.

37. Narmada Bachao Andolan, 1989, *Sardar Sarovar Project: An Economic, Environmental and Human Disaster*, Bombay; B. Amte, 1989, *Cry, the Beloved Narmada*, Maharogi Sewa Samiti, Chandrapur, Maharashtra, India.

38. T. Scudder, 1989, Conservation vs. development: River basin projects in Africa, *Environment*, **31**(2), 4–9, 27–31.

39. L. Master, 1991, Aquatic animals: Endangerment alert, *Nature Conservancy*, March/April; A.D. Tarlock, 1991, From reclamation to reallocation of western water, *Journal of Soil and Water Conservation*, **46**(2), 122–124.

40. Environmental Defense Fund, 1992, The Central Valley Project Improvement Act – General Summary, Oakland, California.

41. M. Falkenmark, 1989, The massive water scarcity now threatening Africa – Why isn't it being addressed?, *Ambio*, **18**(2), 112–118; Population Reference Bureau, 1989, *World Population Data Sheet*, Washington, DC.

42. J.R. Starr and D.C. Stoll, 1987, *U.S. Foreign Policy on Water Resources in the Middle East*, Center for Strategic and International Studies, Washington, DC; J.R. Starr and D.C. Stoll (eds.), 1988, *The Politics of Scarcity: Water in the Middle East*, Westview Press, Boulder, Colorado; population figure calculated from Population Reference Bureau, 1989, *World Population Data Sheet*, Washington, DC.

43. T. Naff, 1991, *The Jordan Basin: Political, economic, and institutional issues*, paper prepared for The World Bank International Workshop on Comprehensive Water Resources Management Policies, Washington, DC, 24–28 June; population statistics from Population Reference Bureau, 1991, *1991 World Population Data Sheet*, Washington, DC; Tahal Consulting Engineers Ltd., 1990, *Israel Water Sector Review: Past Achievements, Current Problems and Future Options*, Tel Aviv.

44. J.E. Nickum, 1987, *Beijing's Rural Water Use*, prepared for East–West Center North China Project, Honolulu; The Chinese Research Team for Water Resources Policy and Management in Beijing–Tianjin Region of China, 1987, *Report on Water Resources Policy and Management for the Beijing–Tianjin Region of China*, Sino–U.S. Cooperative Research Project on Water Resources Policy and Management, Beijing; 1989, Water rules tightened; Fines levied, *China Daily*, 18 May; North China Plain grain output from F.W. Crook, 1988, *Agricultural Statistics of the People's Republic of China, 1949–86*, U.S. Department of Agriculture, Economic Research Service, Washington, DC; L. Hong, 1989, Beijing set to tackle water thirst, *China Daily*, 17 October.

45. East–West Center and State Science and Technology Commission, 1988, *Water Resources Policy and Management for the Beijing–Tianjin Region*, East–West Environment and Policy Institute, Honolulu; The Chinese Research Team for Water Resources Policy and Management in Beijing–Tianjin Region of China, 1987, *Report on Water Resources Policy and Management for the Beijing–Tianjin Region of China*, Sino–U.S. Cooperative Research Project on Water Resources Policy and Management, Beijing.

46. The Chinese Research Team for Water Resources Policy and Management in Beijing–Tianjin Region of China, 1987, *Report on Water Resources Policy and Management for the Beijing–Tianjin Region of China*, Sino–U.S. Cooperative Research Project on Water Resources Policy and Management, Beijing; J.E. Nickum and J. Dixon, 1989, Environmental problems and economic modernization, in C.E. Morrison and R.F. Dernberger (eds.) *Focus: China in the Reform Era, Asia–Pacific Report*, East–West Center, Honolulu; J.E. Nickum, 1987, *Beijing's Rural Water Use*, prepared for East–West Center North China Project, Honolulu.

47. For a good overview of markets, see the interview with S.J. Shupe by N. Zeilig, March 1988, Face to face – Water marketing: An overview, *Journal of the American Water Works Association*, **80**(3), 18–26; 1990 transactions from 1990 Annual Transaction Review: Growing diversity of water agreements, *Water Strategist*, **4**(4), January; *Water Market Update* (now *Water Intelligence Monthly*), various issues, 1988–1990.

48. E. Checchio, 1988, *Water Farming: The Promise and Problems of Water Transfers in Arizona*, University of Arizona, Tucson; G. Thacker, private communication, 1991, U.S. Department of Agriculture extension agent, College of Agriculture, University of Arizona, Tucson.

49. The Chinese Research Team for Water Resources Policy and Management in Beijing–Tianjin Region of China, 1987, *Report on Water Resources Policy and Management for the Beijing–Tianjin Region of China*, Sino–U.S. Cooperative Research Project on Water Resources Policy and Management, Beijing.

50. Efficiency estimate from W.R. Rangeley, 1985, Irrigation and drainage in the world, paper presented at the *International Conference on Food and Water*, 26–30 May, College Station, Texas; see M.T. El-Ashry, J. van Schilfgaarde and S. Shiffman, Salinity pollution from irrigated agriculture, *Journal of Soil and Water Conservation*, **40**(1), 48–52.

51. W.R. Rangeley, 1985, Irrigation and drainage in the world, paper presented at the *International Conference on Food and Water*, 26–30 May, College Station, Texas.

52. For a good overview of irrigation efficiency, see E.G. Kruse and D.F. Heermann, 1977, Implications of irrigation system efficiencies, *Journal of Soil and Water Conservation*, **32**(6), 265–270.

53. D.H. Negri and J.J. Hanchar, 1989, *Water Conservation Through Irrigation Technology*, USDA Economic Research Service, Agriculture Information Bulletin No. 576; California Department of Water Resources, 1984, *Water Conservation in California*, California Resources Agency, Sacramento, California.

54. G. Sloggett, 1982, *Energy and U.S. Agriculture: Irrigation Pumping, 1974–80*, U.S. Department of Agriculture, Washington, DC; California Department of Water Resources, 1991, Tailwater recovery system study released, *Water Conservation News*, April.

55. High Plains Underground Water Conservation District No. 1, 1989, District salutes water savings by area irrigators, *The Cross Section*, newsletter of the High Plains Underground Water Conservation District No. 1, Lubbock, Texas, November; Texas savings and payback from K. Carver, 1992, High Plains Underground Water Conservation District No. 1, personal communications; see also S.M. Masud and R.D. Lacewell, 1990, Energy, water, and economic savings of improved production systems on the Texas High Plains, *American Journal of Alternative Agriculture*, **5**(2), 69–75.

56. For background on sprinkler and other irrigation systems, see K.D. Frederick and J.C. Hanson, 1982, *Water for Western Agriculture*, Resources for the Future, Washington, DC; Saudi Arabian citation from 1985, Saudis convert oil to water and food, *The Groundwater Newsletter*, 28 February 28.

57. Conventional sprinkler efficiency from E.G. Kruse and D.F. Heermann, 1977, Implications of irrigation system efficiencies, *Journal of Soil and Water Conservation*, **32**(6), 265–270, and High Plains Underground Water Conservation District No. 1, 1989, District salutes water savings by area irrigators, *The Cross Section*, newsletter of the High Plains Underground Water Conservation District No. 1, Lubbock, Texas, November; W.M. Lyle and J.P. Bordovsky, 1991, LEPA: Low energy precision application, *Irrigation Journal*, April (including lower end of cost range); S.M. Masud and R.D. Lacewell, 1990, Energy, water, and economic savings of improved production systems on the Texas High Plains, *American Journal of Alternative Agriculture*, **5**(2), 69–75; retrofit cost from K. Carver, 1992, High Plains Underground Water Conservation District No. 1, personal communications.

58. Background and basic features of drip irrigation from K. Shoji, 1977, Drip irrigation, *Scientific American*, **237**(5), 62–68; see also S. Davis and D. Bucks, 1983, Drip irrigation, in C.H. Pair, W.W. Hinz, K.R. Frost, R. Sneed and T.J. Schiltz (eds.) *Irrigation*, The Irrigation Association, Silver Spring, Maryland; estimated water savings from J.S. Abbott, 1984, Micro Irrigation–World Wide Usage, *ICID Bulletin*, January.

59. 1974 estimate from D. Gustafson, 1978, Drip irrigation in the world – State of the art, in *Israqua '78: Proceedings of the International Conference on Water Systems and Applications*, Israel Centre of Waterworks Appliances, Tel Aviv; 1991 data from Dale Bucks, Microirrigation Working Group, International Commission on Irrigation and Drainage (ICID), Beltsville, Md., private communication, June 22, 1992, with irrigated area from U.N. Food and Agriculture Organization (FAO), *1990 Production Yearbook* Rome: 1991, with adjustments from USDA for the United States and Taiwan.

60. M. Keen, 1988, Clearer thoughts flow on irrigation, *Ceres*, Vol. 21, May/June 1988.

61. See California Department of Water Resources, 1984, *Water Conservation in California*, California Resources Agency, Sacramento, California. For a more technical discussion, see E.A. Hiler and T.A. Howell, 1983, Irrigation options to avoid critical stress: An overview, in H.M. Taylor, W.R. Jordan and T.R. Sinclair (eds.) *Limitations to Efficient Water Use in Crop Production*, American Society of Agronomy, Madison, Wisconsin.

62. G. Richardson, J. Tiedeman, K. Crabtree and K. Summ, 1989, Gypsum Blocks "Tell a water tale", in *Journal of Soil and Water Conservation*; for overview of soil moisture monitoring methods see T. Weems, 1991, Survey of moisture measurement instruments, *Irrigation Journal*, Vol. 41, No. 1, January/February; test results and tomato grower estimate from G. Richardson, 1985, *Saving Water from the Ground Up*, INFORM Inc., New York.

63. R. Repetto, 1986, *Skimming the Water: Rent-Seeking and the Performance of Public Irrigation Systems*, World Resources Institute, Washington, DC.

64. R. Repetto, 1986, *Skimming the Water: Rent-Seeking and the Performance of Public Irrigation Systems*, World Resources Institute, Washington, DC; R. Wade, 1988, The management of irrigation systems: How to evoke trust and avoid dilemma, *World Development*, **16**(4), 489–500.

65. E.P. LeVeen and L.B. King, 1985, *Turning off the Tap on Federal Water Subsidies*, vol. 1, Natural Resources Defense Council and California Rural Legal Assistance Foundation, San Francisco; U.S. Department of the Interior, 1988, *1987 Summary Statistics, vol. 1: Water, Land, and Related Data*, U.S. Bureau of Reclamation, Denver; M.R. Moore and C.A. McGuckin, 1988, Program crop production and federal irrigation water, in *Agricultural Resources: Cropland, Water, and Conservation Situation and Outlook Report*, U.S. Department of Agriculture, Economic Research Service, Washington, DC.

66. Conservation and Drought Strategies, 1988, *Water Market Update*, December.

67. B.D. Dhawan, 1988, Indian irrigation: An assessment, *Economic and Political Weekly*, 7 May.

68. See J. Silliman and R. Lenton, 1987, Irrigation and the land-poor, in W.R. Jordan (ed.) W*ater and Water Policy in World Food Supplies*, Texas A & M University, College Station, Texas; United Nations FAO, 1987, *Consultation on Irrigation in Africa*, Rome.

69. P. Miller, 1989, *An Alternative Development Strategy to the Sardar Sarovar Dam*, Testimony before the Subcommittee on Natural Resources, Agricultural Research and Environment, Committee on Science, Space and Technology, U.S. House of Representatives, 24 October.

70. W. Critchley, 1991, *Looking After Our Land: Soil and Water Conservation in Dryland Africa*, Oxfam Publications, Oxford; World Bank, 1987, *Vetiver Grass (Vetiveria zizanioides): A Method of Vegetative Soil and Moisture Conservation*, New Delhi; J.C. Greenfield, 1988, seminar on the vetiver system presented at the World Bank, Washington, DC, 4 August.

71. 1989, Watershed management in state impressive, *Deccan Herald*, 6 January. 1989,

72. Irrigation is key to Bangladesh food needs, *World Water*, May; IFAD quote from 1989, Asia's green revolution: Pause or setback, *Hunger Notes*, May.

73. United Nations FAO, 1987, *Consultation on Irrigation in Africa*, Rome.

74. C.F. Hutchinson, 1989, Will climate change complicate African famine?, *Resources*, **95**, 5–7.

75. T. Scudder, 1989, Conservation vs. development: River basin projects in Africa, *Environment*, **31**(2), 4–9, 27–31.

76. See S. Postel and L. Heise, 1988, *Reforesting the Earth*, Worldwatch Paper 83, Worldwatch Institute, Washington, DC.

77. 1987, Reclaimed municipal wastewater has been found to be as safe as potable well water for irrigating vegetable crops, *Water Newsletter*, 30 April; J. Klein, engineer, private communication, 1989, Monterey Regional Water Pollution Control Agency, Monterey, California; H.I. Shuval, 1987, The development of water reuse in Israel, *Ambio*, **16**(4); Tahal Consulting Engineers Ltd., 1990, *Israel Water Sector Review: Past Achievements, Current Problems and Future Options*, Tel Aviv.

78. A. Charnock, 1988, Plants with a taste for salt, *New Scientist*, **120**(1641), 41–42, 45; B. Forster, 1988, Wheat can take on more than a pinch of salt, *New Scientist*, **120**(1641), 43.

79. 1988, Farm grows salt-tolerant crops, *U.S. Water News*, September; A. Charnock, 1988, Plants with a taste for salt, *New Scientist*, **120**(1641), 41–42, 45; for a comprehensive treatment of salt-tolerant crops, see U.S. National Research Council, 1990, *Saline Agriculture: Salt-Tolerant Plants for Developing Countries*, National Academy Press, Washington, D.C.

# Chapter 6
# Water and energy

Peter H. Gleick

## Introduction

Fresh water and energy, two resources necessary for a reasonable quality of life, are intricately connected. We use energy to help us clean and transport the water we need. We use water to help us produce the energy we need. And as the 21st century nears, we are running up against physical and environmental constraints in our use of both resources. This chapter explores the connections between our demand for and use of energy and water, and suggests that there are strong parallels between both the problems and the solutions to the growing water crisis and conflicts over energy resources. In particular, the arguments during the last two decades over energy prices, equity and efficiency of energy use, technological innovation, and supply versus demand are now being heard in the growing debates over fresh water resources.

Energy is a fundamental requirement for operating modern water supply and purification facilities. Without the input of substantial amounts of energy, major water transfers from water-rich to water-poor regions, the desalination of brackish water or seawater, and massive pumping from ground water aquifers would all be impossible.

The production and use of energy resources, in turn, often requires a significant commitment of water. Water may be required when an energy resource is mined, as a feedstock to alter fuel properties, for the construction, operation, and maintenance of energy generation facilities, for power plant cooling, or for some aspect of waste disposal. Sometimes this water is withdrawn and then returned to a water supply; sometimes it is consumed during operation or contaminated until it is unfit for further use. Even hydroelectric facilities are responsible for the consumptive loss of water that evaporates from reservoir surfaces. Water use in the energy sector can lead to changes in natural hydrological and ecological systems and increase the pressure for inter-basin transfers of water to regions that are water-poor. In some cases, constraints on water availability will limit choices of sites and types of energy facilities.

In the coming years, growing demands for water from competing sectors of society and from growing populations will place new pressures on the amount of water available for energy production. Limitations on the availability of fresh water in some regions of the world may restrict the type and extent of energy development. At the same time, high energy costs or limited energy availability will constrain our ability to provide adequate clean water and sanitation facilities to the thousands of millions of people who lack those basic services. Developing rational water and energy policies will thus increasingly require policy makers to integrate these connections into their decisions.

## Energy for water

The production, transportation, and cleaning of water requires the use of energy. Energy permits us to make use of water that was previously considered either non-potable or unobtainable. We can now remove salts and other contaminants from water using desalination and wastewater treatment techniques, and we pump water from deep underground aquifers or distant sources. The availability and price of energy sets limits on the extent to which unusual sources of water can be tapped. As a result, understanding the links between water supply and quality and energy will help us evaluate constraints on meeting future water needs. This section discusses energy requirements for moving water from one place to another, for pumping ground water, and for desalinating brackish and salt water.

**Energy for moving water.** One of the most important characteristics of the global fresh water cycle, discussed throughout this book, is its grossly uneven spatial and temporal distribution. While water is plentiful on a global average, we often do not get it when we want it, where we want it, or in the form it is needed. The fact that the vast majority of the world's fresh water supply is locked up as ice in Greenland and Antarctica is only one frustrating example. As urban and rural demands grow, we are increasingly faced with the problem of supplying human needs that are often far removed from reliable sources of supply.

Society's first answer to the problem of the grossly uneven distribution of fresh water resources was to build water supply facilities to make up for variations in precipitation or river runoff over time and to move water from regions of surplus to regions of deficit. Legend says that the early kings of Menes, the first of the pharaohs of Egypt, built a masonry dam across the Nile River near Memphis to control the annual flood. By the time of Ramses II in the 14th century BC, an extensive system of irrigation canals and reservoirs had been developed.[1] The ancient Mesopotamians made extensive use of canals to bring water to the city of Babylon. The Hanging Gardens of Babylon, famed as one of the seven wonders of the world, were supplied with water by these systems, and the fertility of Babylonia was a source of envy to the Greeks. Herodotus wrote, "Of all countries, none is more fruitful in grain."[2] Babylon grew corn, barley, wheat, emmer, sesame, flax, fruit trees, vineyards, herbs, and many other crops with water from the Tigris and Euphrates rivers and from reliable ground water aquifers supplied through irrigation canals and qanats, long, sloping tunnels dug from a natural spring to a community or agricultural field. Qanats determined the nature, size, and spread of human settlements in many parts of Iran, Iraq, and northern Africa thousands of years ago.[3]

Another remarkable aqueduct system was built by Sennacherib in 691 BC to bring water from a tributary of the Greater Zab to his capital Nineveh, 80 km away. Jerusalem was supplied in early times by a system thought to have been built under the kings of Judah around 1,000 BC. Parts of this conduit system are still complete.[4]

In Asia, excavations at Harappa and Mohenjo Daro in the Indus valley have revealed ceramic pipes for water supply and brick conduits under the streets for drainage that are thought to have been in operation around 3,000 BC.[5] Over two thousand years ago, the Chinese began construction of the Grand Canal with a 150-km-long canal built to meet the military needs of the Wu Kingdom. The Grand Canal is still in use today and now extends over 1,700 km.[6]

The Romans are also renowned for their aqueducts and water supply systems. The first of the Roman aqueducts was completed around 312 BC, and by the height of the empire, nine major aqueducts supplied Rome with as much water per capita as many parts of the industrialized world today. This water was distributed through an extensive system of lead pipes in the streets and the city was drained by well built sewers.[7]

While significant amounts of water were often provided by these early systems, they were ultimately limited in the amount of water that could be supplied, and where that water could go, by the force of gravity. Water could be transferred from one place to another only as long as the source was uphill of the demand.

Modern civilization has greatly increased its ability to transfer water from one place to another by using energy to pump water over hills and mountains. When the demand for water in a region increases beyond the ability of the region to supply it, new sources of water farther and farther away must be tapped. Throughout the 20th century, large-scale water transfer projects have been developed to permit continued growth in arid and semi-arid regions that would otherwise have already been constrained by natural limits. And new projects are constantly being proposed and evaluated as populations and water requirements increase.

These projects almost always involve a substantial investment of energy. To lift $100 \text{ m}^3$ of water per minute to a height of $100 \text{ m}$ requires over $1.5 \text{ MWe}$ of power if the pumps are 100% efficient. To do this continuously for a year using electricity from a typical oil-fired power plant and pumps that are 50% efficient would require the energy content of 50,000 barrels of oil.

Most long-distance water transfer systems have both pumps for getting water over hills and mountains, and hydroelectric generators to take advantage of the energy in the falling water as it comes down the other side. Whether a system is a net consumer or producer of energy depends on its geographical characteristics. Where additional energy must be supplied, it is typically generated with fossil fuel or nuclear facilities, adding to the environmental costs of the water diversion itself.

Many major water transfer projects have been built or proposed, primarily in industrialized countries. The State Water Project in California, authorized in 1959, now delivers nearly $5 \text{ km}^3$ of water every year from northern California to the drier parts of the state. If the initial plans are fully developed, the project will include 148 pumping plants, 40 power plants, 22 reservoirs and dams, and 1,000 km of aqueducts. The total energy produced by the system's hydroelectric plants will average over 7 million MWh per year, but the energy required by the pumping plants to lift the water over the mountains will exceed 12.4 million MWh per year, making this project a net consumer of energy.

Several other enormous projects that have been proposed would also be net energy consumers. The Texas Water Plan in the south-central United States, originally proposed during a severe drought in the 1950s, would have imported $15–16 \text{ km}^3$ per year from the lower Mississippi River and from the rivers of the more humid eastern part of Texas (e.g., Red, Sabine, Sulphur, Neches).[8] This project would require pumping water up 900 m into west Texas. Overall electrical pumping capacity needed was estimated at nearly 7,000 MWe, which would produce 40 million MWh per year (at a 65% capacity factor). The project has so far been rejected on the grounds that the water would cost far more than irrigators (the major beneficiaries of the water) could afford to pay, and that the environmental impacts of the water transfer would be severe.

Similarly, a proposed transfer of water to the High Plains region of the central United States (Colorado, Kansas, New Mexico, Oklahoma, Texas, and Nebraska) would have replaced a heavy dependence on non-renewable use of the Ogallala aquifer for water. Consumption of ground water from this aquifer exceeds recharge in the region and the costs of additional water extraction are rising rapidly. The High Plains transfer scheme is a set of proposals to move water from the Missouri, Arkansas, White, Red, and Quachita rivers over canals to the High Plains areas. As with the Texas Water Plan, considerable pumping would be required to overcome elevation differences of as much as 1,000 m; estimates range from 7 to 50 million MWh per year for different planned diversions. Operational costs would be quite high because of these huge energy requirements, and this project, too, has never been built because of its high costs.[9]

Several massive water transfer projects have been proposed in China, which, like many other nations, has enormous disparities in regional water supply and demand. The project with the greatest chance of development is the so-called Eastern Route, which would transfer water from the Yangzijiang River west of Shanghai north to the North China Plain near Beijing. The canal would be over 1,100 km long with an average capacity of $14 \text{ km}^3$ per year. Several large pumping plants would be needed in the middle of the project, requiring over 5 million MWh of electricity per year. An interesting characteristic of this project is that it would make use of parts of the ancient Grand Canal.

In the former Soviet Union, enormous Siberian rivers schemes were proposed and cancelled many times, in many forms. One form of the project would have diverted $120 \text{ km}^3$ per year from the Ob, Irtysh, Yenisei, Onega, Pechora, and Dvina rivers toward central Asia and other more populated regions of the country, instead of north into the Arctic Ocean. Five to ten thousand MWe of capacity would have been needed to pump the water over various mountain ranges.[10] Even before the disintegration of the Soviet Union, opposition to the project was high. Now, responsibility for designing, building, and operating such a project has been spread over several new independent nations and new institutions, making its construction even more unlikely.

One of the most grandiose schemes ever conceived was the North American Water and Power Alliance (NAWAPA), proposed in the early 1960s by the Ralph Parsons Company, a construction-engineering firm. NAWAPA would have collected water from the Fraser, Yukon, Peace, Athabasca and other rivers of Alaska, British Columbia, and the Yukon territory and transferred this water throughout Canada, the western and mid-western United States, and to three states in northern Mexico. NAWAPA represents the ultimate fantasy of water engineers, effectively re-plumbing the entire face of western North America with 369 massive projects costing hundreds of billions of dollars. It would have provided over $5,000 \text{ km}^3$ of water storage and eventually transferred $136 \text{ km}^3$ per year.[11] Massive amounts of water would have had to be lifted over the Rocky Mountains, which inconveniently lie in between the water sources and the water demands. The centerpiece of the project is the damming of the Rocky Mountain Trench, an 800-km-long gorge in the Canadian Rockies adjacent to Banff and Jasper National Parks.[12] In short, the project is a dam builder's dream and an environmental and economic nightmare, but the design stands as a monument to what we are willing to consider when water supplies are limited.

Today, even relatively modest projects face growing constraints and opposition. For such projects to succeed, the water to be exported must be considered a real "surplus" for the expected lifetime of the project, which is usually many decades. In addition, the total cost of the water to be delivered, of which the energy cost is often a substantial component, must be less than the cost of any alternative sources of water. This is rarely the case, given the large improvements in water use efficiency that are possible in every sector, at relatively low cost. Postel, in Chapter 5, for example, describes a variety of inexpensive improvements in irrigation techniques and technologies that could free considerable quantities of water for use in other sectors, or elsewhere in the agricultural sector.

Finally, the environmental and ecological costs of such projects are often enormous, given the large volumes of water usually exported from a basin and the extensive construction and hydrologic modifications that must be done. For example, the export of water from northern California to southern California has been implicated in the decimation of several fish species and the loss of important aquatic habitats.[13] The reduction in the flows of water and nutrients at the mouth of the Nile River are implicated in the destruction of the sardine fishery in the eastern Mediterranean Sea.[14] And the complete consumption of the waters of the Colorado River in most years has destroyed the brackish water estuary at the mouth of the river. Some of these effects are described in more detail in Chapter 4.

**Energy for ground water pumping.** The earliest development of irrigated agriculture occurred on the banks of the great rivers of the Middle East: the Nile, the Euphrates, and the Tigris. These developments relied primarily on the natural river flow, which was neither constant over time, nor reliably predictable. In some places, however, reliable and steady flows of ground water were found and used. Qanats, described earlier, made this water available for irrigation, and there is a long history of substantial wells being dug to reach ground water in the desert. Joseph's well in Cairo, made famous by legend and longevity, is dug through more than 100 m of rock. Not until the 20th century, however, when cheap well drilling, pumping technology, and fossil fuels became available, could deep ground water stocks be exploited in a substantial way.

Like many other resources, our ability to extract and use ground water far exceeds, even today, our understanding of the geophysical characteristics of ground water basins. The dynamics of ground water flow and recharge, the limits to regional ground water supply, and the occurrence and migration of contaminants are all still imperfectly understood, and because there has traditionally been little regional competition for ground water resources, legal mechanisms for allocation have rarely been developed or implemented.

The limits to how much water can be extracted from a finite ground water aquifer are economical and environmental. When water is pumped out faster than it is recharged by natural processes, the water level in an aquifer drops and the distance water must be raised to the surface increases. Ultimately, pumping must cease when the energy costs rise to a point that exceeds the value of the water, the quality of the water in the aquifer falls below acceptable levels, or the well runs dry.

Cheap fossil fuels have permitted overpumping of fossil ground water aquifers – ground water basins whose water supplies accumulated over hundreds to thousands of years, or longer. The Ogallala aquifer in the Great Plains region of the United States underlies seven states and spans an area larger than California. In the late 1970s this aquifer supplied over a quarter of the ground water used for irrigation in the United States. By the early 1990s, however, severe depletion in many parts of the aquifer led to rising pumping costs, driving much irrigated agriculture in the region out of production.

Saudi Arabia is also pumping its fossil ground water aquifers far faster than they can be recharged because of the lack of alternative water sources, the availability of cheap energy, and a government decision to subsidize the domestic production of several crops, such as wheat, that could be grown elsewhere at far lower cost. For example, in 1992, the Saudi government paid over $2 billion in subsidies for the domestic production of 4 million tons of wheat – five times what the wheat would have cost on the world market.[15] The Saudis are now a major exporter of wheat, though they should not be considered a long-term reliable supplier, since production depends on ground water reserves that are being rapidly depleted. Ground water overdrafting is also widespread in many parts of India, China, Mexico, northern Africa, and the former Soviet Union. This unsustainable practice reflects both the urgent need for water in many regions and the failure of traditional economics to consider long-term, multi-generational interests when valuing certain non-renewable resources. Ultimately, these resources will be depleted, and future generations will be forced to make the difficult and expensive choices being avoided today.

**Energy for desalination.** Ninety-seven per cent of the water on the planet is too salty to drink or to grow crops. This has led to great interest in devising ways of removing salt from water in the hopes of providing unlimited supplies of fresh water. Despite the lack of technical obstacles to desalination, the high energy cost of these processes continues to make unlimited fresh water supplies an elusive goal. In the energy-rich arid and semi-arid regions of the world with a great discrepancy between water demand and water supply, desalination is an increasingly important option. For poorer countries, desalination continues to be too expensive to pursue on a large scale.

Total global desalting capacity at the end of 1989 exceeded 13.2 million m³ of fresh water per day produced from over 7,500 facilities.[*] Of this total, more than one-quarter of all capacity is in Saudi Arabia, followed by 12% in the United States, 10.5% in Kuwait, and 10% in the United Arab Emirates.[16] While desalination provides a substantial part of the water supply in certain oil-rich Middle Eastern nations, globally, desalination provides just one one-thousandth of total world fresh water use.[17]

The economics of desalination are directly tied to the cost of energy. The theoretical minimum amount of energy required to remove salt from a liter of seawater is 2.8 kJ. The best plants now operating use nearly 30 times this amount, though improvements in technology could reduce this to about 10 times the theoretical minimum.[18] Table 6.1 classifies desalination methods based on the type of energy required for the process. Today, desalinated water in the Middle East costs between $1 and $8 per m³ depending on the technology used (see Tables H-26, H-27, and H-35, in Part II), compared to between $0.01 and $0.05 per m³ paid by farmers in the western United States and about $0.30 per m³ paid by urban users.

Solar energy has been used directly for over a century to distill brackish water. When commercial plate glass began to be produced toward the end

[*]Not including small desalination plants on ships.

**TABLE 6.1** Classification of desalination methods based on energy type

| Energy type | Process |
| --- | --- |
| Thermal | Multiple-effect distillation |
| | Multiple-stage flash distillation |
| | Solar distillation |
| | Supercritical distillation |
| Mechanical | Vapor-compression distillation |
| | Freeze separation |
| | Hydrate separation |
| | Reverse osmosis |
| Electrical | Electrodialysis |
| Chemical | Ion-exchange |
| | Solvent extraction |

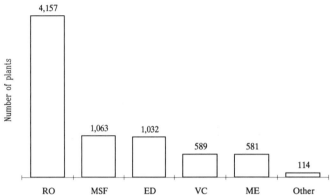

Fig. 6–1. Number of desalination plants worldwide, by process. There were 7,536 desalination plants worldwide at the beginning of 1990 (excluding shipboard units). The most common choice for desalination plants today is reverse osmosis, though multiple-stage flash distillation is still more common in large facilities. More than half of all reverse osmosis plants are in the United States, Saudi Arabia, and Japan. See Table H.32 for data on desalination plants. RO, reverse osmosis; MSF, multiple-stage flash distillation; ED, electrodialysis; VC, vapor-compression; ME, multiple-effect evaporation; "Other" includes freezing, hybrid processes, ultrafiltration, and other unspecified processes.

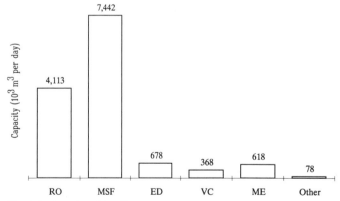

Fig. 6–2. Capacity of desalination plants worldwide, by process. By the beginning of 1990, worldwide desalination capacity exceeded 13 million m³ of fresh water per day for plants capable of producing more than 100 m³ per day. Nearly two-thirds of this capacity uses the multiple-stage flash distillation process; much of the remainder is produced using reverse osmosis techniques. Electrodialysis, vapor-compression, and multiple-effect evaporation each produce less than 1 million m³ per day. See Table H.32. RO, reverse osmosis; MSF, multiple-stage flash distillation; ED, electrodialysis; VC, vapor-compression; ME, multiple-effect evaporation; "Other" includes freezing, hybrid processes, ultrafiltration, and other unspecified processes.

of the 19th century, solar stills began to be developed. One of the first successful ones was built in 1872 in Las Salinas, Chile, which has few alternative sources of fresh water. This still covered 4,500 m², operated for 40 years, and produced just over 20 m³ of fresh water per day.[19] The largest solar desalination plant in operation by the end of 1991 was a 500-m³-per-day plant in the United Arab Emirates, which uses mirror technology to concentrate sunlight.

Some modern desalination facilities are now being run with electricity produced by wind turbines or other solar electric technologies, such as photovoltaics. The world's largest solar desalination plant under construction is a 2,000-m³-per-day system in Libya, designed to be powered by wind turbines.[20] Table H-34 in Part II lists solar desalination plants capable of producing more than 10 m³ of water per day.

Most commercial desalination methods take advantage of inexpensive fossil fuels. The principal techniques for desalinating water involve distillation, where water is evaporated from a saline solution and condensed as fresh water, and reverse osmosis, which separates water and salt ions using selective membranes. Approximately 70% of all desalination capacity uses some form of the distillation process, and most of the rest use membrane technologies. Figures 6.1 and 6.2 show the number and capacity of desalination plants worldwide, by type of desalination process. Tables H-30 to H-34 in Part II provide a broad overview of global desalination capacity.

The majority of distillation plants are installed in Saudi Arabia, Kuwait, and the United Arab Emirates; most reverse osmosis plants and vapor-compression plants are in the United States. Fifty-six per cent of the total installed or contracted capacity is based on multiple-stage flash distillation and 31% is based on reverse osmosis, but the trend over the last decade shows a steady shift toward the construction of reverse osmosis facilities. Sixty-five per cent of all desalination capacity is used to treat seawater and nearly 27% to treat brackish water. See Table H-33 in Part II.

Multiple-stage flash (MSF) distillation delivers high quality fresh water with a salt concentration of only 10 parts per million. Typical MSF systems consist of many evaporation chambers arranged in series, each with successively lower pressures and temperatures that cause sudden (flash) evaporation of hot brine, followed by condensation on tubes in the upper portion of each chamber. At present, distillation techniques require over 200 kJ to desalinate a liter of salt water, though improvements in techniques and increased efficiency of equipment may reduce this to less than 100 kJ/l.[21] (See Table G-22 in Part II for a summary of energy requirements of different desalination techniques.)

Multiple-effect (ME) distillation is one of the oldest and most efficient desalination methods. This approach reuses the heat of vaporization by placing evaporators and condensers in series and is based on the principle that vapor produced by evaporation can be condensed in a way that uses the heat of vaporization to heat brine at a lower temperature and pressure in the following chamber.

Reverse osmosis (RO) uses semipermeable membranes that pass water but retain salts and solids when a pressure difference is maintained across the membranes. The energy requirement for RO depends directly on the concentration of salts in the feedwater, and reverse osmosis facilities are most economical for desalinating brackish water. To desalinate a liter of seawater using RO facilities requires about 90 kJ; to desalinate brackish water requires far less, around 15 kJ/l.[22] The largest RO plant in the world at the beginning of 1990 was located in the United States at Yuma, Arizona. This plant was designed and constructed specifically to fulfill water quality obligations under an international treaty between the U.S. and Mexico on the Colorado River, and has a capacity of about 270,000 m³ per day.[23]

Electrodialysis (ED) depends on the natural ionization of salts in solution and uses membranes that are selectively permeable to ions (either cations or anions). With this method, brackish water is pumped at low pressure between flat, parallel, ion-permeable membranes, some of which allow cations and some of which allow anions to pass. Electric current flows across these parallel channels, pulling ions through the membranes. Like RO, the energy cost of electrodialysis rises with the concentration of the salts in the water. Desalinating brackish water with electrodialysis

requires about 36 kJ/l. Desalinating seawater with this technique requires nearly 150 kJ/l.[24]

Ion-exchange methods use resins to remove undesirable ions in water. For example, cation exchange resins are used in homes and municipal water treatment plants to remove calcium and magnesium ions in "hard" water. The greater the concentration of dissolved solids, the more often the expensive resins have to be replaced, making the entire process economically unattractive compared with RO and ED.

The use of freeze separation takes advantage of the insolubility of salts in ice. Water is frozen out of a saline solution, and the resulting pure ice crystals are then strained from the brine. The most efficient freeze methods use vapor-compression freeze separation systems. Freeze separation requires about 100 kJ to produce a liter of fresh water using present technology. Improvements are expected to be able to reduce this by about 40%.[25]

In the long run, the use of desalination to provide fresh water will be limited by the amount of energy required to purify salt water and by the cost of that energy. Unless major technical advances reduce overall energy requirements, or the price of energy drops substantially, desalination will always be limited to extremely water-poor and energy-rich regions.

## Water for energy

In addition to using energy when we manage water resources, water is required when we produce and use energy. In dry regions, the lack of water for cooling and chemical processes may lead to a decision to locate a power plant near a reliable source of water and to move the fuel instead, or to choose energy sources that require less water. For coal-fired plants, for example, where the weight of the water used for cooling alone is many times the weight of the coal burned, moving the coal to the water has distinct economic advantages.[26]

The amount of water needed to produce energy varies greatly with the type of facility and the characteristics of the fuel cycle. Fossil fuel, nuclear, and geothermal power plants require enormous amounts of water for fuel processing and cooling. Some of this water may be lost to evaporation or contamination; much of it is often returned to a watershed for use by other sectors of society. Solar photovoltaic power systems, wind turbines, and other renewable energy sources often require minimal amounts of water, though some renewable energy technologies are water intensive as well.

Water supply problems have already constrained energy production during periods of extreme shortage. During the severe drought in California between 1987 and 1991, large reductions in hydroelectricity production forced electric utilities there to purchase more fossil fuels than normal at an added cost of approximately $3,000 million to electricity consumers.[27] The decade-long drought in the 1980s in north-eastern Africa caused reductions in hydroelectric generation from the Aswan Dam in Egypt, which supplies nearly half of Egypt's electricity demand.[28] Most recently, Zimbabwe reported in February 1992 that its output of ethanol, which is mixed with gasoline to reduce the country's fuel imports, has been reduced because of the latest severe African drought, which has crippled sugarcane production.[29]

Large conventional fossil fuel power plants cannot be built in many regions of northern and north-eastern Africa because there are no reliable cooling water supplies. In western North America, the development of synthetic fuels from oil shales and tar sands is constrained as much by the limited availability of water as by the marginal economics and severe environmental limitations of these processes.[30] Even small energy developments in semi-arid or arid regions can dramatically affect water supplies. A proposal in the late 1970s for a small coal-gasification system in Southern California using ground water for cooling would have led to a drop in local ground water levels of over 15 cm per year (0.5 feet per year).[31]

Energy use also affects water quality in ways that depend on the characteristics of the energy source and the site. The discharge of waste heat from cooling systems raises the temperature of rivers and lakes, which affects aquatic ecosystems. Wastewaters from mining operations, boilers, and cooling systems may be contaminated with heavy metals, acids, organic materials, and suspended solids. Nuclear fuel production plants, uranium mill tailings ponds, and, under unusual circumstances, nuclear power plants, have all caused radioactive contamination of ground and surface water supplies.

All thermal electric generating facilities, whether they use nuclear, geothermal, fossil fuels, or even some solar sources of heat, convert water or other working fluids into steam or vapor to drive electric generating turbines. This vapor must be condensed in a cooling system in order to be recycled through the turbines. Many different cooling technologies are in use, including once-through circulation, wet and dry cooling towers, cooling ponds, and sprayers.

Once-through cooling has distinct economic advantages where sufficient fresh or salt water is available. In once-through cooling, large volumes of water are withdrawn from a river, lake, or aquifer (or the ocean), circulated through the cooling system, and discharged at a considerably higher temperature. Where water is scarce, or where the discharge of warm water is unacceptable, cooling towers are often used to increase evaporative cooling rates. The consumptive use of water in wet cooling towers is roughly twice that of once-through systems, though overall water withdrawals are considerably less.[32] Closed cycle cooling systems can reduce the total volume of water drawn through the cooling system by nearly 95% compared to the water required for once-through cooling.[33] Nevertheless, in regions without large bodies of water that can be used for once-through cooling, such as arid and semi-arid regions, consumptive use of closed cycle systems can be a much more serious problem.

Closed cycle cooling systems also entail some environmental costs not associated with once-through systems. Facilities that use ocean water for cooling can spread salt-bearing steam across nearby land, damaging agricultural capacity. Cooling towers are capable of causing local fogs and road ice under certain climatic conditions. And closed cycle cooling often involves bulding large cooling towers, which can be visible for miles and are often considered aesthetic liabilities.[34]

Total cooling water use by the electric industry in developed countries is substantial. In the United States in 1990, 270 km$^3$ of water (fresh and saline) were withdrawn for power plant cooling. This is almost half of all water withdrawn for human uses in the United States and nearly 40% of all fresh water withdrawals.[35] In some countries of Europe, such as the Netherlands, France, Germany, and Austria, even greater fractions of total water withdrawals go to power plant cooling.[36] It is important to note that this is water *withdrawn*, not water *consumed*. Three per cent of the water withdrawn by the electric utility sector in the U.S. is actually consumed. Where possible, salt water can be used for some cooling, but over two-thirds of all the water consumed by the United States electric industry for cooling is fresh water. This comprises about 5% of the total U.S. consumptive use of fresh water.[37]

All cooling systems also generate low quality wastewater – called ''blowdown'' when produced by cooling towers – that cannot be returned to the rivers or lakes without treatment. Cooling water returned to rivers and lakes is often at a much higher temperature than the water withdrawn from these water sources. Concern over the ecological impacts of this thermal pollution has led most industrialized nations to set some thermal

**TABLE 6.2**    Consumptive water use for energy production

| Energy technology | Consumptive use ($m^3/10^{12}$J(th) ) | Energy technology | Consumptive use ($m^3/10^{12}$J(th) ) |
|---|---|---|---|
| Nuclear fuel cycle | | Caustic injection | 100 |
| Open pit uranium mining | 20 | Carbon dioxide | 640[c] |
| Underground uranium mining | 0.2 | Oil refining (traditional) | 25–65 |
| Uranium milling | 8–10 | Oil refining (reforming and hydrogenation) | 60–120 |
| Uranium hexafluoride conversion | 4 | Other plant operations | 70[d] |
| Uranium enrichment: gaseous diffusion | 11–13[a] | Natural gas fuel cycle | |
| Uranium enrichment: gas centrifuge | 2 | Onshore gas exploration | Negligible |
| Fuel fabrication | 1 | Onshore gas extraction | Negligible |
| Nuclear fuel reprocessing | 50 | Natural gas processing | 6 |
| Coal fuel cycle | | Gas pipeline operation | 3 |
| Surface mining: no revegetation | 2 | Other plant operations | 100[d] |
| Surface mining: revegetation | 5 | Synthetic fuels | |
| Underground mining | 3–20[b] | Solvent refined and H–coal | 175 |
| Beneficiation | 4 | Lurgi with subbituminous | 125 |
| Slurry pipeline | 40–85 | Lurgi with lignite | 225 |
| Other plant operations | 90[d] | *In situ* gasification | 90–130 |
| Oil fuel cycle | | Coal gasification | 40–95 |
| Onshore oil exploration | 0.01 | Coal liquefaction | 35–70 |
| Onshore oil extraction and production | 3–8 | TOSCO II shale oil retorting | 100 |
| Enhanced oil recovery | 120 | *In situ* retorting of oil shale | 30–60 |
| Water flooding | 600 | Tar sands (Athabasca) | 70–180 |
| Thermal steam injection | 100–180 | Other technologies | |
| Forward combustion/air injection | 50 | Solar active space heat | 265 |
| Micellar polymer | 8,900[c] | Solar passive space heat | Negligible |

*Sources:* N. Buras, 1979, Water constraints on energy related activities, in *Proceedings of the Specialty Conference on Conservation and Utilization of Water and Energy Resources*, American Society of Civil Engineers, 8–11 August.

F.L. Shorney, 1982, Water conservation and reuse at coal-fired power plants, in *Proceedings of the Conference on Water and Energy: Technical and Policy Issues*, American Society of Civil Engineers, pp. 89–95.

C.E. Israelson, V.D. Adams, J.C. Batty, D.B. George, T.C. Hughes, A.J. Seierstad, H.C. Wang and H.P. Kuo, 1980, *Use of Saline Water in Energy Development*, Water Resources Planning Series UWRL/P-80-04, Utah Water Research Laboratory, Utah State University, Logan, Utah.

G.H. Davis and A.L. Velikanov, 1979, *Hydrological Problems Arising from the Development of Energy*, Unesco Technical Papers in Hydrology 17, Paris, France.

G.H. Davis 1985, *Water and Energy Demands and Effects*, Unesco Studies and Reports in Hydrology 42, Paris, France.

US Department of Energy, 1980, *Technology Characterizations: Environmental Information Handbook*, DOE/EV–0072, Washington, DC.

California Department of Water Resources, 1981, *The Availability of Water for Emerging Energy Technologies for the California Region*, US Water Resources Council report, Water Assessment Report Section 13(a), Sacramento, California.

P.H. Gleick, 1992, Environmental consequences of hydroelectric development: The role of facility size and type, *Energy: The International Journal*, Vol. 17, No. 8 pp.735–747 (Pergamon Press Ltd.).

[a]  Excluding water use by additional power plants required for the energy-intensive uranium enrichment process.
[b]  Top end of range reflects once-through system with no recycle.
[c]  Median of a wide range.
[d]  Other plant operations includes plant service, potable water requirements, and boiler make-up water. For coal facilities, this also includes ash handling and flue gas desulfurization process make-up water.

**Essays on fresh water issues**

TABLE 6.3 Consumptive water use for electricity production

| Energy technology | System efficiency[a] (%) | Consumptive use ($m^3/10^3$ kWh) |
|---|---|---|
| Conventional coal combustion | | |
|   Once-through cooling | 35 | 1.2 |
|   Cooling towers | 35 | 2.6 |
| Fluidized bed coal combustion | | |
|   Once-through cooling | 36 | 0.8 |
| Oil and natural gas combustion | | |
|   Once-through cooling | 36 | 1.1 |
|   Cooling towers | 36 | 2.6 |
| Nuclear generation (LWR) | | |
|   Cooling towers | 31 | 3.2 |
| Nuclear generation (HTGR) | | |
|   Cooling towers | 40 | 2.2 |
| Geothermal generation (vapor-dominated) | | |
|   Cooling towers (Geysers, US) | 15 | 6.8 |
|   Once-through cooling (Wairakei, NZ) | 7.5 | 13 |
| Geothermal generation (water-dominated) | | |
|   Cooling towers (Heber, US) | 10 | 15 |
| Wood-fired generation | | |
|   Cooling towers | 32 | 2.3 |
| Renewable energy systems | | |
|   Photovoltaics: residential | | Negligible |
|   Photovoltaics: central utility | | 0.13[b] |
|   Solar thermal: Luz system | | 4 |
|   Wind generation | | Negligible |
|   Ocean thermal | | No fresh water |
| Hydroelectric systems[c] | | |
|   United States (average) | | 17 |
|   California (median of range) | | 5.4 |
|   California (average of range) | | 26 |

*Sources:* US Department of Energy, 1980 *Technology Characterizations: Environmental Information Handbook*, DOE/EV–0072, Washington, DC.
G.H. Davis, 1985, *Water and Energy: Demands and Effects*, UNESCO Studies and Reports in Hydrology 42, Paris, France.
R.L. Ottinger, D.R. Wooley, N.A. Robinson, D.R. Hodas and S.E. Babb, 1990, *Environmental Costs of Electricity*, Pace University Center for Environmental Legal Studies, Oceana Publications, Inc., New York.
P.H. Gleick, 1992, Environmental consequences of hydroelectric development: The role of facility size and type, *Energy: The International Journal*, Vol. 17, No. 8, pp. 735–747 (Permagon Press Ltd.).

[a] Efficiency of conversion of thermal energy to electrical energy.
[b] Maximum water use for array washing and potable water needs.
[c] Assumes all evaporative losses are attributable to the hydroelectric facilities. For reservoirs with significant non-hydroelectric uses, such as recreation and flood

Fig. 6–3. World electrical generation by source, 1989 (values in $10^3$ GWh). The majority of the electricity used in the world is produced from the combustion of oil, coal, and natural gas. Hydroelectric facilities rank second, producing slightly more electricity than nuclear. Geothermal, solar, biomass, wind, and other resources account for the remaining electricity generation. See Table G.3 for data. (□), fossil fuel; (▤), hydroelectric, (▨), nuclear; (■), geothermal and others.

limits to protect the environment.[38] In the future, the increased use of closed systems to minimize thermal pollution will increase overall consumptive water use.

The following sections estimate water consumed per unit energy produced for a wide range of energy facilities and fuels used. The total volume of water withdrawn for use often far exceeds water consumed. Both measures can be important: where total water availability is scarce, large volumes of water may simply not be available on a reliable basis for withdrawal by power plants, even if total consumptive use is low. In all regions, however, the consumptive use of water is a true measure of the quantity of water made unavailable for any other uses in a region. Tables 6.2 and 6.3 summarize water consumed per unit energy and electricity for a variety of commercially available energy sources.

**Hydroelectricity.** The most obvious use of water for the production of energy is in hydroelectric facilities, where the energy in falling water is used directly to turn turbines, which generate electricity. In many areas of the world, water is also used to do mechanical work, such as grinding grain.

At the beginning of the 1990s, there were over 610,000 MWe of installed hydroelectricity capacity worldwide, 24% of total world electrical generating capacity (see Figure 6.3). Hydroelectric potential has been unevenly developed around the world. Over half of all hydro capacity is in North America and Western Europe; only 3% of it is in Africa. Table 6.4 lists hydroelectric capacity and generation by continent for 1990. North America and Europe have developed approximately 60% and 36%

TABLE 6.4 World hydroelectric capacity and generation, 1990

| Continent[a] | Installed hydroelectric capacity ($10^3$ MW) | Per cent of total | Hydroelectric generation ($10^6$ MWh per year) | Per cent of total |
|---|---|---|---|---|
| North America | 156.8 | 26 | 599.6 | 28 |
| Central and South America | 80.3 | 13 | 353.4 | 17 |
| Western Europe | 155.0 | 25 | 444.7 | 21 |
| Eastern Europe | 15.1 | 2 | 26.3 | 1 |
| Soviet Union | 64.4 | 10 | 217.3 | 10 |
| Middle East | 3.1 | 1 | 12.6 | 1 |
| Africa | 18.9 | 3 | 43.2 | 2 |
| Far East and Oceania | 121.3 | 20 | 415.8 | 20 |
| Totals | 614.9 | 100 | 2,112.9 | 100 |

*Source:* US Department of Energy, 1992, *International Energy Annual 1990*, Energy Information Administration, DOE/EIA–0219(90), Washington, DC.

[a] Continental sums use the country assignments of the original source. Since 1990, substantial changes in Eastern Europe and the Soviet Union have occurred.

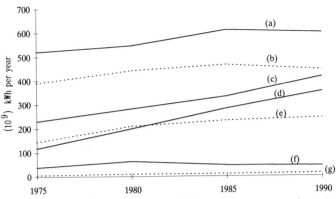

Fig. 6.4. Hydroelectric generation by region, 1975–1990. Electricity generated by hydropower facilities is shown for major regions of the world from 1975 to 1990. North America and western Europe still lead the world in hydroelectric generation in 1990, although their total hydroelectric generation is declining slightly. Hydroelectric generation for Asia and Latin America has increased rapidly in recent years. See Table G.2 for data. (a) North America, (b) western Europe, (c) Asia and Oceania, (d) Central and South America, (e) eastern Europe and the Soviet Union, (f) Africa, (g) the Middle East.

of their large-scale hydropower potential, respectively; Asia and Latin America have harnessed around 10%, and Africa only 5%.[39]

In 1990, hydroelectric dams generated over 2 million GWh of electricity, or just under 7% of the world's primary commercial energy and 20% of global electricity. In South and Central America, 70% of all electricity comes from hydroelectric plants; Canada and the United States together provide 20% of their total electrical demand with hydropower.[40]

Global hydroelectricity production increased more than 20% during the 1980s. In the industrialized nations, however, the development of new hydroelectricity facilities has slowed greatly as the best sites have been developed and as the environmental costs of further construction rise. Indeed, the greatest development of hydroelectric facilities is now occurring in those regions that have seen little development to date. During this same 10-year period, hydroelectric production increased 50% in Asia and more than doubled in parts of Latin America and China.[41] Figure 6.4 shows the trend in hydroelectric generation, by continent, between 1975 and 1990.

Hydropower facilities have a variety of effects on fresh water systems. The creation of a reservoir displaces wildlife and replaces a flowing water ecosystem with a standing water one. The storage of water in a reservoir leads to consumptive water losses from evaporation and seepage. Hydroelectric dams are subject to the risk of catastrophic failure with extensive loss of life and property, an unusual risk associated with few other energy sources, most notably nuclear power. And when hydroelectric facilities are developed on rivers that are shared by two or more nations, political conflicts can arise.[42]

Disputes have arisen over dams throughout the world, including the Farakka Barrage in India, the Itaipú Dam on the Paraná River between Argentina, Brazil, and Paraguay, the Kumgansan hydroelectric dam on the Han River between North and South Korea, the Aswan High Dam in Egypt, which displaced thousands of people living in the Sudan, and many other facilities.[43] Conflicts over water resources and water systems are discussed further in Chapters 8 and 9.

In evaluating and comparing all of these effects, a number of criteria must be evaluated, including the size and type of hydroelectric plant, the temporal and spatial distribution of harm, the possibility of irreversible effects, the coincidence of risks and benefits, and the uncertainty surrounding the nature of the evidence of environmental harm.

Different environmental analysts have quite different interpretations of both the magnitude and extent of the environmental costs of hydroelectric facilities. Some believe that hydropower is a benign alternative source of electrical generation; others have concluded that new large dams may be "arguably the worst electricity option in terms of damage to ecosystems

per unit of electricity."[44] These issues have been widely discussed over the last decade, but many issues remain unresolved.[45]

The greatest consumptive use of water resources from hydroelectric facilities comes from the evaporative loss of water from the surface of reservoirs, though these losses are often left out of environmental assessments. This water represents the loss of a resource that would otherwise be available for downstream human and ecological uses. Conflicts for water among agricultural users, industrial users, commercial users, and ecological support functions are intensified by this water loss. As water has become more precious in different regions, evaporation from artificial reservoirs has received more and more attention.

The evaporation of water is directly related to the surface area of the body of water and varies with the temperature, wind conditions, and humidity of a region. Average annual evaporation from standing water in the United States varies from 0.5 m in north-eastern regions to over 1 m in the desert regions of the south-west. For 700 reservoirs and regulated lakes in 17 western U.S. states with a total effective surface area of approximately 14,000 km$^2$, annual evaporative losses based on variations in location and climate were approximately 15.2 km$^3$, or an average annual evaporation of 1.1 m.[46] Evaporative losses can be much higher, depending on the climate of a region. At the Aswan High Dam on the Nile, about 14 km$^3$ of water (11% of reservoir capacity) are estimated to evaporate annually from a surface area of 5,200 km$^2$.[47] This is equivalent to nearly 3 m of evaporative loss per year.

A recent study estimated evaporative losses from a set of 100 diverse California hydroelectric facilities at between 0.04 and 200 m$^3$ of water per 10$^3$ kWh electricity produced, with a median estimate of 5.4 m$^3$ of water per 10$^3$ kWh†. This is several times larger than the consumptive water use of nuclear or fossil fuel facilities per unit energy produced.[48] Evaporative losses from all U.S. reservoirs are approximately half this amount, because the evaporative losses in the more semi-arid western U.S. are higher than in other parts of the country.

Differences in evaporative losses also result from the type and size of the hydroelectric plant. Table G-19 in Part II breaks down evaporative losses for plants over and under 25 MWe installed capacity and for differences in the ratio of gross static head (GSH) to dam height (DH).** In general, where the GSH exceeds the DH – typical of dams with long penstocks carrying the water downslope to a power plant – the median water losses are smaller than where GSH is smaller than the DH – typical of large dams with powerhouses at their base. When the size of the power plant is considered, plants under 25 MWe lose more water to evaporation than plants larger than 25 MWe per unit energy generated: 14 m$^3$ per 10$^3$ kWh as opposed to 2.5 m$^3$ per 10$^3$ kWh.[49]

Seepage losses from porous foundations underlying hydroelectric reservoirs can also lead to a consumptive use of water. It has been estimated that an average of 5% of the volume of reservoirs is lost annually to seepage,[50] and seepage losses at some facilities have become big political and environmental problems. The Anchor Dam in Wyoming, for example, is built in a location that is so porous that the reservoir has never totally filled in 30 years of operation.[51]

Evaluating data on reservoir storage in the 18 hydrographic regions of the United States reveals that there are approximately 210 km$^3$ of storage in hydroelectric reservoirs and over 2,100 km$^3$ of storage in 49,000 reservoirs. Hydroelectric production from these facilities averages about 290,000 million kWh annually. Assuming seepage losses of about 5% annually from these reservoirs, losses of 36 m$^3$ per 10$^3$ kWh can be expected. For the California reservoirs studied, about 40 m$^3$ of water are lost to seepage for every 10$^3$ kWh of hydroelectricity produced, in good agreement with the overall U.S. estimate.[52]

Seepage and evaporative losses have an important qualitative differ-

---

†Water use numbers are presented here as either m$^3$ per 10$^{12}$ J(th) – the number of cubic meters required per trillion joules of thermal energy – or as m$^3$ per 10$^3$kWh – cubic metres per thousand kilowatthours of electricity produced. The two are not strictly comparable without making assumptions about the efficiency of conversion of thermal to electric energy.

**The gross static head is the vertical distance from the surface of the reservoir to the top of the water in the tailrace, below the dam. It effectively represents the distance the water falls during electricity production.

ence. Water lost to evaporation usually leaves the hydrologic basin and thus is a true loss. Water lost to seepage remains in the basin and may become available downstream or for ground water pumping.

Calculating consumptive water requirements for hydroelectric facilities is complicated by the multiple-use nature of many dams and reservoirs, and by the way they are operated. The mode of operation itself is determined by a set of economic criteria, together with competing upstream and downstream demands for the water. On occasion, the mode of operation is determined by requirements set to maintain certain ecological conditions, such as temperature or flow rates needed to support fish populations. All of these factors need to be considered in evaluating overall water requirements.

**Coal.** Many parts of the coal fuel cycle are water intensive, including coal mining, reclamation of mined land, and coal combustion, which requires substantial water for cooling, ash handling, and waste disposal. Coal mining operations, particularly underground mining, can lead to the contamination of large volumes of water. Water draining from underground coal mines contains minerals and heavy metals, and is usually highly acid. Some 12,000 km of streams in the eastern U.S. are seriously polluted from underground coal mining.[53] Surface mining also causes significant water quality problems by increasing sediment transport in streams and increasing dissolved mineral content if soluble minerals are exposed during mining.

Estimates of average water use in underground coal mining vary from between 3 and 20 $m^3$ of water per $10^{12}$ J(th) of energy in the coal. About 2 $m^3$ per $10^{12}$ J(th) are required to produce coal in surface mines if no revegetation is required. These estimates include water used for disposing of mining wastes.[54] The greater water use in underground mining arises, in part, from water used for suppressing dust for health and safety reasons. No good estimates of the volume of water contaminated in mining are available, per unit energy in mined coal.

Some countries, such as the United States, require land disturbed by surface strip mining to be "reclaimed" after mining, i.e., to be restored to its approximate original contour and vegetation. In such circumstances, water is required to establish vegetation on reclaimed land. The amount of water needed depends on the natural climatic conditions in a region. In the semi-arid western United States, where much of the strip mined coal originates, nearly 3 $m^3$ per $10^{12}$ J(th) are needed to establish vegetation on reclaimed land.[55]

Following mining, coal is often "refined" to separate coals of different quality and to increase the thermal performance of the fuel. Such refining, including washing, beneficiation (which removes non-fuel contaminants), and thermal processing may severely degrade water with organic and inorganic impurities. Any water used for these purposes will require either additional treatment or isolation from the environment. The volumes of water used in this manner are small, typically under 5 $m^3$ per $10^{12}$ J(th).[56]

One method for transporting coal is the slurry pipeline, which moves large quantities of coal suspended in water. Such pipelines require enormous amounts of water (typically equal volumes of coal and water) and in some cases this means exporting water from regions of existing water scarcity, such as the arid western U.S., where coal is often mined.

Several coal slurry pipelines have already been built. The largest is the Black Mesa project in the United States, which transfers 5 million tonnes of coal per year over 400 km from mines in Arizona to the 1,500 MWe Mojave power plant in southern Nevada. Water for suspending the coal is supplied by wells pumping 4 million $m^3$ per year from ground water aquifers (this is equivalent to about 45 $m^3$ per $10^{12}$ J(th) energy in mined coal). Recharge of these aquifers is negligible compared to this rate of withdrawal; hence this is a non-renewable use of water. After the coal is taken out of the slurry at the power plant, some of the water is treated and used for other plant operations, including power plant cooling. At the Mohave plant, about one-seventh of the total cooling demand is supplied using recovered water from the slurry pipeline.[57]

A number of other slurry pipelines have been proposed in recent years, including the Energy Transportation Systems, Inc. (ETSI) line, which was to have delivered coal from South Dakota to Oklahoma and Arkansas. For a variety of economic and environmental reasons, this project has

now been cancelled. Other pipelines would link Colorado and California, Wyoming and Texas, and Virginia and various south-eastern states. The longest pipeline proposed would have carried 25 million tonnes of coal per year over 2,000 km. Total consumptive water use by slurry pipelines has been estimated at between 40 and 85 $m^3$ per $10^{12}$ J(th).[58]

Once it has been mined, processed, and transported to a power plant, coal can be turned into useful energy using many different processes: gasification, liquefaction, and combustion in a variety of direct-fired systems such as fluidized beds and pulverized boilers. The amounts of water used per unit energy produced by these different processes vary considerably, as shown in Table 6.2.

The most typical form of coal-fired power plant burns coal directly to generate steam, which is then used to drive a turbine to produce electricity. The overall efficiency of the system depends on the pressure at the outlet of the turbine. Reducing this pressure to increase efficiency is accomplished with a cooling system, which accounts for the greatest consumption of water in the entire traditional fuel cycle as described above. Only some of the synthetic fuel processes exceed this intensive water use. For coal-fired power plants, consumptive use for cooling ranges from just under 1 $m^3$ per $10^3$ kWh(e) for once-through cooling systems to over 2.5 $m^3$ per $10^3$ kWh(e) for facilities with cooling towers.

Additional water is used in coal facilities for dust suppression, drinking and sanitation, ash handling, as flue gas desulfurization make-up water, and for other plant operations. These demands total approximately 90 $m^3$ per $10^{12}$ J(th).[59]

**Oil and natural gas.** In the past, the production of oil and natural gas has required relatively modest amounts of water. Water is used during the exploration and drilling process, for treating the oil or gas before use, and for human sanitation and drinking water. Overall, between 2 and 8 $m^3$ per $10^{12}$ J(th) of water have historically been required to extract oil, including water for drilling, flooding, and treating. One source indicates that 45 million $m^3$ of fresh water are used annually in the United States for mixing drilling mud to produce about 500 million tonnes of oil (about 2 $m^3$ per $10^{12}$ J(th) of oil produced).[60] Oil production also results in the simultaneous production of large quantities of saline water found with the oil. This water must be disposed of safely. Drilling for natural gas only requires water for preparing drilling fluid.[61]

As the largest fossil fuel reservoirs have been drawn down, however, methods of increasing the percentage of these fuels recovered from wells have been developed. These secondary and tertiary recovery methods increase overall water requirements. In particular, some of the most common and effective oil recovery techniques are water intensive. Secondary recovery uses water flooding to increase the flow of oil to the wells. One-third of all U.S. and Canadian oil production uses water flooding recovery methods. The most widely used tertiary recovery technique is steam injection, where steam is pumped into the depleted oil field and the heat increases oil flow and recovery rates. This technique is used for three-quarters of all U.S. tertiary oil recovery today.[62] Several other enhanced oil recovery methods are in use, with widely varying water requirements ranging from under 100 to 9,000 $m^3$ per $10^{12}$ J(th) of oil recovered,[63] listed in Table 6.2. This water use is entirely consumptive, though in coastal regions salt water may be used for some of these processes.

After oil is extracted, it must be refined into different forms of liquid fuel. Average water withdrawals for traditional refining facilities in industrialized countries are about 325 $m^3$ per $10^{12}$ J(th) of crude oil input; consumptive use ranges from 25 to 65 $m^3$ per $10^{12}$ J(th) of oil input. Of this consumptive loss, 70% is lost in evaporative cooling, 26% is boiler feed water, and the rest goes to other in-plant uses. Recent changes in fuel formulations and improvements in techniques for restructuring organic molecules have increased water requirements, since hydrogen, obtained by dissociating water, is used to upgrade the quality of the product in a process called hydrogenation. Refineries where substantial hydrogenation and reforming take place use between 60 and 120 $m^3$ per $10^{12}$ J(th).[64]

The generation of electricity by burning oil or natural gas also requires water for cooling. Because the thermal efficiency of these plants is comparable to that of coal-fired plants, about the same amount of water is used for cooling – between 1 and 2.6 $m^3$ per $10^3$ kWh(e).

**TABLE 6.5**  Water consumed by oil shale production

| Processes[a] | Water consumption ($m^3/10^{12}$ J(th) ) |
|---|---|
| Processed shale disposal[b] | 40 |
| Shale ore upgrading | 21 |
| Power requirements | 9 |
| Associated urban | 8 |
| Retorting | 7 |
| Mining and crushing | 6 |
| Revegetation | 2 |
| Sanitation | <1 |

*Source:*  G.H. Davis and A.L.Velikanov, 1979, *Hydrological Problems Arising from the Development of Energy: A Preliminary Report*, Unesco Technical Papers in Hydrology 17, Paris, France.

[a]  Assuming an oil shale mine, retort, and upgrading plant producing 16,000 $m^3$ per day of oil shale. Such a plant would require about 57,000 $m^3$ per day of water, or about 95 $m^3$ water per $10^{12}$ J(th) oil shale produced.

[b]  Assuming spent shale contains about 20% moisture content by weight.

**Oil shale and tar sand.**  Large resources of fossil fuels are bound up in shale or tar sands, but the economic and environmental costs of extracting them are too high at present for any significant commercial operations to proceed. If oil shale or tar sands are to become commercially successful, they are likely to be extracted using surface mining or underground mining methods. In situ retorting processes, where the fuel is separated from ore in place, have also been developed. While these methods have some environmental advantages, they are not economically or technically competitive with more traditional extraction processes.

Synthetic fuels production has the potential to consume vast amounts of water; ironically, the greatest deposits of the fuels tend to occur in regions with scarce natural water supplies. Estimates of the volume of consumptive water demand for oil shale and tar sands production range widely depending on the process.[65] The largest uses of water for both oil shales and tar sands come from waste disposal, processing, power generation, and land reclamation. Because of the low oil content of shales and tar sands, extremely large volumes of the raw material are needed in order to produce commercial volumes of oil. For example, to produce just over 400,000 tonnes of oil a day (3 million barrels of oil a day), over 3 million tonnes a day of raw shale would have to be mined and processed, considerably more material than the total daily production of coal in the United States in 1990. And the energy content of this shale oil per unit of raw material mined is far less than the energy content of the coal produced.

The only operating synthetic fuel facility of commercial size using tar sands or shale oil is at Athabasca, Canada, where tar sands are mined and processed. Water consumption at this facility under normal operation is about 8 tonnes of water for every tonne of final product, or 180 $m^3$ per $10^{12}$ J(th). Larger and more efficient facilities could reduce this consumptive use somewhat.[66] An estimate for oil shale suggests that from 2.5 to 4 times as much water is used as oil is produced, the equivalent of 70–100 $m^3$ of water per $10^{12}$ J(th) of oil.[67] Table 6.5 shows the consumption of water for different aspects of the production of oil shale. By far the majority of this water goes for processing the shale, for cooling, and for disposing of the residual ore and slurry.

Huge volumes of waste are produced from both oil shale and tar sands facilities. These wastes must be handled extremely carefully to avoid contamination of downstream water resources. For both energy resources, the waste products are composed of 50% or more water and occupy more volume than the original ore, requiring that large areas of land be set aside for disposal. Additional dry waste material is used in the construction of retaining dikes, as much as 100 m high, for the liquid tailings produced. Recycling processes to remove suspended solids from the water may ultimately permit reuse of some part of the wastewater, but technical problems have prohibited this so far.

Retorting oil shale in place typically requires the same amount of water as refining crude oil – about 1–2.5 volumes of water per volume of petroleum input or 25–70 $m^3$ per $10^{12}$ J(th) of product. The environmental impacts on water resources of oil shale retorting include the generation of inorganic and organic pollutants and the generation of thermal pollution, but no good estimates of the volumes of water contaminated are available.

**Nuclear power.**  Nuclear electricity provides 12% of total global electrical demand, and a considerably higher fraction in several industrialized nations. France, Belgium, and South Korea, for example, all supply more than 50% of their total electrical needs with nuclear plants.[68] As with other large thermal plants, the greatest use of water in the nuclear fuel cycle is for power plant cooling, and many nuclear plants are located in coastal regions to permit the use of salt water. Other aspects of the fuel cycle also require water, and consumptive uses are summarized in Tables 6.2 and 6.3.

Uranium mining requires water for dust control, ore beneficiation, and revegetation of mined surfaces. The quantity of water required is approximately the same as for surface mining of coal, about 20 $m^3$ per $10^{12}$ J(th) energy in the ore. The mining of uranium also causes the mobilization of radioactive minerals that may reach waterways and pose a health hazard. Waste ore from mining and processing activities is often disposed of in evaporation ponds that threaten surface and ground water quality, which in turn can have direct human and ecological health implications.

Milling, refining, and enriching uranium entail the second largest consumptive use of water in the nuclear fuel cycle. An early estimate of water consumption in the nuclear fuel cycle suggested that about 10 $m^3$ per $10^{12}$ J(th) is consumed in the milling stage, almost entirely as evaporation from tailings ponds. Another 1.2 $m^3$ per $10^{12}$ J(th) is consumed during the production of uranium hexafluoride and reprocessing of used fuel.[69] The principal method for enriching uranium is gaseous diffusion, which requires an additional 10–15 $m^3$ per $10^{12}$ J(th). Most of this water is consumed by evaporative cooling. Alternative methods of enrichment, such as centrifuge separation, require considerably less water but are not in widespread use. Water is also required at power plants that provide energy for uranium enrichment, which is extremely energy intensive. Including these water requirements would increase overall consumptive use by an additional 20 $m^3$ per $10^{12}$ J(th).[70]

The current generation of nuclear plants is less efficient than fossil fuel plants because of technological characteristics, restrictions on maximum steam temperatures, and because fossil fuel plants emit substantial waste heat through the flue gases. A typical nuclear plant operating at 31% efficiency requires much more water for cooling than a comparably sized fossil fuel plant, as shown in Table 6.3. High temperature gas reactors (HTGRs) or other high efficiency designs, which may operate at 40% efficiency, would reduce consumptive water use per unit electricity produced to approximately the level of present oil- and gas-fired facilities.

Finally, low-probability, but high-consequence accidents associated with nuclear plants will also have an impact on water resources. The meltdown or burning of a reactor core could result in long-term radioactive poisoning of land and water supplies. The accident at Chernobyl in the Ukraine in 1986, for example, led to the contamination of nearby lakes and ground water aquifers. The extent and severity of this contamination is not yet fully known or reported.

**Geothermal.**  Two forms of geothermal resources are currently considered technologically and economically feasible: vapor-dominated dry-steam systems and liquid-dominated hot-water systems. Considerable geothermal energy potential exists, though total development has been limited. Where the heat resource lies fairly close to the surface, and is sufficiently hot, geothermally-produced electricity can be economically attractive compared to fossil fuels.

Vapor-dominated systems consist of wells drilled into a steam field. Steam is then used to drive a turbine generator to produce electricity. Nearly 2,000 MWe of dry-steam systems are in operation, mostly in the Geysers region of California, in the United States. At the Geysers, no outside source of cooling water has been required because water condensed from the geothermal steam condensate is used for cooling. Where outside cooling water is necessary, between 7 and 13 $m^3$ per $10^3$ kWh(e) output will be required for vapor-dominated systems.[71]

Several forms of liquid-dominated geothermal systems are in use, including flash-steam systems and binary systems. The temperature of the geothermal fluid determines which technology is appropriate for use in producing electricity. At the present time, flash conversion is the simplest and least costly of liquid-dominated systems. In flash-conversion systems, a high temperature geothermal fluid is brought to the surface under pressure, where it "flashes" into steam to drive a turbine. Such systems are in use in Italy, Iceland, Mexico, New Zealand, the Philippines, and the United States. Flash geothermal systems also use geothermal condensate for cooling whenever possible, minimizing outside water requirements.[72]

Binary systems, in which low-temperature (150° C) geothermal fluid is used to vaporize a working fluid, are closed and non-polluting, since all geothermal fluids are reinjected down into the well field. This, however, creates the need for an outside source of cooling water. One estimate is that up to 15 $m^3$ of cooling water are required per $10^3$ kWh(e) output for water-dominated systems, such as at Heber in California.[73] For all plants, some additional water is required for fire protection, facility maintenance, landscaping, and sanitation.

**Solar thermal power stations.** Energy from the sun can be used to heat and vaporize water or another working fluid in order to produce electricity. Among the different designs for such systems are centralized utility solar power towers that use mirrors to focus sunlight onto a boiler, and individual concentrating collectors with tubes of a working fluid located at the collector's focal point. As with other power plants, the working fluid must then be condensed and reused. Estimates of water consumption, which include make-up cooling water and water for washing the mirrors, range considerably depending on the type of facility. Most published estimates of water consumption are low, around 1 $m^3$ per $10^3$ kWh(e). Water consumption at the 10 MWe power tower built in southern California, for example, was estimated at only 0.1 $m^3$ per $10^3$ kWh(e) output, though there are questions about the reliability of this estimate.[74] The consumptive water use of the most advanced, commercially available system, built by the Luz Corporation in southern California, is considerably higher. This project now consists of over 300 MWe, with fields of mirrors directing sunlight onto tubes of a working fluid at each collector's focal point. The Luz plants, which operate reliably, consume over 4 $m^3$ per $10^3$ kWh(e), primarily to operate the condensers and cooling towers.[75]

Water requirements for solar ponds are likely to be extremely high since the ponds will be built in regions with high evaporative loss rates. One estimate for make-up water for cooling and evaporative losses from solar ponds is over 25 $m^3$ per $10^3$ kWh(e).[76]

**Photovoltaics.** Electricity can be generated directly from sunlight using photovoltaic cells. Water use for photovoltaic electricity production is considered negligible.[77] No estimates are available of water consumed in the manufacturing process, but such cells can be made in water-rich regions and shipped anywhere for the production of electricity. Some minor volumes of water may be required for periodic cleaning of photovoltaic arrays.

**Wind energy.** Wind energy facilities require no water for the production of electricity and almost none for the construction and erection of the wind turbines. As with photovoltaic cells, wind turbines can be fabricated anywhere and set up in regions with energy demand and sufficient wind resources.

**Summary.** Water is required for practically every aspect of energy production and use. Because of the wide variations in energy fuel cycles and choices around the world, no overall water requirements for each energy source have been provided here, and great care should be taken in simply summing up water requirements for different aspects of each energy technology. In addition, water use in many portions of the energy cycle is poorly understood or quantified at present. Our knowledge of the severity and extent of water contamination by different fuel cycle activities, for example, is especially limited. Similarly, no data on water use in energy transportation are available, except for coal slurry pipelines and natural gas pipeline operation, and no data on facilities construction are included.

Despite these uncertainties, water requirements for the production and use of some forms of energy are substantial. In water-poor regions, there

may simply be insufficient water to support the cooling needs of conventional fossil fuel power plants, for example, or the emergency cooling requirements of a nuclear plant. In these regions, particular care must be taken in choosing and building energy systems.

## Conclusions

The supply and use of both water and energy resources are intricately connected. The production of useful energy and electricity uses and contaminates fresh water resources. The pumping, cleaning, and transport of fresh water requires the use of energy. In the last few decades, we have begun to run up against physical and environmental limitations in the availability and use of both resources. We can no longer consider the formulation of rational energy policy and water policy to be independent.

Gross inequities in energy use between developed and developing nations will have to be addressed in the near future. Individuals in developed countries of the world – less than one-quarter of the world's population – use an average of 7.5 kW per person. The other three-quarters of the world's population in the developing world use far less, about 1.1 kW per person, for a total global energy use of nearly 14 terawatts ($10^{12}$ W) in 1990, and these averages hide even larger discrepancies between the richest and the poorest people. Yet supplying even this total amount of energy is severely straining the planet's environmental, technological, and managerial resources, without successfully meeting crucial human needs.[78] Even assuming great progress in energy efficiency and the closing of the gap in energy use between the rich and the poor, providing 9,000 million people in the middle of the next century with an average of 3 kW each would require doubling today's global energy use.

Increasing energy use in developing countries through traditional expansion of fossil fuel combustion will lead to severe environmental problems and enormous increases in the consumptive use of water. At the same time, improvements in health, economic conditions, and our overall quality of life will require better access to clean drinking water, sanitation services, and water for other activities.

Where water supplies are plentiful, energy planners have a wide choice of technologies and sites, and decisions about the form and size of energy facilities ultimately rest on other economic, environmental, and social factors. Where energy is plentiful and inexpensive, such as in parts of the Middle East, water planners have greater flexibility in developing water supplies, including the possibility of long-distance transfers and the use of desalination facilities.

Where water or energy is partially or seriously limited in quantity, choices about both energy futures and water supplies will be far more difficult and constrained. In arid and semi-arid regions especially, water availability will play a central role in defining energy choices, and the cost of energy will limit options for providing fresh water.

Where water quality criteria are stringent, a different set of concerns apply. Energy developers must carefully select systems that can meet necessary limits. For example, limits on thermal discharges will require planners to consider alternatives with modest thermal wastes or with closed–cycle cooling systems. Processes that produce significant quantities of hazardous, chemical, or solid wastes will be at a competitive disadvantage with cleaner facilities and in water-short regions will have to meet environmental restrictions and minimize consumptive water use.

There are some bright spots, however. Many of the renewable energy sources, such as photovoltaics and wind generation, require far less water per unit energy produced than do conventional systems. In water-short regions, sources of energy with low water requirements may increasingly be the systems of choice. At the same time, the growing recognition of the serious problems associated with long-distance water transfers and massive ground water pumping have increased the amount of attention given to alternative water supply options, including water trading and marketing, technological improvements in water use efficiency, proper water pricing, and the elimination of subsidies. These approaches have the potential to reduce the pressure on our requirements for both water and energy.

**Notes**

1. H. Rouse and S. Ince, 1957, *History of Hydraulics*, Iowa Institute of Hydraulic Research, State University of Iowa. Other sources say the earliest known dam across a river, the Sadd el Kafara, was built more than 5,000 years ago in the Middle East. See, for example, A. Mascarenhas, 1985, Rationale for establishing a river basin authority in Tanzania, in J. Lundqvist, U. Lohm and M. Falkenmark (eds.) *Strategies for River Basin Management*, D. Reidel, Dordrecht, pp. 317–328.

2. Quoted in M.S. Drower, 1954, Water-supply, irrigation, and agriculture, in C. Singer, E.J. Holmyard and A.R. Hall (eds.) *A History of Technology*, Oxford University Press, New York.

3. For a description, see A.K.S. Lambton, 1989, The origin, diffusion, and functioning of the qanat, in P.Beaumont, M. Bonine and K. McLachian (eds.) *Qanats, Kariz, and Khattara: Traditional Water Systems in the Middle East and North Africa*, University of London, London, pp. 5–34; also T. Naff, University of Pennsylvania, personal communication, 1992.

4. H. Rouse and S. Ince, 1957, *History of Hydraulics*, Iowa Institute of Hydraulic Research, State University of Iowa; M.S. Drower, 1954, Water-supply, irrigation, and agriculture, in C. Singer, E.J. Holmyard and A.R. Hall (eds.) *A History of Technology*, Oxford University Press, New York.

5. H. Rouse and S. Ince, 1957, *History of Hydraulics*, Iowa Institute of Hydraulic Research, State University of Iowa.

6. Liu Changming, Zuo Dakang, Xu Yuexian, 1985, Water transfer in China: The East Route project, in G.N. Golubev and A.K. Biswas (eds.) *Large Scale Water Transfers: Emerging Environmental and Social Experiences*, United Nations Environment Program, Water Resources Series, vol. 7, Tycooly Publishing Ltd., Oxford, UK, pp. 103–118.

7. H. Rouse and S. Ince, 1957, *History of Hydraulics*, Iowa Institute of Hydraulic Research, State University of Iowa; M.S. Drower, 1954, Water-supply, irrigation, and agriculture, in C. Singer, E.J. Holmyard and A.R. Hall (eds.) *A History of Technology*, Oxford University Press, New York.

8. P.P. Micklin, 1985, Inter-basin water transfers in the United States, in G.N. Golubev and A.K. Biswas (eds.) *Large Scale Water Transfers: Emerging Environmental and Social Experiences*, United Nations Environment Program, Water Resources Series, vol. 7, Tycooly Publishing Limited, Oxford, UK, pp. 37–65.

9. P.P. Micklin, 1985, Inter-basin water transfers in the United States, in G.N. Golubev and A.K. Biswas (eds.) *Large Scale Water Transfers: Emerging Environmental and Social Experiences*, United Nations Environment Program, Water Resources Series, vol. 7, Tycooly Publishing Ltd., Oxford, UK, pp. 37–65.

10. R. Clarke, 1991, *Water: The International Crisis*, Earthscan Publications, Ltd., London; G.V. Voropaev and A.L. Velikanov, 1985, Partial southward diversion of northern and Siberian rivers, in G.N. Golubev and A.K. Biswas (eds.) *Large Scale Water Transfers: Emerging Environmental and Social Experiences*, United Nations Environment Program, Water Resources Series, vol. 7, Tycooly Publishing Ltd., Oxford, UK, pp. 67–83.

11. W.R.D. Sewell, 1985, Inter-basin water diversions: Canadian experiences and perspectives, in G.N. Golubev and A.K. Biswas (eds.) *Large Scale Water Transfers: Emerging Environmental and Social Experiences*, United Nations Environment Program, Water Resources Series, vol. 7, Tycooly Publishing Ltd., Oxford, UK, pp. 7–35.

12. United States Senate, 1964, *Western Water Development: A Summary of Water Resources Projects, Plans, and Studies Relating to the Western and Midwestern United States*, Committee on Public Works, Special Subcommittee on Western Water Development, U.S. Government Printing Office, Washington, DC.

13. B. Herbold, A. Jassby and P. Moyle, 1992, *Status and Trends Report on Aquatic Resources in the San Francisco Estuary*, Final Report, San Francisco Estuary Project, Oakland, California.

14. G.F. White, 1988, The environmental effects of the high dam at Aswan, *Environment*, **30**(7), 4–40; B. Blackwelder and P. Carlson, 1986, *Disasters in International Water Development: Fact Sheets on International Water Development Projects*, Environmental Policy Institute, Washington, DC.

15. S. Postel, 1992, *Last Oasis: Facing Water Scarcity*, W.W. Norton and Company, New York.

16. K. Wangnick, 1990, *1990 IDA Worldwide Desalting Plants Inventory Report No. 11*, Wangnick Consulting, Gnarrenburg, Germany.

17. Total global water withdrawals are estimated to be 3,240 km$^3$ per year. The total annual supply of desalinated water is approximately 4.8 km$^3$ per year (see Section H in Part II).

18. E.D. Howe, 1974, *Fundamentals of Water Desalination*, Marcel Dekker, Inc., New York; U.S. Department of the Interior, 1978, *Desalting Plans and Progress: An Evaluation of the State-of-the-Art and Future Research and Development Requirements*, prepared by Fluor Engineering and Con-

19. A.A. Delyannis and E. Delyannis, 1984, Solar Desalination, *Desalination*, **50**, 71–81.

20. K. Wangnick, 1990, *1990 IDA Worldwide Desalting Plants Inventory Report No. 11*, Wangnick Consulting, Gnarrenburg, Germany.

21. U.S. Department of the Interior, 1978, *Desalting Plans and Progress: An Evaluation of the State-of-the-Art and Future Research and Development Requirements*, prepared by Fluor Engineering and Constructors, Inc., Irvine, California. Office of Water Research and Technology, Washington, DC.

22. U.S. Department of the Interior, 1978, *Desalting Plans and Progress: An Evaluation of the State-of-the-Art and Future Research and Development Requirements*, prepared by Fluor Engineering and Constructors, Inc., Irvine, California. Office of Water Research and Technology, Washington, DC.

23. J.D. Birkett, 1985, Alternative water resources in the Middle East, *U.S.–Arab Commerce*, November/December, p. 14.

24. U.S. Department of the Interior, 1978, *Desalting Plans and Progress: An Evaluation of the State-of-the-Art and Future Research and Development Requirements*, prepared by Fluor Engineering and Constructors, Inc., Irvine, California. Office of Water Research and Technology, Washington, DC.

25. U.S. Department of the Interior, 1978, *Desalting Plans and Progress: An Evaluation of the State-of-the-Art and Future Research and Development Requirements*, prepared by Fluor Engineering and Constructors, Inc., Irvine, California. Office of Water Research and Technology, Washington, DC.

26. The energy content of 40 tonnes of coal is about 10$^{12}$ joules (thermal). A coal-fired power plant using once-through cooling consumes nearly 500 tonnes of water for every 10$^{12}$ joules (thermal), excluding water for all other aspects of the coal fuel cycle. Far more water than this would be withdrawn for use. Thus the weight of water consumed by a power plant is approximately ten times the weight of coal required, making it economical to site the plant where there is sufficient water and to transport the fuel. A plant using cooling towers, as would be expected in a semi-arid region, would consume nearly two-and-a-half times this amount of water just for cooling.

27. P.H. Gleick and L. Nash, 1991, *The Societal and Environmental Costs of the Continuing California Drought*, Pacific Institute for Studies in Development, Environment, and Security, Berkeley, California. Report to Congress, August.

28. S. Shalash and M. Abu-Zeid, 1989, Large-scale projects to cope with climatic fluctuations; and M. Abu-Zeid and S. Abdel-Dayem, 1989, Egypt programmes and policy options for facing the low Nile flows, in *Proceedings of the International Seminar on Climatic Fluctuations and Water Management*, 11–14 December, Cairo, Egypt.

29. *The Herald*, 1992, Drought reduces output of ethanol, Harare, Zimbabwe, 24 February, p. 1.

30. J. Harte and M. El-Gasseir, 1978, Energy and water, *Science*, **199**, 623–634.

31. U.S. Water Resources Council, 1981, *The Availability of Water for Emerging Energy Technologies for the California Region*, California Department of Water Resources, Section 13(a) Water Assessment Report, Sacramento, California.

32. J. Harte and M. El-Gasseir, 1978, Energy and water, *Science*, **199**, 623–634; G.H. Davis, 1985, Water and energy: Demand and effects, *Studies and Reports in Hydrology*, 42, United Nations Educational, Scientific, and Cultural Organization (UNESCO), Paris.

33. R.L. Ottinger, D.R. Wooley, N.A. Robinson, D.R. Hodas and S.E. Babb, 1990, *Environmental Costs of Electricity*, Pace University Center for Environmental Legal Studies, Oceana Publications, Inc., New York.

34. R.L. Ottinger, D.R. Wooley, N.A. Robinson, D.R. Hodas and S.E. Babb, 1990, *Environmental Costs of Electricity*, Pace University Center for Environmental Legal Studies, Oceana Publications, Inc., New York; J. Harte and M. El-Gasseir, 1978, Energy and water, *Science*, **199**, 623–634.

35. United States Department of the Interior, 1992, Preliminary data for the publication *Estimated Use of Water in the United States in 1990*, courtesy of Wayne Solley and Howard Perlman of the U.S. Geological Survey, Reston, Virginia.

36. U.S. data on water withdrawals and consumption come from the U.S. Department of the Interior, 1988, *Estimated Use of Water in the United States in 1985*, U.S. Geological Survey Circular 1004, Washington, DC; data for Europe come from OECD, 1989, *Environmental Data Compendium*, Organization for Economic Cooperation and Development, Paris;

and OECD, 1991, *The State of the Environment*, Organization for Economic Cooperation and Development, Paris.

37. U.S. Department of the Interior, 1988, *Estimated Use of Water in the United States in 1985*, U.S. Geological Survey Circular 1004, Washington, DC; United States Department of the Interior, 1992, Preliminary data for the publication *Estimated Use of Water in the United States in 1990*, courtesy of Wayne Solley and Howard Perlman of the U.S. Geological Survey, Reston, Virginia.

38. Temperature limits for drinking water are not usually set. Canada has a secondary (i.e., set for aesthetic reasons) drinking water goal of 15°C. The European Economic Community uses a guide number of 12°C and a maximum of 25°C. Much stricter standards limiting the temperature of cooling water discharges from power plants and industries are set to protect natural aquatic ecosystems in the United States, but these tend to be set on a state or regional level.

39. U.S. Department of Energy, 1992, *International Energy Annual – 1990*, Energy Information Administration, DOE/EIA-0219(90), Washington, DC. Total global installed hydroelectricity capacity as of 1 January 1990 was estimated by this source to be 614,900 MW, out of a total world electrical generating capacity of 2,582,700 MW.

40. U.S. Department of Energy, 1992, *International Energy Annual – 1990*, Energy Information Administration, DOE/EIA-0219(90), Washington, DC.

41. U.S. Department of Energy, 1992, *International Energy Annual – 1990*, Energy Information Administration, DOE/EIA-0219(90), Washington, DC.

42. P.H. Gleick, 1990, Environment, resources, and international security and politics, in E. Arnett (ed.) *Science and International Security: Responding to a Changing World*, American Association for the Advancement of Science, Washington, DC, pp. 501–523.

43. P.H. Gleick, 1992, *Water and Conflict*, Occasional Paper Number 1 of the American Academy of Arts and Sciences, Cambridge, Massachusetts and the University of Toronto (September).

44. OECD, 1988, *Environmental Impacts of Renewable Energy*, Organization for Economic Cooperation and Development, Paris.

45. See, for example, J.P. Holdren, K. Anderson, P.M. Deibler, P.H. Gleick, I. Mintzer and G. Morris, 1983, Health and safety aspects of renewable, geothermal, and fusion energy systems, in C.C. Travis and E.L. Etnier (eds.) *Health Risks of Energy Technologies*, American Association for the Advancement of Science, Westview Press, Inc., Boulder, Colorado; R.L. Ottinger, D.R. Wooley, N.A. Robinson, D.R. Hodas and S.E. Babb, 1990, *Environmental Costs of Electricity*, Pace University Center for Environmental Legal Studies, Oceana Publications, Inc., New York.

46. A.R. Golze (ed.) 1977, *Handbook of Dam Engineering*, Van Nostrand Reinhold, Co., New York.

47. M. Shahin, 1985, *Hydrology of the Nile Basin*, Elsevier Publishing, Amsterdam; G.F. White, 1988, The environmental effects of the high dam at Aswan, *Environment*, **30**(7), 4–40.

48. P.H. Gleick, 1992, Environmental consequences of hydroelectric development: The role of facility size and type, *Energy: The International Journal*, **17**(8), pp. 735–747.

49. P.H. Gleick, 1992, Environmental consequences of hydroelectric development: The role of facility size and type, *Energy: The International Journal*, **17**(8) pp. 735–747.

50. J. Ingersoll, 1978, *Hydroelectric Power in California*, Lawrence Berkeley Laboratory, unpublished manuscript, University of California, Berkeley.

51. After reservoir was nearly filled, Wyoming dam just doesn't hold water, *U.S. Water News*, **8**(3), 4. 1991,

52. P.H. Gleick, 1992, Environmental consequences of hydroelectric development: The role of facility size and type, *Energy: The International Journal*, **17**(8), pp. 735–747.

53. G.H. Davis, 1985, Water and energy: Demand and effects, *Studies and Reports in Hydrology*, 42, United Nations Educational, Scientific, and Cultural Organization (UNESCO), Paris. Despite effort in the late 1970s to reduce this pollution, many of these streams remain severely polluted today due to lack of enforcement of water quality and reclamation regulations and the difficult technical problem of correcting drainage problems from underground mines.

54. G.H. Davis and A.L. Velikanov, 1979, Hydrologic problems arising from the development of energy: A preliminary report, *Technical Papers in Hydrology*, 17, UNESCO, Paris; another estimate, from the Council for Mutual Economic Assistance, 1973, Extended Standards for Water Rates and Amount of Wastewater per Unit of Production for Different Branches of Industry, Stoiizdat, Moscow (in Russian) is for 17 m$^3$ per 10$^{12}$ J(th) for underground coal, which can be reduced to under 5 m$^3$ per 10$^{12}$ J(th) with a water recycling system.

55. For data on water requirements for surface and underground coal mining, with and without revegetation, see N. Buras, 1979, Water constraints on energy related activities, in *Proceedings of the Specialty Conference on Conservation and Utilization of Water and Energy Resources*, American Society of Civil Engineers, 8–11 August; see also, C.E. Israelson, V.D. Adams, J.C. Batty, D.B. George, T.C. Hughes, A.J. Seierstad, H.C. Wang and H.P. Kuo, 1980, *Use of Saline Water in Energy Development*, Water Resources Planning Series UWRL/P-80-04, Utah Water Research Laboratory, Utah State University, Logan, Utah; U.S. Department of Energy, 1980, Technology Characterizations: Environmental Information Handbook, DOE/EV-0072, Washington, DC.

56. U.S. Department of Energy, 1980, *Technology Characterizations: Environmental Information Handbook*, DOE/EV-0072, Washington, DC.

57. G.H. Davis, 1985, Water and energy: Demand and effects, *Studies and Reports in Hydrology*, 42, United Nations Educational, Scientific, and Cultural Organization (UNESCO), Paris.

58. W.R.D. Sewell, 1985, Inter-basin water diversions: Canadian experiences and perspectives, in G.N. Golubev and A.K. Biswas (eds.) *Large Scale Water Transfers: Emerging Environmental and Social Experiences*, United Nations Environment Program, Water Resources Series, vol. 7, Tycooly Publishing Ltd., Oxford, UK, pp. 7–35.

59. F.L. Shorney, 1982, Water conservation and reuse at coal-fired power plants, in *Proceedings of the Conference on Water and Energy: Technical and Policy Issues*, American Society of Civil Engineers, pp. 89–95.

60. G.H. Davis and A.L. Velikanov, 1979, Hydrologic problems arising from the development of energy: A preliminary report, *Technical Papers in Hydrology*, 17, UNESCO, Paris.

61. C.E. Israelson, V.D. Adams, J.C. Batty, D.B. George, T.C. Hughes, A.J. Seierstad, H.C. Wang and H.P. Kuo, 1980, *Use of Saline Water in Energy Development*, Water Resources Planning Series UWRL/P-80-04, Utah Water Research Laboratory, Utah State University, Logan, Utah; U.S. Department of Energy, 1980, *Technology Characterizations: Environmental Information Handbook*, DOE/EV-0072, Washington, DC; G.H. Davis and A.L. Velikanov, 1979, Hydrologic problems arising from the development of energy: A preliminary report, *Technical Papers in Hydrology*, 17, UNESCO, Paris.

62. G.H. Davis, 1985, Water and energy: Demand and effects, *Studies and Reports in Hydrology*, 42, United Nations Educational, Scientific, and Cultural Organization (UNESCO), Paris.

63. U.S. Water Resources Council, 1981, *The Availability of Water for Emerging Energy Technologies for the California Region*, California Department of Water Resources, Section 13(a) Water Assessment Report, Sacramento, California.

64. N. Buras, 1979, Water constraints on energy related activities, in *Proceedings of the Specialty Conference on Conservation and Utilization of Water and Energy Resources*, American Society of Civil Engineers, 8–11 August; C.E. Israelson, V.D. Adams, J.C. Batty, D.B. George, T.C. Hughes, A.J. Seierstad, H.C. Wang and H.P. Kuo, 1980, *Use of Saline Water in Energy Development*, Water Resources Planning Series UWRL/P-80-04, Utah Water Research Laboratory, Utah State University, Logan, Utah; G.H. Davis, 1985, Water and energy: Demand and effects, *Studies and Reports in Hydrology*, 42, United Nations Educational, Scientific, and Cultural Organization (UNESCO), Paris.

65. Considerable information on water requirements for synthetic fuel production can be found in R.F. Probstein and H. Gold, 1978, *Water in Synthetic Fuel Production*, Massachusetts Institute of Technology Press, Cambridge, Massachusetts.

66. R.F. Probstein and H. Gold, 1978, *Water in Synthetic Fuel Production*, Massachusetts Institute of Technology Press, Cambridge, Massachusetts; G.H. Davis, 1985, Water and energy: Demand and effects, *Studies and Reports in Hydrology*, 42, United Nations Educational, Scientific, and Cultural Organization (UNESCO), Paris.

67. G.H. Davis and A.L. Velikanov, 1979, Hydrologic problems arising from the development of energy: A preliminary report, *Technical Papers in Hydrology*, 17, UNESCO, Paris.

68. U.S. Department of Energy, 1992, *International Energy Annual – 1990, Energy Information Administration*, DOE/EIA-0219(90), Washington, DC.

69. U.S. Atomic Energy Commission, 1974, *Environmental Survey of the Uranium Fuel Cycle*, WASH-1248, Washington, DC; U.S. Department of Energy, 1980, Technology Characterizations: Environmental Information Handbook, DOE/EV-0072, Washington, DC.

70. U.S. Atomic Energy Commission, 1972, *Environmental Survey of the Nuclear Fuel Cycle*, Atomic Energy Commission, Washington, DC; G.H. Davis and A.L. Velikanov, 1979, Hydrologic problems arising from the

development of energy: A preliminary report, *Technical Papers in Hydrology*, 17, UNESCO, Paris.

71. D.W. Layton, 1980, *An Assessment of Geothermal Development in the Imperial Valley of California. Vol. I: Environment*, Health and Socioeconomics, U.S. Department of Energy, EOR/EV-0092, UC-66e, Washington, DC; U.S. Water Resources Council, 1980, *Water Assessment for Heber Demonstration Project-Geothermal Binary Electric Power Generation*, Bookman-Edmonston Engineering, Inc., California; C.E. Israelson, V.D. Adams, J.C. Batty, D.B. George, T.C. Hughes, A.J. Seierstad, H.C. Wang and H.P. Kuo, 1980, *Use of Saline Water in Energy Development*, Water Resources Planning Series UWRL/P-80-04, Utah Water Research Laboratory, Utah State University, Logan, Utah; G.H. Davis and A.L. Velikanov, 1979, Hydrologic problems arising from the development of energy: A preliminary report, *Technical Papers in Hydrology*, 17, UNESCO, Paris; G.H. Davis, 1985, Water and energy: Demand and effects, *Studies and Reports in Hydrology*, 42, United Nations Educational, Scientific, and Cultural Organization (UNESCO), Paris.

72. U.S. Water Resources Council, 1981, *The Availability of Water for Emerging Energy Technologies for the California Region*, California Department of Water Resources, Section 13(a) Water Assessment Report, Sacramento, California.

73. D.W. Layton, 1980, *An Assessment of Geothermal Development in the Imperial Valley of California. Vol. I: Environment*, Health and Socioeconomics, U.S. Department of Energy, EOR/EV-0092, UC-66e, Washington, DC; U.S. Water Resources Council, 1980, *Water Assessment for Heber Demonstration Project-Geothermal Binary Electric Power Generation*, Bookman-Edmonston Engineering, Inc., California; C.E. Israelson, V.D. Adams, J.C. Batty, D.B. George, T.C. Hughes, A.J. Seierstad, H.C. Wang and H.P. Kuo, 1980, *Use of Saline Water in Energy Development*, Water Resources Planning Series UWRL/P-80-04, Utah Water Research Laboratory, Utah State University, Logan, Utah; G.H. Davis and A.L. Velikanov, 1979, Hydrologic problems arising from the development of energy: A preliminary report, *Technical Papers in Hydrology*, 17, UNESCO, Paris; G.H. Davis, 1985, Water and energy: Demand and effects, *Studies and Reports in Hydrology*, 42, United Nations Educational, Scientific, and Cultural Organization (UNESCO), Paris.

74. U.S. Department of Energy, 1980, *Technology Characterizations: Environmental Information Handbook*, DOE/EV-0072, Washington, DC.

75. Though the company that built these plants is in financial reorganization, the plants are continuing to operate well. Water estimates were provided to the author by D. Rib and D. Gaskin, personal communication, 1992.

76. California Energy Commission, 1981, *Electricity Tomorrow, 3rd Biennial Report*, Sacramento, California.

77. R.L. Ottinger, D.R. Wooley, N.A. Robinson, D.R. Hodas and S.E. Babb, 1990, *Environmental Costs of Electricity*, Pace University Center for Environmental Legal Studies, Oceana Publications, Inc., New York.

78. For a concise description of the world's energy predicament, see J.P. Holdren, 1990, Energy in transition, *Scientific American*, **263**(3), 157–163.

# Chapter 7
# Water and economic development

Malin Falkenmark and Gunnar Lindh

## Introduction

Earlier chapters have discussed water's role in various sectors of society – agriculture, industry, ecosystems, and energy. Water is indispensable for human health and welfare; its presence or absence can mean life or death, prosperity or poverty; access to it and control over it can lead to political conflict, even war.[1] Water is a necessary commodity in household and municipal activities, and a critical factor in agricultural and industrial production. However, the total amount of water that can actually be used is limited by the geophysical aspects of the water cycle. Some 80 countries, representing 40% of the world's population, already suffer from serious water shortages in some regions or at some times during the year.[2]

In Africa, shortages are related to both underdevelopment of potentially available water resources and to their uneven distribution. In the Americas and the republics of the former Soviet Union, water resources are apparently abundant in relation to demand, although wide regional disparities exist. Europe, endowed with a generally temperate climate and many small rivers with reliable stream flows, can use a relatively high proportion of water runoff to meet its needs for water.[3] However, on a global scale, water bodies are all subject to a variety of pollution loads. In some cases, that load is so great as to render the water unusable, given present technologies.

In this chapter, water is discussed in relation to a nation's economic development. The world is approaching a breaking point in terms of socioeconomic development and its relation to water resources. Rapid population growth in the tropics and subtropics, where the climate is dry, will lead to increasing water scarcity. In a few decades, unrestrained water demand will outstrip the amount that can be sustainably provided in many areas.[4] Conventional ways of handling solid waste and wastewater are also unsustainable. Urban wastewater volume, growing at the same pace as water demand, in many areas can no longer be treated safely and effectively with conventional technologies. Nor are sufficient time and money available for these purposes. The spreading of dry waste in surrounding landscapes is also unsustainable at present loads. As long as rain keeps on falling, pollutants will be dissolved and carried along underground water pathways into ground water and water bodies.

Thus, an analysis of how shrinking per capita water resources may be used most efficiently and sustainably to improve socioeconomic development is urgently needed. To understand the urgency and the growing failure of presently used technologies to support socioeconomic development and concurrently to sustain water resources, it may be instructive to recall the history of how people have used water resources.

## Water as a factor in socioeconomic development: Past experience

Easy access to water is not an end in itself, for any society, but a means to other ends: health, industrial and agricultural production, generation of foreign currency. Water is also essential to human welfare. A regular intake of water is necessary for human metabolic processes, which are extremely vulnerable to disturbances in the water balance. Provision of safe water for households in villages and cities is thus a fundamental component of socioeconomic development, and of the social contract between the governed and the government. Adequate provision of water is also necessary for "societal" metabolism by allowing essential socioeconomic functions, such as industrial and agricultural production.

Adequate water resources have, in the past, been essential as a nation industrializes. In the early phases of industrial development, water demand increases rapidly. Industry needs water for cooling, heating, processing, and transporting, as well as for drinking, air conditioning, and cleaning. There is a wide range of water needed to run industrial processes, and some industries use considerably larger amounts of water than others.

However, water is only one factor necessary for economic development. As Cox puts it, "Water resources management is not an end itself and cannot generally ensure socioeconomic development in the absence of a variety of other conditions on which water depends."[5] What these other factors are has sometimes to do with a nation's stage of economic development. For example, a country where agriculture is still the predominant economic base is likely to exhibit a stronger relationship between water and development than an industrialized country less dependent on agriculture. Moreover, water demand in developed countries may likely be reduced through technological advances in managing industrial processes.

The fundamental importance of easy access to water – by permitting a multitude of water-dependent societal activities – leads to the hypothesis that many of the least developed countries would be found in water-scarce regions in the tropics and subtropics. Figure 7.1 shows that the world's poorest nations (i) are located where water is scarce for part of the year, (ii) experience intermittent drought years, or (iii) experience a high evaporative demand, which prevents much rainfall from being used in human activities since most of it returns to the atmosphere. Figure 7.1 also illustrates that water scarcity can be overcome by access to other of the fundamental factors in development stressed by Cox: knowledge, energy, and money.[6]

**Three developmental phases.** A culture's prevailing use of its water resources goes through three phases. Initially, in pre-industrialized societies, water is abundant and the human influence on it is characterized by relatively few encroachments in the river basin. Under such conditions, water is considered a free gift, easily accessible, merely part of how society functions.

The next phase of development involves more active water exploitation. Dams are built to generate hydroelectric power or for irrigation or both. Inter-basin transfers of water from better endowed regions into nearby dry regions occur, allowing socioeconomic growth in the dry regions and a more efficient regional use of unevenly distributed water resources.

In the third developmental phase, the maximum attainable and acceptable level of stream flow regulation has been reached in many major river basins. All water that can be readily mobilized is put to beneficial use, and demand management has begun. At this stage the costs for further water resources development and management increase rapidly. Further supplies must be made accessible by employing non-conventional techniques.

In all three stages, water is a necessary factor for socioeconomic development. It is part of the social infrastructure, playing a role in both industrial production and in societal consumption for human welfare. Water resources for drinking and economic growth are part of a nation's social

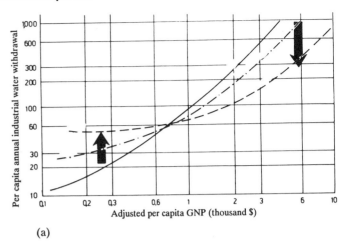

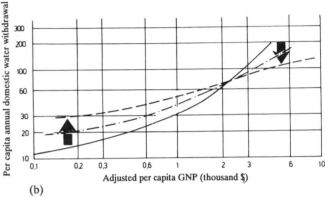

(a)

(b)

Fig. 7–2. (a) Per capita industrial annual water withdrawals as a function of per capita gross national product in the years (———), 1965; (— · — · —), 1980; and (— — —), 2000. (b) Per capita domestic annual water withdrawals as a function of per capita gross national product in the years (———), 1965; (— · — · —), 1980; and (— — —), 2000. From J. Orlóci, K. Szesztay and L. Várkonyi, 1985, *National Infrastructures in the Field of Water Resources*, UNESCO, Paris, France.

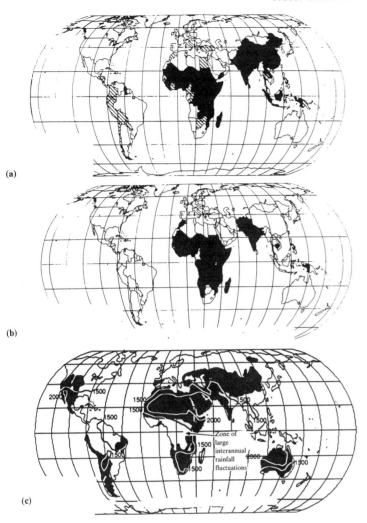

Fig. 7–1. Location of (a) the lowest income countries: (◪), per capita income $500–$1500/p; (■), poorest countries, per capita income 500/p, (b) countries with low "human capability index" (i.e., purchasing power of the poor, life expectancy, and literacy), and (c) regions with dry climate zones: (■), at least part of the year is dry. Isolines indicate high evaporative demand of the atmosphere. From U.S. Simonis, *Human Resources Development: A Neglected Dimension of Development Strategy*; M. Falkenmark and T. Chapman (eds.), 1989, *Comparative Hydrology*, UNESCO, Paris, France.

capital, a necessary foundation for successful national socioeconomic growth and the well-being of the population.

**Water withdrawals and gross national product (GNP).** Given the crucial role played by water in development, does a clear relation exist between GNP and a country's water withdrawals? Statistics do not show such a relation. In part, GNP is not an ideal measure of a nation's overall well-being. It does not include such non-monetary concepts as social well-being. The overall statistical relation could change markedly were such non-monetary social concepts given an estimated financial value.

Additionally, countries are at different phases of development, and have different levels of water reuse and conservation, both of which reduce conventional water demand. Orlóci *et al.* have shown that industrial and urban withdrawals vary considerably as a function of GNP in over fifty countries.[7] Moreover, in making projections for the year 2000, the authors found that water withdrawals increased up to a certain level of GNP, but beyond it, decreased (Figure 7.2a and 7.2b).

A possible explanation for this finding may be that a change from extensive to intensive water use takes place. Also, during industrialization, water-related development is generally characterized by an increase

of the capacity of structures and installations providing water and water-related services. However, during later phases of socioeconomic development, technological alternatives that more efficiently use and save water come gradually into play, slowing down the increase in water withdrawals.

**Water accessibility and well-being.** The study by Orlóci *et al.* – like similar studies – shows the fundamental difficulty in analyzing water's role in socioeconomic development.[8] The quality of water as opposed to its quantity is, for example, not considered in most such studies. In the early stages of a nation's development, the quantity of water is seen as the most pertinent factor. Today, however, polluted water is responsible not only for about 80% of all illnesses, but also for massive disturbances of aquatic ecosystems. Consequently, in a world with a rapidly growing population, water quality management should become as important as water quantity management is today.

**Constraints in water accessibility.** Another global water resources issue is that in many countries all easily accessible water resources have been developed, and the costs of utilizing remaining sources will likely be expensive and will certainly require long-term planning. Population-driven increases in drinking water demand will force governments to find more efficient ways of allocating the water at their disposal.

A more comprehensive and well thought out approach to water use for irrigation may be necessary. Present irrigation methods result in a tremendous misuse of water at the expense of water needs in other sectors of a nation's economy. Moreover, irrigation often waterlogs and salinizes soils. A reassessment of how water is used in agriculture is essential, especially as up to a certain limit increased availability of water for other domestic and municipal uses may increase a nation's standard of living. These issues are discussed in greater detail in Chapter 5.

In permanently or periodically dry areas, non-conventional methods of improving water accessibility may be required. In these areas, quantity is the key water issue. Prerequisites for increasing overall availability are access to money, energy, and knowledge. Two examples are desalination as practised in Saudi Arabia and Kuwait, and solar distillation as found in North Africa.

Technical means may alleviate water scarcity in many ways. For example, water in industrial processes could be recycled, and water could be reused for various purposes. On a modest scale, drought-ridden areas in North America are experimenting in recycling domestic "dirty" water for use in gardens and landscaping. All these methods will permit more water to be used for improving social well-being and facilitating socio-economic development.

**Israel: An illustrative example.** It may be instructive to learn from Israel's water management policy how a high standard of living can be maintained by using a flexible approach to water resources management and use. The Sea of Galilee in the north is one of Israel's significant raw water sources. The Israelis have built a 3-m-wide pipeline from this lake, known as the National Water Carrier, that transports 1 million m$^3$ of water each day across the country to the coast and to the Negev desert in the south. The project is very expensive to operate because the Sea of Galilee is 200 m below sea level, and Israel uses one-fifth of its electricity to pump water to the coastal and desert farms and cities.[9]

Before this project was completed, deep wells were dug in Israel's coastal plain to catch underground water flowing from the West Bank toward the Mediterranean. However, overexploitation of the wells led to sea water intrusion in the aquifer. In its natural state, the water table of the coastal aquifer is 3–5 m above sea level, which is enough to flush out contamination from agriculture or sewage. Overpumping has now left large parts of the aquifer below sea level, and consequently, these wells now yield salty water. Restoring water quality in wells contaminated in this manner is extremely difficult. During early 1991, it was concluded that the country had run out of water, igniting a political debate over water allocation. During the preceding years, over 200 million m$^3$ more water than had gone into the aquifers and the reservoirs had been allocated by the commissioner. A hydrological crisis is now a political crisis.

The present water situation has divided Israeli hydrologists into two camps, but both are optimistic about the country's water potential. The first camp stresses the need to reduce demand, but to remain flexible. This approach could mean, for instance, shifting water from agribusiness and re-employing dislocated workers in industry. Expansionists, who constitute the other camp, say water needed for irrigation may be produced by recycling more wastewater. Moreover, in the Negev desert many plants have been introduced that tolerate salty fossil water from the Nubian sandstone beneath the desert.

This picture of the socioeconomic problems in a country facing a serious water shortage – omitting any discussion of the political conflicts that it creates with Israel's neighbors (see Chapter 8) – is an example of what can be achieved when flexibility in approach is the guiding political as well as hydrological intention. Cox underlines this attitude in stating that "water shortage or any other adverse water resources conditions can be a major constraint on development if not adequately considered within a given development programme." He adds that "it is important to recognize that water resources conditions are complementary to many other development inputs indicating the futility of viewing water management in isolation and mandates a close relationship with many other socio-economic factors essential to the development process." Access to water is, in other words, a necessary but not sufficient condition for socioeconomic development.[10]

## Accessing more of the available fresh water

### Distinction between locally produced and imported water.
Any region's renewable water resources are composed of an endogenous

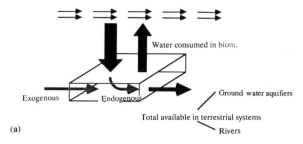

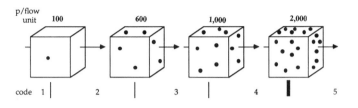

(b)

Fig. 7–3. (a) The overall fresh water availability in a country/region is finite when seen in a long-term perspective. It is basically composed of an endogenous part, i.e., the surplus from regional rainfall (after subtracting the return flow to the atmosphere) that recharges national aquifers and rivers; and the exogenous part imported from upstream countries and entering in aquifers and rivers. (b) Visualization of different levels of water competition. Each cube indicates one flow unit of 1 million m$^3$ of water per year available in aquifers and rivers; each dot represents 100 individuals jointly depending on each flow unit.

part (i.e., precipitation that does not evaporate, feeding aquifers and renewable water courses), and an exogenous part (i.e., water entering a region via rivers and aquifers originating in neighboring countries (see Figure 7.3a)). The exogenous and the endogenous water sum up to the potential water availability. This water can be put to use as it passes through a country. Under optimal conditions, taking into account ecological considerations, all exogenous and endogenous water resources can be mobilized and put to use. For this to be possible, surplus water from the rainy season must be stored in reservoirs and made accessible during the dry season, and water must be redistributed from areas where it is plentiful to areas where it is scarce.

To use exogenous water effectively, international river and lake basins must be more carefully managed, in both political and practical terms. With a rapidly increasing world population, water demand rises daily. The United Nations and others have pointed out that over 200 such basins are shared by two or more countries.[11] Fifty-seven of these basins are in Africa. Claims laid to those common water resources have already caused serious political conflicts among nations. Resolving conflicting claims over so valuable a natural resource means not only establishing a fair division of water quantitatively, but also ensuring that nations polluting the water are held accountable for restoring its quality.

Additionally, on an international scale, the approach to negotiating a fair share of common water resources in international rivers has varied greatly. Most shared river and lake basins fall within developing countries. And those negotiations over how to share international river basins in already-developed nations offer little help as a model for developing countries. McCaffrey, in Chapter 8, discusses these issues in greater detail.

**Adaptation of resource and demand.** Water demands are often concentrated in urban areas, where the gross demands of the entire urban population can be thought of as a point demand. The conventional response to this demand is to transfer water in pipelines and canals from an area with excess water supply to the area where demand is greater.

Water availability is also typically subject to strong seasonal variations, with a wet season or a snowmelt season followed by a dry season. For example, in the semi-arid tropics and subtropics, the minimum reliable part of a river's flow is often quite limited. It may amount to only 10% of the annual average runoff. As water demand for energy production and for crops peaks during the dry season, water storage then becomes a key issue in adapting availability to need. In the dry tropics, river flow may be minimal in the dry season and storage is of great importance to protect agriculture against drought. In temperate regions, energy production dictates storage so that hydropower can be produced during periods of peak energy demand.

**Preconditions for full use of potentially available water.** In the dry tropics and subtropics, where many developing nations are struggling to meet the basic needs of their citizens, a large difference exists between the potential (annual) water availability and the time-stable part of it (the water reliably available during the dry season). In view of the rapidly increasing demands for water in these developing regions, what then are the preconditions for making full use of available water? In a hot climate, where evaporation greatly diminishes availability, water should be stored underground. However, such storage is limited by geology and economics. The water must be easily and inexpensively retrieved when needed without excessive pumping. Geology often does not cooperate in the enterprise – as we saw with the example of Israel in an earlier section, where wells were overpumped, leading to salt water intrusion. Only if availability can be somehow brought into alignment with the demand pattern in time and space can all of a region's water resources really be put to use.

**The dam dilemma.** Wherever underground storage is not feasible, surface water must be dammed and stored in reservoirs. And in such areas, evaporation losses and ecological and societal disruptions are inevitable. Topography determines where reservoirs are located. Water must be stored in deep basins to reduce losses via evaporation. Some countries lack such terrain. One such example is Bangladesh, which must store water from the rainy season in upstream countries in the basins of the Ganges and Brahmaputra rivers. Building dams and creating reservoirs often means entire populations are displaced – only one of many problems created by this storage method. Table G.11, in Part II, presents estimates of people displaced by major hydroelectric and water supply projects.

How water is stored is an environmental issue as well as an economic one in developing nations. However, the emerging debate is muddled. Often, only the negative elements of the debate are focused upon, whereas the beneficial effects (for example, drought-proofing of dryland agriculture) are left out of the discussion. And seldom are distinctions made among the effects of the dam and the reservoir as such, the downstream use of the dry season water made accessible through the dam, and the avoidable and unavoidable effects of the entire process.

**Resilience to climate change.** To further complicate an already complex debate, strong scientific consensus has emerged that the world is entering a period of climate change related to the accumulation of carbon dioxide and other trace gases in the atmosphere, commonly referred to as the greenhouse effect. Although the regional consequences for water availability and quality related to global warming trends cannot yet be predicted with any certainty because of the difficulties of modeling atmospheric circulation, one general expectation is that significant changes in rainfall patterns, soil moisture, and storm frequency and intensities will occur.[12] We cannot say with certainty what may occur. But, clearly, the various economic and political structures in place for handling finite water resources will have to be made more resilient than they presently are if a global warming trend is occurring. See Chapter 9 of this book for further discussion.

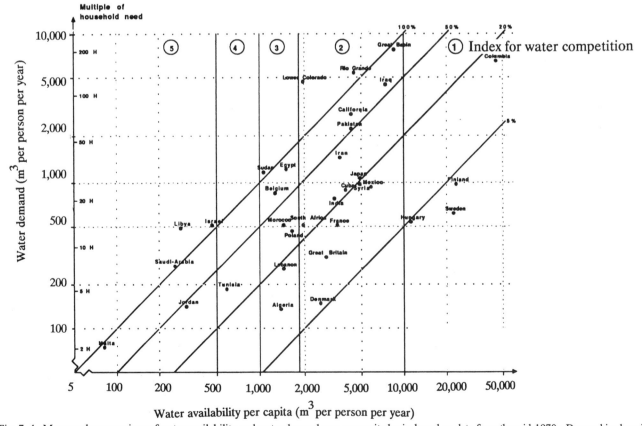

Fig. 7–4. Macroscale comparison of water availability and water demand on a per capita basis, based on data from the mid-1970s. Demand is also given as multiples of a household demand assumed at 100 l per day (H). Crossing lines indicate the degree to which potentially available water has been mobilized for use. Circled numbers refer to the population pressure intervals in Fig. 7.3b. J. Forkasiewicz and J. Margat. 1980. ''World Table of the National Economic Allotment to Water, Resources and Utilization''. Department Hydrogeológie, 79SGN 784HYD, Orléans (in French).

**How much of the available fresh water can realistically be mobilized over the medium term?** Full utilization of all potentially available water resources is not possible without: (i) technical competence to compensate for variations in water availability over time and space, (ii) the presence of geological and topographical features for stored water to be economically and safely retrieved, and (iii) successful management of shared water resources, including a workable political model for sharing exogenous water. In reality, the water a nation can mobilize usually falls far short of water potential: geology may be inadequate for easy retrieval of stored water, technical incompetence and lack of manpower may be issues, and the cooperation and goodwill essential to sharing resources may be entirely lacking. Many countries, therefore, consider 30%–60% of theoretically available water resources to be the practical limit of what they can mobilize.

Economic factors may also limit overall water accessibility. In developed Europe, when over 20% of overall availability was needed to satisfy gross societal water demands, the costs for water resources development structures were increasingly dominant in national economies. Thus, for developing nations with struggling economies and growing populations, 20% is probably a more realistic figure for overall water accessibility in the short to medium term. Higher levels may be achieved in the longer run, once technical capability and managerial skills have improved.

## Overuse and underuse

**Regional water availability.** How scarce is water? Part of the answer comes in calculating present per capita availability and per capita water demand, then recalculating it for a future situation where a better quality of life has been attained. Figure 7.4 relates per capita availability to per capita demand for a number of countries based on statistics from the mid-1970s. The diagram shows that the wettest countries use under 1% of the potentially available water. Sweden mobilizes just under 3% of its water resources to satisfy a demand of 1,700 l per person per day (or 17 H, where H corresponds to a fair level of domestic supply, taken as 100 l per day). Many developing countries have low water demands (below 5 H), but their water availability potential may allow for much higher demands: how high obviously depends on their population growth during the next few decades. Nevertheless, *underuse* of potentially available water is prevalent in many countries and more water could be used to facilitate socioeconomic development in these areas.

At the other extreme, a number of countries or regions are *overusing* their available water resources – water demand exceeds renewability. Libya, for example, is now beginning to pump fossil ground water aquifers in the south. The Lower Colorado and Rio Grande river basins in the United States also have overexploited ground water aquifers. It must be stressed that this discussion of overuse and underuse is based on the *macroscale* perspective. When looking at problems in the more conventional *mesoscale*, both overuse and underuse may exist in parallel, but in different parts of the same country.

**Water resources predicament: Five alternatives.** We distinguish here five alternatives for meeting the water resources predicament (see Figure 7.5). Conventional wisdom with its project-related approaches traditionally focuses on water needs for different purposes. It asks: how much water do we need and how should that amount be made accessible? In our work, we ask: how much water is there and what future demand levels could realistically be met given regional constraints on availability?

The 100% line in Figure 7.5 represents the "ceiling" at different levels of per capita availability. As population growth pushes a country's position to the left in the diagram, the ceiling drops. As the population doubles, the ceiling drops to 50% of the original position. A country's water predicament is also related to the fraction of gross water availability that can be made accessible for use, i.e., the possible mobilization level. Moving upward in the diagram is, in other words, a question not only of technology and economy but also of topography and climate.

Table 7.1 depicts the overall water predicament, the degree of freedom for implementing solutions, and possible policy options for various re-

**TABLE 7.1** Overall water predicament

| Present demand level | Present mobilization level | |
|---|---|---|
| | High | Low |
| High | Alpha | Beta |
| Low | Gamma | Delta |

gions of the world. Nations are entered into one of four categories: alpha, gamma, beta, or delta.

The principal options as population grows in an *alpha country* are demand reduction through rationing and increased efficiency of water use and restructuring of societal activities, such as cultivating low water use crops rather than high water use ones. In such countries, it becomes increasingly important to avoid water pollution as human activities per flow unit of water intensify. A central water authority may be needed to maximize coordination among water-related sectors users and to integrate land use and water resources.

The challenge facing a typical *gamma country* as population grows is tremendous. A gamma country must mobilize a large part of overall water availability through storage above ground and preferably below ground. Integrated land/water management will be essential to secure the best possible use of the very limited amount of water available. Strong efforts will be needed to reduce or avoid water pollution and to increase usability of scarce water.

The typical *beta country* may have difficulties in continuing to meet current high demand levels unless resource development is modernized. The medium-term task in beta countries will be to solve regional imbalances.

The typical *delta country* has the largest degree of flexibility in meeting medium-term demand increases as population grows. But even under such fortunate conditions, a large water availability may be rapidly consumed if a significant fraction of the available water cannot be utilized because of geographical, economic, or political problems, or due to excessive silting of reservoirs.

There is another category of country – the *epsilon country* – which wastes water. In an epsilon nation, water demands may exceed some 100 H per capita. Such a situation is not sustainable when the water supply includes extraction of fossil ground water, as it inevitably does, implying a decreasing water table and increasing pumping costs. In parts of the south-western United States, legislation now forces states to present plans for reducing water demand to a renewable yield. The year 2025 is the latest target year for achieving that goal.

**Predicament of North Africa and the Middle East.** Today, the realization that renewable fresh water availability is finite, while population growth will continuously increase water demand, is starting to cause serious concern in developing nations. Concern is particularly high in arid and semi-arid Arab countries. A recent overview by the Islamic Water Resources Management Network demonstrates the extent of the water crisis in this region that used to be self-reliant in food production. A calculation of future water demands includes water for the production of a basic annual per capita food demand of 375 kg of fruits and vegetables, 35 kg of meats and poultry, and 125 kg of cereal products. Together with 55 m$^3$ per person per year for domestic water needs, and using the best available irrigation practices, up to 1,205 m$^3$ per capita per year or 33 H (as gross water demand) are required.[13]

The study shows that the number of countries unable to meet these demands will increase from 12 at present to 16 by 2000, and to 18 by 2025. At that time, only Iraq, Lebanon, and Mauritania "can emerge from the almost common water deficit situation to face Arab countries." By the year 2000, the study suggests that the regional average available water share will have dropped below the per capita demand. By 2025 the average availability will have dropped to 535 m$^3$ per capita year – less than half of what is considered necessary.

By the year 2000, the region will be facing an overall deficit if food security is the criterion. By 2025 the available water will be far less than that needed for food security. Thus, the region will become increasingly dependent on importing food to feed its populations.

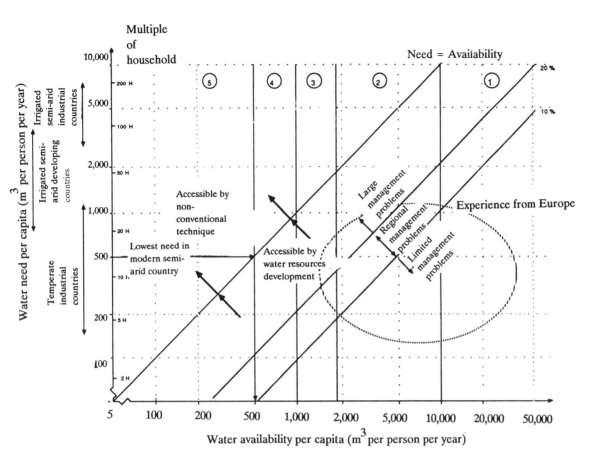

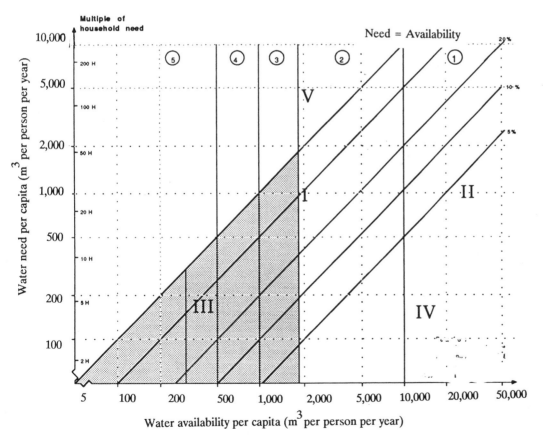

Fig. 7–5. Logarithmic diagram showing water demand possible to supply at different levels of mobilization of the potentially available water. The horizontal axis shows per capita availability (m³ per person and year) – Circled code numbers at the top of the diagrams refer to the water competition levels indicated in Fig. 7.3b. From M. Falkenmark, J. Lundqvist and C. Widstrand, 1989, Macro-scale water scarcity requires micro-scale approaches: Aspects of vulnerability in semi-arid development, *Natural Resources Forum*, **13**(4), 258–267.

The vertical axis shows water demand expressed both as m³ per person and year and as multiples of a household demand H, assumed to be 100 l per person and day. Crossing lines show different mobilization levels of water availability, achieved through water storage, flow control, and other measures of water resources development. Roman numbers refer to water predicament positions (cf Table 7.1)

I   alpha
II  beta
III gamma
IV  delta
V   epsilon

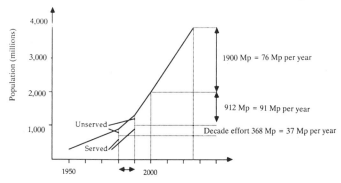

Fig. 7–6. Urban population in developing countries according to present estimates. The dotted lines visualize the relative size of efforts needed to serve the total population with safe water supply by 2000 and 2025 as compared to what was achieved during the International Drinking Water Supply and Sanitation Decade.

## The water future of large urban centers

By the year 2000, half the world's population will be living in urban areas. This growth is the consequence of the urban revolution, equally dramatic in its implications for the history of civilization as have been its predecessors, the agricultural and industrial revolutions.[14] This unprecedented growth in urban areas is especially evident in developing countries (cities in already developed nations grew far more slowly) (see Figure 7.6). Thus, between 1950 and 1980, cities in Latin America, such as Bogota, Mexico City, São Paolo, and Managua have tripled or quadrupled in population, and cities like Nairobi, Dar es Salaam, Lagos, and Kinshasa have increased over sevenfold.[15] Behind this growth is a rural exodus that will, in the 1990s, force cities to cope with 60 million new arrivals per year.[16] This urban explosion has created tremendous problems for city governments, which are increasingly unable to secure the power, resources, and trained staff to provide an adequate quality of life for their citizens.

What are the origins of this move to the cities? Initially, it began during the mercantile era when the world's populations mainly lived in rural areas. During industrialization, a gradual redistribution of the population from the countryside to the cities occurred.[17] Sometimes this migration has been characterized as a "pull" exerted by the urban center, regardless of rural density, or as a "push" from the rural areas.[18] According to this characterization, the first category of migrants "pulled" to urban areas are needed as labor for industrialization. Those people "pushed" into urban areas generally arrive once industrialization is well under way, lured to cities in the hope of creating better lives for themselves.

Urbanization in developed and developing nations has proceeded in different ways. To some extent, these differences are rooted in the colonization process, with its search for cheap raw materials for industries in developed countries. At the same time, rural people in developing countries moved to towns when capital-intensive activities supplanted labor-intensive farming as the nation's developing economic focus. Thus, rural to urban migration in developed countries arose more from the development of urban industries, which, like a magnet, attracted people from rural areas. Rural to urban migration in developing countries is rooted now in the search for a better way of life. In sheer numbers, this current rural to urban migration far exceeds that of earlier waves of rural to urban migration, and industry cannot supply jobs for the excess labor pool as it once did in developed countries. Industrial development in developing countries has also often been based on borrowed capital from developed countries.

Add to this scenario the inability of Third World governments to take over the economic power of earlier colonial rulers and the result has been over-urbanization, meaning a higher level of urbanization relative to the level of industrialization. This imbalance is also the reason why we find the largest urban areas in the Third World. As a contrast we may now observe a counter-urbanization process in already developed nations because of industrial restructuring.

Among the problems related to over-urbanization that city govern-

ments must solve are two that strike at the very existence of a city: how to acquire an adequate supply of water and how to dispose of wastewater. Unfortunately, city governments in developing countries will not learn all that much from similar experiences in already industrialized countries. As we have already suggested, cities in developed countries grew more slowly, giving planners and decision-makers time to work out and implement plans to accommodate the growth. It took Paris about a century – or three generations – to grow from about half a million inhabitants to 3 million.[19] In the Third World, the corresponding growth may take place within less than one generation.

Another difficulty urban governments in developing nations are facing is that the political, institutional, and legal framework for existing governments was mainly designed to deal with rural and agricultural societies. The experience and infrastructure to cope with large and rapidly growing urban populations is quite inadequate.

**The major city predicament.** Some of the largest cities will create special water management problems. Such cities may be important because of their economic, cultural, political, and infrastructural characteristics, rather than simply because of sheer numbers of citizens. And these cities may be more closely related to urban areas in foreign countries than to the country itself. A good example of this predicament is Mexico City, which has grown from 1 million inhabitants in 1920 to 15 million in 1980, and is expected to have 25 million people by the year 2000. It has been said that "the cost of supporting Mexico City may be exceeding its contribution in goods and services: the nation's economic locomotive is becoming a financial drain."

Mexico City's water situation adds directly to its economic problems. Residents pay only 20% of the actual costs of being provided with water. This subsidy may effectively prevent the more efficient use of water for other purposes. The difficulty of supporting the city with water results in part from its geography. In 1982, Mexico City had to pump water from a distance of 100 km and from 1,000 m below the city. By the 1990s, rapid population growth required the city to withdraw additional water from 200 km away and 2,000 m lower. The costs of satisfying this increasing water demand are enormous –equal to roughly half of Mexico's annual interest payments on its external debt. And unfortunately, to some extent the many leaks and breaks in the water system defeat the enormous effort expended. The amount of water lost corresponds to the gross needs of Rome.[20]

Will water scarcity and the escalating costs of managing it ultimately hamper the city's projected growth? Because the urbanized area and the surrounding rural area both draw on the same reserve of natural resources (water, fuelwood, crops, etc.), we must now ask seriously whether the limit of the carrying capacity of the area has been reached. Whatever the case, long-term planning for sustainable development will be a challenge.

The case of Mexico City also adds to our understanding of the role water plays in a nation's socioeconomic development. Excessive demand for any one of the four interrelated factors (water, knowledge, energy, and money) necessary for development –in this case water –may set back development.

**Bringing more water to the city.** Mexico City's situation is common to many large cities. The fundamental problem is the need to expand water supply systems to provide for a growing population. This expansion can be difficult even if the city is situated on or near a river. In an early stage of development, the withdrawal of water from a river facilitates socioeconomic development. Later withdrawals may lead to over-exploitation and a dwindling supply or to degradation of water quality. Currently, many urban areas now must import some water from more remote, water-rich regions. Los Angeles is a primary example. Similar transfers –existing or planned around the world –are often on an enormous scale. Analyzing their benefits and drawbacks can be difficult. Chapter 6 discusses large regional water transfer in more detail.

In addition to overuse of available sources, undersupply, and geographical difficulties that impede provision of water supplies and add greatly to the cost of such provisions, water supply problems for major cities may occur during extended dry periods. For example in 1981 in New York City, fresh water in reservoirs declined critically. A new drought emergency was declared for New York City in December 1991. Similar situations occurred in Europe during dry years in the late 1970s, especially in London and Paris. In Paris, water is usually tapped from underground aquifers located about 150 km away. Some surface water is also taken from the Seine, Marne, and Oise rivers. During some periods of the drought, domestic water had to be brought into Paris by tank cars.

Paris's use of river water is also of concern because of extensive industrial pollution and overextended sewage treatment systems. Even when measures are taken to prevent industrial wastewater disposal, there still remain diffuse sources, mainly from stormwater runoff, that complicate Paris's water supply situation.[21]

These examples raise the question of whether municipal governments are capable of intelligent long-range resource planning. This question was raised as far back as 1956 by Nelson Manfred Blake in his book, *Water for the Cities*.[22] Powledge's comment in 1982 on the question raised here was that "the unfortunate answer, a quarter of a century later is, not yet."[23]

**Disposing of large quantities of wastewater.** The problem of wastewater disposal rivals that of water supply in many urban areas. In both developed and developing countries, the disposal of wastewater and other wastes from large urban centers now threatens natural ecosystems. Waste is expensive and creates the greatest problems where urban populations grow most rapidly, namely in developing countries. Unfortunately, these regions also have some of the world's lowest GNPs, which limits the capital available for wastewater treatment.

Another important concern in wastewater management in developing nations is the extent to which treatment methods developed in industrial countries can be applied. Niemczynowicz writes that modern wastewater technology must be combined with locally applied methods.[24] As an example he cites China, where by the year 2000 more than 2,000 additional wastewater treatment plants will be needed, which must be considered an unrealistic goal to meet in the near future. An alternative solution would use China's long tradition of ecologically sound recycling of wastewater in multilevel biological systems. This combination of old and new methods based on the principles of ecological engineering would certainly help meet China's future wastewater treatment needs.

Treatment issues aside, wastewater could be managed in other ways. One alternative is to use partially treated wastewater for irrigation. Near Mexico City, 90,000 hectares of land are irrigated now and 250,000 hectares will be irrigated in the future using treated wastewater.[25] Eighty per cent of the wastewater from Santiago, Chile is used to irrigate 16,000 hectares of land.

**Toward a more holistic water policy.** Despite modern technology, water demand in growing urban areas cannot be indefinitely satisfied. Even now, environmental damages are being caused by concentrated overexploitation of fresh water and by disposal of wastewater, whether partially treated or untreated. Urban growth consequently is ultimately self-limiting and urban planners and decision-makers must carefully consider how to make long-range, holistic plans that harmonize with the demands for sustainable socioeconomic development at a national as well as a global level.

## Water and ecological security

**The unsustainable American west.** Water's fundamental importance in sustaining life and a culture makes any threat to an area's water supply a threat to its economic life as well. Illustrative examples of nonsustainable water bases for economic activities are the American west and south-west.[26] The areas have arid or semi-arid climates that have attracted large and growing populations, who have historically relied on expensive supply-side projects and subsidized water to maintain irrigated agriculture and a way of life that wastes water.

For example, agriculture in California's Central Valley is almost completely irrigated and supported by large irrigation water subsidies. The right to the subsidized water could, however, be lost if not put to beneficial use. The legal system in the western United States, rooted in the spirit of the rules governing the gold and silver mining in the 19th century

("first in time, first in right" or "use it or lose it") is a tremendous barrier to efficient management of the region's water resources. Moreover, 50% of the irrigation in the Central Valley relies on water drawn from ground water aquifers that are often heavily overdrafted. The system as a whole is experiencing both water quantity and quality problems, soil degradation, and decreasing crop yields. In some areas, ground water is polluted with agricultural chemicals, and irrigation return flows carry heavy salt loads. Much Central Valley land is improperly drained, leading to salinization and waterlogging that has now degraded 20% of all irrigated land in the region. In certain districts, the return flows are contaminated with selenium, leached from minerals under the irrigated soils, and dumped into the Kesterton reservoir –an important bird sanctuary. Subsequent evaporation concentrates the toxic elements within the leachate and kills or injures wildlife.[27]

Elsewhere, the water base for the development of southern Californian cities has been imported via aqueduct from northern and central California and from the Colorado River. These cities are green gardens in a desert climate, with high per capita municipal water use, high consumptive water losses from residential landscaping and watering of lawns, and widespread use of evaporative cooling for power plants.

There are multiple causes for concern in this overreliance on exogenous water. First of all, it has become increasingly difficult to import additional water from the water-rich part of the state. A recent referendum rejected a proposal to export additional water from northern to southern California. Second, some water now imported to southern California from the Colorado River system actually belongs to other interests within the basin that are beginning to draw on this resource. As the Central Arizona Project is completed, California's withdrawals from the Colorado will inevitably drop.

The water situation in this part of the United States is unsustainable as presently operated. Similar conditions appear in the south-western United States, including the multistate region overlying the Ogallala aquifer. A typical case of unsustainable water use is the rapidly growing city of Tucson, Arizona, which used to rely entirely on ground water. But because of heavy ground water overdraft, the Arizona Water Management Act requires the city to return to zero overdraft by the year 2025 at the latest. Water supply is being extended in this area with the Central Arizona Project, which imports water from the Colorado River. This amount is, however, not enough to compensate for the ongoing overdraft. In the present situation, animosity is developing between farmers and city managers. These managers –seeing irrigation as a waste of water –are buying agricultural land with water rights. The alternative option of reducing water demand is seen as highly controversial. Some citizens intend to keep their oases green, and will undoubtedly fight for the water they "need." This attitude makes it very difficult politically to raise water rates to stimulate water conservation; the result is that the politicians are reluctant to take on the issue of reform in the area of water resources and water pricing. Even water recycling is controversial.[28]

**The legal system as barrier.** In the Colorado Basin itself there is competition for water between residents of the upper and lower portions of the basin. Water demand is higher in the lower basin, but upstream users have no incentive to conserve water for use in the lower basin. On the contrary, Coloradans living in the upper basin feel they *have to* develop their water rights before it is too late. All of this wrangling has led to a tendency to develop unjustifiable or inefficient water uses rather than to admit that the water is put to better use in the arid downstream basin. There are even legal barriers to selling water from upstream to downstream users.

Unfortunately, though this region well matches water scarcity conditions typical for a large part of the Third World, it has so far not provided a good model for others to follow. The tremendous challenge for the American west, parts of which use far more than the renewable yield (i.e., the 100% line in Figure 7.4), is to overcome legal, attitudinal, and administrative barriers to effective water use under conditions of scarcity.

**The Mediterranean region.** Another region using water in an unsustainable manner is the Mediterranean.[29] In this region, water may soon seriously limit growth (see Figure 7.7). The Mediterranean landscape as

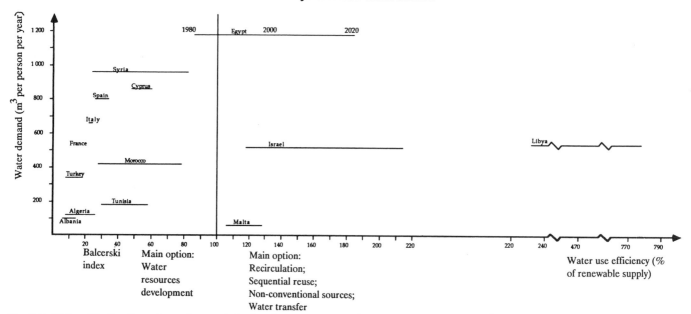

Fig. 7–7. Water utilization for an increasing population based on present per capita water supply levels in some Mediterranean countries. Water utilization is expressed as a percentage of the total water availability in the terrestrial water systems. Horizontal lines show population increase from 1980 to 2025.

a whole is ecologically fragile and seriously endangered by prevailing social and economic trends. "Stress on the coastal regions, unbalance in rural and metropolitan areas, severe water dependence, extreme sensitivity to pollution, vulnerable equilibrium between soil and water, are some of the factors to be considered" in assessing the region's socioeconomic future.[30] Providing water supplies and wastewater management are crucial elements for both the region's economic and social development. Regional soils are extremely vulnerable to erosion with resulting sedimentation of reservoirs and changes in river stability. All these factors call for early, anticipatory water planning rather than waiting for even more serious water shortages and pollution to occur. Wastewater management problems proliferate with the booming urban population in the region, and wastewaters threaten the quality of coastal waters. Agriculture is still expanding in terms of area irrigated and productivity per unit area.

Most of the population in the region is concentrated in the coastal zone, and increasing tourism causes a strong seasonal water demand cycle. Uneven water demands in both space and time greatly increase the cost of making water accessible. Cultural attitudes inhibit raising prices as an incentive for water conservation. "Water is widely seen as a gift of God to be praised rather than paid for."[31]

Distinguishing between charging for water and charging for the service of providing water may ameliorate the situation, however. Where strong cultural opposition exists to charging for water itself, charging for the satisfactory *provision* of water in a distribution system may be possible instead.

**A new conceptual framework.** Having examined, however briefly, some of the impediments to changing how water resources are allocated and used, what then can be said about how better to manage water globally and to protect natural ecosystems? The links between water use and threats to natural aquatic ecosystems, including both water quantity and water quality, are numerous. Ground water pollution is particularly problematic because it is, to a large degree, irreversible. In addition to existing ground water pollution, such as that caused by nitrates released from agricultural fertilization, there are chemical time bombs contained in our water resources from pollutants that have already been released but that have not yet shown up in ground water supplies.[32] Other dangers to our waters include land degradation from salinization and waterlogging, erosion, and desertification (see Chapter 5).

To arrive at an integrated concept of land and water conservation and management, an appropriate philosophical underpinning must be developed. The present concept of "environment" is usually used to describe

that which surrounds a central object. Talk about environment typically focuses on local scales, and often excludes the role of humans. A complementary concept can be usefully proposed and appended to our current notions about what constitutes environment: the concept of the landscape, which focuses on the mesoscale surroundings of all those living there, humans as well as other species. The landscape concept is better adapted for the interprofessional, intersectoral, and interdisciplinary dialogues needed to integrate land and water management.[33] In this conceptualization, humans live in, and are part of, a landscape, which is sustained in part because of the water provided from the atmosphere. To access all the resources in that landscape (water, biomass production, minerals, energy), we manipulate the land. We dig wells, drain, fertilize, clear vegetation, harvest, cut trees, and so forth. Through the intricate interactions among the different components of the ecosystem, secondary, sometimes unintended effects will be produced (see Figure 7.8). Consequently, sus-

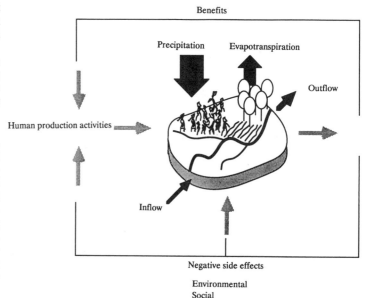

Fig. 7–8. Whereas societal demands for goods and services determine human production activities, the water available in that landscape determines both its productive capacity and the negative side effects of water use.

tainable development of that landscape is equivalent to a careful management of our interaction with it so as to meet three basic criteria for ecological security:

1. Ground water must remain drinkable, land productive, and fish edible.
2. Biological diversity should be conserved.
3. Long-term overdraft of renewable resources should be avoided.

**Develop scientific leadership.** The Canadians have provided an excellent model of sustainable socioeconomic development in relation to water management.[34] Even though water is abundant in Canada, water pollution from both endogenous and external activities occurs. Water is scarce in Canada's prairie region; and conflicting water use interests must be dealt with. A cornerstone of Canadian water policy is reliance on its scientific community. Until recently, hydrological sciences have too often been viewed only as a background necessary for constructing water use facilities or as the knowledge base needed to repair earlier environmental mismanagement. This view has diminished hydrology's potential to anticipate and prevent problems. In the future, hydrology has to become central to water policy decisions. The solution of current and future water problems involves a complex integration of ecological and economic components and long-term, strategic interdisciplinary research.

## Development in water scarce regions

**Avoiding a catastrophic turn.** In the preceding sections of this chapter, the following points have been made:

• The largest population growth is expected in semi-arid tropics and subtropics; i.e., regions where water is already scarce.
• Socioeconomic development in a hot climate entails particularly large water needs, especially for agriculture where the water requirement is controlled by the evaporative demand of the atmosphere.
• Poorer countries will be faced with tremendous economic problems in trying to provide water for food production for a rapidly growing population.
• Rapidly growing cities will need enormous amounts of water that will have to be transferred from increasingly distant sources. Wastewater produced by these cities will have to be dealt with in an acceptable manner, i.e., treated and/or reused.
• No completely satisfactory models for socioeconomic development under conditions of water scarcity exists. The region in the industrialized zone most similar to water conditions experienced by developing nations – the American west – is heavily overexploiting its water availability and must reduce demand.

What then must be done to defuse and rebalance the present crisis-driven approach to water management and conservation in those developing nations where water is scarce? What type of water-related crises do we foresee in the next few decades? What model can be developed for sound socioeconomic development under conditions of water scarcity? How can the growing water scarcity crisis best be met?

**Sound development requires an integrated strategy on a catchment basis.** The scale of the problems detailed in the questions just posed and the short lead time in which we have to address them make it impossible to rely solely on the top-down solutions with which most water experts in developed nations are generally familiar. On the contrary, initiatives originated locally (''bottom-up'') and an interdisciplinary systems approach have to be encouraged and facilitated through various incentives.

It is equally evident that providing safe drinking water to several billion new urban and rural inhabitants in just a few decades cannot be successfully accomplished by municipal authorities. The scale of the problem, in combination with the short time available, makes it impossible to rely on the top-down solutions with which northern experts are most familiar.

One example has been the project-oriented thinking that dominated the activities of the International Drinking Water Supply and Sanitation Decade, which relied to a considerable degree on initiatives by outsiders, as seen from the local perspective, and on government-related activities. The

failures of these and other programs suggest strongly that local involvement is crucial for projects to be successful and sustainable.

In shaping the necessary land and water policies, interregional differences must be thoroughly understood –both by national policy makers, foreign consultants, and international experts – so that the debate addresses the crucial issues. One fundamental distinction concerns the role of controlling demand where water is scarce. Two types of scarcity may be distinguished. First, there is scarcity in the sense that the local population uses ''too much'' water in comparison with its availability, leading to perceived shortages. In this case, demand can be brought down, for instance by raising the cost of undesirable, low-priority water uses. Second, scarcity exists where populations are underusing available water and are suffering from water-borne diseases. This kind of scarcity can be mitigated by developing safe water supplies in adequate per capita amounts, and using appropriate sanitation techniques to protect water quality. In this case the population uses too little water. Controlling demand by prohibitive pricing does nothing to alter this situation. However, pricing water, once a technical system is in place to deliver clean water, would be an obvious way to sustain the system economically.

Too often, professionals used to a technical discussion of water conservation as it would apply to problems in the American west propose the same measure in cases where the problem is not misuse or mismanagement of potentially sustainable water resources but making available adequate amounts of safe water.

**Training of an army of water professionals.** There is a serious lack of locally trained water resources experts in developing countries. The education and training of such people is vitally needed. Training should include study in the many fields related to water resources, planning, and use, including hydrologists, engineers, ground water experts, policy makers, lawyers, ecological economists, ecologists, agricultural engineers, and technical staff. This education must take place *in the region itself* to avoid the professional biases of people trained solely in developed nations.

**Securing rural livelihoods.** According to Leonard,[35] the majority of the rural poor are spread over a limited set of environments:

• degrading rainfed croplands
• eroding hillsides
• drylands
• tropical forests

The task in this case is the simultaneous alleviation of poverty and the cessation of environmental destruction. At present, about 1 billion people live in a state of ''absolute poverty.'' Close to 1 billion more live ''along a subsistence margin that, while not life-threatening, precludes attainment of much beyond the minimal necessities.''[36] At least 80% of the people in these two groupings are said to live in rural areas where they depend almost exclusively on agricultural activities for their daily subsistence (see Figure 7.9).

The sheer magnitude of the task of developing economic security for

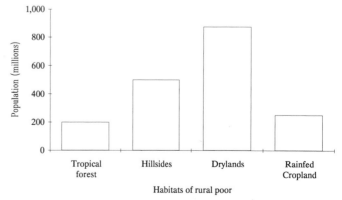

Fig. 7–9. Populations of rural poor living in different habitats. From H.J. Leonard, 1989, *Environment and the Poor: Development Strategies for a Common Agenda*, Transaction Books, New Brunswick.

the rural poor based on self-reliance in food production strongly suggests that the most effective solutions to water resource problems must be from the bottom up and that further water-related soil degradation must be stopped. In the undulating uplands where many of these poor are living, catchment-based cooperatives may be a workable solution. Soil conservation on eroding hillsides is also necessary, not only to provide sustainable agricultural production for the local populace but also to protect the soils in the fertile valley bottoms from further deterioration from the flooding and silt flows generated by the farming methods now used.

**Securing urban livelihoods.** In rapidly growing urban areas, two crucial measures must be implemented in relation to water supply and sanitation. First, leaks in the distribution system must be repaired to reduce losses and to prevent polluted ground water from entering the system. Second, safe water, that is, unpolluted water, must be provided to the population on the urban fringes.

During the 1980s, as part of the massive work of the International Drinking Water Supply and Sanitation Decade, it was estimated that no more than 380 million urban dwellers were in fact supplied with safe drinking water. Although this is a huge absolute number, it pales in comparison to the task before us. During the 1990s alone, another 900 million people will have to be supplied with safe drinking water, and between 2000 and 2025 a further 1.9 billion will require safe drinking water and sanitation. If urban migration cannot be stopped by rural development, the amount of water that will have to be brought to urban areas is enormous. In the next few decades, cities in the developing world will need more than 3,000 m$^3$ per second so that their *new* inhabitants can be supplied at a level of 100 l per day. If an adequate hygienic level cannot be achieved, urban morbidity and mortality rates will rise and social unrest may ensue.

**A business-as-usual approach will fail.** Business-as-usual approaches will fail to meet future water needs in developing nations. Business-as-usual solutions are too large in scale, too expensive to implement, and consume enormous amounts of professional resources. The money, knowledge, and energy are simply not available in the amounts needed to implement the old solutions in developing nations.

**Urban push and urban pull.** We have shown that population growth is one important reason for increasing water scarcity, and that this growth may, in the long run, create volatile social situations. Water crises may develop along other lines as well. The interplay between urban and rural populace is complex and its most salient characteristic is the movement of people from rural areas to urban areas, movement that creates water scarcity and leads to over-urbanization.[37]

What are the implications of over-urbanization for the development and resolution of water scarcity? The extremely rapid urbanization of cities in developing countries leads, sooner or later, to the establishment of slums and squatter settlements. These areas, unsupported by city authorities, do not typically have any connections with existing water supply or wastewater systems. To access drinking water, migrants are forced to buy water from street vendors at prices up to ten times the price paid by citizens supplied by water authorities.

## Conclusions

Water is necessary to sustain life. At the same time, water is an essential resource in industrial production and agriculture, where water security is fundamental if reliable yields are to be produced. Easy access to water is a necessary condition for societal development, and where water is scarce, human health and economic welfare suffer. On a purely human scale, the effort of carrying water in water-poor regions dominates the lives of women in developing countries where piped water supplies are not yet available. Water scarcity can, however, be overcome by skillful use of three other resources: money, knowledge, and energy. With money, knowledge, and energy, fresh water can be produced from the sea by desalination as in Saudi Arabia and elsewhere, or imported from water-rich regions in large transfer systems as in the western and south-western United States.

The early civilizations in dry climate tropics and subtropics had easy access to water. These societies developed along rivers that provided easy water for irrigated agriculture. But such dependence on exogenous sources of water can be threatened by land use and mismanaged water development in a river's upper basin. Increased biomass production upstream may reduce the surplus water feeding the river. This vulnerability, and its political ramifications, are discussed in more detail in Chapter 8.

The fundamental importance of water accessibility for socioeconomic development can be seen in the large numbers of poorer nations in low latitude regions having high evaporative demand, water deficiencies during part of the year, and recurrent droughts. Complicating the water scarcity problem in low-income countries are rapid population growth, which increases the pressure on a scarce and finite water availability, and a need for increased water storage to mobilize more of the available water. The fundamental need to store more water in some regions and to secure supplies is thought provoking in the context of the conventional wisdom among environmental activists against building any dams. We must improve our skills and reevaluate the effects of building storage dams. We must better distinguish between the avoidable and unavoidable effects of dam projects on the one hand, and between the dam project as such and the downstream activities made possible through dam construction, on the other.

Although there are some hydroclimatic similarities between the poverty-stricken countries in the south on the one hand and the American west on the other, there are fundamental differences as well. The latter region, with easier access to energy, knowledge, and money, is now in a situation of both far-reaching overexploitation of ground water and unsustainable transfers of exogenous water. This region must now learn how to bring down excessive demand, or *overuse* of available water. In the south, the issue is instead related to the present *underuse* of water, and people must learn how better to utilize available water so that human health, industrial growth, and food security can be achieved. As a consequence, the pricing of water has two different functions in the two situations: in the American west proper pricing (non-subsidized pricing) can discourage excessive and inefficient use; in the arid and semi-arid developing countries pricing can be used to secure more water in sustainable supply systems.

The issue of exploding populations in cities in developing countries is now a massive problem. This growth is driven by high birth rates and a huge rural exodus. These populations have to be supplied both with food and with water. The scale of the task of supplying the *new* urban populations with safe household water is tremendous. During internationally coordinated efforts in the 1980s, some 380 million people were provided with safe water. Although this was a great feat, the challenge for the period 1990 to 2025 is to provide these services for another 2.8 billion people, twice as many people per decade.

The challenge of securing food for these cities is no less a problem than that of water supply and sanitation. Over-urbanization puts a heavy burden on the surrounding farmland from which grains, fruits, and vegetables have to be provided, all requiring substantial, reliable water supplies, which will ironically compete directly with water demands in the cities themselves.

Municipal authorities will not solve these problems using conventional, "top-down," approaches. A new reliance on bottom-up approaches must therefore emerge for both food production and water supply. The trend toward individual initiatives and entrepreneurship must be facilitated. The alternative is a rapidly growing deterioration of urban water quality, with rising morbidity and mortality rates. Moreover, lack of food security easily translates into urban riots and social unrest.

This does not mean to suggest that governments and regional authorities should stay idle. An essential task facing them is to facilitate, not hinder, the sound development of local initiatives. Legislation and administration have to be put in place to allow them, and special support may be needed where larger scale projects for water resources development will be essential components of broader development activities.

Several regions of the world with rapidly growing populations are moving toward unsustainable conditions due to growing pressures on limited renewable fresh water availability. This will first become evident in North Africa and the Middle East, where decreasing food self-reliance will put increasing pressure on world agricultural markets. A choice must be made whether to increase the massive transfers of water interregionally or to transfer food grown in water-abundant regions. The growing problems in North Africa and the Middle East may soon stimulate a break with traditional business-as-usual approaches and focus instead on population policies. In regions with booming populations pressuring limited fresh water resources, the continued insistence on self-reliance in agricultural production will have to be abandoned.

The policy responses needed to meet this crisis contain a number of key activities:

- Stop population growth in water-scarce countries.
- Facilitate, not hinder, sound development activities.
- Facilitate bottom-up-approaches based on local initiatives, the only viable approach in view of the sheer scale of the efforts called for.
- Train an army of water professionals, knowledgeable in local hydrology.
- Improve efforts to end the rural exodus by stimulating economic security in the rural areas and by preventing further ecological deterioration of rainfed croplands, hillsides, drylands, and tropical forests.

In summary, the business-as-usual approach of addressing developing countries' water problems by top-down methods is in many respects outdated. Combinations of top-down approaches will remain necessary for large- and medium-scale technical projects like the building of new reservoirs or water transfer projects, but bottom-up approaches are an increasingly important complement to these activities. Bottom-up approaches include household-scale water supply and sanitation solutions, efforts at drought-proofing crop production through soil and water conservation measures, and the use of individual initiatives and entrepreneurship to protect irrigation water available in small, local watercourses. Food production in rural areas has to meet three different aims: local food supply, food for the urban market, and food for export. Meeting each goal depends on providing an adequate water supply.

## Notes

1. D. Oodit and U.E. Simonis, 1989, *Water and Development. Water Scarcity and Water Pollution and the Resulting Economic, Social, and Technological Interactions*, Wissenschaft Zentrum für Sozialforschung, Rapport SSII 89-406, Berlin, Germany (in German).
2. UNDTCD, 1988, *Our Common Future: Recommendations Relevant to Water Resources Development Strategies for the 1990s*, Document ACC/ISGW/1988/5, United Nations, Geneva, Switzerland.
3. D. Oodit and U.E. Simonis, 1989, *Water and Development. Water Scarcity and Water Pollution and the Resulting Economic, Social, and Technological Interactions*, Wissenschaft Zentrum für Sozialforschung, Rapport SSII 89-406, Berlin, Germany (in German).
4. M. Falkenmark, J. Lundqvist and C. Widstrand, 1989, Macro-scale water scarcity requires micro-scale approaches: Aspects of vulnerability in semi-arid development, *Natural Resources Forum*, **13**(4), 258–267.
5. W.E. Cox (ed.), 1987, The role of water in socio-economic development, *Studies and Reports in Hydrology*, 46, UNESCO, Paris, France.
6. W.E. Cox (ed.), 1987, The role of water in socio-economic development, *Studies and Reports in Hydrology*, 46, UNESCO, Paris, France.
7. J. Orlóci, K. Szesztay and L. Várkonyi, 1985, *National Infrastructures in the Field of Water Resources*, UNESCO, Paris, France.
8. J. Orlóci, K. Szesztay and L. Várkonyi, 1985, *National Infrastructures in the Field of Water Resources*, UNESCO, Paris, France.
9. F. Pearce, 1991, Wells of conflict on the West Bank, *New Scientist*, **1771**.
10. W.E. Cox (ed.), 1987, The role of water in socio-economic development, *Studies and Reports in Hydrology*, 46, UNESCO, Paris, France.
11. See, for example, United Nations, 1978, *Register of International Rivers, Department of Economic and Social Affairs*, Pergamon Press, Oxford, UK, which lists and describes international river basins and treaties relevant to those basins. The number of international rivers depends on accepted borders and the creation and dissolution of nations. The recent breakup of the Soviet Union into independent entities will increase the number of international river basins. See also, A.K. Biswas, 1991, Presidential Address, *VIIth World Congress on Water Resources*, Rabat, Morocco.
12. Paul E. Waggoner (ed.), 1990, *Climate Change and U.S. Water Resources*, John Wiley and Sons, New York.
13. Islamic Network on Water Resources Development and Management (INWARDAM), 1990, *Water Resources Assessment for Arab League Countries*, no. 10, Amman, Jordan.
14. G. Lindh, 1991, Hydrological and environmental consequences of the rapid growth of urban population, in G. Tsakiris (ed.) *Advances in Water Resources Technology*, A.A. Balkema, Sweden.
15. World Conference on Environment and Development (WCED), 1987, *Our Common Future*, Oxford University Press, London.
16. G. Lindh, 1990, *Towards the Resourceful City*, UNESCO Sources, no. 17, Paris, France, July/August.
17. R.J. Johnston, 1984, *City and Society: An Outline for Urban Geography*, Hutchinson University Library, Hutchinson and Co., London.
18. A.M. Gunn, 1978, *Habitat: Human Settlements in an Urban Age*, Pergamon Press, Oxford, UK.
19. L.R. Brown and J. Jacobson, 1987, Assessing the future of urbanization, in L. Brown (ed.) *State of the World*, W.W. Norton and Co., New York.
20. 1990, *World Resources 1990-91*, Oxford University Press, New York.
21. G. Lindh, 1985, *Water and the City*, UNESCO, Paris, France.
22. Described by Nelson Manfred Blake in his 1956 book, *Water for the Cities: A History of Urban Water Supply Problems in the United States*, Syracuse University Press, Syracuse, New York.
23. F. Powledge, 1982, *Water: The Nature, Uses, and Future of Our Most Precious and Abused Resource*, Farrar Straus Giroux, New York.
24. J. Niemczynowicz, 1991, Environment and urban areas – The necessity of changing views, in *International Conference on New Technologies in Urban Drainage*, Supplements, Dubrovnik, Yugoslavia, pp. 106–118.
25. J. Niemczynowicz, 1992 (in press), Water management and urban development – Call for realistic alternatives for the future, *Impact of Science on Society*, UNESCO, Paris, France.
26. M.T. El-Ashry and D.C. Gibbons, 1986, *Troubled Waters, New Policies for Managing Water in the American West*, World Resources Institute, Study 6, Washington, DC.
27. M.T. El-Ashry and D.C. Gibbons, 1986, *Troubled Waters, New Policies for Managing Water in the American West*, World Resources Institute, Study 6, Washington, DC.
28. M.T. El-Ashry and D.C. Gibbons, 1986, *Troubled Waters, New Policies for Managing Water in the American West*, World Resources Institute, Study 6, Washington, DC.
29. F.N. Correia, 1990, Water resources and waste water management in Mediterranean environments, in *Conference on Management of the Environment in the Mediterranean Basin*, Nicosia, Cyprus, 26–28 April 1990.
30. F.N. Correia, 1990, Water resources and waste water management in Mediterranean environments, in *Conference on Management of the Environment in the Mediterranean Basin*, Nicosia, Cyprus, 26–28 April 1990.
31. F.N. Correia, 1990, Water resources and waste water management in Mediterranean environments, in *Conference on Management of the Environment in the Mediterranean Basin*, Nicosia, Cyprus, 26–28 April 1990.
32. W.M. Stigliani, 1988, *Changes in Valued "Capacities" of Soils and Sediments as Indicators of Nonlinear and Time-Delayed Environmental Effects*, Working Paper WP-88-38, International Institute for Applied Systems Analysis, Laxenburg, Austria.
33. R. Castensson, M. Falkenmark and J.E. Gustafsson (eds.), 1990, *Water Awareness in Societal Planning and Decision-Making*, Swedish Council for Planning and Coordination of Research, FRN Report 90:9, Stockholm, Sweden.
34. H.J. Leonard, 1989, *Environment and the Poor: Development Strategies for a Common Agenda*, Transaction Books, New Brunswick.
35. H.J. Leonard, 1989, *Environment and the Poor: Development Strategies for a Common Agenda*, Transaction Books, New Brunswick.
36. H.J. Leonard, 1989, *Environment and the Poor: Development Strategies for a Common Agenda*, Transaction Books, New Brunswick.
37. R.J. Johnston, 1984, *City and Society: An Outline for Urban Geography*, Hutchinson University Library, Hutchinson and Co., London.

# Chapter 8
# Water, politics, and international law

## Stephen C. McCaffrey

## Introduction

During the second half of the present century, human use of fresh water has increased dramatically,[1] in part because the global population has doubled over the same period.[2] The combination of these phenomena heightens the potential for political conflict over the Earth's finite fresh water resources, especially since a substantial portion of those resources is contained in international drainage basins.[3] In fact, those basins make up nearly 47% of the Earth's land area, excluding Antarctica, and some 60% of the area of both Africa and Latin America.[4] And these figures leave aside entirely an increasingly important source of fresh water, shared aquifers.[5] While there are numerous treaties regulating the utilization of water resources shared by two or more countries, international agreements are either inadequate or lacking entirely in some parts of the world where water is in greatest demand. Some observers predict that "the political tension between certain neighboring countries over the use of international rivers, lakes and aquifers may escalate to the point of war, even before we move into the 21st century."[6]

This chapter will first survey a selection of situations throughout the world involving actual or potential disputes over fresh water. It will then offer an overview of international water law and, finally, will assess the adequacy of international law to regulate water use and to avoid and resolve controversies concerning international fresh water resources.

## Selected international controversies over shared water resources

No region of the world is exempt from controversies between different countries over water. Some of these controversies have the potential to erupt into hostilities while others are routinely resolved in a businesslike way through existing joint institutions. Treaties govern the utilization and management of many international drainage basins, but some of these agreements have not proved effective and others do not include all of the states sharing the basin in question. Where treaty regimes are lacking it is often because of tensions between the states concerned, tensions that are sometimes due precisely to the scarcity of water resources in the region. This section will review an illustrative sampling of situations in different regions of the world that have led to disputes between countries over water resources or have the potential to do so.[7] It will also indicate the extent to which law has played a role in the avoidance or resolution of the situations described. We begin with the Middle East,[8] where a combination of water scarcity and political frictions produce what is perhaps the most volatile situation in the world.

**The Middle East.** For the purposes of this chapter, the Middle East includes not only the countries in the Tigris–Euphrates and Jordan river basins, but also those in the Nile basin in north-east Africa. Because of space constraints, the present discussion can but scratch the surface of the many complex water problems in the region.

**The Jordan.** The Jordan River basin drains parts of the territories of four of the states most involved in hostilities in the region during the past four decades: Israel, Jordan, Lebanon, and Syria. The Jordan is not a long river, extending only 93 km from its sources in Lebanon to the point at which it empties into the Dead Sea. Each of the three streams forming the headwaters of the Jordan is situated in a different country: the Hasbani in

southern Lebanon; the Dan in northern Israel; and the Banias in the Golan Heights region of Syria. Since 1967, however, Israel has controlled the areas in which these streams are located, giving that country complete control of the headwaters of the Jordan.

The upper Jordan River flows south through the Huleh basin – formerly wetlands that Israel drained to form farmland – and into Lake Tiberias (the Sea of Galilee). This lake provides the largest freshwater storage capacity along the Jordan River. From Lake Tiberias the Jordan flows south through the rift valley to its terminus in the Dead Sea. The Jordan has two principal tributaries, both of which join the river from the east between Lake Tiberias and the Dead Sea. By far the most important is the Yarmouk, which rises in Syria and forms the border between Syria and Jordan, then briefly between Israel and Jordan, before it joins the Jordan River. From that point, the Yarmouk contributes approximately 40% of the total flow of the Jordan. The second tributary of the Jordan River is the Zarqa, which is situated entirely in Jordan, and joins the river below the Yarmouk. In all, 77% of the water of the Jordan River system originates in Arab countries.

Two large diversion works have been constructed in the Jordan basin: Israel's National Water Carrier and Jordan's East Ghor Canal. Both projects were initiated after the breakdown of negotiations on a comprehensive water resource development plan for the Jordan basin, known as the Johnston plan after its chief negotiator, Eric Johnston.[9] The National Water Carrier, which was completed in 1964, is a network of pipelines and canals that conveys water from Lake Tiberias along the coastal plain to Tel Aviv and the Negev desert to the south. As such, the project diverts water out of the Jordan River watershed. While Israel does not disclose information concerning the amount of water diverted through the National Water Carrier, considering it a matter of national security, it is widely believed that "this system is capable of transporting the entire average flow of the Jordan water leaving Lake Tiberias."[10]

The East Ghor Canal runs parallel to the Jordan on its east side. Originally intended as the first part of a large plan known as the Greater Yarmouk Project,[11] it carries water from the Yarmouk River to the south for irrigation of the main terrace or Ghor of the Jordan valley. While the canal has been in operation since 1966 and has had a significant impact on agricultural patterns, one specialist concludes that "it has never been able to reach its full potential as Israeli threats have meant that Jordan has never been able to construct adequate reservoir capacity on the Yarmouk."[12]

Competition between states in the region for scarce water resources is such that it has led to hostilities on a number of occasions. One such case that may be cited for purposes of illustration involved a 1964 plan by the Arab states to divert the headwaters of the Jordan into the Yarmouk and to utilize the planned Mukheiba Dam (which has not been built) for storage of the diverted waters. The plan was in part a response to Israel's construction, then nearly complete, of the National Water Carrier. Arab states emphasized that the Carrier would give Israel the capability to support more immigrants, which would disadvantage the Palestinians. It has been estimated that the Arab states' Headwater Diversion Project would have halved the amount of water supplied to the National Water Carrier.[13]

Israel responded that the Carrier did not exceed the limits established by the Johnston Plan and that it had the sovereign right to set its own immigration policies. It also stated that it would consider the diversion of Jordan headwaters as a violation of its sovereign rights. The commencement of work on the project soon led to hostilities, as Israel launched a series of military strikes against the diversion works. "The attacks culminated in April 1967 in air strikes deep inside Syria. The increase in water-related Arab–Israeli hostility was a major factor leading to the 1967 June War."[14]

Among the territories seized by Israel in that war is the area known as the West Bank of the Jordan River. A large aquifer under the limestone mountains in this area provides at least one-fourth of Israel's water. Its importance to that country is such that one of Israel's first acts after the 1967 war was to declare the water resources of the West Bank and Gaza a strategic resource under military control.[15] An official of the Israeli Agriculture Ministry has stated that Israel is so dependent on this water that it could never return control of the West Bank to the Palestinians.[16] The Minister of Agriculture himself has declared, in full-page advertisements in Israeli newspapers, that Israel would be in "mortal danger" if it lost control of the aquifer.[17] The Israeli occupation administration strictly controls Arab water use in the West Bank by licensing artesian wells on a limited basis and restricting pumping from other wells through the use of meters.[18] Palestinians have long charged that the numerous deep Israeli wells have reduced the yield of shallower Arab wells and springs or caused them to dry up.[19] Some 51% of Arab villages in the West Bank do not have running water while all Israeli settlements have that service.[20] Palestinians also claim that Israeli withdrawals from the West Bank aquifer have increased substantially. It has in fact been estimated that Israelis utilize approximately 83% of the water from the West Bank aquifer, either in the occupied territories or in Israel itself.[21] The Israelis' response to the Palestinian charges is that their wells draw upon sources other than those that feed Arab wells and that they have not increased their withdrawals since 1967. It is difficult to assess the accuracy of these claims and responses since Israeli water statistics are state secrets. Thus there is scant public knowledge of the depth or direction of Israeli wells.

Jordan and Syria have developed new plans for a "unity dam" on the Yarmouk which would provide electricity to Syria and water to Jordan – commodities badly needed in the two countries. But they cannot proceed with the project without financial assistance, and the World Bank, following its usual practice, will not provide funding if one of the countries using the river's water opposes the project. In this case one country does. Not surprisingly, that country is Israel. Jordan's current water situation is desperate[22] but it has few options that do not in some way involve Israel. Thus the Middle East Peace Conference – which, at the time of writing, is still in its initial stages – holds special importance for the Hashemite Kingdom.

It must be concluded that international law has played only a minor role in the relations between the countries riparian to the Jordan River system since Israel became a state in 1948.[23] The Johnston Plan could have formed the basis of an agreed regime of water use and was in fact followed for a period by the countries concerned. But the plan was formally rejected by the Arab states, partly because Lake Tiberias, located inside Israel, would have been used for water storage, and partly because of the simple fact that Israel had not yet been recognized by the Arab states. While some of the military actions relating to water resources in the region have been brought before the United Nations Security Council,[24] the use of international watercourses cannot be managed effectively through a crisis-response approach. What is needed is a joint management mechanism, established by agreement, participated in by all riparian states, and perhaps supported by neutral, outside parties, whether states or international organizations. This would have seemed a fanciful idea even as recently as the mid-1980s, but may be a possibility now that superpower relations have warmed, leading to the Middle East Peace Conference in which the world places so much hope.

The Tigris–Euphrates basin. The Tigris and Euphrates river systems are often treated as one basin because they unite in the Shatt-al-Arab waterway shortly before emptying into the Persian Gulf. Both rivers rise in Turkey and flow through or along Syrian territory before entering Iraq. While the Euphrates flows through Syria for a considerable distance, the Tigris never enters Syria entirely but only forms the border of that country with Turkey briefly before flowing into Iraq. Tributaries flowing from the Zagros Mountains in Iran also contribute importantly to the Tigris. The Euphrates is currently the subject of the greatest development efforts, especially in Turkey, but the flow of the Tigris system is approximately 65% greater than that of the Euphrates. Iraq is more heavily dependent upon water from the Tigris and Euphrates than Syria or Turkey, with a majority of the Iraqi population relying upon those rivers for all of their water needs.[25]

Turkey is currently constructing a U.S.$21 billion water project on the Euphrates in south-east Anatolia, called the Greater Anatolia Project (GAP), which has given rise to considerable concern on the part of Syria and Iraq. The GAP project will ultimately consist of 21 dams which will be used for the production of hydroelectric power and for the irrigation of up to 1,500,000 hectares.[26] It has been estimated that when completed in the late 1990s, the project could cause Syria to lose up to 40% of its water from the Euphrates and Iraq as much as 90%.[27] The centerpiece of the project is the gigantic Ataturk dam, the world's ninth largest, which was completed in 1990. In order to begin filling the reservoir behind the dam, Turkey stopped the flow of the Euphrates entirely for one month, from mid-January to mid-February of 1990.

Needless to say, this action disturbed the two lower riparians greatly. Turkey has stated that it will guarantee a flow of 500 m³ per second (CMS) below the project and maintains that increased releases before the January 1990 stoppage of the Euphrates kept the average flow at that level.[28] Iraq, however, is not satisfied with that amount. At a ministerial level meeting of the three states in June of 1990, Iraq requested 700 CMS,[29] basing its claim on what it termed its "acquired rights" to the use of Euphrates waters for irrigation. According to Iraq, these rights derived from use of the waters for that purpose for thousands of years. Indeed, the waters of the Tigris and Euphrates have been used for irrigation in what is now Iraq since the time of ancient Mesopotamia, some 6,000 years ago.

Unlike the situation in the Jordan basin, the relations between the countries in the Tigris–Euphrates basin have not been marked by military conflict over water.[30] There are, in fact, bilateral agreements between Turkey and Iraq and between Syria and Iraq concerning certain aspects of their water relations.[31] Furthermore, since 1983 Syria has been participating in the work of a Joint Technical Regional Rivers Committee that had been established by Iraq and Turkey in 1980, at the invitation of those two countries.[32] Yet water is undeniably a source of tension in the region and a crucial element in the overall political relations between the three countries.

Turkey has accused both Syria and Iraq of providing sanctuary to the Marxist Kurdish Workers' Party (PKK), which is waging a violent and bloody independence campaign against the Turkish government. In October, 1989, then prime minister Turgut Ozal, who has since become Turkey's president, declared that the water of the Euphrates would be cut off unless activities of the PKK were curtailed.[33] But since assuming the presidency, Ozal has stated that Turkey "will never use the control of water to coerce or threaten [our neighbors]."[34] While this should be of some comfort to Turkey's neighbors, Turkish officials have denied any legal obligation to provide water to downstream countries. For example, Turkey's Minister of State, Kamran Inan, has reportedly stated that "we have no international obligations" concerning the Tigris and Euphrates.[35]

Even if coercion or pressure is not an objective, however, the GAP project poses a significant threat to downstream water users. Syria, for example, depends heavily upon the Euphrates for drinking water, irrigation, and industrial uses, and to a lesser extent for electricity. According to one account, "with its population growing at 3.7% per year, Syria would be running short of water by the end of the century even without the GAP. With it, Syria faces a water catastrophe."[36] While Iraq may look to the Tigris as an alternative source of supply, it is sufficiently concerned with the flow of the Euphrates that in 1974 it threatened to bomb the Al Thawra dam in Syria and massed troops along the border, alleging that the dam reduced the river's flow. It is not only water quantity, but also water quality that is at stake. Return flows from the massive irrigation projects contemplated as part of GAP will carry salts as well as fertilizer and pesticide residues back into the Euphrates. This may render the water unfit for drinking, unless treated, and reduce its suitability for irrigation and some industrial uses.

The countries in the region do not appear to have followed the general principle of international law that a state contemplating a project that may harm other riparians must notify and consult with those countries. This failure can have practical as well as political consequences. The World Bank has incorporated the prior notification rule in its policy directive concerning projects on international watercourses.[37] It has been reported that Turkey's refusal to engage in high-level discussions concerning apportionment of Euphrates waters resulted in the Bank's denial of funding for GAP.[38]

The "Peace Pipeline" project. This project[39] is as yet only in the planning stages and, as such, has generated no serious disputes. It is mentioned here because of its novelty and the ways in which it could alter the water picture in the region. The pipeline would involve delivery of water, on a commercial basis, from Turkey to Syria, Jordan, Saudi Arabia, and the Gulf States. The water would be drawn from what Turkey has determined to be an excess supply of some 16.1 million $m^3$ of water per day in the Seyhan and Ceyhan rivers, which rise in that country and empty into the Mediterranean.

There would actually be two pipelines, the Western Pipeline, serving Syria, Jordan, and western Saudi Arabia, and the Gulf Pipeline, serving Kuwait, eastern Saudi Arabia, Bahrain, Qatar, the United Arab Emirates, and Oman. The approximate cost of the two pipelines, in 1986 U.S. dollars, has been estimated to be $8.5 billion for the Western Pipeline and $12.5 billion for the Gulf Pipeline. The major portion of the pipelines would be buried 2 m below ground level or passed through tunnels in mountain areas.

Although Turkey has stressed that the water delivered through the pipeline is intended to supplement, not to replace, existing water supplies in the countries served, those countries have not as yet indicated that they would be willing to participate in the project. Their concerns may stem in part from a reluctance to become dependent upon Turkey for water, in part from fears of terrorist or other sabotage of the pipeline, and in part from cost considerations. In any event, the idea remains under serious and active consideration by Turkey, and could be a factor in the future water equation of the Middle East.

The Nile.  The Nile[40] is the longest river in the world and drains an area of 3,030,300 $km^2$ in nine countries,[41] or nearly one-tenth of the African continent. It is composed principally of the Blue and White Nile Rivers, which join in Khartoum, Sudan. From its source in Lake Tana in the Ethiopian highlands, the Blue Nile flows through a deep gorge to the plains of Sudan and its confluence with the White Nile. The latter river is considerably longer yet contributes only about half as much water as the Blue Nile at Khartoum.[42] The White Nile begins as the Kagera River in Burundi and flows into Lake Victoria, which is generally considered the source of the White Nile proper. Below Khartoum the Nile is joined by only one major tributary, the Atbara, which originates in Ethiopia.

While the waters of the Nile have been used by Egyptians for irrigation for some 5,500 years, the states in the upper Nile basin have not to date made significant use of them. The lack of serious conflict over the use of Nile waters is attributable in part to this fact and in part to Britain's historical control over the river's two principal users, Egypt and Sudan. Britain's control over Egypt lasted from the late 19th century until 1937, and over the Sudan from 1899 until 1956. The absence of hostilities over the Nile waters should by no means be interpreted as a reflection of a lack of interest on the part of the lower riparian states, however. In one early incident that helped shape Egyptian attitudes toward upper basin states, for example, military action nearly ensued between Britain and France in 1898 when a French expedition attempted to gain control of the headwaters of the White Nile. While the parties ultimately negotiated a settlement of the dispute, the incident has been characterized as having "dramatized Egypt's vulnerable dependence on the Nile, and fixed the attitude of Egyptian policy-makers ever since."[43] Especially since the rise of Egyptian nationalism in 1948, the fundamental policy of that country has been that all important Nile works should be situated in Egypt, so that other countries would not take advantage of Egypt's dependence on the river by using it as a political weapon.[44]

While there are a number of agreements concerning the flow of Nile waters and the regime of various parts of the basin,[45] none of them includes all basin states or applies to the basin as a whole. The two agreements that are perhaps best known are the agreements between Egypt and Sudan of 1929 and 1959. The former was concluded in response to a British plan proposed in 1920 for the comprehensive utilization of the Nile, which gave rise to strong opposition in Egypt. The dispute was settled with the assistance of a 1925 report of the multinational Nile Project Commission, which had been established in 1920. The settlement, which took the form of the 1929 agreement, allocated twelve times the amount of water to Egypt that it gave to Sudan and expressly preserved "Egypt's natural and historical rights in the waters of the Nile."[46]

The purpose of the 1959 treaty is well described by its title: "Agreement for the Full Utilization of the Nile Waters." The 1929 treaty had left approximately 32 billion $m^3$ per year (CMY) of Nile water unallocated. The idea of the two lowest riparian states allocating all of the waters of an international watercourse shared by seven other countries is rather striking, and well illustrates the attitude that had been developed, especially by Egypt, over five millennia of use of Nile waters. Negotiation of the agreement was prompted by planning for the Aswan High Dam (Sadd el Aali Dam), which would have a storage capacity of approximately twice the average annual flow of the Nile.[47] The agreement indeed allocates all available waters of the Nile, which have been estimated at 84 billion CMY. Of this total, the agreement allots 55.5 billion CMY to Egypt, an amount that experts believe "will have to be cut most drastically"[48] as upper riparian states develop their sections of the Nile watershed.

Most of the countries in the upper Nile basin were still colonies of the United Kingdom when the 1959 agreement was being negotiated. While Britain attempted to represent the interests of these countries during the negotiations, its efforts "were rebuffed by Egypt, still smarting over the 1956 Suez campaign." Egypt also rejected a British proposal for the creation of an International Water Authority,[49] a joint body in which all Nile basin states would have participated and which would have managed the use and allocation of Nile waters. Instead, the agreement established a Joint Technical Committee between Egypt and Sudan to oversee completion of projects provided for in the agreement and to continue hydrological surveys in the upper reaches of the Nile. The two countries also agreed that they would present a common position in any future negotiations with other Nile riparians, and the agreement envisions that such discussions would be conducted by the Joint Technical Committee.[50]

Negotiations may indeed be necessary in the coming years, as the upper riparian countries begin to develop their water resources. A number of these states have plans for various water projects, some of which would increase, while others would decrease, the quantity of water reaching Egypt. Furthermore, the most efficient solutions to Egypt's water problems involve storage in other countries.[51] It would thus seem to be in Egypt's best interest to support and actively encourage basin-wide planning and development of the water resources of the Nile. Management of Nile water resources could best be accomplished through a joint commission participated in by all basin states, such as the one proposed by Britain in the negotiations leading to the 1959 treaty. Such an approach would promote optimal use of the resource, help to avoid disputes, and provide lower riparian states with a voice in plans concerning the upper Nile basin.

### The Asian subcontinent.

The Indus.  The Indus River,[52] which rises in Tibet and flows some 2,900 km through India and Pakistan to its mouth in the Arabian Sea, has been the subject of controversy since the partition of British India into the dominions of India and Pakistan on 14 August 1947. The Indus drains an area of approximately 906,500 $km^2$ and has an annual flow which is twice that of the Nile and three times that of the Tigris–Euphrates basin. The civilizations of the Indus basin have utilized the waters of the Indus for irrigation for up to 4,500 years, constructing a network of canals for this purpose.[53] Partition left a portion of the basin in each of the new dominions, with Pakistan receiving the greater share of the canal system and irrigated lands in pre-partition Punjab.[54]

Disputes soon arose, with East Punjab (in India) cutting off the water flowing through canals to West Punjab (in Pakistan) in April 1948. The Prime Minister of Pakistan reacted to the stoppage in a telegram to the Prime Minister of India:

The view of the West Punjab Government is that the water

supply cannot be stopped on any account whatsoever and we fully endorse this view. Such stoppage is a most serious matter and affects nearly a million acres of land. It will cause distress to millions and will result in calamitous reduction in production of foodgrains, etc.[55]

East Punjab declared that it would not restore the flow of water in these canals "unless West Punjab recognized that it had no right to the water."[56] But Pakistan and India soon agreed to hold an inter-Dominion conference on the question and East Punjab resumed the water supply on the eve of the meeting "as a gesture of good-will." At the conference, the parties arrived at an agreement, or "Joint Statement," concerning the controversy but just over two weeks later, a fresh dispute erupted regarding new works in Pakistan. India feared these works could result in flooding damage in its territory and threatened to retaliate by diverting certain waters that would normally flow into Pakistan. Pakistan eventually relented but maintained that India's complaint concerned a matter of internal administration of Pakistan's own waters.[57]

These and subsequent disputes ultimately led the World Bank to propose a comprehensive solution on the basis of ideas advanced by one of the principal architects of the Tennessee Valley Authority.[58] The Bank's efforts to mediate the controversy were successful and ultimately resulted, nearly ten years after its original proposal was made, in the 1960 Indus Waters Treaty.[59] The Treaty is a complex instrument whose basic approach was to increase the amount of water available to the two parties and to apportion the water resources of the Indus equitably between them. The latter objective was achieved by allocating the waters of the "Eastern rivers" of the Indus basin (the Sutlej, the Beas, and the Ravi) to India and those of the "Western rivers" (the Indus, the Jhelum, and the Chenab) to Pakistan.

The Bank's settlement efforts succeeded in part because it was in fact possible to increase the amount of available water by the construction of various works, and in part because the Bank was able to secure the participation of various other countries in arrangements to fund those works.[60] While one of the Bank's objectives, and indeed the effect of the treaty, was to achieve an equitable apportionment of Indus waters, the treaty itself expressly provides that nothing it contains is to be "construed by the Parties as in any way establishing any general principle of law or any precedent."[61]

Thus the two countries would presumably revert to their fundamental legal postures in any future water dispute that was not governed by the treaty, such as those concerning ground water.[62] On the basis of positions taken in past disputes, one might predict that India's posture would emphasize territorial sovereignty and freedom of action, while Pakistan's would rely upon historic use and acquired rights. Since international law in this field has undergone considerable development since the late 1950s, however, it is difficult to predict with confidence what positions the two countries might take today. But, as discussed below in connection with the Ganges, India has more recently embraced the idea of sharing water resources with its neighbors. One would certainly hope that, despite continuing political tensions between them – principally involving Kashmir, not water – their experience under the 1960 treaty would encourage India and Pakistan to resolve any future water dispute[63] in a spirit of good neighborliness and cooperation.

**The Ganges.** The Ganges, or Ganga,[64] rises in the Himalayas and flows through India to Bangladesh, where it joins the Brahmaputra to form the Padma, which empties into the Bay of Bengal through a vast delta. Between 1961 and 1975, India constructed a barrage on the Ganges at Farakka, some 18 km upstream from the border with Bangladesh. The dam diverts water through a canal into a distributary of the Ganges which flows to Calcutta, the Bhagirathi-Hooghly River, in order to prevent siltation of that river and Calcutta harbor. This diversion deprives Bangladesh of Ganges water, which according to Bangladesh is especially needed in the dry months of November to May for irrigation, to prevent siltation and resulting flooding of the Bangladesh portion of the Ganges, and to hold back salt water intrusion from the Bay of Bengal.

Bangladesh brought the matter before the United Nations General Assembly, where discussions between 1968 and 1976 elucidated the positions of the parties. While India at first asserted that the Ganges was not

an international river, since 99% of its catchment area lay within India, it nevertheless stated its willingness to discuss the matter with what was then East Pakistan, to assure that country that the Farakka Barrage would not cause it harm. India later not only ceased to deny the internationality of the Ganges but accepted the principle that "each riparian State was entitled to a reasonable and equitable share of the waters of an international river."[65] Bangladesh, for its part, stated that at least in the dry season, it was "entitled to the natural flow of the Ganges in order to satisfy existing human and ecological needs that could not be met in any other way," contrasting its existing uses with India's new use.

The controversy was resolved, for the time being, at least, by the 1977 Agreement on Sharing of the Ganges Waters.[66] The Agreement resulted from a ministerial level meeting between the two states in Dacca that was held pursuant to a joint statement adopted by the UN General Assembly. It provided for the allocation of Ganges waters in accordance with an annexed schedule during the annual dry period. The terms of the Agreement limited its duration to five years, but the parties have continued to follow it on the basis of a "gentlemen's agreement" that it will remain in force on a yearly basis unless terminated by one of them.[67] A Joint Committee of the previously established Joint Rivers Commission[68] oversees the implementation of the Agreement.

Even if the two countries continue to follow the regime of the 1977 Agreement, the problem of low flow of the Ganges during the lean period will not have been solved for Bangladesh. That country has proposed the construction of storage facilities upstream on the Ganges in order to increase available water supplies during these periods, but there is as yet no agreement on such a project.[69]

There is little doubt but that the airing of the controversy in the United Nations assisted Bangladesh in obtaining a resolution – even if provisional and, in its view, incomplete – of the problem. Holding the debate in the spotlight of world public opinion probably also had the effect of causing the parties to moderate their positions. Like the Indus Waters Treaty, the 1977 Agreement provides, in its preamble, that it is not to be taken as establishing any general principles of law or precedent. But the actual settlement of the dispute may be regarded as an equitable one, and the statements made by the parties in the later stages of the debate in the United Nations indicate that they do not disagree fundamentally on the applicable legal principles.

Discussion of the Ganges should not conclude without brief mention of water problems between Nepal, whose rivers contribute the bulk of the Ganges' waters, and India. Indo-Nepali relations in general have for at least 30 years been fraught with problems on many fronts, including trade, transit, and water resources management. One of the few bargaining chips Nepal possesses is its vast hydroelectric power potential. In December of 1991, the two countries concluded a series of agreements on the subjects mentioned above that have long been the source of controversy between them. In the area of water resources, India and Nepal agreed to go forward with hydroelectric, flood control, and other projects that have been pending for several decades. According to press reports, Nepal will receive primarily irrigation and flood control benefits under the agreement, while India will gain access to power from five hydroelectric projects. Critics of the Nepalese prime minister charge that he "sold out" the national interest by allowing India to share Nepal's vast hydroelectric power potential.

Not included as a party to the water agreement, of course, is Bangladesh. It is not yet clear whether the projects contemplated in the agreement will adversely affect that country, but they are likely to influence the flow of the Ganges in a variety of ways: dams can make the flow of a river somewhat more regular; evaporation from reservoirs and irrigation can reduce the amount of water eventually delivered to Bangladesh; return flows from irrigated fields can be polluted by salts, fertilizers, and pesticides; and dams can trap large quantities of silt that would otherwise be washed downstream. Some of these possible effects may prove to be irritants in future relations between the states in the Ganges basin. If so, Bangladesh may claim that the Indo-Nepali projects deprive it of its equitable share of the benefits of Ganges waters or cause it appreciable harm. As discussed further below, the latter argument could be used in support of a claim that no development at all may occur on the upper reaches of the Ganges without the consent of Bangladesh.

**The United States and its neighbors.** The United States shares a number of major lakes and rivers with its neighbors to the north and south, Canada and Mexico. These watercourses include the Great Lakes and the Columbia River to the north, and the Rio Grande and Colorado rivers to the south. Most water-related controversies between the United States and its neighbors in the latter half of the 20th century have been resolved through joint commissions established to deal with water problems along the respective borders: the International Joint Commission between the United States and Canada (IJC);[70] and the International Boundary and Water Commission between the United States and Mexico (IBWC).[71] However, in certain specific controversies, some dating from the end of the last century, each of the three countries has taken extreme positions concerning its legal rights. While such views have not been espoused in the last 30 years, the disputes which produced them demonstrate that shared water resources can arouse strong national sentiments in North America as readily as in other regions of the world.

The first dispute gave rise to what is perhaps the most infamous legal view yet espoused publicly by a state concerning international riparian rights. The controversy stemmed from diversions of water from the Rio Grande in the late 19th century by farmers and ranchers in the U.S. states of Colorado and New Mexico. According to Mexico, these diversions reduced the supply of water to Mexican communities in the vicinity of Ciudad Juarez. The Rio Grande rises in Colorado, flows through New Mexico, then forms the border between the United States state of Texas and Mexico before emptying into the Gulf of Mexico. Mexico protested the diversions in October 1895, declaring that the legal claim of those living on its side of the Rio Grande "to the use of the water of that river is incontestable, being prior to that of the inhabitants of Colorado by hundreds of years."[72]

The U.S. Secretary of State responded that the United States was not obligated to halt the diversions. He relied on a legal opinion prepared by Attorney General Judson Harmon, which has since become known as the "Harmon Doctrine." According to the opinion:

> The fundamental principle of international law is the absolute sovereignty of every nation, as against all others, within its own territory...
>
> All exceptions ... to the full and complete power of a nation within its own territories must be traced up to the consent of the nation itself. They can flow from no other legitimate source.[73]

While nothing in the opinion denied a duty to avoid causing harm to other countries, it has been taken as standing for the proposition that international law allows a state complete freedom of action with regard to international watercourses within its territory, irrespective of any consequence that might ensue in other countries.[74] In fact, the response of the State Department to Mexico on the basis of Harmon's opinion took precisely this position. It declared that:

> the rules of international law imposed upon the United States no duty to deny to its inhabitants the use of the water of that part of the Rio Grande lying wholly within the United States, although such use resulted in reducing the volume of water in the river below the point where it ceased to be entirely within the United States, the supposition of the existence of such a duty being inconsistent with the sovereign jurisdiction of the United States over the national domain.[75]

The United States nonetheless joined Mexico in instructing the International Boundary Commission, which the two countries had established in 1889,[76] to investigate and report on the Rio Grande question.

> The Commissioners' report stated that Mexico had been wrongfully deprived for many years of its equitable rights and they recommended that the matter be settled by a treaty dividing the use of the waters equally, Mexico to waive all claims for indemnity for the past unlawful use of water.[77]

The United States and Mexico substantially followed the recommendations of the report in their agreed resolution of the dispute, which was embodied in the 1906 Convention concerning the Equitable Distribution of the Waters of the Rio Grande for Irrigation Purposes.[78]

The Convention recited that the United States, by entering into the agreement, did not concede any legal basis for claims of landowners in Mexico, or that any general principle or precedent had been established. However, according to scholarly analysis,[79] as well as subsequent positions taken by the U.S. government itself, "it is necessary to distinguish between what States say and what they do."[80]

In this dispute, both countries advanced extreme positions: the Harmon Doctrine, by logical extension, would justify flooding a neighboring state or even shooting into populated areas across a border. The Mexican position of acquired or historic rights – also asserted by Iraq, Egypt, and Pakistan – would prevent any use of the water in the United States having an effect on the established uses in Mexico, thus precluding significant development in the upstream country. On the other hand, the two countries actually agreed to apportion the water in a manner that was acceptable to each of them. This cooperative approach has been followed in other cases, some of which are ongoing at the time of writing.[81]

In a second dispute, this one between Canada and the United States over the Columbia River,[82] the position of the United States was reversed. Here it was a downstream state, at the mercy of the upstream state but for any constraints imposed on the latter by international law. The dispute was ultimately resolved by the 1961 Columbia River Basin Treaty,[83] an agreement which one commentator characterized as having "ended one of the bitterest debates ever waged between Canada and the United States."[84] The Columbia rises in the Columbia Ice Field in British Columbia, Canada, flows across the U.S.–Canadian border into the state of Washington and empties into the Pacific Ocean at the border between the states of Washington and Oregon. It is joined in Canada by the Kootenay River, which originates in Canada, flows into the United States, then returns to Canada and merges with the Columbia.

While it took on a number of complexities before it was resolved, the dispute had its genesis in a 1951 U.S. proposal to construct Libby Dam on the Kootenay River in Montana. The reservoir behind the dam would have flooded 68 km of Canadian territory and raised the water level some 45 m at the border. The United States offered to compensate Canada for flooding the lands and the resulting dislocations, but not for the power benefits that the raised water level in Canada would have provided, both at Libby Dam itself and further downstream. Canada insisted on a share of the power benefits and indicated that it might divert the Kootenay into the Columbia, an action that would deny Libby and other Kootenay dams the diverted waters.[85] Canada later announced the possibility of diverting up to 18 billion $m^3$ (15 million acre-feet) annually from the Columbia into the Fraser River basin,[86] which for present purposes may be deemed to be situated entirely in Canada.

The two states ultimately agreed upon a comprehensive and integrated plan for the development of the resources of the Columbia River basin. Canada agreed to construct large storage dams, which would benefit the United States by enhancing downstream power generation and protecting against floods. The United States agreed to provide Canada with one-half the additional power generated by the Canadian projects and to pay Canada for flood control benefits.[87]

During the course of the dispute, the United States took positions that were very similar to those espoused by Mexico, and rejected by the U.S., in the earlier Rio Grande controversy. For example, in contesting the Canadian proposal to divert Columbia River waters into the Fraser River basin, the United States relied in part on the doctrine of "prior appropriation," under which "the appropriator who is first in time is first in right." The United States referred in this connection to the substantial investments that it had made in hydroelectric plants on the lower Columbia River and indicated that the proposed diversion would result in serious injury to these downstream interests.[88] In addition, it argued that Canada's reliance upon article II of the 1909 Boundary Waters Treaty[89] between the two countries was misplaced, because "the reservation of sovereign rights in article II is based on the Harmon Doctrine, which is not part of international law"[90] – thus repudiating the extreme position it had taken in its dispute with Mexico.

While Canada had indeed argued that article II of the 1909 treaty embodied the Harmon Doctrine, in the end both sides agreed that this self-centered policy "was more in tune with the pre-industrial revolution era ... than with the close economic, social, and political ties that characterize our present, rapidly shrinking world." The Harmon Doctrine thus "suffered an ignominious rout," while the principle of equitable apportionment of the benefits of an international watercourse "gained enormously in prestige and acceptance."[91]

**The Paraná River.**[92]   A dispute arose in the early 1970s between Brazil and Argentina over plans by Brazil and Paraguay to construct one of the world's largest dams[93] across the Paraná River at Itaipú, where the Paraná forms the border between Brazil and Paraguay. Argentina was concerned that the Itaipú project would adversely affect a dam it planned to construct, also with Paraguay, farther downstream on the Paraná where the river forms the border between Argentina and Paraguay. Argentina maintained that Brazil had an obligation under international law to inform it of the technical details of the Itaipú project and to consult with it so that Brazil might take Argentina's concerns into account.

Brazil vigorously denied that such obligations of prior notification and consultation existed. In so doing, it effectively espoused a view similar to the Harmon Doctrine in that it denied that it had any obligations toward the downstream state, Argentina, with regard to its planned project. Brazil did not go so far as to declare that it had complete freedom to do as it wished with the river while it was within Brazilian territory, irrespective of any harmful consequences Argentina might suffer. But since it seems implausible that an immense dam, once constructed, would be removed, the most that Argentina could realistically claim in the event that it were harmed would be some form of compensation. The end result of Brazil's position would thus appear to be tantamount to an involuntary expropriation by Brazil of Argentina's natural resources: Brazil would be entitled to proceed with its project regardless of apprehended harm in Argentina, but Brazil might have to compensate Argentina for any harm the completed project caused.

Brazil clung steadfastly to its position when the same general issue arose in other contexts in United Nations meetings, presumably so that it would not be undercutting its stance on the Itaipú project. Despite wide acceptance of the general principle of prior notification,[94] the power of Brazil's advocacy was such that it managed to dilute considerably the effect of the instruments under discussion at those meetings.[95] But, after one false start,[96] Brazil and Argentina ultimately reached agreement on the coordination of the projects they were planning. The agreement not only sets forth technical requirements concerning such matters as the maximum normal operational level of the water of Argentina's dam and minimum flow variations of the Brazilian project, but also provides for prior notification and the provision of technical information concerning the filling of reservoirs. It further provides that the entities responsible for operating the two projects are to "establish adequate procedures of operational coordination for the attainment of reciprocal benefits, including the exchange of hydrological information."[97] Thus, Argentina and Brazil were ultimately able to resolve the dispute in an amicable fashion. It is interesting to note that the very principles that were at issue in the dispute – prior notification and consultation – were enshrined in the agreement that resolved it.

**Summary.**   In the first part of the chapter we have taken a brief look at controversies concerning a number of international watercourses: the Jordan, the Tigris–Euphrates, the Nile, the Indus, the Ganges, the Rio Grande, the Columbia, and the Paraná. These rivers were selected because disputes concerning them well illustrate attitudes taken by states with regard to shared water resources. While the list could easily be lengthened were it not for space constraints, further examples would not add in material ways to the illustrations of state behavior already presented.

Some of the controversies considered have been settled while others are ongoing and show no signs of an imminent resolution. Four of the eight basins examined are now governed by treaties dealing with the problem that gave rise to the dispute. Significant[98] hostilities over water have erupted only in the Jordan basin. But this is too facile a way in which to measure the seriousness of a dispute over shared water resources: existing political tensions unrelated to water may result in recourse to violence over questions that would have been resolved peacefully between other countries. In all of the cases reviewed, the disputes were matters of high national interest, engaging the attention of government officials, diplomats, and scholars alike. Several of the controversies were

brought before the United Nations, usually with good effect. And several of the rivers will probably receive increasing international attention as expanding populations, climate change, and upstream development efforts combine to place further demands on already scarce water resources.

The next section will present an overview of the general principles of international law governing the use and protection of international water resources that are relevant to the present study. This survey should be read against the background of the illustrations of state practice reviewed earlier. These illustrations demonstrate that even the most serious and complex controversies can be resolved in an equitable manner, provided the states concerned have the political will to do so and are willing to make the necessary accommodations, as in the Indus controversy. On the other hand, problems for which there are readily available solutions may continue to fester because the countries concerned refuse to compromise with each other, as happened over the Jordan. In either case, the history of controversies concerning international water resources suggests that law will almost certainly be invoked by at least one, and probably all of the parties to any dispute over these resources. The second section will examine whether any firm norms exist in this field, or whether it is open to states to base their claims on any ground they wish.

## International water law

International law, being a decentralized system which relies for its enforcement principally on self-help (within certain limits) and the opinion of the world community, lacks such features as compulsory jurisdiction and centralized enforcement that are characteristic of domestic legal systems. Nonetheless, as observed by a prominent American legal scholar, "almost all nations observe almost all principles of international law and almost all of their obligations almost all of the time."[99] This is not difficult to understand. Observance of international law is in states' self-interest. History has shown time and time again that serious violations of international law entail costs, and that these are often not easily borne. But for this system of enforcement – comparatively crude though it may be – to function at all effectively, rules of international law must possess a determinacy, a clarity, which leaves little doubt about their content.[100] In short, states must know with a reasonable degree of certainty what it is that is expected of them.

What evidence is there that there are rules of international law concerning shared water resources that relate to the kinds of problems discussed above and satisfy the requirement of determinacy? Most rules of international law derive from one of two categories of sources: treaties or international custom. Treaty-based rules are relatively easy to ascertain, although there is always the possibility of differing interpretations of individual provisions. Norms of customary international law[101] are somewhat more difficult to establish, but efforts at "codification" of those rules by organizations of high repute[102] greatly assist the process. This part will present an overview of treaties and international custom relating to international watercourses. All that can be attempted in the limited scope of this chapter is a rather one-dimensional snapshot; references will be provided to more detailed treatments for the interested reader.

**Treaties.**   A striking aspect of treaties concerning international watercourses is their sheer number. A systematic index prepared by the Food and Agriculture Organization (FAO) of the United Nations contains more than 2,000 instruments relating to international watercourses.[103] Another remarkable feature of these agreements is that some date from as early as the first and second centuries of the present millennium.[104] Most agreements concerning shared water resources are bilateral, and relate to specific rivers that form or cross boundaries, or lakes that straddle them. The number of multilateral agreements, while small in comparison to that of bilateral treaties, is growing steadily.[105] Virtually all of these multilateral agreements concern specific watercourses or drainage basins. Only two purport to lay down principles of general applicability, and neither of those agreements is of great contemporary significance.[106]

But the fact that most water treaties relate to specific watercourses does not mean that these agreements are of no use in the search for general principles of law in this field. Similar provisions occurring in numerous agreements between different countries provide indications as to general expectations that states have in common regarding the use of international watercourses. These expectations influence behavior, which may ripen into a general practice accepted by states as law.[107]

While it is difficult to generalize, it can be said with some confidence that most treaties that allocate water quantity or quality between the states concerned, or establish management mechanisms, reflect the principle of equitable utilization. That is, the objective of these agreements appears from the effect of their provisions, if not from their actual language, to be the apportionment of the uses and benefits of the watercourse in question in an *equitable* manner.[108] What is equitable must be determined in each individual case, in light of all relevant facts and circumstances. But the case-by-case nature of the determination does not affect the generality of the principle. Indeed, as discussed below, attempts by groups of experts to derive general rules from treaties and other forms of state practice have concluded that equitable utilization is a – and perhaps *the* – fundamental rule in the field.[109]

The other basic rule reflected in treaty practice, though perhaps not so widely, is that a state may not, through its actions affecting an international watercourse, significantly harm other states.[110] Experts also recognize the fundamental nature of this rule,[111] though some maintain that it is subordinate to the principle of equitable utilization.[112] A survey of the work of various groups of these experts will assist in the ascertainment of rules of customary international law.

**General (customary) international law.** It will be recalled that many, if not most, of the controversies discussed above were triggered because one state perceived it had been, or might be, harmed by the actions of another with regard to a shared watercourse. The "harm" involved was sometimes the actual or prospective deprivation of what the "harmed" state perceived to be its equitable share of the uses and benefits of the water. This kind of reaction usually derives from a perception that the conduct in question is at variance with the state's expectations, which are often based upon customary norms.

Yet it is difficult to prove the existence of these norms. This is especially so when, as in this case, the object of the practice – watercourses – is unique in each instance. In view of this difficulty, those seeking to ascertain rules of customary international law often have recourse to secondary sources. One of the most widely relied upon sources of evidence of customary international law is the work of recognized experts. These individuals or groups conduct empirical studies on the basis of which they form conclusions regarding customary norms. Three international organizations of high repute have prepared sets of draft rules in this field, rules which to a large extent are based on state practice. While they are not conclusive evidence of norms of customary international law, the drafts are highly authoritative. They will be reviewed briefly as a convenient method of illustrating the basic rules of customary international law operative in the field of international watercourses.

The Institut de Droit International (Institute of International Law) is a non-official body established in 1873 and composed of some 120 elected members. It adopts resolutions which purport to state existing rules of international law and sometimes proposes such rules. The Institut's work has been relied upon on a number of occasions by international tribunals and by states in diplomatic exchanges. The Institut has adopted two resolutions concerning shared water resources. They are the 1961 Salzburg Resolution on the Use of International Non-Maritime Waters[113] and the 1979 Athens Resolution on the Pollution of Rivers and Lakes and International Law.[114]

The former resolution declares that a state's right to make use of shared waters "is limited by the right of use by the other States concerned with the same river or watershed." It then provides that any dispute as to the extent of the respective states' rights "shall be settled on the basis of equity, taking into consideration the respective needs of the States, as well as any other circumstances relevant to any particular case."[115] This provision thus affirms the principle of equitable utilization. The resolution goes on to provide for advance notice of new uses and negotiations in the event of objections to such uses.

The 1979 Resolution declares that states must "ensure" that activities within their borders "cause no pollution in the waters of international rivers and lakes beyond their boundaries." This obligation is, however, moderated in a subsequent article, which provides that it may be fulfilled by preventing new forms of pollution and increases in existing levels of pollution, and by abating existing pollution as soon as practicable. The resolution also contains detailed provisions concerning forms of cooperation between states sharing the same basin, such as exchange of data concerning pollution, prior notification of potential polluting activities, consultation concerning pollution problems, and the establishment of international commissions competent to deal with basin-wide pollution problems.[116] Taken together, the two resolutions of the Institut demonstrate that a select body of international legal authorities of high repute support a flexible approach to international water problems, emphasizing the need for regular communication and the establishment of mechanisms within which experts from the countries concerned can work together to anticipate and solve problems on the technical level.

Like the Institut de Droit International, the International Law Association (ILA) was founded in 1873. The membership of the ILA is, however, much larger, numbering around a thousand. The ILA also adopts resolutions setting forth rules and recommendations concerning international law. Among its best known products is the set of articles adopted in 1966, known as the Helsinki Rules on the Uses of the Waters of International Rivers.[117] It is clear from the Helsinki Rules and subsequent work of the ILA that this organization regards the dominant principle in the field of international water resources to be that of equitable utilization or apportionment. Article IV of the Helsinki Rules provides that states are entitled "to a reasonable and equitable share in the beneficial uses of the waters of an international drainage basin." The commentary to the ILA's Montreal Rules on Water Pollution in an International Drainage Basin, adopted in 1982,[118] states that "the principle of equitable utilization ... is the foundation on which the Helsinki Rules are built."[119] And the commentary to the Helsinki Rules themselves makes clear that an existing use may have to give way to a new use in order to achieve an equitable apportionment of shared water resources; compensation would, however, have to be paid for the impairment or discontinuance of the existing use.[120]

Thus, far from being prohibited, harm to an existing use is treated by the ILA as only one factor to be taken into account in achieving an equitable apportionment of the waters. In fact, the very example used in the Helsinki Rules commentary is strikingly similar to existing situations discussed earlier in this chapter: "State A, a lower co-basin State, has for many years, used the waters of an international river for irrigation purposes. State B upstream now wishes to utilize the waters for hydro-electric power production." According to the ILA's commentary, while an "existing reasonable use is entitled to significant weight as a factor," it may be outweighed by other factors such as "the existence of alternative sources of agricultural products, the conservation benefits to the co-basin States, the employment of a wasteful and antiquated method of utilisation and its potential for replacement by a less wasteful method within the financial ability of State A and the potential value of the proposed use—all dictate modification and accommodation."[121] Thus the principle of equitable apportionment might require that states such as Iraq and Egypt modify their irrigation methods, or even sharply curtail their agricultural uses in exchange for some form of compensation, in order to permit upstream states such as Turkey and Ethiopia to develop the hydroelectric potential of shared water resources.

The final group of experts whose work in this field should be mentioned is the International Law Commission (ILC) of the United Nations.[122] This body is composed of 34 experts on international law, from as many countries, who are elected by the UN General Assembly and serve in their individual capacities, not as representatives of governments. The ILC, like the two organizations previously discussed, prepares drafts setting forth existing and proposed rules of international law. In the case of the ILC, however, the subjects on which it works are determined by the UN General Assembly.

At its 1991 session, the ILC completed the "first reading," or provisional adoption of a set of 32 draft articles on the Law of the Non-Navigational Uses of International Watercourses.[123] Among the "general principles" set forth in the draft are those of equitable utilization (article 5), the obligation not to cause harm to other riparian states (article 7), and the obligation to exchange hydrologic and other relevant data and information on a regular basis (article 9). The articles contain a variety of additional provisions, including a detailed set of procedures concerning new uses of international watercourses (Part III), and articles on protection of ecosystems and water quality (Part IV), joint management (article 26), protection of installations (article 27), and equal access to judicial and administrative procedures (article 32).

On the important issue of whether one state may develop its water resources later than another, even though that development would result in "harm" to the latter, the Commission's answer is, in effect, in the negative. That is, under the ILC's articles the obligation not to cause harm to other states prevails over that of equitable utilization in the event the two come into conflict. In the words of the Commission's commentary to article 7 (Obligation Not to Cause Appreciable Harm):

A watercourse State's right to utilize an international watercourse in an equitable and reasonable manner has its limit in the duty of that State not to cause appreciable harm to other watercourse States. In other words – *prima facie*, at least – utilization of an international watercourse is not equitable if it causes other watercourse States appreciable harm.[124]

Thus the ILC differs from the ILA – and, by implication, the Institut de Droit International – on this crucial issue. Neither the articles themselves nor the commentary specifically address this apparent lack of flexibility. The ILC seems to take refuge in the hope that the states concerned will be able to reach an agreement in cases such as those of the Tigris–Euphrates and Nile basins:

The Commission recognizes ... that in some instances the attainment of equitable and reasonable utilization will depend upon the toleration by one or more watercourse States of a measure of harm. In these cases, the necessary accommodations would be arrived at through specific agreements. Thus a watercourse State may not justify a use that causes appreciable harm to another watercourse State on the ground that the use is "equitable," in the absence of agreement between the watercourse States concerned.[125]

This may be an appropriate solution in cases involving pollution or where the actually or potentially harmed state is weaker than the other state.[126] As to the former, it is becoming less acceptable to speak of an "equitable" right to pollute, or an "equitable" share of the absorptive capacity of a river, lake or aquifer. As to the relative power of the states concerned, it would be much more difficult for a state to establish conclusively, without the aid of a neutral third party, that another state had exceeded its equitable share than it would be to prove that the first state had been harmed. The latter is a comparatively simple factual question, whereas the former requires, as has been seen, the balancing of all relevant factors and circumstances. While a relatively powerful state could bring economic or other forms of pressure to bear on the other state in order to convince it to cease its "inequitable" behavior, a comparatively weaker state would not have these means at its disposal. Yet proof that it had been harmed would be relatively straightforward.

In other cases, however, the ILC's solution is at least questionable. It would appear to encourage a "race to the river," and to reward the winner. That is, a state that develops first would thereby establish an entitlement to the amount of water used or to the particular use made of the watercourse, and other riparian states would not be permitted to trench upon that amount or adversely affect that use without the agreement of the former state. If that state refused to agree, the only *legal* avenues open to the other states would appear to be (i) to attempt to construe the term "harm" in a relative rather than an absolute sense, or (ii) to attempt to convince the first state to reach a practical resolution of the conflict (with or without the aid of a neutral third party), probably according to equitable principles. The latter avenue might lead to an acceptable settlement if the states wishing to develop their water resources were otherwise in a strong negotiating position *vis-à-vis* the potentially harmed state.

The former avenue would appear to be a difficult one to negotiate, especially under the ILC's articles, which use the term "harm" in its factual sense.[127] But at least two lines of argument might be possible. One would emphasize that the law only takes into account "legally significant" harm, and ignores that which is *de minimis* (trifling or insignificant). It would maintain that harm which is reasonably tolerable and

which results from a reasonable or necessary use by another state should not be regarded as being legally significant, i.e., that such "harm" should not prevent it from making an otherwise reasonable use of the watercourse. A second line of argument might follow the Coase theorem of social cost, which reveals that the problem of harm in cases such as the instant one is of a reciprocal nature: to avoid harm to one state (e.g., the earlier-developing downstream state) is to inflict harm on another (the later-developing upstream state) by preventing it from developing its water resources. In Coase's words, "The problem is to avoid the more serious harm."[128] Or as the Helsinki Rules would have it, the problem is to accommodate the conflicting uses, even if the only way of doing so is by allowing the "harmful" but reasonable use to proceed, while requiring that compensation (whether in the form of money, electricity, or some other medium) be paid for the harm caused. But this line of argument reveals the basic flaw with the entire approach, namely, that it is vulnerable to the charge that the "no harm" rule would actually become nothing less than the principle of equitable utilization in a very transparent disguise.

## Conclusion

Several conclusions are suggested by the foregoing review of controversies concerning shared water resources and applicable rules of international law. First, it seems apparent that downstream countries, whose topography generally lends itself more readily to agriculture than that of countries at the headwaters of international rivers, make intensive use of their water resources before their upstream neighbors. This is true, for example, of Iraq in relation to Turkey, Egypt in relation to Sudan and the countries of the upper Nile basin, and Mexico in relation to the United States. Second, efforts or even plans by later-developing upstream countries to begin to utilize their water resources often give rise to strenuous objections by their downstream neighbors. The downstream countries often take the position that they have acquired rights to the quantity (and perhaps quality) of the water they have used in the past. According to this theory, any use by the upstream state that would interfere with these rights would be unlawful. They base this claim, in effect, on (i) the obligation not to cause harm to other riparian states, and (ii) their historic use of the water without objection by upper riparians, coupled with their reliance on this "acquiescence." Third, the later-developing upstream countries typically take the position that they have not heretofore had the need or the capability to develop their water resources, or that the technology for such development simply did not exist while the lower riparian country was developing its dependence on the water (as in the case of hydroelectric facilities). They rely upon the principle of equitable utilization and point out that a strict rule against causing harm to other riparian states forecloses, or at least unreasonably limits, development. Fourth, these controversies are typically resolved (i) when the states concerned otherwise enjoy good relations with each other (as in the case of Canada and the United States), (ii) where one of the states is clearly more powerful than the other but wishes to end the dispute (as in the case of the United States *vis-à-vis* Mexico) or (iii) where it is otherwise in the mutual interest of the states concerned to do so (as in the case of India and Pakistan, due largely to the intervention of the World Bank). In other cases, the dispute often remains unresolved.

The role of law in these situations differs from case to case, but it must be concluded that it has generally been only one factor influencing the outcome of major international water controversies. On the other hand, states have rarely shown a disposition to defy generally accepted principles of international law and, indeed, usually rely on those principles in their diplomatic exchanges. Further, the more concrete and generally accepted the applicable legal principles become, the more likely it is that they will play a major role in the resolution of international water controversies. In this connection, the work of organizations such as those discussed above is most welcome and will without question contribute to the resolution of existing disputes and the avoidance of future controversies.

There is little doubt that the key to both peaceful relations with regard to shared water resources and sound management thereof is ongoing communication between the states concerned, preferably on the technical level, concerning everything from hydrologic and meteorological data to basin-wide development plans. Experience has shown that such communication can most effectively occur through some form of joint mechanism such as a commission composed of experts from all basin states. Unfortunately, political frictions often prevent the formation of such bodies precisely in the cases where they are needed most. It may well be, however, that the end of the Cold War will usher in a new era of cooperation, even where tensions have been highest, since in many cases the political conflicts preventing the resolution of water controversies were perpetuated by external support of the disputants by the superpowers. Let us hope that this concluding note is not unduly optimistic.

## Notes

1. From 1950 to 1980, global water withdrawals grew from approximately 1,000 km$^3$ to around 3,500 km$^3$. This rate of growth is itself increasing. L. Riviere, 1989, Threats to the world's water, *Scientific American*, **261**(3) 80. See also R. Linden, 1990, The last drops, *Time*, 20 August, p. 58. For an historical discussion of the development of different uses of international watercourses, see United Nations, Department of Economic and Social Affairs, 1970, *Integrated River Basin Development*, Report of a Panel of Experts, UN Document E/3066/Rev. 1, p. 2.

2. The world population nearly doubled between 1950 (2,500 million) and 1985 (4,800 million); World Commission on Environment and Development (Brundtland Commission), 1987, *Our Common Future*, Oxford University Press, Oxford, UK. It had reached 5,300 million by mid-1990 and was projected to climb to 6,300 million by 2000 according to the Encyclopedia Britannica, 1991, *Britannica Book of the Year (1991)*, p. 278.

3. "[A] high proportion of the world's great basins are international ..." United Nations, Department of Economic and Social Affairs, 1970, *Integrated River Basin Development*, Report of a Panel of Experts, UN Document E/3066/Rev. 1. Map 2 of the publication just referred to shows some 165 international drainage basins. Comparing a list of the world's longest rivers from National Geographic Society, 1984, *Great Rivers of the World*, Washington, DC with Map 2 shows that thirteen of the world's twenty longest rivers (65%) are international.

4. A. Biswas, 1991, Water for sustainable development. A global perspective, *Development and Cooperation*, 5, 17–20 (German Foundation for International Development, Berlin).

5. Concerning shared aquifers (ground water) and the law governing them, see generally the following: R. Hayton and A. Utton, 1989, Transboundary groundwaters: The Bellagio draft treaty, *Natural Resources Journal*, **29**, 663; International Law Association, 1986, Report of the Committee on International Water Resources Law, in ILA, 1986, *Report of the Sixty-second Conference*, Seoul, 1986; S. McCaffrey, 1991, *Seventh Report on the Law of the Non-Navigational Uses of International Watercourses*, ILA, 43d Session, UN Document A/CN.4/436 and Corrs. 1, 2 and 3, 14 May, pp. 13–34.

6. A. Biswas, 1991, Water for sustainable development. A global perspective, *Development and Cooperation*, 5, 17–20 (German Foundation for International Development, Berlin).

7. For an early discussion of particular interstate controversies, see H.A. Smith, 1931, *The Economic Uses of International Rivers*, P.S. King and Son, Ltd., London. A number of the 14 cases discussed by Smith are between states that are members of federal systems.

8. See generally T. Naff and R. Matson, 1984, *Water in the Middle East: Conflict or Cooperation?*, Westview Press, Boulder, Colorado.

9. Eric Johnston was appointed by President Eisenhower on 16 October 1953 as special ambassador to negotiate a comprehensive plan for the development and utilization of the waters of the Jordan River system. See T. Naff and R. Matson, 1984, *Water in the Middle East: Conflict or Cooperation?*, Westview Press, Boulder, Colorado; P. Beaumont, 1991, Transboundary water disputes in the Middle East, paper delivered at the Conference on Transboundary Waters in the Middle East: Prospects for Regional Cooperation, Ankara, Turkey, 2–3 September 1991. See generally, for example, United Nations, C.T. Main, Inc., 1953, *The United Development of the Water Resources of the Jordan Valley Region*, Gordon Clapp, Boston, Massachusetts; K. Doherty, 1965, Jordan waters conflict, *International Conciliation*, Carnegie Endowment for International Peace, New York.

10. P. Beaumont, 1991, Transboundary water disputes in the Middle East, paper delivered at the Conference on Transboundary Waters in the Middle East: Prospects for Regional Cooperation, Ankara, Turkey, 2–3 September 1991.

11. The project is described by T. Naff and R. Matson, 1984, *Water in the Middle East: Conflict or Cooperation?*, Westview Press, Boulder, Colorado.

12. P. Beaumont, 1991, Transboundary water disputes in the Middle East, paper delivered at the Conference on Transboundary Waters in the Middle East: Prospects for Regional Cooperation, Ankara, Turkey, 2–3 September 1991.

13. P. Beaumont, 1991, Transboundary water disputes in the Middle East, paper delivered at the Conference on Transboundary Waters in the Middle East: Prospects for Regional Cooperation, Ankara, Turkey, 2–3 September 1991.

14. P. Beaumont, 1991, Transboundary water disputes in the Middle East, paper delivered at the Conference on Transboundary Waters in the Middle East: Prospects for Regional Cooperation, Ankara, Turkey, 2–3 September 1991. Other water-related hostilities in the region are listed by T. Naff and R. Matson, 1984, *Water in the Middle East: Conflict or Cooperation?*, Westview Press, Boulder, Colorado, Table 3, "Water-related ceasefire violations in Jordan River system from 1951 to 1967." Among additional incidents were those on 23 June and 10 August 1969, when Israel attacked and effectively disabled the East Ghor Canal in response to PLO activities which it had been unsuccessful in halting.

15. T. Walker, 1991, Another Middle East issue of life and death, *Financial Times*, 8 October, p. 40.

16. National Public Radio, "All Things Considered," 2 December 1991, statement of Martin Sherman, assistant to Agriculture Minister Refael Eitan (John Nielsen reporting).

17. National Public Radio, "All Things Considered," 3 December 1991, statement of Refael Eitan.

18. T. Naff and R. Matson, 1984, *Water in the Middle East: Conflict or Cooperation?*, Westview Press, Boulder, Colorado.

19. See, for example, the National Public Radio report cited in note 16, especially the statement of Nadir al-Hatib, a geologist with the Palestinian Hydrology Group. T. Walker, 1991, Another Middle East issue of life and death, *Financial Times*, 8 October, p. 40.

20. T. Walker, 1991, Another Middle East issue of life and death, *Financial Times*, 8 October, p. 40. According to Abed el Rahman Tamimi, director of the Palestinian Hydrology Group, "In some areas [Israeli] settlements have swimming pools, while Palestinian villages next door have a shortage of drinking water."

21. T. Walker, 1991, Another Middle East issue of life and death, *Financial Times*, 8 October, p. 40.

22. "Jordan now rations water, and everyone is affected. Water is allocated at specific times, and many families often run out. Those who can afford it buy more from trucks stationed throughout the city at 10 times the normal price." National Public Radio, "All Things Considered," 4 December 1991 (Joyce Davis reporting).

23. Compare J. Dellapenna, 1989, Water in the Jordan Valley: The potential and limits of law, *The Palestine Yearbook of International Law*, **5**, 15.

24. For example, T. Naff and R. Matson, 1984, refer to five separate occasions between 1951 and 1967 on which disputes concerning water resources were taken to the Security Council, *Water in the Middle East: Conflict or Cooperation?*, Westview Press, Boulder, Colorado.

25. P. Beaumont, 1991, Transboundary water disputes in the Middle East, paper delivered at the Conference on Transboundary Waters in the Middle East: Prospects for Regional Cooperation, Ankara, Turkey, 2–3 September 1991.

26. D. Caponera, 1991, Legal and institutional concepts of cooperation, paper delivered at the Conference on Transboundary Waters in the Middle East: Prospects for Regional Cooperation, Ankara, Turkey, 2–3 September 1991.

27. G. Moffett, III, 1990, Downstream fears feed tensions, *The Christian Science Monitor*, 13 March, p. 4, citing Thomas Naff of the University of Pennsylvania.

28. Statement of the Minister of State of Turkey, Kamran Inan, at the Conference on Transboundary Waters in the Middle East: Prospects for Regional Cooperation, Ankara, Turkey, 3 September 1991. See also G. Moffett, III, 1990, Downstream fears feed tensions, *The Christian Science Monitor*, 13 March, p. 4.

29. P. Beaumont, 1991, Transboundary water disputes in the Middle East, paper delivered at the Conference on Transboundary Waters in the Middle East: Prospects for Regional Cooperation, Ankara, Turkey, 2–3 September 1991. Iraq and Syria are reported to have reached their own agreement on 16 April 1990 on the apportionment between them of Euphrates water. Under this agreement, Syria is to obtain 42% and Iraq 58% of the amount of water flowing from Turkey into Syria. The agreement further provides for resolution of details concerning the monthly division of water by a technical committee. British Broadcasting Corporation, Summary of World Broadcasts/The Monitoring Report, 1 May 1990 (LEXIS).

30. The conflict between Iran and Iraq over the Shatt al Arab has been charac-

terized as "more a symptom than [a] cause of hostility between the powers on its opposite banks," and "a reflection of the struggle of Iraq and Iran for regional supremacy." T. Naff and R. Matson, 1984, *Water in the Middle East: Conflict or Cooperation?*, Westview Press, Boulder, Colorado. Thus it is not treated here as a conflict over water, *per se.*

31. See the Protocol on Flow Regulation of the Tigris and Euphrates Rivers and of their Tributaries, annexed to the Treaty of Friendship and Good Neighbourly Relations between Iraq and Turkey, Ankara, 29 March 1946; United Nations, 1963, *Legislative Texts and Treaty Provisions Concerning the Utilization of International Rivers for Other Purposes than Navigation,* p. 376, UN Document ST/LEG/SER.B/12; and the 1978 technical agreement between Iraq and Syria providing for the "exchange of hydrologic data and other relevant information" mentioned in D. Caponera, 1991, Legal and institutional concepts of cooperation, paper delivered at the Conference on Transboundary Waters in the Middle East: Prospects for Regional Cooperation, Ankara, Turkey, 2–3 September 1991.

32. D. Caponera, 1991, Legal and institutional concepts of cooperation, paper delivered at the Conference on Transboundary Waters in the Middle East: Prospects for Regional Cooperation, Ankara, Turkey, 2–3 September 1991.

33. *The Economist* reports that as of December 1989, the campaign had claimed 2,000 lives; Mesopotamia: Send for the dowsers, 16 December 1989, p. 56.

34. *Financial Times*, 15 January 1990, Section I, p. 4. See also the statement of the Minister of State of Turkey, Kamran Inan, at the Conference on Transboundary Waters in the Middle East: Prospects for Regional Cooperation, Ankara, Turkey, 3 September 1991. At that meeting, Minister Inan stated that Turkey would never use water as a means of political pressure and noted that it had declined to do so during the Gulf War. In addition, during the controversy over cutting off the flow of the Euphrates to begin filling the Ataturk reservoir, a delegation of Turkish foreign ministry officials visited Iraq, Syria, Egypt, and Saudi Arabia to assure those countries that Turkey did not intend to use its control of the Euphrates for political advantage. *San Francisco Chronicle*, 13 January 1990, p. A9, column 1, p. A13.

35. G. Moffett, III, 1990, Downstream fears feed tensions, *The Christian Science Monitor*, 13 March, p. 4. Inan referred instead to a "gentleman's agreement" with Syria and Iraq "to do our best not to harm our neighbors."

36. *The Economist*, 1989, Mesopotamia: Send for the dowsers, 16 December, p. 56.

37. Operational Directive 7.50: Projects on International Waterways, 18 September 18. See the discussion of the directive by Raj Krishna, 1991, International Law Adviser, Operations, World Bank, at a seminar on The Non-Navigational Uses of International Watercourses, in *American Society of International Law, Proceedings, Eighty-Fourth Annual Meeting,* 28–31 March, p. 232.

38. *The Economist*, 1989, Mesopotamia: Send for the dowsers, 16 December, p. 56. The Bank also denied funding for Syria's Ghab Project on the Orontes River (Al Asi) in the 1950s on the ground that Turkey, a lower riparian on that river, objected. This was the first international water project considered by the Bank. D. Caponera, 1991, Legal and institutional concepts of cooperation, paper delivered at the Conference on Transboundary Waters in the Middle East: Prospects for Regional Cooperation, Ankara, Turkey, 2–3 September 1991. The Orontes rises in Lebanon and flows through Syria and into Turkey before emptying into the Mediterranean Sea.

39. The information in this section is drawn from a paper entitled "Peace pipeline project," distributed at a Roundtable on Transboundary Watercourses held at Bilkent University, Ankara, Turkey, 26–27 November 1990.

40. See generally A. Garretson, 1967, The Nile basin, in A. Garretson, R. Hayton and C. Olmstead (eds.) *The Law of International Drainage Basins,* Oceana, Dobbs Ferry, p. 256; and T. Naff and R. Matson, 1984, *Water in the Middle East: Conflict or Cooperation?*, Westview Press, Boulder, Colorado.

41. National Geographic Society, 1984, *Great Rivers of the World*, Washington, DC. The Nile is 6,700 km long from its sources in Burundi and Rwanda to its mouth in the Mediterranean Sea. The nine nations of the Nile Basin are Rwanda, Burundi, Zaire, Tanzania, Kenya, Uganda, Ethiopia, Sudan, and Egypt.

42. According to official Egyptian estimates, the White Nile contributes 2/7, the Blue Nile 4/7 and the Atbara 1/7 of Nile waters. D. Caponera, 1991, Legal and institutional concepts of cooperation, paper delivered at the Conference on Transboundary Waters in the Middle East: Prospects for Regional Cooperation, Ankara, Turkey, 2–3 September 1991.

43. T. Naff and R. Matson, 1984, *Water in the Middle East: Conflict or Cooperation?*, Westview Press, Boulder, Colorado.

44. See, for example, D. Caponera, 1991, Legal and institutional concepts of cooperation, paper delivered at the Conference on Transboundary Waters in the Middle East: Prospects for Regional Cooperation, Ankara, Turkey, 2–3 September 1991. This is not to say that no important projects have been implemented outside Egypt, however. For example, as early as the 1920s, a plan to irrigate the Gezira area of Sudan, south of Khartoum, was developed by the British in response to the worldwide cotton shortage of that time. This plan, which later became known as the Gezira Cotton Scheme, involved construction of a dam at Sennar on the Blue Nile. The Gezira Scheme was inaugurated in 1926 after completion of the Sennar Dam and continues to operate to the present day; T. Naff and R. Matson, 1984, *Water in the Middle East: Conflict or Cooperation?*, Westview Press, Boulder, Colorado. It has been characterized as "the most important irrigation project in the Sudan" and is one which the Sudanese government would like to enlarge.

45. D. Caponera, 1991, lists eleven such agreements, ranging in time from the Protocol of 15 April 1891 between the UK and Italy in which Italy agrees, *inter alia*, not to alter the flow of the Atbara, to the Treaty of 24 August 1977 concerning the Establishment of the Organization for the Management and Development of the Kagera River Basin; Legal and institutional concepts of cooperation, paper delivered at the Conference on Transboundary Waters in the Middle East: Prospects for Regional Cooperation, Ankara, Turkey, 2–3 September 1991.

46. The 1929 agreement: Exchange of Notes between the UK and Egypt in regard to the Use of the Waters of the River Nile for Irrigation Purposes, Cairo, 7 May 1929, League of Nations Treaty Series (hereafter LNTS), no. 93, 44. The 1959 agreement: Agreement between the United Arab Republic and the Republic of Sudan for the Full Utilization of the Nile Waters, Cairo, 8 November 1959, *Egyptian Review of International Law* (Cairo), **15**, 321.

47. P. Beaumont, 1991, Transboundary water disputes in the Middle East, paper delivered at the Conference on Transboundary Waters in the Middle East: Prospects for Regional Cooperation, Ankara, Turkey, 2–3 September 1991. Construction of the High Dam "was begun in 1961 and completed in 1968, but the entire project did not become fully operational until 1975," T. Naff and R. Matson, 1984, *Water in the Middle East: Conflict or Cooperation?*, Westview Press, Boulder, Colorado.

48. P. Beaumont, 1991, Transboundary water disputes in the Middle East, paper delivered at the Conference on Transboundary Waters in the Middle East: Prospects for Regional Cooperation, Ankara, Turkey, 2–3 September 1991, citing D. Whittington and K.E. Haynes, 1985, Nile water for whom? Emerging conflicts in water allocation for agricultural expansion in Egypt and Sudan, in P. Beaumont and K.S. McLachlan (eds.) *Agricultural Development in the Middle East*, John Wiley and Son, Chicester, UK.

49. T. Naff and R. Matson, 1984, *Water in the Middle East: Conflict or Cooperation?*, Westview Press, Boulder, Colorado.

50. The 1959 agreement: Agreement between the United Arab Republic and the Republic of Sudan for the Full Utilization of the Nile Waters, Cairo, 8 November 1959, *Egyptian Review of International Law* (Cairo), **15**, 321.

51. See, for example, T. Naff and R. Matson, 1984, *Water in the Middle East: Conflict or Cooperation?*, Westview Press, Boulder, Colorado, especially Table 7, "Upper Nile water projects," p. 139; D. Caponera, 1991, Legal and institutional concepts of cooperation, and P. Beaumont, 1991, Transboundary water disputes in the Middle East, papers delivered at the Conference on Transboundary Waters in the Middle East: Prospects for Regional Cooperation, Ankara, Turkey, 2–3 September 1991. The increased flow to Egypt from the planned projects on the upper Nile has been estimated by Caponera at approximately 18 km$^3$ per year. The increase would come principally from canals that would circumvent the vast swamps in the Sudd region; considerable water is currently lost from the swamps through evaporation. Decreases in flow could result from upper Nile basin irrigation projects, particularly in Tanzania and Kenya, using water from Lake Victoria. Ethiopia has considerable hydroelectric potential, but Beaumont points out that there is a major siltation problem on the upper Blue Nile, meaning that any works would have a short life expectancy.

52. See generally R. Baxter, 1967, The Indus basin, in A. Garretson, R. Hayton and C. Olmstead (eds.) *The Law of International Drainage Basins*, Oceana, Dobbs Ferry, Chapter 9; and M. Whiteman, 1964, Indus, *Digest of International Law*, vol. 3, Chapter VII, section 20, Department of State, Washington, DC.

53. While no evidence remains of early irrigation works and canals, a system of canals had been constructed in the Punjab region by the mid-19th century and was considerably expanded during the period of British rule.

R. Baxter, 1967, The Indus basin, in A. Garretson, R. Hayton and C. Olmstead (eds.) *The Law of International Drainage Basins*, Oceana, Dobbs Ferry, Chapter 9. See also G.C. Taylor, Jr., 1965, Water, history, and the Indus plain, *Natural History*, **74**, 40.

54. A Boundary Commission had been constituted to determine the location of the border in Bengal and the Punjab according to the Muslim or non-Muslim character of the areas in question. The boundary drawn by the Commission left ten of the thirteen canals in pre-partition Punjab in Pakistan and two in India, and divided the remaining canal between the two countries. The head of the Commission noted that the canal system was "vital to the life of the Punjab" but had been "developed only under the conception of a single administration." He expressed the hope that some form of joint control over the canal system could be established. R. Baxter, 1967, The Indus basin, in A. Garretson, R. Hayton and C. Olmstead (eds.) *The Law of International Drainage Basins*, Oceana, Dobbs Ferry, Chapter 9.

55. Telegram of 15 April 1948, from Prime Minister Liaquat Ali Khan of Pakistan to Prime Minister Nehru of India, as quoted in R. Baxter, 1967, The Indus basin, in A. Garretson, R. Hayton and C. Olmstead (eds.) *The Law of International Drainage Basins*, Oceana, Dobbs Ferry, Chapter 9. "Without *water for irrigation* [West Punjab] would be desert, 20,000,000 acres would dry up in a week, tens of millions would starve. No army, with bombs and shellfire, could devastate a land as thoroughly as Pakistan could be devistated by the simple expelent of India's permanently shutting off the sources of water that keep the fields and the people of Pakistan alive." D. Lilienthal. 1951. "Another Korea in the Making?" *Collier's* Aug. 4, 1951, p. 58. col. 1 (emphasis in original).

56. F.J. Fowler, 1955, The Indo-Pakistan water dispute, *Year Book of World Affairs, 1955*, vol. IX, London, p. 101.

57. R. Baxter, 1967, The Indus basin, in A. Garretson, R. Hayton and C. Olmstead (eds.) *The Law of International Drainage Basins*, Oceana, Dobbs Ferry, Chapter 9. The flow of water through the canals to Pakistan appears to have been cut off for most, if not all, of the month of April, 1949. The conference was held in New Delhi on 3 May 1948, and the Inter-Dominion Agreement between India and Pakistan on the Canal Water Dispute between East and West Punjab was signed on 4 May 1948. The agreement is published in United Nations Treaty Series (hereafter UNTS), no. 54, p. 45, and relevant portions are quoted in M. Whiteman, 1964, Indus, *Digest of International Law*, vol. 3, Chapter VII, section 20, Department of State, Washington, DC.

58. The ideas were those of David Lilienthal. D. Lilienthal, 1951, Another "Korea" in the making? *Collier's*, 5, July/September, 22. The proposal of the Bank was made by its President, Eugene R. Black, in "The World Economic Situation," address before the Board of Governors of the International Bank for Reconstruction and Development (IBRD, or World Bank), Department of State, 1952, *Bulletin*, **XXVII**(690), 15 September 1952, p. 385.

59. Signed at Karachi, 19 September 1960, UNTS, no. 419, p. 126.

60. Indus Basin Development Fund Agreement, Australia, Canada, Germany, New Zealand, Pakistan, the United Kingdom, the United States, and the World Bank, signed at Karachi, 19 September 1960, UST, no. 12, p. 19; UNTS, no. 444, p. 259.

61. Indus Waters Treaty, 1960, UNTS, no. 419, p. 126, article 11, paragraph 2.

62. While ground water that interacts with surface water should be included in any agreement concerning international watercourses, the Indus Waters Treaty seems to be understood to exclude ground water. See, for example, Office of the Commissioner for Indus Waters, Pakistan, The Permanent Indus Commission, in United Nations, 1983, *Experiences in the Development and Management of International River and Lake Basins*, Natural Resources/Water Series No. 10, UN Document ST/ESA/120, UN, New York.

63. A recent article by an official of the Pakistani Water and Power Development Authority points to several areas in which Pakistan believes India is not fulfilling its obligations under the 1960 Treaty: sharing of data; new hydroelectric projects on western rivers; and India's construction of Wuller Barrage and Storage Works on the Jhelum Main without having informed Pakistan, and creation of a live storage capacity behind the barrage far in excess of treaty allowances. M.Y. Khan, 1990, Boundary water conflict between India and Pakistan, *Water International*, **15**, 195; abstracted in *Transboundary Resources Report*, **5**(1), Spring 1991, International Transboundary Resources Center, Albuquerque. India would undoubtedly contest at least some of these charges, but no published account of such replies has been discovered.

64. For a fuller discussion of the controversy dealt with in this section, see J.

Lammers, 1984, *Pollution of International Watercourses*, Martinus Nijhoff, Boston.

65. United Nations, 1968, *Official Records of the General Assembly*, 23rd Session, Plenary Meetings, 1682nd meeting, paragraph 177; 1976, 31st session, Special Political Committee, 21st meeting, paragraph 15; and 1969, 24th session, Plenary Meetings, 1776th meeting, paragraph 285.

66. 1978, International Legal Materials (hereafter ILM), no. 17, p. 103. The Agreement entered into force on 5 November 1977 for a period of five years.

67. As reported by Mr Suresh Chadurgedi, Legal Adviser, Mission of India to the United Nations. Telephone conversation of 21 October 1991.

68. The Joint Rivers Commission was established in March 1972 to advise the governments of India and Bangladesh on the development and allocation of shared water resources. The Commission consists of four members from each country, at least two of whom are to be engineers. Ministry of Power, Water Resources and Flood Control, Bangladesh, 1983, International rivers – The experience of Bangladesh, in United Nations, 1983, *Experiences in the Development and Management of International River and Lake Basins*, Natural Resources/Water Series No. 10, UN Document ST/ESA/120, UN, New York.

69. India had proposed linking the Ganges with the Brahmaputra and storing water in the latter river system. United Nations, 1976, *Official Records of the General Assembly*, 31st session, Special Political Committee, 21st meeting, paragraph 15.

70. See generally International Joint Commission, 1983, The International Joint Commission: Canada–United States, in United Nations, 1983, *Experiences in the Development and Management of International River and Lake Basins*, Natural Resources/Water Series No. 10, UN Document ST/ESA/120, UN, New York.

71. See generally J. Friedkin, 1983, The International Boundary and Water Commission: Mexico–United States, in United Nations, 1983, *Experiences in the Development and Management of International River and Lake Basins*, Natural Resources/Water Series No. 10, UN Document ST/ESA/120, UN, New York.

72. Letter of 21 October 1895 from the Mexican Minister, Matias Romero, to the United States Secretary of State, Richard Olney.

73. United States of America, 1898, *Official Opinions of the Attorneys General of the United States*, vol. XXI, PP. 281–282, Washington, DC, quoting, remarkably, from an opinion of Chief Justice Marshall of the U.S. Supreme Court in a sovereign immunity case.

74. See, for example, R. Johnson, 1967, The Columbia basin, in A. Garretson, R. Hayton and C. Olmstead (eds.) *The Law of International Drainage Basins*, Oceana, Dobbs Ferry: "Briefly, this doctrine [the Harmon Doctrine] suggests that a sovereign nation can do as it pleases with the portion of an international river found within its borders, regardless of the impact on the downstream nation."

75. J. Moore, 1906, *A Digest of International Law*, **1**, 653–654, U.S. Government, Washington, DC.

76. Boundary Convention between the United States and Mexico, 1 March 1889, C. Parry (ed.), 1978, *The Consolidated Treaty Series*, vol. 172, p. 21.

77. United States of America, Memorandum of the State Department of 21 April 1958, *Legal Aspects of the Use of Systems of International Waters with Reference to Columbia-Kootenay River System under Customary International Law and the Treaty of 1909*, 85th Congress, 2nd Session, Senate document no. 118, Washington, DC, p. 64.

78. Signed at Washington on 21 May 1906, C. Parry (ed.), 1978, *The Consolidated Treaty Series*, vol. 201, p. 225.

79. See, for example, A. D'Amato, 1970, *The Concept of Custom in International Law*, Cornell University Press, Ithaca.

80. United States of America, Memorandum of the State Department of 21 April 1958, *Legal Aspects of the Use of Systems of International Waters with Reference to Columbia-Kootenay River System under Customary International Law and the Treaty of 1909*, 85th Congress, 2nd Session, Senate document no. 118, Washington, DC, p. 64.

81. For example, the Colorado River salinity problem, caused by diversions and irrigation in the south-western United States, a problem that has persisted despite the 1973 Agreement concerning the Permanent and Definitive Solution to the International Problem of the Salinity of the Colorado River, UST, no. 24, 1968, reprinted in ILM, vol. 12, p. 1105, 1973. On this problem, see A. Kneese, 1988, Environmental stress and political conflicts: Salinity in the Colorado River, discussion paper for the conference on Environmental Stress and Political Conflicts, Royal Swedish Academy of Sciences, Stockholm, Sweden, 13–15 December 1988; and R. Fradkin, 1981, *A River No More*, University of Arizona Press, Tucson. While the United States is sending salty water to Mexico, Mexico is sending sewage

to the U.S. See, for example, C. Metzner, Transboundary sewage problems: Tijuana/San Diego – New River/Imperial Valley, *Transboundary Resources Report*, **2**(1), International Transboundary Resources Center, Albuquerque; California in brief: Emergency declared on sewage discharge, *LA Times*, 16 March 1991, p. A23; P. McDonnell, 1990, U.S., Mexico sign pact on sewage plant: Border agreement calls for building $200 million facility to handle sewage flowing from Tijuana into San Diego, *LA Times*, 3 July 1990, p. A25; and W. Branigan, 1989, Pollution under scrutiny at U.S.-Mexican border, *Washington Post*, 24 October 1989, p. A27.

82. See generally M. Whiteman, 1964, Indus, *Digest of International Law*, vol. 3, Chapter VII, section 20, Department of State, Washington, DC; and R. Johnson, 1967, The Columbia basin, in A. Garretson, R. Hayton and C. Olmstead (eds.) *The Law of International Drainage Basins*, Oceana, Dobbs Ferry.

83. Treaty relating to Cooperative Development of the Water Resources of the Columbia River Basin, signed at Washington, 17 January 1961, and exchanges of notes at Washington, 22 January 1964, and at Ottawa, 16 September 1964, UST no. 15, p 1555; UNTS, no. 542, p. 244.

84. R. Johnson, 1967, The Columbia basin, in A. Garretson, R. Hayton and C. Olmstead (eds.) *The Law of International Drainage Basins*, Oceana, Dobbs Ferry.

85. R. Johnson, 1967, The Columbia basin, in A. Garretson, R. Hayton and C. Olmstead (eds.) *The Law of International Drainage Basins*, Oceana, Dobbs Ferry.

86. M. Whiteman, 1964, Indus, *Digest of International Law*, vol. 3, Chapter VII, section 20, Department of State, Washington, DC.

87. R. Johnson, 1967, The Columbia basin, in A. Garretson, R. Hayton and C. Olmstead (eds.) *The Law of International Drainage Basins*, Oceana, Dobbs Ferry.

88. M. Whiteman, 1964, Indus, *Digest of International Law*, vol. 3, Chapter VII, section 20, Department of State, Washington, DC, quoting from a statement of the Chairman of the United States Section of the International Joint Commission, Governor Len Jordan. See also R. Johnson, 1967, The Columbia basin, in A. Garretson, R. Hayton and C. Olmstead (eds.) *The Law of International Drainage Basins*, Oceana, Dobbs Ferry, summarizing U.S. and Canadian arguments.

89. Treaty relating to the Boundary Waters and Questions Arising Along the Boundary between the United States and Canada, Signed at Washington, 11 January 1909, C. Bevans, 1968–1974, *Treaties and Other International Agreements of the United States of America 1776–1949*, vol. 12, p. 319. In essence, article II reserves to each of the two countries "the exclusive jurisdiction and control over the use and diversion ... of all waters on its own side of the line which in their natural channels would flow across the boundary or into boundary waters." It also provides that any diversion of these waters causing injury on the other side of the boundary "shall give rise to the same rights and entitle the injured parties to the same legal remedies as if such injury took place in the country where such diversion or interference occurs." Canada argued that the latter provision made clear that the intent of the treaty was to permit the contemplated diversions but to permit private parties injured thereby to seek redress through the courts.

90. As summarized by L. Bloomfield, and G. Fitzgerald, 1958, *Boundary Waters Problems of Canada and the United States, The International Joint Commission 1912–1958*, Carswell, Toronto.

91. R. Johnson, 1967, The Columbia basin, in A. Garretson, R. Hayton and C. Olmstead (eds.) *The Law of International Drainage Basins*, Oceana, Dobbs Ferry.

92. See generally G. Cano, 1976, Argentina, Brazil and the De la Plata River basin: A summary review of their legal relationship, *Natural Resources Journal*, **16**, Albuquerque, New Mexico; C. Caubet, 1983, Le Barrage d'Itaipu et le droit international fluvial, doctoral thesis, University of Toulouse; and P. Dupuy, 1978, La gestion concertée des ressources naturelles: à propos du différend entre le Brésil et L'Argentine relatif au barrage d'Itaipu, *Annuaire français de droit international*, **VI**, XXIV, Paris.

93. According to one account, the Itaipú dam "will be the mightiest hydroelectric project in the world." It is longer than 40 city blocks and the reservoir it created stretches for 200 km. National Geographic Society, 1984, *Great Rivers of the World*, Washington, DC.

94. In addition to the UNEP Principles concerning Shared Natural Resources, referred to in the following footnote, see, for example, article 3 of the Charter of Economic Rights and Duties of States, 12 December 1974, UN General Assembly Resolution 3281 (XXIX), 29 UN GAOR Supplement, no. 31, p. 50, UN Document A/9631, 1975, reprinted in vol. 14 ILM, **14**, p. 251, 1975.

95. This was true of both Principle 20 of the Draft Stockholm Declaration on the Human Environment, adopted by the UN Conference on the Human Environment, Stockholm, 16 June 1972; and the UNEP Draft Principles of Conduct in the Field of the Environment for the Guidance of States in the Conservation and Harmonious Utilization of Natural Resources Shared by Two or More States, Decision 6/14 of the Governing Council of UNEP, 19 May 1978. Concerning the former, see, for example, A. Beesley, 1973, The Canadian approach to international environmental law, *Canadian Yearbook of International Law*, p. 3. Concerning the latter, see, for example, 1983 Note presented by Mr Constantin A. Stavropoulos, UN Document A/CN.4/L.353, *Yearbook of the International Law Commission, 1985*, vol. II, part One, UN, New York.

96. On 29 September 1972 Argentina and Brazil concluded an agreement providing that they would cooperate by providing technical data regarding works to be undertaken within their jurisdiction in order to prevent any appreciable harm which might be caused in the human environment of neighboring areas. The text of the agreement was later adopted as UN General Assembly resolution 2995 (XXVII) of 15 December 1972, but Argentina denounced the agreement on 10 June 1973, *inter alia*, because of disagreements with Brazil over which country would be the judge of whether planned works might cause harm to a neighboring country.

97. Agreement between Argentina, Brazil, and Paraguay on Paraná River projects of 19 October 1979. English translation reproduced in ILM, no. 19, p. 615, 1980.

98. The 1898 incident concerning the headwaters of the Nile (see note 37) is not considered significant for this purpose because hostilities did not, in fact, ensue.

99. L. Henkin, 1979, *How Nations Behave*, 2nd ed., Columbia University Press, New York.

100. See generally T. Frank, 1988, Legitimacy in the international system, *American Journal of International Law*, **82**, 705; and P. Weil, 1983, Towards relative normativity in international law?, *American Journal of International Law*, **77**, 413.

101. See generally A. D'Amato, 1970, *The Concept of Custom in International Law*, Cornell University Press, Ithaca; and M.B. Akehurst, 1977, Custom as a source of international law, in [1974–1975] *British Yearbook of International Law*, Oxford, UK, p. 40.

102. Examples of such organizations are the International Law Commission of the United Nations, the International Law Association and the Institute of International Law. As will be seen below, each of these organizations has produced drafts concerning the law of international watercourses which at least to some extent codify customary international law on that subject.

103. Food and Agriculture Organization of the United Nations, 1978 (as updated), *Systematic Index of International Water Resources Treaties, Declarations, Acts and Cases by Basin*, Legislative Study no. 15, FAO, Rome. A compilation of treaty provisions relating to non-navigational water uses published by the United Nations in 1963 contains excerpts from 253 agreements. United Nations, 1963, *Legislative Texts and Treaty Provisions Concerning the Utilization of International Rivers for Other Purposes than Navigation*, UN Document ST/LEG/SER.B.B./12, UN, New York. This compilation is updated to 1974 in United Nations, 1976, *Yearbook [1974] of the International Law Commission*, vol. II, part 2, UN, New York.

104. For example, the FAO, 1978, *Systematic Index of International Water Resources Treaties, Declarations, Acts and Cases by Basin*, Legislative Study no. 15, FAO, Rome, contains a grant of freedom of navigation to a monastery of the year 805 and a bilateral treaty concerning the Weser River basin of 2 October 1221.

105. There are some fourteen multilateral agreements concerning international watercourses, excluding the Barcelona Convention (see note 106), which deals with navigation. All but one of these (the 1923 Geneva Convention on hydraulic power, referred to in note 106) were concluded since 1950 and nine were concluded, amended, or supplemented since 1970. The rivers, lakes, and basins covered, some of which are the subject of more than one agreement, are as follows: the Danube River, Lake Constance, the Mosel River, the Rhine River, the Niger basin, the Lake Chad basin, the La Plata river basin, the Senegal River, the Kagera River Basin, the River Gambia, the Amazon River, and the Zambezi River system. These agreements, as well as various non-binding instruments concerning transboundary freshwaters, are conveniently listed in UN Document A/CONF.151/PC/79, 1 July 1991, at pp. 4–5.

106. See the Convention relating to the Development of Hydraulic Power Affecting More than One State, and Protocol of Signature, Geneva, 9 December 1923, LNTS, no. 36, p. 77; and the Convention and Statute on the Regime of Navigable Waterways of International Concern, Barcelona, 20 April 1921, LNTS, no 7, p. 37. The former convention has not been applied in practice, while the latter concerns navigation and applies only to international rivers declared to be "navigable waterways of international concern," United Nations, 1976, *Yearbook [1974] of the International Law*

*Commission*, vol. II, part 2, UN, New York. After this chapter was written a multilateral treaty on shared water resources was adopted by states in the ECE (U.N. Economic Commission for Europe) region. See the U.N. Convention on the Protection and Use of Transboundary Watercourses and International Lakes, done at Helsinki, March 17, 1992, reprinted in International Legal Materials, vol. 31, p. 1312 (1992).

107.   One of the sources of international law that the UN International Court of Justice is to apply in deciding cases submitted to it is ''international custom, as evidence of a general practice accepted as law,'' Statute of the International Court of Justice, article 38, paragraph 1 (b). Charter of the United Nations with the Statute of the International Court of Justice annexed thereto, San Francisco, 26 June 1945, UNTS, no. 1, p. 16. Of course, one difficulty with extrapolating general, community-wide rules from bilateral water treaties is that the parties to those treaties rarely have relevant practice with other states – in contrast, for example, with the field of diplomatic law.

108.   See generally the survey, by region, of treaty provisions concerning contiguous and successive watercourses, respectively, in S. McCaffrey, 1986, *Second Report on the Law of the Non-Navigational Uses of International Watercourses*, 38th Session of the International Law Commission, in United Nations, 1976, *Yearbook [1974] of the International Law Commission*, vol. II, part 1, p. 88, UN, New York. Illustrative treaties are the 1906 Convention concerning the Equitable Distribution of the Waters of the Rio Grande, 1980 (see note 78); the 1960 Indus Waters Treaty (see note 59); the 1961 Columbia River Treaty (see note 77); the Convention relating to the Status of the Senegal River, 11 March 1972, United Nations, 1984, *Treaties concerning the Utilization of International Watercourses for Other Purposes than Navigation: Africa*, Natural Resources/Water Series no. 13, p. 16; the Convention between Yugoslavia and Austria concerning Water Economy Questions relating to the Drava, 25 May 1954, UNTS, no. 227, p. 111; and the Treaty of the Rio de la Plata Basin, 23 April 1969, UNTS, no. 875, p. 3.

109.   See, for example, article 5 of the draft articles on international watercourses adopted by the UN International Law Commission, 1991, *Report of the International Law Commission on the Work of Its Forty-Third Session*, p. 163; and article IV of the Helsinki Rules on the Uses of the Waters of International Rivers, International Law Association, 1966, *Report of the Fifty-Second Conference, Helsinki, 1966*. According to the Helsinki Rules, the principle of equitable utilization is the paramount and guiding principle with regard to the uses of international rivers.

110.   This principle is reflected in many of the treaty provisions concerning successive watercourses cited in S. McCaffrey, 1986, *Second Report on the Law of the Non-Navigational Uses of International Watercourses*, 38th

Session of the International Law Commission, in United Nations, 1976, *Yearbook [1974] of the International Law Commission*, vol. II, part 1, p. 88, UN, New York.

111.   See, for example, article 7 of the UN International Law Commission, 1991, *Report of the International Law Commission on the Work of Its Forty-Third Session*. See also article 1 of the International Law Association, 1982, Montreal Rules on Water Pollution in an International Drainage Basin, 1982, *Report of the Sixtieth Conference*, Montreal, p. 535.

112.   The International Law Association subscribes to this view. See note 109. In contrast, the UN International Law Commission has taken the position that the obligation not to harm other states is the controlling norm. See its 1988 *Report of the International Law Commission on the Work of Its Fortieth Session*, p. 84.

113.   *Annuaire de l'Institut de Droit International*, 1961, vol. 49-II, Salzburg Session, September, Basel, pp. 381–384 (hereafter referred to as the 1961 Resolution).

114.   *Annuaire de l'Institut de Droit International*, 1980, vol. 58-I, Athens Session, September, Basel/Munich, p. 197 (hereafter referred to as the 1979 Resolution).

115.   1961 Resolution.

116.   1979 Resolution.

117.   See note 109.

118.   See note 111.

119.   See International Law Association's Montreal Rules, 1982 (note 111).

120.   International Law Association (Helsinki Rules Report), 1966, see note 109.

121.   International Law Association (Helsinki Rules Report), 1966, see note 109.

122.   On the ILC, see generally I. Sinclair, 1987, *The International Law Commission*, Grotius, Cambridge.

123.   The full set of articles is contained in the Commission's report to the General Assembly, cited in note 109. The articles have been sent to member governments of the UN for their comments, which are due by 1 January 1993.

124.   ILC draft articles, 1991, see note 109.

125.   ILC draft articles, 1991, see note 109.

126.   These points are discussed in S. McCaffrey, 1989, The law of international watercourses: Some recent developments and unanswered questions, *Denver Journal of International Law and Policy*, **17**(3), 505.

127.   See paragraph 5 of the commentary to article 7 of the ILC's draft, 1991 (see note 109).

128.   R. Coase, 1960, The problem of social cost, *Journal of Law and Economics*, **3**, 1.

# Chapter 9
# Water in the 21st century

**Peter H. Gleick**

## Introduction

As the 20th century comes to a close, arguments about the continued viability of our patterns of development, industrialization, and resource use are intensifying. For the first time, the goal of attaining an equitable and ''sustainable'' society has been raised, and despite difficult and unresolved questions about the meaning of these terms, the debate has reached the entire international community. Fresh water resources play an integral role in this debate because they play an integral role in determining our overall quality of life.

Many of the fresh water problems that now challenge us are discussed in detail in the earlier chapters of the book. But as we look to the future, a new set of problems looms large, including rapidly growing populations in developing countries, uncertainties about the impacts of global climatic changes, and possible conflicts over shared fresh water resources. The purpose of this chapter is to explore these new problems and to offer some ideas for how we might respond to them.

First we must admit that ''we don't know what we don't know.'' Our crystal ball is murky, and we can see only dimly what the future might bring. On the positive side, new technologies may offer solutions to some of our problems. Renewable energy resources could come of age and eliminate our excessive dependence on fossil fuels and nuclear power. The end of the Cold War could, if we chose, permit us to devote renewed attention and resources to more pressing human needs. And massive environmental problems such as global climate change might not materialize as quickly or as severely as some predict.

On the other side, however, there could be unpleasant surprises. Natural ecosystems, which provide vital goods and services to humans, may completely collapse if further stressed. New holes in our protective ozone layer may appear over populated areas. Nonlinear atmospheric behavior may lead to abrupt and unpredictable changes in climatic conditions. New or existing diseases may spread rapidly and uncontrollably in parts of the world, as is AIDS in Africa and Asia, and as is cholera in Latin America.

Lack of perfect foresight does not mean that we have to be completely unprepared. Problems are more easily solved for a population of 6,000 million people than for a population twice that size. Protecting natural ecosystems that provide us with essential goods and services or that harbor populations of endangered species makes sense today for many reasons. We want and we need, food and water, clean air and benevolent climates, safe energy, and medicines and materials to support the generations to come. Yet much of our current resource use is characterized by enormous waste. Improving the efficiency with which we use resources increases the number of people whose needs can be supplied, and reduces the environmental costs of satisfying those needs. By looking for solutions that can be tried today, we may be able to avoid the worst of tomorrow's problems.

## Population pressures

Population plays a fundamental role in questions about future water availability, use, and quality. The population of the earth has more than doubled in just the last 40 years from 2,500 million in 1950 to over 5,300 million in 1990, and it is increasing faster than ever before. By the turn of the century, the total population will exceed 6,000 million people, an increase of 20% in only 10 years. New medium population estimates from the United Nations suggest that if present trends continue, global population will reach 10,000 million by 2050 and ultimately stabilize at just under 12,000 million people after 2100. This estimate is 10% higher than the previous United Nations medium estimate because of observed increases in the upper limit of life expectancy and the projected rate of changes in fertility.[1]

Total populations and growth rates are very different in developing and developed countries. The population of the major industrialized countries in 1990 was about 1,200 million people; the other 77% of the population lived in developing countries. The rate of growth in developed countries is well under 1% per year; in developing countries it exceeds 2% per year, and in some parts of Africa, Asia, and the Middle East it exceeds 3% per year. As a result, over 90% of all future population increases will occur in the developing world. This means that almost all new births will be in regions where access to clean water and sanitation services, adequate health care and education, and other fundamental requirements for a satisfactory quality of life are severely lacking.

Our failure to provide these requirements already has appalling consequences. Over one-third of all deaths in developing countries in recent years were children under five years old and under-five mortality rates in Africa are roughly fifteen times the rates in North America.[2] In almost every developing country, diarrhea is the primary cause of death among these children; the direct result of inadequate sanitation and water supply. Other water-related diseases, including malaria, schistosomiasis, onchocerciasis, and cholera are also responsible for hundreds of millions of cases of disease, and millions of deaths every year.

In 1980, the United Nations launched the International Drinking Water Supply and Sanitation Decade, with the goal of providing sanitation services and access to safe drinking water to those without them. Yet a decade later, after enormous effort, expense, and progress, 1,800 million people are still without access to sanitation services and nearly 1,300 million lack access to clean water, in large part because population growth wiped out progress achieved in these areas. And these overall numbers hide some ugly regional problems. For example, the total populations in urban areas needing both clean water and sanitation grew substantially over the decade, reflecting the massive migrations to large, urban centers in developing countries and our inability to provide for them.[3] The situation looks even bleaker for the next decade; the United Nations estimates that by the year 2000, an additional 900 million people will be born into regions where these services are not available.[4]

Because the total amount of fresh water is fixed, growing population will continuously reduce the amount available per person. Some regions of the world are already unable to provide adequate supplies of water to their population, and more will run up against this constraint in the future. Many countries already use a substantial portion of the total water supply available to them. Table 9.1 lists 21 countries whose present water withdrawals exceed one-third of their total renewable supply of water. Nine of these, all in the Middle East, withdraw more than 100% of their annual renewable supply and are already forced to import additional fresh water, pump ground water at a non-renewable rate, or desalinate seawater at great expense.

Growing populations will worsen problems with absolute water availability. In 18 countries of the world in 1990, absolute per capita water availability falls below 1,000 m$^3$ per year, a level some suggest is an approximate minimum necessary for an adequate quality of life in a moderately developed country.[5] By the year 2025, over 30 countries will be unable to provide 1,000 m$^3$ per person per year, simply because of popu-

**TABLE 9.1** Ratio of water demand to supply by country

| Country | Water withdrawals as a percentage of internal renewable supplies and river flows from other countries[a] |
|---|---|
| Libya | 374 |
| Qatar | 174 |
| United Arab Emirates | 140 |
| Yemen | 135 |
| Jordan | 110 |
| Israel | 110 |
| Saudi Arabia | 106 |
| Kuwait | >100 |
| Bahrain | >100 |
| Egypt | 97 |
| Malta | 92 |
| Belgium | 72 |
| Cyprus | 60 |
| Tunisia | 53 |
| Afghanistan | 52 |
| Pakistan | 51 |
| Barbados | 51 |
| Iraq | 43 |
| Madagascar | 41 |
| Iran | 39 |
| Morocco | 37 |

*Source:* P.H. Gleick, 1992, Effects of climate change on shared fresh water resources, in I.M. Mintzer (ed.) *Confronting Climate Change: Risks, Implications and Responses*, Cambridge University Press, Cambridge, pp. 127–140.

[a] These data are for the late 1980s. Nine countries use more than 100% of available renewable supply. This means that these countries partly depend on imports of fresh water, ground water, or desalination of brackish or salt water.

lation growth. Table 9.2 lists per capita water availability in 1990 and in 2025 for those countries with less than 1,000 m$^3$ per person per year of fresh water. Note the large numbers of countries in Africa and Asia with low per capita water availability. In contrast, few countries in North and Central America and in Europe are near these limits.

Water use of 500 m$^3$ per person per year might suffice in a semi-arid society with extremely sophisticated water management, as in Israel, but even here water resources scarcity is already causing political and social stresses. Currently, 12 countries cannot provide this level of supply according to Table 9.2, and by 2025 this will increase to 19 countries. Unless water of adequate quality and quantity is provided to the world's population, the level of disease and misery will not only remain high, but will increase with the growing population.

The world's population cannot continue to grow indefinitely. It must be stabilized as quickly as possible while efforts are also made to reduce the enormous suffering already experienced by hundreds of millions of the world's people. The problem of population must be tackled directly, through improved family planning, increased economic development in poorer regions, increased availability of contraception, and education about birth control. The education of women, in particular, appears to be one of the most important steps that can be taken to improve the quality of health care, reduce population growth rates, and provide clean drinking water. Throughout the world, poor rural women spend 60–90 hours per week gathering wood, collecting water, preparing food, and caring for children. In some regions, women must spend 4–5 hours each day carrying water from distant sources. Education of women can improve child health and often leads to improved availability of water and sanitation.[6] At the same time, educating women about family planning and public health appears to be effective in reducing population growth rates. The ideological reluctance to tackle these issues in some parts of the world needs to be overcome.

## Climate changes

Global climatic changes – the so-called greenhouse effect – further complicate future water resources planning. Human activities over the last

**TABLE 9.2** Per capita water availability today and in 2025, selected countries

| Country | Per capita water availability, 1990 (m$^3$ per person per year) | Projected per capita water availability, 2025 (m$^3$ per person per year) |
|---|---|---|
| *Africa* | | |
| Algeria | 750 | 380 |
| Burundi | 660 | 280 |
| Cape Verde | 500 | 220 |
| Comoros | 2,040 | 790 |
| Djibouti | 750 | 270 |
| Egypt | 1,070 | 620 |
| Ethiopia | 2,360 | 980 |
| Kenya | 590 | 190 |
| Lesotho | 2,220 | 930 |
| Libya | 160 | 60 |
| Morocco | 1,200 | 680 |
| Nigeria | 2,660 | 1,000 |
| Rwanda | 880 | 350 |
| Somalia | 1,510 | 610 |
| South Africa | 1,420 | 790 |
| Tanzania | 2,780 | 900 |
| Tunisia | 530 | 330 |
| *North and Central America* | | |
| Barbados | 170 | 170 |
| Haiti | 1,690 | 960 |
| *South America* | | |
| Peru | 1,790 | 980 |
| *Asia/Middle East* | | |
| Cyprus | 1,290 | 1,000 |
| Iran | 2,080 | 960 |
| Israel | 470 | 310 |
| Jordan | 260 | 80 |
| Kuwait | <10 | <10 |
| Lebanon | 1,600 | 960 |
| Oman | 1,330 | 470 |
| Qatar | 50 | 20 |
| Saudi Arabia | 160 | 50 |
| Singapore | 220 | 190 |
| United Arab Emirates | 190 | 110 |
| Yemen | 240 | 80 |
| *Europe* | | |
| Malta | 80 | 80 |

*Source:* P.H. Gleick, 1992, Effects of climate change on shared fresh water resources, in I.M. Mintzer (ed.) *Confronting Climate Change: Risks, Implications and Responses*, Cambridge University Press, Cambridge, pp. 127–140.

century have led to increases in the atmospheric concentration of trace gases* that trap heat in the atmosphere. As the concentrations of these gases rise, the behavior of the earth's climate will be affected in ways that are only partly understood, but that will include higher temperatures, changes in precipitation patterns and sea level, and alterations in the frequency and intensity of major storms.

Greenhouse gases come in large part from the combustion of fossil fuels and from deforestation, which removes a sink for carbon dioxide and releases $CO_2$ into the atmosphere when the forests are burned. While many actions can be taken to reduce the emissions of these gases, growing populations and growing energy use will make it exceedingly difficult to prevent some climatic changes from occurring. And political disputes over the responsibility for and consequences of these changes will complicate international responses.[7]

These climatic changes will, in turn, greatly affect fresh water re-

*These gases, sometimes call ''greenhouse gases,'' include carbon dioxide ($CO_2$), nitrous oxide ($N_2O$), methane ($CH_4$), chlorofluorocarbons (CFCs), and ozone ($O_3$).

sources. Higher temperatures will increase evaporation, change snowfall and snowmelt patterns, and lead to alterations in water demand. Changes in rainfall will affect water availability in rivers and lakes, hydroelectricity generation, and agricultural productivity. Rising oceans will contaminate coastal fresh water aquifers.

Many uncertainties remain about climate changes and their impacts on water resources. This results in part from limitations in the ability of large-scale climate models to incorporate and reproduce important aspects of the hydrologic cycle. Many important hydrologic processes, such as the dynamics of clouds and rain-generating storms, occur on spatial scales far smaller than can yet be modeled. At the same time, the hydrologic processes included in the models are far simpler than those in the real world. We thus know far less than we would like about how the global water cycle is likely to change. Despite these limitations, research in the last few years has revealed some important things about how hydrology and water supplies may be affected by climatic changes.

**Changes in precipitation and temperature.** Temperature and precipitation are key factors in determining the distribution of natural ecosystems, what crops we grow in what places, characteristics of human habitations and energy use, and how much water we use for different aspects of human development. The best current estimate is that doubling the concentration of greenhouse gases in the atmosphere, which is expected to occur within the next several decades given current trends, would cause the global average temperature of the earth to increase by about 3°C.[8] This increase will occur at a rate far faster than any comparable change seen over the past 10,000 years, and will raise the earth's temperature higher than that experienced over the past two to three million years. Today's atmospheric carbon dioxide concentration is already higher than it has been for 160,000 years.[9] As temperatures rise, the evaporation of water from land and water surfaces will increase, as will global average precipitation. Most recent estimates are that global precipitation will increase by between 3% and 15% for an equivalent doubling of atmospheric carbon dioxide concentration. The greater the warming, the larger these increases.[10]

Regional changes will differ substantially from global ones. For example, air temperatures over land will increase faster than over water, warming in the winter is predicted to be greater in the northern latitudes than toward the equator, and temperature increases in southern Europe and central North America are predicted to be higher than the global average.[11] Precipitation is expected to fall more consistently and intensely throughout the year at high latitudes and in the tropics, and in mid-latitudes in winter. In certain zones, such as 35–55°N, precipitation may increase by as much as 10%–20%,[12] though summer rainfall may decrease over much of the northern mid-latitude continents. There are few consistent and large-scale changes in precipitation predicted for subtropical arid regions, but even small changes in these arid zones can lead to large changes in ecological and human systems.[13]

**Changes in soil moisture.** Another important water resource variable is the water held in soils, which determines what plants can grow in different regions and how much additional water needs to be applied with irrigation to grow a crop. Soil moisture is often overlooked in climate impact studies, in part because of the difficulty of accurately modeling the stocks and flows of water in the ground. An increase in precipitation does not necessarily mean a wetter land surface or more soil moisture because increases in evaporation may exceed the increases in precipitation, leading to a net drying of the land surface.[14] Indeed, one important recent finding was that the incidence of droughts in the United States, measured by an index that looks at soil wetness conditions, may dramatically increase as temperatures go up despite an accompanying increase in precipitation, because of the increased evaporative losses.[15] A similar result has been observed in detailed hydrologic modeling of river basins where large increases in precipitation may be necessary in order to maintain present river runoff levels as temperatures and evaporative losses rise.[16]

Most climate models also suggest large-scale drying of the Earth's surface over northern mid-latitude continents in summer owing to higher temperatures and either insufficient precipitation increases or actual re-

ductions in rainfall. For example, a review of the soil–moisture results from five different climate models showed substantial agreement on decreased soil moisture in the central United States.[17] Drying in these regions would affect both agricultural production and other water demands.

Climate models typically exclude the direct effect of $CO_2$ concentrations on vegetation and the role of other factors in altering evapotranspiration. Greenhouse warming may alter temperature, cloudiness, wind conditions, humidity, plant growth rates, rooting, leaf area, and so on. Higher $CO_2$ levels have been shown in laboratory and greenhouse studies to enhance plant growth and alter the efficiency with which water is used by certain plants, but the effect on plants in the real world is still uncertain because of many complicating factors and feedbacks. Recent work suggests that evapotranspiration in some plant communities is more responsive to changes in air temperature, the resistance of plant stomata, and net radiation, and less responsive to leaf area index, vapor pressure, and windspeed.[18] More careful study of these many complicating factors is necessary.

**Changes in snowfall and snowmelt.** One of the most important hydrologic impacts of climatic change will be changes in snowfall and snowmelt behavior. In basins with substantial snowfall and snowmelt at some time during the year, snowpack acts as a large storage reservoir that redistributes runoff over time. Temperature increases in these basins will have three effects: they will increase the ratio of rain to snow in cold months, decrease the overall length of the snow season, and increase the rate and intensity of snowmelt. As a result of these effects, average winter runoff and average peak runoff will increase, peak runoff will occur earlier in the year, and there will be a faster and more intense drying out of soil moisture. This mechanism was identified as a leading cause of summer soil moisture drying in climate model results,[19] and has also been observed in regional hydrologic modeling results.[20]

One consequence of these effects may be a substantial increase in the risk of floods. In many mid- and high-latitude river basins, the worst floods occur during snowmelt runoff periods or when rain falls on snow. One of the greatest worries about higher temperatures is, therefore, the increased probability *and* intensity of flood flows. When combined with possible precipitation increases in many regions, flooding becomes a critical concern.

Changes in snow dynamics will also complicate water management in places where reservoirs are managed both to control winter floods and to provide water during the dry summer season, such as in the western United States. Determining when to begin storing water for coming dry periods without increasing flood risks under conditions of global warming is a challenge water managers will have to begin to address.[21]

**Changes in storm frequency and intensity.** Another vitally important question is what will happen to the variability of the climate, i.e., the frequency and intensity of extremes. Extreme weather events, including typhoons, hurricanes, and monsoons, bring both death and life to many regions. Of all natural disasters, floods, tropical cyclones, and typhoons are the largest killers. Ninety-five per cent of these deaths occur in developing countries. Single typhoons in southern Asia have killed hundreds of thousands of people in Bangladesh through flooding of low-lying lands. Floods along the Huanghe River in China have killed more people than any other single feature of the earth. In 1887 between 900,000 and 2 million people died from flooding or subsequent starvation; in 1931, between 1 million and 3.7 million people died.[22]

At the same time, intense monsoons bring vitally needed rain to India during the growing seasons, and without this rain, the reliability of crop production would decrease. Although not enough research has been done on this complex issue, there are some indications that the variability of the hydrologic cycle increases when mean precipitation increases and vice versa.[23] In one model study the total global precipitation increased but the area over which it fell decreased, implying more intense local storms and, hence, runoff.[24]

Other changes in variability are also likely, though we have little confidence that we can predict what they will be. There is some indication of possible reductions in day-to-day and interannual variability of storms in the mid-latitudes. At the same time, there is evidence from both model simulations and empirical studies that the frequency, intensity, and area of tropical disturbances may increase.[25] Far more modeling and analytical efforts are needed in this area.

**Changes in runoff.** Fresh water runoff in rivers is of vital importance to human water supplies and natural ecosystems. Any changes in either the timing or magnitude of runoff because of climate change will have widespread ramifications for water supply, energy generation, human health, commercial and industrial development, and environmental conditions. Less runoff in some places will increase pressures on remaining water supplies. More runoff could help water-short areas or cause severe flooding.

Estimates of surface runoff in climate models are not generated directly. Runoff is typically derived from the difference between precipitation and evaporation at the land surface and has no direction of flow, no discrete river basins, and no realistic surface runoff processes. As a result, runoff from these models does not always agree with more detailed regional model results, even in the direction of possible changes.

More realistic estimates of changes in runoff have been produced using detailed regional hydrologic models of specific river basins. By using expected changes in temperature and precipitation, these models permit the incorporation of realistic small-scale hydrology and suggest that some significant changes in the timing and magnitude of runoff are likely to result from quite plausible changes in climatic variables. In basins where demands for water are close to the limit of reliable supplies, such changes will have enormous implications for water supply and planning; where the river basins are shared by two or more nations, such changes could have political implications.

**Water management under uncertainty.** Perhaps the most important effect of climatic change on water resources will be a great increase in the overall uncertainty associated with the management and supply of fresh water resources. Rainfall, runoff, and storms are all natural events with a substantial random component to them – in the language of hydrologists they are stochastic. In many ways, therefore, the science of hydrology is the science of estimating the probabilities of certain types of events. But these estimates are almost always done assuming that climate is stationary, i.e., that the patterns of climate we have seen in the recent past will continue into the future. Indeed, hydrologists have few analytical tools with which they can incorporate future changes of uncertain magnitude.

Recent studies on the effects of future climatic changes suggest that the present methods of water allocation and reservoir operation may worsen water supply and quality problems.[26] Yet no organizations or agencies responsible for shared international river management have yet indicated a willingness to consider changing operating rules to improve their ability to handle possible climatic changes.[27] Adding to this problem is the fact that many water data are still classified as secret by national governments, complicating the rational management of internationally shared fresh water.

Future climatic changes effectively make obsolete all our old assumptions about the behavior of water supply. Perhaps the greatest *certainty* about future climatic changes is that the future will *not* look like the past. We may not know precisely what it *will* look like, but changes are coming, and by the turn of the century, many of these changes may already be apparent.

## Water wars

History is replete with examples of competition and disputes over shared fresh water resources, and there are reasons to believe that tensions over water will increase as more and more people compete for a fixed water supply, as improving standards of living increase the demand for fresh water, and as future global climatic changes make supply and quality more problematic and uncertain. Many rivers and sources of fresh water are shared by two or more nations. This geographical fact has led to the geopolitical reality of disputes over shared international rivers, including the Nile, Jordan, and Euphrates in the Middle East, the Indus, Ganges, and Brahmaputra in southern Asia, and the Colorado, Rio Grande, and

Paraná in the Americas. As growing populations demand more water for agriculture and economic development, strains on limited water resources will grow, and international disputes in water-short regions will worsen. While regional and international legal mechanisms can reduce water-related tensions, as described by McCaffrey in Chapter 8, these mechanisms have never received the support or attention necessary to resolve many conflicts over water. Indeed, there is growing evidence that existing international water law may be unable to handle the strains of ongoing and future problems.

Not all water-resources disputes will lead to violent conflict; indeed most lead to negotiations, discussions, and non-violent resolutions. But in certain regions of the world, such as the Middle East and southern Asia, water is an increasingly scarce resource important for economic and agricultural development. In these regions, the probability of violence, due at least in part to water disputes, is increasing. The following sections identify several classes of water-related disputes and present brief historical examples of each.[28]

**Water resources as military goals.** Although non-renewable resources such as oil and other minerals are more typically the focus of traditional international security analyses,[29] water often provides a source of economic or political strength, and many nations have a large fraction

**TABLE 9.3**   Dependence on imported surface water, selected countries

| Country | Percentage of total flow originating outside of border | Ratio of external water supply to internal supply[a] |
|---|---|---|
| Egypt | 97 | 32.3 |
| Hungary | 95 | 17.9 |
| Mauritania | 95 | 17.5 |
| Botswana | 94 | 16.9 |
| Bulgaria | 91 | 10.4 |
| Netherlands | 89 | 7.9 |
| Gambia | 86 | 6.4 |
| Cambodia | 82 | 4.6 |
| Romania | 82 | 4.6 |
| Luxembourg | 80 | 4.0 |
| Syria | 79 | 3.7 |
| Congo | 77 | 3.4 |
| Sudan | 77 | 3.3 |
| Paraguay | 70 | 2.3 |
| Czechoslovakia | 69 | 2.2 |
| Niger | 68 | 2.1 |
| Iraq | 66 | 1.9 |
| Albania | 53 | 1.1 |
| Uruguay | 52 | 1.1 |
| Germany | 51 | 1.0 |
| Portugal | 48 | 0.9 |
| Yugoslavia | 43 | 0.8 |
| Bangladesh | 42 | 0.7 |
| Thailand | 39 | 0.6 |
| Austria | 38 | 0.6 |
| Pakistan | 36 | 0.6 |
| Jordan | 36 | 0.6 |
| Venezuela | 35 | 0.5 |
| Senegal | 34 | 0.5 |
| Belgium | 33 | 0.5 |
| Israel[b] | 21 | 0.3 |

*Source:* P.H. Gleick, 1992, Effects of climate change on shared fresh water resources, in I.M. Mintzer (ed.) *Confronting Climate Change: Risks, Implications and Responses*, Cambridge University Press, Cambridge, pp. 127–140.

[a] Using national average annual flows. "External" represents river runoff originating outside national borders; "internal" includes average flows of rivers and aquifers from precipitation within the country.
[b] Although only 21% of Israel's water comes from outside current borders, a significant fraction of Israel's fresh water supply comes from disputed lands, complicating the calculation of the origin of surface water supplies. This percentage would be affected by a political settlement of the Middle East conflict.

of their total water supply originating outside of their borders and under the control of other nations. Over 30 nations receive more than one-third of their surface water from outside their national borders, as shown in Table 9.3.[30] Under these conditions, water may provide a justification for going to war or water resources can be the object of military conquest.

Perhaps the best example of a region where fresh water supplies have had clear strategic implications is the Middle East. The Middle East, with its age-old ideological, religious, and geographical disputes, is extremely arid. Even those parts of the Middle East with relatively extensive water resources, such as the Nile, Tigris, and Euphrates river valleys, are coming under increasing population, irrigation, and energy pressures. And every major river in the region crosses international borders.

As far back as the 6th century BC, Ashurbanipal, King of Assyria from 669 to 626 BC, seized water wells as part of his strategy of desert warfare against Arabia.[31] In modern times, we have seen disputes over the Jordan, Nile, Euphrates, and Litani rivers, and over the control and use of ground water resources in the West Bank. One outcome of the 1967 Arab–Israeli War, for example, was the occupation of much of the headwaters of the Jordan River by Israel and the loss to Jordan of a significant fraction of its available water supply. Approximately 40% of the ground water upon which Israel now depends – and more than 33% of its sustainable annual water yield – originates in the occupied territories.[32]

Similarly, Egypt's dependence on the waters of the Nile River is vulnerable to the actions of its upstream neighbors. Ninety-seven per cent of the waters of the Nile originate outside of Egypt's border in the other eight nations of the basin.[†] A treaty signed by Egypt and the Sudan in 1959 resolved many important issues, but none of the other basin states are parties to the treaty. Additional water development in these other nations, which has long been feared by Egypt and sought by Ethiopia and others, would greatly increase tensions over water.

**Water resource systems as military targets.** In political conflicts that escalate to military aggression, water resource systems have regularly been the targets of war. When Sennacherib of Assyria destroyed Babylon in 689 BC as retribution for the death of his son, he pulled down temples and palaces and destroyed the water supply canals to the city.[33] In recent times, hydroelectric dams were bombed during World War II and the Korean War. Irrigation systems in North Vietnam were bombed by the United States in the late 1960s. When Syria tried to stop Israel from building its National Water Carrier in the early 1950s, fighting broke out across the demilitarized zone, and when Syria tried to divert the headwaters of the Jordan in the mid-1960s, Israel used force, including air strikes against the diversion facilities to prevent their construction and operation. These military actions contributed to the tensions that led to the 1967 war.[34]

Most recently, dams, desalination plants, and water conveyance systems were targeted by both sides during the Persian Gulf War. Most of Kuwait's extensive desalination capacity was destroyed by the retreating Iraqis, and in mid-1992, the Iraqis were still suffering severe problems rebuilding Baghdad's modern water supply and sanitation system, which had been intentionally destroyed.[35] As water supplies and delivery systems become increasingly valuable in water-scarce regions, their value as military targets also increases.

**Water resources as military means.** The usual tools of conflict are military weapons of destruction, though the use of water and water resources systems as both offensive and defensive weapons also has a long history. King Nebuchadnezzar of Babylon (605 to 562 BC) described the defense of the city:

> To strengthen the defenses of Babylon, I had a mighty dike of earth thrown up, above the other, from the banks of the Tigris to that of the Euphrates 5 bern long and I surrounded the city with a great expanse of water, with waves on it like the sea, for an extent of 20 bern.[36]

In the last few years, non-military tools have been increasingly used to achieve military ends, including resource ''weapons'' and embargoes. In some of these instances, the resource manipulated was water. Although fresh water resources are renewable, in practice they are finite, poorly distributed, and often subject to substantial control by one nation or regional group. In such circumstances, the temptation to use water for political or military purposes may prove irresistible.

For example, a pro-apartheid council in South Africa cut off water in 1990 to the Wesselton township of 50,000 blacks following a protest over miserable sanitation and living conditions.[37] In 1986, North Korea announced plans to build the Kumgansan hydroelectric dam on a tributary of the Han River upstream of Seoul, South Korea. South Korean military analysts predicted that the deliberate destruction of the dam by the North would raise the level of the Han River by over 50 m as it passes through Seoul – enough to destroy most of the city. A formal request to halt construction was made to the North Korean government and South Korea built a series of levees and check dams above Seoul to try to mitigate possible impacts.[38]

Similarly, it is increasingly difficult to keep the management of the Euphrates River non-political. This river flows from the mountains of southern Turkey through Syria to Iraq, before emptying into the Persian Gulf. Both Syria and Iraq rely heavily on the Euphrates River for drinking water, irrigation, industrial uses, and hydroelectricity, and view any upstream development with concern. In 1974, Iraq threatened to bomb the Al Thawra dam in Syria and massed troops along the border, alleging that the flow of water to Iraq had been reduced by the dam. In mid-1990, Turkish President Ozal threatened to use the newly-completed Ataturk Dam to restrict water flow to Syria to force it to withdraw support for Kurdish rebels operating in southern Turkey. While this threat was later disavowed, Syrian officials argue that Turkey has already used its power over the headwaters of the Euphrates for political goals and could do so again.[39]

The ability of Turkey to shut off the flow of the Euphrates was noted by political and military strategists at the beginning of the Persian Gulf conflict. In the early days of the war, there were behind-the-scenes discussions at the United Nations about using Turkish dams on the Euphrates River to deprive Iraq of a significant fraction of its fresh water supply in response to its invasion of Kuwait.[40] While no such action was ever taken, the threat of the ''water weapon'' was again made clear.

**Resource inequities and water developments as causes of conflict.** There are growing tensions between rich and poor nations due to the inequitable distribution and use of both renewable resources (such as water) and non-renewable resources (such as minerals and fossil fuels). Previous chapters have described the great maldistribution of fresh water resources. Unlike rare metals, water is quite expensive to move in large quantities from one place to another. And unlike oil, water has no substitutes.

Great improvements in the efficiency of water use can be made throughout the world, as can trade-offs between water-consumptive and water-efficient sectors, but these actions only push back the day when limits on water availability will be reached, they do not eliminate it. Indeed, some limits to supply are already being reached in fast-growing semi-arid nations and regions, despite efforts to reduce wasteful use and to redirect priorities. As a result, the type and extent of industrial development in some countries will be limited because of constraints on the availability of fresh water.

In most cases, resource deficiencies will lead to more poverty, shortened lives, and misery rather than to direct violent conflict. But in some cases, inequitable use of resources could increase the likelihood of regional and international disputes, create refugees that cross borders, and decrease the ability of a nation to resist economic and military activities by neighboring countries. For example, while arable land suitable for irrigation often exists, political or physical constraints may hinder any expansion of irrigation. In northern Africa, the Sudan is considered one of the few nations with great potential for increased irrigation: there is sufficient arable land and there is, in theory, sufficient water in the Nile. In reality, however, withdrawing the water from the Nile would require renegotiating or abrogating the treaty signed in 1959 with Egypt.[41]

Violent conflicts have arisen over water allocations within India, such as in early 1992 following a court decision to allocate the waters of the

[†]The nine riparian nations of the Nile are Egypt, the Sudan, Ethiopia, Kenya, Rwanda, Burundi, Uganda, Tanzania, and Zaire.

Cauvery River between Karnataka and Tamil Nadu. The Cauvery River originates in Karnataka and flows through Tamil Nadu before entering the Bay of Bengal. The greatest use of water is in Tamil Nadu. Over fifty people were reported killed in riots following this court decision.[42]

The impacts of water development schemes such as irrigation facilities, hydroelectric developments, and flood control reservoirs can also cause controversy and international disputes. Major water developments may displace large local populations, affect downstream water users, change control over local resources, and cause economic dislocations. These impacts may, in turn, lead to disputes among ethnic or economic groups, between urban and rural populations, and across borders.

For example, the construction of the Aswan High Dam by Egypt led to flooding and dislocation of populations in the Sudan; the construction of the Farakka Barrage on the Ganges in India affected water conditions and availability in Bangladesh; the construction of several major irrigation projects in the south-western United States led to the serious degradation of Colorado River water quality delivered to Mexico and an intense political dispute that was ultimately resolved through diplomatic negotiations.[43]

## Conclusions: An agenda for change

Fresh water touches every aspect of life, from the health of aquatic ecosystems to human health, from the need to grow food for the Earth's growing population to the need to provide energy. Only in the last few years have the complexities and connections among these issues become more apparent.

As population grows, and as the level of economic development increases, human needs for water will grow. To meet these needs we already modify all aspects of the hydrologic cycle. We move vast quantities of water from one region to another, build huge reservoirs that store water for dry periods, seed clouds to squeeze more moisture out of them, divert entire rivers from their beds, harness the power of falling water for electricity, pump great quantities of water from ancient underground aquifers, and strip the salt from sea water.

We must now acknowledge that we are failing to meet many of even the most basic human needs. Biological and chemical contamination of surface and ground water is a growing problem. Vast and growing numbers of people lack clean drinking water and even rudimentary sanitation services. Water-related diseases such as malaria, typhoid, and cholera kill millions of people every year, and the spread of some diseases is out of control. Aquatic ecosystems are threatened everywhere by water developments and diversions. And human needs in other sectors drain the economic and human resources needed for new efforts to reduce these problems.

Enormous resources, human and otherwise, can be brought to bear on these problems, if we choose to do so. Our water resources systems can be made more flexible, more equitable, and more efficient, which would help us overcome many of the problems we already see coming. Better information can be obtained and exchanged on alternative water management approaches. The international community can reaffirm its commitment to improving water supply and quality in the developing world. The following sections offer some brief suggestions for achieving these things.

**Water resources data collection and sharing.** We remain remarkably ignorant about the state of many basic water resources conditions in this day of satellite observation, geographical information systems, and increasing sophistication and dispersion of computers. Efforts must be intensified to gather fundamental water data, organize it into usable and accessible forms, and disseminate it to all who need it. Regional data collection and sharing is an important part of the rational management of any resource. Basic water resources data must not be considered classified or withheld from other nations. Unless nations share hydrologic data, no satisfactory agreements on allocations, responses during shortages, flood management, or long-range planning can be reached. International organizations, such as those under the umbrella of the United Nations or scientific associations, have a major role to play in encouraging the collecting and open sharing of water resources data. A longer discussion of

water resources data issues can be found in the ''About the Data'' section at the beginning of Part II.

**The economics of water.** Water is a scarce resource in many parts of the world, yet it is still often treated as a free good, available at no charge to whoever can pump it from the ground or remove it from a river or lake. We pay to move water from one place to another, and perhaps to treat it following use, but we tend to ignore the so-called ''opportunity cost'' of putting water to one use at the expense of another.[44] Equally important, water can be polluted for free and the costs of waste disposal are rarely taken into account. Nor do we measure the environmental costs of our water use. Thus the fish, birds, or other species dependent on water have no say in what, for them, is often a life or death issue. Until water is priced at its true value, there will be few incentives for wise and efficient water use.

In some regions of the world, clean water of sufficient quantity is no longer available to all who wish it. Traditional economic approaches, be they Marxian, Keynesian, Malthusian, or whatever, often fail to account for natural resource issues and environmental values, making the proper allocation of scarce water unlikely. Among the more innovative tools traditional economists are now trying to apply to water management are revised pricing of water, the reduction or elimination of traditional subsidies, more open markets for trading water or the rights to water, and either providing money for guaranteeing water to natural systems or setting aside some water for ecosystems through legislative actions. While these actions may prove effective in most developed economies, water is a public good and privatization can solve only part of the problems of water distribution and quality. In the long run, the limitations of traditional economics in solving environmental and resource problems must generate some new thinking about the proper sustainable allocation of scarce environmental resources in the real world.[45]

**Water management, laws, and customs.** Institutional arrangements for water allocation tend toward the arcane and complex; water management systems are often inflexible and inefficient. Legal rights to water and water customs vary widely around the world. Governments may put the control of water issues under ministries of the environment, commerce, agriculture, development, finance, fish and wildlife, health, or tourism. Different aspects of water resources protection and use are often divided up into different agencies. Sometimes central government control is retained; sometimes local governments manage local water resources.

Water laws may allocate the right to water to the occupant of land bordering a stream, to the first user of a source of water, or through a system of permits or licenses. Where water is scarce, government control of water allocation is typical. The system of land tenure may constrain water allocations and rights.

All of these factors complicate the rational management and use of water resources in a world with growing populations and uncertainties about long-term supplies and demands. Yet these systems of local, regional, or international laws and customs make up the framework into which all future decisions must be made. Greater attention must be paid to increasing the flexibility of these arrangements to handle the unknown and unexpected and to reduce the risks of conflicts over water.

**Water use efficiency.** Much of the debate in the late 1970s and 1980s about the efficiency of energy use is relevant in the debate over water. Just as the price of energy fails to reflect the true cost of obtaining and using energy, so the price of water fails to include many of the most important costs of its use. As a result, we use water inefficiently. Enormous improvements in the efficiency of water use are possible. In many places, less than half the water diverted to irrigation actually benefits a crop, and agriculture typically accounts for 80% or more of a region's total water use. Carefully designed irrigation systems can raise the efficiency of water use to 95%, freeing vast quantities of water for new irrigation projects or for use in other sectors.

Similar improvements in efficiency are possible in other sectors. In industrialized nations, current residential and commercial water use can be cut substantially with modest investments. New water-efficient technologies are now available for the residential and commercial sectors, and new buildings equipped with such systems can often save water with no increases in marginal costs. In California, a six-year drought led to voluntary and mandatory municipal restrictions on water use, tightened standards for new construction, and changes in industrial processes. These actions often led to reductions in water use of 20% or more at little economic cost.[46]

**New water developments.** Massive water developments have been built throughout the world in the last several decades. Rivers have been dammed and diverted. Land has been flooded and cut with canals. Lakes have been drained and filled. In many places, these projects have had very beneficial effects. In some places, unanticipated problems have worsened human and ecological conditions.

In the last decade, the development of major new water projects has slowed greatly. In industrialized nations, new developments are constrained by growing environmental opposition, the fact that the best sites in many places have already been developed, and high costs. In developing nations, new projects are increasingly constrained by high costs and funding problems, competing demands for water, and local opposition from groups that will be affected by either the project itself or the changes in water quality and availability it would entail.

Some new developments may be necessary, particularly in the poorer regions of the world where flood control, irrigation supply, or hydroelectricity are desperately needed. The inevitability of such developments, however, does not mean that they should be pursued in the same style as earlier projects; indeed, there is a growing realization that national governments and major international funders have a responsibility to review more carefully the wide range of direct and indirect environmental and social impacts such projects entail, to mitigate those impacts wherever possible, and to ensure that the benefits from the project are distributed equitably.

**Traditional solutions.** We have much to re-learn from traditional water management experience. Small-scale, indigenous systems can often be more effective at meeting community needs without the large, unexpected impacts of large-scale developments, and community-level participation in water supply development and management often leads to other economical, educational, or health benefits as well. This section cannot provide detailed solutions to water supply problems in arid regions – such solutions must vary depending on the availability of water resources, the cultural and social characteristics of the local population, soil and climatic conditions, and many other factors. But many approaches are possible.

Crop choices are being reconsidered to more closely match water and soil conditions, rather than crop export requirements. Micro-irrigation for gardens, as well as for commercial cropping, promises benefits for families and communities. In the Negev Desert of the Middle East, rainwater harvesting techniques developed over 2,000 years ago are being revived and applied to developing countries throughout Africa. The use of terracing and contour farming in some regions can improve the size and reliability of crop yields. The sustainability of the lifestyles of pastoral nomads in the Sahel is increasingly being recognized as a source of insight and ideas for other communities. Small-scale dams, if properly designed and sited, may often avoid some of the worst impacts of major developments, and still provide irrigation, power, or water supply benefits.[47]

Unfortunately, such traditional approaches are often ignored by the international development community and governments. They are excluded in surveys of water systems, they do not get investment credits from international aid programs, they are denied the support of information and educational services, and they lack the glamour and high profile of big projects. In the end, we must explore and encourage these small-scale options to increase our knowledge, expand our bag of tricks, and improve the standard of living among the poorest populations of the world.

**Reducing the risks of water-related conflicts.** The disturbing trend toward the use of force in resource-related disputes, the apparent willingness of some nations to use water supply systems as targets and tools of war, and growing disparities in water availability and demand among nations makes it urgent that we work to reduce the probability and consequences of water-related conflicts. International law and international institutions can play a leading role in this area. There have already been some attempts to develop international law protecting environmental resources, but almost all of these focus on attempting to limit environmental damages from conflict and war; virtually no effort has been made to address the equally important problem of preventing conflicts over resource disparities or environmental damage.

For example, the Environmental Modification Convention of 1977, negotiated under the auspices of the United Nations, states:

> Each State Party to this Convention undertakes not to engage in military or any other hostile use of environmental modification techniques having widespread, long-lasting or severe effects as the means of destruction, damage or injury to any other State Party. (Article I.1)

In 1982 the United Nations General Assembly promulgated the World Charter for Nature, supported by over 110 nations, which states:

> Nature shall be secured against degradation caused by warfare or other hostile activities (Article V)

and

> Military actions damaging to nature shall be avoided (Article XX).

The 1977 Bern Geneva Convention on the Protection of Victims of International Armed Conflicts (additional to the Geneva Convention of 1949), states:

> It is prohibited to employ methods or means of warfare which are intended, or may be expected, to cause widespread, long-term and severe damage to the natural environment. (Article XXXV.3)

and

> Care shall be taken in warfare to protect the natural environment against widespread, long-term and severe damage. This protection includes a prohibition of the use of methods or means of warfare which are intended or may be expected to cause such damage to the natural environment and thereby to prejudice the health or survival of the population. (Article LV.1)

These kinds of agreements and statements typically lack any enforcement mechanisms and they carry little weight in the international arena when politics, economics, and other factors are considered more important.

In the last few decades, however, international organizations have attempted to derive more general principles and new concepts governing shared fresh water resources. Developing such principles is difficult because of the intricacies of interstate politics, differing national practices, and reluctance of nations to take a long-term perspective at the expense of short-term flexibility. Ultimately, if nations can recognize the benefits of enforceable accords, the successful negotiation and implementation of international agreements will greatly reduce the risks of water-related conflict.

**A sustainable goal.** Sustaining an adequate quality of life for the world's people, both those now alive and the generations to come, requires matching our use of resources to our goals and desires. One of those resources, fundamental to life, is fresh water. Better understanding of the connections between water and all of our other concerns will lead to better understanding of proper and successful approaches to water management. The day when water-related diseases are eliminated, when natural ecosystems are guaranteed the water necessary for their survival, when conflicts over water are avoided, and when adequate clean water is available to everyone on Earth is a day we should strive to see. May it come in our lifetime.

## Notes

1.   United Nations Population Division, 1991 *Long-Range World Population Projections: Two Centuries of Population Growth, 1950–2150*, United Nations, New York.

2.   United Nations, 1988, *Mortality of Children Under Age 5: World Estimates and Projection 1950–2025*, United Nations, New York.

3.   J. Christmas and C. de Rooy, 1991, The decade and beyond: At a glance, *Water International* **16**(3), 127–134, International Water Resources Association, Urbana, Illinois.

4.   B. Grover and D. Howarth, 1991, Evolving international collaborative arrangements for water supply and sanitation, *Water International* **16**(3), 145–152, International Water Resources Association, Urbana, Illinois.

5.   See, for example, M. Falkenmark and G. Lindh, in Chapter 7, this volume; also M. Falkenmark, 1986, Fresh water – Time for a modified approach, *Ambio* **15**(4), 192–200.

6.   United Nations, 1991, *The World's Women: 1970 to 1990*, United Nations, New York; also I. Dankelman and J. Davidson, 1988, *Women and Environment in the Third World: Alliance for the Future*, Earthscan Publications, Ltd.; the International Union for the Conservation of Nature (IUCN), Sustainable Development Series, London.

7.   For an overview of the impacts of climatic changes, see the report of Working Group II of the Intergovernmental Panel on Climatic Changes (IPCC), J. Jager and H.L. Ferguson (eds.), 1991, Climate change: Science, impacts and policy, in *Proceedings of the Second World Climate Conference*, Cambridge University Press, Cambridge. The growing disputes over responses to climatic change can be seen in the contentious discussions over the climate treaty signed in June 1992 at the United Nations Conference on Environment and Development, in Rio de Janeiro, Brazil.

8.   IPCC, 1990, *Climate Change: The IPCC Scientific Assessment*, Cambridge University Press, Cambridge.

9.   IPCC, 1990, *Climate Change: The IPCC Scientific Assessment*, Cambridge University Press, Cambridge; M.C. MacCracken, M.I. Budyko, A.D. Hecht and Y.I. Izrael (eds.), 1990, *Prospects for Future Climate*, Lewis Publishers, Inc., Chelsea, Michigan; F.R. Rysberman and R.J. Swart (eds.), 1990, *Targets and Indicators of Climatic Change*, vol. II of the Report of the Advisory Group on Greenhouse Gases, Stockholm Environment Institute, Stockholm.

10.  IPCC, 1990, *Climate Change: The IPCC Scientific Assessment*, Cambridge University Press, Cambridge.

11.  For more detail on regional climate changes, see IPCC, 1990, *Climate Change: The IPCC Scientific Assessment*, Cambridge University Press, Cambridge, and J.T. Houghton, 1991, Scientific Assessment of Climate Change: Summary of the IPCC Working Group I Report, in J. Jager and H.L. Ferguson (eds.), Climate change: Science, impacts and policy, in *Proceedings of the Second World Climate Conference*, Cambridge University Press, Cambridge.

12.  IPCC, 1990, *Climate Change: The IPCC Scientific Assessment*, Cambridge University Press, Cambridge.

13.  IPCC, 1990, *Climate Change: The IPCC Scientific Assessment*, Cambridge University Press, Cambridge.

14.  IPCC, 1990, *Climate Change: The IPCC Scientific Assessment*, Cambridge University Press, Cambridge.

15.  D. Rind, R. Goldberg, J. Hansen, C. Rosenzweig and R. Ruedy, 1990, Potential evapotranspiration and the likelihood of future drought, *Journal of Geophysical Review*, **95**, 9983–10004.

16.  See, for example, P.H. Gleick, 1987, Regional hydrologic consequences of increases in atmospheric $CO_2$ and other trace gases, *Climatic Change*, **10**, 137–161; also L. Nash and P.H. Gleick, 1991, The sensitivity of streamflow in the Colorado basin to climatic changes, *Journal of Hydrology*, **125**, 221–241.

17.  W. Kellogg and Z.-Ci Zhao, 1988, Sensitivity of soil moisture to doubling of carbon dioxide in climate model experiments: Part I: North America, *Journal of Climate* **1**, 348–366.

18.  N.J. Rosenberg, B.A. Kimball, P. Martin and C.F. Cooper, 1990, From climate and $CO_2$ enrichment to evapotranspiration, in P.E. Waggoner (ed.) *Climate Change and U.S. Water Resources*, John Wiley and Sons, Inc., New York, pp. 151–176.

19.  S. Manabe and R.T. Wetherald, 1986, Reduction in summer soil wetness induced by an increase in atmospheric carbon dioxide, *Science*, **232**, 626–628; C.A. Wilson and J.F.B. Mitchell, 1987, A doubled $CO_2$ climate sensitivity experiment with a global climate model including a simple ocean, *Journal of Geophysical Research*, **92**, 13315–13343.

20.  P.H. Gleick, 1987, Regional hydrologic consequences of increases in atmospheric $CO_2$ and other trace gases, *Climatic Change*, **10**, 137–161; D.P. Lettenmaier, T.Y. Gan and D.R. Dawdy, 1988, Interpretation of Hydrologic Effects of Climate Change in the Sacramento–San Joaquin River Basin, California, Water Resources Technological Report no. 110, University of Washington, Seattle; F. Bultot, A. Coppens, G.L. Dupriez, D. Gellens and F. Meulenberghs, 1988, Repercussions of a $CO_2$ doubling on the water cycle and on the water balance: A case study for Belgium, *Journal of Hydrology* **99**, 319–347; L. Nash and P.H. Gleick, 1991, The sensitivity of streamflow in the Colorado basin to climatic changes, *Journal of Hydrology*, **125**, 221–241; I.A. Shiklomanov, 1987, Effects of climatic changes on Soviet rivers, in *International Symposium of the XIXth General Assembly*, International Union of Geodesy and Geophysics (IUGG), 9–22 August 1987, Vancouver, British Columbia.

21.  P.E. Waggoner (ed.), 1991, *Climate Change and U.S. Water Resources*, John Wiley and Sons, Inc., New York.

22.  A. Wijkman and L. Timberlake, 1984, *Natural Disasters: Acts of God or Acts of Man?*, Earthscan Publishers, Ltd., London; A.T. McDonald and D. Kay, 1988, *Water Resources: Issues and Strategies*, Longman Scientific and Technical Publishers, Ltd., London.

23.  S.H. Schneider, P.H. Gleick and L.O. Mearns, 1990, Prospects for climate change, in P.E. Waggoner (ed.) *Climate Change and U.S. Water Resources*, John Wiley and Sons, Inc., New York, pp. 41–73.

24.  A. Noda and T. Tokioka, 1989, The effect of doubling the $CO_2$ concentration on convective and non-convective precipitation in a general circulation model with a simple mixed layer ocean, *Journal of the Meteorological Society of Japan*, **67**, 1055–1067.

25.  K.A. Emanuel, 1987, The dependence of hurricane intensity on climate, *Nature*, **326**, 483–485; see also IPCC, 1990, *Climate Change: The IPCC Scientific Assessment*, Cambridge University Press, Cambridge.

26.  P.E. Waggoner (ed.), 1991, *Climate Change and U.S. Water Resources*, John Wiley and Sons, Inc., New York.

27.  G. Goldenman, 1989, *International River Agreements in the Context of Climatic Change*, Pacific Institute for Studies in Development, Environment, and Security, Berkeley, California; L. Nash and P.H. Gleick, 1991, The sensitivity of streamflow in the Colorado basin to climatic changes, *Journal of Hydrology*, **125**, 221–241; L. Nash and P.H. Gleick, (1993) The Sensitivity of Streamflow and Water Supply in the Colorado Basin to Climatic Changes, U.S. Environmental Protection Agency, Washington, DC; P.H. Gleick, 1992, Effects of climate change on shared fresh water resources, in I.M. Mintzer (ed.) *Confronting Climate Change: Risks, Implications and Responses*, Cambridge University Press, Cambridge.

28.  For a more detailed analysis, see P.H. Gleick, 1992, *Water and Conflict*, American Academy of Arts and Sciences Occasional Paper, No. 1, Cambridge, Massachusetts, with the University of Toronto (September).

29.  See, for example, T. Naff and R. Matson, 1984, *Water in the Middle East, Conflict or Cooperation?*, Westview Press, Boulder, Colorado; M.R. Lowi, 1990, The Politics of Water Under Conditions of Scarcity and Conflict: The Jordan River and Riparian States, PhD dissertation, Department of Politics, Princeton University, Princeton, New Jersey.

30.  These data precede the breakup of the Soviet Union and Yugoslavia. Many major rivers in these regions cross the borders of the newly formed political states. When the political status of these regions becomes clearer, it will be possible to recalculate the number of nations receiving significant fractions of water from sources originating outside of their political boundaries.

31.  M.S. Drower, 1954, Water-supply, irrigation, and agriculture, in C. Singer, E.J. Holmyard and A.R. Hall (ed.) *A History of Technology*. Oxford University Press, New York.

32.  M.R. Lowi, 1990, The Politics of Water Under Conditions of Scarcity and Conflict: The Jordan River and Riparian States, PhD dissertation, Department of Politics, Princeton University, Princeton, New Jersey.

33.  M.S. Drower, 1954, Water-supply, irrigation, and agriculture, in C. Singer, E.J. Holmyard and A.R. Hall (ed.) *A History of Technology*. Oxford University Press, New York.

34.  See, for example, T. Naff and R. Matson, 1984, *Water in the Middle East, Conflict or Cooperation?*, Westview Press, Boulder, Colorado.

35.  1992, Iraq's water systems still in shambles, *U.S. Water News*, **8**(10), p. 2.

36.  Quoted in M.S. Drower, 1954, Water-supply, irrigation, and agriculture, in C. Singer, E.J. Holmyard and A.R. Hall (ed.) *A History of Technology*. Oxford University Press, New York.

37.  R. Pinder, 1990, 50,000 Blacks deprived of water, Reuters Press/*San Francisco Chronicle*, 24 October.

38.  S. Chira, 1986, North Korea dam worries the South, *The New York Times*, 30 November, p. 3; N. Koch, 1987, North Korean dam seen as potential "water bomb", *Washington Post/San Francisco Chronicle*, 30 September. North Korea denied any military intentions, but construction on the dam was halted in the late 1980s.

39.  A. Cowell, 1990, Water rights: Plenty of mud to sling, *The New York Times*, 7 February, p. A4.

40.  These closed-door discussions were described to the author by the ambassador of a member nation of the UN Security Council under the condition that he remain anonymous. Further evidence for this comes from the statement of Kamran Inan, the Minister of State of Turkey, at the Conference on Transboundary Waters in the Middle East: Prospects for Regional Co-

operation, Ankara, Turkey, 3 September 1991. At that meeting, Minister Inan stated that Turkey would never use water as a means of political pressure and noted that it had declined to do so during the Gulf War. See also a *New York Times* op-ed piece, ''The Spigot Strategy'', by P. Schweizer, 11 November 1990.

41.   P.H. Gleick, 1990, Climate changes, international rivers, and international security: The Nile and the Colorado, in R. Redford and T.J. Minger (eds) *Greenhouse Glasnost*, The Ecco Press, New York, pp. 147–165; P.H. Gleick, 1991, The vulnerability of runoff in the Nile basin to climatic changes, *The Environmental Professional*, **13**, 66–73.

42.   M. Moench, 1992, personal communication with author.

43.   P.H. Gleick, 1988, The effects of future climatic changes on international water resources: The Colorado River, the United States, and Mexico, *Policy Sciences*, **21**, 23–39.

44.   D.C. Gibbons, 1986, *The Economic Value of Water*, Resources for the Future, Inc., Washington, DC.

45.   For some recent new ideas, see H.E. Daly and J.B. Cobb, 1989, *For the Common Good: Redirecting the Economy Toward Community, the Environment, and a Sustainable Future*, Beacon Press, Boston; R.C. Darge, R.B. Norgaard, M. Olson and R. Somerville, 1991, Economic growth, sustainability, and the environment, *Contemporary Policy Issues*, **9**(1), 1–23; and H.E. Daly, 1991, Towards an environmental macroeconomics, *Land Economics*, **67**(2), 255–259.

46.   P.H. Gleick and L. Nash, 1991, *The Societal and Environmental Costs of the Continuing California Drought*, Pacific Institute for Studies in Development, Environment, and Security, Berkeley, California (Report to Congress), August.

47.   For brief descriptions of many of these techniques and where they have successfully been applied, see S. Postel, 1992, *The Last Oasis: Facing Water Scarcity*, W.W. Norton and Co., New York; R. Clarke, 1991, *Water: The International Crisis*, Earthscan Publications, Ltd., London.

# Part II
# **Fresh water data**

# About the data

## Peter H. Gleick

The following section presents diverse fresh water resources data from around the world. These data provide quantitative and qualitative information on a wide variety of critical fresh water questions, including the global water balance, water use and supply, water quality and health, the characteristics of aquatic ecosystems, agricultural water use around the world, water requirements for energy production and use, and many other issues.

Good water data are hard to come by. Data are often not collected regularly and systematically. They may not be reported in useful or consistent forms. If collected, they may never be compiled or distributed. Standards and techniques of measurement differ from region to region, or worse, from year to year. Units of measure vary around the world – different quantitative measures may have the same name, or a single measure may be called different things in different places. Data are collected by individuals with differing skills, goals, and intents. Some data are collected objectively; other data are collected to support particular ideological or political biases. The more we know about a data set and how it was collected, the better equipped we will be to evaluate it and use it.

Anyone with an interest in water data, or any natural resources data for that matter, needs to be aware of the limitations of the following data sets. There are gaping holes in this collection, such as in the areas of ground water, water quality, and aquatic ecosystems. We have chosen the data here on the basis of their quality and scope. We have also chosen *not* to present some available data because of the problems described below and because of space considerations. Throughout this section we try to include precise information on sources and citations to permit the interested reader to further pursue details that may not be found here. Comments, corrections, and additions are welcome.

## Data format

Because of the widely varying scope and quality of the data in these tables, we include with each table a standard introduction that includes the title, a description, a discussion of data limitations, and sources.

We have made every effort to track down original sources of data, rather than rely on secondary sources, but in some cases we cite data compiled by others. In these instances, we have made an effort to explicitly note both the original and secondary sources. Data are often recycled from one source to another, with inadequate or incomplete citations. Over time, it can become extremely difficult to find the original source, the original assumptions, and the limitations of the measurement. For example, all estimates for the sediment discharge of the Irrawaddy River are based on one set of measurements made by the British in the 1870s and quoted over and over since then.

The data in this book are presented in standard international metric units. Original non-metric data were converted to the international standard. While this increases the risk of errors during the conversion process, it is important to have a standard set of units for comparative purposes. To assist those familiar with other units, we have included a detailed set of data conversions in Section J, ''Units, Data Conversions, and Constants.''

We use exponent format, rather than the prefixes of ''thousand,'' ''million,'' ''billion,'' ''milliard,'' and so forth. This should reduce some of the confusion caused by certain ambiguous terms, such as ''billion,'' which means $10^9$ in some parts of the world, and $10^{12}$ in others.

## Errors in the data

There are likely to be errors in these data, either inadvertently propagated from errors in the original sources, or introduced in transcription and conversion. In some cases, we have found and corrected typographical errors in the original sources, in an attempt to *prevent* error propagation. Anyone interested in using these or other water resources data should exercise extreme caution: check original sources where possible, compare different data sets, and beware of false precision (described later).

## Difference between ''measured'' data and ''derived'' data

In the past, almost all geophysical hydrologic data were measured directly by field researchers. This is no longer the case. Three factors are leading to increasing dependence on derived or computed data: first, the growing demand for comprehensive data on water resources and on human use of water; second, the costs of high-quality field work; and third, the increasing sophistication (and decreasing cost) of high-quality, high-speed computers.

Where measured data are unavailable, researchers often attempt to derive approximate values using other available data. For example, while few countries specifically report the percentage of water used in the agricultural sector, some researchers compute approximate national agricultural water use by making assumptions about crop types and planting patterns, typical crop water consumption rates, regional climatology, and irrigation methods.

There are many problems and pitfalls with such computed data. The implicit assumptions used in computer modeling are rarely described, much less understood, by those who later use these data. While such derived data are often valuable, they are only as good as the quality of the input data and model assumptions used, and must *not* be considered a substitute for actual measurements if they can be found.

In the last decade, some efforts have been made to improve the collection of water data. Under the auspices of different international organizations, such as the United Nations (UN) and the International Association of Hydrological Sciences (IAHS), standards and repositories for certain types of data have been set up. For example, a systematic attempt to gather global water quality data worldwide is being made by the Global Environmental Monitoring System (GEMS) launched in 1977 by UNEP and the World Health Organization. Unfortunately, several problems affect the quality of these data: monitoring data are based on voluntary country contributions and are affected by limited financial resources; the task is immense given the huge numbers of sites that need to be monitored; and there are problems with setting and following consistent monitoring standards. Similarly, a new center for compiling global runoff data has been formed at the Federal Institute of Hydrology in Koblenz, Germany, which may help standardize river runoff data collection and availability over the next decade. These efforts to promulgate consistent data standards and to set up other fresh water data repositories should be accelerated.

## Time-series data and time of estimate

Three kinds of temporal variables appear in the following tables: single estimates made at a specific time, estimates of a fixed variable made at *different* times by different researchers, and *changes* in fresh water data measured over a period of time. Where possible, we present time-series data. Time-series data on water resources can be extremely valuable and often reveal interesting information about environmental trends and about the political, economic, or societal conditions surrounding resources use. For example, single annual estimates of the size of the Aral Sea or of the extent of wetlands hide the dramatic year-to-year changes affecting these resources because of human interference with the natural water balance. In these cases, only time-series data should be used.

In some cases, we present data on specific water resources variables made at different times. For example, in Section A we show six estimates of the global water balance estimated over a twenty-year period. Readers should be cautioned, however, that although later estimates of certain types of data may be better than earlier ones due to improvements in measuring techniques or newer information, this is not always the case. We have tried to present the best information available; when there is no obvious reason for picking one estimate over another, we often present several estimates of the same thing to permit direct comparisons.

## Uneven regional data coverage and uneven data quality

One of the greatest shortcomings a data collection like this reveals is the great disparity in the availability of data from the developed and developing parts of the world. Although we have made every effort to include data from around the globe, the quantity and quality of data from the United States, Canada, and western Europe are far better than for the rest of the world. For many nations, data on any aspect of fresh water are almost completely lacking. This reflects the unhappy fact that collecting, compiling, and publishing information on natural resources are luxuries that cannot be afforded by the poor.

Where data are available, their quality and scope vary widely. There are few standards for collecting and reporting water resource data, data are often not available for the same year or location, and measurement methods vary from place to place. The situation is worst in the developing world. Few developing countries have regular reporting systems, even for issues as fundamental as access to drinking water supply and to sanitation facilities. Despite major efforts during the International Drinking Water Supply and Sanitation Decade (1981–1990) launched by the 1980 United Nations General Assembly, we are still missing large blocks of information about access to clean water and sanitation. In some regions, data are available for less than half the population, typically excluding poorer rural areas almost entirely.

In the area of water quality, there is a pronounced lack of adequate data for tropical regions because the vast majority of water quality research has been conducted in the temperate zones. With the exception of the most common problems, this has resulted in a poor understanding of tropical zone water quality, and in some cases inappropriate policies. Furthermore, while the emphasis on newer problems – such as the prevalence of organic chemicals in ground water – is focused on the industrialized world, it should not be inferred that these problems are not also present in Africa, Asia, and Latin America, but only that they have not been sufficiently studied in those regions.

## Variability, uncertainty, and illusory precision and accuracy

Another pervasive shortcoming in the literature of fresh water resources is insufficient attention given to the variability and uncertainties associated with hydrologic data. "Variability" arises from the stochastic nature of the hydrologic process and from changes in human factors (e.g., different technologies, different economic conditions, changing resource needs). "Uncertainties" refer to incomplete knowledge that may or may not be reducible with additional effort. For water resources data to be useful for researchers, policy makers, and the public, information should be provided on both the nature of the variability and the extent of the uncertainty in the data.

There are many examples of variability and uncertainty presented here. Given the variability of water quality, for example, a single, discrete set of measurements often reveals very little. Accurate estimates of regional water quality require that several samples be taken at different locations and times; and that these samples be selected to ensure that they are representative of the waterbody as a whole. While a well-mixed lake may require only a few samples to describe water quality, a contaminated heterogeneous aquifer may require hundreds of samples to obtain even a rough approximation of overall water quality. In these days of remote sensing and direct satellite observations, there is still surprising disagreement about such fundamental pieces of information as the area of individual watersheds and the lengths and average runoff of the world's major rivers. Many different references provide these data, but the references disagree about the values of these variables to a remarkable extent. In part these disagreements result from different analytical methods and different assumptions, such as the geographical fractal scale used for measuring lengths and different periods used to define average runoff. Unfortunately, these assumptions and methods are rarely described in the original sources, making comparisons extremely difficult. We have often chosen to present here *several* data sets for the same issue rather than choose a single one from among different sets. In some cases, we present several data sets side-by-side to facilitate comparisons. When in doubt about a data set, readers are urged to find original references and investigate original assumptions.

Related to the issue of uncertainty is that of illusory precision. A striking and pervasive shortcoming in the hydrologic literature is insufficient attention to the magnitudes of the variations associated with the data. Hydrologic data are often reported in the literature with far more precision than our understanding warrants. Yet rounding data off, which reduces the implied precision of the data, can also hide real differences. While we have attempted to prevent the proliferation of insignificant figures here, we also feel compelled to report, verbatim, data as they are presented in the original sources. In no cases have we added significant figures unless better data have actually become available; in some cases we have trimmed the number of significant figures where the number given in the original was not justified. When in doubt, we report the same number as in the original source, and we always indicate when we have rounded numbers.

Finally, beware of false "accuracy": different sources may agree about a value or set of values, yet they may *all* be incorrect. The sources may not be independent, they may be based on similar errors, or they may simply be the best available, though still incorrect, numbers.

## Aggregation of data

Hydrologic data are often processed in some way to make them presentable or to prove a point. Long-term measurements may be presented as single mean values, regional data may be blended, figures from multiple sources may be combined, missing data may be "estimated," and so on. There are advantages and disadvantages of processing data. The combination of large amounts of data into single values may, under some circumstances, be extremely useful by illuminating some things not immediately obvious. On the other hand, such aggregation also conceals important differences revealed by the raw data.

River runoff, for example, is usually reported as an average annual volume, or an average instantaneous flow. Such averages, however, conceal information about the statistical nature of the runoff distribution, including the nature and severity of both droughts and floods. In addition, averages are all too often reported without specifying either the time period over which the average was taken, or the range of extreme values that are often more important for making policy decisions.

Data from different sources are often combined in order to fill holes in our knowledge. When this occurs, the separate assumptions and idiosyncrasies of each data set are usually lost, as are the details of the uncertainties and measurement techniques. For water quality, different analytical and sample-handling techniques can yield results that differ by an order of magnitude or more, depending on the quality of the equipment, the presence of other contaminants, and the skill of the analyst. Thus the comparison of water quality measurements from different studies, regions, or times will often be inappropriate and will lead to misleading conclusions. Similarly, data are often aggregated on a regional or national basis, which can hide enormous regional differences. For example, average ground water withdrawal and use data hide the fact that independent aquifers only kilometers apart may experience very different levels of use or overdraft.

## Inconsistent definitions, standards, and boundaries

Different organizations, countries, and regions use different standards and definitions for measuring, evaluating, and reporting water data. For example, there are many different terms used to describe undeveloped hydroelectric potential: "theoretical capacity," "exploitable potential," "economic potential," "maximum technical potential." Yet these terms are regularly used without clear definitions of what precisely is being measured, making comparisons among sources an exercise in frustration.

The inconsistency of water quality measurement techniques is another serious problem. While some standards of measurement have been accepted for the most common water quality parameters (such as fecal coliforms and dissolved oxygen), the same is not true for constituents such as metals and organic chemicals. Accurate measurement requires the resources and training to conduct the appropriate analytical tests, and the development of consistent definitions and standards of analysis.

Definitions often change over time for practical or political reasons. Unless these changes are made obvious and apparent, they can mistakenly lead to the appearance of a change where no change has occurred. For example, changes in the definition of "safe drinking water" can lead to a sizeable improvement – on paper – in the number of people supplied by water systems, without any actual improvement in fresh water supply or the quality of life. We provide definitions where possible, and we cite original sources throughout to permit further investigation of assumptions.

Another "boundary" problem is that of the international political system. National borders are rarely drawn using natural boundaries. Thus, a political change can lead to dramatic changes in natural resource definitions. The breakup of the Soviet Union has led to the creation of several new independent states and, as of 1993, a loose coalition of the remaining former Soviet republics. Thus, the water availability and use information for the former Soviet Union must be recomputed for these new political entities. Similarly, the number of "international rivers," defined as rivers shared by two or more nations, has increased significantly in the last few year as rivers that were completely contained in what was the Soviet Union and other countries of Eastern & Central Europe now cross newly-formed international borders and must be governed and shared by newly-formed nations.

## International politics and hydrologic data

Too often, political considerations play a role in the collection and dissemination of hydrologic data. Because political borders rarely coincide with hydrologic boundaries, and because water is an increasing source of tensions among nations, some countries restrict the release of environmental and resource data on "national security" grounds. For example, India and Israel, to mention only two countries, do not release complete information on runoff or water availability and use for some of the surface and ground water resources they share with neighboring nations.

Another problem is that changes are constantly occurring in the names and borders of nations. Keeping track of name changes alone is a daunting task for geographers. Similarly, the recent enormous political changes in Europe and the former Soviet Union are giving map-makers nightmares. In our country tables, we have reported most data for the "Soviet Union" or "USSR," since the data predate the Soviet breakup. But we have also been able include some new data on water resources and use in the separate Soviet republics, many of which are now independent nations.

## Summary

What is in a number? Each number has a value, such as 2.5 rather than 6.7. Each number has a unit that defines its character, such as a volume, a flow, or a level of contamination. Each has a level of precision, 2.5 rather than 2.531. But each also represents more than just a physical or scientific fact. It represents something real, something tangible about the world in which we live, something that may be far more complicated than it appears on the surface. For example, data on the per capita water availability in a country is a measure of what is theoretically available, not what is actually supplied to each human being living there, which is often far less. Data on potential hydroelectric resources in a country may suggest enormous untapped energy reserves, but tapping them may be possible only at the cost of strangling our fisheries, or turning our Grand Canyons into flat-water reservoirs. Data on the fraction of the population without access to basic sanitation and clean water gives a sense of the staggering raw numbers of people without the most basic human services. But there are other meanings hidden in this single number as well: it means 1,300 million people cannot turn on a tap in their household and drink clean water, as the readers of this book can. It means that tens or hundreds of millions of (primarily) women and children must spend hours each day searching for water for their basic needs. And it means that millions die every year of preventable water-related diseases. Readers should keep these things in mind as they review and use the data here.

# A. Global and regional fresh water resources

**TABLE A.1**  Global stocks of water ($10^3$ km$^3$)

| Stock | Nace (1967) | UNESCO (1974) | L'vovich (1974) | Baumgartner and Reichel (1975) | Berner and Berner (1987) | WRI (1988) |
|---|---|---|---|---|---|---|
| Fresh water lakes | 125 | 91 | 280[e] | 225[f] | 125[e] | 100 |
| Saline lakes and inland seas | 104 | 85.4 | [e] | [f] | [e] | 105 |
| Rivers[a] | 1.25 | 2.12 | 1.2 | [f] | 1.7 | 1.7 |
| Soil moisture | 67[b] | 16.5 | 85 | [g] | 65 | 70 |
| Ground water | 8,350[c] | 23,400[k] | 60,000[h] | 8,062[g] | 9,500[c] | 8,200 |
| Ice caps and glaciers | 29,200 | 24,064 | 24,000 | 27,820 | 29,000 | 27,500 |
| Underground permafrost ice | [d] | 300 | [d] | [d] | [d] | [d] |
| Swamp water | [d] | 11.47 | [d] | [d] | [d] | [d] |
| Biota | [d] | 1.12 | [d] | [d] | 0.6 | 1.1 |
| Total Inland Water | 37,800 | | | | | |
| Atmospheric Water | 13 | 12.9 | 14 | 13 | 13 | 13 |
| Ocean Water | 1,320,000 | 1,338,000 | 1,370,323 | 1,348,000 | 1,370,000 | 1,350,000 |
| Total Stocks[j] | 1,360,000 | 1,385,985 | 1,454,193 | 1,384,120 | 1,408,700 | [i] |

[a]Average instantaneous volume.
[b]Includes vadose water (subsurface water above the water table level).
[c]To depth of 4 km.
[d]These values included in other categories or not included.
[e]Fresh and saline water.
[f]All lake and river water included in the value for fresh water lakes.
[g]Soil water included in ground water value.
[h]Refers to volume of water in upper 5 km of earth's crust, excluding chemically bound water.
[i]No total given in original source.
[j]Total given in original source. May not add up to sum of individual stocks.
[k]Of this total, 10,530 are fresh water.

**TABLE A.1**  Global stocks of water

## DESCRIPTION

Six estimates of the global stocks of fresh and saline water are presented here. Each source uses slightly different assumptions and categories, as described in the notes. All data are in km$^3$ × $10^3$. The number of significant figures and the values presented here as total stocks are the same as in the original sources, even if the sum of the individual categories is different.

## LIMITATIONS

These stocks are difficult to accurately estimate, and the data are grouped in different ways by the authors. For some categories, such as ground water, the total volume varies considerably due to different assumptions. Not all estimates are independently obtained, and care should be exercised in choosing one over another.

## SOURCES

R. Nace, 1967, Are we running out of water? U.S. Geological Survey Circular no. 536, Washington, DC.
USSR Committee for the International Hydrological Decade, 1978, *World Water Balance and Water Resources of the Earth*, Studies and Reports in Hydrology, no. 25. UNESCO, Paris and Gidrometeoizdat, Leningrad (English translation of a 1974 USSR publication).
M.I. L'vovich, 1979, *World Water Resources and Their Future*, Copyright by American Geophysical Union, Washington, DC (English translation of a 1974 USSR publication edited by R. Nace). By permission.
A. Baumgartner and E. Reichel, 1975, *The World Water Balance: Mean Annual Global, Continental and Maritime Precipitation, Evaporation and Runoff*, Elsevier, Amsterdam.
World Resources Institute, 1988, *World Resources 1988–89*, World Resources Institute and the International Institute for Environment and Development in collaboration with the United Nations Environment Programme, Basic Books, New York. By permission.
E.K. Berner and R.A. Berner, 1987, *The Global Water Cycle: Geochemistry and Environment*, pp. 13–14, reprinted/adapted by permission of Prentice-Hall, Englewood Cliffs, New Jersey.

**TABLE A.2** Flows of the global water cycle (km$^3$/yr)

| | Nace (1967) | L'vovich (1974) | Berner and Berner (1987) | WRI range (1988) | Shiklomanov (1992) |
|---|---|---|---|---|---|
| Evaporation from land | 70,000 | 72,500 | 72,900 | 63,000–73,000 | 72,000 |
| Precipitation on land | 100,000 | 113,500 | 110,300 | 99,000–19,000 | 119,000 |
| Evaporation from the ocean | 350,000 | 452,600 | 423,100 | 383,000–505,000 | 505,000 |
| Precipitation on the ocean | 320,000 | 411,600 | 385,700 | 320,000–458,000 | 458,000 |
| Runoff from land to the ocean | 39,600 | 41,000 | 37,400 | 33,500–47,000 | 47,000 |
| Rivers | | | | 27,000–45,000 | |
| Direct ground water runoff | 1,600 | | | 0–12,000 | |
| Glacier runoff (water and ice) | | | | 1,700–4,500 | |

**TABLE A.2** Flows of the global water cycle

## DESCRIPTION

Five estimates of total global evaporation, precipitation, and runoff in km$^3$ per year are given here. Data include estimates of evaporation and precipitation to and from the oceans and land, and runoff from the land, glaciers, and ground water to the oceans.

## LIMITATIONS

These flows are extremely difficult to measure accurately, and the differences among the numbers cited here are probably smaller than their overall uncertainties. The Berner and Berner (1987) and World Resources Institute (1988) estimates compile data from several primary and secondary sources.

## SOURCES

R. Nace, 1967, Are we running out of water? U.S. Geological Survey Circular no. 536, Washington, DC.

M.I. L'vovich, 1979, *World Water Resources and Their Future*, Copyright by American Geophysical Union, Washington, DC (English translation of a 1974 USSR publication edited by R. Nace). By permission.

E.K. Berner and R.A. Berner, 1987, *The Global Water Cycle: Geochemistry and Environment*, pp. 13–14, reprinted/adapted by permission of Prentice-Hall, Englewood Cliffs, New Jersey.

World Resources Institute, 1988, *World Resources 1988-89*, World Resources Institute and the International Institute for Environment and Development in collaboration with the United Nations Environment Programme, Basic Books, New York. By permission.

I.A. Shiklomanov, Chapter 2, this volume.

**TABLE A.3** Inflow of surface water and ground water to the world oceans

| Ocean | Ocean area ($10^6$ km²) | Inflow of surface waters from the continents to the oceans (km³/yr) | | | | | | Total surface water inflow | Total inflow of surface and ground water (km³/yr) |
|---|---|---|---|---|---|---|---|---|---|
| | | Eurasia | Africa | North America | South America | Australia and Oceania | Antarctica | | |
| Pacific | 178.7 | 7,450 | | 2,540 | 1,330 | 1,800 | 975 | 14,100 | 14,800 |
| Including southern section | 25.3 | | | | 700 | 100 | 975 | 1,780 | 1,820 |
| Atlantic | 91.7 | 2,300 | 3,360 | 3,260 | 10,370 | | 570 | 19,860 | 20,760 |
| Including southern section | 15.5 | | | | | | 570 | 570 | 570 |
| Indian | 76.2 | 3,560 | 748 | | | 568 | 765 | 5,640 | 6,150 |
| Including southern section | 28.5 | | | | | 70 | 765 | 835 | 845 |
| Arctic | 14.7 | 3,100 | | 2,040 | | | | 5,140 | 5,220 |
| World | 361.3 | 16,400 | 4,110 | 7,840 | 11,700 | 2,370 | 2,310 | 44,700 | 47,000 |
| Including southern section | 69.3 | | | | 700 | 170 | 2,310 | 3,180 | 3,230 |

**TABLE A.3** Inflow of surface water and ground water to the world oceans

**DESCRIPTION**

Fresh water flows from rivers and ground water aquifers into the oceans of the world are given here in km³ per year. This table offers a breakdown of fresh water flowing into each major world ocean, from each continent.

**LIMITATIONS**

These data are rough estimates of total global runoff. The range of uncertainty in these estimates is not described in the original source. The inflow of ground water not drained by rivers has been estimated based on coastline length. All data are presented and rounded as in the original source.

**SOURCE**

USSR Committee for the International Hydrological Decade, 1978, *World Water Balance and Water Resources of the Earth*, Studies and Reports in Hydrology, no. 25. UNESCO, Paris and Gidrometeoizdat, Leningrad (English translation of a 1974 USSR publication).

**TABLE A.4**   Areas of external and internal runoff, by continent

| | Total area of continent with islands ($10^6$ km$^2$) | Areas of external runoff | | | | Areas of internal runoff ($10^6$ km$^2$) |
|---|---|---|---|---|---|---|
| | | Arctic Ocean | Atlantic Ocean | Indian Ocean | Pacific Ocean | |
| Europe | 10.5 | 1.5 | 6.8 | — | — | 2.2 |
| Asia | 43.5 | 11.7 | 0.6 | 7.0 | 11.9 | 12.3 |
| Africa | 30.1 | — | 14.9 | 5.6 | — | 9.6 |
| North America | 24.2 | 9.2 | 9.2 | — | 5.0 | 0.8 |
| South America | 17.8 | — | 15.2 | — | 1.2 | 1.4 |
| Australia | 8.9 | — | — | 3.3 | 1.7 | 3.9 |
| Antarctica | 14.0 | — | 4.0 | 5.0 | 5.0 | — |
| Total land area including islands | 149 | 22.4 | 50.7 | 20.9 | 24.8 | 30.2 |
| As a % of total area | 100 | 15 | 34 | 14 | 17 | 20 |

**TABLE A.4**   Areas of external and internal runoff, by continent

**DESCRIPTION**

The drainage area of each continent is listed, divided into areas producing internal runoff and areas producing runoff that flows into four major oceans. All areas are measured in million km$^2$.

**SOURCE**

USSR Committee for the International Hydrological Decade, 1978, *World Water Balance and Water Resources of the Earth*, Studies and Reports in Hydrology, no. 25. UNESCO, Paris and Gidrometeoizdat, Leningrad (English translation of a 1974 USSR publication).

**TABLE A.5**   Annual average water balance of the continents

| | Europe[a] | Asia | Africa | North America[b] | South America | Australia[c] | Total Land area[d] |
|---|---|---|---|---|---|---|---|
| | | | | (km$^3$/yr) | | | |
| Precipitation | 7,165 | 32,690 | 20,780 | 13,910 | 29,355 | 6,405 | 110,305 |
| Ground water runoff | 1,065 | 3,410 | 1,465 | 1,740 | 3,740 | 465 | 11,885 |
| Surface water runoff | 2,045 | 9,780 | 2,760 | 4,220 | 6,640 | 1,500 | 26,945 |
| Total surface wetting | 5,120 | 22,910 | 18,020 | 9,690 | 22,715 | 4,905 | 83,360 |
| Evapotranspiration | 4,055 | 19,500 | 16,555 | 7,950 | 18,975 | 4,440 | 71,475 |
| | | | | (mm/yr) | | | |
| Precipitation | 734 | 726 | 686 | 670 | 1,648 | 736 | 834 |
| Ground water runoff | 109 | 76 | 48 | 84 | 210 | 54 | 90 |
| Surface water runoff | 210 | 217 | 91 | 203 | 378 | 172 | 204 |
| Total surface wetting | 524 | 509 | 595 | 467 | 1,275 | 564 | 630 |
| Evapotranspiration | 415 | 433 | 547 | 383 | 1,065 | 510 | 540 |
| | | | | (10$^6$ km$^2$) | | | |
| Land Area | 9.8 | 45.0 | 30.3 | 20.7 | 17.8 | 8.7 | 132.3 |

[a]Europe includes Iceland
[b]North America includes Central America and excludes the Canadian archipelago.
[c]Australia includes New Zealand, New Guinea, and Tasmania.
[d]Total land excludes Antarctica, Greenland, and the Canadian archipelago.

**TABLE A.5**   Annual average water balance of the continents

**DESCRIPTION**

Precipitation, surface runoff, ground water runoff, and evapotranspiration are given for each continent in km$^3$ per year and in mm per year. Land area is shown in million km$^2$. Continental areas exclude Antarctica, Greenland, the Canadian archipelago, and many smaller islands.

**LIMITATIONS**

The water balance data are presented with greater accuracy than actual measurements permit. Neither the uncertainty ranges for the data nor the time periods for which the averages are calculated are given in the original source. The author discusses the general quality of the data for different regions.

**SOURCE**

M.I. L'vovich, 1979, *World Water Resources and Their Future*, Copyright by American Geophysical Union, Washington, DC (English translation of a 1974 USSR publication edited by R. Nace). By permission.

**TABLE A.6**   Six estimates of the annual average continental water balance (mm/yr)

| Source | North America | | | South America | | | Europe | | | Asia | | | Africa | | | Australasia | | |
|---|---|---|---|---|---|---|---|---|---|---|---|---|---|---|---|---|---|---|
| | P | E | R | P | E | R | P | E | R | P | E | R | P | E | R | P | E | R |
| a | 645 | 403 | 242 | 1,564 | 946 | 618 | 657 | 375 | 282 | 696 | 420 | 276 | 696 | 582 | 114 | 803 | 534 | 269 |
| b | 660 | 396 | 264 | 1,350 | 864 | 486 | 580 | 348 | 242 | 610 | 390 | 220 | 660 | 502 | 158 | NA | NA | NA |
| c | 756 | 418 | 339 | 1,600 | 910 | 685 | 790 | 507 | 283 | 740 | 416 | 324 | 740 | 587 | 153 | 791 | 511 | 280 |
| d | 670 | 383 | 287 | 1,648 | 1,065 | 583 | 734 | 415 | 319 | 726 | 433 | 293 | 686 | 547 | 139 | 736 | 510 | 229 |
| e | 563 | 427 | 136 | 1,543 | 984 | 559 | 584 | 454 | 130 | 846 | 553 | 293 | 639 | 495 | 144 | 861 | 575 | 286 |
| f | 800 | 470 | 330 | 1,600 | 940 | 660 | 770 | 490 | 280 | 630 | 370 | 260 | 720 | 580 | 140 | NA | NA | NA |

P, precipitation; E, evapotranspiration; R, runoff; NA, not available

[a]Baumgartner and Reichel (1975)
[b]AWRC (1976)
[c]UNESCO (1978)
[d]L'vovich (1979)
[e]Willmott et al. (1985)
[f]Budyko (1986)

**TABLE A.6**   Six estimates of the annual average continental water balance

## DESCRIPTION

Many studies estimate continental water balances. Included here are three that represent comprehensive assessments of the world water balance (Baumgartner and Reichel, 1975; USSR Committee, 1978; L'vovich, 1979), one database that includes calculations based on a water balance analysis using observed temperature and precipitation (Willmott et al., 1985), and two less detailed continental breakdowns (AWRC, 1976; Budyko, 1986). Precipitation (P), evaporation (E), and runoff (R) are expressed in mm per year as annual averages of water spread to a uniform depth across the land area of each continent. Reference b in the original includes estimates for Australia alone.

## LIMITATIONS

The differences among these numbers reflect limitations in the quality of the data, inconsistent methods of calculations among the sources, and different continental area assumptions. For example, the annual averages are computed over different periods of time, though no details about these time periods are provided. There is no evidence of consistent systematic bias in the continental values, though significant differences appear among the estimates. See the original sources for detailed assumptions.

## SOURCES

G. Thomas and A. Henderson-Sellers, 1992, Global and continental water balance in a GCM, *Climatic Change*, **20**(4), 251–276. By permission of Kluwer Academic Publishers.

A. Baumgartner and E. Reichel, 1975, *The World Water Balance: Mean Annual Global, Continental and Maritime Precipitation, Evaporation and Runoff*, Elsevier, Amsterdam.

Australian Water Resources Council (AWRC), 1976, *Review of Australia's Water Resources 1975*, Australian Government Publishing Service, Canberra.

USSR Committee for the International Hydrological Decade, 1978, *World Water Balance and Water Resources of the Earth*, Studies and Reports in Hydrology, no. 25. UNESCO, Paris and Gidrometeoizdat, Leningrad (English translation of a 1974 USSR publication).

M.I. L'vovich, 1979, *World Water Resources and Their Future*, Copyright by the American Geophysical Union, Washington, DC (English translation of a 1974 USSR publication edited by R. Nace). By permission.

C.J. Willmott, C.M. Rowe and Y. Mintz, 1985, Climatology of the terrestrial seasonal water cycle, *Journal of Climatology*, **5**, 589–606.

M.I. Budyko, 1986, *The Evolution of the Earth's Biosphere*. Reidel, Dordrecht.

**TABLE A.7**   Continental areas and mean annual runoff, three estimates

| | Area ($10^3$ km$^2$) | | Runoff (km$^3$/yr) | | |
|---|---|---|---|---|---|
| | Baumgartner and Reichel (1975) | UNESCO (1978) | Baumgartner and Reichel (1975) | UNESCO (1978)[e] | Shiklomanov (1992)[a] |
| Europe | 10,025 | 10,500 | 2,826 | 3,210 | 2,970 |
| Asia[b] | 44,133 | 43,475 | 12,205 | 14,410 | 14,100 |
| Africa | 29,785 | 30,120 | 3,409 | 4,570 | 4,600 |
| Australia[c] | 8,895 | 8,950 | 2,394 | 2,388 | 2,510 |
| North America[d] | 24,120 | 24,200 | 5,840 | 8,200 | 8,180 |
| South America | 17,884 | 17,800 | 11,039 | 11,760 | 12,200 |
| Antarctica | 14,062 | 13,980 | 1,987 | 2,230 | 2,310 |
| World | 148,904 | 149,000[f] | 39,700 | 46,800[f] | 46,870 |

[a]Shiklomanov (1992) uses the continental areas of UNESCO (1978).
[b]Includes Taiwan, the Philippines, Japan, Indonesia, and some islands of Oceania.
[c]Includes New Guinea, Tasmania, New Zealand, and some islands of Oceania.
[d]Includes Central America to the Panama Canal, Greenland, and the Canadian archipelago
[e]Average runoff measured from 1918 to 1967.
[f]Totals given in original source. May not add to totals of column due to rounding.

**TABLE A.7**   Continental areas and mean annual runoff, three estimates

**DESCRIPTION**

Even rudimentary hydrologic data are often uncertain because of problems with definitions, techniques of measurement, or other assumptions. Shown here for the continents are two estimates of total areas, in km$^2$ × 10$^3$, and three estimates of mean annual runoff, in km$^3$ per year.

**LIMITATIONS**

Some of the differences in runoff values and continental areas are attributable to the use of different periods of measurement or assumptions about what land areas to include in each continent, but some of the differences arise from basic uncertainties about actual runoff values, which are not always measured directly. USSR Committee

(1978) provides additional information on maximum and minimum runoff values and on the coefficients of variation.

**SOURCES**

A. Baumgartner and E. Reichel, 1975, *The World Water Balance: Mean Annual Global, Continental and Maritime Precipitation, Evaporation and Runoff,* Elsevier, Amsterdam.
USSR Committee for the International Hydrological Decade, 1978, *World Water Balance and Water Resources of the Earth,* Studies and Reports in Hydrology, no. 25. UNESCO, Paris and Gidrometeoizdat, Leningrad (English translation of a 1974 USSR publication).
I.A. Shiklomanov, 1992, Chapter 2, this volume.

**TABLE A.8**   Water reserves in surface ice *(page 127)*

**DESCRIPTION**

Shown here are the areas of surface ice, in km$^2$, and the volumes of water reserves in ice, in km$^3$, for the glaciated regions of the world.

**LIMITATIONS**

Data on the thickness and water content of glaciers are only approximate. At the time these estimates were made (the mid-1970s), they were based on measurements obtained through borings or seismic soundings. These data were then applied to other glaciers with similar morphological characteristics. The quality of glacial water volume estimates depends on the number and quality of the soundings and borings. The total amounts of water contained in the ice cover of the earth depend mainly on the volume of ice in Antarctica and Greenland.

The accuracy of the determination of the amount of water in the ice of Antarctica, for example, is estimated to be within ±3 million km$^3$. This is somewhat higher than the total reserves of water in the ice of all remaining areas. Future satellite surveys should greatly improve the accuracy of these numbers.

**SOURCE**

USSR Committee for the International Hydrological Decade, 1978, *World Water Balance and Water Resources of the Earth,* Studies and Reports in Hydrology, no. 25. UNESCO, Paris and Gidrometeoizdat, Leningrad (English translation of a 1974 USSR publication).

**TABLE A.8**   Water reserves in surface ice

| | Area of ice (km$^2$) | | Water reserves (km$^3$) | |
|---|---|---|---|---|
| *Antarctica* | 13,980,000 | | 21,600,000 | |
| *Greenland* | 1,802,400 | | 2,340,000 | |
| *Arctic islands* | 226,090 | | 83,500 | |
| Franz Josef Land | | 13,735 | | 2,530 |
| Novaya Zemlya | | 24,420 | | 9,200 |
| Severnaya Zemlya | | 17,470 | | 4,620 |
| Spitsbergen (Western) | | 21,240 | | 18,690 |
| Small islands | | 400 | | 60 |
| Canadian Arctic archipelago | | 148,825 | | 48,400 |
| *Europe* | 21,415 | | 4,090 | |
| Iceland | | 11,785 | | 3,000 |
| Scandinavia | | 5,000 | | 645 |
| Alps | | 3,200 | | 350 |
| Caucasus | | 1,430 | | 95 |
| *Asia* | 109,085 | | 15,630 | |
| Pamir Alai | | 11,255 | | 1,725 |
| Tien Shan | | 7,115 | | 735 |
| Dzungarian Alatau, Altai, Sayan Mountains | | 1,635 | | 140 |
| Eastern Siberia | | 400 | | 30 |
| Kamchatka, Koryak Range | | 1,510 | | 80 |
| Hindu Kush | | 6,200 | | 930 |
| Karakoram Range | | 15,670 | | 2,180 |
| Himalayas | | 33,150 | | 4,990 |
| Tibet | | 32,150 | | 4,820 |
| *North America* | 67,522 | | 14,062 | |
| Alaska (Pacific Coast) | | 52,000 | | 12,200 |
| Inland Alaska | | 15,000 | | 1,800 |
| United States | | 510 | | 60 |
| Mexico | | 12 | | 2 |
| *South America* | 25,000 | | 6,750 | |
| Andes of Patagonia | | 17,900 | | 4,050 |
| All other South America[a] | | 7,100 | | 2,700 |
| *Africa*[b] | 22.5 | | 3 | |
| *New Zealand* | 1,000 | | 100 | |
| *New Guinea* | 14.5 | | 7 | |
| Total | 16,232,549 | | 24,064,142 | |

[a]Venezuela, Colombia, Andes of Ecuador, Andes of Peru, Andes of Chile and Argentina, Tierra del Fuego.

[b]Kenya, Kilimanjaro, Ruwenzori Mountains.

## TABLE A.9   Distribution of ice cover by region

| Glacier region | Area of ice-covered surface ($10^3$ km$^2$) |
|---|---|
| North polar regions | 2,000 |
| Temperate countries of the Northern Hemisphere | 190 |
| Tropical countries | 0.1 |
| Temperate countries of the Southern Hemisphere | 26 |
| Antarctica | 14,000 |
| Total for the earth | 16,200[a] |

[a]Total given in the original table; does not add to total of entries because of rounding.

## TABLE A.9   Distribution of ice cover by region

### TABLE A.9   Distribution of ice cover by region

**DESCRIPTION**

The area of ice cover, in km$^2 \times 10^3$, is shown for each of five regions, and for the earth as a whole.

**LIMITATIONS**

Data on the distribution, thickness, and water content of glaciers are only approximate and depend on the accuracy, number, and extent of ice surveys. Satellite measurements now underway may greatly increase the accuracy of these estimates as well as provide information on fluctuations in ice cover over time.

**SOURCE**

USSR Committee for the International Hydrological Decade, 1978, *World Water Balance and Water Resources of the Earth*, Studies and Reports in Hydrology, no. 25. UNESCO, Paris and Gidrometeoizdat, Leningrad (English translation of a 1974 USSR publication).

## TABLE A.10   Global fresh water resources, by country *(pages 129–133)*

**DESCRIPTION**

Global fresh water resources are listed by continent and country for all internal renewable water resources, river inflows, and outflows. All quantities are given in km$^3$ per year with the exception of per capita renewable resources, measured in m$^3 \times 10^3$ per year per person, using 1990 population estimates. Annual internal renewable resources are average fresh water resources renewably available over a year from precipitation falling within a country's borders. Data on several small countries not listed individually in the table are included in the continental and world totals.

**LIMITATIONS**

These detailed country data should be viewed with healthy skepticism. Many countries do not measure or report internal water resources data, so some of these entries were produced using other, less reliable methods. The World Resources Institute, which compiled the table originally, notes "Margat compiles water resources ... data from published documents, including national, United Nations, and professional literature. Data for small countries and countries in arid and semi-arid zones are less reliable than are those for larger and wetter countries. Belyaev compiles data on water resources ... from the world's literature and estimates resources ... from models using other data, such as area under irrigated agriculture, livestock populations, and precipitation, when necessary." Caution should be used when comparing data from different countries because of differences in the sources, methods of data collection, and periods used for measurement, which are not described in detail in the original source. The annual average figures hide large seasonal, interannual, and long-term variations. When no data are provided for annual river flows to or from other countries, the annual internal renewable water resources figure may include these flows.

**SOURCE**

World Resources Institute, 1992, *World Resources 1992–93*, Oxford University Press, New York. By permission.

**TABLE A.10** Global fresh water resources, by country

| Region[b] | 1990 population (10^6 people) | Annual internal renewable water resources | | Annual river flows | |
|---|---|---|---|---|---|
| | | Total (km³/yr) | 1990 (10³ m³/yr per capita) | From other countries (km³/yr) | To other countries (km³/yr) |
| *World* | 5,292.20 | 40,673.00[a] | 7.69 | | |
| *Africa* | 642.11 | 4,184.00[a] | 6.46 | | |
| Algeria | 24.96 | 18.90 | 0.75 | 0.20 | 0.70 |
| Angola | 10.02 | 158.00[a] | 15.77 | | |
| Benin | 4.63 | 26.00 | 5.48 | | |
| Botswana | 1.30 | 1.00 | 0.78 | 17.00 | |
| Burkina Faso | 9.00 | 28.00[a] | 3.11 | | |
| Burundi | 5.47 | 3.60[a] | 0.66 | | |
| Cameroon | 11.83 | 208.00 | 18.50 | | |
| Cape Verde | 0.37 | 0.20 | 0.53 | 0.00 | 0.00 |
| Central African Republic | 3.04 | 141.00[a] | 48.40 | | |
| Chad | 5.68 | 38.40[a] | 6.76 | | |
| Comoros | 0.55 | 1.02[a] | 1.97 | 0.00 | 0.00 |
| Congo | 2.27 | 181.00[a] | 90.77 | 621.00 | |
| Côte d'Ivoire | 12.00 | 74.00 | 5.87 | | |
| Djibouti | 0.41 | 0.30 | 0.74 | 0.00 | |
| Egypt | 52.43 | 1.80 | 0.03 | 56.50 | 0.00 |
| Equatorial Guinea | 0.35 | 30.00[a] | 68.18 | | |
| Ethiopia | 49.24 | 110.00 | 2.35 | | |
| Gabon | 1.17 | 164.00[a] | 140.05 | | |
| Gambia | 0.86 | 3.00 | 3.50 | 19.00 | |
| Ghana | 15.03 | 53.00 | 3.53 | | |
| Guinea | 5.76 | 226.00[a] | 32.87 | | |
| Guinea-Bissau | 0.96 | 31.00[a] | 31.41 | | |
| Kenya | 24.03 | 14.80[a] | 0.59 | | |
| Lesotho | 1.77 | 4.00[a] | 2.25 | | |
| Liberia | 2.58 | 232.00[a] | 90.84 | | |
| Libya | 4.55 | 0.70 | 0.15 | 0.00 | 0.00 |
| Madagascar | 12.00 | 40.00 | 3.34 | 0.00 | 0.00 |
| Malawi | 8.75 | 9.00[a] | 1.07 | | |
| Mali | 9.21 | 62.00[a] | 6.62 | | |
| Mauritania | 2.02 | 0.40 | 0.20 | 7.00 | |
| Mauritius | 1.08 | 2.20 | 1.99 | 0.00 | 0.00 |
| Morocco | 25.06 | 30.00 | 1.19 | 0.00 | 0.30 |
| Mozambique | 15.66 | 58.00[a] | 3.70 | | |

*continued*

| | | Annual internal renewable water resources | | Annual river flows | |
|---|---|---|---|---|---|
| Region[b] | 1990 population ($10^6$ people) | Total ($km^3/yr$) | 1990 ($10^3$ $m^3/yr$ per capita) | From other countries ($km^3/yr$) | To other countries ($km^3/yr$) |
| Namibia | 1.78 | 9.00[a] | | | |
| Niger | 7.73 | 14.00[a] | 1.97 | 30.00 | |
| Nigeria | 108.54 | 261.00[a] | 2.31 | 47.00 | |
| Rwanda | 7.24 | 6.30[a] | 0.87 | | |
| Senegal | 7.33 | 23.20[a] | 3.15 | 12.00 | |
| Sierra Leone | 4.15 | 160.00[a] | 38.54 | | |
| Somalia | 7.50 | 11.50 | 1.52 | 0.00 | |
| South Africa | 35.28 | 50.00 | 1.42 | | |
| Sudan | 25.20 | 30.00 | 1.19 | 100.00 | 56.50 |
| Swaziland | 0.79 | 6.96[a] | 8.82 | | |
| Tanzania | 27.32 | 76.00[a] | 2.78 | | |
| Togo | 3.53 | 11.50 | 3.33 | | |
| Tunisia | 8.18 | 3.75 | 0.46 | 0.60 | 0.00 |
| Uganda | 18.79 | 66.00[a] | 3.58 | | |
| Zaire | 35.57 | 1,019.00[a] | 28.31 | | |
| Zambia | 8.45 | 96.00[a] | 11.35 | | |
| Zimbabwe | 9.71 | 23.00[a] | 2.37 | | |
| | | | | | |
| *North and Central America* | 427.23 | 6,945.00[a] | 16.26 | | |
| Barbados | 0.26 | 0.05 | 0.20 | 0.00 | 0.00 |
| Belize | 0.19 | 16.00 | | | |
| Canada | 26.52 | 2,901.00 | 109.37 | | |
| Costa Rica | 3.02 | 95.00 | 31.51 | | |
| Cuba | 10.61 | 34.50 | 3.34 | 0.00 | 0.00 |
| Dominican Republic | 7.17 | 20.00 | 2.79 | | |
| El Salvador | 5.25 | 18.95 | 3.61 | | |
| Guatemala | 9.20 | 116.00 | 12.61 | | |
| Haiti | 6.51 | 11.00 | 1.69 | | |
| Honduras | 5.14 | 102.00 | 19.85 | | |
| Jamaica | 2.46 | 8.30 | 3.29 | 0.00 | 0.00 |
| Mexico | 88.60 | 357.40 | 4.03 | | |
| Nicaragua | 3.87 | 175.00 | 45.21 | | |
| Panama | 2.42 | 144.00 | 59.55 | | |
| Trinidad and Tobago | 1.28 | 5.10[a] | 3.98 | 0.00 | 0.00 |
| United States | 249.22 | 2,478.00 | 9.94 | | |

*continued*

| Region[b] | 1990 population (10^6 people) | Annual internal renewable water resources | | Annual river flows | |
|---|---|---|---|---|---|
| | | Total (km³/yr) | 1990 (10³ m³/yr per capita) | From other countries (km³/yr) | To other countries (km³/yr) |
| *South America* | 296.72 | 10,377.00[a] | 34.96 | | |
| Argentina | 32.32 | 694.00 | 21.47 | 300.00 | |
| Bolivia | 7.31 | 300.00[a] | 41.02 | | |
| Brazil | 150.37 | 5,190.00 | 34.52 | 1,760.00 | |
| Chile | 13.17 | 468.00[a] | 35.53 | | |
| Colombia | 32.98 | 1,070.00 | 33.63 | | |
| Ecuador | 10.59 | 314.00 | 29.12 | | |
| Guyana | 0.80 | 241.00[a] | 231.73 | | |
| Paraguay | 4.28 | 94.00[a] | 21.98 | 220.00 | |
| Peru | 21.55 | 40.00 | 1.79 | | |
| Suriname | 0.42 | 200.00[a] | 496.28 | | |
| Uruguay | 3.09 | 59.00[a] | 18.86 | 65.00 | |
| Venezuela | 19.74 | 856.00 | 43.37 | 461.00 | |
| *Asia* | 3,112.70 | 10,485.00 | 3.37 | | |
| Afghanistan | 16.56 | 50.00 | 3.02 | | |
| Bahrain | 0.52 | 0.00 | 0.00 | | |
| Bangladesh | 115.59 | 1,357.00 | 11.74 | 1,000.00 | |
| Bhutan | 1.52 | 95.00[a] | 62.66 | | |
| Cambodia | 8.25 | 88.10 | 10.68 | 410.00 | |
| China | 1,139.06 | 2,800.00 | 2.47 | 0.00 | |
| Cyprus | 0.70 | 0.90 | 1.28 | 0.00 | 0.00 |
| India | 853.09 | 1,850.00 | 2.17 | 235.00 | |
| Indonesia | 184.28 | 2,530.00 | 14.02 | | |
| Iran | 54.61 | 117.50 | 2.08 | | |
| Iraq | 18.92 | 34.00[a] | 1.80 | 66.00 | |
| Israel | 4.60 | 1.70 | 0.37 | 0.45 | 0.00 |
| Japan | 123.46 | 547.00 | 4.43 | 0.00 | 0.00 |
| Jordan | 4.01 | 0.70 | 0.16 | 0.40 | |
| Korea DPR | 21.77 | 67.00[a] | 2.92 | | |
| Korea Rep | 42.79 | 63.00 | 1.45 | | |
| Kuwait | 2.04 | 0.00 | 0.00 | 0.00 | |
| Laos | 4.14 | 270.00 | 66.32 | | |
| Lebanon | 2.70 | 4.80 | 1.62 | 0.00 | 0.86 |
| Malaysia | 17.89 | 456.00 | 26.30 | | |
| Mongolia | 2.19 | 24.60 | 11.05 | | |

*continued*

| Region[b] | 1990 population ($10^6$ people) | Annual internal renewable water resources | | Annual river flows | |
| | | Total ($km^3/yr$) | 1990 ($10^3$ $m^3/yr$ per capita) | From other countries ($km^3/yr$) | To other countries ($km^3/yr$) |
| --- | --- | --- | --- | --- | --- |
| Myanmar | 41.68 | 1,082.00 | 25.96 | | |
| Nepal | 19.14 | 170.00 | 8.88 | | |
| Oman | 1.50 | 2.00 | 1.36 | 0.00 | |
| Pakistan | 122.63 | 298.00 | 2.43 | 170.00 | |
| Philippines | 62.41 | 323.00 | 5.18 | 0.00 | 0.00 |
| Qatar | 0.37 | 0.02 | 0.06 | 0.00 | |
| Saudi Arabia | 14.13 | 2.20 | 0.16 | 0.00 | |
| Singapore | 2.72 | 0.60 | 0.22 | 0.00 | 0.00 |
| Sri Lanka | 17.22 | 43.20 | 2.51 | 0.00 | 0.00 |
| Syria | 12.53 | 7.60 | 0.61 | 27.90 | 30.00 |
| Thailand | 55.70 | 110.00 | 1.97 | 69.00 | |
| Turkey | 55.87 | 196.00 | 3.52 | 7.00 | 69.00 |
| United Arab Emirates | 1.59 | 0.30 | 0.19 | 0.00 | |
| Vietnam | 66.69 | 376.00[a] | 5.60 | | |
| Yemen Arab Rep | 9.20 | 1.00 | 0.12 | 0.00 | |
| Yemen Dem Rep | 2.49 | 1.50 | 0.60 | 0.00 | |
| | | | | | |
| *Europe* | 498.37 | 2,321.00[a] | 4.66 | | |
| Albania | 3.25 | 10.00 | 3.08 | 11.30 | |
| Austria | 7.58 | 56.30 | 7.51 | 34.00 | |
| Belgium | 9.85 | 8.40 | 0.85 | 4.10 | |
| Bulgaria | 9.01 | 18.00 | 2.00 | 187.00 | |
| Czechoslovakia | 15.67 | 28.00 | 1.79 | 62.60 | |
| Denmark | 5.14 | 11.00 | 2.15 | 2.00 | |
| Finland | 4.98 | 110.00 | 22.11 | 3.00 | |
| France | 56.14 | 170.00 | 3.03 | 15.00 | 20.50 |
| German DR | 16.25 | 17.00 | 1.02 | 17.00 | |
| Germany FR | 61.32 | 79.00 | 1.30 | 82.00 | |
| Greece | 10.05 | 45.15 | 4.49 | 13.50 | 3.00 |
| Hungary | 10.55 | 6.00 | 0.57 | 109.00 | |
| Iceland | 0.25 | 170.00 | 671.94 | 0.00 | 0.00 |
| Ireland | 3.72 | 50.00 | 13.44 | 0.00 | |
| Italy | 57.06 | 179.40 | 3.13 | 7.60 | 0.00 |
| Luxembourg | 0.37 | 1.00 | 2.72 | 4.00 | |
| Malta | 0.35 | 0.03 | 0.07 | 0.00 | 0.00 |
| Netherlands | 14.95 | 10.00 | 0.68 | 80.00 | |
| Norway | 4.21 | 405.00 | 96.15 | 8.00 | |

*continued*

| Region[b] | 1990 population (10^6 people) | Annual internal renewable water resources | | Annual river flows | |
|---|---|---|---|---|---|
| | | Total (km³/yr) | 1990 (10³ m³/yr per capita) | From other countries (km³/yr) | To other countries (km³/yr) |
| Poland | 38.42 | 49.40 | 1.29 | 6.80 | |
| Portugal | 10.29 | 34.00 | 3.31 | 31.60 | |
| Romania | 23.27 | 37.00 | 1.59 | 171.00 | |
| Spain | 39.19 | 110.30 | 2.80 | 1.00 | 17.00 |
| Sweden | 8.44 | 176.00 | 21.11 | 4.00 | |
| Switzerland | 6.61 | 42.50 | 6.52 | 7.50 | |
| United Kingdom | 57.24 | 120.00 | 2.11 | 0.00 | |
| Yugoslavia | 23.81 | 150.00 | 6.29 | 115.00 | 200.00 |
| | | | | | |
| *USSR* | 288.60 | 4,384.00 | 15.22 | 300.00 | |
| | | | | | |
| *Oceania* | 26.48 | 2,011.00[a] | 75.96 | | |
| Australia | 16.87 | 343.00 | 20.48 | 0.00 | 0.00 |
| Fiji | 0.76 | 28.55[a] | 38.12 | 0.00 | 0.00 |
| New Zealand | 3.39 | 397.00 | 117.49 | 0.00 | 0.00 |
| Papua New Guinea | 3.87 | 801.00[a] | 199.70 | | |
| Solomon Islands | 0.32 | 44.70[a] | 149.00 | 0.00 | 0.00 |

Figures may not add to totals due to independent rounding.
[a]Estimates from Belyaev, Institute of Geography, USSR.
[b]Data on several small countries not shown in the table are included in world and regional totals.

*continued*

**TABLE A.11**   Renewable water resources for the European Economic Community

| Country | Average flow formed in country (km³/yr) | Average flow imported from neighboring country | | Total average flow available (km³/yr) | Average flow imported (%) | Average fresh water resources per capita[a] (m³/yr/person) |
|---|---|---|---|---|---|---|
| | | EEC | Non-EEC | | | |
| Belgium | 8.4 | 8.5 | 0 | 16.9 | 50 | 1,715 |
| Denmark | 11 | 2 | 0 | 13 | 15 | 2,548 |
| France | 170 | 3 | 12 | 185 | 8 | 3,367 |
| Germany FR | 79 | 9 | 84 | 172 | 54 | 2,817 |
| Greece | 45.15 | 0 | 13.5 | 58.65 | 23 | 5,902 |
| Ireland | 50 | 0 | 0 | 50 | 0 | 14,045 |
| Italy | 179.4 | 1 | 6.6 | 187 | 4 | 3,284 |
| Luxembourg | 1 | 4 | 0 | 5 | 80 | 13,661 |
| Netherlands | 10 | 80 | 0 | 90 | 89 | 6,213 |
| Portugal | 34 | 31.6 | 0 | 65.5 | 48 | 6,433 |
| Spain | 110.3 | 1 | 0 | 113.3 | 1 | 2,874 |
| United Kingdom | 117.8 | 2.2 | 0 | 120 | 2 | 2,122 |

[a]Assuming a 1985 population.

**TABLE A.11**   Renewable water resources for the European Economic Community

## DESCRIPTION
Total runoff generated from internal rainfall and from rivers that flow over their borders is reported for the twelve nations of the European Economic Community (EEC) in km³ per year. Runoff imports are reported from both EEC and non-EEC neighbors. Per capita fresh water availability is reported in m³ × 10³ per year per person, using 1985 population figures. Also presented are the mean annual rainfall for each country, in mm per year, and the percentage of total fresh water flow that originates outside of a country's borders.

## LIMITATIONS
These data are long-term average flows. No estimate of the natural variability of water availability is presented, though this is extremely important for water planning purposes. No time periods are specified for the long-term averages.

## SOURCE
J. Margat, 1989, The sharing of common water resources in the European Community (EEC), *Water International*, **14**, 59–91, International Water Resources Association, Illinois. By permission.

**TABLE A.12** Summary of the characteristics of principal aquifer types

| Aquifer type | Lithology | Ground water flow regime | Aquifer properties | | | Natural flow rates (m/day) | Residence times |
|---|---|---|---|---|---|---|---|
| | | | Porosity (percent) | Permeability (m/day) | Specific yield (percent) | | |
| Shallow alluvium | Gravel | Intergranular | 25–35 | 100–1,000 | 12–25 | 2–10 | Could be very short; a few months to years, depending on volume |
| | Sand | Intergranular | 30–42 | 1–50 | 10–25 | 0.05–1 | |
| | Silt | Intergranular | 40–45 | 0.0005–0.1 | 5–10 | 0.001–0.1 | |
| Deep sedimentary formations | Sands and silts | Intergranular | 30–40 | 0.1–5 | 2–10 | 0.001–0.01 | Many thousands of years |
| Sandstone | Cemented quartz grains | Intergranular and fissure | 10–30 | 0.1–10 | 8–20 | 0.001–0.1 | Tens to hundreds of years |
| Limestone | Cemented carbonate | Mainly fissure | 5–30 | 0.1–50 | 5–15 | 0.001–1 | Tens to hundreds of years |
| Karstic limestone | Cemented carbonate | Fissures and channels | 5–25 | 100–10,000 | 5–15 | 10–2,000 | A few hours to days |
| Volcanic rock basalt | Fine grained crystalline | Fissure | 2–15 | 0.1–1,000 | 1–5 | 1–500 | Very wide range; can be very short |
| tuff | Cemented grains | Intergranular and fissure | 15–30 | 0.1–5 | 10–20 | 0.001–1 | Wide range |
| Igneous and metamorphic rocks (granites and gneisses): Fresh | Crystalline | Fissure | 0.1–2 | $10^{-7}$–$5\times10^{-5}$ | 1–5 | $10^{-6}$–$10^{-2}$ | Thousands of years, but can be rapid where fractured |
| Weathered | Disaggregated crystalline | Intergranular and fissure | 10–20 | 0.1–2 | 1–5 | 0.001–0.1 | Tens to hundreds of years |

**TABLE A.12** Summary of the characteristics of principal aquifer types

**DESCRIPTION**

Aquifers are rock beds capable of storing water and allowing it to pass through at a variety of rates. Aquifers have different properties that govern their capacity to hold and transport water. This table summarizes some of the principal qualities with respect to aquifer lithology or rock type, the physical path through which ground water can pass, and the residence times, the amount of time any given molecule of water is likely to remain in the aquifer. Residence times can range from hours to millennia. The table also quantifies the porosity of the rock,

permeability, specific yield, and natural flow rates. Porosity is defined as the percentage of empty spaces in a given volume of rock. Specific yield is the proportion of ground water that is available for drainage under forces of gravity.

**LIMITATIONS**

Only general aquifer characteristics are presented here. Individual ground water aquifers will have region-specific characteristics

**SOURCE**

P.J. Chilton, 1990, cited in M. Meybeck, D.V. Chapman and R. Helmer, *Global Freshwater Quality: A First Assessment*, Global Environmental Monitoring System (GEMS), World Health Organization and United Nations Environment Programme, Basil Blackwell, Oxford. By permission.

**TABLE A.13**    Discharge to the oceans of ground water and subsurface dissolved solids

| Oceans, continents, and major islands | Ground water discharge | | Subsurface dissolved solids discharge | |
|---|---|---|---|---|
| | Areal ($l/sec/km^2$) | Volume ($km^3/yr$) | Areal (tonnes/yr/$km^2$) | Volume ($10^6$ tonnes/yr) |
| *Pacific Ocean* | | | | |
| Australia | 1.1 | 7.1 | 24.9 | 5.0 |
| Asia | 4.8 | 254.3 | 98.2 | 165.2 |
| North America | 5.4 | 124.6 | 50.1 | 36.7 |
| South America | 11.5 | 199.6 | 64.1 | 35.5 |
| Major islands | 13.0 | 714.7 | 159.8 | 278.1 |
| Total | | 1,300.3 | | 520.5 |
| *Atlantic Ocean* | | | | |
| Africa | 3.9 | 208.7 | 99.9 | 169.2 |
| Europe | 4.2 | 71.2 | 47.8 | 25.8 |
| North America | 4.6 | 219.4 | 74.6 | 77.7 |
| South America | 3.0 | 185.3 | 40.2 | 42.9 |
| Major islands | 4.4 | 77.7 | 76.0 | |
| *Mediterranean Sea* | | | | |
| Africa | 0.4 | 5.1 | 24.4 | 9.9 |
| Asia | 2.4 | 8.3 | 110.3 | 11.9 |
| Europe | 4.0 | 33.9 | 68.4 | 18.4 |
| Major islands | 2.8 | 5.7 | 34.9 | 2.3 |
| Total | | 815.3 | | 470.0 |
| *Indian Ocean* | | | | |
| Australia | 0.2 | 16.4 | 28.4 | 66.7 |
| Africa | 0.6 | 22.1 | 38.7 | 49.0 |
| Asia | 1.7 | 65.3 | 97.2 | 119.2 |
| Major islands | 5.1 | 115.6 | 84.7 | 60.6 |
| Total | | 219.4 | | 295.5 |
| *Arctic Ocean* | | | | |
| Europe | 5.6 | 47.5 | 26.6 | 7.2 |
| World total | | 2,382.5 | | 1,293.5 |

**TABLE A.13**  Discharge to the oceans of ground water and subsurface dissolved solids *(page 136)*

**DESCRIPTION**

An estimate of ground water discharged to each ocean from all major land areas is given here, measured in l/s/km$^2$ of land area and in km$^3$ per year. Also shown is the discharge of dissolved solids measured in tonnes per year per km$^2$ and in million tonnes per year.

**LIMITATIONS**

Ground water discharge volumes and the concentrations of dissolved solids are difficult to measure and these values should be considered

approximations. No ranges of uncertainty or the annual variability of these discharges are presented in the original source.

**SOURCE**

I.S. Zekster and R.G. Dzhamalov, 1988, *Role of Ground Water in the Hydrological Cycle and in Continental Water Balance*, IHP III Project 2.3, Technical Documents in Hydrology, International Hydrological Programme, UNESCO and USSR National Committee for the International Hydrological Programme, Water Problems Institute of the USSR Academy of Sciences, Paris.

**TABLE A.14**  Ground water recharge estimates for India

| State | Net recharge or replenishable ground water resource (km$^3$/yr) | |
| | GOI (1960) | CGWB (1991) |
|---|---|---|
| Andhra Pradesh | 21.0 | 43.4 |
| Bihar | 27.0 | 33.8 |
| Delhi | 0.4 | 0.5 |
| Gujarat | 12.6 | 22.6 |
| Haryana | 4.3 | 8.5 |
| Himachal Pradesh | 1.1 | 0.4 |
| Jammu and Kashmir | 4.9 | 4.4 |
| Kerala | 6.7 | 8.1 |
| Madhya Pradesh | 32.9 | 59.7 |
| Maharashtra | 15.5 | 38.8 |
| Orissa | 19.7 | 23.3 |
| Punjab | 8.5 | 18.0 |
| Rajasthan | 4.2 | 16.2 |
| Uttar Pradesh | 43.8 | 80.5 |
| West Bengal | 19.9 | 20.7 |
| Other areas | 47.1 | 72.6 |
| Total | 269.6 | 451.5 |

**TABLE A.14**  Ground water recharge estimates for India

**DESCRIPTION**

Two estimates for net ground water recharge rates for several states in India are given in km$^3$ per year. These estimates were made using assumed values for infiltration, canal seepage, evapotranspiration losses, and subsurface runoff. The two estimates differ substantially, reflecting both improved measurements in the time between estimates and the uncertainty in total ground water recharge rates. The data cited here in the GOI (1972) document reflect recharge estimates made in 1960.

**LIMITATIONS**

"Other areas" include states such as Tamil Nadu where boundaries have changed between estimates. Moench (1991) states, "While [the 1960] estimate was admittedly very rough, it is not clear how much better the current estimates are," since many of the general assumptions of the earlier analysis were used in generating most of the currently accepted figures. Estimates of ground water recharge rates throughout the world often have a political component because of the desire to maximize the use of ground water for irrigation purposes.

**SOURCES**

M. Moench, 1991, *Sustainability, Efficiency, and Equity in Groundwater Development: Issues in India and Comparison with the Western US*, Pacific Institute for Studies in Development, Environment, and Security. Berkeley, California. By permission.

GOI, 1972, *Report of the Irrigation Commission, 1972*, Ministry of Irrigation and Power, New Delhi.

Central Ground Water Board (CGWB) figures, given as "Replenishable ground water resources," are from a personal communication with B.P. Sinha, Chief Hydrologist at the Central Ground Water Board.

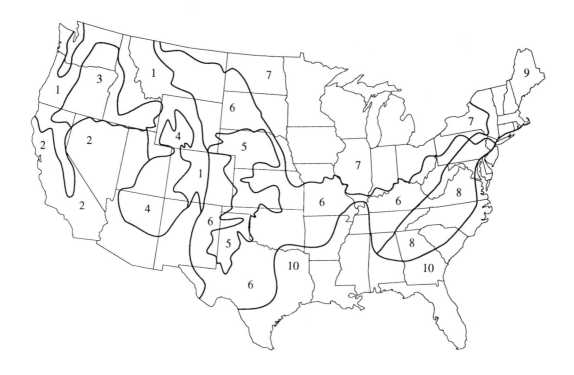

Fig. A-15   Ground water resources in the continental United States by geologic region. See Table A.15 for details.

**TABLE A.15**   Ground water resources in geologic regions of the United States *(page 139)*

**DESCRIPTION**

This table and the accompanying map present a summary of the morphology of twelve United States ground water regions and the resources they contain. Each section describes the geological characteristics and the nature of ground water resources of each region. Alaska and Hawaii are not shown on the map.

**LIMITATIONS**

The coarse grouping of geographical types and the poor spatial resolu-tion of the figure hides smaller ground water basins of regional importance.

**SOURCE**

American Institute of Professional Geologists, 1983, *Ground Water Issues and Answers*, Arvada, Colorado. By permission.

**TABLE A.15**  Ground water resources in geologic regions of the United States

**1. Western Mountains**

Underlain by hard, dense rocks; weathered rock locally yields modest supplies, as does alluvium in intermontane valleys. Large supplies are rare.

**2. Alluvial Basins**

Large depressed areas flanked by highlands and filled with erosional debris. Alluvial fill functions as an ideal aquifer, absorbing water readily from streams issuing from highlands and yielding large supplies to wells. Supports large-scale irrigated agriculture and provides municipal water for many cities.

**3. Columbia Lava Plateau**

Underlain by thousands of feet of basaltic lava flows, interbedded with alluvial and lake sediments. Lava rocks are highly permeable because of lava tubes, shrinkage cracks, and interflow rubble zones. Yields large supplies of water for irrigation and municipal use.

**4. Colorado Plateaus and Wyoming Basins**

Underlain by gently dipping sediments, mainly poorly permeable sandstone and shale. Most productive aquifers are sandstone, furnishing small supplies for stock and domestic use. Prospects poor for large-scale ground water developments, but such supplies are found at a few favorable localities.

**5. High Plains**

Underlain by alluvium of the Ogallala Formation, as much as 450 feet thick, which yields large supplies to wells, mainly for irrigation. Opportunity for recharge from streams is small, because rainfall is low and large streams have cut below the base of alluvium. Water table is gradually declining in much of the area due to overdraft.

**6. Unglaciated Central Region**

Complex area of plains and plateaus, underlain by consolidated sedimentary rocks. Alluvium of stream valleys provides large supplies for industry and cities. Most productive aquifers in much of the region are dolomitic limestones and sandstones of low-to-moderate yield.

**7. Glaciated Central Region**

Similar to Unglaciated Central Region, except that area is mantled by glacial deposits as much as 900 feet thick. These contain lenses and beds of well-sorted sand and gravel, which yield large supplies of water for industrial and municipal use.

**8. Unglaciated Appalachians**

Mountainous area underlain mainly by consolidated sedimentary rocks of small-to-moderate water yield. Locally, limestones yield large supplies of water.

**9. Glaciated Appalachians**

Glacial deposits mantle steep areas and underlie valleys and lowlands. Yields from bedrocks are generally small to moderate. Principal ground water sources are sand and gravel of glacial outwash plains, or channel fillings in stratified drift.

**10. Atlantic and Gulf Coastal Plains**

A huge, seaward-thickening wedge of sedimentary rocks consisting mainly of clay, sand, marl, and limestone. Thickness along coast increases southward from 300 to 30,000 feet. Large supplies of ground water can be obtained almost anywhere, although salt-water encroachment is a problem locally.

**11. Alaska**

Most of the area has been glaciated, and large supplies of ground water can be obtained from glacial sand and gravel. Permafrost is present in northern Alaska, restricting the availability of ground water.

**12. Hawaii**

Entire island chain is composed of basaltic lava flows, which are highly permeable and yield water readily to wells and tunnels. Fresh water body forms a lens floating upon seawater, so extraction must be carefully managed to avoid seawater intrusion.

**TABLE A.16**   Surface flow and useful resources of ground water in the former Soviet republics

| | Average surface flows (km³/yr) | Usable ground water resources (km³/yr) |
|---|---|---|
| Russia | 4,270 | 228 |
| Ukraine | 210 | 21 |
| Byelorussia | 56 | 18 |
| Uzbekistan | 108 | 22 |
| Kazakhstan | 125 | 44 |
| Georgia | 61 | 4 |
| Azerbaidzhan | 28 | 5 |
| Lithuania | 23 | 1 |
| Moldavia | 13 | 1 |
| Latvia | 32 | 2 |
| Kirghizia | 49 | 13 |
| Tadjikistan | 95 | 6 |
| Armenia | 8 | 5 |
| Turkmenistan | 71 | 2 |
| Estonia | 16 | 2 |

**TABLE A.16**   Surface flow and useful resources of ground water in the former Soviet republics

**DESCRIPTION**

Data are presented here for average surface water flows and ground water resources of the fifteen former Soviet republics in km³ per year. For ground water, data refer to the volume of ground water that can potentially be withdrawn on an annually renewable basis. This value is a rough estimate and requires further study, according to the original source.

**LIMITATIONS**

For the surface flows, no information is presented on water that flows across borders and whether the values shown are internally renewable resources or total water availability, including flows from neighboring regions. Ground water aquifers do not follow political borders, so the values described here are unlikely to be precise for each region. No data are provided on the period of observation for the average surface flows.

**SOURCE**

Goscomstat, USSR, 1989, *Protection of the Environment and Rational Utilization of Natural Resources in the USSR*, Statistical Handbook, Government Committee on Statistics, Moscow (in Russian).

**TABLE A.17**   Stream flow, ground water, and rainfall resources in the Arab region

| Country | Streamflow (km$^3$/yr) | Ground water (km$^3$/yr) | Rainfall (km$^3$/yr) | Rainfall (mm/yr) |
|---|---|---|---|---|
| Algeria | 13.00 | 4.20 | 192.5 | 81 |
| Bahrain | - | 0.09 | 0.1 | 80 |
| Djibouti | - | - | 4.0 | 160 |
| Egypt | 55.50 | 3.42 | 15.3 | 15 |
| Iraq | 43.20 | - | 99.9 | 230 |
| Jordan | 0.90 | 0.41 | 6.7 | 73 |
| Kuwait | - | 0.16 | 2.4 | 135 |
| Lebanon | 4.38 | 0.60 | 6.8 | 657 |
| Libya | 0.22 | 4.40 | 49.0 | 28 |
| Mauritania | 1.00 | 0.88 | 157.2 | 153 |
| Morocco | 23.00 | 5.00 | 82.4 | 185 |
| North Yemen | 2.10 | 1.00 | 46.1 | 236 |
| Oman | 1.37 | 0.56 | 15.0 | 71 |
| Qatar | - | 0.05 | 0.8 | 73 |
| Saudi Arabia | 2.20 | 2.35 | 126.2 | 59 |
| Somalia | 8.20 | 0.30 | 190.6 | 298 |
| South Yemen | 1.70 | 0.40 | 21.1 | 73 |
| Sudan | 20.50 | 0.28 | 1,094.4 | 437 |
| Syria | 22.12 | 3.67 | 52.7 | 287 |
| Tunisia | 2.63 | 1.73 | 39.8 | 242 |
| United Arab Emirates | 0.10 | 0.39 | 7.4 | 89 |

**TABLE A.17**   Stream flow, ground water, and rainfall resources in the Arab region

**DESCRIPTION**

Stream flow and ground water availability are presented for 21 countries of the Middle East and North Africa, in km$^3$ per year. Average annual rainfall is presented in km$^3$ per year and mm per year. Data for Israel were not included in the original source. Available stream flow includes both perennial and intermittent streams. Estimates of annual ground water availability are for annually renewable recharge, which is considerably less than the total water volumes in storage.

**LIMITATIONS**

Estimates of ground water availability are highly uncertain. The stream flow, ground water, and rainfall data represent long-term averages, but no information on the period of measurement is included in the original

source. Actual precipitation and stream flow are extremely variable in this region, both annually and seasonally. The stream flow and ground water availability figures do not represent water withdrawn for use. Many countries of the Arabian Peninsula, such as Kuwait, Saudi Arabia, Bahrain, and Qatar use desalination facilities to supply additional fresh water. These non-conventional water resources are not included here.

**SOURCE**

M. Shahin, 1989, Review and assessment of water resources in the Arab region, *Water International*, **14**(4), 206–219, International Water Resources Association, Illinois. By permission.

**TABLE A.18**   Abundances of chemical elements in seawater

| Element | | Concentration (mg/l) | Some probable dissolved species | Total amount in the oceans (tonnes) |
|---|---|---|---|---|
| Chlorine | Cl | $1.95 \times 10^4$ | $Cl^-$ | $2.57 \times 10^{16}$ |
| Sodium | Na | $1.077 \times 10^4$ | $Na^+$ | $1.42 \times 10^{16}$ |
| Magnesium | Mg | $1.290 \times 10^3$ | $Mg^{2+}$ | $1.71 \times 10^{15}$ |
| Sulfur | S | $9.05 \times 10^2$ | $SO_4^{2-}$, $NaSO_4^-$ | $1.2 \times 10^{15}$ |
| Calcium | Ca | $4.12 \times 10^2$ | $Ca^{2+}$ | $5.45 \times 10^{14}$ |
| Potassium | K | $3.80 \times 10^2$ | $K^+$ | $5.02 \times 10^{14}$ |
| Bromine | Br | 67 | $Br^-$ | $8.86 \times 10^{13}$ |
| Carbon | C | 28 | $HCO_3^-$, $CO_3^{2-}$, $CO_2$ | $3.7 \times 10^{13}$ |
| Nitrogen | N | 11.5 | $N_2$ gas, $NO_3^-$, $NH_4^+$ | $1.5 \times 10^{13}$ |
| Strontium | Sr | 8 | $Sr^{2+}$ | $1.06 \times 10^{13}$ |
| Oxygen | O | 6 | $O_2$ gas | $7.93 \times 10^{12}$ |
| Boron | B | 4.4 | $B(OH)_3$, $B(OH)_4^-$, $H_2BO_3^-$ | $5.82 \times 10^{12}$ |
| Silicon | Si | 2 | $Si(OH)_4$ | $2.64 \times 10^{12}$ |
| Fluorine | F | 1.3 | $F^-$, $MgF^+$ | $1.72 \times 10^{12}$ |
| Argon | Ar | 0.43 | Ar gas | $5.68 \times 10^{11}$ |
| Lithium | Li | 0.18 | $Li^+$ | $2.38 \times 10^{11}$ |
| Rubidium | Rb | 0.12 | $Rb^+$ | $1.59 \times 10^{11}$ |
| Phosphorus | P | $6 \times 10^{-2}$ | $HPO_4^{2-}$, $PO_4^{3-}$, $H_2PO_4^-$ | $7.93 \times 10^{10}$ |
| Iodine | I | $6 \times 10^{-2}$ | $IO_3^-$, $I^-$ | $7.93 \times 10^{10}$ |
| Barium | Ba | $2 \times 10^{-2}$ | $Ba^{2+}$ | $2.64 \times 10^{10}$ |
| Molybdenum | Mo | $1 \times 10^{-2}$ | $MoO_4^{2-}$ | $1.32 \times 10^{10}$ |
| Arsenic | As | $3.7 \times 10^{-3}$ | $HAsO_4^{2-}$, $H^2AsO_4^-$ | $4.89 \times 10^9$ |
| Uranium | U | $3.2 \times 10^{-3}$ | $UO_2(CO_3)_2^{4-}$ | $4.23 \times 10^9$ |
| Vanadium | V | $2.5 \times 10^{-3}$ | $H_2VO_4^-$, $HVO_4^{2-}$ | $3.31 \times 10^9$ |
| Aluminum | Al | $2 \times 10^{-3}$ | $Al(OH)_4^-$ | $2.64 \times 10^9$ |
| Iron | Fe | $2 \times 10^{-3}$ | $Fe(OH)_2^+$, $Fe(OH)_4^-$ | $2.64 \times 10^9$ |
| Nickel | Ni | $1.7 \times 10^{-3}$ | $Ni^{2+}$ | $2.25 \times 10^9$ |
| Titanium | Ti | $1 \times 10^{-3}$ | $Ti(OH)_4$ | $1.32 \times 10^9$ |
| Zinc | Zn | $5 \times 10^{-4}$ | $ZnOH^+$, $Zn^{2+}$, $ZnCO_3$ | $6.61 \times 10^8$ |
| Cesium | Cs | $4 \times 10^{-4}$ | $Cs^+$ | $5.29 \times 10^8$ |
| Chromium | Cr | $3 \times 10^{-4}$ | $Cr(OH)_3$, $CrO_4^{2-}$ | $3.97 \times 10^8$ |
| Antimony | Sb | $2.4 \times 10^{-4}$ | $Sb(OH)_6^-$ | $3.17 \times 10^8$ |
| Manganese | Mn | $2 \times 10^{-4}$ | $Mn^{2+}$, $MnCl^+$ | $2.64 \times 10^8$ |
| Krypton | Kr | $2 \times 10^{-4}$ | Kr gas | $2.64 \times 10^8$ |
| Selenium | Se | $2 \times 10^{-4}$ | $SeO_3^{2-}$ | $2.64 \times 10^8$ |
| Neon | Ne | $1.2 \times 10^{-4}$ | Ne gas | $1.59 \times 10^8$ |

*continued*

| Element | | Concentration (mg/l) | Some probable dissolved species | Total amount in the oceans (tonnes) |
|---|---|---|---|---|
| Cadmium | Cd | $1 \times 10^{-4}$ | $CdCl_2$ | $1.32 \times 10^8$ |
| Copper | Cu | $1 \times 10^{-4}$ | $CuCO_3, CuOH^+$ | $1.32 \times 10^8$ |
| Tungsten | W | $1 \times 10^{-4}$ | $WO_4^{2-}$ | $1.32 \times 10^8$ |
| Germanium | Ge | $5 \times 10^{-5}$ | $Ge(OH)_4$ | $6.61 \times 10^7$ |
| Xenon | Xe | $5 \times 10^{-5}$ | Xe gas | $6.61 \times 10^7$ |
| Mercury | Hg | $3 \times 10^{-5}$ | $HgCl_4^{2-}, HgCl_2$ | $3.97 \times 10^7$ |
| Zirconium | Zr | $3 \times 10^{-5}$ | $Zr(OH)_4$ | $3.97 \times 10^7$ |
| Bismuth | Bi | $2 \times 10^{-5}$ | $BiO^+, Bi(OH)_2^+$ | $2.64 \times 10^7$ |
| Niobium | Nb | $1 \times 10^{-5}$ | Not known | $1.32 \times 10^7$ |
| Tin | Sn | $1 \times 10^{-5}$ | $SnO(OH)_3^-$ | $1.32 \times 10^7$ |
| Thallium | Tl | $1 \times 10^{-5}$ | $Tl^+$ | $1.32 \times 10^7$ |
| Thorium | Th | $1 \times 10^{-5}$ | $Th(OH)_4$ | $1.32 \times 10^7$ |
| Hafnium | Hf | $7 \times 10^{-6}$ | Not known | $9.25 \times 10^6$ |
| Heiium | He | $6.8 \times 10^{-6}$ | He gas | $8.99 \times 10^6$ |
| Beryllium | Be | $5.6 \times 10^{-6}$ | $BeOH^+$ | $7.40 \times 10^6$ |
| Gold | Au | $4 \times 10^{-6}$ | $AuCl_2^-$ | $5.29 \times 10^6$ |
| Rhenium | Re | $4 \times 10^{-6}$ | $ReO_4^-$ | $5.29 \times 10^6$ |
| Cobalt | Co | $3 \times 10^{-6}$ | $Co^{2+}$ | $3.97 \times 10^6$ |
| Lanthanum | La | $3 \times 10^{-6}$ | $La(OH)_3$ | $3.97 \times 10^6$ |
| Neodymium | Nd | $3 \times 10^{-6}$ | $Nd(OH)_3$ | $3.97 \times 10^6$ |
| Silver | Ag | $2 \times 10^{-6}$ | $AgCl_2^-$ | $2.64 \times 10^6$ |
| Tantalum | Ta | $2 \times 10^{-6}$ | Not known | $2.64 \times 10^6$ |
| Gallium | Ga | $2 \times 10^{-6}$ | $Ga(OH)_4^-$ | $2.64 \times 10^6$ |
| Yttrium | Y | $1.3 \times 10^{-6}$ | $Y(OH)_3$ | $1.73 \times 10^6$ |
| Cerium | Ce | $1 \times 10^{-6}$ | $Ce(OH)_3$ | $1.32 \times 10^6$ |
| Dysprosium | Dy | $9 \times 10^{-7}$ | $Dy(OH)_3$ | $1.19 \times 10^6$ |
| Erbium | Er | $8 \times 10^{-7}$ | $Er(OH)_3$ | $1.06 \times 10^6$ |
| Ytterbium | Yb | $8 \times 10^{-7}$ | $Yb(OH)_3$ | $1.06 \times 10^6$ |
| Gadolinium | Gd | $7 \times 10^{-7}$ | $Gd(OH)_3$ | $9.25 \times 10^5$ |
| Praseodymium | Pr | $6 \times 10^{-7}$ | $Pr(OH)_3$ | $7.93 \times 10^5$ |
| Scandium | Sc | $6 \times 10^{-7}$ | $Sc(OH)_3$ | $7.93 \times 10^5$ |
| Lead | Pb | $5 \times 10^{-7}$ | $PbCO_3, Pb(CO_3)_2^{2-}$ | $6.61 \times 10^5$ |
| Holmium | Ho | $2 \times 10^{-7}$ | $Ho(OH)_3$ | $2.64 \times 10^5$ |
| Lutetium | Lu | $2 \times 10^{-7}$ | $Lu(OH)$ | $2.64 \times 10^5$ |
| Thulium | Tm | $2 \times 10^{-7}$ | $Tm(OH)_3$ | $2.64 \times 10^5$ |
| Indium | In | $1 \times 10^{-7}$ | $In(OH)_2^+$ | $1.32 \times 10^5$ |
| Terbium | Tb | $1 \times 10^{-7}$ | $Tb(OH)_3$ | $1.32 \times 10^5$ |
| Tellurium | Te | $1 \times 10^{-7}$ | $Te(OH)_6$ | $1.32 \times 10^5$ |

*continued*

| Element | | Concentration (mg/l) | Some probable dissolved species | Total amount in the oceans (tonnes) |
|---|---|---|---|---|
| Samarium | Sm | $5 \times 10^{-8}$ | $Sm(OH)_3$ | $6.61 \times 10^4$ |
| Europium | Eu | $1 \times 10^{-8}$ | $Eu(OH)_3$ | $1.32 \times 10^4$ |
| Radium | Ra | $7 \times 10^{-11}$ | $Ra^{2+}$ | 92.5 |
| Protactinium | Pa | $5 \times 10^{-11}$ | Not known | 66.1 |
| Radon | Rn | $6 \times 10^{-16}$ | Rn gas | $7.93 \times 10^{-4}$ |
| Polonium | Po | | $PoO_3^{2-}, Po(OH)_2$? | |

**TABLE A.18**   Abundances of chemical elements in seawater *(pages 142–144)*

**DESCRIPTION**

Chemical elements are listed in order of their concentrations in seawater. Some common or probable dissolved ions are also given. Concentrations are by weight, mg/l. The total amount of the element in the oceans is also provided, in metric tonnes.

**LIMITATIONS**

This table does not represent the last word on seawater composition. Even for the more abundant constituents, compilations from different sources differ in detail. For the rarer elements, many of the entries will be subject to revision, as analytical methods improve. Seawater composition also varies around the world, depending on local mineral sources and sinks.

**SOURCE**

The Open University, 1989, *Seawater: Its Composition, Properties and Behaviour*, Pergamon Press, Oxford, UK. By permission.

# B. Rivers, lakes, and waterfalls

**TABLE B.1** Drainage area and average runoff for selected rivers, four estimates *(pages 145–148)*

## DESCRIPTION

Four different estimates (Baumgartner and Reichel, 1975 (16 rivers); Czaya, 1981 (55 rivers); Szestay, 1982 (30 rivers); Meybeck, 1988 (47 rivers)) of the watershed area and average runoff of major rivers of the world are presented in the following four tables. All drainage areas are given in $km^2 \times 10^3$ and all long-term average runoff in $m^3/s$. Rivers are sorted by runoff.

## LIMITATIONS

Four different estimates of river drainage areas and average runoff are presented here because of the lack of agreement among sources. Various values can be computed for drainage area depending upon the definition of where a river commences. In comparing the numbers cited by different authors, the reader should consider this choice, as one author may include all tributaries of a major river, while another may define tributaries as separate river basins.

The differences among the runoff estimates result from the use of different measurement periods, methods, and measurement locations, but complete information on these variables is not presented in the original sources, making comparisons difficult. For a better example of appropriate presentation of data, see Table B.2, by Probst and Tardy (1987). Mean annual flow of a river is a poor measure of water availability, since river flow fluctuates enormously seasonally and interannually. Some of these rivers are heavily controlled by human activities, and the authors do not typically discuss whether these figures account for such activities. For the Meybeck data, runoff for the Nile, Amu Dar'ya, and Huanghe are measured prior to major irrigation schemes. See the original sources for more information on the rivers and assumptions.

## SOURCES

A. Baumgartner and E. Reichel, 1975, *The World Water Balance: Mean Annual Global, Continental and Maritime Precipitation, Evaporation and Runoff*, Elsevier, Amsterdam.

E. Czaya, 1981, *Rivers of the World*, Van Nostrand Reinhold, New York. By permission.

K. Szestay, 1982, River basin development and water management, *Water Quality Bulletin*, **7**, 155–162. (Cited in M. Meybeck, D.V. Chapman and R. Helmer, 1989, *Global Freshwater Quality: A First Assessment*, Global Environmental Monitoring System (GEMS), World Health Organization and United Nations Environment Programme. Basil Blackwell, Oxford.

M. Meybeck, 1988, How to establish and use world budgets of riverine materials, in A. Lerman and M. Meybeck (eds.) *Physical and Chemical Weathering in Geochemical Cycles*. Reprinted by permission, Kluwer Academic Publishers, Reidel Press, Dordrecht.

**TABLE B.1 I** Drainage area and average runoff for selected rivers (Baumgartner and Reichel)

| River | Drainage area ($10^3$ km$^2$) | Average runoff (m$^3$/sec) |
|---|---|---|
| Amazon | 7,180 | 190,000 |
| Zaire-Congo | 3,822 | 42,000 |
| Yangzijiang | 1,970 | 35,000 |
| Orinoco | 1,086 | 29,000 |
| Brahmaputra | 589 | 20,000 |
| Rio de la Plata | 2,650 | 19,500 |
| Yenisei | 2,599 | 17,800 |
| Mississippi | 3,224 | 17,700 |
| Lena | 2,430 | 16,300 |
| Mekong | 795 | 15,900 |
| Ganges | 1,073 | 15,500 |
| Irrawaddy | 431 | 14,000 |
| Ob | 2,950 | 12,500 |
| Xijiang | 435 | 11,500 |
| Amur | 1,843 | 11,000 |
| St. Lawrence | 1,030 | 10,400 |

**TABLE B.1 II**   Drainage area and average runoff for selected rivers (Czaya)

| River | Drainage area ($10^3$ km$^2$) | Average runoff (m$^3$/sec) | River | Drainage area ($10^3$ km$^2$) | Average runoff (m$^3$/sec) |
|---|---|---|---|---|---|
| Amazon (with Toçantins) | 7,180 | 180,000 | Nelson | 1,072 | 2,300 |
| Zaire-Congo | 3,822 | 42,000 | Nile | 2,881 | 1,584 |
| Yangzijiang | 1,970 | 35,000 | Salween | 280 | 1,500 |
| Orinoco | 1,086 | 28,000 | Huanghe | 745 | 1,365 |
| Brahmaputra | 938 | 20,000 | Amu Darya | 227 | 1,300 |
| Yenisei | 2,605 | 19,600 | Olenek | 220 | 1,210 |
| Rio de la Plata | 2,650 | 19,500 | Dnieper | 503 | 1,160 |
| Mississippi-Missouri | 3,221 | 17,545 | Shatt-al-Arab | 808 | 856 |
| Lena | 2,490 | 16,400 | Syr Darya | 462 | 430 |
| Mekong | 795 | 15,900 | Murray-Darling | 1,072 | 391 |
| Ganges | 1,073 | 15,000 | Ural | 220 | 347[b] |
| Irrawaddy | 431 | 14,000 | Colorado[a] | 629 | 168 |
| Ob-Irtysh | 2,975 | 12,600 | Rio Grande | 570 | 82 |
| Amur | 1,855 | 12,500 | Tarim-Khotan | 1,000 | * |
| Xijiang | 435 | 11,000 | Orange | 850 | * |
| St. Lawrence | 1,030 | 10,400 | Okavango | 800 | * |
| Volga | 1,380 | 8,000 | Chari | 700 | * |
| Rio Magdelena | 240 | 8,000 | Helmand | 500 | * |
| Mackenzie-Peace | 1,805 | 7,500 | | | |
| Yukon | 855 | 7,000 | | | |
| Columbia | 669 | 6,650 | | | |
| Danube | 805 | 6,450 | | | |
| Niger | 2,092 | 5,700 | | | |
| Pechora | 327 | 4,060 | | | |
| Godavari | 313 | 3,980 | | | |
| Song-Koi | * | 3,900 | | | |
| Indus | 960 | 3,850 | | | |
| Kolyma | 644 | 3,800 | | | |
| Fraser | 225 | 3,750 | | | |
| Northern Dvina | 360 | 3,560 | | | |
| São Francisco | 610 | 3,300 | | | |
| Khatanga | 422 | 3,280 | | | |
| Grijalva-Usumacinta | * | 3,265 | | | |
| Mahanadi | 133 | 2,940 | | | |
| Pyassina | 182 | 2,600 | | | |
| Neva | 281 | 2,530 | | | |
| Zambezi | 1,330 | 2,500 | | | |

[a]At the United States-Mexico border
[b]Near Kushum.
*Czaya omitted these figures from the original table

**TABLE B.1 III** Drainage area and average runoff for selected rivers (Szestay)

| River | Drainage area ($10^3$ km$^2$) | Average runoff (m$^3$/sec) |
|---|---|---|
| Amazon | 5,578 | 212,000 |
| Zaire-Congo | 4,015 | 40,000 |
| Yangzijiang | 1,943 | 22,000 |
| Brahmaputra | 935 | 20,000 |
| Ganges | 1,060 | 19,000 |
| Mississippi | 3,222 | 17,300 |
| Orinoco | 881 | 17,000 |
| Paraná | 2,305 | 14,900 |
| Irrawaddy | 430 | 13,600 |
| Ob-Irtysh | 2,430 | 12,000 |
| Mekong | 803 | 11,000 |
| Toçantins | 907 | 10,000 |
| Yukon | 932 | 9,100 |
| Magdalena | 241 | 7,500 |
| Zambezi | 1,295 | 7,000 |
| Danube | 817 | 6,200 |
| Niger | 1,114 | 6,100 |
| Indus | 927 | 5,600 |
| Huanghe | 673 | 3,300 |
| Nile | 2,980 | 2,800 |
| São Francisco | 673 | 2,800 |
| Rhine | 145 | 2,200 |
| Rhone | 96 | 1,700 |
| Tigris-Euphrates | 541 | 1,500 |
| Po | 70 | 1,400 |
| Vistula | 197 | 1,100 |
| Senegal | 338 | 700 |
| Colorado | 629 | 580 |
| Orange | 640 | 350 |
| Rio Grande | 352 | 120 |

**TABLE B.1 IV** Drainage area and average runoff for selected rivers (Meybeck)

| River | Drainage area ($10^3$ km$^2$) | Average runoff (m$^3$/sec) |
|---|---|---|
| Amazon | 6,300 | 175,000 |
| Negro[a] | 755 | 45,300 |
| Zaire-Congo | 4,000 | 39,200 |
| Madeira[a] | 1,380 | 32,000 |
| Orinoco | 950 | 30,000 |
| Yangzijiang | 1,950 | 28,000 |
| Brahmaputra | 580 | 19,300 |
| Mississippi | 3,267 | 18,400 |
| Mekong | 795 | 18,300 |
| Paraná | 2,800 | 18,000 |
| Yenisei | 2,600 | 17,200 |
| Lena | 2,430 | 16,300 |
| Irrawaddy | 430 | 13,400 |
| Ob | 2,500 | 12,350 |
| Ganges | 975 | 11,600 |
| Amur | 1,850 | 11,000 |
| Toçantins | 900 | 11,000 |
| Maranon[a] | 407 | 11,000 |
| St. Lawrence | 1,025 | 10,700 |
| Ucayali[a] | 400 | 9,700 |
| Mackenzie | 1,800 | 9,600 |
| Volga | 1,350 | 8,400 |
| Columbia | 670 | 7,960 |
| Xijiang | 1,350 | 7,800 |
| Xingu[a] | 540 | 7,800 |
| Magdalena | 240 | 7,500 |
| Tapajos[a] | 500 | 7,200 |
| Zambezi | 1,340 | 7,100 |
| Indus | 950 | 6,700 |
| Danube | 805 | 6,430 |
| Yukon | 770 | 6,200 |
| Niger | 1,125 | 6,100 |
| Uruguay | 350 | 5,000 |
| Nelson | 1,150 | 3,500 |
| Kolyma | 645 | 3,150 |
| São Francisco | 470 | 2,900 |
| Nile | 3,000 | 2,830 |

*continued*

| River | Drainage area ($10^3$ km$^2$) | Average runoff (m$^3$/sec) |
|---|---|---|
| Shatt-al-Arab | 410 | 1,750 |
| Dnieper | 500 | 1,650 |
| Huanghe | 745 | 1,480 |
| Amu Darya | 450 | 1,450 |
| Chari | 600 | 1,320 |
| Don | 420 | 870 |
| Murray | 1,070 | 740 |
| Colorado | 635 | 640 |
| Orange | 800 | 300 |
| Rio Grande | 670 | 100 |

[a]Tributaries of the Amazon.

**TABLE B.2**   Characteristics of fifty selected rivers *(pages 149–150)*

**DESCRIPTION**

Fifty of the world's major rivers are described in this table, compiled by Probst and Tardy in 1987. As with the previous tables on world rivers, this one includes data on the drainage area of the basin in km$^2$ × 10$^3$ and the long-term average runoff ("mean interannual modules") in m$^3$/s. Unlike the earlier tables, however, this one includes valuable detailed information on the location of the gauging stations where the runoff data were collected and the period of the data. This additional information permits the reader to better judge data quality and comparability with other data sets. River basin latitude is also included.

**LIMITATIONS**

Some major rivers and areas of the world are not represented in this table, including the Orinoco, the Mackenzie, the Yukon, the Huanghe, the Yangzijiang, and the Ganges. These omissions are due to a decision by the authors to include only rivers with high-quality data sets exceeding fifty years in length. Apparently, these major rivers failed to meet the necessary criteria.

**SOURCE**

J.L. Probst and Y. Tardy, 1987, Long range streamflow and world continental runoff fluctuations since the beginning of this century, *Journal of Hydrology*, **94**, 289–311. By permission.

**TABLE B.2**  Characteristics of fifty selected rivers

| River | Gauging station | Country | Area ($10^3$ km$^2$) | Period of data | Mean interannual module (m$^3$/sec) | Latitudes of basin |
|---|---|---|---|---|---|---|
| Garonne | Mas d'Agenais | France | 52.0 | 1832–1978[a] | 632 | 40–50° N |
| Loire | Montjean | France | 110.0 | 1863–1974 | 814 | 40–50° N |
| Seine | Paris | France | 44.3 | 1928–1982 | 273 | 40–50° N |
| Rhone | Lyon | France | 50.2 | 1900–1982 | 1,048 | 40–50° N |
| Rhine | Rees | W. Europe | 159.7 | 1936–1975 | 2,250 | 40–50° N |
| Ebre | Tortosa | Spain | 84.2 | 1913–1974[a] | 1,374 | 40–50° N |
| Guadalquivir | Alcala del Rio | Spain | 49.9 | 1913–1972[a] | 508 | 30–40° N |
| Po | Pontelagoscuro | Italy | 70.1 | 1918–1975 | 1,488 | 40–50° N |
| Gota | Vanesborg | Sweden | 46.8 | 1807–1983 | 532 | 56–60° N |
| Vuoksi | Imatra | Finland | 61.3 | 1847–1975 | 590 | 60–64° N |
| Elbe | Decin | Czechoslovakia | 51.1 | 1851–1975 | 300 | 50–55° N |
| Vistule | Tczew | Poland | 193.8 | 1901–1978[a] | 1,041 | 50–55° N |
| Oder | Gozdowice | Poland | 109.3 | 1901–1978[a] | 532 | 50–55° N |
| Danube | Orsova | Hungary | 578.3 | 1840–1975 | 5,447 | 40–50° N |
| Dvina | Ust Pinega | W. USSR | 348.0 | 1882–1969 | 3,368 | 60–65° N |
| Don | Razdorskaya | W. USSR | 378.0 | 1891–1975[a] | 793 | 50–55° N |
| Volga | Volograd | W. USSR | 1,350.0 | 1879–1975[a] | 8,137 | 45–55° N |
| Neman | Smalininkai | W. USSR | 81.2 | 1812–1975[a] | 540 | 52–56° N |
| Ural | Kushum | W. USSR | 190.0 | 1915–1975 | 308 | 45–55° N |
| Kolyma | Sredne-Kolymesk | E. USSR | 361.0 | 1927–1975[a] | 2,200 | 65–75° N |
| Amur | Komsomolsk | E. USSR | 1,730.0 | 1933–1975 | 9,987 | 45–55° N |
| Ob | Salekhard | E. USSR | 2,950.0 | 1930–1975 | 12,454 | 50–70° N |
| Yenisei | Igarka | E. USSR | 2,440.0 | 1936–1975 | 17,805 | 50–75° N |
| Lena | Kusur | E. USSR | 2,430.0 | 1935–1975 | 16,480 | 50–75° N |
| Mekong | Mukdahan | Thailand | 391.0 | 1925–1968 | 8,318 | 10–30° N |
| Godavari | Dowlaishawaran | India | 299.3 | 1902–1979[a] | 3,065 | 18–20° N |
| Nile | Aswan | Egypt | 1,500.0 | 1800–1976 | 2,830 | 0–30° N |
| Senegal | Bakel | Senegal | 218.0 | 1903–1983 | 733 | 10–15° N |
| Niger | Koulikoro | Niger | 120.0 | 1908–1983 | 1,487 | 5–15° N |
| Chari | N'Djamena | Chad | 600.0 | 1932–1979 | 1,185 | 6–13° N |
| Congo | Kinshasa | Zaire | 3,500.0 | 1902–1983 | 41,134 | 6° N–10° S |
| Zambezi | Matundo-Cais | Mozambique | 540.0 | 1925–1983 | 2,429 | 10–12° S |
| Limpopo | Chokule | South Africa | 342.0 | 1920–1975 | 54 | 20–26° S |
| Orange | Dimo 3 | South Africa | 37.1 | 1914–1977 | 163 | 26–32° S |
| St. Lawrence | Ogdensburg | Canada | 764.0 | 1861–1983 | 6,892 | 43–47° N |
| Red River | Emerson | Canada | 104.0 | 1913–1975 | 92 | 47–50° N |
| Assiniboine | Headingly | Canada | 162.0 | 1914–1975 | 51 | 47–50° N |
| N. Saskatchewan | Prince Albert | Canada | 119.5 | 1912–1975 | 245 | 50–53° N |

*continued*

| River | Gauging station | Country | Area ($10^3$ km²) | Period of data | Mean interannual module (m³/sec) | Latitudes of basin |
|-------|-----------------|---------|-------------------|----------------|-----------------------------------|---------------------|
| S. Saskatchewan | Saskatoon | Canada | 139.5 | 1912–1975 | 274 | 50–53° N |
| Susquehanna | Harrisburg | Canada | 62.4 | 1912–1975 | 968 | 40–42° N |
| Niagara | Queenston | Canada | 665.0 | 1860–1975 | 5,776 | 42–44° N |
| Colorado | Lees Ferry | United States | 279.5 | 1911–1983 | 457 | 35–45° N |
| Mississippi | Alton III | United States | 444.2 | 1928–1983 | 2,856 | 38–48° N |
| Missouri | Herman | United States | 1,368.0 | 1898–1983 | 2,274 | 38–48° N |
| Ohio | Metropolis III | United States | 525.7 | 1928–1983 | 7,409 | 35–42° N |
| Columbia | The Dalles | United States | 614.0 | 1879–1982 | 5,452 | 45–55° N |
| Snake | Clarkston | United States | 267.3 | 1916–1972 | 1,415 | 43–46° N |
| Paraná | Guaira | Brazil | 806.0 | 1921–1975 | 8,506 | 5–25° S |
| São Francisco | Juazeiro | Brazil | 490.8 | 1921–1975 | 2,730 | 10–20° S |
| Amazon | Obidos | Brazil | 4,688.0 | 1928–1975[a] | 149,950 | 5° N–15° S |

[a]Some lack periods in data series.

**TABLE B.3**  Lengths and basin countries of major rivers, four estimates *(pages 151–153)*

**DESCRIPTION**

Although the length of a river is a fundamental geographical measure, there is rarely agreement on its actual value. Presented here are four different estimates of the lengths of 63 major rivers of the world, in km. Also listed are the basin countries of each river.

**LIMITATIONS**

River length is a fractal measure, varying with the geographical scale used, the accuracy of the measurement technique, and the assumptions about what tributaries to include in each river. Not every estimate here was made independently, and original sources are cited irregularly. Length estimates differ radically for some rivers such as the Ogooué, Oranje, Senegal, and Zambesi in Africa, the Amur, Huanghe, Ob, and Yenisei in Asia, the Dvina in Europe, the St. Lawrence in North America, Coopers Creek and the Fly in Oceania, and the Negro and Orinoco in South America. In these cases, the original sources are unclear about which tributaries are included. Where possible, specific tributaries included in the length estimates are identified in the notes following the table. Check with the original sources for details on their assumptions.

The basin countries and their political boundaries change constantly due to political volatility. Most basin country names listed here are consistent with those from the United Nations listing of international rivers (United Nations, 1978). At the time of writing, for example, the Soviet Union is fragmenting into smaller political units and new countries are forming. This will greatly increase the number of international rivers and basin countries for several rivers on this list.

**SOURCES**

*Encyclopaedia Britannica*, 1988, 15th edn. Encyclopaedia Britannica, Inc., New York. (Cited in F. van der Leeden, F.L. Troise and D.K. Todd (eds.), 1990, *The Water Encyclopedia*, Lewis Publishers, Michigan.) By permission.

V. Showers, 1989, *World Facts and Figures*, 3rd edn. John Wiley and Sons, New York. By permission.

USSR Committee for the International Hydrological Decade, 1978, *World Water Balance and Water Resources of the Earth*, Studies and Reports in Hydrology, no. 25. UNESCO, Paris and Gidrometeoizdat, Leningrad (English translation of a 1974 USSR publication).

E. Czaya, 1981, *Rivers of the World*, Van Nostrand Reinhold, New York. By permission.

United Nations, 1978, *Register of International Rivers*, Pergamon Press, Oxford, UK. By permission.

**TABLE B.3**  Lengths and basin countries of selected major rivers, four estimates

| | Basin countries | Length in kilometers, four estimates | | | |
|---|---|---|---|---|---|
| | | UNESCO (1974) | Czaya (1981) | TWE (1990) | WFF (1989) |
| *Africa* | | | | | |
| Chari | Chad, Cameroon | 1,400 | | 1,400 | 1,450 |
| Limpopo | Mozambique, Zimbabwe, South Africa, Botswana | 1,600 | | 1,800 | 1,590 |
| Niger | Mali, Nigeria, Niger, Algeria, Guinea, Chad, Cameroon, Burkina Faso, Benin, Côte d'Ivoire | 4,160 | 4,030 | 4,200 | 4,100 |
| Nile | Sudan, Ethiopia, Egypt, Uganda, Burundi, Tanzania, Kenya, Zaire, Rwanda | 6,670 | 6,484 | 6,650 | 6,670 |
| Ogooue | Gabon, Congo, Cameroon, Equatorial Guinea | 850 | | | 1,210 |
| Okavango | Botswana, Namibia, Angola, Zimbabwe | 1,800 | | 1,600 | 1,610 |
| Orange | South Africa, Namibia, Botswana, Lesotho | 1,860 | 1,860 | 2,100 | 2,250 |
| Senegal | Guinea, Mali, Mauritania, Senegal | 1,430 | | 1,641 | 1,700 |
| Shebelli | Ethiopia, Somalia, Kenya | 1,600[a] | | | 1,930 |
| Volta | Burkina Faso, Ghana, Togo, Côte d'Ivoire, Benin, Mali | 1,600 | | | 1,600 |
| Zaire-Congo | Zaire, Central African Republic, Angola, Zambia, Tanzania, Cameroon, Congo, Burundi, Rwanda | 4,370 | 4,700 | 4,700 | 4,630 |
| Zambezi | Zambia, Angola, Zimbabwe, Tanzania, Mozambique, Malawi, Botswana, Namibia | 2,660 | 2,660 | 3,500 | 2,650 |
| *Asia* | | | | | |
| Amu Darya[b] | USSR | 2,820 | 2,620 | 2,540 | 2,540 |
| Amur | USSR, China, Mongolia | | 4,510 | 2,824 | 5,780 |
| Brahmaputra | China, India, Bangladesh, Bhutan, Nepal[c] | 3,000[c] | 2,900 | 2,900 | 2,840 |
| Ganges | [c] | | 2,700 | 2,510 | 2,510 |
| Huanghe | China | 4,670 | 4,845 | 5,464 | 4,840 |
| Indus | Pakistan, India, Afghanistan, China | 3,180 | 3,180 | 2,900 | 2,880 |
| Irrawaddy | Burma, India, China | 2,300 | 2,150 | 1,992 | 1,990 |
| Lena | USSR | 4,400 | 4,270 | 4,400 | 4,400 |
| Mekong | China, Laos, Burma, Thailand, Cambodia, Vietnam | 4,500 | 4,500 | 4,350 | 4,180 |
| Ob | USSR, China | 3,650 | 5,570 | 3,650 | 3,180 |
| Syr Darya[e] | USSR | | 3,078 | 3,019 | 3,020 |
| Tigris-Euphrates | Iraq, Iran, Turkey, Syria | 2,760 | 2,900 | 2,900 | 2,430[f] |

*continued*

| | Basin countries | Length in kilometers, four estimates | | | |
|---|---|---|---|---|---|
| | | UNESCO (1974) | Czaya (1981) | TWE (1990) | WFF (1989) |
| Xijiang | China, Vietnam | | | 1,957 | 1,960[d] |
| Yangzijiang | China | 5,520 | 5,800 | 6,300 | 5,980 |
| Yenisei | USSR, Mongolia | 3,490 | 5,550 | 4,102 | 5,870 |
| *Europe* | | | | | |
| Danube | Romania, Yugoslavia, Hungary, Albania, Italy, Austria, Czechoslovakia, Germany, USSR, Poland, Bulgaria, Switzerland | 2,860 | 2,850 | 2,850 | 2,860 |
| Dnieper | USSR | 2,200 | 2,285 | 2,200 | 2,200 |
| Don | USSR | 1,870 | | 1,870 | 1,870 |
| Duero | Portugal, Spain | 925 | 776 | | 890 |
| Ebro | Spain, France | 930 | 927 | | 910 |
| Elbe | Germany, Czechoslovakia, Austria, Poland | 1,110 | 1,165 | | 1,160 |
| Loire | France | 1,110 | 1,020 | | 1,020 |
| Mesen | USSR | 966 | 857 | | |
| Neman | USSR, Poland | 937 | 990 | | 940 |
| Neva | USSR | 74 | 74 | | 1,000[g] |
| Northern Dvina | USSR | 744 | 1,302 | 1,302 | 1,860 |
| Oder | Poland, Germany | 907 | 912 | | 950 |
| Pechora | USSR | 1,810 | 1,809 | 1,809 | 1,810 |
| Po | Italy, Switzerland | 650 | 676 | | 620 |
| Rhine | Germany, Switzerland, France, Liechtenstein, Netherlands, Austria, Luxembourg, Belgium | 1,360 | | 1,392 | 1,320 |
| Rhone | France, Switzerland | 810 | 812 | | 810 |
| Seine | France | 780 | 776 | | 780 |
| Volga | USSR | 3,350 | 3,688 | 3,530 | 3,530 |
| Wisla | Poland, USSR, Czechoslovakia | 1,090 | 1,095 | | 1,200 |
| *North and Central America* | | | | | |
| Colorado | United States, Mexico | 2,180 | 3,200 | 2,333 | 2,330 |
| Columbia | Canada, United States | 1,950 | 2,250 | 2,000 | 2,240[h] |

*continued*

| Basin countries | Length in kilometers, four estimates | | | |
|---|---|---|---|---|
| | UNESCO (1974) | Czaya (1981) | TWE (1990) | WFF (1989) |
| Mackenzie — Canada | 4,240 | 4,250 | 4,241 | 4,240 |
| Mississippi-Missouri — United States, Canada[i] | 5,985[i] | 6,019[i] | 3,779 | 5,970[i] |
| Nelson — Canada, United States | 2,600 | 2,575 | 2,575 | 2,570 |
| Rio Grande — United States, Mexico | 2,880 | 2,870 | 3,034 | 3,030 |
| St. Lawrence — Canada, United States | 3,060 | | 4,000 | 3,320 |
| Yukon — Canada, United States | 3,000 | 3,185 | 3,018 | 3,180 |
| *Oceania* | | | | |
| Cooper Creek — Australia | 2,000 | 2,000 | 882 | |
| Fly — Papua New Guinea | 620 | | | 1,290 |
| Murray-Darling — Australia | 3,490 | 2,570 | 2,739[j] | 3,750 |
| *South America* | | | | |
| Amazon — Brazil, Peru, Bolivia, Columbia, Ecuador, Venezuela, Guyana | 6,280 | 6,516 | 6,400 | 6,570 |
| La Plata[k] — Brazil, Argentina, Paraguay, Bolivia, Uruguay | 4,700 | 4,700 | | 4,880 |
| Madeira — Brazil, Bolivia | | | 3,350 | 3,200 |
| Maranon — Peru | | | 1,905 | 1,410 |
| Negro — Columbia, Venezuela, Brazil | 1,000 | | 2,253 | 1,210 |
| Orinoco — Venezuela, Columbia | 2,740 | 2,500 | 2,736 | 2,140 |

[a] Length of Juba with Shebelli.
[b] Including Pyandzh to outflow in Aral Sea.
[c] The Ganges–Brahamputra systems combined drain land in India, Bhutan, Nepal, Bangladesh, and China.
[d] With Zhu.
[e] Including Arabelsu to outflow in Aral Sea.
[f] Euphrates alone. The Tigris is shown as having a length of 1,850 km to its outflow at the Shatt-al-Arab.
[g] With Volkov and Lovat.
[h] With Snake.
[i] Including Missouri. None of Missouri basin is in Canada.
[j] Darling River to inflow in Murray.
[k] With Paraná and Grande.

**TABLE B.4**    River transport of suspended sediment load to the oceans, by climatic region

| Morphoclimatic region | Sample river basins used | Average transport of suspended sediments (tonnes/km$^2$/yr) | All basins Total area (10$^6$ km$^2$) | All basins Total load (10$^6$ tonnes/yr) |
|---|---|---|---|---|
| Dry tundra and taiga | Mackenzie, Yukon; Kara, Laptev, Bering Seas watersheds | 16 | 20 | 320 |
| Humid taiga | Barentz watershed; Finland | 25 | 3.15 | 79 |
| Wet taiga | Lerman and Meybeck estimate | 350 | 0.2 | 70 |
| Semi-arid | Black, Aral, Caspian watersheds[a] | 360 | 2.6 | 940 |
| Temperate | Huanghe | 1,500 | 0.75 | 1,080 |
| Dry temperate | Baltic watershed, Shatt-al-Arab, Indus, Mississippi, Danube | 190 | 6.85 | 1,300 |
| Humid temperate | Alps watersheds; Columbia, Yangzijiang | 365 | 7.7 | 2,820 |
| Wet temperate (highlands) | Ganges, Brahmaputra | 850 | 4.5 | 3,820 |
| Wet tropics (lowlands) | Orinoco, middle and lower Amazon watersheds | 56 | 7.05 | 395 |
| Wet tropics (highlands) | Mekong, Magdalena, Ucayali, Maranon | 440 | 7.8 | 3,430 |
| Wet tropics (mixed) | Zaire, Chao Phrya, Burma rivers | 115 | 8.95 | 1,030 |
| Savanna | Niger, Chari[a], Paraná, Blue Nile, Zambezi | 90 | 13.1 | 1,180 |
| Arid | Murray, Orange, Nile, Rio Grande | 75 | 15.5 | 1,160 |
| Desert | | 0 | 1.7 | 0 |

| Percent of total | Total area | Total load |
|---|---|---|
| Cold regions | 23.4 | 2.7 |
| Temperate | 22.4 | 56.5 |
| Tropic | 37.0 | 34.2 |
| Arid | 17.2 | 6.6 |

[a]These watersheds, draining to closed interior regions, were used to determine the average transport of suspended sediments for comparable basins draining to the oceans.

**TABLE B.4**    River transport of suspended sediment load to the oceans, by climatic region

**DESCRIPTION**

Average suspended sediment loads have been determined for 13 morphoclimatic zones, defined as regions with a set of consistent climatic conditions. Sample river basins have been chosen from each region and their average transport of suspended sediment computed in tonnes per km$^2$ per year. These sample basins were then used to compute total sediment discharge, in million tonnes per year, for each climatic zone. Total zonal areas are also given, in km$^2$ × 10$^6$.

**LIMITATIONS**

Total sediment load for each region is assumed to have the same average load as the documented rivers used. This is unlikely to be precise, but the lack of accurate data for every river basin necessitates this simplification. If better monitoring and measuring are implemented, more accurate estimates can be made. The average suspended sediment data here hide the variability in sediment loads in each river. For the Indus, Niger, Nile, and Rio Grande rivers, data used predate dam construction.

**SOURCE**

A. Lerman and M. Meybeck, 1988, *Physical and Chemical Weathering in Geochemical Cycles*. Reprinted by permission, Kluwer Academic Publishers, Reidel Press, Dordrecht.

**TABLE B.5**  Calculated total suspended sediment discharge from the continents

|  | Sediment yield (tonnes/km²/yr) | Drainage area (10⁶ km²) | Sediment discharge (10⁶ tonnes/yr) |
|---|---|---|---|
| *North America* | | | |
| St. Lawrence | 4 | 1.03 | 4 |
| U.S. Atlantic Coast | 17 | 0.74 | 13 |
| Gulf Coast | 59 | 4.5 | 256 |
| Colorado | 0.2 | 0.63 | 0.1 |
| Columbia | 12 | 0.69 | 8 |
| Rest of Western United States | 193 | 0.32 | 62 |
| Canada West Coast | 91 | 0.67 | 61 |
| S. Alaska (glacial) | 1,000 | 0.34 | 340 |
| S. Alaska (nonglacial) | 76 | 1.37 | 104 |
| N. Alaska | 120 | 0.35 | 42 |
| Mackenzie | 55 | 1.81 | 100 |
| N. NE Canada | 8 | 3.73 | 30 |
| Subtotals | | 15.42[a] | 1,020 |
| *Central America* | | | |
| Mexico | 140 | 1.5 | 210 |
| Remainder | 400 | 0.58 | 232 |
| Subtotals | | 2.08 | 442 |
| *South America* | | | |
| Northwest | 500 | 0.3 | 150 |
| Magdalena | 900 | 0.24 | 220 |
| Northern | 150 | 7.79 | 1,218 |
| Eastern | 9.4 | 3.00 | 28 |
| Southern | 32 | 4.38 | 154 |
| Western and South | 10 | 1.77 | 18 |
| Subtotals | | 17.9[a] | 1,788 |
| *Europe* | | | |
| Western | 12 | 2.60 | 31 |
| Alpine | 120 | 0.55 | 66 |
| Black Sea | 72 | 1.86 | 133 |
| Subtotals | | 4.61[a] | 230 |
| *Eurasian Arctic* | | | |
| West of 140′ E | 6 | 9.9 | 59 |
| East of 140′ E | 20 | 1.27 | 25 |
| Subtotals | | 11.17 | 84 |
| *Asia* | | | |
| Northeast | 28 | 3.2 | 100 |
| NE China-Korea | 658 | 1.00 | 658 |

*continued*

| | Sediment yield (tonnes/km²/yr) | Drainage area (10⁶ km²) | Sediment discharge (10⁶ tonnes/yr) |
|---|---|---|---|
| Yellow (Huanghe) | 1,400 | 0.77 | 1,080 |
| Rest of China | 250 | 3.72 | 930 |
| SE Asia and Himalayas (exclusive of Indus) | 796 | 3.93 | 3,128 |
| India | 154 | 1.86 | 286 |
| Indus | | | 100 |
| Asia Minor | 50 | 1.35 | 67 |
| Subtotals | | 16.88[a] | 6,349 |
| *Africa* | | | |
| Northwest | 100 | 1.10 | 110 |
| West | 16.5 | 6.86 | 113 |
| Southwest | 17 | 1.02 | 17 |
| East | 80 | 3.00 | 240 |
| Zambesi | 17 | 1.20 | 20 |
| (Tana)[b] | (1,000) | | (30) |
| Nile | 0 | 2.16 | 0 |
| Subtotals | | 15.34 | 500 |
| *Australia* | | | |
| East, North | 28 | 2.20 | 62 |
| *Oceanic Islands* | 1,000 | 3.00 | 3,000 |
| Totals | 116 | 86.40[a] | 13,505[a] |

[a]Total given in original table.
[b]Local anomaly due to overgrazing of pasture land.

**TABLE B.5**   Calculated total suspended sediment discharge from the continents *(pages 155–156)*

**DESCRIPTION**

Rivers discharge considerable sediment into the oceans annually. This table divides the world into major drainage basin regions and presents sediment yields in tonnes per km² per year, regional drainage areas in km² × 10⁶, and total sediment discharges in million tonnes per year.

**LIMITATIONS**

Reliable sediment data are hard to obtain for many parts of the world. The data here are considered to be better than earlier information due to improvements in measurement techniques, more regional information, and changes in land management practices not included in earlier studies. The regional sums presented here hide more detailed information on sediment yields from particular rivers. Some of these data are available in the next table and in the original source, which carefully describes assumptions and primary references. Only limited data are available for many regions of the world, including Chile and southern Peru, South America, south-east Asia, rivers draining the Himalayas, and parts of Africa. Some of the data in each column in the original table do not sum to the subtotals presented. These are identified in the table notes.

**SOURCE**

J.D. Milliman and R.H. Meade, 1983, World-wide delivery of river sediment to the oceans, *Journal of Geology*, **91**(1), 1–21. By permission.

**TABLE B.6** Drainage area, runoff, and suspended sediment discharges for major rivers of the world

| | Drainage area ($10^3$ km$^2$) | Runoff (km$^3$/yr) | Sediment discharge ($10^3$ tonnes/yr) |
|---|---|---|---|
| *North America* | | | |
| St. Lawrence (Canada) | 1,030 | 447 | 4,000 |
| Hudson (United States) | 20 | 12 | 1,000 |
| Mississippi (United States) (including Atchafalaya) | 3,270 | 580 | 210,000 |
| Brazos (United States) | 110 | 7 | 16,000 |
| Colorado (Mexico) | 640 | 20 | 100 |
| Eel (United States) | 8 | | 14,000 |
| Columbia (United States) | 670 | 251 | 8,000 |
| Fraser (Canada) | 220 | 112 | 20,000 |
| Yukon (United States) | 840 | 195 | 60,000 |
| Copper (United States) | 60 | 39 | 70,000 |
| Susitna (United States) | 50 | 40 | 25,000 |
| Mackenzie (Canada) | 1,810 | 306 | 100,000 |
| *South America* | | | |
| Chira (Peru) | 20 | 5 | 4,000-75,000 |
| Magdalena (Colombia) | 240 | 237 | 220,000 |
| Orinoco (Venezuela) | 990 | 1,110 | 210,000 |
| Amazon (Brazil) | 6,150 | 6,300 | 900,000 |
| São Francisco (Brazil) | 640 | 97 | 6,000 |
| La Plata (Argentina) | 2,830 | 470 | 92,000 |
| Negro (Argentina) | 100 | 30 | 13,000 |
| *Europe* | | | |
| Rhone (France) | 90 | 49 | 10,000 |
| Po (Italy) | 70 | 46 | 15,000 |
| Danube (Romania) | 810 | 206 | 67,000 |
| Drini (Albania) | 10 | | |
| *Eurasian Arctic* | | | |
| Yana (USSR) | 220 | 29 | 3,000 |
| Ob (USSR) | 2,500 | 385 | 16,000 |
| Yenisei (USSR) | 2,580 | 560 | 13,000 |
| Severnay Dvina (USSR) | 350 | 106 | 4,500 |
| Lena (USSR) | 2,500 | 514 | 12,000 |
| Kolyma (USSR) | 640 | 71 | 6,000 |
| Indigirka (USSR) | 360 | 55 | 14,000 |
| *Asia* | | | |
| Amur (USSR) | 1,850 | 325 | 52,000 |

*continued*

| | Drainage area ($10^3$ km$^2$) | Runoff (km$^3$/yr) | Sediment discharge ($10^3$ tonnes/yr) |
|---|---|---|---|
| Liaohe (China) | 170 | 6 | 41,000 |
| Daling (China) | 20 | 1 | 36,000 |
| Haihe (China) | 50 | 2 | 81,000 |
| Huanghe (China) | 770 | 49 | 1,080,000 |
| Yangzijiang (China) | 1,940 | 900 | 478,000 |
| Huaihe (China) | 260 | | 14,000 |
| Zhujiang (China) | 440 | 302 | 69,000 |
| Hungho (Vietnam) | 120 | 123 | 130,000 |
| Mekong (Vietnam) | 790 | 470 | 160,000 |
| Irrawaddy (Burma) | 430 | 428 | 265,000 |
| Ganges-Brahmaputra (Bangladesh) | 1,480 | 971 | 1,670,000 |
| Mehandi (India) | 130 | 67 | 2,000 |
| Damodar (India) | 20 | 10 | |
| Godavari (India) | 310 | 84 | 96,000 |
| Indus (Pakistan) | 970 | 238 | 100,000 |
| Tigris-Euphrates (Iraq) | 1,050 | 46 | |

### Africa

| | | | |
|---|---|---|---|
| Nile (Egypt) | 2,960 | 30 | 0[a] |
| Niger (Nigeria) | 1,210 | 192 | 40,000 |
| Zaire (Zaire) | 3,820 | 1,250 | 43,000 |
| Orange (S. Africa) | 1,020 | 11 | 17,000 |
| Zambesi (Mozambique) | 1,200 | 223 | 20,000 |
| Limpopo (Mozambique) | 410 | 5 | 33,000 |
| Rufiji (Tanzania) | 180 | 9 | 17,000 |
| Tana (Kenya) | 32 | | 32,000 |

### Oceania

| | | | |
|---|---|---|---|
| Murray (Australia) | 1,060 | 22 | 30,000 |
| Waiapu (New Zealand) | | | 28,000 |
| Haast (New Zealand) | 1 | 6 | 13,000 |
| Fly (New Guinea) | 61 | 77 | 30,000 |
| Purari (New Guinea) | 31 | 77 | 80,000 |
| Choshui (Taiwan) | 3 | 6 | 66,000 |
| Kaoping (Taiwan) | 3 | 9 | 39,000 |
| Tsengwen (Taiwan) | 1 | 2 | 28,000 |
| Hualien (Taiwan) | 2 | 4 | 19,000 |
| Peinan (Taiwan) | 2 | 4 | 17,000 |
| Hsiukuluan (Taiwan) | 2 | 4 | 16,000 |

[a]New estimate due to effect of Aswan High Dam.

**TABLE B.6** Drainage area, runoff, and suspended sediment discharges for major rivers of the world *(pages 157–158)*

**DESCRIPTION**

Major rivers are listed by continent and country where the river discharges. Data include the drainage area in km$^2$ × 10$^3$, water discharge in km$^3$ per year, and sediment discharge in thousand tonnes per year.

**LIMITATIONS**

Incomplete information is available for some rivers and not all rivers are listed. Water discharges and sediment discharges shown do not consistently reflect the impact of human activities. For example, the Colorado River flow data do not include withdrawals, which greatly reduce actual flows to Mexico, but the sediment discharge values do reflect the impacts of dams, which retain sediment. The sediment dis-

charge of the Nile into the Mediterranean as estimated in this source is zero, because of the complete trapping of sediments behind the Aswan High Dam. Runoff numbers are long-term avarages, but no information on the duration of the record is given in the original.

**SOURCE**

J.D. Milliman and R.H. Meade, 1983, World-wide delivery of river sediment to the oceans, *Journal of Geology*, **91**(1), 1–21. By permission.

**TABLE B.7** Drainage area, runoff, and sediment yields for fourteen major rivers

| River | Country of river mouth | Catchment area (10$^6$ km$^2$) | Runoff (cm/yr) | Sediment (tonnes/km$^2$/yr) | Yield (ppm) |
|---|---|---|---|---|---|
| Haihe | China | 0.05 | 4 | 1,620 | 40,500 |
| Huanghe | China | 0.77 | 6 | 1,403 | 22,041 |
| Yangzijiang | China | 1.94 | 46 | 246 | 531 |
| Mekong | Vietnam | 0.79 | 59 | 203 | 340 |
| Ganges-Brahmaputra | Bangladesh | 1.48 | 66 | 1,128 | 1,720 |
| Indus | Pakistan | 0.97 | 25 | 454 | 1,849 |
| Tigris-Euphrates | Iraq | 1.05 | 4 | 50 | 1,152 |
| Amur | USSR | 1.85 | 18 | 28 | 160 |
| Niger | Nigeria | 1.21 | 16 | 33 | 208 |
| Nile[a] | Egypt | 2.96 | 1 | 38 | 3,700 |
| Zaire | Zaire | 3.82 | 33 | 11 | 34 |
| Mississippi | United States | 3.27 | 18 | 107 | 602 |
| Amazon | Brazil | 6.15 | 102 | 146 | 143 |
| Orinoco | Venezuela | 0.99 | 111 | 212 | 191 |

[a]Nile before Aswan High Dam.

**TABLE B.7** Drainage area, runoff, and sediment yields for fourteen major rivers

**DESCRIPTION**

The sediment yields of fourteen major rivers are given here in tonnes per km$^2$ per year and in parts per million (ppm). Also shown are the catchment areas in km$^2$ × 10$^6$, the runoff in cm per year distributed over the catchment area, and the country where the mouth of the river discharges. The table shows the wide range of sediment yields – up to a factor of a thousand – for the world's major rivers.

**LIMITATIONS**

The limitations described in Table B.6 apply for these data as well. In particular, the quality of these data vary due to differences in measuring

techniques and periods of sampling, which are not described in the original. For example, the sediment yield of the Nile is now considered to be zero, due to the Aswan High Dam. Some typographical errors in the original have been corrected here, such as the catchment area of the Huanghe.

**SOURCE**

M. Abu Zeid and A.K. Biswas, 1990, Impacts of agriculture on water quality, *Water International*, **15**(3), 160–167. International Water Resources Association, Illinois. By permission.

**TABLE B.8** Quality of data base for twenty-one of the largest river-sediment discharges to the ocean

| River | Average sediment discharge ($10^6$ tonnes/yr) | Adequacy of data base |
|---|---|---|
| Ganges-Brahmaputra | 1,670 | Inadequate |
| Huanghe | 1,080 | Good |
| Amazon | 900 | Inadequate |
| Yangzijiang | 478 | Good |
| Irrawaddy | 285 | Inadequate |
| Magdalena | 220 | Inadequate |
| Mississippi | 210 | Good |
| Orinoco | 210 | Sufficient |
| Hunghe (Red) | 160 | Inadequate |
| Mekong | 160 | Sufficient |
| Indus | 100 | Sufficient |
| Mackenzie | 100 | Poor to Fair |
| Godavari | 96 | Inadequate |
| La Plata | 92 | Inadequate to Sufficient |
| Haiho | 81 | Good |
| Purari | 80 | Inadequate to Sufficient |
| Copper | 70 | Sufficient |
| Zhujiang | 69 | Sufficient to Good |
| Danube | 67 | Good |
| Choshui | 66 | Sufficient |
| Yukon | 60 | Sufficient |

**TABLE B.8** Quality of data base for twenty-one of the largest river-sediment discharges to the ocean

**DESCRIPTION**

The 21 rivers with the greatest sediment delivery to the ocean are listed here. Average sediment discharge is given in million tonnes per year and the adequacy of the database is rated for each river. Of the 21 largest rivers that contribute nearly 50% of the total sediment to the oceans, only five (the Huanghe, Yangzijiang, Mississippi, Haiho, and Danube) can be considered adequately documented.

**LIMITATIONS**

According to the authors, "The data upon which our estimates are based have a number of serious potential errors, which need to be taken into account when considering either local or world-wide budgets. The most important factor is the widely variable quality of data, which is a result of differences in measurement techniques, in lengths of observation, and in sampling procedures.... Moreover, many rivers are poorly studied or unmeasured during large floods, when sediment discharge may be particularly important."

Data from developed countries are considered to be better than data from less developed countries. Unfortunately, the rivers with the greatest sediment discharges to the ocean tend to be in developing regions of the world. The authors also note the difficulty in obtaining original data, and the problem of recycled data. One of the worst examples, described in the original source, is the estimate for the Irrawaddy, quoted by many researchers, but which in fact is based on measurements made by the British in the 1870s.

**SOURCE**

J.D. Milliman and R.H. Meade, 1983, World-wide delivery of river sediment to the oceans, *Journal of Geology*, **91**(1), 1–21. By permission.

**TABLE B.9**　Sediment yields to selected reservoirs in China

| Reservoir | River | Catchment area ($km^2$) | Storage volume ($10^6$ $m^3$) | Sedimentation | | |
|---|---|---|---|---|---|---|
| | | | | ($10^6$ $m^3$) | Years[a] | (Percent of storage) |
| Sanmenxia | Huanghe | 688,421 | 9,700 | 3,391 | 7.5 | 35 |
| Qingtongxia | Huanghe | 285,000 | 627 | 527 | 5 | 84 |
| Yanguoxia | Huanghe | 182,800 | 220 | 150 | 4 | 68 |
| Liujiaxia | Huanghe | 172,000 | 5,720 | 522 | 8 | 11 |
| Danjiangkou | Hanshui | 95,217 | 16,000 | 625 | 15 | 4 |
| Guanting | Yongdinghe | 47,600 | 2,270 | 553 | 24 | 24 |
| Hongshan | Laohe | 24,486 | 2,560 | 440 | 15 | 17 |
| Gangnan | Hutuohe | 15,900 | 1,558 | 185 | 17 | 12 |
| Xingqiao | Hongliuhe | 1,327 | 200 | 156 | 14 | 71 |

[a]Period over which sedimentation occurred.

**TABLE B.9**　Sediment yields to selected reservoirs in China

**DESCRIPTION**

A major concern about artificial reservoirs is the loss of storage volume that occurs when sediments carried in by rivers settle to the reservoir bottom. In China, where some rivers carry enormous sediment loads, many reservoirs may lose a significant fraction of their total storage volume in the span of only a few years. This table lists sedimentation volumes in $m^3 \times 10^6$ over different periods of time for nine reservoirs in China. Also given are total reservoir storage volumes in $m^3 \times 10^6$ and the catchment areas over which sediments are generated, in $km^2$. For some of these reservoirs, their useful storage may be lost to sedimentation long before the end of the useful life of the dam itself.

**LIMITATIONS**

Sedimentation rates vary considerably from year to year depending on rates and volumes of runoff, and human actions within a drainage basin. Some limited remedial actions can be taken to reduce sedimentation and increase the useful life of a reservoir. The data presented here do not reflect these actions. Thus, estimates of when a reservoir's storage volume will be fully lost to sedimentation may be inaccurate if these data are used.

**SOURCE**

M. Abu Zeid and A.K. Biswas, 1990, Impacts of agriculture on water quality, *Water International*, **15**(3), 160–167. International Water Resources Association, Illinois. By permission.

**TABLE B.10**　Major lakes of the world　*(pages 162–165)*

**DESCRIPTION**

The principal geophysical characteristics of the world's major lakes are given here, including lake surface and drainage basin area in $km^2$, maximum lake depth in meters, and lake volume in $km^3$. Both fresh water and saline lakes are included. Because lakes can fluctuate in size and volume due to climatic variability and human impacts, ranges are often given here for surface area and lake depth. Some of the lakes shown are ephemeral, such as Lake Torrens in Australia. This lake may have significant amounts of water in it only once in a hundred years, but it can also be the second largest lake in Australia as measured by surface area. The geologic type of lake is also included, from Czaya (1983). Drainage basin areas exclude the surface area of the lake. The lakes are sorted by continent and surface area.

**LIMITATIONS**

Incomplete or contradictary data are available for some lakes, even those that have been extensively studied. For example, estimates of the maximum depth of Lake Superior, the largest lake in North America, vary by a third; maximum depth estimates for Great Bear and Great Slave Lakes in Canada vary by a factor of three or more. Some ranges are provided for lake volumes, surface areas, and depth due to natural variations over time. Lake names also vary with time, political conditions in a region, and language.

**SOURCES**

USSR Committee for the International Hydrological Decade, 1978, *World Water Balance and Water Resources of the Earth*, Studies and Reports in Hydrology, no. 25. UNESCO, Paris and Gidrometeoizdat, Leningrad (English translation of a 1974 USSR publication).

World Resources Institute, 1988, *World Resources 1988–89*, World Resources Institute and the International Institute for Environment and Development in collaboration with the United Nations Environment Programme, Basic Books, New York. The World Resources Institue notes that much of their lake data come from the Lake Biwa Research Institute, 1984, *Data Book of World Lakes*, National Institute for Research Advancement, Tokyo; and C.E. Herdendorf, 1984, *Inventory of the Morphometric Limnologic Characteristics of the Large Lakes of the World*, Ohio Sea Grant Program, Ohio State University, Columbus. By permission.

E. Czaya, 1983, *Rivers of the World*, Van Nostrand Reinhold, New York. By permission.

V. Showers, 1989, *World Facts and Figures*, John Wiley and Sons, New York. By permission.

**TABLE B.10** Major lakes of the world

| | Countries | Surface area (km²) | Maximum depth (m) | Drainage basin area (km²) | Volume (km³) | Geologic type | Sources |
|---|---|---|---|---|---|---|---|
| *Africa* | | | | | | | |
| Victoria | Tanzania, Kenya, Uganda | 62,940-69,000 | 80-92 | 184,000 | 2,700 | Tect | 1,2,3 |
| Tanganyika | Tanzania, Zaire, Zambia, Burundi, Rwanda | 32,000-34,000 | 1,435-1,470 | 263,000 | 18,900 | Tect | 1,2,3 |
| Nyasa (Malawi) | Malawi, Mozambique, Tanzania | 22,490-30,900 | 706 | 65,000 | 6,140-7,725 | Tect | 1,2,3 |
| Chad | Chad, Niger, Nigeria, Cameroon | 7,000-26,000 | 4-12 | 2,426,730 | 44 | Tect/aeo | 1,2,3 |
| Bangweulu | Zambia | 4,000-15,000 | 5 | | 5 | | 3 |
| Rudolph | Kenya, Ethiopia, Sudan, Uganda | 6,400-8,660 | 73 | | | | 3,4 |
| Nai Ndombe (Leopold II) | Zaire | 2,070-8,210 | 6 | | | | 3,4 |
| Mobutu Sese Seko (Albert) | Uganda, Zaire | 5,300-5,590 | 48-57 | | 64 | Tect | 1,3,4 |
| Kariba | Zambia, Zimbabwe | 5,400 | | 663,000 | 160 | | 2 |
| Mweru | Zambia, Zaire | 4,350-5,100 | 9-15 | | 32 | Tect | 1,3,4 |
| Rukwa[b] | Tanzania | 2,850-4,500 | shallow | | | | 3,4 |
| Kyoga | Uganda | 4,430 | 8 | | | | 4 |
| Tana | Ethiopia | 3,150-3,600 | 9-14 | | 28 | | 3,4 |
| Kivu | Zaire, Rwanda | 2,220-2,650 | 480-780 | | 569 | Tect/vol | 1,3,4 |
| Rutanzige (Edward) | Zaire, Uganda | 2,150-2,500 | 117-131 | | 78 | | 3,4 |
| *Asia* | | | | | | | |
| Aral[a,c] | USSR (Kazakhstan, Uzbekistan) | 64,100 | 68 | 1,618,000 | 1,020 | Tect | 1,2,3,4 |
| Baikal | USSR (Russia) | 31,500 | 1,620-1,741 | 560,000 | 23,000 | Tect | 1,2,3 |
| Tonle Sap[b] | Cambodia | 2,700-30,000 | 12 | | 40 | | 3,4 |
| Balkhash[a] | USSR (Kazakhstan) | 17,000-19,000 | 26 | 176,500 | 112 | Tect | 1,2,3,4 |
| Dongtinghu[b] | China | 3,100-12,000 | 10 | 259,430 | 18 | | 2,3,4 |
| Issyk-kul | USSR (Kirghizia) | 6,200 | 702 | | 1,730 | Tect | 1,3,4 |
| Rezaiyeh (Urmia)[a] | Iran | 3,900-5,930 | 16 | | 45 | Tect | 1,3 |
| Zaisan | USSR | 5,510 | 9 | | 53 | | 3 |
| Poyanghu | China | 2,700-5,160 | 20 | | | Fluv | 1,3 |
| Taimyr | USSR (Russia) | 4,560 | 26 | | 13 | | 3 |

*continued*

| | Countries | Surface area (km²) | Maximum depth (m) | Drainage basin area (km²) | Volume (km³) | Geologic type | Sources |
|---|---|---|---|---|---|---|---|
| Chanka | USSR, China | 3,030–4,403 | 11 | | 19 | Tect | 1,3,4 |
| Chovsgol Nuur | Mongolia | 2,620 | 246–270 | | 480 | — | 3,4 |
| Taihu | China | 2,210 | 5 | | 4 | | 2,4 |
| Helmand[b] | Afghanistan, USSR | 2,080 | | 350,000 | 8 | | 2 |
| Toba | Indonesia | 1,100–1,129 | 450 | | 1,258 | Vol/tect | 1,2 |
| Dead Sea[a] | Israel, Jordan | 940–1,020 | 356–433 | 3,440 | 188 | Tect | 1,3,4 |
| Songkhla | Thailand | 987 | | 8,000 | 2 | | 2 |
| Laguna de Bay | Philippines | 890 | | 3,820 | 3 | | 2 |
| Biwa | Japan | 674–690 | 96–103 | 3,174 | 28 | | 2,3,4 |
| Matana | Indonesia | 164–190 | 590 | | | Tect | 1,4 |
| Sarez | USSR (Tadzhikistan) | 88 | 505 | | | Nat dam | 1 |
| Shikotsu | Japan | 78 | 363 | | | Vol | 1 |
| Tazawa | Japan | 26 | 425 | | | Vol | 1 |
| *Europe* | | | | | | | |
| Caspian[a] | USSR, Iran | 374,000–378,400 | 995–1,025 | 3,625,000 | 78,200 | Tect | 1,2,3,4 |
| Ladoga | USSR (Russia) | 17,700–18,400 | 230 | | 908 | Tect/glac | 1,2,4 |
| Onega | USSR (Russia) | 9,600–9,720 | 127 | | 295 | Tect/glac | 1,3,4 |
| Vanern | Sweden | 5,546–5,648 | 89–100 | | 153–180 | Tect/glac | 1,2,3,4 |
| Saimaa | Finland | 1,760–4,400 | 58–82 | | 36 | Glac | 1,3,4 |
| Vattern | Sweden | 1,856–1,910 | 119–128 | 4,503 | 72–74 | | 2,3,4 |
| Beloye (White) | USSR (Russia) | 1,290 | 20 | | 5 | | 3,4 |
| Malaren | Sweden | 1,096–1,140 | 64 | 22,600 | 10–14 | | 2,3 |
| Geneva | France, Switzerland | 581 | 310 | 7,975 | 90 | Glac | 1,2,3 |
| Constance | Switzerland, Austria, Germany | 476–540 | 252 | 10,446 | 48 | Glac | 1,2,3,4 |
| Garda | Italy | 370 | 346 | | 50 | Glac | 1,2,3 |
| Mjosa | Norway | 363–370 | 434–449 | | 56 | Glac | 1,3,4 |
| Ohrid | Albania, Yugoslvia | 349–370 | 256–286 | | 61 | Tect/kar | 1,3,4 |
| Neusiedler See | Austria, Hungary | 323 | 2 | | | | 3 |

| | Countries | Surface area (km$^2$) | Maximum depth (m) | Drainage basin area (km$^2$) | Volume (km$^3$) | Geologic type | Sources |
|---|---|---|---|---|---|---|---|
| Prespansko | Greece, Albania, Yugoslavia | 270–290 | 54 | | 4 | | 3,4 |
| Maggiore | Italy, Switzerland | 214 | 372 | 6,387 | 38 | Glac | 1,2,3,4 |
| Como | Italy | 146 | 410 | | | Glac | 1 |
| Tinnvatn | Norway | 54 | 438 | | | Fjord | 1 |
| Hornindalsvatn | Norway | 51 | 514 | | | Fjord | 1 |
| Salsvatan | Norway | 45–49 | 445–464 | | | Fjord | 1,4 |
| *North and Central America* | | | | | | | |
| Superior | Canada, United States | 82,100–83,300 | 307–406 | 127,700 | 11,600–12,230 | Glac | 1,2,3,4 |
| Huron | Canada, United States | 59,500–59,800 | 223–229 | 133,900 | 3,537–3,580 | Glac | 1,2,3,4 |
| Michigan | United States | 57,016–58,100 | 265–285 | 118,100 | 4,680–4,920 | Glac | 1,2,3,4 |
| Great Bear | Canada | 30,200–31,792 | 137–445 | 146,000 | 1,010–2,381 | Glac | 1,2,3,4 |
| Great Slave | Canada | 27,200–28,570 | 156–614 | 971,000 | 1,070–2,088 | Glac | 1,2,3,4 |
| Erie | Canada, United States | 25,657–25,720 | 64 | 58,800 | 483–545 | Glac | 1,2,3,4 |
| Winnipeg | Canada | 24,387–24,600 | 19–28 | 984,200 | 127–371 | Tect | 1,2,3,4 |
| Ontario | Canada, United States | 18,760–19,480 | 225–273 | 70,700 | 1,637–1,710 | Glac | 1,2,3,4 |
| Nicaragua | Nicaragua | 8,150–8,430 | 70 | | 108 | | 2,3 |
| Athabasca | Canada | 7,935–8,080 | 60–124 | 158,000 | 110–204 | Glac | 1,2,3,4 |
| Great Salt[a] | United States | 2,800–6,000 | 14–16 | | 19 | Glac | 1,3,4 |
| Winnipegosis | Canada | 5,370–5,470 | 12 | | 16 | Glac | 1,3,4 |
| Manitoba | Canada | 4,660–4,720 | 28 | | 17 | Glac | 1,3,4 |
| Managua | Nicaragua | 1,490 | 80 | | | | 3 |
| Cedar | Canada | 1,353 | | 339,000 | | | 2 |
| Chapala | Mexico | 1,080–1,140 | 10–13 | | 10 | | 3,4 |
| Tahoe | United States | 499 | 486–501 | | | Tect | 1 |
| Chelan | United States | 140–150 | 458–489 | | | Glac | 1,4 |
| Atitlan | Guatemala | 130–136 | 320–341 | | | Vol | 1,4 |
| Washington | United States | 88 | | 1,274 | 3 | | 2 |
| Crater | United States | 55 | 608 | | | Vol | 1 |

*continued*

*South America*

| Countries | | Surface area (km²) | Maximum depth (m) | Drainage basin area (km²) | Volume (km³) | Geologic type | Sources |
|---|---|---|---|---|---|---|---|
| Maracaibo[a] | Venezuela | 13,010–14,343 | 35–250 | 90,200 | 280 | Lag | 1,2,3,4 |
| Titicaca | Peru, Bolivia | 8,030–8,110 | 230–304 | 60,800 | 710–827 | Tect | 1,2,3,4 |
| Poopo[a] | Bolivia | 1,340–2,530 | 3 | | 2 | | 3,4 |
| Buenos Aires | Chile, Argentina | 2,240–2,400 | | | | | 3,4 |
| Lago Argentina | Argentina | 1,400 | 300 | | | Glac | 1,3 |
| Valencia | Venezuela | 350–370 | 40 | | | | 3,4 |
| *Oceania* | | | | | | | |
| Eyre[a] | Australia | 0–40,000 | 0–20 | 1,122,250 | 0–23 | Tect | 1,2,3,4 |
| Torrens[a] | Australia | 0–30,000 | Shallow | | | Tect | 1,3,4 |
| Amadeus[a] | Australia | 8,000 | | | | | 3 |
| Gairdner[a] | Australia | 0–7,000 | Shallow | | | Tect | 1,3,4 |
| Taupo | New Zealand | 611 | 159 | | | | 3 |
| Te Anau | New Zealand | 340–352 | 270–276 | | | Glac | 1,3,4 |
| Wakatipu | New Zealand | 293 | 371–378 | | | Glac | 1,3 |
| George | Australia | 145 | 3 | | 0.3 | | 3 |
| Manapouri | New Zealand | 130–145 | 445 | | | Glac | 1,3 |

Tect, tectonic; Aeo, aeolian, Vol, volcanic; Fjord, fjord; Kar, karst; Fluv, fluvial; Glac, glacial; Lag, lagoon; Nat dam, natural dam.
[a]Saline water.
[b]Subject to large fluctuations.
[c]Aral Sea volume, area, and depth have dropped enormously from these values in recent years due to consumptive use of the rivers that flow into it.

Sources:
1. UNESCO (1978).
2. World Resources Institute (1988).
3. Czaya (1983).
4. Showers (1989).

**TABLE B.11**   Number and area of lakes in the former Soviet Union

| Lake area (hectares) | Number of lakes | Total water area (hectares) | Percent of total lake area |
|---|---|---|---|
| Less than 100 | 2,844,890 | 15,922,500 | 32.8 |
| 100-999 | 36,660 | 8,647,000 | 17.8 |
| 1,000-4,999 | 2,145 | 3,966,700 | 8.2 |
| 5,000-9,999 | 228 | 1,555,800 | 3.2 |
| 10,000-100,000 | 154 | 4,115,500 | 8.5 |
| Over 100,000 | 27 | 14,362,500 | 29.6 |
| Total | 2,884,104 | 48,570,000 | 100.1[a] |

[a]Does not add to 100% due to rounding.

**TABLE B.11**   Number and area of lakes in the former Soviet Union

**DESCRIPTION**

The total number and area of lakes in the former Soviet Union are shown here as a function of lake area. For each category, the number of lakes, the total water area in hectares, and the percentage of total lake area are given. There are nearly 3 million lakes smaller than 100 hectares, though the 27 lakes over 100,000 hectares each have almost the same total area.

**LIMITATIONS**

Lake areas fluctuate with water inflow, climate, and water withdrawals. These data presumably represent an average value, but this is not specified in the original.

**SOURCE**

R. Berka, 1989, *Inland Capture Fisheries of the USSR*, FAO Fisheries Technical Paper no. 311, Food and Agricultural Organization of the United Nations, Rome.

**TABLE B.12**   Distribution of large lakes in five republics of the former Soviet Union

| | Total area of large lakes (hectares) | Proportion (%) |
|---|---|---|
| Russia | 8,852,400 | 73.1 |
| Kazakhstan | 2,343,000 | 19.3 |
| Kirghizia | 620,000 | 5.1 |
| Estonia | 157,000 | 1.3 |
| Armenia | 137,000 | 1.1 |
| Total | 12,109,400 | 99.9[a] |

[a]Does not add to 100% due to rounding.

**TABLE B.12**   Distribution of large lakes in five republics of the former Soviet Union

**DESCRIPTION**

The distribution of lakes of the former Soviet Union larger than 10,000 hectares is shown here for five former republics. The area of the largest lakes is given in hectares. Russia, with by far the largest total land area of the republics, also contains the majority of all large lakes.

**LIMITATIONS**

Lake areas fluctuate with water inflow, climate, and water withdrawals. These data presumably represent an average value, but this is not specified in the original. Large lakes in other republics are not listed here.

**SOURCE**

R. Berka, 1989, *Inland Capture Fisheries of the USSR*, FAO Fisheries Technical Paper no. 311, Food and Agricultural Organization of the United Nations, Rome.

**TABLE B.13**  Waterfalls with the greatest flow

| Waterfall | Country | River | Average flow (m³/sec) | Height (m) |
|---|---|---|---|---|
| Kohne | Cambodia-Laos | Mekong | 11,610 | 21 |
| Sete Quedas[a] | Brazil-Paraguay | Paraná | 8,260 | 65 |
| Niagara | Canada-United States | Niagara | 5,830 | 57 |
| Grande | Argentina-Uruguay | Uruguay | 4,500 | 23 |
| Paulo Afonso | Brazil | São Francisco | 2,890 | 80 |
| Urubupunga | Brazil | Paraná | 2,750 | 9 |
| Iguacu | Argentina-Brazil | Iguacu | 1,700 | 70 |
| Maribondo | Brazil | Grande | 1,500 | 35 |
| Churchill | Canada | Churchill | 1,390 | 75 |
| Kabalega | Uganda | Nile | 1,200 | 40 |
| Victoria | Zambia-Zimbabwe | Zambezi | 1,090 | 92 |
| Kaveri | India | Kaveri | 934 | 98 |
| Rhine | Switzerland | Rhine | 700 | 21 |
| Kaieteur | Guyana | Potaro | 650 | 226 |

[a]The Sete Quedas waterfall was totally submerged after the completion of the Itaipú Dam in 1982.

**TABLE B.13**  Waterfalls with the greatest flow

**DESCRIPTION**
The fourteen waterfalls with the greatest average flow, in m³/s, are shown here, together with their height in meters, and their location and river.

**LIMITATIONS**
The world's second highest waterfall, Sete Quedas on the Paraná River, was innundated by the reservoir formed behind the Itaipú Dam in 1982. The average flow figures are long-term averages, but the periods of measurement are not listed in the original source.

**SOURCE**
V. Showers, 1989, *World Facts and Figures*, John Wiley and Sons, New York. By permission.

**TABLE B.14** Highest waterfalls

| Waterfall | Country | River | Height (m) |
|---|---|---|---|
| Angel[a] | Venezuela | Churun | 807 |
| Monge | Norway | Rauma | 774 |
| Itatinga | Brazil | Itatinga | 628 |
| Ormeli | Norway | | 563 |
| Kahiwa[b] | United States | Wailau Stream | 533 |
| Tusse | Norway | Tussa Lake | 533 |
| Pilao | Brazil | Itajai | 524 |
| Montoya | Venezuela | Porah-Pi | 505 |
| Ribbon | United States | Ribbon Creek | 491 |
| Great | Guyana | Kamarang | 488 |
| Vestre Mardals | Norway | Eikesdals Lake | 468 |
| Della | Canada | Tofino Creek | 440 |
| Yosemite[a] | United States | Yosemite Creek | 436 |
| Gavarnie[a] | France | Pau | 422 |
| Tugela[a] | South Africa | Tugela | 411 |
| Verma | Norway | Verma | 381 |
| Austerbo | Norway | | 380 |
| Papalaua[b] | United States | Kawainui Stream | 366 |
| Takakkaw[a] | Canada | Yoho tributary | 366 |
| Silver Strand | United States | Meadow Brook | 357 |
| Ahui | French Polynesia | | 350 |
| Giessbach | Switzerland | Giessbach Creek | 350 |
| Mawsmai | India | Sohryngkew | 350 |
| Honokohau[b] | United States | Honokohau Stream | 341 |
| Kaloba | Zaire | Lofoi | 340 |
| Kalapis | Malaysia | Kalapis | 335 |
| Wollomomb[a] | Australia | Wollomombi | 334 |
| Cuquenan | Guyana-Venezuela | Cuquenan | 317 |
| Basaseachic | Mexico | Basaseachic Creek | 311 |
| Hiilawe | United States | Wailoa Stream | 305 |
| Mtarazi | Mozambique-Zimbabwe | Mtarazi | 305 |
| Thylliejlongwa | India | Umngi | 304 |
| Belmore: three falls | Australia | Barrengarry Creek | 300 |
| Candelas | Colombia | Cusiana | 300 |
| Cannabullen | Australia | Cannabullen Creek | 300 |
| Horseshoe | Australia | Govetts Leap Creek | 300 |
| Rembesdals[a] | Norway | Rembesdals Lake | 300 |
| Skykkjedals[a] | Norway | Skykkjua | 300 |
| Tyssestrengene[a] | Norway | Tysso | 300 |

*continued*

| Waterfall | Country | River | Height (m) |
|---|---|---|---|
| Staubbach | Switzerland | Staubbach Creek | 299 |
| Austre Mardals[a] | Norway | Eikesdals Lake | 297 |
| Wallaman | Australia | Stony Creek | 296 |
| Farina del Diavolo | Italy | | 280 |
| Pungwe | Zimbabwe | Pungwe | 277 |
| Vettis[a] | Norway | Morkedola | 275 |
| Elizabeth Grant | Australia | Tully | 274 |
| Twin | Canada | Twin Falls Creek | 274 |
| Valur | Norway | Veig | 272 |
| Molli | Norway | Molles | 269 |
| Sewerd | Peru | Cutibireni | 267 |
| Austerkrok[a] | Norway | Austerkrok | 257 |
| Tiboku | Guyana | Semang | 256 |
| Gersoppa[a] | India | Sharavati | 253 |
| Helena | New Zealand | Helena | 253 |
| Hunlen | Canada | Atnarko Creek | 253 |

[a]Highest fall
[b]Total fall

**TABLE B.14**  Highest waterfalls *(pages 168–169)*

**DESCRIPTION**
The world's highest waterfalls are ranked here, in meters. Their location and river are also included. All waterfalls higher than 250 m are listed.

**LIMITATIONS**
Only a subset of the world's waterfalls is listed here. The names used are in most cases the local names. For waterfalls with multiple drops, sometimes the total fall is listed and sometimes the highest fall is listed. These are noted in the table.

**SOURCE**
V. Showers, 1989, *World Facts and Figures*, John Wiley and Sons, New York. By permission.

# C. Sanitation and water-related disease

**TABLE C.1**   Worldwide access to safe drinking water *(pages 171–178)*

## DESCRIPTION

One of the most important issues in fresh water resources is access to clean water for drinking and sanitation. The United Nations devoted the decade of the 1980s to the goal of providing safe drinking water and access to sanitation to the thousands of millions of people without these resources. During this decade, significant funds were made available for new and improved facilities. This table presents the percentage of the urban and rural populations in selected countries with access to safe drinking water. The data are for 1970, 1975, 1980, and 1985. Countries included here are those responding to World Health Organization (WHO) questionnaires, and represent about 70% of the world's population. Almost all of the rest of the people in developing countries live in China, which did not report here. The total percentage of the population in each country with access to safe drinking water is also given, as is the urban population as a percentage of the total population, for 1970, 1980, and 1985.

## LIMITATIONS

The definition of ''access to safe drinking water'' has varied over time. As used here by the WHO, safe drinking water includes treated surface water and untreated water from protected springs, boreholes, and wells. The WHO defines access to safe drinking water in urban areas as piped water to housing units or to public standpipes within 200 m. In rural areas, reasonable access implies that fetching water does not take up a disproportionate part of the day. This latter definition is ambiguous and measured in different ways in different countries and regions. Definitions of rural and urban are supplied by national governments. Incomplete information is given in particular categories, and no data are available in this table for some major countries, such as China, the United States, Australia, Canada, South Africa, and Japan. Population growth in many cities during the 1980s increased the need for services at a rate faster than could be supplied by United Nations efforts. For information on total numbers of people served and unserved by adequate access to safe drinking water in 1980 and 1990, see Table C.3.

## SOURCES

United Nations Environment Programme, 1989, *Environmental Data Report*, GEMS Monitoring and Assessment Research Centre, Basil Blackwell, Oxford. By permission.

World Health Organization data, cited by the World Resources Institute, 1988, *World Resources 1988–89*, World Resources Institute and the International Institute for Environment and Development in collaboration with the United Nations Environment Programme, Basic Books, New York.

**TABLE C.1** Worldwide access to safe drinking water

Percentage of population with access to safe drinking water

| Region | Urban population as percentage of total population | | | Urban | | | | Rural | | | | Total | | | |
|---|---|---|---|---|---|---|---|---|---|---|---|---|---|---|---|
| | 1970 | 1980 | 1985 | 1970 | 1975 | 1980 | 1985 | 1970 | 1975 | 1980 | 1985 | 1970 | 1975 | 1980 | 1985 |
| World | 37.1 | 39.6 | 41.0 | | | | | | | | | | | | |
| *Africa* | | | | | | | | | | | | | | | |
| Algeria | 39.5 | 41.2 | 42.6 | 84 | 100 | | 85 | | 61 | | 55 | | 77 | | 68 |
| Angola | 15.0 | 21.0 | 24.5 | | | 85 | 87 | | | 10 | 15 | | | 26 | 33 |
| Benin | 16.0 | 28.2 | 35.2 | 83 | 100 | 26 | 80 | 20 | 20 | 15 | 34 | 29 | 34 | 18 | 50 |
| Botswana | 8.4 | 15.3 | 19.2 | 71 | 95 | | 84 | 26 | 39 | | 46 | 29 | 45 | | 53 |
| Burkina Faso | 5.7 | 7.0 | 7.9 | 35 | 50 | 27 | 43 | 10 | 23 | 31 | 69 | 12 | 25 | 31 | 67 |
| Burundi | 2.2 | 4.1 | 5.6 | 77 | | 90 | 98 | | | 20 | 21 | | | 23 | 25 |
| Cameroon | 20.3 | 34.7 | 42.4 | 77 | | | 43 | 21 | | | 24 | 32 | | | 32 |
| Cape Verde | 5.6 | 5.1 | 5.3 | | | 100 | 83 | | | 21 | 50 | | | 25 | 52 |
| Central African Republic | 30.4 | 38.2 | 42.4 | | | | 13 | | | | | | | | |
| Chad | 11.4 | 20.8 | 27.0 | 47 | 43 | | | 24 | 23 | | | 27 | 26 | | |
| Comoros | 11.3 | 23.2 | 25.2 | | | | | | | | | | | | |
| Congo | 34.8 | 37.3 | 39.5 | 63 | 81 | 42[a] | | 6 | 9 | 7[a] | | 27 | 38 | 20[a] | |
| Côte d'Ivoire | 27.4 | 37.1 | 42.0 | 98 | | | | 29 | | | | 44 | | | |
| Djibouti | 62.0 | 73.7 | 77.7 | | | 50 | 50 | | | 20 | 20 | | | 43 | 45 |
| Egypt | 43.5 | 44.7 | 46.4 | 94 | | 88 | | 93 | | 64 | | 93 | | 84 | |
| Equatorial Guinea | 39.0 | 53.7 | 59.7 | | | 47[a] | | | | | | | | | |
| Ethiopia | 8.6 | 10.5 | 11.6 | 61 | 58 | | 69 | | 1 | | 9 | 6 | 8 | | 16 |
| Gabon | 25.6 | 35.8 | 40.9 | | | | | | | | | | | | |
| Gambia | 15.0 | 18.1 | 20.1 | 97 | 86 | 85 | 97 | 3 | | | 50 | 12 | | | 59 |
| Ghana | 29.0 | 30.7 | 31.5 | 86 | 86 | 72 | 93 | 14 | 14 | 33 | 39 | 35 | 35 | 45 | 56 |
| Guinea | 13.8 | 19.1 | 22.2 | 68 | 69 | 69 | 41 | | | 2 | 12 | | 14 | 15 | 18 |
| Guinea-Bissau | 18.1 | 23.8 | 27.1 | | | 18 | 17 | | | 8 | 22 | | | 10 | 21 |

*continued*

Percentage of population with access to safe drinking water

| Region | Urban population as percentage of total population | | | Percentage of population with access to safe drinking water | | | | | | | | | | | |
|---|---|---|---|---|---|---|---|---|---|---|---|---|---|---|---|
| | 1970 | 1980 | 1985 | Urban | | | | Rural | | | | Total | | | |
| | | | | 1970 | 1975 | 1980 | 1985 | 1970 | 1975 | 1980 | 1985 | 1970 | 1975 | 1980 | 1985 |
| Kenya | 10.3 | 16.1 | 19.7 | 100 | 100 | 85 | 65 | 2 | 4 | 15 | 30 | 15 | 17 | 26 | |
| Lesotho | 8.6 | 13.6 | 16.7 | 100 | 65 | 37 | | 1 | 14 | 11 | 30 | 3 | 17 | 15 | 36 |
| Liberia | 26.0 | 34.9 | 39.5 | 100 | | 100 | 100 | 6 | | | 23 | 15 | | | 53 |
| Libya | 35.8 | 56.6 | 64.5 | 100 | 100 | 100 | | 42 | 82 | 90 | 17 | 58 | 87 | 96 | |
| Madagascar | 14.1 | 18.9 | 21.8 | 67 | 76 | 80 | 81 | 1 | 14 | 7 | 17 | 11 | 25 | 21 | 31 |
| Malawi | 6.0 | 9.7 | 12.0 | | | 77 | 97 | | | 37 | 50 | | | 41 | 56 |
| Mali | 14.3 | 17.3 | 18.0 | 29 | | 37 | 46 | 10 | | | 10 | | | | 16 |
| Mauritania | 13.9 | 26.9 | 34.6 | 98 | | 80 | 73 | | | 85 | | 17 | | 84 | |
| Mauritius | 42.0 | 42.9 | 42.2 | 100 | 100 | 100 | 100 | 29 | 22 | 98 | 100 | 61 | 60 | 99 | 100 |
| Morocco | 34.6 | 41.3 | 44.8 | 92 | | 100 | 100 | 28 | | | 25 | 51 | | | 59 |
| Mozambique | 5.7 | 13.1 | 19.4 | | | | 38 | | | | 9 | | | | 15 |
| Namibia | | | | | | | | | | | | | | | |
| Niger | 8.5 | 13.2 | 16.2 | 37 | 36 | 41 | 35 | 19 | 26 | 32 | 49 | 20 | 27 | 33 | 47 |
| Nigeria | 16.4 | 20.4 | 23.0 | | | | 100 | | | | 20 | | | | 38 |
| Reunion | | | | | | | | | | | | | | | |
| Rwanda | 3.2 | 5.0 | 6.2 | 81 | 84 | 48 | 79 | 66 | 68 | 55 | 48 | 67 | 68 | 55 | 50 |
| São Tomé and Principe | | | | | | | | | | | 45 | | | | 45 |
| Senegal | 33.4 | 34.9 | 36.4 | 87 | 56 | 77 | 79 | | | 25 | 38 | | | 43 | 53 |
| Seychelles | | | | | | | | | | | 95 | | | | 95 |
| Sierra Leone | 18.1 | 24.5 | 28.3 | 75 | | 50 | 68 | 1 | | 2 | 7 | 12 | | 14 | 24 |
| Somalia | 23.1 | 30.2 | 34.1 | 17 | 77 | | 58 | 14 | 22 | | 22 | 15 | 38 | | 34 |
| South Africa | 47.9 | 53.2 | 55.9 | | | | | | | | | | | | |
| Sudan | 16.4 | 19.7 | 20.6 | 61 | 96 | 100[a] | | 13 | 43 | 31[a] | | 19 | 50 | 51[a] | |
| Swaziland | 9.7 | 19.8 | 26.3 | | 83 | | 100 | | 29 | | 7 | | 37 | | 31 |
| Tanzania | 6.9 | 16.5 | 22.3 | 61 | 88 | 70 | 90 | 9 | 36 | 31 | 42 | 13 | 39 | 38 | 53 |
| Togo | 13.1 | 18.8 | 22.1 | 100 | 49 | 70 | 100 | 5 | 10 | 31 | 41 | 17 | 16 | 38 | 54 |

continued

Percentage of population with access to safe drinking water

| Region | Urban population as percentage of total population | | | Urban | | | | Rural | | | | Total | | | |
|---|---|---|---|---|---|---|---|---|---|---|---|---|---|---|---|
| | 1970 | 1980 | 1985 | 1970 | 1975 | 1980 | 1985 | 1970 | 1975 | 1980 | 1985 | 1970 | 1975 | 1980 | 1985 |
| Tunisia | 43.5 | 52.3 | 56.8 | 92 | 93 | 100 | 100 | 17 | | 17 | 31 | 49 | | 60 | 70 |
| Uganda | 7.8 | 8.7 | 9.5 | 88 | 100 | | 37 | 17 | 29 | | 18 | 22 | 35 | | 20 |
| Zaire | 30.3 | 34.2 | 36.6 | 33 | 38 | | 52 | 4 | 12 | | 21 | 11 | 19 | | 32 |
| Zambia | 30.4 | 42.8 | 49.5 | 70 | 86 | | 76 | 22 | 16 | | 41 | 37 | 42 | | 58 |
| Zimbabwe | 16.9 | 21.9 | 24.6 | | | | | | | | 32 | | | | |
| *North and Central America* | | | | | | | | | | | | | | | |
| Bahamas | | | | 100 | 100 | 100 | 100 | 12 | 13 | | | 65 | 65 | 100 | 100 |
| Barbados | 37.1 | 40.1 | 42.2 | 95 | 100 | 99 | 100 | 100 | 100 | 98 | 99 | 98 | 100 | 99 | 99 |
| Belize | | | | | | 99 | 100 | | | 36 | 26 | | | 68 | 64 |
| Canada | 75.7 | 75.7 | 75.9 | | | | | | | | | | | | |
| Cayman Islands | | | | | | 100 | 98 | | | | | | | | |
| Costa Rica | 39.7 | 46.0 | 49.8 | 98 | 100 | 100 | 100 | 59 | 56 | 82 | 83 | 74 | 72 | 90 | 91 |
| Cuba | 60.2 | 68.1 | 71.8 | 82 | 96 | | | 15 | | | | 56 | 55 | 60 | 62 |
| Dominican Republic | 40.3 | 50.5 | 55.7 | 72 | 88 | 85 | 85 | 14 | 27 | 34 | 33 | 37 | 55 | 50 | 51 |
| El Salvador | 39.4 | 39.3 | 39.1 | 71 | 89 | 67 | 68 | 20 | 28 | 40 | 40 | 40 | 53 | 50 | 51 |
| Guadeloupe | | | | | | | | | | | | | | | |
| Guatemala | 35.7 | 38.5 | 40.0 | 88 | 85 | 90 | 72 | 12 | 14 | 18 | 14 | 38 | 39 | 46 | 37 |
| Haiti | 19.8 | 24.6 | 27.2 | | 46 | 51 | 59 | | 3 | 8 | 30 | | 12 | 19 | 38 |
| Honduras | 28.9 | 36.1 | 40.0 | 99 | 99 | 93 | 56 | 10 | 13 | 40 | 45 | 34 | 41 | 59 | 49 |
| Jamaica | 41.6 | 49.8 | 53.8 | 100 | 100 | 55 | 99 | 48 | 79 | 46 | 93 | 62 | 86 | 51 | 96 |
| Martinique | | | | | | | | | | | | | | | |
| Mexico | 59.0 | 66.4 | 69.6 | 71 | 70 | 90 | 99 | 29 | 49 | 40 | 47 | 54 | 62 | 73 | 83 |
| Nicaragua | 47.0 | 53.4 | 56.6 | 58 | 100 | 67 | 76 | 16 | 14 | 6 | 11 | 35 | 56 | 39 | 48 |
| Panama | 47.6 | 50.5 | 52.4 | 100 | 100 | 100 | 100 | 41 | 54 | 62 | 64 | 69 | 77 | 81 | 82 |
| Puerto Rico | | | | | | | | | | | | | | | |

continued

| Region | Urban population as percentage of total population | | | Percentage of population with access to safe drinking water | | | | | | | | | | | |
|---|---|---|---|---|---|---|---|---|---|---|---|---|---|---|---|
| | | | | Urban | | | | Rural | | | | Total | | | |
| | 1970 | 1980 | 1985 | 1970 | 1975 | 1980 | 1985 | 1970 | 1975 | 1980 | 1985 | 1970 | 1975 | 1980 | 1985 |
| St. Lucia | | | | | | | | | | | | | | | |
| St. Vincent | | | | | | | | | | | | | | | |
| Trinidad and Tobago | 38.8 | 56.9 | 63.9 | 100 | 79 | 100 | 100 | 95 | 100 | 93 | 95 | 96 | 93 | 97 | 98 |
| Turks and Caicos Islands | | | | | | | 87 | | | | 68 | | | | 77 |
| United States | 73.6 | 73.7 | 73.9 | | | | | | | | | | | | |
| *South America* | | | | | | | | | | | | | | | |
| Argentina | 78.4 | 82.7 | 84.6 | 69 | 76 | 61 | 63 | 12 | 26 | 17 | 17 | 56 | 66 | 54 | 56 |
| Bolivia | 40.8 | 44.3 | 47.8 | 92 | 81 | 69 | 75 | 2 | 6 | 10 | 13 | 33 | 34 | 36 | 43 |
| Brazil | 55.8 | 67.5 | 72.7 | 78 | 87 | 83 | 85 | 28 | | 51 | 56 | 55 | 70 | 72 | 77 |
| Chile | 75.2 | 81.1 | 83.6 | 67 | 78 | 100 | 98 | 13 | 28 | 17 | 29 | 56 | 70 | 84 | 87 |
| Colombia | 57.2 | 64.2 | 67.4 | 88 | 86 | 93 | 100 | 28 | 33 | 73 | 76 | 63 | 64 | 86 | |
| Ecuador | 39.5 | 47.3 | 52.3 | 76 | 67 | 79 | 81 | 7 | 8 | 20 | 31 | 34 | 36 | 50 | 57 |
| Guyana | 29.4 | 30.5 | 32.2 | 100 | 100 | 100 | 100 | 63 | 75 | 60 | 65 | 75 | 84 | 72 | 76 |
| Paraguay | 37.1 | 41.7 | 44.4 | 22 | 25 | 39 | 53 | 5 | 5 | 9 | 8 | 11 | 13 | 21 | 28 |
| Peru | 57.4 | 64.5 | 67.4 | 58 | 72 | 68 | 73 | 8 | 15 | 18 | 17 | 35 | 47 | 50 | 55 |
| Suriname | 45.9 | 44.8 | 45.7 | | | 100 | 71 | | | 79 | 94 | | | 88 | 83 |
| Uruguay | 82.1 | 83.8 | 84.6 | 100 | 100 | 96 | 95 | 59 | 87 | 2 | 27 | 92 | 98 | 81 | 85 |
| Venezuela | 76.2 | 83.7 | 86.6 | 92 | | 93 | 93 | 38 | | 53 | 65 | 75 | | 86 | 89 |
| *Asia* | | | | | | | | | | | | | | | |
| Afghanistan | 11.0 | 15.6 | 18.5 | 18 | 40 | 28 | 38 | 1 | 5 | 8 | 17 | 3 | 9 | 8 | 17 |
| Bahrain | 78.2 | 80.5 | 81.7 | 100 | 100 | | 100 | 94 | 100 | | 100 | 99 | 100 | | 100 |
| Bangladesh | 7.6 | 10.4 | 11.9 | 13 | 22 | 26 | 24 | 47 | 61 | 40 | 49 | 45 | 56 | 39 | 46 |
| Bhutan | 3.1 | 3.9 | 4.5 | | | 50 | | | | 5 | 19 | | | 7 | |

*continued*

| Region | Urban population as percentage of total population | | | Percentage of population with access to safe drinking water | | | | | | | | | | | |
| | | | | Urban | | | | Rural | | | | Total | | | |
| | 1970 | 1980 | 1985 | 1970 | 1975 | 1980 | 1985 | 1970 | 1975 | 1980 | 1985 | 1970 | 1975 | 1980 | 1985 |
|---|---|---|---|---|---|---|---|---|---|---|---|---|---|---|---|
| Brunei Darus | | | | | | 100 | | | | 95 | | | | | |
| Cambodia | 11.7 | 10.3 | 10.8 | | | | | | | | | | | | |
| China | 20.1 | 20.4 | 20.6 | | | | | | | | | | | | |
| Cyprus | 40.8 | 46.3 | 49.5 | 100 | 94 | | 100 | 92 | 96 | | 100 | 95 | 95 | | 100 |
| Hong Kong | | | | | | 100 | | | | 95 | | | | | |
| India | 19.8 | 23.4 | 25.5 | 60 | 80 | 77 | 76 | 6 | 18 | 31 | 50 | 17 | 31 | 42 | 56 |
| Indonesia | 17.1 | 22.2 | 25.3 | 10 | 41 | 35 | 43 | 1 | 4 | 19 | 36 | 3 | 11 | 23 | 38 |
| Iran | 41.0 | 49.1 | 51.9 | 68 | 76 | 82[a] | | 11 | 30 | 50[a] | | 35 | 51 | 66[a] | |
| Iraq | 56.2 | 66.4 | 70.6 | 83 | 100 | | 100 | 7 | 11 | 54 | 54 | 51 | 66 | | 86 |
| Israel | 84.2 | 88.6 | 90.3 | | | | | | | | | | | | |
| Japan | 71.2 | 76.2 | 76.5 | | | | | | | | | | | | |
| Jordan | 50.6 | 60.1 | 64.4 | 98 | | 100 | 100 | 59 | | 65 | 88 | 77 | | 86 | 96 |
| Korea DPR | 50.1 | 59.7 | 63.8 | | | | | | | | | | | | |
| Korea Rep | 40.7 | 56.9 | 65.3 | 84 | 95 | 86 | 90 | 38 | 33 | 61 | 48 | 58 | 66 | 75 | 75 |
| Kuwait | 77.8 | 90.2 | 93.5 | 60 | 100 | 86[a] | 97 | | | 100[a] | | 51 | 89 | 87[a] | |
| Laos | 9.6 | 13.4 | 15.9 | 97 | 100 | 28[a] | | 39 | 32 | 20[a] | | 48 | 41 | 21[a] | |
| Lebanon | 59.4 | 74.8 | 80.1 | | | | | | | | | | | | |
| Malaysia | 27.0 | 34.2 | 38.2 | 100 | 100 | 90 | 96 | 1 | 6 | 49 | 76 | 29 | 34 | 63 | 84 |
| Maldives | | | | | | 11 | 58 | | | 3 | 12 | | | 2 | 21 |
| Mongolia | 45.1 | 51.1 | 50.8 | | | | | | | | | | | | |
| Myanmar | 22.8 | 23.9 | 23.9 | 35 | 31 | 38 | 36 | 13 | 14 | 15 | 24 | 18 | 17 | 21 | 27 |
| Nepal | 3.9 | 6.1 | 7.7 | 53 | 85 | 83 | 70 | | 5 | 7 | 25 | 2 | 8 | 11 | 28 |
| Oman | 5.1 | 7.3 | 8.8 | | 100 | | 90 | | 48 | | 49 | | 52 | | 53 |
| Pakistan | 24.9 | 28.1 | 29.8 | 77 | 75 | 72 | 83 | 4 | 5 | 20 | 27 | 21 | 25 | 35 | 44 |
| Philippines | 33.0 | 37.4 | 39.6 | 67 | 82 | 49 | 49 | 20 | 31 | 43 | 54 | 36 | 50 | 45 | 52 |
| Qatar | 80.3 | 86.0 | 88.0 | 100 | 100 | 76[a] | | 75 | 83 | 43[a] | | 95 | 97 | 71[a] | |

continued

Percentage of population with access to safe drinking water

| Region | Urban population as percentage of total population | | | Urban | | | | Rural | | | | Total | | | |
|---|---|---|---|---|---|---|---|---|---|---|---|---|---|---|---|
| | 1970 | 1980 | 1985 | 1970 | 1975 | 1980 | 1985 | 1970 | 1975 | 1980 | 1985 | 1970 | 1975 | 1980 | 1985 |
| Saudi Arabia | 48.7 | 65.9 | 72.4 | 100 | 97 | 92 | 100 | 37 | 56 | 87 | 88 | 49 | 64 | 90 | 94 |
| Singapore | 100.0 | 100.0 | 100.0 | | | 100 | 100 | | | | | | | 100[a] | 100[a] |
| Sri Lanka | 21.9 | 21.6 | 21.1 | 46 | 36 | 65 | 82 | 14 | 13 | 18 | 29 | 21 | 19 | 28 | 40 |
| Syria | 43.3 | 47.4 | 49.5 | 98 | | 98 | | 50 | | 54 | | 71 | | 74 | |
| Thailand | 13.3 | 17.3 | 19.8 | 60 | 69 | 65 | 56 | 10 | 16 | 63 | 66 | 17 | 25 | 63 | 64 |
| Turkey | 38.4 | 43.8 | 45.9 | | | 95[a] | | | | 62[a] | | | | 76[a] | |
| United Arab Emirates | 42.3 | 81.2 | 77.8 | | | 95 | | | | 81 | | | | 92 | |
| Vietnam | 18.3 | 19.3 | 20.3 | | | | 70 | | 32 | 39 | | | | | 45 |
| Yemen Arab Rep | 7.5 | 15.3 | 20.0 | 45 | | 100 | 100 | 2 | | 18 | 25 | 4 | | 31 | 40 |
| Yemen Dem Rep | 32.1 | 36.9 | 39.9 | 88 | | 85 | | 43 | | 25 | | 57 | | 52 | |
| *Europe* | | | | | | | | | | | | | | | |
| Albania | 33.5 | 33.4 | 34.0 | | | | | | | | | | | | |
| Austria | 51.7 | 54.6 | 56.1 | | | | | | | | | | | | |
| Belgium | 94.3 | 95.4 | 96.3 | | | | 94[a] | | | | 91[a] | | | | 94[a] |
| Bulgaria | 52.3 | 62.5 | 66.5 | | | | | | | | | | | | |
| Czechoslovakia | 55.2 | 62.3 | 65.3 | | | | 100[a] | | | | 100[a] | | | | 100[a] |
| Denmark | 79.7 | 84.3 | 85.9 | | | | | | | | | | | | |
| Finland | 50.3 | 59.6 | 64.0 | | | | 100[a] | | | | 99[a] | | | | 100[a] |
| France | 71.0 | 73.2 | 73.4 | | | | 100[a] | | | | 100[a] | | | | 100[a] |
| German DR | 73.7 | 76.2 | 77.0 | | | | 100[a] | | | | 100[a] | | | | 100[a] |
| Germany FR | 81.3 | 84.4 | 85.5 | | | | | | | | | | | | |
| Greece | 52.5 | 57.7 | 60.1 | | | | | | | | | | | | |
| Hungary | 45.6 | 53.5 | 56.2 | | | | 100[a] | | | | 98[a] | | | | 99[a] |
| Iceland | 84.9 | 88.2 | 89.4 | | | | | | | | | | | | |
| Ireland | 51.7 | 55.3 | 57.0 | | | | | | | | | | | | |

*continued*

Percentage of population with access to safe drinking water

| Region | Urban population as percentage of total population | | | Total | | | | Rural | | | | Urban | | | |
|---|---|---|---|---|---|---|---|---|---|---|---|---|---|---|---|
| | 1970 | 1980 | 1985 | 1970 | 1975 | 1980 | 1985 | 1970 | 1975 | 1980 | 1985 | 1970 | 1975 | 1980 | 1985 |
| Italy | 64.3 | 66.5 | 67.4 | | | | | | | | | | | | |
| Luxembourg | 67.8 | 77.6 | 81.0 | | | | | | | | | | | | |
| Malta | 77.5 | 83.1 | 85.3 | | | | | | | | | | | | |
| Monaco | | | | | | | 100[a] | | | | | | | | 100[a] |
| Netherlands | 86.1 | 88.4 | 88.4 | | | | | | | | 99[a] | | | | 100[a] |
| Norway | 64.8 | 70.5 | 72.8 | | | | 89[a] | | | | 73[a] | | | | 100[a] |
| Poland | 52.3 | 58.2 | 61.0 | | | | | | | | | | | | 100[a] |
| Portugal | 26.2 | 29.5 | 31.2 | | | | 46[a] | | | | 22[a] | | | | 100[a] |
| Romania | 41.8 | 48.1 | 49.0 | | | | | | | | | | | | |
| Spain | 66.0 | 72.8 | 75.8 | | | | 95[a] | | | | 81[a] | | | | 100[a] |
| Sweden | 81.1 | 83.1 | 83.4 | | | | | | | | | | | | |
| Switzerland | 54.5 | 57.0 | 58.2 | | | | | | | | | | | | 100[a] |
| United Kingdom | 88.5 | 90.7 | 91.5 | | | | 100[a] | | | | 100[a] | | | | 100[a] |
| Yugoslavia | 34.8 | 42.3 | 46.3 | | | | 74[a] | | | | 60[a] | | | | 91[a] |
| *USSR* | 56.7 | 63.1 | 65.6 | | | | | | | | | | | | |
| *Oceania* | | | | | | | | | | | | | | | |
| Australia | 85.2 | 85.8 | 85.5 | | | | | | | | | | | | |
| Cook Islands | | | | | | | 92 | | | | 88 | | | 100 | 99 |
| Fiji | 34.8 | 38.7 | 41.2 | 37 | 69 | 77 | | 15 | 56 | 66 | | 78 | 89 | 94 | |
| Kiribati | | | | | | | | | | 25 | | | | 93 | |
| New Zealand | 81.1 | 83.3 | 83.7 | | | | | | | | | | | | |
| Niue | | | | | | | | | | | 100[a] | | | | |
| Papua New Guinea | 9.8 | 13.0 | 14.3 | 70 | 20 | 16 | 26 | 72 | 19 | 10 | 15 | 44 | 30 | 55 | 95 |
| Samoa | | | | 17 | 43 | | | | 23 | 94 | | 86 | 100 | 97 | |

*continued*

Percentage of population with access to safe drinking water

| Region | Urban population as percentage of total population | | | Urban | | | | Rural | | | | Total | | | |
|---|---|---|---|---|---|---|---|---|---|---|---|---|---|---|---|
| | 1970 | 1980 | 1985 | 1970 | 1975 | 1980 | 1985 | 1970 | 1975 | 1980 | 1985 | 1970 | 1975 | 1980 | 1985 |
| Solomon Islands | 8.9 | 9.2 | 9.7 | | | | | | | | | | | | |
| Tokelau | | | | | | 96 | | | | 45 | 100[a] | | | | |
| Tonga | | | | 100 | 100 | 86 | 99 | 53 | 71 | 70 | 99 | 63 | 83 | 17 | 99 |
| Tuvalu | | | | | | | 100[a] | | | | 100[a] | | | | |
| Vanuatu | | | | | | 65 | 95 | | | 53 | 54 | | | | 64 |
| Western Samoa | | | | | | 97[a] | 75[a] | | | 94[a] | 67[a] | | | | 69 |

[a]Missing UNEP (1989) data supplemented with numbers from World Resources (1988).

**TABLE C.2**  Worldwide access to sanitation services *(pages 180–186)*

## DESCRIPTION

One of the most important issues in fresh water resources is access to clean water for drinking and sanitation. The United Nations devoted the decade of the 1980s to the goal of providing safe drinking water and access to sanitation to the thousands of millions of people without these resources. During this decade, significant funds were made available for new and improved facilities. This table presents the percentage of the urban and rural populations in selected countries with access to sanitation services. The data are for 1970, 1975, 1980, and 1985. Countries included here are those responding to World Health Organization (WHO) questionnaires, and represent about 70% of the world's population. Almost all of the rest of the people in developing countries live in China, which did not report data here. The total percentage of the population in each country with access to sanitation services is also given, as is the urban population as a percentage of the total population, for 1970, 1980, and 1985.

## LIMITATIONS

The definition of "access to sanitation services" is subject to interpretation and involves minimal standards, as usually measured by industrialized nations. As used here by the WHO, urban areas with access to sanitation services are defined as urban populations served by connections to public sewers or household systems such as pit privies, pour-flush latrines, septic tanks, and communal toilets. Rural populations with access to sanitation services are defined as those with adequate disposal such as pit privies, pour-flush latrines, septic tanks, and communal toilets. These definitions are ambiguous and measured in different ways in different countries and regions. Definitions of rural and urban are supplied by national governments. Incomplete information is given in particular categories, and no data are available in this table for some major countries, such as China, the United States, Australia, Canada, South Africa, and Japan. Population growth in many cities during the 1980s increased the need for sanitation services at a rate faster than could be supplied by United Nations efforts. For information on numbers of people served and unserved by adequate sanitation systems in 1980 and 1990, see Table C.3.

## SOURCES

United Nations Environment Programme, 1989, *Environmental Data Report*, GEMS Monitoring and Assessment Research Centre, Basil Blackwell, Oxford. By permission.

World Health Organization data, cited by the World Resources Institute, 1988, *World Resources 1988–89*, World Resources Institute and the International Institute for Environment and Development in collaboration with the United Nations Environment Programme, Basic Books, New York.

**TABLE C.2**  Worldwide access to sanitation services

| Region | Urban population as percentage of total population | | | Percentage of population with access to sanitation services | | | | | | | | | | | |
| | 1970 | 1980 | 1985 | Urban | | | | Rural | | | | Total | | | |
| | | | | 1970 | 1975 | 1980 | 1985 | 1970 | 1975 | 1980 | 1985 | 1970 | 1975 | 1980 | 1985 |
| World | 37.1 | 39.6 | 41.0 | | | | | | | | | | | | |
| *Africa* | | | | | | | | | | | | | | | |
| Algeria | 39.5 | 41.2 | 42.6 | 13 | 100 | | 80 | 6 | 50 | | 40 | 9 | 67 | | 57 |
| Angola | 15.0 | 21.0 | 24.5 | | | 40 | 29 | | | 15 | 16 | | | 20 | 19 |
| Benin | 16.0 | 28.2 | 35.2 | 83 | | 48 | 58 | 1 | | 4 | 20 | 14 | | 16 | 33 |
| Botswana | 8.4 | 15.3 | 19.2 | | | | 93 | | | | 28 | | | | 40[a] |
| Burkina Faso | 5.7 | 7.0 | 7.9 | 49 | 47 | 38 | 44 | | | 5 | 6 | 4 | 4 | 7 | 9 |
| Burundi | 2.2 | 4.1 | 5.6 | 96 | 40 | 84 | | | 35 | 56 | | | 35 | 58 | |
| Cameroon | 20.3 | 34.7 | 42.4 | | | | 100 | | | | 1 | | | | 43 |
| Cape Verde | 5.6 | 5.1 | 5.3 | | | 34 | 32 | | | 10 | 9 | | | 11 | 10 |
| Central African Republic | 30.4 | 38.2 | 42.4 | 64 | 100 | | | 96 | 100 | | | 72 | 100 | | |
| Chad | 11.4 | 20.8 | 27.0 | 7 | 9 | | | | 1 | | | 1 | 1 | | |
| Comoros | 11.3 | 23.2 | 25.2 | | | | | | | | | | | | |
| Congo | 34.8 | 37.3 | 39.5 | 8 | 10 | | | 6 | 9 | | | 6 | 9 | | |
| Côte d'Ivoire | 27.4 | 37.1 | 42.0 | 23 | | | | 5 | | | | 5 | | | |
| Djibouti | 62.0 | 73.7 | 77.7 | | | 43 | 78 | | | 20 | 17 | | | 39 | 64 |
| Egypt | 43.5 | 44.7 | 46.4 | | | | | | | 10 | | | | | |
| Equatorial Guinea | 39.0 | 53.7 | 59.7 | | | | | | | | | | | | |
| Ethiopia | 8.6 | 10.5 | 11.6 | 67 | 56 | | 96 | 8 | 8 | | 96[a] | 14 | 14 | | |
| Gabon | 25.6 | 35.8 | 40.9 | | | | | | | | | | | | |
| Gambia | 15.0 | 18.1 | 20.1 | | | | | | | | | | | | |
| Ghana | 29.0 | 30.7 | 31.5 | 92 | 95 | 47 | 51 | 40 | 40 | 17 | 16 | 55 | 56 | 26 | 30 |
| Guinea | 13.8 | 19.1 | 22.2 | 70 | | 54 | | 2 | | 1 | | 13 | | 11 | |
| Guinea-Bissau | 18.1 | 23.8 | 27.1 | | | 21 | 29 | | | 13 | 18 | | | 15 | 21 |
| Kenya | 10.3 | 16.1 | 19.7 | 85 | 98 | 89 | | 45 | 48 | 19 | | 50 | 55 | 30 | |

*continued*

| Region | Urban population as percentage of total population | | | Percentage of population with access to sanitation services | | | | | | | | | | | |
| | | | | Urban | | | | Rural | | | | Total | | | |
| | 1970 | 1980 | 1985 | 1970 | 1975 | 1980 | 1985 | 1970 | 1975 | 1980 | 1985 | 1970 | 1975 | 1980 | 1985 |
|---|---|---|---|---|---|---|---|---|---|---|---|---|---|---|---|
| Lesotho | 8.6 | 13.6 | 16.7 | 44 | 51 | 13 | 22 | 10 | 12 | 14 | 14 | 11 | 13 | 14 | 15 |
| Liberia | 26.0 | 34.9 | 39.5 | 100 | | | 6[b] | 9 | | | 2 | 19 | | | |
| Libya | 35.8 | 56.6 | 64.5 | 100 | 100 | 100 | | 54 | 69 | 72 | | 67 | 79 | 88 | |
| Madagascar | 14.1 | 18.9 | 21.8 | 88 | | 9 | 55[b] | | 9 | | | | | | |
| Malawi | 6.0 | 9.7 | 12.0 | | | 100 | | | | 81 | | | | 83 | |
| Mali | 14.3 | 17.3 | 18.0 | 63 | | 79 | 90 | | | | 3 | 8 | | | 19 |
| Mauritania | 13.9 | 26.9 | 34.6 | 100 | | 5 | 8 | | | | | 7 | | | |
| Mauritius | 42.0 | 42.9 | 42.2 | 51 | 63 | 100 | 100 | 99 | 100 | 90 | 86 | 77 | 82 | 94 | 92 |
| Morocco | 34.6 | 41.3 | 44.8 | 75 | | 100 | 62[a,b] | 4 | | | 16 | 29 | | | |
| Mozambique | 5.7 | 13.1 | 19.4 | | | | 53 | | | | 12 | | | | 20 |
| Namibia | | | | | | | | | | | | | | | |
| Niger | 8.5 | 13.2 | 16.2 | 10 | 30 | 36 | | | 1 | 3 | 5 | 1 | 3 | 7 | |
| Nigeria | 16.4 | 20.4 | 23.0 | | | | | | | | | | | | |
| Reunion | | | | | | | | | | | | | | | |
| Rwanda | 3.2 | 5.0 | 6.2 | 83 | 87 | 60 | 77 | 52 | 56 | 50 | 55 | 53 | 57 | 51 | 56 |
| São Tome and Principe | | | | | | | | | | | 15 | | | | 15 |
| Senegal | 33.4 | 34.9 | 36.4 | | 87 | 100 | 87 | | | 2 | | | | 36 | |
| Seychelles | | | | | | | | | | | | | | | |
| Sierra Leone | 18.1 | 24.5 | 28.3 | | | 31 | 60 | | | 6 | 10 | | | 12 | 24 |
| Somalia | 23.1 | 30.2 | 34.1 | | 77 | | 44 | | 35 | | 5 | | 47 | | 18 |
| South Africa | 47.9 | 53.2 | 55.9 | | | | | | | | | | | | |
| Sudan | 16.4 | 19.7 | 20.6 | 100 | 100 | 73[a] | 73 | 4 | 10 | | | 16 | 22 | | |
| Swaziland | 9.7 | 19.8 | 26.3 | | 99 | | 100 | | 25 | | 25 | | 36 | | 45 |
| Tanzania | 6.9 | 16.5 | 22.3 | | 88 | 100 | 93 | | 14 | | 58 | | 17 | 66 | 55 |
| Togo | 13.1 | 18.8 | 22.1 | 4 | 36 | 24 | 31 | | 12 | 10 | 9 | 1 | 15 | 13 | 14 |
| Tunisia | 43.5 | 52.3 | 56.8 | 100 | | 100 | 84 | 34 | | | 16 | 62 | | | 55 |
| Uganda | 7.8 | 8.7 | 9.5 | 84 | 82 | | 32 | 76 | 95 | | 30 | 76 | 94 | | 30 |

*continued*

Percentage of population with access to sanitation services

| Region | Urban population as percentage of total population | | | Urban | | | | Rural | | | | Total | | | |
|---|---|---|---|---|---|---|---|---|---|---|---|---|---|---|---|
| | 1970 | 1980 | 1985 | 1970 | 1975 | 1980 | 1985 | 1970 | 1975 | 1980 | 1985 | 1970 | 1975 | 1980 | 1985 |
| Zaire | 30.3 | 34.2 | 36.6 | 5 | 65 | | | 5 | 6 | | 9 | 5 | 22 | | |
| Zambia | 30.4 | 42.8 | 49.5 | 12 | 87 | | 76 | 18 | 16 | | 34 | 16 | 42 | | 55 |
| Zimbabwe | 16.9 | 21.9 | 24.6 | | | | | | | | 15 | | | | |
| *North and Central America* | | | | | | | | | | | | | | | |
| Bahamas | | | | 100 | 100 | 88 | 100 | 13 | 13 | | | 66 | 65 | 88 | 100 |
| Barbados | 37.1 | 40.1 | 42.2 | 100 | 100 | | 100 | 100 | 100 | | | 100 | 100 | 88 | |
| Belize | | | | | | 62 | 87 | | | 75 | 45 | | | 69 | 66 |
| Canada | 75.7 | 75.7 | 75.9 | | | | | | | | | | | | |
| Cayman Islands | | | | | | 94[a] | 96[a] | | | | 94 | 96 | | | |
| Costa Rica | 39.7 | 46.0 | 49.8 | 66 | 94 | 99 | 99 | 43 | 93 | 84 | 89 | 52 | 93 | 91 | 95 |
| Cuba | 60.2 | 68.1 | 71.8 | 57 | 100 | | | | | | | | | | |
| Dominican Republic | 40.3 | 50.5 | 55.7 | 63 | 74 | 25 | 41 | 54 | 16 | 4 | 10 | 58 | 42 | 15 | 23 |
| El Salvador | 39.4 | 39.3 | 39.1 | 66 | 71 | 48 | 82 | 18 | 17 | 26 | 43 | 37 | 39 | 35 | 58 |
| Guadeloupe | | | | | | | | | | | | | | | |
| Guatemala | 35.7 | 38.5 | 40.0 | | | 45 | 41 | 11 | 16 | 20 | 12 | | | 30 | 24 |
| Haiti | 19.8 | 24.6 | 27.2 | | | 42 | 42 | 43 | 1 | 10 | 13 | | | 19 | 21 |
| Honduras | 28.9 | 36.1 | 40.0 | 64 | 53 | 49 | 24 | 9 | 13 | 26 | 34 | 24 | 26 | 35 | 30 |
| Jamaica | 41.6 | 49.8 | 53.8 | 100 | 100 | | 92 | 92 | 91 | | 90 | 94 | 94 | | 91 |
| Martinique | | | | | | | | | | | | | | | |
| Mexico | 59.0 | 66.4 | 69.6 | | | 77 | 77 | 13 | 14 | 12 | 13 | | | 55 | 58 |
| Nicaragua | 47.0 | 53.4 | 56.6 | | | 34 | 35 | 8 | 24 | | 16 | | | | 27 |
| Panama | 47.6 | 50.5 | 52.4 | 87 | 78 | 83 | 99 | 69 | 76 | 59 | 61 | 78 | 77 | 71 | 81 |
| Puerto Rico | | | | | | | | | | | | | | | |
| St. Lucia | | | | | | | | | | | | | | | |
| St. Vincent | | | | | | | | | | | | | | | |
| Trinidad and Tobago | 38.8 | 56.9 | 63.9 | | | | | | | | | | | | |

*continued*

Percentage of population with access to sanitation services

| Region | Urban population as percentage of total population | | | Urban | | | | Rural | | | | Total | | | |
|---|---|---|---|---|---|---|---|---|---|---|---|---|---|---|---|
| | 1970 | 1980 | 1985 | 1970 | 1975 | 1980 | 1985 | 1970 | 1975 | 1980 | 1985 | 1970 | 1975 | 1980 | 1985 |
| Turks and Caicos Islands | | | | 51 | 83 | 96 | 100 | 96 | 97 | 88 | 95 | 81 | 92 | 93 | 98 |
| United States | 73.6 | 73.7 | 73.9 | | | | | | | | | | | | |
| *South America* | | | | | | | | | | | | | | | |
| Argentina | 78.4 | 82.7 | 84.6 | 87 | 100 | 80 | 75 | 79 | 83 | 35 | 35 | 85 | 97 | | 69 |
| Bolivia | 40.8 | 44.3 | 47.8 | 25 | | 37 | 33 | 4 | 9 | 4 | 10 | 12 | | 18 | 21 |
| Brazil | 55.8 | 67.5 | 72.7 | 85 | | | 86 | 24 | | 1 | 1 | 58 | | | 63 |
| Chile | 75.2 | 81.1 | 83.6 | 33 | 36 | 100 | 100 | 10 | 11 | 10 | 4 | 29 | 32 | 83 | 84 |
| Colombia | 57.2 | 64.2 | 67.4 | 75 | 73 | 93 | 96 | 8 | 13 | 4 | 13 | 47 | 48 | 61 | |
| Ecuador | 39.5 | 47.3 | 52.3 | | | 73 | 98 | | 7 | 17 | 29 | | | 43 | 65 |
| Guyana | 29.4 | 30.5 | 32.2 | 95 | 99 | 73 | 100 | 92 | 94 | 80 | 79 | 93 | 96 | 78 | 86 |
| Paraguay | 37.1 | 41.7 | 44.4 | 16 | 28 | 95 | 89 | | | 80 | 83 | 6 | 10 | 86 | 85 |
| Peru | 57.4 | 64.5 | 67.4 | 52 | | 57 | 67 | 16 | | 0[a] | 12 | 36 | | 36 | 49 |
| Suriname | 45.9 | 44.8 | 45.7 | | | 100 | 78 | | | 79 | 48 | | | 88 | 62 |
| Uruguay | 82.1 | 83.8 | 84.6 | 97 | 97 | 59 | 59 | 13 | 17 | 6 | 59 | 82 | 83 | 51 | 59 |
| Venezuela | 76.2 | 83.7 | 86.6 | | | 60 | 57 | 45 | | 12 | 5 | | | 52 | 50 |
| *Asia* | | | | | | | | | | | | | | | |
| Afghanistan | 11.0 | 15.6 | 18.5 | 69 | 63 | | 5 | 16 | 15 | | | 21 | 21 | | |
| Bahrain | 78.2 | 80.5 | 81.7 | | | | 100 | | | | 100 | | | | 100 |
| Bangladesh | 7.6 | 10.4 | 11.9 | 87 | 40 | 21 | 24 | | | 1 | 3 | 6 | 5 | 3 | 5 |
| Bhutan | 3.1 | 3.9 | 4.5 | | | | | | | | | | | | |
| Brunei Darus | | | | | | | | | | | | | | | |
| Cambodia | 11.7 | 10.3 | 10.8 | | | 100 | | | | 76 | | | | | |
| China | 20.1 | 20.4 | 20.6 | | | | | | | | | | | | |
| Cyprus | 40.8 | 46.3 | 49.5 | 100 | 94 | 100 | 100 | 92 | 95 | | 100 | 95 | 95 | | 100 |
| Hong Kong | | | | | | | | | | | | | | | |

*continued*

Percentage of population with access to sanitation services

| Region | Urban population as percentage of total population | | | Urban | | | | Rural | | | | Total | | | |
|---|---|---|---|---|---|---|---|---|---|---|---|---|---|---|---|
| | 1970 | 1980 | 1985 | 1970 | 1975 | 1980 | 1985 | 1970 | 1975 | 1980 | 1985 | 1970 | 1975 | 1980 | 1985 |
| India | 19.8 | 23.4 | 25.5 | 85 | 87 | 27 | 31 | 1 | 2 | 1 | 2 | 18 | 20 | 7 | 9 |
| Indonesia | 17.1 | 22.2 | 25.3 | 50 | 60 | 29 | 33 | 4 | 5 | 21 | 38 | 12 | 15 | 23 | 37 |
| Iran | 41.0 | 49.1 | 51.9 | 100 | 100 | 96[a] | | 48 | 59 | 43[a] | | 70 | 78 | 69[a] | |
| Iraq | 56.2 | 66.4 | 70.6 | 82 | 75 | | 100 | | 1 | | 11 | 47 | 47 | | 74 |
| Israel | 84.2 | 88.6 | 90.3 | | | | | | | | | | | | |
| Japan | 71.2 | 76.2 | 76.5 | | | | | | | | | | | | |
| Jordan | 50.6 | 60.1 | 64.4 | | | 94 | 92 | | | 34 | | | | 70 | |
| Korea DPR | 50.1 | 59.7 | 63.8 | 59 | | | | | | | | | | | |
| Korea Rep | 40.7 | 56.9 | 65.3 | | 80 | 100 | 100 | | 50 | 100 | 100 | 25 | 64 | 100 | 100 |
| Kuwait | 77.8 | 90.2 | 93.5 | | | 100[a] | 100 | | | 100[a] | | | | 100[a] | |
| Laos | 9.6 | 13.4 | 15.9 | | 10 | 13[a] | | | 2 | 4[a] | | | 3 | 5[a] | |
| Lebanon | 59.4 | 74.8 | 80.1 | | | | | | | | | | | | |
| Malaysia | 27.0 | 34.2 | 38.2 | 100 | 100 | 100 | 100 | 43 | 43 | 55 | 60 | 59 | 60 | 70 | 75 |
| Maldives | | | | | 21 | 60 | 100 | | | 1 | 2 | | 3 | 13 | 22 |
| Mongolia | 45.1 | 51.1 | 50.8 | | | | | | | | | | | | |
| Myanmar | 22.8 | 23.9 | 23.9 | 45 | 38 | 38 | 33 | 32 | 32 | 15 | 21 | 35 | 33 | 20 | 24 |
| Nepal | 3.9 | 6.1 | 7.7 | 14 | 14 | 16 | 17 | | 1 | 1 | 1 | 1 | 1 | 1 | |
| Oman | 5.1 | 7.3 | 8.8 | 100 | 100 | | 88 | | 5 | | 25 | 3 | 12 | | 31 |
| Pakistan | 24.9 | 28.1 | 29.8 | 12 | 21 | 42 | 51 | | | 2 | 6 | 3 | 6 | 13 | 19 |
| Philippines | 33.0 | 37.4 | 39.6 | 90 | 76 | 81 | 83 | 40 | 44 | 67 | 56 | 57 | 56 | 75 | 67 |
| Qatar | 80.3 | 86.0 | 88.0 | 100 | 100 | | | 16 | 100 | | | 83 | 100 | | |
| Saudi Arabia | 48.7 | 65.9 | 72.4 | 67 | 91 | 81 | 100 | 11 | 35 | 50 | 33 | 21 | 47 | 70 | 82 |
| Singapore | | | | | | 80 | 99 | | | | | | | 80[a] | 99[a] |
| Sri Lanka | 21.9 | 21.6 | 21.1 | 76 | 68 | 80 | 65 | 61 | 55 | 63 | 39 | 64 | 59 | 67 | 44 |
| Syria | 43.3 | 47.4 | 49.5 | | | 74 | | | | 28 | | | | 50 | |
| Thailand | 13.3 | 17.3 | 19.8 | 65 | 58 | 64 | 78 | 8 | 36 | 41 | 46 | 17 | 40 | 45 | 52 |
| Turkey | 38.4 | 43.8 | 45.9 | | | 56[a] | | | | | | | | | |

*continued*

## Percentage of population with access to sanitation services

| Region | Urban population as percentage of total population | | | Urban | | | | Rural | | | | Total | | | |
|---|---|---|---|---|---|---|---|---|---|---|---|---|---|---|---|
| | 1970 | 1980 | 1985 | 1970 | 1975 | 1980 | 1985 | 1970 | 1975 | 1980 | 1985 | 1970 | 1975 | 1980 | 1985 |
| United Arab Emirates | 42.3 | 81.2 | 77.8 | 100 | | 93[a] | | | | 22[a] | | | | 80[a] | |
| Vietnam | 18.3 | 19.3 | 20.3 | | | | | | 2 | 55 | | 26 | | | |
| Yemen Arab Rep | 7.5 | 15.3 | 20.0 | | | 60 | 83 | | | | | | | 35 | |
| Yemen Dem Rep | 32.1 | 36.9 | 39.9 | | | 70 | | | | 15 | | | | | |
| *Europe* | | | | | | | | | | | | | | | |
| Albania | 33.5 | 33.4 | 34.0 | | | | | | | | | | | | |
| Austria | 51.7 | 54.6 | 56.1 | | | | | | | | | | | | |
| Belgium | 94.3 | 95.4 | 96.3 | | | | 100[a] | | | | 80[a] | | | | 99[a] |
| Bulgaria | 52.3 | 62.5 | 66.5 | | | | | | | | | | | | |
| Czechoslovakia | 55.2 | 62.3 | 65.3 | | | | 100[a] | | | | 100[a] | | | | 100[a] |
| Denmark | 79.7 | 84.3 | 85.9 | | | | | | | | | | | | |
| Finland | 50.3 | 59.6 | 64.0 | | | | 99[a] | | | | 99[a] | | | | 99[a] |
| France | 71.0 | 73.2 | 73.4 | | | | 71[a] | | | | 71[a] | | | | 71[a] |
| German DR | 73.7 | 76.2 | 77.0 | | | | | | | | | | | | |
| Germany FR | 81.3 | 84.4 | 85.5 | | | | | | | | | | | | |
| Greece | 52.5 | 57.7 | 60.1 | | | | 100[a] | | | | 98[a] | | | | 99[a] |
| Hungary | 45.6 | 53.5 | 56.2 | | | | | | | | | | | | |
| Iceland | 84.9 | 88.2 | 89.4 | | | | | | | | | | | | |
| Ireland | 51.7 | 55.3 | 57.0 | | | | | | | | | | | | |
| Italy | 64.3 | 66.5 | 67.4 | | | | | | | | | | | | |
| Luxembourg | 67.8 | 77.6 | 81.0 | | | | | | | | | | | | |
| Malta | 77.5 | 83.1 | 85.3 | | | | | | | | | | | | |
| Monaco | | | | | | | | | | | | | | | |
| Netherlands | 86.1 | 88.4 | 88.4 | | | | 100[a] | | | | 100[a] | | | | 100[a] |
| Norway | 64.8 | 70.5 | 72.8 | | | | | | | | | | | | |
| Poland | 52.3 | 58.2 | 61.0 | | | | 100[a] | | | | 42[a] | | | | 77[a] |

*continued*

Percentage of population with access to sanitation services

| Region | Urban population as percentage of total population | | | Urban | | | | Rural | | | | Total | | | |
|---|---|---|---|---|---|---|---|---|---|---|---|---|---|---|---|
| | 1970 | 1980 | 1985 | 1970 | 1975 | 1980 | 1985 | 1970 | 1975 | 1980 | 1985 | 1970 | 1975 | 1980 | 1985 |
| Portugal | 26.2 | 29.5 | 31.2 | | | | 100[a] | | | | 11[a] | | | | 39[a] |
| Romania | 41.8 | 48.1 | 49.0 | | | | | | | | | | | | |
| Spain | 66.0 | 72.8 | 75.8 | | | | 73[a] | | | | 69[a] | | | | 72[a] |
| Sweden | 81.1 | 83.1 | 83.4 | | | | | | | | | | | | |
| Switzerland | 54.5 | 57.0 | 58.2 | | | | 100[a] | | | | | | | | |
| United Kingdom | 88.5 | 90.7 | 91.5 | | | | 100[a] | | | | 100[a] | | | | 100[a] |
| Yugoslavia | 34.8 | 42.3 | 46.3 | | | | | | | | | | | | |
| USSR | 56.7 | 63.1 | 65.6 | | | | | | | | | | | | |
| Oceania | | | | | | | | | | | | | | | |
| Australia | 85.2 | 85.8 | 85.5 | | | | | | | | | | | | |
| Cook Islands | | | | | | 100 | 100 | | | 76 | 99 | | | | 99 |
| Fiji | 34.8 | 38.7 | 41.2 | 100 | 100 | 85 | | 87 | 93 | 60 | | 91 | 96 | 70 | |
| Kiribati | | | | | | | | | | | | | | | |
| New Zealand | 81.1 | 83.3 | 83.7 | | | | | | | | | | | | |
| Niue | | | | | | | | | | | 100[a] | | | | |
| Papua New Guinea | 9.8 | 13.0 | 14.3 | 100 | 100 | 96 | 99 | 5 | 5 | 3 | 35 | 14 | 18 | 15 | 44 |
| Samoa | | | | 100 | 100 | 86 | | 80 | 99 | 83 | | 84 | 99 | | |
| Solomon Islands | 8.9 | 9.2 | 9.7 | | 80 | | | 21 | | | | | | | |
| Tokelau | | | | | | | | | | 41[a] | | | | | |
| Tonga | | | | 100 | 100 | 97 | 99 | 100 | 100 | 94 | 40 | 100 | 100 | 19 | 52 |
| Tuvalu | | | | | | 100[a] | 81[a] | | | 80[a] | 73a | | | | |
| Vanuatu | | | | | | 95 | 86 | | | 68 | 25 | | | | 40 |
| Western Samoa | | | | | | 86[a] | 88[a] | | | 83[a] | 83a | | | | 84a |

[a]Missing UNEP (1989) data supplemented with numbers from World Resources (1988).
[b]Data refer to sewage only.

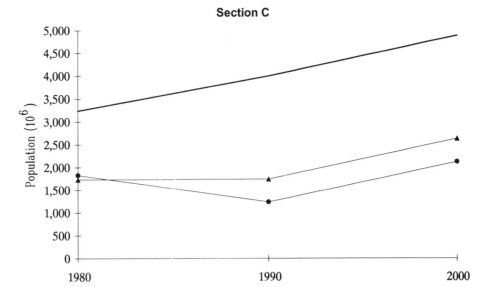

Fig. C–3. Population lacking (•) safe drinking water and (▲) sanitation services, and (——) total population, for developing countries, 1980–2000. In the year 2000, it is estimated that roughly half of all people living in developing parts of the world will lack access to safe water for drinking, and will be subject to water-related diseases stemming from inadequate sanitation. See Tables C.1–C.4 for more data.

**TABLE C.3**  Water supply and sanitation coverage for developing regions, 1980 and 1990 *(page 188)*

**DESCRIPTION**

Improving access to safe drinking water and sanitation services was the goal of the International Water Supply and Sanitation Decade, sponsored by the United Nations in the 1980s. Shown here is one of the first estimates of the accomplishments of that decade, comparing changes in access to fresh water supply and sanitation services for the urban and rural sectors in regions of developing countries, for the years 1980 and 1990. Data are given for total urban and rural populations by developing region, numbers of people served and unserved by water supply and sanitation services, and the percentage coverage for the beginning and the end of the decade. Four-fifths of the world's population is covered by this table; and nearly 100% of the population of developing countries.

**LIMITATIONS**

The data in this table were initially prepared in 1988 and updated in 1989 to provide estimates for 1990. Unlike Tables C.1 and C.2, this table includes China, which reports high rural sanitation coverage. The total percentage rural sanitation coverage in the rest of the world will be much lower than shown if China is excluded. Definitions of access to safe drinking water and sanitation services are somewhat ambiguous, as described in Tables C.1 and C.2. Definitions of urban and rural are provided by national governments. Population growth in many urban and rural areas during the 1980s increased the need for services at a rate faster than could be supplied by United Nations efforts.

**SOURCE**

J. Christmas and C. de Rooy, 1991, The decade and beyond: at a glance, *Water International*, **16**(3), 127–134, International Water Resources Association, Illinois. By permission.

**TABLE C.3**   Water supply and sanitation coverage for developing regions, 1980 and 1990

| Region/sector | 1980 | | | | 1990 | | | |
|---|---|---|---|---|---|---|---|---|
| | Population $(10^6)$ | Percent coverage | Number served $(10^6)$ | Number unserved $(10^6)$ | Population $(10^6)$ | Percent coverage | Number served $(10^6)$ | Number unserved $(10^6)$ |
| *Africa* | | | | | | | | |
| Urban water | 119.77 | 83 | 99.41 | 20.36 | 202.54 | 87 | 176.21 | 26.33 |
| Rural water | 332.83 | 33 | 109.83 | 223.00 | 409.64 | 42 | 172.06 | 237.59 |
| Urban sanitation | 119.77 | 65 | 77.85 | 41.92 | 202.54 | 78 | 160.01 | 42.53 |
| Rural sanitation | 332.83 | 18 | 59.91 | 272.92 | 409.64 | 26 | 106.51 | 303.13 |
| *Latin America and the Caribbean* | | | | | | | | |
| Urban water | 236.72 | 82 | 194.11 | 42.61 | 324.08 | 87 | 281.95 | 42.13 |
| Rural water | 124.91 | 47 | 58.71 | 66.20 | 123.87 | 62 | 76.80 | 47.07 |
| Urban sanitation | 236.72 | 78 | 184.64 | 52.08 | 324.08 | 79 | 256.02 | 68.06 |
| Rural sanitation | 124.91 | 22 | 27.48 | 97.43 | 123.87 | 37 | 45.83 | 78.04 |
| *Asia and the Pacific* | | | | | | | | |
| Urban water | 549.44 | 73 | 401.09 | 148.35 | 761.18 | 77 | 586.11 | 175.07 |
| Rural water | 1,823.30 | 28 | 510.52 | 1,312.78 | 2,099.40 | 67 | 1,406.60 | 692.80 |
| Urban sanitation | 549.44 | 65 | 357.14 | 192.30 | 761.18 | 65 | 494.77 | 266.41 |
| Rural sanitation | 1,823.30 | 42 | 765.79 | 1,057.51 | 2,099.40 | 54 | 1,133.68 | 965.72 |
| *Western Asia (Middle East)* | | | | | | | | |
| Urban water | 27.54 | 95 | 26.16 | 1.38 | 44.42 | 100 | 44.25 | 0.17 |
| Rural water | 21.95 | 51 | 11.19 | 10.76 | 25.60 | 56 | 14.34 | 11.26 |
| Urban sanitation | 27.54 | 79 | 21.76 | 5.78 | 44.42 | 100 | 44.42 | 0.00 |
| Rural sanitation | 21.95 | 34 | 7.46 | 14.49 | 25.60 | 34 | 8.70 | 16.90 |
| *Totals for these regions* | | | | | | | | |
| Urban water | 933.47 | 77 | 720.77 | 212.70 | 1,332.22 | 82 | 1,088.52 | 243.70 |
| Rural water | 2,302.99 | 30 | 690.25 | 1,612.74 | 2,658.51 | 63 | 1,669.79 | 988.72 |
| Urban sanitation | 933.47 | 69 | 641.39 | 292.08 | 1,332.23 | 72 | 955.22 | 377.00 |
| Rural sanitation | 2,302.99 | 37 | 860.64 | 1,442.35 | 2,658.51 | 49 | 1,294.72 | 1,363.79 |

**TABLE C.4** Developing country needs for urban
and rural water supply and sanitation, 1990 and 2000

| | Population not served in 1990 ($10^6$) | Expected population increase 1990–2000 ($10^6$) | Total additional population requiring service by 2000 ($10^6$) |
|---|---|---|---|
| **Water Supply** | | | |
| Urban | 243 | 570 | 813 |
| Rural | 989 | 312 | 1,301 |
| Total | 1,232 | 882 | 2,114 |
| **Sanitation** | | | |
| Urban | 377 | 570 | 947 |
| Rural | 1,364 | 312 | 1,676 |
| Total | 1,741 | 882 | 2,623 |

**TABLE C.4** Developing country needs for urban
and rural water supply and sanitation, 1990 and 2000

**DESCRIPTION**

The United Nations made a substantial effort during the International Water Supply and Sanitation Decade of the 1980s to provide drinking water and sanitation services to populations lacking those services. Despite these efforts, the growth in developing country populations almost entirely wiped out the gains, and nearly as many people lack those services today as at the beginning of the 1980s. Shown here are the populations presently unserved and the expected additional growth between 1990 and 2000 in rural and urban populations that will have to be served by new drinking water and sanitation systems. In total, an additional 880 million more people will have to be supplied with these services, on top of the 1,741 million currently lacking them. This table covers four-fifths of the world's population and approximately 100% of the population of developing countries.

**LIMITATIONS**

These data present the drinking water and sanitation service needs in developing countries only and use United Nations population estimates for 2000. The level of service is typically defined by the World Meteorological Organization. As used here by the World Health Organization (WHO), safe drinking water includes treated surface water and untreated water from protected springs, boreholes, and wells. The WHO defines access to safe drinking water in urban areas as piped water to housing units or to public standpipes within 200 m. In rural areas, reasonable access implies that fetching water does not take up a disproportionate part of the day. Urban areas with access to sanitation services are defined as urban populations served by connections to public sewers or household systems such as pit privies, pour-flush latrines, septic tanks, and communal toilets. Rural populations with access to sanitation services are defined as those with adequate disposal such as pit privies, pour-flush latrines, septic tanks, and communal toilets.

**SOURCES**

B. Grover and D. Howarth, 1991, Evolving international collaborative arrangements for water supply and sanitation. *Water International*, **16**(3), 145–152, International Water Resources Association, Illinois. By permission.

Data from Report A/45/327 of the Secretary General of the Economic and Social Council to the United Nations General Assembly, July 1990.

**TABLE C.5**  Constraints in improving water supply and sanitation services in developing countries

| Constraints | Number of countries indicating constraint | | | Ranking Index[a] |
|---|---|---|---|---|
| | Very Severe | Severe | Moderate | |
| Insufficiency of trained personnel (professional) | 16 | 40 | 27 | 155 |
| Funding limitations | 21 | 31 | 30 | 155 |
| Insufficiency of trained personnel (subprofessional) | 16 | 38 | 29 | 153 |
| Operation and maintenance[b] | 16 | 36 | 23 | 143 |
| Logistics[b] | 11 | 35 | 23 | 126 |
| Inadequate cost-recovery framework | 11 | 34 | 22 | 123 |
| Inappropriate institutional framework | 6 | 30 | 35 | 113 |
| Insufficient health education efforts | 7 | 24 | 43 | 112 |
| Intermittent water service | 10 | 19 | 32 | 100 |
| Lack of planning and design criteria | 6 | 17 | 41 | 93 |
| Noninvolvement of communities | 6 | 15 | 44 | 92 |
| Inadequate or outmoded legal framework | 10 | 14 | 34 | 92 |
| Inappropriate technology | 5 | 18 | 33 | 84 |
| Insufficient knowledge of water resources | 1 | 20 | 39 | 82 |
| Inadequate water resources | 5 | 11 | 40 | 77 |
| Lack of definite government policy for sector | 4 | 10 | 44 | 76 |
| Import restrictions | 5 | 12 | 21 | 60 |

[a]Ranking index: (Number very severe × 3) + (Number severe × 2) + (Number moderate).
[b]"Logistics" is ranked ahead of "operations and maintenance" in the group of least developed countries.


**TABLE C.5**  Constraints in improving water supply and sanitation services in developing countries

## DESCRIPTION
Many factors interfere with the improvement of water supply and the provision of sanitation services, including lack of trained personnel, shortages of capital, and inadequate political or community infrastructure. This table identifies the many shortcomings that need to be overcome for 87 developing countries. Each problem is ranked as moderate, severe, or very severe in importance, and an overall ranking index is calculated for each constraint. According to these measures, insufficient trained professionals and insufficient funding are the greatest constraints. These data come from country questionnaires returned to the World Health Organization.

## LIMITATIONS
Improving water supply and sanitation services in every country or region is limited by very particular constraints, and the list provided here is simply an overview of all limiting factors. The ranking index is a simple measure of overall severity, and can hide differences among the countries. Nevertheless, it does give a sense of where additional effort should be devoted by the international community.

## SOURCE
World Health Organization, 1984, *The International Drinking Water Supply and Sanitation Decade: Review of National Baseline Data (as at 31 December 1980)*, WHO Offset Publication no. 85, World Health Organization, Geneva. By permission.

**TABLE C.6**  Fraction of population served by wastewater treatment plants, for selected countries, 1970 to 1987

| | Primary only | | | | | Primary and secondary or tertiary | | | | | Total served | | | | |
|---|---|---|---|---|---|---|---|---|---|---|---|---|---|---|---|
| | 1970 | 1975 | 1980 | 1985 | 1987 | 1970 | 1975 | 1980 | 1985 | 1987 | 1970 | 1975 | 1980 | 1985 | 1987 |
| Canada [k,c] | | | 13.0 | 10.0 | 14.7 | | | 43.0 | 47.0 | 51.5 | | | 56.0 | 57.0 | 62.2 |
| United States [c,d,l] | | 23.0 | 17.0 | 15.0 | | | 44.0 | 53.0 | 59.0 | | | 67.0 | 70.0 | 74.0 | |
| Japan [d,m] | | | | | | 16.0 | 23.0 | 30.0 | 36.0 | 39.0 | 16.0 | 23.0 | 30.0 | 36.0 | 39.0 |
| New Zealand [n] | | 9.0 | 10.0 | 8.0 | | | 47.0 | 49.0 | 80.0 | | 52.0 | 56.0 | 59.0 | 88.0 | |
| Austria | 12.0 | | 10.0 | 7.0 | 5.0 | 5.0 | | 28.0 | 58.0 | 62.0 | 16.0 | | 38.0 | 65.0 | 67.0 |
| Belgium [b] | 0.0 | 0.0 | 0.0 | | | 3.8 | 5.5 | 22.9 | | | 3.8 | 5.5 | 22.9 | | |
| Denmark [e,a] | 31.9 | 29.0 | | 20.0 | 8.0 | 22.4 | 41.6 | 63.0 | 70.0 | 90.0 | 54.3 | 70.6 | | 90.0 | 98.0 |
| Finland | | 3.0 | 2.0 | 0.0 | 0.0 | 22.0 | 47.0 | 63.0 | 72.0 | 74.0 | 27.0 | 50.0 | 65.0 | 72.0 | 74.0 |
| France [d] | | | | | | | | | | | 19.0 | | 43.6 | 49.7 | |
| Germany [g] | 20.5 | 18.4 | 10.2 | 7.5 | | 41.3 | 56.4 | 71.6 | 79.0 | | 61.8 | 74.8 | 81.8 | 86.5 | |
| Greece | | | 0.0 | | | | | 0.5 | | | | | 0.5 | | |
| Ireland | | | 0.2 | | | | | 11.0 | | | | | 11.2 | | |
| Italy [h] | 8.0 | | | | | 6.0 | | | | | 14.0 | | 30.0 | | |
| Luxembourg | 23.0 | | 16.0 | 14.0 | | 5.0 | | 65.0 | 69.0 | 83.0 | 28.0 | | 81.0 | 83.0 | 90.0 |
| Netherlands | | 8.0 | 7.0 | 7.0 | 7.0 | | 37.0 | 61.0 | 78.0 | 83.0 | | 45.0 | 68.0 | 85.0 | 90.0 |
| Norway [a,f] | 1.0 | 2.0 | 3.0 | 6.0 | 6.0 | 21.0 | 25.0 | 31.0 | 36.0 | 37.0 | 21.0 | 27.0 | 34.0 | 42.0 | 43.0 |
| Portugal [a,o] | 1.0 | 2.0 | 3.0 | 3.5 | | 2.1 | 4.0 | 7.0 | 9.0 | | 3.1 | 6.0 | 10.0 | 12.5 | |
| Spain | | 7.0 | 8.8 | 13.2 | | | 7.3 | 9.1 | 15.8 | | | 14.3 | 17.9 | 29.0 | |
| Sweden [i] | 23.0 | 4.0 | 1.0 | 1.0 | 1.0 | 55.0 | 94.0 | 98.0 | 99.0 | 99.0 | 78.0 | 98.0 | 99.0 | 100.0 | 100.0 |
| Switzerland | 0.0 | 0.0 | 0.0 | 0.0 | 0.0 | 35.0 | 55.0 | 70.0 | 81.0 | 85.0 | 35.0 | 55.0 | 70.0 | 81.0 | 85.0 |
| Turkey | | | 1.6 | 1.6 | | | | 0.4 | 1.8 | | | | 2.0 | 3.3 | |
| United Kingdom [j] | | | 6.0 | 6.0 | 6.0 | | | 76.0 | 77.0 | 78.0 | | | 82.0 | 83.0 | 84.0 |
| North America | | | | | | | | | | | 42.0 | 65.0 | 69.0 | 72.0 | 77.0 |
| OECD Europe [o] | | | | | | | | | | | 29.0 | 39.0 | 46.0 | 51.0 | 55.0 |
| European Economic Community (EEC) [o] | | | | | | | | | | | 32.0 | 42.0 | 50.0 | 56.0 | 61.0 |
| OECD Total [o] | | | | | | | | | | | 32.0 | 45.0 | 51.0 | 56.0 | 60.0 |

continued

a 1985 data refer to 1983.
b 1980 data refer to 1979.
c 1975 data refer to 1976.
d 1985 data refer to 1984.
e 1975 data refer to 1977.
f 1987 data refer to 1988.
g 1970, 1980, 1985 data refer to 1969, 1979, 1983.
h 1970 data refer to 1971.
i Urban population only (85% of total). Primary: sedimentation; secondary: chemical or biological treatments; tertiary: chemical, biological, and complementary treatments.
j England and Wales only.
k Secondary: usually includes private treatment and oxidation ponds; tertiary: refers to secondary treatment with phosphorous precipitation.
l 1980 and 1985 data for the second category include 1% and 2% of non-discharge treatment. 1980 and 1984 data were determined by using different methods than previous data, and therefore may not be comparable.
Primary: may include some biological treatment; Secondary: preliminary and biological treatments together.
m Data for the second category may include data for primary treatment only.
n 1970 data are Secretariat estimates.
o Secretariat estimates.

**TABLE C.6** Fraction of population served by wastewater treatment plants, for selected countries, 1970 to 1987 (*pages 191–192*)

## DESCRIPTION

The percentages of the population of selected industrialized countries served by primary, secondary, or tertiary wastewater treatment plants are given here for 1970, 1975, 1980, 1985, and 1987. These data fill in several gaps in Table C.2, but are not strictly comparable because of different standards of measure, different methods of data collection, and different estimate dates.

## LIMITATIONS

The difficulty of finding consistent methods of measurement and wastewater treatment definitions is clearly shown in this table. For the 22 countries listed here, 15 different footnotes are required to clarify the actual date of measurement, the differences in the definitions used, and the regions included. The levels of treatment described here tend to be higher than the minimum levels used by the World Health Organization for

developing countries in Table C.2. There are still many gaps in the data.

## SOURCE

Organization for Economic Co-operation and Development, 1989, *OECD Environmental Data – Compendium 1989*, Organization for Economic Co-operation and Development, Paris. By permission.

**TABLE C.7** United States population served by wastewater treatment systems, by level of treatment, 1960 to 1988 (million people)

| Level of Treatment | 1960 | 1978 | 1982 | 1984 | 1986 | 1988 |
|---|---|---|---|---|---|---|
| Not served | 70.0 | 66.0 | 62.0 | 65.7 | 67.8 | 69.9 |
| No discharge | n.d. | n.d. | n.d. | 5.5 | 5.7 | 6.1 |
| Raw sewage discharge | n.d. | n.d. | n.d. | 1.3 | 1.6 | 1.5 |
| Less than secondary treatment | 36.0 | n.d. | 37.0 | 33.7 | 28.8 | 26.5 |
| Secondary treatment | n.d. | 56.0 | 63.0 | 70.7 | 72.3 | 78.0 |
| Greater than secondary treatment | 4.0 | 49.0 | 53.0 | 59.5 | 54.9 | 65.7 |

Data for 1960 are not strictly comparable to other data because of different methods of data collection.
n.d. is no data.

**TABLE C.7** United States population served by wastewater treatment systems, by level of treatment, 1960 to 1988

**DESCRIPTION**

The populations in the United States served by different levels of wastewater treatment are shown here for 1960, 1978, 1982, 1984, 1986, and 1988. The levels of treatment include no treatment, no discharge, raw sewage discharge, less than secondary treatment, secondary treatment, and greater than secondary treatment. The population "not served" represent those with onsite septic systems; "no discharge" means other wastewater is not discharged to public wastewater treatment facilities. Populations are given in millions of people. Note that the population not served by wastewater treatment facilities has risen since the late 1970s and is approximately equal to the population not served in 1960. The total fraction of the population receiving treatment, however, has risen substantially over that period, as has the average level of treatment.

**LIMITATIONS**

The data for 1960 were collected in a different way than the later data and may not be strictly comparable. There are no data for some of the categories of treatment in the earlier years.

**SOURCE**

Council on Environmental Quality, 1991, *Environmental Quality 1990: The 21st Annual Report of the Council on Environmental Quality*, Washington, DC.

**TABLE C.8** Estimated unit costs for various wastewater treatment processes

| Description of treatment process | Plant capacity of 3,800 m³/day (1983 $/m³) | Plant capacity of 38,000 m³/day (1983 $/m³) |
|---|---|---|
| Aerated lagoons plus spreading basins (high-rate infiltration system) | 0.19 | 0.07 |
| Aerated lagoons plus irrigation (low-rate infiltration system) | 0.30 | 0.23 |
| Biological treatment by extended aeration (oxidation process) | 0.16 | 0.11 |
| Secondary treatment[a] | 0.39 | 0.21 |
| Secondary treatment plus nitrification | 0.51 | 0.30 |
| Secondary treatment plus nitrogen reduction | 0.64 | 0.31 |
| Secondary treatment plus phosphorus reduction | 0.66 | 0.29 |
| Secondary treatment plus nitrogen and phosphorus reduction | 0.83 | 0.36 |
| Secondary treatment plus filtration and carbon adsorption | 0.66 | 0.27 |
| Advanced wastewater treatment (AWT)[b] | 1.08 | 0.48 |
| AWT plus reverse osmosis | 1.80 | 0.80 |
| Physical-chemical[c] | 0.95 | 0.41 |

[a]Secondary treatment is biological oxidation by activated sludge, trickling filters, or rotating biological contactors with the appropriate pretreatment and clarifiers. This cost is an average of the three.
[b]AWT is secondary treatment plus nitrogen and phosphorus reduction, filtration, and carbon adsorption.
[c]Physical-chemical treatment consists of precipitation with lime, stripping towers (ammonia removal), filtration, carbon adsorption, and disinfection or precipitation with ferric chloride, filtration, carbon adsorption, selective ion exchange, and disinfection.

**TABLE C.8** Estimated unit costs for various wastewater treatment process

**DESCRIPTION**

The estimated costs of various wastewater treatment processes are given here in 1983 $U.S. per m³ for waste treatment plants operating at 3,800 and 38,000 m³ per day capacity. Costs include amortization of capital costs, equipment, operation, and maintenance.

**LIMITATIONS**

Some treatment processes have been combined for clarity, and the original 1977 costs were converted to 1983 prices by the United Nations assuming an annual average increase of 6%. Costs include operation and maintenance and amortizing the capital cost at 7% per year for 20 years. Labor, energy, chemicals, and equipment costs were computed assuming conditions in the United States.

**SOURCE**

United Nations, 1985, *The Use of Non-conventional Water Resources in Developing Countries*, Natural Resources/Water Series no. 14, Department of Technical Co-operation for Development, New York.

**TABLE C.9**  Average annual onsite wastewater collection and treatment cost, per household (1978 $/household/yr)

| | Total | Onsite | Collection | Treatment |
|---|---|---|---|---|
| Low Cost, Below $100/yr | | | | |
| Pour flush toilet | 19 | 19 | | |
| Pit privy | 28 | 28 | | |
| Communal toilet | 34 | 34 | | |
| Vacuum truck cartage | 38 | 17 | 14 | 7 |
| Composting toilets | 55 | 47 | | 8 |
| Medium Cost, $100 - 200/yr | | | | |
| Sewered aquaprivy | 160 | 90 | 40 | 30 |
| Aquaprivy with drain field | 170 | 170 | | |
| Japanese vacuum truck | 190 | 130 | 30 | 30 |
| High cost, Above $300/yr | | | | |
| Septic tanks | 370 | 330 | 30 | 10 |
| Sewerage | 400 | 200 | 80 | 120 |

**TABLE C.9**  Average annual onsite wastewater collection and treatment cost, per household

**DESCRIPTION**

A wide variety of wastewater collection and treatment alternatives are available, with widely varying costs. Shown here is a comparison of the total annual costs per houshold in 1978 $U.S. for low-, medium-, and high-cost alternative systems. Listed in the table are onsite costs, collection costs, and treatment costs, including onsite and offsite capital and recurrent costs. Where the costs of water are high, choosing systems with low requirements for flushing will have economic benefits. The options in this table can provide approximately the same health benefits at widely varying costs.

**LIMITATIONS**

These data are average figures for developing countries in the late 1970s and should be updated. Total costs will vary considerably depending on the site and the cost of labor and equipment. No ranges were given in the original source. Different systems may have comparable health benefits, but very different social and economic status.

**SOURCE**

F.L. Golladay, 1982, Cost-effective water supply and sanitation projects for developing countries, in E.J. Schiller and R.L. Droste (eds.) *Water Supply and Sanitation in Developing Countries*, Ann Arbor Science, Ann Arbor, Michigan, pp. 221–249.

**TABLE C.10**   Annual cost per household of alternative water-supply systems (1978 $/household/yr)

|  | Amortization of investment | Operation and maintenance | Total cost |
|---|---|---|---|
| Standpipes (one per 400 persons) | 8.64 | 1.86 | 10.50 |
| Yard hydrants for all | 20.88 | 3.78 | 24.66 |
| Kitchen tap and shower | 32.40 | 6.60 | 39.00 |
| Multiple housetaps | 53.28 | 11.70 | 64.98 |

**TABLE C.10**   Annual cost per household of alternative water-supply systems

**DESCRIPTION**

The annual costs of four methods of providing water for households are given here, in 1978 $U.S. normalized per household. Costs include amortization of initial capital investment, and operation and maintenance. These estimates are based on a study of a community of 6,000 people expected to grow to a population of 10,800 over 20 years. Family size was assumed to be six people, and a discount rate of 10% per year was used.

**LIMITATIONS**

These data are average figures for developing countries in the late 1970s and should be updated. Total costs will vary considerably depending on the site and the cost of labor and equipment. No ranges were given in the original source. Different systems may have comparable health benefits, but very different social and economic status.

**SOURCE**

F.L. Golladay, 1982, Cost-effective water supply and sanitation projects for developing countries, in E.J. Schiller and R.L. Droste (eds.) *Water Supply and Sanitation in Developing Countries*, Ann Arbor Science, Ann Arbor, Michigan, pp. 221–249.

**TABLE C.11**   Environmental classification of water-related infections *(page 197)*

**DESCRIPTION**

Many diseases come from organisms that reproduce or spend part of their life-cycle in fresh water – so called "waterborne diseases." This table categorizes the diseases by type and means of infection.

**LIMITATIONS**

This table is limited to diseases caused by organisms, and excludes diseases caused by water quality problems, such as chemical contamination. It also omits those diseases spread through the consumption of unhealthy fish, shellfish, and other organisms that are associated with fresh water and may carry and pass on disease.

**SOURCES**

P.G. Bourne (ed.), 1984, *Water and Sanitation*, Academic Press, Orlando.

**TABLE C.11**  Environmental classifications of water-related infections

| Category | Infection | Pathogenic agent |
|---|---|---|
| Fecal-oral (waterborne or water-washed) | Diarrheas and dysenteries | |
| | Amebic dysentery | P |
| | Balantidiasis | P |
| | *Campylobacter* enteritis | B |
| | Cholera | B |
| | *Escherichia coli* diarrhea | B |
| | Giardiasis | P |
| | Rotavirus diarrhea | V |
| | Salmonellosis | B |
| | Shigellosis (bacillary dysentery) | B |
| | Yersiniosis | B |
| | Enteric fevers | |
| | Typhoid | B |
| | Paratyphoid | B |
| | Poliomyelitis | V |
| | Leptospirosis | S |
| | Ascariasis | H |
| | Trichuriasis | H |
| Water-washed | | |
| Skin and eye infections | Infectious skin diseases | M |
| | Infectious eye diseases | M |
| Other | Louse-borne typhus | R |
| | Louse-borne relapsing fever | S |
| Water-based | | |
| Penetrating skin | Schistosomiasis | H |
| Ingested | Guinea worm | H |
| | Clonorchiasis | H |
| | Diphyllobothriasis | H |
| | Fasciolopsiasis | H |
| | Paragonimiasis | H |
| | Others | H |
| Water-related insect vector | | |
| Biting near water | Sleeping sickness | P |
| Breeding in water | Filariasis | H |
| | Malaria | P |
| | River blindness | H |
| | Mosquito-borne viruses | |
| | Yellow fever | V |
| | Dengue | V |
| | Others | V |

B, bacterium; P, protozoan; S, spirochete; M, miscellaneous; H, helminth; R, rickettsia; V, virus.

**TABLE C.12**   Environmental classification of excreted infections

| Category and epidemiological features[a] | Infection | Environmental transmission focus | Major control measure |
|---|---|---|---|
| I.   Nonlatent; low infective dose | Amebiasis<br>Balantidiasis<br>Enterobiasis<br>Enteroviral infections[b]<br>Giardiasis<br>Hymenolepiasis<br>Infectious hepatitus<br>Rotavirus infection | Personal<br>Domestic | Domestic water supply<br>Health education<br>Improved housing<br>Provision of toilets |
| II.   Nonlatent; medium or high infective dose; moderately persistent; able to multiply | *Campylobacter* infection<br>Cholera<br>Pathogenic *Escherichia coli* infection[c]<br>Salmonellosis<br>Shigellosis<br>Typhoid<br>Yersiniosis | Personal<br>Domestic<br>Water<br>Crop | Domestic water supply<br>Health education<br>Improved housing<br>Provision of toilets<br>Treatment of excreta prior to discharge or reuse |
| III.   Latent and persistent; no intermediate host | Ascariasis<br>Hookworm infection[d]<br>Strongyloidiasis<br>Trichuriasis | Yard<br>Field<br>Crop | Provisions of toilets<br>Treatment of excreta prior to land application |
| IV.   Latent and persistent; cow or pig as intermediate host | Taeniasis | Yard<br>Field<br>Fodder | Provision of toilets<br>Treatment of excreta prior to land application<br>Cooking, meat inspection |
| V.   Latent and persistent; aquatic intermediate host(s) | Clonorchiasis<br>Diphyllobothriasis<br>Fascioliasis<br>Fasciolopsiasis<br>Gastrodiscoidiasis<br>Heterophyiasis<br>Metagonimiasis<br>Opisthorchiasis<br>Paragonimiasis<br>Schistosomiasis | Water | Provision of toilets<br>Treatment of excreta prior to discharge<br>Control of animal reservoirs<br>Control of intermediate hosts<br>Cooking of water plants and fish<br>Reducing water contact |
| VI.   Spread by excreta-related insects | Bancroftian filariasis (transmitted by *Culex pipiens*)<br>All the infections in I–IV able to be transmitted mechanically by flies and cockroaches | Various fecally contaminated sites in which insects breed | Identification and elimination of suitable insect breeding sites |

[a] See Table C.13 for data on additional epidemiological features by pathogen.
[b] Includes polio-, echo-, and coxsackievirus infections.
[c] Includes enterotoxigenic, enteroinvasive, and enteropathogenic *E. coli* infections.
[d] *Ancylostoma duodenale* and *Necator americanus.*

**TABLE C.12**   Environmental classification of excreted infections

**DESCRIPTION**

Inadequate sanitation is a major contributor to serious human health problems, particularly in developing areas. Microorganisms excreted by human hosts can easily be transmitted to a new human carrier through contact with contaminated waters, or by way of intermediate hosts. Effective treatment of excreta drastically reduces the opportunity for transmission, and simultaneously, the incidence of the disease. This table qualitatively discusses the major types of infections associated with inadequate sanitation, the methods of transmission, and important measures that can be taken to control each type of infection.

**SOURCE**

R.G. Feachem, D.J. Bradley, H. Garelick and D.D. Mara, 1983, *Sanitation and Disease: Health Aspects of Excreta and Wastewater Management*, The International Bank for Reconstruction and Development (World Bank), John Wiley and Sons, New York. By permission.

**TABLE C.13** Basic epidemiological features of excreted pathogens, by environmental category

| Pathogen | Excreted load[a] | Latency[b] | Persistence[c] | Multiplication outside human host | Median infective dose (ID$^{50}$) | Significant immunity? | Major nonhuman reservoir? | Intermediate host |
|---|---|---|---|---|---|---|---|---|
| *Category I* | | | | | | | | |
| Enteroviruses[d] | $10^7$ | 0 | 3 months | No | L | Yes | No | None |
| Hepatitis A virus | $10^6$(?) | 0 | ? | No | L(?) | Yes | No | None |
| Rotavirus | $10^6$(?) | 0 | ? | No | L(?) | Yes | No(?) | None |
| *Balantidium coli* | ? | 0 | ? | No | L(?) | No(?) | Yes | None |
| *Entamoeba histolytica* | $10^5$ | 0 | 25 days | No | L | No(?) | No | None |
| *Giardia lamblia* | $10^5$ | 0 | 25 days | No | L | No(?) | Yes | None |
| *Enterobius vermicularis* | Not usually found in feces | 0 | 7 days | No | L | No | No | None |
| *Hymenolepis nana* | ? | 0 | 1 month | No | L | Yes(?) | No(?) | None |
| *Category II* | | | | | | | | |
| *Campylobacter fetus* ssp. *jejuni* | $10^7$ | 0 | 7 days | Yes[e] | H(?) | ? | Yes | None |
| Pathogenic *Escherichia coli*[f] | $10^8$ | 0 | 3 months | Yes | H | Yes(?) | No(?) | None |
| *Salmonella* | | | | | | | | |
| S. typhi | $10^8$ | 0 | 2 months | Yes[e] | H | Yes | No | None |
| Other salmonellae | $10^8$ | 0 | 3 months | Yes[e] | H | No | Yes | None |
| *Shigella* | $10^7$ | 0 | 1 month | Yes[e] | M | No | No | None |
| *Vibrio cholerae* | $10^7$ | 0 | 1 month(?) | Yes | H | Yes(?) | No | None |
| *Yersinia enterocolitica* | $10^5$ | 0 | 3 months | Yes | H(?) | No | Yes | None |
| *Category III* | | | | | | | | |
| *Ascaris lumbricoides* | $10^4$ | 10 days | 1 year | No | L | No | No | None |
| Hookworms[g] | $10^2$ | 7 days | 3 months | No | L | No | No | None |
| *Strongyloides stercoralis* | 10 | 3 days | 3 weeks (free-living stage much longer) | Yes | L | Yes | No | None |
| *Trichuris trichiura* | $10^3$ | 20 days | 9 months | No | L | No | No | None |
| *Category IV* | | | | | | | | |
| *Taenia saginata* and *T. solium*[h] | $10^4$ | 2 months | 9 months | No | L | No | No | Cow (*T. Saginata*) or pig (*T. solium*) |

*continued*

| Pathogen | Excreted load[a] | Latency[b] | Persistence[c] | Multiplication outside human host | Median infective dose ($ID_{50}$) | Significant immunity? | Major nonhuman reservoir? | Intermediate host |
|---|---|---|---|---|---|---|---|---|
| *Category V* | | | | | | | | |
| *Clonorchis sinensis*[i] | $10^2$ | 6 weeks | Life of fish | Yes[j] | L | No | Yes | Snail and fish |
| *Diphyllobothrium latum*[i] | $10^4$ | 2 months | Life of fish | No | L | No | Yes | Copepod and fish |
| *Fasciola hepatica*[h] | ? | 2 months | 4 months | Yes[j] | L | No | Yes | Snail and aquatic plant |
| *Fasciolopsis buski*[h] | $10^3$ | 2 months | ? | Yes[j] | L | No | Yes | Snail and aquatic plant |
| *Gastrodiscoides hominis*[h] | ? | 2 months(?) | ? | Yes[j] | L | No | Yes | Snail and aquatic plant |
| *Heterophyes heterophyes*[i] | ? | 6 weeks | Life of fish | Yes[j] | L | No | Yes | Snail and fish |
| *Metagonimus yokogawai*[i] | ? | 6 weeks(?) | Life of fish | Yes[j] | L | No | Yes | Snail and fish |
| *Paragonimus westermani*[i] | ? | 4 months | Life of crab | Yes[j] | L | No | Yes | Snail and crab or crayfish |
| *Schistosoma* | | | | | | | | |
| *S. haematobium*[h] | 4 per milliliter of urine | 5 weeks | 2 days | Yes[j] | L | Yes | No | Snail |
| *S. japonicum*[h] | 40 | 7 weeks | 2 days | Yes[j] | L | Yes | Yes | Snail |
| *S. mansoni*[h] | 40 | 4 weeks | 2 days | Yes[j] | L | ? | No | Snail |
| *Leptospira* spp.[k] | Urine(?) | 0 | 7 days | No | L | Yes(?) | Yes | None |

L, low ($<10^2$); M, medium ($\approx10^4$); H, high ($>10^6$).
?, Uncertain.

[a]Typical average number of organisms per gram of feces (except for *Schistosoma haematobium* and *Leptospira*, which occur in urine).

[b]Typical minimum time from excretion to infectivity.

[c]Estimated maximum life of infective stage at 20°–30°C.

[d]Includes polio-, echo-, and coxsackieviruses.

[e]Multiplication takes place predominantly on food.

[f]Includes enterotoxigenic, enteroinvasive, and enteropathogenic *E. coli.*

[g]*Ancylostoma duodenale* and *Necator americanus.*

[h]Latency is minimum time from excretion by man to potential reinfection of man. Persistence here refers to maximum survival time of final infective stage. Life cycle involves one intermediate host.

[i]Latency and persistence as for *Taenia.* Life cycle involves two intermediate hosts.

[j]Multiplication takes place in intermediate snail host.

[k]For the reasons given in original source, *Leptospira* spp. do not fit any of the categories defined in the table.

**TABLE C.13** Basic epidemiological features of excreted pathogens, by environmental category *(pages 199–200)*

**DESCRIPTION**

Quantitative background for the five categories of waterborne diseases described in the previous table are given here. Among the data presented are the average number of organisms excreted, the latency period, the peristence of an organism in the environment, whether or not an organism can multiply outside of a human host, the median dose required to cause an infection, whether immunity can result from infection, and the possible non-human hosts. Inadequate sanitation is a major contributor to serious human health problems, particularly in developing areas. Microorganisms excreted by human hosts can easily be transmitted to a new human carrier through contact with contaminated waters, or by way of intermediate hosts. Effective treatment of excreta drastically reduces the opportunity for transmission, and, simultaneously, the incidence of the disease.

**LIMITATIONS**

Complete information is not available on the life-cycles of some waterborne disease related organisms, on their persistence in the environment, and on their infective doses. Developing appropriate control strategies often requires this information.

**SOURCE**

R.G. Feachem, D.J. Bradley, H. Garelick and D.D. Mara, 1983, *Sanitation and Disease: Health Aspects of Excreta and Wastewater Management*, The International Bank for Reconstruction and Development (World Bank), John Wiley and Sons, New York. By permission.

**TABLE C.14** Water-related disease mechanisms and prevention

| Transmission mechanism | Disease | Preventive strategy |
| --- | --- | --- |
| Waterborne | Cholera, typhoid, bacillary dysentery, infectious hepatitis | Improve water quality, prevent casual use of polluted sources, educate people about cause and effect |
| Water-washed | Trachoma, scabia, dysentery, louse-borne fever | Improve water quality, water accessibility, hygenic education |
| Water-based | Schistosomiasis (bilharziasis), Guinea worm | Decrease necessity for water contact, snail control, educate about life cycle |
| Water-related insect vector | Malaria, sleeping sickness, onchocerciasis | Improve surface water management, destroy breeding habitat, decrease necessity to visit breeding site |

**TABLE C.14** Water-related disease mechanisms and prevention

**DESCRIPTION**

Many severe infectious diseases are commonly transmitted through some aspect of the water cycle. Diseases caused by oral ingestion of an infective agent in contaminated water are termed waterborne diseases. Waterborne diseases include cholera, typhoid, dysentery, and hepatitis. "Water-washed" diseases include skin sepsis, scabies, and some diseases of the gastrointestinal tract. "Water-based" diseases are transmitted by organisms that rely on humans for some portion of their life-cycle. The best known examples include schistosomiasis and guinea worm. The last category of "water-related" diseases result from insects that serve as disease vectors by transmitting disease-causing microorganisms and who breed or live in water. Diseases in this category include malaria, sleeping sickness, and river blindness (onchocerciasis). This table summarizes water-related diseases and general preventative strategies for reducing disease incidence, including improved water quality and sanitation, hygiene education, insect and parasite control, and elimination of breeding habitat for disease vectors.

**SOURCE**

R.C. Cembrowicz, 1984, Technically, socially and economically appropriate technologies for drinking water supply in small communities, in U. Neis and A. Bittner (eds). *Water Reuse: Selected Reports in Water Reuse in Urban and Rural Areas*, Institute of Water Resources Engineering, Karlsruhe, Institute for Scientific Co-operation, Tubingen.

**TABLE C.15**   Water- and sanitation-related infections and their control

| Infection | Importance of Alternate Control Methods | | | | | | | Public health importance |
|---|---|---|---|---|---|---|---|---|
| | Water quality | Water availability | Excreta disposal | Excreta treatment | Personal/ domestic cleanliness | Drainage/ sullage disposal | Food hygiene | |
| Diarrheal diseases and enteric fevers | | | | | | | | |
|    Viral agents | 2 | 3 | 2 | 1 | 3 | 0 | 2 | 3 |
|    Bacterial agents | 3 | 3 | 2 | 1 | 3 | 0 | 3 | 3 |
|    Protozoal agents | 1 | 3 | 2 | 1 | 3 | 0 | 2 | 2 |
| Poliomyelitis and hepatitis A | 1 | 3 | 2 | 1 | 3 | 0 | 1 | 3 |
| Worms with no intermediate host | | | | | | | | |
|    Ascaris and Trichuris | 0 | 1 | 3 | 2 | 1 | 1 | 2 | 2 |
|    Hookworms | 0 | 1 | 3 | 2 | 1 | 0 | 1 | 3 |
| Beef and pork tapeworms | 0 | 0 | 3 | 3 | 0 | 0 | 3 | 2 |
| Worms with intermediate aquatic stages | | | | | | | | |
|    Schistosomiasis | 1 | 1 | 3 | 2 | 1 | 0 | 0 | 3 |
|    Guinea worm | 3 | 0 | 0 | 0 | 0 | 0 | 0 | 2 |
|    Worms with two aquatic stages | 0 | 0 | 2 | 2 | 0 | 0 | 3 | 1 |
| Skin, eye, and louse-borne infections | 0 | 3 | 0 | 0 | 3 | 0 | 0 | 2 |
| Infections spread by water-related insects | | | | | | | | |
|    Malaria | 0 | 0 | 0 | 0 | 0 | 1 | 0 | 3 |
|    Yellow fever and dengue | 0 | 0 | 0 | 0 | 0 | 1 | 0 | 3 |
|    Bancroftian filariasis | 0 | 0 | 3 | 0 | 0 | 3 | 0 | 3 |

0, no importance; 1, little importance; 2, moderate importance; 3, great importance.

**TABLE C.15**   Water- and sanitation-related infections and their control

**DESCRIPTION**

Presented here for the major water- and sanitation-related infections are a variety of alternate control methods that can slow the spread of diseases prevalent in many developing countries. A rough quantitative measure of their effectiveness is also given. Control methods include water quality, water availability, excreta disposal, excreta treatment, personal and domestic cleanliness, drainage and sullage disposal, and food hygiene. Each control method is rated in order of importance, from 0 to 3, with 0 being of no importance and 3 being of great importance.

**LIMITATIONS**

These measures are only rough indicators of the effectiveness of alternative control measures. Combinations of control measures are usually required to combat waterborne disease outbreaks successfully.

**SOURCE**

R.G. Feachem, 1984, The health dimension of the decade, *World Water '83: the World Problem*, Proceedings of a conference organized by the Institution of Civil Engineers, 12–15 July, Thomas Telford Ltd, London.

**TABLE C.16**  Water-related disease prevalence in Africa, Asia, and Latin America, 1977–78

| Infection | Infections ($10^3$/yr) | Deaths ($10^3$/yr) | Cases of disease ($10^3$/yr) | Relative Disability | Method of transmission |
|---|---|---|---|---|---|
| Diarrheas | 3–5,000,000 | 5–10,000 | 3–5,000,000 | 2 | A |
| Malaria | 800,000 | 1,200 | 150,000 | 2 | D |
| Schistosomiasis (bilharzia) | 200,000 | 500–1,000 | 20,000 | 3–4 | C |
| Hookworm | 7–900,000 | 50–60 | 1,500 | 4 | E |
| Amebiasis | 400,000 | 30 | 1,500 | 3 | A |
| Ascariasis (roundworm) | 800,000–1,000,000 | 20 | 1,000 | 3 | B |
| Polio | 80,000 | 10–20 | 2,000 | 2 | A |
| Typhoid | 1,000 | 25 | 500 | 2 | A |
| African trypanosomiasis | 1,000 | 5 | 10 | 1 | D |
| Leprosy | – | Very low | 12,000 | 2–3 | B |
| Trichuriasis (whipworm) | 500,000 | Low | 100 | 3 | B |
| Onchocerciasis (river blindness and skin disease) | 30,000 | 20–50 | 200–500 | 1–2 | D |

Method of transmission:
A, Disease transported through drinking water or through oral contact with contaminated objects.
B, Infection spread through poor sanitary conditions and bad hygiene.
C, Any physical contact with water containing the disease transmitting vector brings infection.
D, Transmitted by infected insects that breed in or near especially stagnant water.
E, Infecting organisms grow in feces when sanitation is inadequate.

Level of disability:
1, Bedridden.
2, Limited ability to function.
3, Ambulatory.
4, Minor.

**TABLE C.16**  Water-related disease prevalence in Africa, Asia, and Latin America, 1977–78

**DESCRIPTION**

Information on common water-related infections is presented here for Africa, Asia, and Latin America for the period 1977–1978. For twelve important diseases, data are provided on the number of cases of infections, deaths, and diseases in thousand cases per year. Also given is detail on the level of relative disability, on a scale of 4 (minor) to 1 (bedridden), and the method of transmission.

**LIMITATIONS**

These data represent gross regional disease prevalences for a very specific time period – 1977–1978. Actual outbreaks are highly variable in time. For example, cholera was, until the early 1990s, not found in Latin America. A recent dramatic upsurge in the disease, however, is not reflected in these short period data. See Tables C.22 and C.23 for additional information on the recent cholera outbreak, and Table C.20 for time-series information on malaria.

**SOURCE**

Adapted from information appearing in J.A. Walsh and K.S. Warren, 1979, Selective primary health care: An interim strategy for disease control in developing countries, *The New England Journal of Medicine*, **301**(18), 969.

**TABLE C.17**   Incidence of waterborne disease, 1980

| Country | (Cases per 100,000 population) | Country | (Cases per 100,000 population) |
|---|---|---|---|
| Cook Islands | 27,000 | New Caledonia | 1,200 |
| Sierra Leone | 15,400 | Kiribati | 1,080 |
| Tuvalu | 14,560 | Philippines | 974 |
| Colombia | 8,200 | Sri Lanka | 962 |
| Mauritania | 8,200 | Vietnam | 956 |
| Madagascar | 7,690 | Pacific Islands | 940 |
| Cape Verde | 7,160 | Cayman Islands | 843 |
| Bolivia | 7,000 | India | 800 |
| Nicaragua | 6,190 | Peru | 705 |
| Rwanda | 5,400 | Argentina | 700 |
| Angola | 4,690 | Brunei | 391 |
| Mali | 4,600 | Senegal | 360 |
| Nepal | 4,500 | Pakistan | 315 |
| Jordan | 4,000 | Djibouti | 193 |
| Guinea-Bissau | 3,880 | Syria | 149 |
| El Salvador | 3,600 | Chile | 144 |
| Tonga | 2,900 | Indonesia | 116 |
| Panama | 2,788 | Malawi | 92 |
| Tunisia | 2,740 | Egypt | 78 |
| United Arab Emirates | 2,600 | Morocco | 60 |
| Niger | 2,500 | Uruguay | 60 |
| Yemen | 2,341 | Hong Kong | 54 |
| Upper Volta | 2,200 | Macao | 42 |
| Burundi | 2,000 | Turkey | 33 |
| Malaysia | 2,000 | Lesotho | 30 |
| | | Singapore | 10 |

**TABLE C.17**   Incidence of waterborne disease, 1980

**DESCRIPTION**

Listed here are the number of cases of waterborne disease in selected countries per 100,000 population, sorted by incidence, for 1980. Waterborne diseases include cholera, typhoid fever, infectious hepatitis, shigellosis, schistosomiasis, dracunculiasis, onchocerciasis, malaria, and the more common, and more deadly, diarrheal diseases.

**LIMITATIONS**

The incidence of waterborne diseases varies greatly from year to year and from region to region. This table does not differentiate among different types and severities of waterborne diseases, nor does it given any indication of trends, since data for only a single year are reported. Data on the incidence of waterborne disease are irregularly and inconsistently reported by individual countries to the World Health Organization, and there are likely to be countries with more diseases than those reported here.

**SOURCE**

World Health Organization, 1984, *The International Drinking Water Supply and Sanitation Decade: Review of National Baseline Data*, WHO Offset Publication no. 85, World Health Organization, Geneva. By permission.

**TABLE C.18** Incidence and effects of selected diseases in developing countries, late 1980s

| Disease | Incidence[a] ($10^6$/cases/yr) | Estimated[a] deaths/yr |
|---|---|---|
| Diarrhea | 875[b] | 4,600,000 |
| Ascariasis | 900 | 20,000 |
| Guinea worm | 4 | c |
| Schistosomiasis | 200 | c |
| Hookworm | 800 | c |
| Trachoma | 500 | d |

[a]Excluding China
[b]Estimated cases per year.
[c]Effect is usually debilitation rather than death.
[d]Major disability is blindness.

**TABLE C.18** Incidence and effects of selected diseases in developing countries, late 1980s

**DESCRIPTION**

Despite efforts to improve water quality and the availability of sanitation services in recent decades, waterborne diseases are still widespread, particularly in the developing world. For six waterborne diseases, this table gives the number of cases per year, in millions, and the estimated deaths per year in developing countries in the late 1980s.

**LIMITATIONS**

Incidence and deaths from these diseases in China are not included in the data given. This is not a complete list of water-related illnesses. These numbers are approximate and are based on individual country reports to the World Health Organization. Actual number of cases will vary from year to year and region to region.

**SOURCE**

S.A. Esry, J.B. Potash, L. Roberts and C. Shiff, 1990, *Health Benefits from Improvements in Water Supply and Sanitation: Survey and Analysis of the Literature on Selected Diseases*, WASH Technical Report no. 66, Water and Sanitation for Health (WASH) Project, Office of Health, Bureau for Science and Technology, United States Agency for International Development, Arlington, Virginia.

**TABLE C.19** Onchocerciasis morbidity, 1983 *(page 206)*

**DESCRIPTION**

Onchocerciasis, or river blindness, is a major health risk in many tropical countries, and is endemic to Africa, and more limited parts of the Americas and Arabia. Data on onchocerciasis morbidity is presented here for 1983, by country, and on the severity of the disease. Roughly 85 million people, mostly rural, are at risk of contracting the disease, and there were over 17 million people infected with the filarial worm, *Onchocerca volvulus*, worldwide around 1983. Also shown are the numbers of people blinded by the disease. The worm that causes the illness is carried by the blackfly, which breeds in rapidly flowing rivers. The disease causes skin dermatitis, lymphadenitis, and eye lesions that can lead to blindness.

**LIMITATIONS**

These estimates may underestimate the prevalence of the disease. For many countries, figures have been estimated from surveys without accurate information on the disease distribution. Significant progress has been made in eliminating the disease by controlling blackflies. The World Health Organization estimates that the prevalence of blindness has fallen by 50% from the early 1970s. As a result, more recent data might show a smaller infected population than these data.

**SOURCES**

Data from World Health Organization, 1987, *WHO Expert Committee on Onchocerciasis – Third Report*, Technical Report Series no. 752, World Health Organization, Geneva.

Cited in United Nations Environment Programme, 1989, *Environmental Data Report 1989/90*, GEMS Monitoring and Assessment Research Centre, Basil Blackwell, Oxford. By permission.

**TABLE C.19**  Onchocerciasis morbidity, 1983

| | Total population ($10^3$) | Number at risk in onchocerciasis endemic areas ($10^3$) | Number infected ($10^3$)[a] | Number blind as result of onchocerciasis ($10^3$) |
|---|---|---|---|---|
| *Africa/Middle East* | 320,340 | 80,333 | 18,583 | 325.1 |
| Angola | 8,340 | 2,000 | 100 | 2.0 |
| Benin | 3,720 | 600 | 300 | 7.8 |
| Burkina Faso | 6,610 | 0 | 160 | 9.0 |
| Burundi | 4,420 | 60 | 12 | 0.4 |
| Cameroon | 9,160 | 4,935 | 1,200 | 20.0 |
| Central African Republic | 2,450 | 1,800 | 390 | 19.0 |
| Chad | 4,790 | 560 | 127 | 11.0 |
| Congo | 1,650 | 100 | 20 | 0.5 |
| Côte d'Ivoire | 9,160 | 100 | 200 | 10.4 |
| Equatorial Guinea | 380 | 50 | 4 | 0.1 |
| Ethiopia | 33,680 | 7,300 | 1,380 | 20.7 |
| Gabon | 1,130 | 100 | 60 | 3.0 |
| Ghana | 12,700 | 1,200 | 1,200 | 1.2 |
| Guinea | 5,180 | 2,100 | 560 | 20.0 |
| Guinea Bissau | 860 | 132 | 30 | 1.4 |
| Liberia | 2,060 | 1,300 | 600 | 2.6 |
| Malawi | 6,430 | 458 | 120 | 1.0 |
| Mali | 7,530 | 860 | 360 | 15.0 |
| Niger | 5,770 | 0 | 5 | 2.4 |
| Nigeria | 89,020 | 38,920 | 6,962 | 113.0 |
| Senegal | 6,320 | 198 | 44 | 1.5 |
| Sierra Leone | 3,470 | 1,000 | 300 | 10.0 |
| Sudan | 20,360 | 2,000 | 520 | 8.4 |
| Togo | 2,760 | 1,000 | 160 | 8.5 |
| Uganda | 14,630 | 200 | 30 | 0.8 |
| Tanzania | 20,380 | 1,300 | 325 | 7.5 |
| Yemen | 6,230 | 60 | 20 | |
| Zaire | 31,150 | 12,000 | 3,394 | 27.9 |
| *Americas* | 265,850 | 5,251 | 98 | 1.4 |
| Brazil | 129,660 | 8 | 6 | 0.1 |
| Colombia | 27,520 | 732 | | 0.0 |
| Ecuador | 9,250 | 11 | 7 | 0.5 |
| Guatemala | 7,930 | 441 | 40 | 0.6 |
| Mexico | 75,100 | 191 | 25 | 0.1 |
| Venezuela | 16,390 | 3,868 | 20 | 0.1 |
| Total | 586,190 | 85,584 | 18,681 | 326.5 |

[a] Infected with *Onchocerca volvulus*.

**TABLE C.20** Continental trends in malaria morbidity, 1965 to 1987

| Region | | 1965 | 1970 | 1975 | 1980 | 1981 | 1982 | 1983 | 1984 | 1985 | 1986 | 1987 |
|---|---|---|---|---|---|---|---|---|---|---|---|---|
| World | ($10^3$ cases/year) | 4,883.4 | 6,971.7 | 10,845.1 | 15,954.8 | 14,577.5 | 12,449.2 | 8,405.4 | 9,676.6 | 7,736.5 | 7,315.0 | 5,939.6 |
| | Number of countries reporting | 115 | 121 | 131 | 143 | 150 | 143 | 129 | 128 | 120 | 112 | 105 |
| Africa | ($10^3$ cases/year) | 3,805.5 | 4,831.9 | 3,861.3 | 7,893.7 | 6,770.8 | 6,053.7 | 2,731.4 | 4,161.6 | 2,909.3 | 2,364.7 | 839.9 |
| | Number of countries reporting | 35 | 32 | 32 | 44 | 44 | 41 | 23 | 26 | 20 | 14 | 10 |
| North America | ($10^3$ cases/year) | 91.3 | 199.5 | 197.5 | 316.0 | 325.2 | 364.7 | 319.2 | 348.3 | 287.3 | 269.1 | 226.5 |
| | Number of countries reporting | 16 | 16 | 20 | 20 | 25 | 24 | 25 | 24 | 25 | 24 | 21 |
| South America | ($10^3$ cases/year) | 152.3 | 144.6 | 159.3 | 286.8 | 312.9 | 352.9 | 511.7 | 583.3 | 605.9 | 681.5 | 783.0 |
| | No. of countries reporting | 12 | 12 | 11 | 11 | 12 | 12 | 13 | 12 | 11 | 12 | 12 |
| Asia | ($10^3$ cases/year) | 816.1 | 1,751.2 | 6,562.8 | 7,293.9 | 6,973.8 | 5,482.9 | 4,600.7 | 4,326.7 | 3,678.7 | 3,827.0 | 3,892.0 |
| | Number of countries reporting | 29 | 31 | 34 | 38 | 37 | 36 | 37 | 36 | 35 | 35 | 33 |
| Europe | ($10^3$ cases/year) | 0.4 | 1 | 2.6 | 3.5 | 3.6 | 3.7 | 3.8 | 4.2 | 5.9 | 6.4 | 5.6 |
| | Number of countries reporting | 18 | 21 | 24 | 24 | 26 | 25 | 26 | 25 | 24 | 24 | 26 |
| USSR | ($10^3$ cases/year) | 0.4 | 0.5 | 0.3 | 0.4 | 0.3 | 0.4 | 0.5 | 0.4 | 0.5 | 0.5 | 1.7 |
| Oceania | ($10^3$ cases/year) | 17.4 | 43 | 61.3 | 160.5 | 190.9 | 190.9 | 238.1 | 251.1 | 248.9 | 165.7 | 190.8 |
| | Number of countries reporting | 4 | 8 | 9 | 5 | 5 | 4 | 4 | 4 | 4 | 2 | 2 |

**TABLE C.20**  Continental trends in malaria morbidity, 1965 to 1987 *(page 207)*

## DESCRIPTION

Listed here are the number of countries reporting the incidence of malaria for each continent during eleven different years from 1965 to 1987, and the number of new cases reported each year, in thousands. Malaria is transmitted to human populations by mosquitoes carrying the parasite *Plasmodium*, particularly *P. falciparum*. Water serves as the breeding grounds for the mosquitoes. According to the World Health Organization, more than 40% of the global population is at risk of contracting malaria. Thirty-one per cent live in areas in which the disease was once diminished, but has now staged a comeback, and an additional 10% live in areas in which the disease thrives. An estimated 110 million people currently suffer from malaria, while a total of 280 million carry the parasite. Annual deaths from malaria are estimated at around 1 million, and the disease can also be extremely debilitating for years. Although each episode can be treated with medication, no cure exists that can safely remove the parasite from the human liver. Both the vector mosquito and the parasite itself have become increasingly resistant to the insecticides and drugs used to eradicate them.

The figures show malarial cases reported and the number of countries reporting for major continental areas.

## LIMITATIONS

Extreme caution should be used in interpreting these numbers and evaluating trends, for one very important reason: the number of countries reporting varies enormously from region to region and year to year. For example, in 1980 nearly 16 million new cases of malaria were reported worldwide, while in 1987, only 6 million new cases were reported. For the same two years, however, the number of countries reporting dropped from 143 to 105. The greatest changes appear in countries in Africa. Thirty to forty countries in Africa used to report malaria numbers regularly, but in recent years this has dropped to under fifteen.

## SOURCES

United Nations Environment Programme, 1989, *Environmental Data Report*, 2nd edn., Basil Blackwell, Oxford. By permission.

World Health Organization, 1991, World malaria situation in 1989 – Part I, *Weekly Epidemiological Record*, **66**, 157–163, World Health Organization, Geneva. By permission.

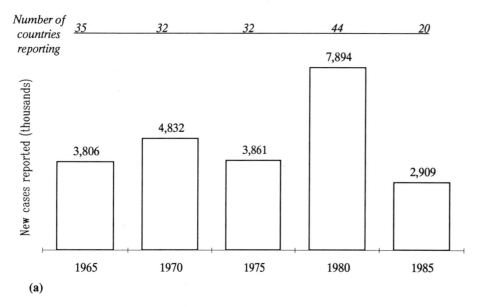

(a)

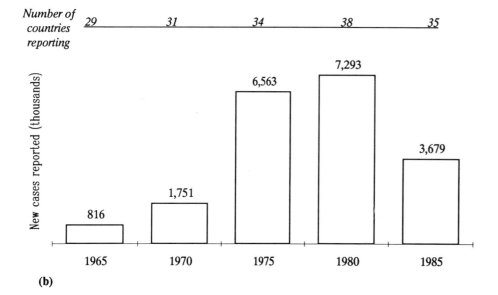

(b)

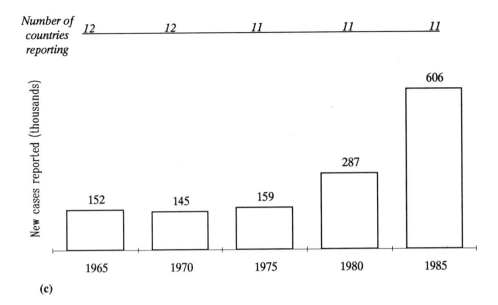

(c)

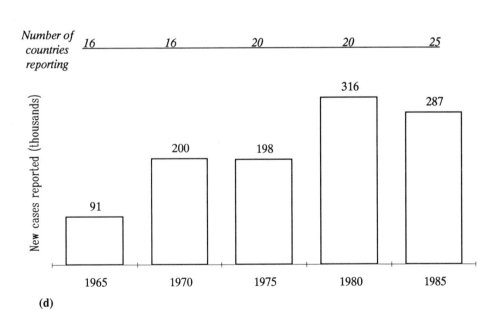

(d)

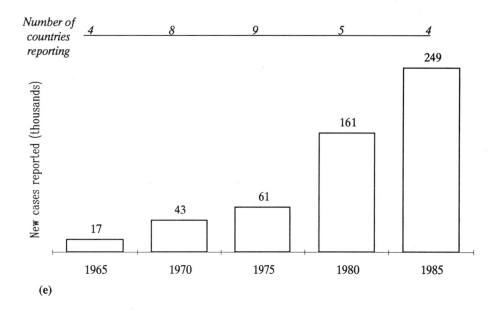

(e)

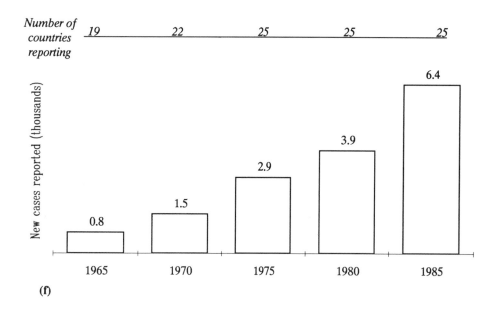

(f)

Fig. C–20. Malaria cases reported by continent, 1965 to 1985. Hundreds of millions of people suffer from malaria, with several million deaths each year. These graphs show, for each continent, the number of new cases per year reported to the World Health Organization from 1965 to 1985, and the number of countries reporting. (a) Africa, (b) Asia, (c) South America, (d) North America, (e) Oceania, (f) Europe. While the total number of malarial cases in Europe and Oceania are low, malaria is increasing in these regions. For Asia and Africa, severe problems with consistent reporting and the quality of the data prevent us from drawing any conclusions from these figures. See Table C.20 for complete data, including specific country figures.

**TABLE C.21**  National trends in malaria morbidity, 1965 to 1987 ($10^3$ cases)

| Region | 1965 | 1970 | 1975 | 1980 | 1981 | 1982 | 1983 | 1984 | 1985 | 1986 | 1987 |
|---|---|---|---|---|---|---|---|---|---|---|---|
| World | 4,883.4 | 6,971.7 | 10,845.1 | 15,954.8 | 14,577.5 | 12,449.2 | 8,405.4 | 9,675.6 | 7,736.5 | 7,315.0 | 5,939.6 |
|  | 115 | 121 | 131 | 143 | 150 | 143 | 129 | 128 | 120 | 112 | 105 |
| Africa[a] | 3,805.5 | 4,831.9 | 3,861.3 | 7,893.7 | 6,770.8 | 6,053.7 | 2,731.4 | 4,161.6 | 2,909.3 | 2,364.7 | 839.9 |
| Countries reporting | 35 | 32 | 32 | 44 | 44 | 41 | 23 | 26 | 20 | 14 | 10 |
| Angola | 160.6 | 162.1 |  | 797.7 | 242.9 |  | 818.0 | 912.0 | 443.0 | 449.9 |  |
| Benin | 94.2 | 107.4 | 103.6 | 131.9 | 122.2 | 96.9 |  | 160.2 | 207.2 | 151.1 |  |
| Burkina Faso | 228.7 | 532.8 |  | 211.7 | 326.2 | 306.0 |  |  |  |  |  |
| Burundi |  |  |  | 79.2 | 169.9 | 159.1 |  |  |  |  |  |
| Cameroon | 214.3 | 222.0 | 48.0 | 614.4 |  | 39.0 |  | 818.0 | 799.79 |  |  |
| Central African Republic | 20.0 | 75.0 | 55.9 | 139.1 | 89.1 | 107.0 | 79.7 | 35.1 |  |  |  |
| Chad | 107.3 | 127.5 | 167.3 |  |  | 10.4 | 24.7 | 53.1 | 51.2 | 32.2 |  |
| Congo | 104.7 | 150.0 | 175.7 | 77.8 | 81.9 | 46.2 | 51.4 | 67.2 |  |  |  |
| Côte d'Ivoire | 375.6 |  |  |  |  |  |  |  |  |  |  |
| Ethiopia |  |  |  | 97.6 | 256.4 | 110.5 | 29.4 | 59.6 | 101.6 |  |  |
| Ghana | 116.9 |  |  |  | 56.2 | 341.0 |  |  |  |  |  |
| Guinea |  | 60.9 | 64.3 |  | 23.4 | 46.2 | 103.6 |  |  |  |  |
| Guinea-Bissau | 58.1 | 26.9 |  | 148.1 | 161.5 | 30.0 |  | 160.5 |  |  |  |
| Kenya |  |  | 663.1 | 1,050.2 | 324.6 | 685.0 |  |  |  |  |  |
| Liberia | 5.1 |  | 18.1 | 318.3 | 302.8 | 282.9 |  |  |  |  |  |
| Madagascar | 13.3 | 88.9 |  | 349.1 | 440.0 | 403.5 | 448.0 | 422.6 | 348.7 | 380.1 |  |
| Malawi | 11.5 |  |  | 28.5 | 192.8 |  |  |  |  |  |  |
| Mali | 346.0 | 517.7 | 408.9 | 149.3 | 145.8 | 42.6 |  |  |  |  |  |
| Niger | 101.6 | 158.6 | 114.3 | 385.0 | 395.0 | 407.1 |  | 547.8 | 753.3 | 727.0 |  |
| Nigeria |  | 628.5 | 1,083.3 | 1,375.2 | 1,268.8 | 832.9 |  |  |  |  |  |
| Rwanda |  | 34.3 | 66.4 | 123.9 | 173.4 | 137.3 |  |  |  |  |  |
| Senegal | 280.2 | 472.5 | 483.8 | 568.0 | 498.9 |  |  |  |  |  |  |

| Region | 1965 | 1970 | 1975 | 1980 | 1981 | 1982 | 1983 | 1984 | 1985 | 1986 | 1987 |
|---|---|---|---|---|---|---|---|---|---|---|---|
| Sudan | 638.6 | 110.1 | | 2.9 | 2.2 | 3.4 | | 7.7 | 8.5 | 18.1 | 17.4 |
| Tanzania | 463.4 | 46.7 | | 214.8 | 374.4 | 414.2 | 322.5 | | | | |
| Togo | 241.5 | 258.2 | 264.6 | 280.1 | 301.0 | 274.7 | 553.9 | 745.3 | | 570.0 | 779.2 |
| Uganda | | | 31.8 | 52.1 | 70.5 | 42.9 | | | | | |
| Zaire | | 901.2 | | 295.7 | 391.2 | | 196.8 | | | | |
| Zambia | | | 1.1 | 116.1 | 126.9 | 1,108.1 | | | | | |
| *North and Central America*[b] | 91.3 | 199.5 | 197.5 | 316.0 | 325.2 | 364.7 | 319.2 | 348.3 | 287.3 | 269.1 | 226.5 |
| Countries reporting | 16 | 16 | 20 | 20 | 25 | 24 | 25 | 24 | 25 | 24 | 21 |
| El Salvador | 34.1 | 45.4 | 83.1 | 95.8 | 93.2 | 86.2 | 65.4 | 66.9 | 44.5 | 24.0 | 12.8 |
| Guatemala | 14.5 | 11.0 | 5.0 | 62.6 | 68.0 | 77.4 | 64.0 | 74.1 | 55.0 | 42.6 | 57.7 |
| Haiti | 10.3 | 10.7 | 24.7 | 53.5 | 46.7 | 65.4 | 53.9 | 69.9 | 16.7 | 14.4 | 12.1 |
| Honduras | 6.9 | 34.5 | 30.3 | 43.0 | 49.4 | 57.5 | 37.5 | 27.3 | 33.8 | 29.1 | 19.1 |
| Mexico | 10.1 | 57.3 | 27.9 | 25.7 | 42.1 | 52.1 | 74.2 | 85.5 | 116.0 | 130.9 | 99.6 |
| Nicaragua | 10.3 | 27.3 | 24.7 | 25.5 | 17.4 | 15.6 | 12.9 | 15.7 | 15.1 | 20.3 | 17.0 |
| *South America*[b] | 152.3 | 144.6 | 159.3 | 286.8 | 312.9 | 352.9 | 511.7 | 583.3 | 605.9 | 681.5 | 783 |
| Countries reporting | 12 | 12 | 11 | 11 | 12 | 12 | 13 | 12 | 11 | 12 | 12 |
| Bolivia | 0.9 | 6.8 | 6.6 | 16.6 | 9.8 | 6.7 | 14.4 | 16.3 | 14.4 | 21.0 | 24.9 |
| Brazil | 110.3 | 54.6 | 88.6 | 176.2 | 205.5 | 221.9 | 297.7 | 378.3 | 401.9 | 443.6 | 508.9 |
| Colombia | 18.9 | 32.3 | 32.7 | 57.3 | 61.0 | 78.6 | 105.4 | 55.3 | 55.8 | 89.3 | 90.0 |
| Ecuador | 4.2 | 28.4 | 6.6 | 8.7 | 12.7 | 14.6 | 51.6 | 78.6 | 69.0 | 51.4 | 63.5 |
| Guyana | — | — | 1.1 | 3.2 | 2.1 | 1.7 | 2.1 | 3.0 | 7.9 | 16.4 | 34.1 |
| Paraguay | 6.7 | 1.4 | 0.2 | 0.1 | 0.1 | 0.1 | — | 0.6 | 4.6 | 4.3 | 3.6 |
| Peru | 1.9 | 4.5 | 14.3 | 15.0 | 14.8 | 20.5 | 28.6 | 33.7 | 35.0 | 36.9 | 39.1 |
| Suriname | | | | | | | | | 1.6 | | |
| Venezuela | 5.4 | 15.3 | 6.0 | 3.9 | 3.4 | 4.3 | 8.4 | 12.2 | 14.3 | 14.4 | 13.0 |

*continued*

| Region | 1965 | 1970 | 1975 | 1980 | 1981 | 1982 | 1983 | 1984 | 1985 | 1986 | 1987 |
|---|---|---|---|---|---|---|---|---|---|---|---|
| *Asia*[b] | 816.1 | 1,751.2 | 6,562.8 | 7,293.9 | 6,973.8 | 5,482.9 | 4,600.7 | 4,326.7 | 3,678.7 | 3,827.0 | 3,892.0 |
| Countries reporting | 29 | 31 | 34 | 38 | 37 | 36 | 37 | 36 | 35 | 35 | 33 |
| Afghanistan | 2.1 | 37.0 | 80.6 | 47.3 | 87.4 | 110.3 | 127.3 | 176.7 | 227.8 | 422.4 | 428.1 |
| Bangladesh | 0.6 | 8.2 | 31.2 | 67.7 | 45.9 | 38.2 | 40.3 | 31.8 | 30.3 | 41.4 | 35.8 |
| Bhutan | 0.5 | 0.6 | 7.9 | 3.6 | 4.2 | 5.6 | 5.2 | 18.4 | 16.0 | 19.9 | 13.1 |
| Cambodia | 19.7 | 3.7 | 30.9 | 3.1 | 16.5 | 61.5 | 41.6 |  |  |  |  |
| China | 99.7 | 694.6 |  | 3,300.3 | 3,059.6 | 2,041.4 | 1,377.6 | 903.8 | 563.8 | 363.7 | 210.6 |
| India |  |  | 2,898.1 | 2,898.1 | 2,701.1 | 2,182.3 | 2,018.6 | 2,184.4 | 1,864.4 | 1,765.6 | 1,611.2 |
| Indonesia | 8.9 | 117.1 | 176.7 | 176.7 | 124.6 | 84.3 | 133.6 | 86.1 | 47.7 | 20.1 | 19.3 |
| Iran | 10.6 | 24.3 | 32.6 | 32.6 | 29.7 | 42.8 | 45.9 | 30.8 | 26.4 | 32.3 |  |
| Iraq | 6.1 | 14.2 | 2.8 | 2.8 | 2.6 | 3.3 | 2.4 | 3.3 | 4.8 | 3.0 | 3.7 |
| Korea |  | 15.9 |  |  |  |  |  |  |  |  |  |
| Laos | 175.5 | 29.2 | 12.6 | 12.6 |  |  | 5.6 | 7.2 | 21.2 | 21.7 | 35.0 |
| Malaysia | 321.4[c] | 40.8 | 60.9 | 44.2 | 59.4 | 43.9 | 22.2 | 32.1 | 49.5 | 44.1 | 36.7 |
| Myanmar | 5.8 | 8.1 | 11.9 | 30.9 | 42.0 | 42.0 | 47.7 | 60.5 | 65.3 | 33.4 | 61.7 |
| Nepal | 4.6 | 2.9 | 12.4 | 14.1 | 16.1 | 16.9 | 16.7 | 29.4 | 42.3 | 36.5 | 26.7 |
| Oman |  |  | 0.4 | 1.3 | 2.2 | 30.6 | 34.9 | 16.6 | 16.4 | 16.7 | 15.5 |
| Pakistan |  | 108.8 | 238.3 | 17.7 | 37.9 | 56.4 | 51.6 | 7.4 | 77.6 | 90.4 | 64.3 |
| Philippines | 29.0 | 35.5 | 72.7 | 105.7 | 97.5 | 97.5 | 90.3 | 107.3 | 103.1 | 102.6 | 154.1 |
| Saudi Arabia |  | 6.1 | 1.8 | 6.5 | 5.5 | 15.2 | 17.9 | 11.1 | 16.2 | 13.0 | 17.7 |
| Sri Lanka | 0.3 | 468.2 | 400.8 | 47.9 | 47.4 | 38.6 | 127.3 | 149.5 | 117.8 | 412.5 | 676.6 |
| Thailand | 108.0 | 123.7 | 267.5 | 395.4 | 473.2 | 420.8 | 243.9 | 306.6 | 275.4 | 252.1 | 321.5 |
| Turkey | 4.6 | 1.3 | 9.8 | 34.1 | 55.5 | 60.0[c] | 66.7 | 26.8 | 24.6 | 37.9 | 20.1 |
| United Arab Emirates |  |  |  | 8.6 | 7.6 | 6.2 | 4.8 | 3.5 | 2.6 | 3.1 |  |
| Vietnam | 17.0 | 8.6 |  | 31.8 | 40.3 | 50.9 | 63.6 | 58.8 | 78.4 | 87.4 | 130.7 |
| Yemen Arab Rep |  |  | 2.8 | 5.8 | 10.0 | 20.6 | 2.2 | 1.3 | 1.2 | 1.9 | 2.6 |
| Yemen Dem Rep |  |  | 13.1 | 2.1 | 4.1 | 7.6 | 9.3 | 3.6 | 3.2 | 3.6 | 5.5 |

*continued*

| Region | 1965 | 1970 | 1975 | 1980 | 1981 | 1982 | 1983 | 1984 | 1985 | 1986 | 1987 |
|---|---|---|---|---|---|---|---|---|---|---|---|
| *Europe*[d] | 0.4 | 1.0 | 2.6 | 3.5 | 3.6 | 3.7 | 3.8 | 4.2 | 5.9 | 6.4 | 5.6 |
| Countries reporting | 18 | 21 | 24 | 24 | 26 | 25 | 26 | 25 | 24 | 24 | 26 |
| Austria | — | — | — | — | 0.1 | 0.1 | 0.1 | 0.1 | 0.1 | 0.1 | 0.1 |
| Belgium | | — | — | 0.1 | — | — | 0.1 | 0.1 | 0.2 | 0.3 | 0.3 |
| Bulgaria | | — | — | 0.1 | 0.4 | 0.4 | 0.2 | 0.3 | | 0.1 | 0.1 |
| Denmark | — | — | 0.1 | 0.1 | 0.1 | 0.1 | 0.1 | 0.1 | 0.1 | 0.2 | 0.1 |
| France | | — | 0.1 | 0.1 | 0.1 | 0.1 | 0.1 | 0.1 | 1.6 | 1.0 | 1.0 |
| German DR | | | — | — | — | 0.1 | — | 0.1 | 0.1 | | 0.1 |
| Germany FR | — | 0.1 | 0.2 | 0.6 | 0.4 | 0.5 | 0.4 | 0.5 | 0.5 | 1.1 | 0.7 |
| Greece | 0.1 | — | — | 0.1 | 0.1 | 0.1 | — | 0.1 | — | — | 0.1 |
| Italy | — | — | 0.1 | 0.2 | 0.2 | 0.2 | 0.2 | 0.2 | 0.2 | 0.2 | 0.3 |
| Netherlands | — | — | 0.1 | 0.1 | 0.1 | 0.1 | 0.1 | 0.1 | 0.1 | 0.2 | 0.2 |
| Norway | — | — | — | — | — | — | 0.1 | 0.1 | 0.1 | 0.1 | — |
| Portugal | 0.1 | 0.5 | 0.1 | — | — | — | — | — | 0.1 | 0.1 | 0.1 |
| Spain | — | — | 0.1 | 0.1 | 0.1 | 0.1 | 0.1 | 0.1 | 0.1 | 0.2 | 0.2 |
| Sweden | — | — | 0.1 | 0.1 | 0.1 | 0.1 | 0.1 | 0.1 | 0.1 | 0.1 | 0.2 |
| Switzerland | | — | 0.1 | 0.1 | 0.1 | 0.1 | 0.2 | 0.2 | 0.2 | 0.2 | 0.2 |
| UK | — | 0.2 | 0.8 | 1.7 | 1.6 | 1.5 | 1.7 | 1.9 | 2.2 | 2.3 | 1.8 |
| Yugoslavia | — | — | — | 0.1 | 0.1 | 0.1 | 0.1 | 0.1 | 0.1 | 0.1 | 0.1 |
| *USSR* | 0.4 | 0.5 | 0.3 | 0.4 | 0.3 | 0.4 | 0.5 | 0.4 | 0.5 | 0.5 | 1.7 |
| *Oceania*[d] | 17.4 | 43.0 | 61.3 | 160.5 | 190.9 | 190.9 | 238.1 | 251.1 | 248.9 | 165.7 | 190.8 |
| Countries reporting | 4 | 8 | 9 | 5 | 5 | 4 | 4 | 4 | 4 | 2 | 2 |
| Australia | 0.1 | 0.2 | 0.3 | 0.6 | 0.5 | 0.6 | 0.6 | 0.6 | 0.7 | | |
| Guam | — | 0.1 | — | | | | | | | | |

*continued*

| Region | 1965 | 1970 | 1975 | 1980 | 1981 | 1982 | 1983 | 1984 | 1985 | 1986 | 1987 |
|---|---|---|---|---|---|---|---|---|---|---|---|
| Papua New Guinea | | 30.3 | 56.6 | 121.6 | 122.9 | 109.3 | 126.9 | 150.3 | 182.5 | 143.3 | 164.2 |
| Solomon Islands | 15.7 | 9.8 | 3.6 | 35.0 | 61.1 | 69.8 | 84.5 | 72.1 | 40.8 | | |
| Vanuatu | 1.6 | 2.6 | 0.9 | 3.2 | 6.3 | 11.2 | 26.0 | 28.0 | 24.9 | 22.4 | 26.6 |

Data are the number of new cases reported for that year.
The information provided does not cover the total population at risk in some countries.
A dash signifies less than sixty cases.
The countries comprising each region account for more than 90% of the total regional incidence for most years, with the exception of European countries (average more than 80%).
[a]Data for individual countries in Africa are included where an annual incidence of more than 100,000 cases has been reported during the study period. Mainly clinically diagnosed cases in this region (i.e., not confirmed cases).
[b]Data for individual countries in this region are included where an annual incidence of more than 10,000 cases has been reported during the study period.
[c]Estimated, provisional, or incomplete.
[d]Data for individual countries in this region are included where an annual incidence of more than sixty cases has been reported during the study period.

**TABLE C.21**   National trends in malaria morbidity, 1965 to 1987 (*pages 211–215*)

### DESCRIPTION

World Health Organization (WHO) data are given here on the number of new cases of malaria for reporting countries and regions of the world, in thousands of cases, during eleven different years from 1965 to 1987, and the number of reporting countries for each year. Malaria is transmitted to human populations by mosquitoes carrying the parasite *Plasmodium*, particularly *P. falciparum*. Water serves as the breeding grounds for the mosquitoes. According to the WHO, more than 40% of the global population is at risk of contracting malaria. Thirty-one per cent live in areas in which the disease was once diminished, but has now staged a comeback, and an additional 10% live in areas in which the disease thrives. An estimated 110 million people currently suffer from malaria, while a total of 280 million carry the parasite. Annual deaths from malaria are estimated at approximately 1 million, and the disease can also be extremely debilitating for years. Although each episode can be treated with medication, no cure exists that can safely remove the parasite from the human liver. Both the vector mosquito and the parasite itself have become increasingly resistant to the insecticides and drugs used to eradicate them.

### LIMITATIONS

Extreme caution should be used in interpreting these numbers and evaluating trends because the number of countries reporting varies enormously from region to region and year to year. There are many gaps in national malaria reports and data for several major countries are missing in the later years of this table. For example, figures for Kenya, Nigeria, Ghana, Liberia, Zambia, and many other African nations are not given for the late 1980s. Between 1980 and 1987, the number of countries worldwide reporting malaria data to the WHO dropped from 143 to 105. For African countries, annual incidence was only listed if it exceeded 100,000 cases per year. For Asia and the Americas, 10,000 was the cutoff point, while for Europe and Oceania, the cutoff is 60.

### SOURCES

United Nations Environment Programme, 1989, *Environmental Data Report*, 2nd edn., Basil Blackwell, Oxford.
World Health Organization, 1991, World malaria situation in 1989 – Part I, *Weekly Epidemiological Record*, **66**, 157–163, World Health Organization, Geneva. By permission.

**TABLE C.22**  Cholera cases and deaths, by continent, 1979 to 1991

Cases reported

| Continent | 1979 | 1980 | 1981 | 1982 | 1983 | 1984 | 1985 | 1986 | 1987 | 1988 | 1989 | 1990 | 1991 |
|---|---|---|---|---|---|---|---|---|---|---|---|---|---|
| Africa | 21,586 | 18,742 | 19,415 | 46,924 | 37,383 | 17,504 | 31,884 | 35,585 | 31,324 | 23,583 | 35,951 | 39,211 | 153,367 |
| Asia | 35,732 | 24,015 | 32,343 | 17,991 | 27,877 | 11,809 | 13,389 | 17,131 | 17,558 | 20,871 | 18,007 | 30,979 | 49,791 |
| Europe | 289 | 16 | 46 | 21 | 12 | 11 | 6 | 52 | 14 | 14 | 11 | 349 | 316 |
| North America | 1 | 13 | 21 | 0 | 3 | 1 | 4 | 21 | 7 | 10 | 1 | 7 | 28 |
| Oceania | 64 | 3 | 6 | 2,217 | 326 | 20 | 7 | 3 | 2 | 1 | 0 | 40 | 0 |
| South America | 0 | 0 | 0 | 0 | 0 | 0 | 0 | 0 | 0 | 0 | 0 | 0 | 391,192 |
| Total | 57,672 | 42,789 | 51,831 | 67,153 | 65,601 | 29,345 | 45,290 | 52,792 | 48,905 | 44,479 | 53,970 | 70,586 | 594,694 |

Deaths reported

| Continent | 1979 | 1980 | 1981 | 1982 | 1983 | 1984 | 1985 | 1986 | 1987 | 1988 | 1989 | 1990 | 1991 |
|---|---|---|---|---|---|---|---|---|---|---|---|---|---|
| Africa | 1,869 | 1,185 | 1,581 | 2,988 | 1,903 | 1,711 | 3,837 | 3,490 | 2,658 | 1,500 | 1,445 | 2,288 | 13,998 |
| Asia | 1,602 | 769 | 860 | 833 | 765 | 119 | 276 | 477 | 238 | 378 | 224 | 628 | 1,286 |
| Europe | 8 | 0 | 0 | 0 | 0 | 0 | 1 | 0 | 0 | 0 | 0 | 2 | 9 |
| North America | 0 | 0 | 0 | 0 | 0 | 0 | 0 | 0 | 0 | 0 | 0 | 0 | 0 |
| Oceania | 0 | 0 | 0 | 17 | 1 | 0 | 0 | 0 | 0 | 0 | 0 | 1 | 0 |
| South America | 0 | 0 | 0 | 0 | 0 | 0 | 0 | 0 | 0 | 0 | 0 | 0 | 4,002 |
| Total | 3,479 | 1,954 | 2,441 | 3,838 | 2,669 | 1,830 | 4,114 | 3,967 | 2,896 | 1,878 | 1,669 | 2,919 | 19,295 |

**TABLE C.22** Cholera cases and deaths, by continent, 1979 to 1991 *(page 216)*

**DESCRIPTION**

The official number of cases and deaths from cholera reported to the World Health Organization (WHO) are shown here, from 1979 to 1991, by continent. Cholera affects the poorest populations and is linked to tainted water supplies, inadequate sanitation, and lack of access to basic health care. To control the spread of cholera and limit its effects requires better detection and reporting of cases, prompt and efficient treatment of diarrhea, and measures to provide safe water, sewage treatment, and food handling. This table reveals the extraordinary outbreak of cholera in 1991 in Latin America, and the great increase in both cases and deaths in Asia. All figures were compiled in the summer of 1991, except for the 1991 data themselves, which include all 1991 cases reported through 28 May 1992.

**LIMITATIONS**

These data are only the cases of cholera identified and reported to the WHO. No estimate is available of the extent of undiagnosed or unreported cases.

**SOURCE**

June 1992, Dr S.J. Siméant, World Health Organization, Geneva, personal communication.

**TABLE C.23** Cholera cases and deaths worldwide, 1991 *(page 218)*

**DESCRIPTION**

The official number of cases and deaths from cholera reported to the World Health Organization (WHO) for 1991 are shown here, by country and continent. The vast majority of the cases were reported from Asia, Africa, and the 1991 outbreak in Latin America. Cholera is a disease linked to tainted water supplies, inadequate sanitation, and lack of access to basic health care. Cholera affects those living under the poorest conditions. To control the spread of cholera and limit its effects requires better detection and reporting of cases, prompt and efficient treatment of diarrhea, and measures to provide safe water, sewage treatment, and food handling. The data include all 1991 cases reported through 28 May 1992.

**LIMITATIONS**

These data are only the cases and deaths identified and reported to the WHO. No estimate is available for the extent of undiagnosed or unreported cases.

**SOURCE**

June 1992, Dr S.J. Siméant, World Health Organization, Geneva, personal communication.

**TABLE C.23**   Cholera cases and deaths
worldwide, 1991

| Country/area | Cases | | Deaths | Country/area | Cases | | Deaths |
|---|---|---|---|---|---|---|---|
| *Africa* | 153,367 | | 13,998 | *Americas* | 391,220 | | 4,002 |
| Angola | 8,590 | | 582 | Bolivia | 206 | | 12 |
| Benin | 7,474 | | 259 | Brazil | 1,567 | | 26 |
| Burkina Faso | 537 | | 61 | Canada | 2 | (2[c]) | 0 |
| Burundi | 3 | | 0 | Chile | 41 | | 2 |
| Cameroon | 4,026 | | 491 | Colombia | 11,979 | | 207 |
| Chad | 13,915 | | 1,344 | Ecuador | 46,320 | | 697 |
| Côte d'Ivoire | 604 | | 116 | El Salvador | 947 | | 34 |
| Ghana | 13,172 | | 409 | French Guiana | 1 | (1[c]) | 0 |
| Liberia | 132 | | 40 | Guatemala | 3,674 | | 50 |
| Malawi | 8,088 | | 245 | Honduras | 11 | | 0 |
| Mozambique | 7,847 | | 328 | Mexico | 2,690 | | 34 |
| Niger | 3,238 | | 367 | Nicaragua | 1 | | 0 |
| Nigeria | 59,478 | | 7,654 | Panama | 1,178 | | 29 |
| Rwanda | 679 | | 35 | Peru | 322,562 | | 2,909 |
| São Tome and Principe | 3 | | 1 | United States | 26 | (7[c]) | 0 |
| South Africa | 10 | | 0 | Venezuela | 15 | (11[c]) | 2 |
| Tanzania | 5,676 | | 572 | | | | |
| Togo | 2,396 | | 81 | *Asia* | 49,791 | | 1,286 |
| Uganda | 279 | | 28 | Bhutan | 422 | | 19 |
| Zaire | 4,066 | | 294 | Cambodia | 770 | | 97 |
| Zambia | 13,154 | | 1,091 | China | 205 | | 0 |
| | | | | Hong Kong | 5 | (2[c]) | 0 |
| *Europe* | 316 | | 9 | India | 6,993 | | 149 |
| France | 7 | (7[c]) | 0 | Indonesia | 6,202[a] | | 55 |
| Romania | 226 | | 9 | Iran | 1,880 | | 32 |
| Russia | 3 | (2[c]) | 0 | Iraq | 877 | | 6 |
| Spain | 1 | (1[c]) | 0 | Japan | 90 | (65[c]) | 1 |
| Ukraine | 75 | | 0 | Malaysia | 506 | | 6 |
| United Kingdom | 4 | (4[c]) | 0 | Myanmar | 924 | | 39 |
| | | | | Nepal | 30,648[b] | | |
| | | | | Korea, Rep. | 113 | | 4 |
| | | | | Singapore | 34 | (6[c]) | 2 |
| | | | | Sri Lanka | 70 | | 2 |
| | | | | Vietnam | 52 | | 1 |
| | | | | World total | 594,694 | | 19,295 |

[a]Provisional figure.
[b]Estimated cases and deaths based on laboratory-confirmed samples.
[c]Imported cases.

**TABLE C.24**  Reported cases of dracunculiasis, 1986 to 1990

| Country | 1986 | 1987 | 1988 | 1989 | 1990 |
|---|---|---|---|---|---|
| Benin | n/a | 400 | 33,962 | 7,172 | 37,414[b] |
| Burkina Faso | 2,558[a] | 1,957 | 1,266 | 45,004[b] | 42,187[b] |
| Cameroon | 86 | n/a | 752[b] | 871[b] | 742[b] |
| Central African Republic | 0 | 1,322 | n/a | n/a | 10 |
| Chad | 314 | n/a | n/a | n/a | n/a |
| Côte d'Ivoire | 1,177 | 1,272 | 1,370 | 1,555 | 1,360 |
| Ethiopia | 3,385 | 2,302 | 1,487 | 3,565 | n/a |
| Gambia | 0 | 0 | n/a | n/a | 0[b] |
| Ghana | 4,717 | 18,398 | 71,767 | 179,556[b] | 117,034[b] |
| Guinea | 0 | 0 | n/a | 1 | 0[b] |
| India | 23,070[b] | 17,031[b] | 12,023[b] | 7,881[b] | 4,798[b] |
| Kenya | n/a | n/a | n/a | 5[b] | 6[b] |
| Mali | 5,640 | 435 | 564 | 1,111 | 884 |
| Mauritania | n/a | 227 | 608 | 447 | 8,036[b] |
| Niger | n/a | 699 | n/a | 288 | n/a |
| Nigeria | 2,821 | 216,484 | 653,492[b] | 640,008[b] | 394,082[b] |
| Pakistan | n/a | 2,400 | 1,110[b] | 534[b] | 160[b] |
| Senegal | 128 | 132 | 138 | n/a | 38 |
| Sudan | 822 | 399 | 542 | n/a | n/a |
| Togo | 1,325 | n/a | 178 | 2,749 | 3,042[b] |
| Uganda | n/a | n/a | 1,960 | 1,309 | 4,704 |

[a]From passive reporting and/or area-limit searches unless otherwise indicated.
[b]Data based on national survey or active case search.
n/a is data not available.

**TABLE C.24**  Reported cases of dracunculiasis, 1986 to 1990

**DESCRIPTION**

The number of cases of dracunculiasis, or guinea worm, reported to the World Health Organization (WHO) from 1986 to 1990 are shown here. Dracunculiasis affects the poorest populations and is linked to lack of secure and safe supplies of drinking water and lack of access to basic health care. To control the spread of this disease and limit its effects requires better detection and reporting of cases and access to safe water. There is an international campaign now underway to eradicate this disease under the auspices of the WHO.

from the time-series trends presented here. For example, the great increases in cases in Burkina Faso, Mauritania, Uganda, and elsewhere in 1989 and 1990 typically result from the completion of the first national searches for this disease and better reporting. In Ghana, the reduction from 1989 to 1990 represents real progress in detecting and eliminating the disease. Similarly, in India and Pakistan, major programs to eradicate the disease are having clear effects, shown in the declines from the mid-1980s.

**LIMITATIONS**

These data are only the cases of dracunculiasis identified and reported to the WHO. No estimate is available of the extent of undiagnosed or unreported cases. Extreme care should be taken in drawing conclusions

**SOURCES**

World Health Organization, 1991, Dracunculiasis: Global surveillance summary, 1990, *Weekly Epidemiological Record*, **66**(31), 225–230. By permission.

**Fresh water data**

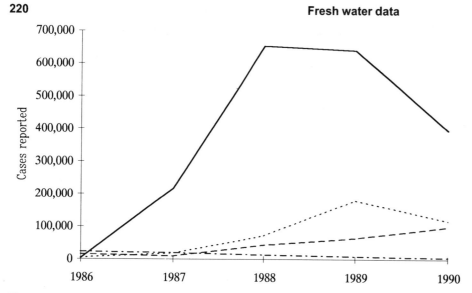

Fig. C–24. Cases of dracunculiasis (guinea worm) reported worldwide, 1986 to 1990. (———), Nigeria; (– – –), Ghana; (——), other Africa; (— · — · —), other. Although this figure suggests an enormous outbreak of the disease in recent years, it is thought that this surge represents more accurate reporting, rather than any real change in the extent of the disease. See Table C.24.

**TABLE C.25** Waterborne disease outbreaks in the United States, by type of water-supply system, 1971 to 1988

| Year | Community | Noncommunity | Individual | Total outbreaks | Total cases |
|------|-----------|--------------|------------|-----------------|-------------|
| 1971 | 8 | 8 | 4 | 20 | 5,184 |
| 1972 | 9 | 19 | 2 | 30 | 1,650 |
| 1973 | 6 | 16 | 3 | 25 | 1,762 |
| 1974 | 11 | 9 | 5 | 25 | 8,356 |
| 1975 | 6 | 16 | 2 | 24 | 10,879 |
| 1976 | 9 | 23 | 3 | 35 | 5,068 |
| 1977 | 14 | 18 | 2 | 34 | 3,860 |
| 1978 | 10 | 19 | 3 | 32 | 11,435 |
| 1979 | 24 | 13 | 8 | 45 | 9,841 |
| 1980 | 26 | 20 | 7 | 53 | 20,045 |
| 1981 | 14 | 18 | 4 | 36 | 4,537 |
| 1982 | 26 | 15 | 3 | 44 | 3,588 |
| 1983 | 30 | 9 | 4 | 43 | 21,036 |
| 1984 | 12 | 5 | 10 | 27 | 1,800 |
| 1985 | 7 | 14 | 1 | 22 | 1,946 |
| 1986 | 10 | 10 | 2 | 22 | 1,569 |
| 1987 | 8 | 6 | 1 | 15 | 22,149 |
| 1988 | 4 | 8 | 1 | 13 | 2,128 |
| Total | 234 | 246 | 65 | 545 | 136,833 |
| Percent | 43 | 45 | 12 | 100 | |

**TABLE C.25** Waterborne disease outbreaks in the United States, by type of water-supply system, 1971 to 1988

**DESCRIPTION**

While the United States is not subject to the concentration of waterborne disease per capita found in developing countries, as many as 22,000 cases have been reported for any given year after 1971. Most of these cases are the milder forms of gastrointestinal illnesses and *Giardia lamblia*. Other common cases include chemical problems, shigella, salmonella poisoning, *Cryptosporidium*, Norwalk, and *Campylobacter*. These can be spread in community (municipal), noncommunity (semi-public), or individual water systems, and in locations including lakes, restaurants, hotels, community centers, and private homes. They can be brought about through use of untreated surface water or ground water, problems with treatment or distribution systems, or in other ways. This table summarizes all waterborne disease outbreaks in the United States over the period from 1971 to 1988, identifying the type of water supply system involved, the total number of outbreaks, and the total number of cases.

**LIMITATIONS**

No detail is given here on the types of diseases involved in these outbreaks, though the vast majority are associated with *Giardia*, the *Cryptosporidium* protozoan, and the Norwalk virus. No data are given here on the severity of the outbreaks. Excluded from these data are cases of waterborne diseases imported from other countries.

**SOURCES**

W.C. Levine and G.F. Craun, 1990, Waterborne disease outbreaks, 1986–1988, *CDC Surveillance Summaries: Morbidity and Mortality Report*, **39**(SS-1), 1–13, U.S. Department of Health and Human Services, Public Health Service Centers for Disease Control, Massachusetts Medical Society, Waltham, Massachusetts.

**TABLE C.26**   Health effects of some major water developments

| Project | Date of completion | Health effect | Prevalence | |
|---|---|---|---|---|
| | | | Preproject | Postproject |
| Vo Ha Dam (Ghana) | 1966 | Schistosomiasis | 3% | 70% |
| Sugar estate irrigation (Tanzania) | 1968 | Schistosomiasis | Low | 85% |
| Kainji Dam (Nigeria) | 1969 | Schistosomiasis | Low | 30–70% |
| Aswan High Dam (Egypt, Sudan) | 1969 | Schistosomiasis | 10% | 100% |
| Ubolratana Dam (Thailand) | 1970 | Helminths | No data | 52–90% |
| Malumfashi irrigation project (Nigeria) | 1978 | Schistosomiasis | Low | 65% |
| Srinagarind Dam | 1978 | Malaria | 16% | 25% |
| Gezira irrigation project (Sudan) | 1979 | Schistosomiasis | Low | 70% |

**TABLE C.26**   Health effects of some major water developments

**DESCRIPTION**

Water projects can lead to difficult and costly problems for populations living near them. Among these problems are waterborne and water-related diseases. Increased irrigation or the creation of reservoirs often provides habitat suitable for mosquitoes and the related problems of malaria. Reservoirs can be a breeding ground for the tsetse fly and sleeping sickness. Introduction of year-round irrigation improves conditions for the snail that transmits schistosomiasis. Some positive impacts can occur; for example, water development can reduce the incidence of river blindness (onchocerciasis). This table gives examples of large dam or irrigation projects completed during the 1960s and 1970s in which the incidence of some waterborne diseases increased significantly after project completion. Estimates of the pre- and post-project disease prevalence in the neighboring population is given in percentages.

**LIMITATIONS**

Only a few examples of the health effects of water developments are given here. Estimates of pre- and post-project prevalence are rough, because of the inadequacy of surveys and the inaccuracy of reporting methods. Disease prevalence varies with time, and only average values are reported here. No discussion is provided in the sources about the areal extent of the affected population, the total population at risk, or the period used to estimate the health effects.

**SOURCES**

Data from World Health Organization, 1985, *Environmental Health Impacts Assessment of Irrigated Agricultural Development Projects: Guidelines and Recommendations*, Environmental Health Services, World Health Organization, Copenhagen.

As cited in J.W. Moore, 1985, *Balancing the Needs of Water Use*, Springer-Verlag, New York. By permission.

**TABLE C.27** Proportion of preventable water-related diseases in East Africa

| Diagnosis | Expected reduction if water supply is excellent (%) |
|---|---|
| Guinea worm | 100 |
| Typhoid | 80 |
| Leptospirosis | 80 |
| Urinary schistosomiasis | 80 |
| Trypanosomiasis gambiense | 80 |
| Scabies | 80 |
| Yaws | 70 |
| Inflammatory eye diseases | 70 |
| Trachoma | 60 |
| Schitosomiasis, unspecified | 60 |
| Tinea | 50 |
| Bacillary dysentery | 50 |
| Gastroenteritis, 4 wk to 2 yr | 50 |
| Skin and subcutaneous infections | 50 |
| Amebiasis | 50 |
| Dysentery, unspecified | 50 |
| Diarrhea of the newborn | 50 |
| Gastroenteritis, over 2 yr | 50 |
| Ascariasis | 40 |
| Intestinal schistosomiasis | 40 |
| Louse-borne relapsing fever | 40 |
| Otitis externa | 40 |
| Paratyphoid and other Salmonella | 40 |
| Chronic shin (leg) ulcer | 40 |
| Louse-borne typhus | 40 |
| Trypanosomiasis, unspecified | 10 |
| Dental caries | 10 |

**TABLE C.27** Proportion of preventable water-related diseases in East Africa

**DESCRIPTION**

Some water-related diseases can be controlled or eliminated by improving the quality of the water supply. One estimate of the possible reductions in certain water-related diseases if the water supply is excellent is provided here, for East Africa.

**LIMITATIONS**

Many water-related diseases cannot be eliminated with clean drinking water alone; improvements in sanitation services and the elimination of disease vectors are also necessary. These data are rough estimates for one region of Africa only. No definition of "excellent" is provided. A more recent study with more information on the ranges of expected improvements is described in Table C.28.

**SOURCES**

G.F. White, D.J. Bradley and A.U. White, 1972, *Drawers of Water*, University of Chicago Press, Chicago.

As cited in R.C. Cembrowicz, 1984, Technically, socially and economically appropriate technologies for drinking water supply in small communities, in U. Neis and A. Bittner (eds.) *Water Reuse: Selected Reports in Water Reuse in Urban and Rural Areas*, Institute of Water Resources Engineering, Karlsruhe, Institute for Scientific Co-operation, Tubingen.

**TABLE C.28**   Expected reduction in morbidity from improved water supply and sanitation

| | All Studies | | | Better Studies | | |
|---|---|---|---|---|---|---|
| | Number of Studies | Median (%) | Range (%) | Number of Studies | Median (%) | Range (%) |
| Diarrheal diseases | 55 | 26 | 0–100 | 20 | 29 | 0–68 |
| Ascariasis | 11 | 28 | 0–70 | 4 | 29 | 15–70 |
| Guinea worm | 7 | 76 | 37–98 | 2 | 78 | 75–81 |
| Hookworm | 9 | 4 | 0–100 | | | |
| Schistosomiasis | 4 | 73 | 59–87 | 3 | 77 | 59–87 |
| Trachoma | 13 | 50 | 0–91 | 7 | 27 | 0–79 |

**TABLE C.28**   Expected reduction in morbidity from improved water supply and sanitation

### DESCRIPTION

Despite efforts to improve water quality and the availability of sanitation services in recent decades, waterborne diseases are still widespread, particularly in the developing world. Many studies have been done on the health benefits that might result from improving access to clean drinking water and sanitation services. This table summarizes many of those studies, describing for all studies, and for a subset of the "better" ones, the median reduction and the range of reduction in morbidity that would occur for six important water-related diseases.

### LIMITATIONS

The authors made an assessment of the quality of the many analyses done on possible improvements in health resulting from improvements in water supply and sanitation. This assessment reflects their expert judgement. With the exception of trachoma, the median values of the "better" studies are not significantly different from the median of all studies, though the overall ranges are often considerably smaller. The effects of water supply improvements on the incidence of many other water-related diseases are not evaluated here.

### SOURCE

S.A. Esry, J.B. Potash, L. Roberts and C. Shiff, 1990, *Health Benefits from Improvements in Water Supply and Sanitation: Survey and Analysis of the Literature on Selected Diseases*, WASH Technical Report no. 66, Water and Sanitation for Health (WASH) Project, Office of Health, Bureau for Science and Technology, United States Agency for International Development, Arlington, Virginia.

# D. Water quality and contamination

**TABLE D.1**  Drinking water standards of the United States, Canada, EEC, and WHO *(pages 225–230)*

## DESCRIPTION

Drinking water quality standards set by the World Health Organization (WHO), Canada, the European Economic Community (EEC), and the U.S. Environmental Protection Agency (EPA) are compiled here. The table provides information on the "safe" concentration of a specific contaminant, the relative stringency of water quality standards set by different organizations, and the number of contaminant-specific standards. Standards provided are primary standards required for the protection of human health unless otherwise indicated. In the United States, primary standards, "maximum contaminant levels," are the maximum permissible levels of a contaminant in water that enters the distribution system of public water supply. These standards are legally enforceable. Canadian "maximum acceptable concentrations" are guidelines (generally non-enforceable) for water quality. Both the WHO and the EEC standards are considered recommended, but unenforceable, guidelines. Secondary standards, indicated by a parenthetical "S", are adopted for aesthetic reasons; a parenthetical "P" indicates where EPA or Canadian standards have been proposed but not yet adopted. Most standards are measured in either milligrams or micrograms per liter. Radioactive constituents are measured in becquerels per liter. Water quality criteria are divided into categories, including microbiological, particulate matter, organic pollution indicators, nitrogenous compounds, salinity and specific ions, inorganic micropollutants, organic micropollutants, pesticides, and radioactive constituents.

## LIMITATIONS

Water quality standards are in flux. Standards are being proposed and revised at a rapid rate. This table has been simplified to provide a single number for each contaminant in each region. In many cases, both primary and secondary standards exist for the same contaminant. Moreover, standards may not apply uniformly to all drinking water supplies. In many cases the absence of a standard reflects economic and technical considerations rather than the absence of a threat to human health. The distinction between primary and secondary standards is only made for the U.S. and Canadian data. U.S. secondary standards are non-enforceable, are maintained to protect public welfare, are applied at the point of delivery to the consumer, and involve protection of the taste, odor, or appearance of drinking water. For Canada, "proposed" or "interim" standards are created when there is insufficient data to derive a final maximum acceptable concentrations. These standards are periodically reviewed. Canadian secondary standards are equivalent to "aesthetic objectives." Within the U.S., stricter standards may have been adopted by individual states. The existence of a standard does not imply regular monitoring.

## SOURCES

World Health Organization, 1984, *Guidelines for Drinking-Water Quality*, vol. 1, World Health Organization, Geneva.

United States Environmental Protection Agency, 1992, *Drinking Water Standards and Health Advisory Table*, U.S. Environmental Protection Agency Drinking Water and Ground Water Protection Branch, San Francisco. Updated by personal communication with Bruce Macler, EPA, 1992.

Minister of National Health and Welfare, 1989, *Guidelines for Canadian Drinking Water Quality*, 4th edn., Canadian Government Publishing Center, Ottawa. Updated by personal communication with David Green, MNHW, 1992 (with 5th edn. changes).

Personal communication with European Economic Community scientific attache, Washington, DC, 1992.

**TABLE D.1** Drinking water standards of the United States, Canada, EEC, and WHO.

| Category<br>Variable | Unit | WHO<br>(1984) | Canada<br>(1992) | EEC<br>(1980) | U.S. EPA<br>(1992) |
|---|---|---|---|---|---|
| *Microbiological criteria* | | | | | |
| Total coliforms[a] | Per 100 ml | 0 to 10 | 0 | 0 | 0 |
| Fecal coliforms | Per 100 ml | 0 | 0 | | |
| *Particulate matter* | | | | | |
| Turbidity | NTU | <1 to 5 | <1 to 5 | 0–4 | 1 NTU monthly |
| | JTU | | | 4 | 5 NTU two-day consecutive average |
| *Pollution indicators* | | | | | |
| Hardness | CaCO$_3$ mg/l | 500 | 500 (S) | | |
| pH range | pH | 6.5-8.5 | 6.5-8.5 (S) | 6.5-8.5 | 6.5-8.5 (S) |
| Phosphate | mg/l | | | | |
| Total dissolved solids | mg/l | 1,000 | 500 (S) | | 500 (S) |
| *Aesthetic indicators* | | | | | |
| Color | Color Units | 15 | 15 (S) | 20 | 15 (S) |
| Foaming agents | mg/l | | | | 0.5 (S) |
| Odor (Threshold) | Odor Threshold numbers | | Inoffensive | | 3 (S) |
| Temperature | Degrees C | | 15 (S) | 25 | |
| *Inorganic pollutants* | | | | | |
| Aluminum | mg/l | 0.2 | | 0.2 | 0.02 (S) |
| Antimony | mg/l | | | 0.01 | 0.006 |
| Arsenic | mg/l | 0.05 | 0.025 | 0.05 | 0.05 |
| Asbestos | Long fibers | | | | 7 ×10$^6$ long fibers |
| Barium | mg/l | | 1.0 (P) | 0.1 (GL) | 2 |
| Beryllium | mg/l | | | | 0.004 |
| Boron | mg/l | | 5.0 | 1 (GL) | |
| Cadmium | mg/l | 0.005 | 0.005 | 0.005 | 0.005 |
| Chloride | mg/l | 250 | 250 (S) | 25 (GL) | 250 (S) |
| Chromium | mg/l | 0.05 | 0.05 | 0.05 | 0.1 |
| Cobalt | mg/l | | | | |
| Copper | mg/l | 1 | 1.0 (S) | 3 | 1.3[e] |
| Cyanide | mg/l | 0.1 | 0.2 | 0.05 | 0.2 |
| Fluoride | mg/l | 1.5 | 1.5 | 1.5[i] | 4 |
| Hydrogen sulfide | mg/l | | | | |

*continued*

| Category Variable | Unit | WHO (1984) | Canada (1992) | EEC (1980) | U.S. EPA (1992) |
|---|---|---|---|---|---|
| Iron | mg/l | 0.3 | 0.3 (P) | 0.2 | 0.3 (S) |
| Lead | mg/l | 0.05 | 0.01 (P) | 0.05 | 0.015[f] |
| Manganese | mg/l | 0.1 | 0.05 (P) | 0.05 | 0.05 (S) |
| Mercury | mg/l | 0.001 | 0.001 | 0.001 | 0.002 |
| Nickel | mg/l | | | 0.05 | 0.1 |
| Nitrate-N | mg/l | 10 | 10 (P) | 50 | 10 |
| Nitrite-N | mg/l | | 1.0 (P) | 0.1 | 1 |
| Selenium | mg/l | 0.01 | 0.01 | 0.01 | 0.05 |
| Silver | mg/l | | | 0.01 | 0.05 |
| Sodium | mg/l | 200 | 200 (P) | 150 | |
| Sulfate | mg/l | 400 | 500 (P) | 250 | 250 (S) |
| Thalium | mg/l | | | | 0.0005 |
| Zinc | mg/l | 5 | 5 (P) | 5 | 5 (S) |
| | | | | | |
| *Organic micropollutants* | | | | | |
| Adipates | µg/l | | | | 500 (P) |
| Alachlor | µg/l | | | | 2 |
| Aldicarb | µg/l | | 9 | | 3 |
| Aldicarb Sulfone | µg/l | | | | 2 |
| Aldicarb Sulfoxide | µg/l | | | | 4 |
| Aldrin/dieldrin | µg/l | 0.03 | 0.7 | | |
| Atrazine | µg/l | | 60 (S) | | 3 |
| Azinphos-methyl | µg/l | | 20 | | |
| Bendiocarb | µg/l | | 40 | | |
| Benzene | µg/l | 10[d] | 5 | | 5 |
| Benzo(a)pyrene (PAHs) | µg/l | 0.01[d] | 0.01 | | 0.2 |
| Bromodichloromethane (THM)[c] | µg/l | | | | 100 |
| Bromoform (THM)[c] | µg/l | | | | 100 |
| Bromoxynil | µg/l | | 5 (P) | | |
| Butyl benzylphtalate | µg/l | | | | 4 (P) |
| Carbaryl | µg/l | | 90 | | |
| Carbofuran | µg/l | | 90 | | 40 |
| Carbon tetrachloride | µg/l | 3[d] | 5 | | 5 |
| Chlordane | µg/l | 0.3 | 7 | | 2 |
| Chlorobenzene | µg/l | | | | 100 |
| Chloroform (THM)[c] | µg/l | 30[d] | | | 100 |
| Chlorpyrifos | µg/l | | 90 | | |
| Cyanazine | µg/l | | 10 (P) | | |

*continued*

| Category Variable | Unit | WHO (1984) | Canada (1992) | EEC (1980) | U.S. EPA (1992) |
|---|---|---|---|---|---|
| Dalapon | µg/l | | | | 200 |
| DDT | µg/l | 1 | 30 | | |
| Diadipate | µg/l | | | | 400 |
| Diazion | µg/l | | 20 | | |
| Dibromochloromethane (THM)[c] | µg/l | | | | 100 |
| 1,2-Dibromo-3-chloropropane | µg/l | | | | 0.2 |
| Dibutylphthalate | µg/l | | | | 4 (P) |
| Dicamba | µg/l | | 120 | | |
| 1,2-Dichlorobenzene | µg/l | | 200 | | 600 |
| 1,4-Dichlorobenzene | µg/l | | 5 | | 75 |
| 1,2-Dichloroethane | µg/l | 10[d] | 5 (P) | | 5 |
| 1,1,-Dichloroethene | µg/l | 0.3[d] | | | 7 |
| cis-1,2-Dichloroethene | µg/l | | | | 70 |
| trans-1,2-Dichloroethene | µg/l | | | | 100 |
| Dichloromethane | µg/l | | 50 | | |
| 2,4-Dichlorophenoxyacetic acid | µg/l | 100 | 100 | | 70 |
| 2,4-Dichlorophenol | µg/l | | 900 | | |
| 1,2-Dichloropropane | µg/l | | | | 5 |
| Diclofop-methyl | µg/l | | 9 | | |
| Dimethoate | µg/l | | 20 (P) | | |
| Dinoseb | µg/l | | 0.01 | | 7 |
| Dioxin | µg/l | | | | $3 \times 10^{-5}$ |
| Diphthalate (PAE) | µg/l | | | | 4 |
| Diquat | µg/l | | 70 | | 20 |
| Diuron | µg/l | | 150 | | |
| Endothall | µg/l | | | | 100 |
| Endrin | µg/l | | | | 0.2 g (2.0) |
| Ethylbenzene | µg/l | | 2.4 (S) | | 700 |
| Ethylene Dibromide | µg/l | | | | 0.05 |
| Glyphosphate | µg/l | | 280 (P) | | 700 |
| Heptachlor | µg/l | 0.1[h] | 3[h] | | 0.4 |
| Heptachlor epoxide | µg/l | | | | 0.2 |
| Hexachlorobenzene (HCB) | µg/l | 0.01[d] | | | 1 |
| Hexachlorocyclopentadiene (HEX) | µg/l | | | | 50 |
| Lindane | µg/l | 3 | 4 | | 0.2 |
| Malathion | µg/l | | 190 | | |
| Metolachlor | µg/l | | 50 (P) | | |
| Methoxychlor | µg/l | 30 | 900 | | 40 |

*continued*

| Category Variable | Unit | WHO (1984) | Canada (1992) | EEC (1980) | U.S. EPA (1992) |
|---|---|---|---|---|---|
| Methylene Chloride | μg/l | | | | 5 |
| Metribuzin | μg/l | | 80 | | |
| Monochlorobenzene | μg/l | | 80 | | |
| Nitrilotriacetic acid (NTA) | μg/l | | 400 | | |
| Oxamyl | μg/l | | | | 200 |
| PCBs | μg/l | | | | 0.5 |
| Paraquat | μg/l | | 10 (P) | | |
| Parathion | μg/l | | 50 | | |
| Pentachlorophenol | μg/l | 10 | 60 | | 1 |
| Phenols | μg/l | | | 0.5 | |
| Phorate | μg/l | | 2 (P) | | |
| Picloram | μg/l | | 190 (P) | | 500 |
| Polynuclear aromatic hydrocarbons | μg/l | | | | 0.2 |
| Simazine | μg/l | | 10 (P) | | 4 |
| Styrene | μg/l | | | | 100 |
| Temephos | μg/l | | 280 | | |
| Tetrachloroethene | μg/l | 10[d] | | | 5 |
| Terbufos | μg/l | | 1 (P) | | |
| 2,3,4,6-Tetrachlorophenol | μg/l | | 100 | | |
| Toluene | μg/l | | 24 (S) | | 1,000 |
| Toxaphene | μg/l | | | | 3 |
| Triallate | μg/l | | 230 | | |
| 1,2,4-Trichlorobenzene | μg/l | | | | 70 |
| 1,1,1-Trichloroethane | μg/l | | | | 200 |
| 1,1,2-Trichloroethane | μg/l | | | | 5 |
| Trichloroethene | μg/l | 30[d] | 50 | | 5 |
| 2,4,6-Trichlorophenol | μg/l | 10[d] | 5 | | |
| 2,4,5-Trichlorophenoxyproprionic acid | μg/l | | 280 | | 50 |
| Trifluralin | μg/l | | 45 (P) | | |
| Trihalomethanes (THM)[c] | μg/l | | 350 | 1 | 100 |
| Vinyl Chloride | μg/l | | | | 2 |
| Xylenes (sum of isomers) | μg/l | | 300 (S) | | 10,000 |
| | | | | | |
| *Radioactive constituents* | | | | | |
| Gross alpha activity | Bq/l | 0.1 | | | 0.56 (15pCi/l) |
| Gross beta activity | Bq/l | 1 | | | 4 mrem/yr[b] |
| Cesium 137 | Bq/l | | 50 | | |
| Iodine 131 | Bq/l | | 10 | | |

*continued*

| Category Variable | Unit | WHO (1984) | Canada (1992) | EEC (1980) | U.S. EPA (1992) |
|---|---|---|---|---|---|
| Radium 226, 228 | Bq/l | | 1 | | 0.19 (5 pCi/l) |
| Radon | Bq/l | | | | 11.1 (300 pCi/l) (P) |
| Strontium 90 | Bq/l | | 10 | | |
| Tritium | Bq/l | | 40,000 | | |
| Uranium | mg/l | | 0.1 | | 0.02 |

[a]For systems analyzing at least 40 samples per month, the MCL in no more than 5% of the monthly samples may be total coliform-positive. For systems analyzing less than 40 samples per month, the MCL in no more than one sample per month may be total coliform-positive.

[b]4 mrem/yr is a maximum dosage; a water concentration standard will be created in the future.

[c]Total trihalomethanes (U.S. EPA) MCL includes four compounds: chloroform, bromodichloromethane, dibromochlormethane, and bromoform.

[d]These guideline values were computed from a conservative hypothetical mathematical model which cannot be experimentally verified and values should therefore be interpreted differently. Uncertainties involved may amount to two orders of magnitude (i.e., from 0.1 to 10 times the number).

[e]Treatment technique triggered at action level of 1.3 ppm.

[f]Treatment technique triggered at action level of 0.015 ppm.

[g]Endrin was changed on May 11, 1992, to a final MCL of 2.0 ppb; this will take effect after 18 months.

[h]Heptachlor data include heptachlor epoxide.

[i]At 8–12° C; 0.7 at 25-30°C.

Blank entries are used if the original source had a blank, no guideline set, or not available.

NTU, nephelometric turbidity units; JTU, Jackson turbidity units; MCL, maximum contamination level; MAC, maximum acceptable concentration; Canada (S), aesthetic objective; U.S. EPA (S), secondary MCL; Canada (P), interim MAC or proposed MAC; U.S. EPA (P), Proposed MCL. EEC standards labeled GL (guide level) are targets for which member countries are to aim; all other EEC standards are MAC's.

**TABLE D.2**   Sources of individual ions in rainwater

| | Origin | | |
| Ion | Marine input | Terrestrial inputs | Pollution inputs |
|---|---|---|---|
| $Na^-$ | Sea salt | Soil dust | Burning vegetation |
| $Mg^{2-}$ | Sea salt | Soil dust | Burning vegetation |
| $K^-$ | Sea salt | Biogenic aerosols<br>Soil dust | Burning vegetation<br>Fertilizer |
| $Ca^{2-}$ | Sea salt | Soil dust | Cement manufacture<br>Fuel burning<br>Burning vegetation |
| $H^-$ | Gas reaction | Gas reaction | Fuel burning to form gases |
| $Cl^-$ | Sea salt<br><br>Gas release from sea salt | — | Industrial HCl |
| $SO_4^{2-}$ | Sea salt<br>Marine gases<br>(DMS) | $H_2S$ from biological decay<br>Volcanoes<br>Soil dust<br>(biogenic aerosols) | Burning of fossil fuels to $SO_2$<br>Forest burning |
| $NO_3^-$ | $N_2$ plus lightning | $NO_2$ from biological decay<br>$N_2$ plus lightning | Gaseous auto emissions<br>Combustion of fossil fuels<br>Forest burning<br>Nitrogen fertilizers |
| $NH_4^-$ | — | $NH_3$ from bacterial decay | Ammonia fertilizers<br>Decomposition of human and animal wastes<br>Combustion |
| $PO_4^{3-}$ | — | Soil dust<br>Biogenic aerosols<br>Absorbed on sea salt | Burning vegetation<br>Fertilizer |
| $HCO_3^-$ | $CO_2$ in air | $CO_2$ in air<br>Soil dust | — |
| $SiO_2$, Al, Fe | — | Soil dust | Land clearing |

**TABLE D.2**   Sources of individual ions in rainwater

**DESCRIPTION**

The common ions found in rainwater are listed here with their origins: marine, terrestrial, and anthropogenic. The principal sources of each are also shown.

**LIMITATIONS**

No quantitative information is given on the responsibility of each source for the concentration in rainwater. Actual ionic concentrations vary greatly from region to region, depending on sources, atmospheric conditions, and other factors.

**SOURCES**

E.K. Berner and R.A. Berner, 1987, *The Global Water Cycle: Geochemistry and Environment*, p. 77. Reprinted/adapted by permission of Prentice-Hall, Englewood Cliffs, New Jersey.

**TABLE D.3**  Principal sources and causes of salinity

| Source | Rivers | Lakes and Reservoirs | Ground Water |
|---|---|---|---|
| *Natural* | | | |
| Evaporation | | XX | XX |
| Dissolution of minerals | X | | XX |
| Airborne sea salt | X | X | X |
| Juvenile water | X | | X |
| Connate water | | | X |
| | | | |
| *Anthropogenic* | | | |
| Irrigated agriculture | | | |
| a) Waterlogging and salinization | | | XXXX |
| b) Irrigation return flows | XX | X | X |
| c) Excessive river water withdrawals | XX | XXX | |
| d) Overpumping ground water | | | XX |
| Saline intrusion | | | XXXX |
| Mining activities | XX | X | XX |
| Disposal of oilfield brines | | | XX |
| Upconing of connate water | | | X |
| Highway deicing | X | | X |
| Landfill leachates | X | | XX |
| Leaking sewers | | | X |

Blank spaces are unimportant as a source of salinity.
X, XX, XXX, XXXX are general assessments of the relative importance of sources, increasing in importance.

**TABLE D.3**  Principal sources and causes of salinity

**DESCRIPTION**

The contamination of fresh water by salt or salty water is a major water quality concern. This table qualitatively evaluates the natural and anthropogenic sources of salinity contamination for rivers, lakes and reservoirs, and ground water. Connate water is highly mineralized sea water trapped as ground water. Upconing of connate water is the intrusion of salty ground water into fresh water aquifers.

**LIMITATIONS**

The relative severity of each source is given qualitatively here. No quantitative measure of importance is provided.

**SOURCE**

M. Meybeck, D. Chapman and R. Helmer, 1989, *Global Freshwater Quality: A First Assessment*, World Health Organization and United Nations Environment Programme, Basil Blackwell, Oxford. By permission.

**TABLE D.4** Average concentration of dissolved constituents in river water

| | Concentration (mg/l) | | | | | | | | |
|---|---|---|---|---|---|---|---|---|---|
| | $Ca^{2+}$ | $Mg^{2+}$ | $Na^+$ | $K^+$ | $Cl^-$ | $SO_4^{2-}$ | $HCO_3^-$ | $SiO_2$ | Total |
| Global average[a] | 15 | 4.1 | 6.3 | 2.3 | 7.8 | 11.2 | 58.4 | 13.1 | 118 |
| Global average[b] | 13.4 | 3.3 | 5.1 | 1.3 | 5.7 | 8.3 | 52 | 10.4 | 100 |
| Northwest Territories | 3.3 | 0.7 | 0.6 | 0.4 | 1.9 | 1.9 | 10.1 | 0.4 | 19 |
| Guyana | 2.6 | 1.1 | 2.5 | 0.7 | 3.9 | 2.0 | 12.2 | 10.9 | 36 |
| Philippines | 31 | 6.6 | 10.4 | 1.7 | 3.9 | 13.6 | 131 | 30 | 228 |
| Mackenzie | 33 | 10.4 | 7.0 | 1.1 | 8.9 | 36 | 111 | 3.0 | 210 |
| Colorado | 83 | 24 | 95 | 5.0 | 82 | 270 | 135 | 9.3 | 703 |

Total refers to total dissolved material.
[a]Estimated by Livingstone (1963).
[b]Estimated by Meybeck (1979); all other estimates are by Meybeck (1981).

**TABLE D.4** Average concentration of dissolved constituents in river water

**DESCRIPTION**

Certain ions are commonly found in river water. Shown here are two estimates for the global average concentration of eight common ions, and average estimates for five regions or rivers. All concentrations are in mg/l. The total values given refer to total dissolved material.

**LIMITATIONS**

Average ion concentrations vary over time and space, depending on a wide variety of flow and watershed conditions. Data for only eight ions are shown. No ranges are given here, nor is the period over which these concentrations were measured. The five regions or rivers listed are representative low and high values, but are not necessarily the most extreme.

**SOURCES**

This summary table is from A. Lerman and M.D. Meybeck, 1988, *Physical and Chemical Weathering in Geochemical Cycles*, Kluwer Academic Publishers, Reidel Press, Dordrecht. By permission.

The original data come from:

D.A. Livingstone, 1963, *Chemical Composition of Rivers and Lakes*, U.S. Geological Survey Professional Paper no. 440-G, Washington, DC.

M. Meybeck, 1979, Concentrations des eaux fluviales en elements majeurs et apports en solution aux oceans, *Revue Geogr. Phys.*, 21, 215–246.

M. Meybeck, 1981, Pathways of major elements from land to ocean through rivers, in J.M Martin, J.B. Burton and D. Eisma (eds.) *River Inputs to Ocean Systems*, United Nations Press, New York.

**TABLE D.5** Acidification of precipitation worldwide *(page 234–238)*

**DESCRIPTION**

The problem of acidification of precipitation is increasing worldwide as population growth, urbanization, and fossil fuel use rise. In the mid-1980s, the World Meteorological Organization began coordinating an extensive network of precipitation chemistry monitoring stations known as the Background Air Pollution Monitoring Network (BAPMoN) in cooperation with the U.S. National Oceanic and Atmospheric Administration (NOAA) and the Environmental Protection Agency (EPA). Presented here are data on the pH of precipitation from BAPMoN stations around the world for 1975–1976 through 1983–1984. All parameters are expressed in precipitation weighted means. All stations are situated away from major urban areas and point sources of pollution. Each station monitors, at a minimum, atmospheric turbidity, precipitation chemistry, and suspended particulate matter. Some stations monitor a wider range of pollutants. Baseline stations are in italics.

**LIMITATIONS**

The data here have not been rigorously and scientifically edited, hence they should not be used for continental trend analysis, according to those who manage the data. Some regional trends may be evident, however. Some rejection and editing procedures have not been applied to this data set; for example, differences in the frequency of measurements and the type of precipitation collector have not always been taken into account.

**SOURCES**

Cited in United Nations Environment Programme, 1989, *Environmental Data Report 1989/90*, GEMS Monitoring and Assessment Research Centre, Basil Blackwell, Oxford. By permission.

Data from WMO Collaborative Centre on BAPMoN Data, US Environmental Protection Agency, Research Triangle Park, North Carolina.

**TABLE D.5**  Acidification of precipitation worldwide

| Region | Station | pH | | | | |
|--------|---------|---------|---------|---------|---------|---------|
| | | 1975–76 | 1977–78 | 1979–80 | 1981–82 | 1983–84 |
| *Africa* | | | | | | |
| Tunisia | Thala | | | 6.40[a] | | |
| Zambia | Mfuwe | | | — | | |
| *North America* | | | | | | |
| Canada | Edson | 4.90[a] | 5.10 | 5.20 | 5.50 | 5.40[a] |
| | Hay River | | | 6.10[a] | 6.10 | 6.40[a] |
| | Kelowna | | 5.50 | 5.40 | 5.20 | 5.00[a] |
| | Maniwaki | 4.30 | 4.30 | 4.20 | 4.30 | 4.50[a] |
| | Mould Bay | | 6.50 | — | 5.10 | |
| | Mount Forest | 4.40 | 4.60 | 4.70 | 4.50 | 4.50[a] |
| | Pickle Lake | | 5.30 | 5.20 | 5.00 | |
| | *Sable Island* | 4.80 | 4.70 | 4.70[a] | 4.80[a] | |
| | Wynyard | — | 6.40 | 6.50[a] | 6.40 | 5.90[a] |
| El Salvador | Cerro Verde[d] | 5.00 | 4.90[a] | 5.10[a] | | |
| Greenland | Godhavn[d] | 5.30[a] | 5.10 | 4.90 | 5.10[a] | |
| Panama | La Yeguada | | | | 5.40[a] | — |
| United States | Atlantic Co. NJ | 4.30 | 4.20 | 4.30[a] | | |
| | Bishop CA | — | 5.40[a] | 6.10 | 5.70 | |
| | Caribou ME | 4.60 | 4.30 | 4.50 | | |
| | Glacier N.P. MT | 5.50 | 5.60 | | | |
| | Grand Canyon N.P. AZ | | | | | |
| | Huron SD | 6.30 | 6.60[a] | 6.10 | 5.50 | 5.40[a] |
| | Macon Co. NC | 4.40[a] | 4.30 | 4.50 | | |
| | *Mauna Loa HI* | 4.80[a] | — | 4.70[a] | 5.40 | 5.10[a] |
| | Mercer Co. NJ | | | 4.20[a] | 4.30 | 4.40[a] |
| | Meridian MS | 4.60 | 4.60 | 4.80 | 4.70 | 4.80[a] |
| | Pendleton OR | 5.40 | 5.30 | 5.30 | | |

*continued*

| Region | Station | pH | | | | |
|--------|---------|---------|---------|---------|---------|---------|
| | | 1975–76 | 1977–78 | 1979–80 | 1981–82 | 1983–84 |
| | Raleigh NC | 4.40 | 4.30 | 4.40 | | |
| | Salem IL | 4.40 | 4.20 | 4.40 | 4.40 | |
| | San Angelo TX | | 5.40[a] | | | |
| | Tahlequah OK | 5.00 | 5.00 | 5.00 | | |
| | Victoria TX | 5.20 | 4.90 | 5.10 | 5.00 | 5.00[a] |
| *South America* | | | | | | |
| Columbia | Los Gaviotos | | | | | 4.90 |
| *Asia* | | | | | | |
| China | Shangdianzi | | | | | 6.30[a] |
| India | Allahabad[c] | | 7.00[a] | 7.50[a] | 6.90[a] | 6.80 |
| | Jodphur[c] | | 8.30 | — | 7.30 | 7.30[a] |
| | Kodaikanal[c] | | — | 6.00[a] | 6.30 | 6.00 |
| | Minicoy[c] | | 6.90[a] | 6.60 | 6.60 | 6.20 |
| | Mohanbari[c] | — | | — | 6.50 | 6.60 |
| | Nagpur[c] | | | 6.30[a] | 6.40 | 6.40 |
| | Port Blair[c] | 6.30[a] | 6.50[a] | — | — | 6.00 |
| | Pune[c] | 7.30 | 7.10 | 6.80 | 6.60[a] | 6.00 |
| | Srinagar[c] | | 7.80[a] | 7.10 | 6.80 | 7.40 |
| | Visakhapatnam[c] | | 7.10 | 6.50 | 6.30 | 6.30 |
| Indonesia | BMG (Jakarta) | | | | 5.70[a] | 4.60[a] |
| Japan | *Ryori* | 5.40[a] | 4.90 | 5.00 | 4.70 | 4.50 |
| Malaysia | Tanah Rata | 4.80[a] | 5.00 | 5.00 | 5.20 | 5.30 |
| Turkey | Camkoru | | | | — | — |
| *Europe* | | | | | | |
| Austria | Retz | 4.50 | 4.80 | 4.60 | 4.80 | 5.40[a] |
| | Wiel | 4.30 | 4.40 | 4.50 | 4.80[a] | 4.80[a] |

*continued*

| Region | Station | pH | | | | |
|---|---|---|---|---|---|---|
| | | 1975–76 | 1977–78 | 1979–80 | 1981–82 | 1983–84 |
| Bulgaria | Rojen (Guletchitza) | | | 5.40 | | 4.40 |
| Czechoslovakia | Chopok[c,d] | | 4.30 | 4.20 | 4.30 | 4.40 |
| | Svratouch[c,d] | 4.20 | 4.20 | 4.30 | 4.40 | 4.40 |
| Denmark | Anholt | — | 4.30 | — | | |
| | Edje Loran (Faeroe Is.)[d] | 4.70 | 4.50 | 4.60 | 4.80 | 4.50 |
| | Tange | | | | 4.40 | — |
| Finland | Jokioinen | 4.20 | 4.30 | 4.40 | 4.30 | 4.30 |
| | Sodankyla | 4.60 | 4.60 | 4.60 | 4.50 | 4.60 |
| France | Abbeville | | 4.50 | 4.50 | 4.70 | 4.70 |
| | Carpentras | | 4.90 | 4.90 | 5.10 | 5.00 |
| | Chateau Chinon | | — | 4.80 | 4.90 | 4.70 |
| | Gourdon | | 4.70 | 4.90 | 5.20 | 5.30 |
| | Phalsbourg | | 4.40 | 4.30 | 4.70 | 4.50 |
| | Rostrenen | | 4.90 | 5.00 | 5.20 | 4.90 |
| German DR | Neuglobsov[c,d] | 4.00 | 3.90 | 4.20[b] | 4.20 | 4.20 |
| Germany FR | Brotjacklriegel[c,d] | 4.30 | 4.30 | 4.10 | 4.50 | 4.40[a] |
| | Deuselbach[c,d] | 4.20[a] | 4.20 | 4.20 | 4.20 | 4.20[a] |
| | Langenbrugge[c,d] | 4.10 | 3.90[b] | 3.80 | 4.10 | 4.20[a] |
| | Schauinsland[c,d] | 4.10 | 4.40 | 4.50 | 4.60 | 4.60[a] |
| | Wank | | | | 4.20[a] | |
| | Zugspitze | | | | | 4.80 |
| Hungary | K-Puszta (Kecskemet)[c] | 4.00 | 4.70 | 4.90 | 5.50 | 5.90 |
| Iceland | Irafoss | | | | 5.00 | 5.10 |
| Ireland | Valentia Observatory[c] | 5.30 | 5.10 | 5.30 | 5.40 | 5.20 |
| Italy | *Monte Cimone* | 5.00 | 4.90 | 4.90 | 4.40 | 4.40 |
| | Santa Maria di Leuca | 5.20 | 4.70 | 5.00 | 4.60 | 4.40 |
| | Trapani | 5.80[a] | 5.70 | 6.00 | 4.70 | 4.60 |
| | Verona | 4.40 | 4.40 | 4.40 | 4.20 | 4.40 |

*continued*

| Region | Station | pH | | | | |
|---|---|---|---|---|---|---|
| | | 1975–76 | 1977–78 | 1979–80 | 1981–82 | 1983–84 |
| | Viterbo | 4.80 | 4.60 | 4.60 | 4.60 | 4.70 |
| Netherlands | Witteveen[d] | 4.40 | 4.50 | 4.10[a] | 4.50 | 4.60 |
| Norway | Ås | 4.30 | 4.40 | 4.20 | 4.60[a] | |
| | Birkenes | 4.20 | 4.20 | 4.10 | 4.20[a] | |
| | Kise | 4.40 | 4.40 | 4.50 | 4.40[a] | |
| Poland | Sniezka | | | | 4.00 | 4.10 |
| | Suwalki | 4.40[a] | 4.60 | 4.50 | 4.40 | 4.30 |
| Portugal | Braganca | | | 6.00 | 5.90 | 5.60[a] |
| | Faro | | | 6.50 | — | — |
| Romania | Fundata | | | 5.30[a] | 4.40 | 4.90 |
| | Paring | | | 4.90[a] | 4.10 | 4.30 |
| | Rarau | | | 4.40[a] | 4.40 | 4.30 |
| | Semenic | | | 4.40[a] | 4.00 | 4.30 |
| | Stina de Vale | | | 5.40 | 4.50 | 4.40 |
| | Turia | | | 4.70[a] | 5.00 | 4.60 |
| Sweden | Bredkalen | 4.90 | 4.90 | 4.80 | 4.70 | 4.50 |
| | Velen | 4.30 | 4.40 | 4.30 | 4.40 | 4.30 |
| Switzerland | Jungfraujoch | | 5.70[a] | 5.30 | 5.90 | 5.90 |
| | Payerne | | 5.20[a] | 4.80 | 4.80 | 4.50 |
| United Kingdom | Eskadalemuir (Scotland) | | 4.80[a] | 4.40 | 4.50 | 4.30[a] |
| Yugoslavia | Ivan Sedlo | | | | 5.70[a] | 5.70 |
| | Lazarapole | 4.80[a] | 4.90 | 4.90 | 4.90[a] | 5.80 |
| | Puntijarka | 5.20 | 4.80 | 4.90 | | |
| *USSR* | Irkutsk | 5.50[a] | 6.50[a] | 6.20[a] | 6.10 | |
| | Kurgan | 5.50 | 6.00 | 5.50[a] | 5.40 | |
| | Novopyatigorsk | 6.20 | — | 6.30[a] | 6.20 | |
| | Siktivkar | 5.60 | 5.60 | 5.00[a] | 5.60 | |
| | Turukhansk | 5.20[a] | 5.60 | 5.80[a] | 5.70 | |

*continued*

| Region | Station | pH | | | | |
|--------|---------|--------|---------|---------|---------|---------|
| | | 1975–76 | 1977–78 | 1979–80 | 1981–82 | 1983–84 |
| *Oceania* | | | | | | |
| American Samoa | *Cape Matatula* | 5.50 | 5.30 | 5.30 | 5.40 | 5.50[a] |
| Australia | *Cape Grim*[d] | | | | 5.90[a] | 6.10 |
| | Coffs Harbour | | | | | 5.10 |
| New Zealand | Lauder | | | | | 5.80[a] |

A dash indicates that data are insufficient to calculate annual average (i.e., less than 8 months).
Baseline stations are in italics.
[a]Data for one year only.
[b]Average calculated from monthly values measured by different methods.
[c]Change in measurement method for sulfate occurs during monitoring period.
[d]Change in measurement method for nitrate occurs during monitoring period.

**TABLE D.6** pH of rainwater in selected cities in China, early 1980s

| Location | City | Year | pH |
|----------|------|------|-----|
| North | Beijing | 1981 | 6.80 |
| North | Tianjin | 1981 | 6.26 |
| North | Lanzhou | Jan.-Aug. 1982 | 6.85 |
| South | Nanjing | June-Nov. 1981 | 6.38 |
| South | Hangzhou | Sept.-Dec. 1981 | 5.10 |
| South | Wuhan | Jan.-July 1983 | 6.44 |
| South | Fuzhou | May 1982 | 4.49 |
| South | Nanning | June-Nov. 1981 | 5.74 |
| South | Yibin | 1982 | 4.87 |
| South | Chongqing | 1982 | 4.14 |
| South | Guiyang | 1982 | 4.02 |

**TABLE D.6** pH of rainwater in selected cities in China, early 1980s

**DESCRIPTION**

Acid rain is a growing concern, particularly in heavily industrialized areas. For China, massive amounts of uncontrolled coal burning have caused major air pollution problems in urban areas, including high levels of sulfur dioxide and suspended particulate matter. Measurements of the pH of rainfall taken in the early 1980s also indicate that acid rain is prevalent in some urban areas. Presented here are some examples of the range of pH in rainfall found in Chinese cities in the early 1980s.

**LIMITATIONS**

The values given here are average values for multiple measurements, but no detail is given on the number of measurements taken in any city, or on the measurement methods used. The acidity of precipitation varies with the time of year, the atmospheric conditions, and distance from the sources of sulfate and nitrate ions, and is not well represented by single estimates such as those given in this source.

**SOURCE**

D. Zhao and B. Sun, 1986, Air pollution and acid rain in China, *Ambio*, **15**(1), 2–5. By permission.

**TABLE D.7** Emissions factors for the release of trace metals to soil and water *(pages 240–241)*

**DESCRIPTION**

All industries discharge trace elements into soil or water. Shown here is an inventory of the principal industrial and commercial users of water and producers of solid wastes, and emissions factors of trace metals into water (in ng/l) and soils (in μg/g). Ranges are given for all emissions factors. These ranges come from a critical survey of the published literature.

**LIMITATIONS**

Where the authors thought the emissions factors estimates in the literature to be too high, they reported the lower end of the concentration range. It is not possible to place an error range on the calculated inventories, according to the authors, though they believe that the exact global values of the metals released into the environment "are unlikely to differ from those used [in this table] by more than a factor of two or so."

**SOURCES**

J.O. Nriagu and J.M. Pacyna, 1988, Quantitative assessment of worldwide contamination of air, water and soils by trace metals, *Nature*, **333**, 134–139. By permission.

**TABLE D.7** Emissions factors for the release of trace metals to soil and water

| | As | Cd | Cr | Cu | Hg | Mn | Mo | Ni | Pb | Sb | Se | V | Zn |
|---|---|---|---|---|---|---|---|---|---|---|---|---|---|
| **WATER (ng/l)** | | | | | | | | | | | | | |
| Domestic wastewater | 0.02–0.09 | 0.002–0.02 | 0.09–0.4 | 0.05–0.2 | 0–0.002 | 0.2–0.9 | 0–0.03 | 0.1–0.6 | 0.01–0.08 | 0–0.03 | 0–0.05 | 0–0.3 | 0.1–0.5 |
| Central | 0.02–0.12 | 0.005–0.02 | 0.1–0.7 | 0.07–0.5 | 0–0.007 | 0.5–1.5 | 0–0.03 | 0.2–0.8 | 0.01–0.08 | 0–0.03 | 0–0.05 | 0–0.3 | 0.1–0.6 |
| Noncentral | 0.4–0.12 | 0.001–0.04 | 0.5–1.4 | 0.6–3.8 | 0–0.6 | 0.8–3.0 | 0.01–0.2 | 0.5–3.0 | 0.04–0.2 | 0–0.06 | 1.0–5.0 | 0–0.1 | 1.0–5.0 |
| Steam electric | | | | | | | | | | | | | |
| Base metal mining and dressing | 0.04–1.5 | 0.001–0.6 | 0.008–1.4 | 0.2–18 | 0–0.3 | 1.5–23 | 0.005–1.1 | 0.02–1.0 | 0.5–5.0 | 0.08–0.7 | 0.5–5.0 | 0–0.02 | 0.04–12 |
| Smelting and refining | | | | | | | | | | | | | |
| Iron and steel | | | | | | 2.0–5.2 | | | 0.2–0.4 | | | | 0.8–3.5 |
| Nonferrous metals | 0.5–6.4 | 0.004–1.8 | 1.5–10 | 1.2–8.5 | 0.001–0.002 | 1.0–7.5 | 0.003–0.2 | 1.0–12 | 0.5–3.0 | 0.04–3.8 | 1.5–10 | 0–0.6 | 1.0–10 |
| Manufacturing processes | | | | | | | | | | | | | |
| Metals | 0.01–0.06 | 0.02–0.07 | 0.6–2.3 | 0.4–1.5 | 0–0.03 | 0.1–0.8 | 0.02–0.2 | 0.008–0.3 | 0.1–0.9 | 0.1–0.6 | 0–0.2 | 0–0.03 | 1.0–5.5 |
| Chemicals | 0.12–1.4 | 0.02–0.5 | 0.5–4.8 | 0.2–3.6 | 0.004–0.3 | 0.4–3.0 | 0–0.6 | 0.2–1.2 | 0.08–0.6 | 0.004–0.5 | 0–0.07 | 0.04–1.0 | |
| Pulp and paper | 0–0.3 | — | 0.004–0.5 | 0.01–0.13 | — | 0.01–0.15 | — | 0.02–0.04 | 0.004–0.3 | 0.001–0.09 | 0.002–0.3 | — | 0.03–0.5 |
| Petroleum products | 0.002–0.2 | 0–0.04 | 0.001–0.7 | 0.002–0.2 | 0–0.08 | — | — | 0.004–0.2 | 0.003–0.4 | — | 0.002–0.3 | — | 0.01–0.8 |
| **SOILS (µg/g)** | | | | | | | | | | | | | |
| Agricultural and food wastes | 0–0.4 | 0–0.2 | 0.3–6.0 | 0.2–2.5 | 0–0.1 | 1.0–7.5 | 0.6–2.0 | 0.4–3.0 | 0.1–1.8 | 0–0.6 | 0–0.5 | 0.2–1.5 | 0.8–10 |
| Animal wastes, manure | 0.6–2.2 | 0.1–0.6 | 5.0–30 | 7.0–40 | 0–0.1 | 25–70 | 2–12 | 1.5–18 | 1.6–10 | 0–0.43 | 0.2–0.7 | 1.0–5.5 | 75–160 |
| Logging and other wood wastes | 0–0.3 | 0–0.6 | 0.2–1.6 | 0.3–4.7 | 0–0.2 | 1.6–9.5 | 0–0.3 | 0.2–2.1 | 0.6–7.5 | 0–0.5 | 0–0.3 | 0.1–0.9 | 1.2–15 |
| Urban refuse | 0.2–1.6 | 2.0–17 | 15–75 | 30–90 | 0–0.6 | 55–320 | 0.5–10 | 5.0–23 | 40–150 | 0.3–5.0 | 0.1–1.5 | 0.2–1.2 | 80–220 |

continued

|  | As | Cd | Cr | Cu | Hg | Mn | Mo | Ni | Pb | Sb | Se | V | Zn |
|---|---|---|---|---|---|---|---|---|---|---|---|---|---|
| Municipal sewage sludge | 0.3–12 | 1.0–20 | 8–550 | 240–1,030 | 0.5–9.0 | 220–540 | 4.0–16 | 25–110 | 140–480 | 2.1–10 | 0.3–6.9 | 11–73 | 900–2,800 |
| Miscellaneous organic wastes including excreta | 0–0.25 | 0–0.06 | 0.04–2.3 | 0.2–2.9 | 0–0.02 | 0.4–3.0 | 0.3–1.9 | 0.8–15 | 0.08–7.6 | 0–0.5 | 0–0.4 | 0.5–3.6 | 0.6–10 |
| Solid wastes, metal manufacturing | 0.03–0.6 | 0–0.2 | 1.7–6.3 | 2.5–20 | 0–0.1 | 1.1–13 | 0.03–0.4 | 2.2–6.5 | 11–28 | 0–0.4 | 0.03–0.5 | 0.1–0.6 | 7.0–50 |
| Coal fly ash and bottom ash[a] | 1.8–10 | 0.4–3.6 | 40–120 | 25–90 | 0.1–1.3 | 134–445 | 4.1–20 | 15–75 | 12–65 | 0.7–6.0 | 1.1–16 | 3.0–18 | 30–130 |
| Fertilizer | 0–0.1 | 0.2–15 | 0.2–2.3 | 0.3–3.5 | 0–0.02 | 0.8–5.0 | 0–0.1 | 1.2–3.3 | 2.5–14 | 0–0.03 | 0.1–0.6 | 0.2–0.8 | 1.6–6.5 |
| Peat | 0.1–1.3 | 0–0.3 | 0.1–0.5 | 0.4–5.2 | 0–0.05 | 14–45 | 0.4–2.0 | 0.6–9.4 | 1.2–6.8 | 0.1–1.2 | 0–1.1 | 0.2–4.5 | 0.4–9.4 |

[a]The concentrations given are for the coal rather than for fly ash or bottom ash.

**TABLE D.8** Water pollution from industrial effluents *(pages 243–247)*

## DESCRIPTION

Industry is often a major source of waste discharges into rivers and lakes. These discharges cause water quality problems, which vary widely regionally, depending on the type of industrial process, the amount of water used, and other chemicals and microorganisms present in the effluent. Shown here for many industrial groups are the waste volumes in $m^3$ per unit of product, and the kilograms of waste produced per unit of product for six forms of pollutants. Limited data on pH of water effluents is also given. Units of product vary from ''head'' for livestock production to tonnes or $m^3$ of product.

## LIMITATIONS

These industrial waste loads are average values for industries and hide considerable variations within industries. Actual waste loads from a given plant may vary considerably from the industry average. No detail is given here on the range of outputs within industrial groups.

## SOURCE

World Health Organization, 1982, *Rapid Assessment of Sources of Air, Water, and Land Pollution*, WHO Offset Publication no. 62, World Health Organization. Geneva.

*continued*

# TABLE D.8   Water pollution from industrial effluents

| Industry and Process | Unit | pH | Waste volume ($m^3$/unit) | 5-day biological oxygen demand (kg/unit) | Chemical oxygen demand (kg/unit) | Suspended solids (kg/unit) | Total dissolved solids (kg/unit) | Oil (kg/unit) | Nitrogen (kg/unit) |
|---|---|---|---|---|---|---|---|---|---|
| *Agricultural and livestock production* | | | | | | | | | |
| Beef feedlot | Heads | | 20.2 | 250 | | 1,716 | | | 80.3 |
| Swine feedlot | Heads | | 1.6 | 28.4 | | 183 | | | 8.4 |
| Broiler feedlot | Heads | | 0.04 | 1.4 | | 14.6 | | | 0.51 |
| Lamb feedlot | Heads | | 1.8 | 36.6 | | 201 | | | 8.4 |
| Turkey feedlot | Heads | | 0.04 | 15 | | 14.6 | | | 0.51 |
| Duck feedlot | Heads | | 0.04 | 1.4 | | 14.6 | | | 0.51 |
| Dairy farms | Heads | | | 539 | | | | | |
| Layer houses | Heads | | | 4.6 | | | | | |
| | | | | | | | | | |
| *Food manufacturing* | | | | | | | | | |
| Slaughterhouse | t of LWK | | 5.3 | 6.4 | | 5.2 | | 2.8 | 1.58 |
| (Add if blood not recovered) | t of LWK | | | 11 | | | | | |
| (Add if paunch not recovered) | t of LWK | | | 4.7 | | | | | |
| Packinghouse | t of LWK | | 9.3 | 6.3 | | 3 | | 2.3 | 1.59 |
| Poultry processing | $10^3$ birds | | 37.5 | 11.9 | 22.4 | 12.7 | 15 | 5.6 | |
| Dairy processing | t of milk | | 2.4 | 5.3 | | 2.2 | 3.3 | | |
| Canning fruits/vegetables | t of product | | 11.3 | 12.5 | | 4.3 | | | |
| Canning fish | t of product | | 23 | 7.9 | 16 | 9.2 | | 4.5 | 0.64 |
| Olive oil extraction | t of product | 9–5 | 0.5 | 7.5 | 59 | 33 | | | |
| Vegetable oil refining | t of product | | 57.5 | 12.9 | 21 | 16.4 | 882 | 6.5 | |
| Grain mill | t of product | | 0.6 | 1.1 | | 1.6 | | | |
| Sugar cane factory | t of product | | 28.6 | 2.6 | | 3.9 | | | |
| Beet sugar factory | t of product | | 23.4 | 20 | | 75 | | | |
| Starch/glucose factory | t of product | | 33 | 13.4 | 21.8 | 9.7 | 42.3 | 1.2 | |
| Yeast products | t of product | | 150 | 1,125 | | 18.7 | 2,250 | 127.5 | |
| | | | | | | | | | |
| *Beverage industry* | | | | | | | | | |
| Alcohol distilleries | t of product | | 63 | 220 | | 257 | 385 | | |
| Malt/malt liquor manufacturing | $m^3$ of beer | | 4.5 | 1.1 | | 0.2 | | | |
| Beer fermenting | $m^3$ of beer | | 10 | 7.5 | | 14.5 | | | |
| Total for beer production | $m^3$ of beer | | 14.5 | 8.6 | | 14.7 | | | |
| Wine production | $m^3$ of wine | | 4.8 | 0.26 | | | | | |
| Soft drinks factory | t of product | | 7.1 | 2.5 | | 1.3 | | | |

*continued*

| Industry and Process | Unit | pH | Waste volume ($m^3$/unit) | 5-day biological oxygen demand (kg/unit) | Chemical oxygen demand (kg/unit) | Suspended solids (kg/unit) | Total dissolved solids (kg/unit) | Oil (kg/unit) | Nitrogen (kg/unit) |
|---|---|---|---|---|---|---|---|---|---|
| *Manufacture of textiles* | | | | | | | | | |
| Wool (scouring included) | t of product | 2–10 | 544 | 314 | 1,140 | 196 | 481 | 191 | |
| Wool (no scouring) | t of product | 2–10 | 537 | 87 | 347 | 43 | 365 | | |
| Cotton | t of product | 8–11 | 317 | 155 | | 70 | 205 | | |
| Rayon | t of product | | 42 | 30 | 52 | 55 | 100 | | |
| Acetate | t of product | | 75 | 45 | 78 | 40 | 100 | | |
| Nylon | t of product | | 125 | 45 | 78 | 30 | 100 | | |
| Acrylic | t of product | | 210 | 125 | 216 | 87 | 100 | | |
| Polyester | t of product | | 100 | 185 | 320 | 95 | 150 | | |
| *Manufacture of leather* | | | | | | | | | |
| Leather tanneries | t of hides | 1–13 | 52 | 89 | 258 | 138 | 351 | 20 | 15 |
| *Wood products* | | | | | | | | | |
| Plywood manufacturing | $10^3$ $m^2$ | 10.5 | 4.1 | | 7.3 | 1.1 | 5.1 | | 0.24 |
| Fiberboard manufacturing | t of product | | 20 | 125 | | 20 | | | |
| *Manufacture of pulp, paper, and paperboard* | | | | | | | | | |
| Sulfate (kraft) pulp | t of product | | 61.3 | 31 | | 18 | 166 | | |
| Sulfite pulp | t of product | | 92.4 | 130 | | 26 | 258 | | |
| Semichemical pulp | t of product | | 47 | 27 | | 12.5 | 134 | | |
| Paper mills | t of product | | 54 | 8 | | 23 | 37 | | |
| Paper mills (with water recovery) | t of product | | 22 | 6.4 | | 15.2 | 30 | | |
| Paper mills (improved water recovery) | t of product | | 12.5 | 4 | | 11.5 | 15 | | |
| *Manufacture of industrial chemicals* | | | | | | | | | |
| Basic inorganic chemicals | | | | | | | | | |
| Hydrochloric acid | t of product | | [a] | Negl. | Negl. | Negl. | Negl. | Negl. | Negl. |
| Sulfuric acid | t of product | | 1.62 | Negl. | Negl. | Negl. | Negl. | Negl. | Negl. |
| Nitric acid | t of product | | [a] | Negl. | Negl. | Negl. | Negl. | Negl. | Negl. |
| Phosphoric acid (without pond) | t of $P_2O_5$ | | 670 | | | 3,772 | | | 6 |
| Phosphoric acid (with pond) | t of $P_2O_5$ | 1–1.6 | 2.8 | | | | | | 0.15 |
| Phosphoric acid (thermal process) | t of $P_2O_5$ | | 4.6 | | | | | | |
| Ammonia | t of product | | 2.1 | 0.2 | 0.26 | | | 10 | 0.12 |
| Hydrofluoric acid | t of product | | 11.0 | Negl. | Negl. | 2,711 | | | |
| Chrome pigments | t of product | | | | | 70.4 | | | |

*continued*

| Industry and Process | Unit | pH | Waste volume (m³/unit) | 5-day biological oxygen demand (kg/unit) | Chemical oxygen demand (kg/unit) | Suspended solids (kg/unit) | Total dissolved solids (kg/unit) | Oil (kg/unit) | Nitrogen (kg/unit) |
|---|---|---|---|---|---|---|---|---|---|
| Basic organic chemicals | | | | | | | | | |
| Group 1 (see footnotes) | t of product | | 8.3 | 0.11 | 2 | | | | |
| Group 2 (see footnotes) | t of product | | 12.7 | 0.35 | 11 | | | | |
| Group 3 (see footnotes) | t of product | | 12.6 | 63 | 193 | | | | |
| Group 4 (see footnotes) | t of product | | 450 | 136 | 2,500 | | | | |
| Fertilizers | | (Major effluents are those from the production of phosphoric acid and sulfuric acid) | | | | | | | |
| Pesticides | | | | | | | | | |
| DDT | t of product | | 5.3 | | | | | | |
| Chlorinated hydrocarbon herbicides | t of product | 0.5 | 3.6 | 22.7 | 30 | 9 | 365 | | |
| Carbamate | t of product | 7-10 | | 0 | | 0 | | | |
| Parathion | t of product | 2 | | 0 | | 0 | | | |
| Synthetic resins, plastics and fibers | | | | | | | | | |
| Rayon fibers | t of product | | 471 | 68.4 | 355 | 193 | 2,447 | | |
| Vulcanizable elastomers (synthetic rubber) | t of product | | 19.6 | 2.6 | 20 | 12 | | 1.2 | |
| Polyolefins (polythelenes) | t of product | | 0 | | | | | | |
| Polystyrene resins and copolymer | t of product | | 5.7 | Negl. | Negl. | Negl. | | | |
| Vinyl resins (PVC) | t of product | | 12.5 | 10 | | 1.5 | | | |
| Phenolic resins | t of product | 6.4 | 4 | 47.3 | | 1.6 | 0.5 | | |
| Acrylic resins (bulk polymer) | t of product | | 0 | | | | | | |
| Acrylic resins (emulsion polymer) | t of product | | 0.5 | 1.5 | | | | | |
| Paints, Varnishes, and Lacquers | t of product | | Neg. pollution | | | | | | |
| Manufacture of drugs and medicines | | | | | | | | | |
| Erythromycin | t of product | 7.2 | 4,000 | 13,800 | | 5,600 | | | |
| Streptomycin | t of product | 8.5 | 4,000 | 7,400 | | | | | |
| Tetracyclin | t of product | 9.4 | 4,000 | 5,200 | | 1,776 | | | |
| Penicillin | t of product | 4.5 | 4,000 | 12,800 | | | | | |
| Aureomycin | t of product | 8 | 4,000 | 14,280 | | | | | |
| Soap and cleaning preparations | | | | | | | | | |
| Soap by kettle boiling | t of product | | 4.5 | 6 | 10 | 4 | | 0.9 | |
| Soap from fatty acids | t of product | | 3.1 | 13.5 | 29.5 | 23 | | 3.5 | |
| Detergents | t of product | | 2.8 | 0.4 | 1.2 | 0.7 | | 0.4 | |
| Glycerine refining | t of product | | 10 (1,120) | 20 | 40 | 4 | | 2 | |
| Liquid detergents | t of product | | | 5.3 | 7.9 | 0.6 | | | |
| Animal glue (from fleshing) | t of product | | 421 | 2,500 | 4,800 | 4,280 | | | |
| Animal glue (from hides) | t of product | | 457 | 580 | 1,420 | 1,920 | | | |

*continued*

| Industry and Process | Unit | pH | Waste volume (m³/unit) | 5-day biological oxygen demand (kg/unit) | Chemical oxygen demand (kg/unit) | Suspended solids (kg/unit) | Total dissolved solids (kg/unit) | Oil (kg/unit) | Nitrogen (kg/unit) |
|---|---|---|---|---|---|---|---|---|---|
| Animal glue (from chrome stock) | t of product | | 426 | 280 | 650 | 400 | | | |
| Petroleum refining | | | | | | | | | |
|    Typical topping refineries | $10^3$ m³ feedstock | | 66 | 3.4 | 37 | 11.7 | | 8.3 | 1.2 |
|    Old topping refineries | $10^3$ m³ feedstock | | | 190 | | | | 115 | 24 |
|    Low-cracking refineries | $10^3$ m³ feedstock | | 79 | 71.5 | 200 | 27 | | 27 | 10 |
|    High-cracking refineries | $10^3$ m³ feedstock | | 93 | 72.9 | 217 | 18.2 | | 31.4 | 28.3 |
|    Lubrication refineries | $10^3$ m³ feedstock | | 117 | 217 | 543 | 71.5 | | 120 | 24.1 |
|    Petrochemical refineries | $10^3$ m³ feedstock | | 108 | 171.6 | 463 | 48.6 | | 52.9 | 34.3 |
|    Integrated refineries | $10^3$ m³ feedstock | | 234 | 197 | 328 | 58 | | 75 | 20.5 |
| Asphalt products | | | | | | | | | |
|    Asphalt roofing products | t of product | | 50 | 8 | | 40 | | | |
| Tires and inner tube factory | t of product | | 37 | | 0.78 | 1 | 12 | 0.12 | |
| *Non-metallic mineral industry* | | | | | | | | | |
|    Glass and glass products | t of product | 9 | 45.9 | | 4.6 | 0.7 | 8.0 | | |
| Cement | | | | | | | | | |
|    Wet process | t of product | | 5.1 | | | 0.9 | 6.6 | | |
|    Dry process | t of product | | 5.1 | | | Negl. | 0.3 | | |
| *Basic metal industry* | | | | | | | | | |
| Iron and steel industry | | | | | | | | | |
|    Metallurgical coke oven[b] | t of product | | 0.42 | 0.58 | | 0.44 | | 0.075 | 0.95 |
|    Blast furnace | t of product | | 14.4 | | | 15.8 | | | 0.09 |
|    BOF steel furnace | t of product | | 2.3 | | | 3.5 | | | 0.01 |
|    Open hearth steel furnace | t of product | | 2.41 | | | 4.93 | | | |
|    Electric arc steel furnace | t of product | | 0.8 | | | 11.7 | | | |
|    Steel and grey iron foundries | t of product | | 1.6 | | | 0.3 | | 0.25 | |
| Nonferrous metal basic industry | t of product | | | | | | | | |
|    Aluminum | t of product | | | | 2.9 | 4.47 | 2.2 | 0.46 | |
| *Fabricated metal products* | | | | | | | | | |
|    Household appliances | t of iron sheet used | | 55 | 19.3 | 82 | 8.3 | 23 | 3.4 | |
| Electroplating | | | | | | | | | |
|    Cu | t of anodes | | 1,403 | | | | | | |

*continued*

| Industry and Process | Unit | pH | Waste volume (m³/unit) | 5-day biological oxygen demand (kg/unit) | Chemical oxygen demand (kg/unit) | Suspended solids (kg/unit) | Total dissolved solids (kg/unit) | Oil (kg/unit) | Nitrogen (kg/unit) |
|---|---|---|---|---|---|---|---|---|---|
| Ni | t of anodes | | 1,519 | | | | | | |
| $Cr_2O_3$ | t of anodes | | 36,300 | | | | | | |
| Zn | t of anodes | | 1,815 | | | | | | |
| Cd | t of anodes | | 883 | | | | | | |
| Sn | t of anodes | | 1,125 | | | | | | |
| Copper deposit | m² of deposit | | 94 | | | | | | |
| Nickel deposit | m² of deposit | | 103 | | | | | | |
| Chromium deposit | m² of deposit | | 95 | | | | | | |
| Zinc deposit | m² of deposit | | 93 | | | | | | |
| Motor vehicles manufacturing | t of iron sheets painted | | 55 | 19.3 | 82 | 8.3 | 22.6 | 3.4 | |
| *Electricity and gas* | | | | | | | | | |
| Power plants | 10³ MWh | | 129 | 2.2 | 17 | 286 | 110 | 0.15 | |
| Manufacture of gas from coke ovens[b] | t of coke | | 0.42 | 0.58 | | 0.44 | | 0.075 | 0.095 |
| | 10³ m³ gas | | 0.63 | 0.81 | | 0.66 | | 0.11 | 1.4 |

LWK, live weight killed.
[a]Only cooling water.
[b]The waste factors for metallurgic coke ovens are based on the assumption that the condensates and sludges produced are disposed of as solid wastes. If, however, the wastes discharged are liquid, the liquid pollution and waste loads involved should also be considered.

Basic Organic Chemicals

| Group 1 | Group 2 | Group 3 | Group 4 |
|---|---|---|---|
| Cyclohexane | Ethylene | Acetaldehyde | Organic dyes |
| Ethyl benzene | Propylene | Acetic acid | Azoic dyes and |
| BTX aromatics | Methanol | Acrylic acid | components |
| Vinyl chloride | Acetone | Aniline | |
| (Produced by | Acetaldehyde | Bisphenol A | |
| addition of HCl | Vinyl acetate | Caprolactom | |
| to acetylene) | Butadiene | Coal tar | |
| | Acetylene | Phenol | |
| | Ethylene | Acrylates | |
| | Ethylene oxide | p-Cresol | |
| | Formaldehyde | Oxo chemicals | |
| | Ethylene dichloride | Ethylene glycol | |
| | Styrene | Tetraethyl lead | |
| | Vinyl chloride | Terephthalic acid | |
| | (Produced by | Dimethyl | |
| | cracking | terephthalate | |
| | ethylene | Methyl metacrylates | |
| | dichloride) | | |

**TABLE D.9**　Raw waste loads in selected industries
(based on production rates in kg/tonne product)

| Type of industry | 5-day biological oxygen demand | Suspended solids | Oil and grease | Chemical oxygen demand | Ammonia Nitrogen | Phenols |
|---|---|---|---|---|---|---|
| Canned and preserved fruits and vegetables | 5.13 | 6.33 | | 12.8 | | |
| Southern (nonbreaded) shrimp | | 253.30 | 80.0 | | | |
| Alaskan bottom fish processing | | 11.30 | 0.6 | | | |
| Corn wet milling | 9.02 | 8.93 | | 22.6 | | |
| Corn dry milling | 0.71 | 0.63 | | 1.8 | | |
| Bulgur wheat flour mills | 0.10 | 0.10 | | 0.3 | | |
| Parboiled rice | 0.93 | 0.53 | | 2.3 | | |
| Ready-to-eat cereal | 2.67 | 2.67 | | 6.7 | | |
| Wheat starch gluten | 13.30 | 13.30 | | 33.3 | | |
| Simple slaughterhouse[a] | 0.80 | 1.33 | 0.4 | 2.0 | | |
| Dairy products | 0.90 | 1.35 | | 2.3 | | |
| Crystalline cane sugar | 5.73 | 1.2 | | 14.3 | | |
| Edible oils | 22.30 | 19.50 | 14.0 | 55.8 | | |
| Brewery | 10.20 | 4.73 | | 11.2 | | |
| Soft drinks | 3.15 | 4.33 | | 7.9 | | |
| Coffee | 625 | 50 | | 1,562 | | |
| Petroleum refining (topping) | 0.094 | 0.080 | 0.029 | 0.47 | 0.010 | 0.0006 |
| Petroleum refining (cracking) | 0.126 | 0.080 | 0.048 | 0.35 | 0.026 | 0.0006 |
| Petroleum storage and washing | | | 0.5 | | | |
| Petrochemicals | 0.144 | 0.116 | 0.047 | 0.85 | 0.084 | 0.0009 |
| Manufacturing soap flakes and powders | 0.067 | 0.067 | 0.067 | 0.33 | | |
| Manufacturing bar soap | 2.27 | 3.87 | 0.27 | 5.67 | | |
| Tires and inner tubes | | 0.43 | 0.11 | | | |
| Emulsion crumb rubber | 2.67 | 4.33 | | 53.3 | | |
| Solution crumb rubber | 2.67 | 4.33 | 1.07 | 24.3 | | |
| Latex rubber | 2.27 | 3.67 | 0.93 | 45.7 | | |
| Leather tanning and finishing[b] | 26.67 | 33.3 | 5.0 | 66.7 | | |
| Pulp, paper, and paperboard[c] | 18.67 | 40 | | 46.7 | | |
| Cement manufacturing (leaching) | 2.67 | | | 6.7 | | |
| Explosives | 1.46 | 29.3 | | 3.87 | | |
| Textiles printing and dyeing[d] | 22.7 | 58 | | 282.0 | | 0.40 |
| Paint and lacquer | 0.13 | 0.20 | | 0.33 | | |
| Plywood (kg/m$^3$ of plywood) | 0.62 | | | 1.56 | | 0.70 |
| Veneer (kg/m$^3$ of hardwood) | 3.64 | | | 9.1 | | |
| Iron and steel | | 0.24 | 0.073 | | 0.61 | 0.01 |
| Primary aluminum smelting[e] | | 10.0 | | | | |

*continued*

| Type of industry | 5-day biological oxygen demand | Suspended solids | Oil and grease | Chemical oxygen demand | Ammonia Nitrogen | Phenols |
|---|---|---|---|---|---|---|
| Phosphate manufacturing | | 3.33 | | | | |
| Sulfuric acid | | 0.3 | 0.045 | | | |
| Ammonium sulfate | | | | | 2.5 | |
| Painting and galvanizing | | 1.26 | | | | |
| Fertilizers | | 3.33 | | | | |
| Pharmaceuticals | 21.3 | 47.3 | | 53.3 | | |
| Batteries | 6.24 | 1,560 | | 15.6 | | |

[a]Simple slaughterhouse (kg/tonnes live killed weight).
[b]Leather tanning and finishing (hair pulp with chrome tanning).
[c]Pulp, paper, and paperboard (unbleached kraft).
[d]Textiles printing and dyeing (assume cloth weighs 0.15 kg/m$^2$).
[e]Primary aluminum smelting (by Hall-Heroult process).

**TABLE D.9** Raw waste loads in selected industries *(pages 248–249)*

**DESCRIPTION**

Industry is often a major source of waste discharges into rivers and lakes. These discharges cause water quality problems, which vary widely regionally, depending on the type of industrial process, the amount of water used, and other chemicals and microorganisms present in the effluent. Shown here for a wide variety of industries are the kilograms of waste produced per tonne of product for six pollutants. This example of raw waste load was taken from a study of West and Central Africa.

**LIMITATIONS**

These industrial waste loads were developed from data from West and Central Africa and may not be representative of industrial waste data from other regions. Actual waste loads from a given plant may vary considerably from the industry average. No detail is given here on the range of outputs within industrial groups.

**SOURCE**

Data from United Nations Environment Programme, 1982, *Survey of Marine Pollutants from Industrial Sources in the West and Central Africa Region*, UNEP Regional Seas Reports and Studies no. 2, United Nations Environment Programme, Nairobi.

Cited in M. Meybeck, D. Chapman and R. Helmer, 1989, *Global Freshwater Quality: A First Assessment*, World Health Organization and United Nations Environment Programme, Basil Blackwell, Oxford.

**TABLE D.10**　Presence of heavy metals in waste streams of selected industries

| Industry | As | Cd | Cr | Cu | Pb | Hg | Se | Zn | Cyanides |
|---|---|---|---|---|---|---|---|---|---|
| Mining | X | X | X | X | X | X | X | X | X |
| Paint and dye | | X | X | X | X | X | X | | X |
| Pesticides | X | | | | X | X | | X | X |
| Electrical and electronic | | | | X | X | X | X | | X |
| Printing and duplicating | X | | X | X | X | | X | | |
| Electroplating and metal finishing (plating) | | X | X | X | | | | X | X |
| Chemical manufacturing | | | X | X | | X | | | |
| Explosives | X | | | X | X | X | | | |
| Rubber and plastics | | | | | | X | | X | X |
| Battery | | X | | | X | X | | X | |
| Pharmaceutical | X | | | | | X | | | |
| Textile | | | X | X | | | | | |
| Petroleum and coal | X | | | | X | | | | |
| Pulp and paper | | | | | | X | | | |
| Leather | | | X | | | | | | |

**TABLE D.10**　Presence of heavy metals in waste streams of selected industries

**DESCRIPTION**

The presence or absence of certain heavy metals in the waste streams of selected industries is shown here. Heavy metals surveyed are arsenic (As), cadmium (Cd), chromium (Cr), copper (Cu), lead (Pb), mercury (Hg), selenium (Se), and zinc (Zn). Fifteen industrial groups are surveyed here. The presence of forms of cyanides in the waste streams of these industries are also shown.

**LIMITATIONS**

Other industrial groups may have heavy metals, but were not surveyed by this source. Cyanides are not metals but are often used in metal and other industries and are found in their waste streams. No quantitative assessment of the extent of metals contamination in these waste streams is given here.

**SOURCE**

K.W. Bucksteeg, 1984, Reuse of municipal sewage for agricultural purposes: A solution to water economy problems in arid zones, in E. Neis and A. Bittner (eds.) *Water Reuse: Selected Reports on Water Reuse in Urban and Rural Areas*, Tubingen, Germany.

**TABLE D.11** Characteristics of wastewater from the production of organic chemicals

| Product | 5-day biological oxygen demand (mg/l) | Chemical oxygen demand (mg/l) | Suspended matter (mg/l) |
|---|---|---|---|
| Acetaldehyde | 15,000–25,000 | 40,000–60,000 | 150–300 |
| Acrylates | 1,000–2,000 | 2,000–3,200 | 50–100 |
| Acrylonitrile | 200–500 | 600–1,200 | 80–150 |
| Butadiene and styrene | 4,000–8,000 | 800–1,500 | 200–500 |
| Esters | 5,000–12,000 | 10,000–20,000 | 20–100 |
| Ethylene and propylene | 400–600 | 800–1,200 | 20–40 |
| Isocyanates | 300–600 | 900–1,600 | 40–75 |
| Ketones | 10,000–20,000 | 20,000–40,000 | 50–100 |
| Methacrylic acid | | 7,000–12,000 | 6,000–12,000 |
| Methyl and ethyl parathion | 2,000–3,500 | 4,000–6,000 | 50–100 |
| Organic acids | 300–600 | 5,000–15,000 | 100–200 |
| Organic phosphate compounds | 500–1,000 | 1,500–3,000 | 200–400 |
| Phthalic acid anhydride and maleic acid anhydride | | 150–300 | 20–50 |
| Raw materials for the pigment industry | 200–400 | 1,000–2,000 | 80–200 |

**TABLE D.11**  Characteristics of wastewater from the production of organic chemicals

**DESCRIPTION**

The production of organic chemicals is accompanied by the production of wastewater. Ranges of five-day biological oxygen demand, the chemical oxygen demand, and the concentration of suspended matter are given here for fourteen categories of organic chemicals, in mg/l.

**LIMITATIONS**

There is considerable variation in waste loads from the production of different organic chemicals. This variation is reflected in the ranges presented here, but it is not clear from the original source whether these ranges reflect the maximum and minimum values in an industry or the variability in effluents produced by a particular set of chemical plants monitored.

**SOURCES**

M. Meybeck, D. Chapman and R. Helmer, 1989, *Global Freshwater Quality: A First Assessment*, World Health Organization and United Nations Environment Programme. Basil Blackwell, Oxford. By permission.

Adapted from S.E. Joergensen, 1979, *Industrial Waste Water Management*, Elsevier, Amsterdam.

**TABLE D.12**   Water treatment systems capable of removing common contaminants

| Common Contaminant | Treatment Systems | | | | |
| | Water Softeners | | | | |
| | Ion Exchange | | | | |
| | Cation | Anion | Activated carbon filter | Reverse osmosis | Distillation |
| --- | --- | --- | --- | --- | --- |
| Arsenic | — | Yes | — | Yes | Yes |
| Barium | Yes | — | — | Yes | Yes |
| Fluoride | — | — | — | Yes | Yes |
| Lead | — | — | — | Yes | Yes |
| Mercury (inorganic) | — | — | Yes | Yes | Yes |
| Mercury (organic) | — | — | Yes | — | — |
| Nitrate | — | Yes | — | Yes | Yes |
| Volatile Synthetic Organic Chemicals | — | — | Yes | — | — |
| Total Trihalomethanes | — | — | Yes | — | Yes |
| Radium | Yes | — | — | Yes | — |
| Coliform Bacteria | — | — | — | Yes | Yes |

**TABLE D.12**   Water treatment systems capable of removing common contaminants

**DESCRIPTION**

Wealthier water consumers may improve the quality of their drinking water through use of water treatment systems. This table compares the most commonly used systems for the effectiveness of five treatment systems in removing common contaminants, including arsenic, barium, fluoride, lead, mercury, various organic chemicals, bacteria, and radium.

**LIMITATIONS**

Only a few of the most common water contaminants are listed here. No quantitative measure of the ability of these treatment systems to remove these contaminants, or comparative information on the relative cost of treatment, is given.

**SOURCES**

1991, *Facets of Freshwater Newsletter*, **21**(5), Freshwater Foundation, Wayzata, Minnesota.

**TABLE D.13**  Typical effluent changes in a waste-treatment plant

| | 5-day biological oxygen demand (mg/l) | Suspended solids (mg/l) | Total coliform (count/ml) |
|---|---|---|---|
| Raw sewage | 220 | 220 | $1 \times 10^8$ |
| At end of primary settlement | 200 | 90 | $5 \times 10^6$ |
| At end of secondary settlement | 30[a] | 30[a] | $4 \times 10^6$ |
| At end of tertiary treatment | 12–15 | 12–15 | $2 \times 10^6$ |
| After chlorination of secondary effluent | | | $2 \times 10^4$ |

[a]U.S. Environmental Protection Agency definition of secondary treatment (various sources).

**TABLE D.13**  Typical effluent changes in a waste-treatment plant

**DESCRIPTION**

The goal of wastewater treatment plants is to reduce the concentration of water pollutants, particularly biological oxygen demand (BOD), suspended solids, and of total fecal coliform count. Assuming an initial concentration in raw sewage of 220 mg/l for BOD and suspended solids, this table shows the reductions in these pollutants after different levels of treatment. BOD and suspended solids are measured in mg/l; total coliform count is given in counts per ml.

**LIMITATIONS**

These changes are described as ''typical'' in the original source, but no information is given on how these data were measured and whether they represent average or median values for all plants or for a subset of waste treatment plants.

**SOURCE**

J.J. Sharp, 1990, The use of ocean outfalls for effluent disposal in small communities and developing countries, *Water International*, **15**(1), 35–43. International Water Resources Association, Illinois. By permission.
J.J. Sharp, 1992, Personal communication.

**TABLE D.14**  Impaired waters in the United States, by causes and sources of pollution *(page 254)*

**DESCRIPTION**

Water quality in estuaries, rivers, and lakes in the United States is assessed by the U.S. Environmental Protection Agency on a regular basis. Considerable information is provided in the original source for each state or region on lake and estuary areas and river miles assessed and monitored, on the extent of pollution, and on the sources and causes of pollution for 1988. Some additional information on the severity of pollution is also given in the original source. Summarized here are the fraction of impaired lake, river, and estuary waters, categorized by causes of impairment and the source of pollution.

**LIMITATIONS**

In all, only about 30% of U.S. river miles, 40% of lake areas, and 70% of estuary areas were assessed for this study. In some cases, significant areas have been missed, though the U.S. Environmental Protection Agency states that ''it is likely that unassessed waters are not as polluted as assessed waters.'' For example, no data on estuaries in California or the Delaware River Basin are included. Lake information for California, Maine, Michigan, Missouri, Nebraska, Ohio, Tennessee, and Wisconsin are missing. River data for Arizona, California, the Delaware River Basin, Louisiana, Maine, Michigan, Nebraska, New Mexico, Oregon, and Tennessee are missing.

**SOURCE**

United States Environmental Protection Agency, 1990, *National Water Quality Inventory*, 1988 Report to Congress, Office of Water. EPA 440-4-90-003. Washington DC.

**TABLE D.14** Impaired waters in the United States, by causes and sources of pollution

| | Percent of rivers, lakes, and estuaries surveyed determined to be impaired | | |
| Causes | River length | Lake area | Estuary area |
| --- | --- | --- | --- |
| Siltation | 42.4 | 25.4 | 6.7 |
| Nutrients | 26.6 | 48.8 | 49.6 |
| Pathogens | 18.6 | 8.6 | 48.1 |
| Organic enrichment | 14.6 | 25.3 | 29.0 |
| Metals | 10.8 | 7.4 | 9.5 |
| Pesticides | 10.3 | 5.3 | 1.0 |
| Suspended solids | 6.2 | 7.5 | |
| Salinity | 6.1 | 14.3 | |
| Flow alteration | 5.8 | 3.3 | |
| Habitat modification | 5.7 | 11.3 | |
| pH | 5.1 | 5.1 | 0.4 |
| Priority organics | | 8.2 | 4.1 |
| Oil and grease | | | 23.4 |
| Unknown toxins | | | 5.1 |
| Other inorganics | | | 0.4 |
| Ammonia | | | 0.2 |

| | Percent of rivers, lakes, and estuaries surveyed determined to be impaired | | |
| Sources | River length | Lake area | Estuary area |
| --- | --- | --- | --- |
| Agriculture | 55.2 | 58.2 | 18.6 |
| Municipal | 16.3 | 15.1 | 53.1 |
| Resource extraction | 13.0 | 4.2 | 34.2 |
| Hydrologic/habitat modification | 12.9 | 33.1 | 4.8 |
| Storm sewers/runoff | 8.8 | 27.7 | 28.5 |
| Silviculture | 8.6 | 0.9 | 1.6 |
| Industrial | 8.5 | 7.7 | 12.1 |
| Construction | 6.3 | 3.3 | 12.5 |
| Land disposal | 4.4 | 26.5 | 27.4 |
| Combined sewers | 3.7 | 0.3 | 10.3 |

**TABLE D.15** Activities contributing to ground water contamination in the United States

| Activity | States citing | Estimated sites | Contaminants frequently cited as result of activity | Remarks |
|---|---|---|---|---|
| *Waste disposal* | | | | |
| Septic systems | 41 | $22 \times 10^6$ | Bacteria, viruses, nitrate, phosphate, chloride, and organic compounds such as trichloroethylene | Between 3 and 5.5 billion cubic meters per year discharged to shallowest aquifers. |
| Landfills (active) | 51 | 16,400 | Dissolved solids, iron, manganese, trace metals, acids, organic compounds, and pesticides | Traditional disposal method for municipal and industrial solid waste. Unknown number of abandoned landfills. |
| Surface impoundments | 32 | 191,800 | Brines, acidic mine wastes, feedlot wastes, trace metals, and organic compounds | Used to store oil/gas brines (125,100 sites), mine wastes (19,800), agricultural wastes (17,200), industrial liquid wastes (16,200), other wastes (11,100). |
| Injection wells | 10 | 280,800 | Dissolved solids, bacteria, sodium, chloride, nitrate, phosphate, organic compounds, pesticides, and acids | Wells used for injecting waste below drinking water sources (550), oil/gas brine disposal (161,400), solution mining (22,700), injecting waste into or above drinking water sources (14), and storm-water disposal, agricultural drainage, heat pumps (96,100). |
| Land application of wastes | 12 | 19,000 land application units | Bacteria, nitrate, phosphate, trace metals, and organic compounds | Waste disposal from municipal waste-treatment plants (11,000), industry (5,600), oil/gas production (730), petroleum and woodpreserving wastes (250), other (620). |
| *Storage and handling of materials* | | | | |
| Underground storage tanks | 39 | $2.4 - 4.8 \times 10^6$ | Benzene, toluene, xylene, and petroleum products | Useful life of steel tanks, 15–20 years. About 25–30% of petroleum tanks may leak. |
| Above-ground storage tanks | 16 | Unknown | Organic compounds, acids, metals, and petroleum products | Spills/overflows may contaminate ground water. |
| Material handling and transfers | 29 | 10,000–16,000 spills per year | Petroleum products, aluminum, iron, sulfate, and trace metals | Includes coal storage piles, bulk chemical storage, containers, and accidential spills. |
| *Mining activities* | | | | |
| Mining and spoil disposal— coal mines | 23 | 15,000 active; 67,000 inactive | Acids, iron, manganese, sulfate, uranium, thorium, radium, molybdenum, selenium, and trace metals | Leachates from spoil piles of coal, metal, and nonmetallic mineral mining contain a variety of contaminants. Coal mines are sources of acid drainage. |
| *Oil and gas activities* | | | | |
| Wells | 20 | 550,000 production; $1.2 \times 10^6$ abandoned | Brines | Contamination from improperly plugged wells and oil brine stored in ponds or injected underground. |

*continued*

**TABLE D.15** Activities contributing to ground water contamination in the United States *(pages 255–256)*

| Activity | States citing | Estimated sites | Contaminants frequently cited as result of activity | Remarks |
|---|---|---|---|---|
| *Agricultural activities* | | | | |
| Fertilizer and pesticide applications | 44 | $147 \times 10^6$ hectares | Nitrate, phosphate, and pesticides | Fertilizer applied 1982–83, 38.4 million tonnes per year; active ingredients of pesticides applied 1982, 300,000 tonnes. |
| Irrigation practices | 22 | 376,000 wells; $20 \times 10^6$ hectares irrigated | Dissolved solids, nitrate, phosphate, and pesticides | Salts, fertilizers, pesticides can concentrate in ground water. Improperly plugged abandoned wells contamination source. |
| Animal feedlots | 17 | 1,900 | Nitrate, phosphate, and bacteria | Primarily in the Corn Belt and High Plains states. |
| *Urban activities* | | | | |
| Runoff | 15 | $19 \times 10^6$ hectares urban land | Bacteria, hydrocarbons, dissolved solids, lead, cadmium, and trace metals | Infiltration from detention basins, drainage wells, pits, shafts can reach ground water. Karst areas particularly vulnerable. |
| Deicing chemical storage and use | 14 | Not reported | Sodium chloride, ferric ferrocyanide, sodium, ferrocyanide, phosphate, and trace metals | Winter 1983, 8.5 million tonnes dry salts/abrasives, 30,000 cubic meters of liquid salts applied. |
| *Other* | | | | |
| Saline intrusion or upconing | 29 | Not reported | Dissolved solids and brines | Present in coastal areas and in many inland areas. |

**DESCRIPTION**

This table summarizes the number of states reporting ground water contamination, the number of sites involved, the activity that lead to the contamination, and the categories of contaminants found. Details are provided on each form of contamination. Between 1950 and 1985 the use of fresh ground water in the United States more than doubled, and now accounts for nearly 25% of all withdrawals for offstream uses. While the majority of this ground water is being drawn for self-supplied agricultural irrigation purposes, in 1985, ground water also supplied 40% of all publicly-supplied water, 98% of self-supplied domestic water, and over half of all self-supplied livestock, commercial, and mining water. Studies of water quality in the past found few examples of tainted ground water, but increasing numbers of individual aquifers are now being identified as containing contaminants that could render the water useless for assorted end users.

**LIMITATIONS**

These data provide an overview of sources of potential ground water contamination, though limited detailed information on actual levels or extent of contamination is provided. Ground water quality is poorly monitored in many regions and these data are unlikely to be complete.

**SOURCE**

U.S. Geological Survey, 1988, *National Water Summary 1986: Hydrologic Events and Groundwater Quality*, U.S. Geological Survey, Water Supply Paper no. 2325, Washington, DC.

**TABLE D.16** Estimated phosphorus loadings to the Great Lakes, 1976 to 1986 (tonnes/yr)

| Year | Lake Superior | Lake Michigan | Lake Huron | Lake Erie | Lake Ontario |
|---|---|---|---|---|---|
| 1976 | 3,550 | 6,656 | 4,802 | 18,480 | 12,695 |
| 1977 | 3,661 | 4,666 | 3,763 | 14,576 | 8,935 |
| 1978 | 5,990 | 6,245 | 5,255 | 19,431 | 9,547 |
| 1979 | 6,619 | 7,659 | 4,881 | 11,941 | 8,988 |
| 1980 | 6,412 | 6,574 | 5,307 | 14,855 | 8,579 |
| 1981 | 3,412 | 4,091 | 3,481 | 10,452 | 7,437 |
| 1982 | 3,160 | 4,084 | 4,689 | 12,349 | 8,891 |
| 1983 | 3,407 | 4,515 | 3,978 | 9,880 | 6,779 |
| 1984 | 3,642 | 3,611 | 3,452 | 12,874 | 7,948 |
| 1985 | 2,864 | 3,956 | 5,758 | 11,216 | 7,083 |
| 1986 | 3,059 | 4,981 | 4,210 | 11,118 | 9,561 |
| Target Loads[a] | 3,400 | 5,600 | 4,360 | 11,000 | 7,000 |

[a]The 1978 Great Lakes Water Quality Agreement set target loadings for each lake (in tonnes per year).

**TABLE D.16** Estimated phosphorus loadings to the Great Lakes, 1976 to 1986

**DESCRIPTION**
Inputs of phosphorus to the Great Lakes are given here annually from 1976 to 1986, in tonnes per year. In 1978, target limits for phosphorus were set for all five Great Lakes. These limits have been exceeded regularly for Lake Erie and Lake Ontario.

**SOURCES**
Great Lakes Water Quality Board, 1989, *Great Lakes Water Quality*, Surveillance Subcommittee Report to the International Joint Commission, United States and Canada, Windsor, Ontario.
As cited in Council on Environmental Quality, 1991, *Environmental Quality: the 21st Annual Report of the Council on Environmental Quality*, Washington, DC.

**TABLE D.17** Selected Hazardous Substances at 951 National Priorities Sites *(page 258)*

**DESCRIPTION**
The United States Public Health Services lists sites severely contaminated by hazardous substances. Shown here for major hazardous materials are the number of sites contaminated and an assessment of how many have experienced migration of substances into ground water, surface water, soil, air, food, or sediments, where they may more easily affect human and ecosystem health. Among the substances listed are heavy metals, volatile organic compounds, polychlorinated biphenyls, hydrocarbons, pesticides, and dioxin.

**LIMITATIONS**
Only those substances that are considered high priority by the Agency

for Toxic Substances and Disease Registry (ATSDR) are included here. Some sites severely contaminated may not yet have been identified, and would not be listed in this table.

**SOURCES**
United States Public Health Service, 1989, *ATSDR Biennial Report to Congress October 17, 1986–September 30, 1988*, Agency for Toxic Substances and Disease Registry (ATSDR), Atlanta, Georgia.
As cited in National Academy of Sciences, 1991, *Environmental Epidemiology: Public Health and Hazardous Wastes*, vol. 1, National Research Council, National Academy Press, Washington, DC.

**TABLE D.17**   Selected Hazardous Substances at 951 National Priorities Sites
(Number and Percentage of Sites and Documented Migration of Substances into Specific Media)

| Substance | ATSDR priority group | Number | Percent | Sites with migration | Ground water | Surface water | Soil | Air | Food | Sediment |
|---|---|---|---|---|---|---|---|---|---|---|
| *Metallic Elements* | | 564 | 59 | 327 | 234 | 138 | 122 | 37 | 50 | 114 |
| Lead | 1 | 404 | 43 | 224 | 159 | 84 | 88 | 28 | 39 | 84 |
| Chromium | 1 | 329 | 35 | 142 | 93 | 55 | 48 | 12 | 15 | 46 |
| Arsenic | 1 | 262 | 28 | 36 | 92 | 46 | 54 | 16 | 19 | 50 |
| Cadmium | 1 | 232 | 24 | 112 | 72 | 49 | 45 | 18 | 21 | 44 |
| Mercury | 2 | 129 | 14 | 58 | 29 | 24 | 20 | 6 | 10 | 19 |
| Nickel | 1 | 126 | 13 | 55 | 30 | 24 | 15 | 3 | 8 | 21 |
| Beryllium | 1 | 21 | 2 | 9 | 2 | 3 | 1 | 0 | 0 | 3 |
| *Volatile Organic Compounds (VOCs)* | | 518 | 54 | 268 | 236 | 88 | 81 | 71 | 31 | 58 |
| Trichloroethylene | 1 | 402 | 42 | 231 | 204 | 63 | 41 | 44 | 19 | 27 |
| Benzene | 1 | 323 | 34 | 139 | 115 | 41 | 27 | 29 | 9 | 24 |
| Tetrachloroethylene | 1 | 267 | 28 | 125 | 116 | 28 | 22 | 34 | 11 | 17 |
| Toluene | 2 | 256 | 27 | 101 | 78 | 26 | 29 | 26 | 6 | 20 |
| Vinyl chloride | 1 | 187 | 20 | 87 | 80 | 16 | 14 | 18 | 7 | 9 |
| Methylene chloride | 1 | 183 | 19 | 81 | 61 | 21 | 16 | 17 | 4 | 9 |
| Chloroform | 1 | 142 | 15 | 74 | 61 | 20 | 8 | 7 | 3 | 9 |
| 1,4-Dichlorobenzene | 1 | 31 | 3 | 7 | 6 | 0 | 1 | 1 | 0 | 1 |
| *Polychlorinated Biphenyls (PCBs)* | 1 | 162 | 17 | 86 | 43 | 25 | 40 | 11 | 25 | 39 |
| *Polycyclic Aromatic Hydrocarbons (PAHs)* | | 187 | 20 | 75 | 32 | 22 | 31 | 4 | 6 | 38 |
| Benzo(a)pyrene | 1 | 56 | 6 | 18 | 6 | 6 | 9 | 0 | 2 | 8 |
| Benzo(a)anthracene | 1 | 32 | 3 | 10 | 3 | 4 | 8 | 0 | 1 | 6 |
| Benzo(a)fluoroanthene | 1 | 25 | 3 | 10 | 1 | 3 | 5 | 0 | 0 | 4 |
| Chrysene | 1 | 23 | 2 | 6 | 2 | 1 | 3 | 0 | 1 | 4 |
| Dibenzo(a,h)anthracene | 1 | 4 | 1 | 1 | 0 | 0 | 1 | 0 | 0 | 0 |
| *Phthalates* | | 106 | 11 | 35 | 22 | 13 | 17 | 5 | 5 | 16 |
| Bis(2-ethylhexyl)phthalate | 1 | 88 | 9 | 35 | 22 | 13 | 16 | 3 | 5 | 16 |
| *Pesticides* | | 82 | 9 | 25 | 13 | 8 | 17 | 6 | 7 | 12 |
| Dieldrin/aldrin | 1 | 29 | 3 | 13 | 8 | 2 | 6 | 3 | 2 | 3 |
| Heptachlor/heptachlor epoxide | 1 | 15 | 2 | 4 | 2 | 0 | 1 | 0 | 0 | 1 |
| *Dioxins* | | 47 | 5 | 21 | 8 | 7 | 16 | 2 | 7 | 11 |
| 2,3,7,8-Tetrachlorodibenzo-p-dioxin | 1 | 19 | 2 | 15 | 5 | 3 | 8 | 3 | 7 | 10 |
| *Other* | | | | | | | | | | |
| Cyanide | 1 | 74 | 8 | 23 | 13 | 9 | 7 | 3 | 2 | 8 |
| N-Nitrosodiphenylamine | 1 | 8 | 1 | 4 | 2 | 1 | 2 | 0 | 1 | 2 |

# E. Water and agriculture

**TABLE E.1**  Cropland use by region and country, 1989 *(pages 260–264)*

## DESCRIPTION

The total area of land suitable for crop production is given here by continent and country, for 1989. Cropland area is then broken down into arable land and land devoted to permanent crops, and to irrigated and rainfed cropland. All areas are given in thousand hectares. Arable land refers to land under temporary crops, temporary meadows for pasture, land under market and kitchen gardens, and land fallow less than five years. Abandoned land from shifting cultivation is not included in this category. Permanent cropland refers to land that need not be replanted after each harvest, such as cocoa, coffee, and rubber. Land devoted to tree production is excluded from this category. Total cropland is assumed to be equal to arable land plus permanent cropland. Rainfed area is assumed to equal total cropland area minus the area of irrigated cropland; the percentages are figured with respect to total cropland.

## LIMITATIONS

These numbers are a combination of Food and Agriculture Organization estimates and other country estimates, with different degrees of accuracy. Arable land area listed for Australia includes 27 million hectares of cultivated grassland. For Germany, plots of 1 hectare or more are included, as are smaller parcels that produce a market value above a fixed standard. For Portugal, the data include 800,000 hectares of temporary crops grown in association with permanent crops and forest. With respect to the irrigated area data, for Cuba, data refer to the state sector only. For Denmark and Romania, data refer to land provided with irrigation facilities. For Hungary, data exclude complementary farm plots and individual farms. For Japan, the Republic of Korea, and Sri Lanka, the irrigated land data refer to irrigated rice only. As noted in the description, not all of these data are measured: rainfed area is simply assumed to equal total cropland minus irrigated cropland. Similarly, total cropland is assumed to equal arable land area plus area devoted to permanent cropland.

## SOURCES

Food and Agriculture Organization, 1990, *FAO Production Yearbook 1990*, FAO Statistics Series no. 99, vol. 44, prepared by the Statistics Division of the Economic and Social Policy Department, on the basis of information available as of April 1991, Food and Agriculture Organization of the United Nations, Rome. By permission.

**TABLE E.1**  Cropland use by region and country, 1989
($10^3$ hectares)

| | Total cropland | Arable land | Permanent cropland | Irrigated cropland | Rainfed cropland | Percent irrigated | Percent rainfed |
|---|---|---|---|---|---|---|---|
| *World* | 1,476,709 | 1,373,311 | 103,398 | 232,828 | 1,243,881 | 15.8 | 84.2 |
| *Africa* | 186,995 | 168,162 | 18,833 | 11,186 | 175,809 | 6.0 | 94.0 |
| Algeria | 7,605 | 7,070 | 535 | 336 | 7,269 | 4.4 | 95.6 |
| Angola | 3,600 | 3,050 | 550 | | | | |
| Benin | 1,860 | 1,410 | 450 | 6 | 1,854 | 0.3 | 99.7 |
| Botswana | 1,380 | 1,380 | | 2 | 1,378 | 0.1 | 99.9 |
| Burkina Faso | 3,564 | 3,551 | 13 | 16 | 3,548 | 0.4 | 99.6 |
| Burundi | 1,336 | 1,120 | 216 | 72 | 1,264 | 5.4 | 94.6 |
| Cameroon | 7,008 | 5,940 | 1,068 | 28 | 6,980 | 0.4 | 99.6 |
| Cape Verde | 39 | 37 | 2 | 2 | 37 | 5.1 | 94.9 |
| Central African Republic | 2,006 | 1,920 | 86 | | | | |
| Chad | 3,205 | 3,200 | 5 | 10 | 3,195 | 0.3 | 99.7 |
| Comoros | 100 | 78 | 22 | | | | |
| Congo | 168 | 144 | 24 | 4 | 164 | 2.4 | 97.6 |
| Côte D'Ivoire | 3,660 | 2,420 | 1,240 | 62 | 3,598 | 1.7 | 98.3 |
| Egypt | 2,585 | 2,310 | 275 | 2,585 | 0 | 100.0 | 0.0 |
| Equatorial Guinea | 230 | 130 | 100 | | | | |
| Ethiopia | 13,930 | 13,200 | 730 | 162 | 13,768 | 1.2 | 98.8 |
| Gabon | 452 | 290 | 162 | | | | |
| Gambia | 178 | 178 | | 12 | 166 | 6.7 | 93.3 |
| Ghana | 2,720 | 1,140 | 1,580 | 8 | 2,712 | 0.3 | 99.7 |
| Guinea | 728 | 610 | 118 | 24 | 704 | 3.3 | 96.7 |
| Guinea-Bissau | 335 | 300 | 35 | | | | |
| Kenya | 2,428 | 1,930 | 498 | 52 | 2,376 | 2.1 | 97.9 |
| Lesotho | 320 | 320 | | | | | |
| Liberia | 373 | 128 | 245 | 2 | 371 | 0.5 | 99.5 |
| Libya | 2,150 | 1,805 | 345 | 242 | 1,908 | 11.3 | 88.7 |
| Madagascar | 3,092 | 2,570 | 522 | 900 | 2,192 | 29.1 | 70.9 |
| Malawi | 2,409 | 2,380 | 29 | 20 | 2,389 | 0.8 | 99.2 |
| Mali | 2,093 | 2,090 | 3 | 205 | 1,888 | 9.8 | 90.2 |
| Mauritania | 199 | 196 | 3 | 12 | 187 | 6.0 | 94.0 |
| Mauritius | 106 | 100 | 6 | 17 | 89 | 16.0 | 84.0 |
| Morocco | 9,241 | 8,661 | 580 | 1,265 | 7,976 | 13.7 | 86.3 |
| Mozambique | 3,100 | 2,870 | 230 | 115 | 2,985 | 3.7 | 96.3 |
| Namibia | 662 | 660 | 2 | 4 | 658 | 0.6 | 99.4 |
| Niger | 3,605 | 3,605 | | 32 | 3,573 | 0.9 | 99.1 |
| Nigeria | 31,335 | 28,800 | 2,535 | 865 | 30,470 | 2.8 | 97.2 |
| Reunion | 52 | 47 | 5 | 6 | 46 | 11.5 | 88.5 |
| Rwanda | 1,153 | 849 | 304 | 4 | 1,149 | 0.3 | 99.7 |

*continued*

| | Total cropland | Arable land | Permanent cropland | Irrigated cropland | Rainfed cropland | Percent irrigated | Percent rainfed |
|---|---|---|---|---|---|---|---|
| St. Helena | 2 | 2 | 2 | | | | |
| Sao Tome and Principe | 37 | 2 | 35 | | | | |
| Senegal | 5,226 | 5,220 | 6 | 180 | 5,046 | 3.4 | 96.6 |
| Seychelles | 6 | 1 | 5 | | | | |
| Sierra Leone | 1,801 | 1,655 | 146 | 34 | 1,767 | 1.9 | 98.1 |
| Somalia | 1,039 | 1,022 | 17 | 116 | 923 | 11.2 | 88.8 |
| South Africa | 13,174 | 12,360 | 814 | 1,128 | 12,046 | 8.6 | 91.4 |
| Sudan | 12,510 | 12,450 | 60 | 1,890 | 10,620 | 15.1 | 84.9 |
| Swaziland | 164 | 160 | 4 | 62 | 102 | 37.8 | 62.2 |
| Tanzania | 5,250 | 4,160 | 1,090 | 153 | 5,097 | 2.9 | 97.1 |
| Togo | 1,444 | 1,375 | 69 | 7 | 1,437 | 0.5 | 99.5 |
| Tunisia | 4,700 | 3,034 | 1,666 | 275 | 4,425 | 5.9 | 94.1 |
| Uganda | 6,705 | 5,000 | 1,705 | 9 | 6,696 | 0.1 | 99.9 |
| Zaire | 7,850 | 7,250 | 600 | 10 | 7,840 | 0.1 | 99.9 |
| Zambia | 5,268 | 5,260 | 8 | 32 | 5,236 | 0.6 | 99.4 |
| Zimbabwe | 2,810 | 2,720 | 90 | 220 | 2,590 | 7.8 | 92.2 |
| *North and Central America* | 273,834 | 266,981 | 6,853 | 25,920 | 247,914 | 9.5 | 90.5 |
| Antigua and Barbuda | 8 | 8 | | | | | |
| Bahamas | 10 | 8 | 2 | | | | |
| Barbados | 33 | 33 | | | | | |
| Belize | 56 | 44 | 12 | 2 | 54 | 3.6 | 96.4 |
| British Virgin Islands | 4 | 3 | 1 | | | | |
| Canada | 45,960 | 45,880 | 80 | 840 | 45,120 | 1.8 | 98.2 |
| Costa Rica | 528 | 285 | 243 | 118 | 410 | 22.3 | 77.7 |
| Cuba | 3,329 | 2,607 | 722 | 896 | 2,433 | 26.9 | 73.1 |
| Dominica | 17 | 7 | 10 | | | | |
| Dominican Republic | 1,446 | 1,000 | 446 | 225 | 1,221 | 15.6 | 84.4 |
| El Salvador | 733 | 565 | 168 | 120 | 613 | 16.4 | 83.6 |
| Grenada | 13 | 5 | 8 | | | | |
| Guadeloupe | 29 | 21 | 8 | 3 | 26 | 10.3 | 89.7 |
| Guatemala | 1,875 | 1,390 | 485 | 78 | 1,797 | 4.2 | 95.8 |
| Haiti | 905 | 555 | 350 | 75 | 830 | 8.3 | 91.7 |
| Honduras | 1,810 | 1,600 | 210 | 90 | 1,720 | 5.0 | 95.0 |
| Jamaica | 269 | 207 | 62 | 35 | 234 | 13.0 | 87.0 |
| Martinique | 20 | 10 | 10 | 6 | 14 | 30.0 | 70.0 |
| Mexico | 24,710 | 23,150 | 1,560 | 5,150 | 19,560 | 20.8 | 79.2 |
| Montserrat | 2 | 2 | | | | | |
| Netherland Antilles | 8 | 8 | | | | | |
| Nicaragua | 1,273 | 1,100 | 173 | 85 | 1,188 | 6.7 | 93.3 |
| Panama | 577 | 442 | 135 | 32 | 545 | 5.5 | 94.5 |
| Puerto Rico | 128 | 68 | 60 | 39 | 89 | 30.5 | 69.5 |

*continued*

| | Total cropland | Arable land | Permanent cropland | Irrigated cropland | Rainfed cropland | Percent irrigated | Percent rainfed |
|---|---|---|---|---|---|---|---|
| St. Kitts-Nevis | 14 | 8 | 6 | | | | |
| St. Lucia | 18 | 5 | 13 | 1 | 17 | 5.6 | 94.4 |
| St. Pierre and Miquelon | 3 | 3 | | | | | |
| St. Vincent | 11 | 4 | 7 | 1 | 10 | 9.1 | 90.9 |
| Trinidad and Tobago | 120 | 74 | 46 | 22 | 98 | 18.3 | 81.7 |
| Turks and Caicos Islands | 1 | 1 | | | | | |
| United States | 189,915 | 187,881 | 2,034 | 18,102 | 171,813 | 9.5 | 90.5 |
| U.S. Virgin Islands | 7 | 5 | 2 | | | | |
| | | | | | | | |
| *South America* | 142,134 | 116,102 | 26,032 | 8,835 | 133,299 | 6.2 | 93.8 |
| Argentina | 35,750 | 26,000 | 9,750 | 1,760 | 33,990 | 4.9 | 95.1 |
| Bolivia | 3,460 | 3,270 | 190 | 165 | 3,295 | 4.8 | 95.2 |
| Brazil | 78,650 | 66,500 | 12,150 | 2,700 | 75,950 | 3.4 | 96.6 |
| Chile | 4,525 | 4,276 | 249 | 1,265 | 3,260 | 28.0 | 72.0 |
| Colombia | 5,380 | 3,870 | 1,510 | 515 | 4,865 | 9.6 | 90.4 |
| Ecuador | 2,653 | 1,683 | 970 | 550 | 2,103 | 20.7 | 79.3 |
| French Guiana | 8 | 6 | 2 | | | | |
| Guyana | 495 | 480 | 15 | 130 | 365 | 26.3 | 73.7 |
| Paraguay | 2,216 | 2,100 | 116 | 67 | 2,149 | 3.0 | 97.0 |
| Peru | 3,730 | 3,400 | 330 | 1,250 | 2,480 | 33.5 | 66.5 |
| Suriname | 68 | 57 | 11 | 59 | 9 | 86.8 | 13.2 |
| Uruguay | 1,304 | 1,260 | 44 | 110 | 1,194 | 8.4 | 91.6 |
| Venezuela | 3,895 | 3,200 | 695 | 264 | 3,631 | 6.8 | 93.2 |
| | | | | | | | |
| *Asia* | 452,634 | 420,334 | 32,300 | 146,422 | 306,212 | 32.3 | 67.7 |
| Afghanistan | 8,054 | 7,910 | 144 | 2,660 | 5,394 | 33.0 | 67.0 |
| Bahrain | 2 | 1 | 1 | 1 | 1 | 50.0 | 50.0 |
| Bangladesh | 9,292 | 9,020 | 272 | 2,738 | 6,554 | 29.5 | 70.5 |
| Bhutan | 131 | 112 | 19 | 34 | 97 | 26.0 | 74.0 |
| Brunei Darus | 7 | 3 | 4 | 1 | 6 | 14.3 | 85.7 |
| Cambodia | 3,056 | 2,910 | 146 | 92 | 2,964 | 3.0 | 97.0 |
| China | 96,115 | 92,825 | 3,290 | 45,349 | 50,766 | 47.2 | 52.8 |
| Cyprus | 156 | 104 | 52 | 35 | 121 | 22.4 | 77.6 |
| East Timor | 80 | 70 | 10 | | | | |
| Gaza Strip | 18 | 6 | 12 | | | | |
| Hong Kong | 7 | 6 | 1 | 2 | 5 | 28.6 | 71.4 |
| India | 168,990 | 165,315 | 3,675 | 43,039 | 125,951 | 25.5 | 74.5 |
| Indonesia | 21,260 | 15,800 | 5,460 | 7,550 | 13,710 | 35.5 | 64.5 |
| Iran | 14,830 | 14,100 | 730 | 5,750 | 9,080 | 38.8 | 61.2 |
| Iraq | 5,450 | 5,250 | 200 | 2,550 | 2,900 | 46.8 | 53.2 |
| Israel | 433 | 344 | 89 | 214 | 219 | 49.4 | 50.6 |
| Japan | 4,637 | 4,150 | 487 | 2,868 | 1,769 | 61.9 | 38.1 |

*continued*

| | Total cropland | Arable land | Permanent cropland | Irrigated cropland | Rainfed cropland | Percent irrigated | Percent rainfed |
|---|---|---|---|---|---|---|---|
| Jordan | 376 | 310 | 66 | 57 | 319 | 15.2 | 84.8 |
| Korea DPR | 2,000 | 1,700 | 300 | 1,400 | 600 | 70.0 | 30.0 |
| Korea Rep | 2,127 | 1,984 | 143 | 1,353 | 774 | 63.6 | 36.4 |
| Kuwait | 4 | 4 | | 2 | 2 | 50.0 | 50.0 |
| Laos | 901 | 880 | 21 | 120 | 781 | 13.3 | 86.7 |
| Lebanon | 301 | 208 | 93 | 86 | 215 | 28.6 | 71.4 |
| Malaysia | 4,880 | 1,040 | 3,840 | 342 | 4,538 | 7.0 | 93.0 |
| Maldives | 3 | 3 | | | | | |
| Mongolia | 1,375 | 1,374 | 1 | 77 | 1,298 | 5.6 | 94.4 |
| Myanmar | 10,034 | 9,538 | 496 | 1,018 | 9,016 | 10.1 | 89.9 |
| Nepal | 2,641 | 2,612 | 29 | 943 | 1,698 | 35.7 | 64.3 |
| Oman | 48 | 16 | 32 | 41 | 7 | 85.4 | 14.6 |
| Pakistan | 20,730 | 20,285 | 445 | 16,220 | 4,510 | 78.2 | 21.8 |
| Philippines | 7,970 | 4,550 | 3,420 | 1,620 | 6,350 | 20.3 | 79.7 |
| Qatar | 5 | 5 | | | | | |
| Saudi Arabia | 1,185 | 1,110 | 75 | 435 | 750 | 36.7 | 63.3 |
| Singapore | 1 | 1 | | | | | |
| Sri Lanka | 1,901 | 926 | 975 | 560 | 1,341 | 29.5 | 70.5 |
| Syria | 5,503 | 4,889 | 614 | 670 | 4,833 | 12.2 | 87.8 |
| Thailand | 22,126 | 19,000 | 3,126 | 4,230 | 17,896 | 19.1 | 80.9 |
| Turkey | 27,885 | 24,868 | 3,017 | 2,220 | 25,665 | 8.0 | 92.0 |
| United Arab Emirates | 39 | 29 | 10 | 5 | 34 | 12.8 | 87.2 |
| Vietnam | 6,600 | 5,700 | 900 | 1,830 | 4,770 | 27.7 | 72.3 |
| Yemen | 1,481 | 1,376 | 105 | 310 | 1,171 | 20.9 | 79.1 |
| | | | | | | | |
| *Europe* | 139,865 | 126,014 | 13,851 | 17,240 | 122,625 | 12.3 | 87.7 |
| Albania | 707 | 582 | 125 | 423 | 284 | 59.8 | 40.2 |
| Andorra | 1 | 1 | | | | | |
| Austria | 1,533 | 1,459 | 74 | 4 | 1,529 | 0.3 | 99.7 |
| Belgium-Luxembourg | 822 | 806 | 16 | 1 | 821 | 0.1 | 99.9 |
| Bulgaria | 4,146 | 3,848 | 298 | 1,253 | 2,893 | 30.2 | 69.8 |
| Czechoslovakia | 5,108 | 4,976 | 132 | 310 | 4,798 | 6.1 | 93.9 |
| Denmark | 2,555 | 2,550 | 5 | 430 | 2,125 | 16.8 | 83.2 |
| Faeroe Islands | 3 | 3 | | | | | |
| Finland | 2,453 | 2,453 | | 62 | 2,391 | 2.5 | 97.5 |
| France | 19,119 | 17,899 | 1,220 | 1,160 | 17,959 | 6.1 | 93.9 |
| German DR | 4,913 | 4,676 | 237 | 150 | 4,763 | 3.1 | 96.9 |
| Germany FR | 7,478 | 7,273 | 205 | 330 | 7,148 | 4.4 | 95.6 |
| Greece | 3,924 | 2,875 | 1,049 | 1,190 | 2,734 | 30.3 | 69.7 |
| Hungary | 5,287 | 5,052 | 235 | 175 | 5,112 | 3.3 | 96.7 |
| Iceland | 8 | 8 | | | | | |
| Ireland | 953 | 950 | 3 | | | | |

*continued*

| | Total cropland | Arable land | Permanent cropland | Irrigated cropland | Rainfed cropland | Percent irrigated | Percent rainfed |
|---|---|---|---|---|---|---|---|
| Italy | 12,033 | 9,043 | 2,990 | 3,100 | 8,933 | 25.8 | 74.2 |
| Liechtenstein | 4 | 4 | | | | | |
| Malta | 13 | 12 | 1 | 1 | 12 | 7.7 | 92.3 |
| Netherlands | 934 | 905 | 29 | 550 | 384 | 58.9 | 41.1 |
| Norway | 878 | 878 | | 95 | 783 | 10.8 | 89.2 |
| Poland | 14,759 | 14,414 | 345 | 100 | 14,659 | 0.7 | 99.3 |
| Portugal | 3,771 | 2,906 | 865 | 634 | 3,137 | 16.8 | 83.2 |
| Romania | 10,350 | 9,902 | 448 | 3,450 | 6,900 | 33.3 | 66.7 |
| San Marino | 1 | 1 | | | | | |
| Spain | 20,345 | 15,570 | 4,775 | 3,360 | 16,985 | 16.5 | 83.5 |
| Sweden | 2,853 | 2,853 | | 112 | 2,741 | 3.9 | 96.1 |
| Switzerland | 412 | 391 | 21 | 25 | 387 | 6.1 | 93.9 |
| United Kingdom | 6,736 | 6,685 | 51 | 157 | 6,579 | 2.3 | 97.7 |
| Yugoslavia | 7,766 | 7,039 | 727 | 168 | 7,598 | 2.2 | 97.8 |
| | | | | | | | |
| *Oceania* | 50,617 | 49,618 | 999 | 2,161 | 48,456 | 4.3 | 95.7 |
| American Samoa | 4 | 2 | 2 | | | | |
| Australia | 48,934 | 48,761 | 173 | 1,880 | 47,054 | 3.8 | 96.2 |
| Cook Islands | 6 | 1 | 5 | | | | |
| Fiji | 240 | 152 | 88 | 1 | 239 | 0.4 | 99.6 |
| French Polynesia | 27 | 5 | 22 | | | | |
| Guam | 12 | 6 | 6 | | | | |
| Kiribati | 37 | 37 | | | | | |
| New Caledonia | 20 | 10 | 10 | | | | |
| New Zealand | 507 | 485 | 22 | 280 | 227 | 55.2 | 44.8 |
| Niue | 7 | 5 | 2 | | | | |
| Pacific Islands | 59 | 25 | 34 | | | | |
| Papua New Guinea | 388 | 33 | 355 | | | | |
| Samoa | 122 | 55 | 67 | | | | |
| Solomon Islands | 57 | 40 | 17 | | | | |
| Tonga | 48 | 17 | 31 | | | | |
| Vanuatu | 144 | 20 | 124 | | | | |
| Wallis and Futuna | 5 | 1 | 4 | | | | |
| | | | | | | | |
| *USSR* | 230,630 | 226,100 | 4,530 | 21,064 | 209,566 | 9.1 | 90.9 |

**TABLE E.2**   Global irrigated area, 1900 to 1989

| Year | Irrigated area[a] (10³ hectares) | Period | Percent change over period | Annual average percent change over period |
|------|------|------|------|------|
| 1900 | 48,000[b] | 1900–1950 | 96 | 1.9 |
| 1950 | 94,000[b] | 1950–1959 | 43 | 4.3 |
| 1959 | 134,000[b] | 1960–1969 | 21 | 2.1 |
| 1960 | 136,500[b] | 1970–1979 | 24 | 2.4 |
| 1961 | 139,441 | 1980–1989 | 11 | 1.1 |
| 1962 | 142,156 | | | |
| 1963 | 144,865 | | | |
| 1964 | 147,528 | | | |
| 1965 | 150,549 | | | |
| 1966 | 153,934 | | | |
| 1967 | 156,888 | | | |
| 1968 | 160,462 | | | |
| 1969 | 164,689 | | | |
| 1970 | 168,284 | | | |
| 1971 | 172,069 | | | |
| 1972 | 175,516 | | | |
| 1973 | 181,351 | | | |
| 1974 | 185,109 | | | |
| 1975 | 190,600 | | | |
| 1976 | 196,028 | | | |
| 1977 | 200,526 | | | |
| 1978 | 205,884 | | | |
| 1979 | 209,040 | | | |
| 1980 | 211,336 | | | |
| 1981 | 214,626 | | | |
| 1982 | 216,139 | | | |
| 1983 | 216,798 | | | |
| 1984 | 222,883 | | | |
| 1985 | 225,266 | | | |
| 1986 | 226,993 | | | |
| 1987 | 227,791 | | | |
| 1988 | 230,287 | | | |
| 1989 | 234,888 | | | |

[a]Data from the Food and Agriculture Organization of the United Nations, with adjusted figures for the United States from the U.S. Department of Agriculture
[b]Estimated.

## TABLE E.2  Global irrigated area, 1900 to 1989 *(page 265)*

### DESCRIPTION

Estimates of global irrigated area are given for 1900, 1950, and 1959 through 1989, in thousand hectares. The percentage changes in irrigated area over these time periods are also given. Irrigated areas are approximate for 1900 through 1960. Also computed here is the average annual change in irrigated area over several time periods, showing the recent drop in the rate of increase of irrigated area to about 1% per year from a high of over 4% annually during the 1950s.

### LIMITATIONS

The numbers for irrigated area are a combination of Food and Agriculture Organization estimates and other country estimates, with different degrees of accuracy. The FAO data are then adjusted using U.S. data from the Department of Agriculture.

### SOURCES

Basic data from Food and Agriculture Organization, 1990, *FAO Production Yearbook 1990*, FAO Statistics Series, no. 99, vol. 44, prepared by the Statistics Division of the Economic and Social Policy Department, on the basis of information available as of April 1991, Food and Agriculture Organization of the United Nations, Rome. By permission.

Updated areas from S. Postel, 1992, *Last Oasis: Facing Water Scarcity*, W.W. Norton and Co., New York.

## TABLE E.3  Irrigated area by continent

| | Irrigated area ($10^3$ hectares) | | | | Percent change in irrigated area 1974–1989 | Per capita irrigated area (hectares/person) 1989 |
|---|---|---|---|---|---|---|
| | 1974 | 1979 | 1984 | 1989 | | |
| World | 185,225 | 209,248 | 220,988 | 232,828 | 25.7 | 0.045 |
| Africa | 9,354 | 9,836 | 10,573 | 11,186 | 19.6 | 0.018 |
| North and Central America | 22,599 | 27,567 | 25,380 | 25,920 | 14.7 | 0.061 |
| South America | 6,344 | 7,218 | 8,036 | 8,835 | 39.3 | 0.030 |
| Asia | 119,070 | 131,576 | 140,064 | 146,422 | 23.0 | 0.048 |
| Europe | 12,576 | 14,394 | 15,584 | 17,240 | 37.1 | 0.035 |
| Oceania | 1,607 | 1,657 | 1,866 | 2,161 | 34.5 | 0.083 |
| USSR | 13,675 | 17,000 | 19,485 | 21,064 | 54.0 | 0.074 |

## TABLE E.3  Irrigated area by continent

### DESCRIPTION

The total area of irrigated land changes is given here by continent, for 1974, 1979, 1984, and 1989, in thousand hectares. The percentage change in irrigated area over this time period is also given for each continent. Per capita irrigated area, in hectares per person, is shown using 1989 population data and 1989 irrigated areas. Total population numbers used in this computation come from the United Nations Population Division estimates, adjusted by the Food and Agriculture Organization to incorporate more recently available official population data, particularly for developing countries whose demographic statistics are considered to be reliable.

### LIMITATIONS

The numbers on irrigated area are a combination of Food and Agriculture Organization estimates and other country estimates, with different degrees of accuracy. Table E.4 gives specific information on individual country data. Irrigated area per capita is based on total population for the country, and is not a measure of the actual distribution of people of the surface of the land.

### SOURCE

Food and Agriculture Organization, 1990, *FAO Production Yearbook 1990*, FAO Statistics Series no. 99, vol. 44, prepared by the Statistics Division of the Economic and Social Policy Department, on the basis of information available as of April 1991, Food and Agriculture Organization of the United Nations, Rome. By permission.

**TABLE E.4**  Irrigated area by region and country

| | Irrigated area ($10^3$ hectares) | | | | Total population ($10^3$) 1989 | Agricultural population ($10^3$) 1989 | Total population per irrigated hectare 1989 |
|---|---|---|---|---|---|---|---|
| | 1974 | 1979 | 1984 | 1989 | | | |
| *World* | 185,225 | 209,248 | 220,988 | 232,828 | 5,204,120 | 2,369,139 | 22.4 |
| *Africa* | 9,354 | 9,836 | 10,573 | 11,186 | 623,106 | 380,710 | 55.7 |
| Algeria | 242 | 252 | 298 | 336 | 24,299 | 5,936 | 72.3 |
| Benin | 3 | 5 | 6 | 6 | 4,491 | 2,798 | 748.5 |
| Botswana | 1 | 2 | 2 | 2 | 1,257 | 801 | 628.5 |
| Burkina Faso | 8 | 10 | 12 | 16 | 8,753 | 7,408 | 547.1 |
| Burundi | 44 | 54 | 65 | 72 | 5,313 | 4,855 | 73.8 |
| Cameroon | 8 | 14 | 18 | 28 | 11,444 | 7,094 | 408.7 |
| Cape Verde | 2 | 2 | 2 | 2 | 359 | 159 | 179.5 |
| Chad | 6 | 6 | 10 | 10 | 5,537 | 4,186 | 553.7 |
| Congo | 2 | 3 | 4 | 4 | 2,199 | 1,313 | 549.8 |
| Côte d'Ivoire | 32 | 42 | 52 | 62 | 11,552 | 6,548 | 186.3 |
| Egypt | 2,843 | 2,447 | 2,493 | 2,585 | 51,233 | 21,005 | 19.8 |
| Ethiopia | 158 | 160 | 162 | 162 | 47,870 | 35,975 | 295.5 |
| Gambia | 8 | 10 | 12 | 12 | 837 | 680 | 69.8 |
| Ghana | 7 | 7 | 7 | 8 | 14,561 | 7,398 | 1,820.1 |
| Guinea | 6 | 8 | 16 | 24 | 5,584 | 4,183 | 232.7 |
| Kenya | 40 | 40 | 40 | 52 | 23,185 | 17,944 | 445.9 |
| Liberia | 2 | 2 | 2 | 2 | 2,494 | 1,753 | 1,247.0 |
| Libya | 195 | 220 | 232 | 242 | 4,387 | 616 | 18.1 |
| Madagascar | 426 | 610 | 790 | 900 | 11,625 | 8,955 | 12.9 |
| Malawi | 11 | 18 | 18 | 20 | 8,448 | 6,422 | 422.4 |
| Mali | 100 | 145 | 180 | 205 | 8,930 | 7,265 | 43.6 |
| Mauritania | 10 | 11 | 12 | 12 | 1,969 | 1,277 | 164.1 |
| Mauritius | 15 | 16 | 17 | 17 | 1,069 | 248 | 62.9 |
| Morocco | 1,032 | 1,185 | 1,240 | 1,265 | 24,427 | 9,095 | 19.3 |
| Mozambique | 34 | 60 | 86 | 115 | 15,244 | 12,491 | 132.6 |
| Namibia | 4 | 4 | 4 | 4 | 1,724 | 618 | 431.0 |
| Niger | 18 | 21 | 30 | 32 | 7,493 | 6,573 | 234.2 |
| Nigeria | 810 | 820 | 843 | 865 | 105,019 | 68,405 | 121.4 |
| Reunion | 5 | 5 | 5 | 6 | 587 | 69 | 97.8 |
| Rwanda | 4 | 4 | 4 | 4 | 6,992 | 6,393 | 1,748.0 |
| Senegal | 150 | 170 | 175 | 180 | 7,126 | 5,607 | 39.6 |
| Sierra Leone | 13 | 19 | 26 | 34 | 4,046 | 2,551 | 119.0 |
| Somalia | 100 | 105 | 110 | 116 | 7,280 | 5,206 | 62.8 |

*continued*

| | Irrigated area (10³ hectares) | | | | Total population (10³) 1989 | Agricultural population (10³) 1989 | Total population per irrigated hectare 1989 |
|---|---|---|---|---|---|---|---|
| | 1974 | 1979 | 1984 | 1989 | | | |
| South Africa | 1,017 | 1,100 | 1,128 | 1,128 | 34,507 | 5,414 | 30.6 |
| Sudan | 1,685 | 1,760 | 1,833 | 1,890 | 24,496 | 15,042 | 13.0 |
| Swaziland | 56 | 58 | 60 | 62 | 761 | 511 | 12.3 |
| Tanzania | 50 | 110 | 126 | 153 | 26,331 | 21,037 | 172.1 |
| Togo | 6 | 6 | 7 | 7 | 3,422 | 2,393 | 488.9 |
| Tunisia | 115 | 145 | 220 | 275 | 7,996 | 2,015 | 29.1 |
| Uganda | 4 | 6 | 8 | 9 | 18,111 | 14,745 | 2,012.3 |
| Zaire | | 5 | 8 | 10 | 34,450 | 22,852 | 3,445.0 |
| Zambia | 17 | 19 | 25 | 32 | 8,142 | 5,650 | 254.4 |
| Zimbabwe | 65 | 150 | 185 | 220 | 9,406 | 6,457 | 42.8 |
| *North and Central America* | 22,599 | 27,567 | 25,380 | 25,920 | 422,668 | 56,240 | 16.3 |
| Belize | 1 | 1 | 2 | 2 | 183 | | 91.5 |
| Canada | 485 | 565 | 715 | 840 | 26,253 | 907 | 31.3 |
| Costa Rica | 31 | 56 | 108 | 118 | 2,941 | 730 | 24.9 |
| Cuba | 550 | 740 | 824 | 896 | 10,496 | 2,069 | 11.7 |
| Dominican Republic | 135 | 160 | 190 | 225 | 7,019 | 2,583 | 31.2 |
| El Salvador | 33 | 102 | 110 | 120 | 5,138 | 1,943 | 42.8 |
| Guadeloupe | 2 | 2 | 2 | 3 | 341 | 34 | 113.7 |
| Guatemala | 60 | 66 | 75 | 78 | 8,935 | 4,628 | 114.6 |
| Haiti | 70 | 70 | 70 | 75 | 6,380 | 3,867 | 85.1 |
| Honduras | 75 | 80 | 85 | 90 | 4,982 | 2,865 | 55.4 |
| Jamaica | 30 | 32 | 34 | 35 | 2,427 | 732 | 69.3 |
| Martinique | 2 | 4 | 6 | 6 | 339 | 28 | 56.5 |
| Mexico | 4,293 | 4,940 | 4,882 | 5,150 | 86,737 | 26,530 | 16.8 |
| Nicaragua | 60 | 78 | 83 | 85 | 3,745 | 1,460 | 44.1 |
| Panama | 23 | 28 | 30 | 32 | 2,370 | 597 | 74.1 |
| Puerto Rico | 39 | 39 | 39 | 39 | 3,437 | 115 | 88.1 |
| St. Lucia | 1 | 1 | 1 | 1 | 147 | | 147.0 |
| St. Vincent | 1 | 1 | 1 | 1 | 115 | | 115.0 |
| Trinidad and Tobago | 18 | 20 | 21 | 22 | 1,261 | 97 | 57.3 |
| United States | 16,690 | 20,582 | 18,102 | 18,102 | 248,082 | 6,798 | 13.7 |
| *South America* | 6,344 | 7,218 | 8,036 | 8,835 | 290,988 | 69,832 | 32.9 |
| Argentina | 1,410 | 1,560 | 1,660 | 1,760 | 31,930 | 3,387 | 18.1 |
| Bolivia | 110 | 130 | 155 | 165 | 7,112 | 2,989 | 43.1 |
| Brazil | 1,200 | 1,700 | 2,200 | 2,700 | 147,404 | 36,729 | 54.6 |

*continued*

| | Irrigated area (10³ hectares) | | | | Total population (10³) 1989 | Agricultural population (10³) 1989 | Total population per irrigated hectare 1989 |
|---|---|---|---|---|---|---|---|
| | 1974 | 1979 | 1984 | 1989 | | | |
| Chile | 1,240 | 1,252 | 1,257 | 1,265 | 12,961 | 1,709 | 10.2 |
| Colombia | 290 | 380 | 452 | 515 | 32,351 | 9,136 | 62.8 |
| Ecuador | 506 | 520 | 537 | 550 | 10,327 | 3,231 | 18.8 |
| Guyana | 120 | 122 | 127 | 130 | 795 | 181 | 6.1 |
| Paraguay | 50 | 55 | 62 | 67 | 4,157 | 2,002 | 62.0 |
| Peru | 1,120 | 1,155 | 1,200 | 1,250 | 21,118 | 7,854 | 16.9 |
| Suriname | 32 | 40 | 45 | 59 | 413 | 69 | 7.0 |
| Uruguay | 55 | 70 | 93 | 110 | 3,078 | 425 | 28.0 |
| Venezuela | 211 | 234 | 248 | 264 | 19,246 | 2,093 | 72.9 |
| *Asia* | 119,070 | 131,576 | 140,064 | 146,422 | 3,054,998 | 1,774,368 | 20.9 |
| Afghanistan | 2,440 | 2,640 | 2,660 | 2,660 | 15,827 | 8,769 | 6.0 |
| Bahrain | 1 | 1 | 1 | 1 | 499 | 9 | 499.0 |
| Bangladesh | 1,299 | 1,495 | 1,920 | 2,738 | 112,559 | 77,887 | 41.1 |
| Bhutan | 22 | 26 | 30 | 34 | 1,483 | 1,348 | 43.6 |
| Brunei Darus | | | 1 | 1 | 258 | | 258.0 |
| Cambodia | 89 | 89 | 89 | 92 | 8,035 | 5,661 | 87.3 |
| China | 41,755 | 45,439 | 44,873 | 45,349 | 1,122,179 | 765,178 | 24.7 |
| Cyprus | 30 | 30 | 30 | 35 | 694 | 146 | 19.8 |
| Hong Kong | 6 | 4 | 3 | 2 | 5,761 | 73 | 2,880.5 |
| India | 32,550 | 38,060 | 41,955 | 43,039 | 835,718 | 527,611 | 19.4 |
| Indonesia | 4,840 | 5,360 | 6,254 | 7,550 | 180,840 | 81,889 | 24.0 |
| Iran | 6,000 | 5,280 | 5,730 | 5,750 | 53,314 | 14,679 | 9.3 |
| Iraq | 1,550 | 1,730 | 1,750 | 2,550 | 18,282 | 3,879 | 7.2 |
| Israel | 176 | 207 | 227 | 214 | 4,510 | 199 | 21.1 |
| Japan | 3,209 | 3,081 | 2,952 | 2,868 | 123,116 | 7,972 | 42.9 |
| Jordan | 35 | 37 | 48 | 57 | 3,167 | 194 | 55.6 |
| Korea DPR | 800 | 1,090 | 1,240 | 1,400 | 21,372 | 7,358 | 15.3 |
| Korea Rep | 1,269 | 1,311 | 1,320 | 1,353 | 42,426 | 9,903 | 31.4 |
| Kuwait | 1 | 1 | 2 | 2 | 1,977 | | 988.5 |
| Laos | 35 | 90 | 118 | 120 | 4,017 | 2,892 | 33.5 |
| Lebanon | 86 | 86 | 86 | 86 | 2,678 | 247 | 31.1 |
| Malaysia | 302 | 319 | 335 | 342 | 17,438 | 5,432 | 51.0 |
| Mongolia | 20 | 32 | 55 | 77 | 2,131 | 667 | 27.7 |
| Myanmar | 971 | 1,044 | 1,064 | 1,018 | 40,812 | 19,363 | 40.1 |
| Nepal | 175 | 460 | 720 | 943 | 18,686 | 17,162 | 19.8 |
| Oman | 33 | 38 | 41 | 41 | 1,448 | 594 | 35.3 |

*continued*

| | Irrigated area (10³ hectares) | | | | Total population (10³) 1989 | Agricultural population (10³) 1989 | Total population per irrigated hectare 1989 |
|---|---|---|---|---|---|---|---|
| | 1974 | 1979 | 1984 | 1989 | | | |
| Pakistan | 13,343 | 14,280 | 15,720 | 16,220 | 118,675 | 63,444 | 7.3 |
| Philippines | 990 | 1,167 | 1,408 | 1,620 | 60,924 | 28,634 | 37.6 |
| Saudi Arabia | 375 | 385 | 410 | 435 | 13,599 | 5,431 | 31.3 |
| Sri Lanka | 477 | 523 | 550 | 560 | 17,003 | 8,817 | 30.4 |
| Syria | 578 | 539 | 618 | 670 | 12,082 | 2,979 | 18.0 |
| Thailand | 2,378 | 2,836 | 3,659 | 4,230 | 54,923 | 33,596 | 13.0 |
| Turkey | 1,970 | 2,050 | 2,140 | 2,220 | 54,761 | 24,697 | 24.7 |
| United Arab Emirates | 5 | 5 | 5 | 5 | 1,549 | 41 | 309.8 |
| Vietnam | 980 | 1,555 | 1,750 | 1,830 | 65,270 | 39,988 | 35.7 |
| Yemen | 280 | 286 | 300 | 310 | 11,272 | 6,401 | 36.4 |
| *Europe* | 12,576 | 14,394 | 15,584 | 17,240 | 498,781 | 44,415 | 28.9 |
| Albania | 320 | 366 | 393 | 423 | 3,189 | 1,565 | 7.5 |
| Austria | 4 | 4 | 4 | 4 | 7,618 | 390 | 1,904.5 |
| Belgium-Luxembourg | 1 | 1 | 1 | 1 | 10,271 | 194 | 10,271.0 |
| Bulgaria | 1,101 | 1,185 | 1,210 | 1,253 | 8,991 | 1,106 | 7.2 |
| Czechoslovakia | 134 | 138 | 193 | 310 | 15,638 | 1,514 | 50.4 |
| Denmark | 160 | 372 | 405 | 430 | 5,132 | 251 | 11.9 |
| Finland | 33 | 60 | 60 | 62 | 4,962 | 424 | 80.0 |
| France | 790 | 875 | 1,020 | 1,160 | 56,160 | 2,886 | 48.4 |
| Germany DR | 140 | 145 | 150 | 150 | 16,614 | 1,385 | 110.8 |
| Germany FR | 300 | 314 | 320 | 330 | 61,990 | 1,992 | 187.8 |
| Greece | 848 | 928 | 1,030 | 1,190 | 10,028 | 2,198 | 8.4 |
| Hungary | 308 | 252 | 234 | 175 | 10,577 | 1,324 | 60.4 |
| Italy | 2,680 | 2,840 | 2,970 | 3,100 | 57,541 | 3,688 | 18.6 |
| Malta | 1 | 1 | 1 | 1 | 351 | 14 | 351.0 |
| Netherlands | 420 | 470 | 520 | 550 | 14,842 | 570 | 27.0 |
| Norway | 40 | 70 | 86 | 95 | 4,228 | 244 | 44.5 |
| Poland | 205 | 114 | 100 | 100 | 37,854 | 7,139 | 378.5 |
| Portugal | 625 | 630 | 632 | 634 | 10,319 | 1,843 | 16.3 |
| Romania | 1,396 | 2,253 | 2,612 | 3,450 | 23,145 | 4,288 | 6.7 |
| Spain | 2,783 | 2,997 | 3,215 | 3,360 | 39,090 | 4,247 | 11.6 |
| Sweden | 47 | 65 | 90 | 112 | 8,498 | 381 | 75.9 |
| Switzerland | 25 | 25 | 25 | 25 | 6,647 | 241 | 265.9 |
| United Kingdom | 82 | 135 | 152 | 157 | 57,436 | 1,178 | 365.8 |
| Yugoslavia | 133 | 154 | 161 | 168 | 23,690 | 4,842 | 141.0 |

*continued*

| | Irrigated area (10³ hectares) | | | | Total population (10³) 1989 | Agricultural population (10³) 1989 | Total population per irrigated hectare 1989 |
|---|---|---|---|---|---|---|---|
| | 1974 | 1979 | 1984 | 1989 | | | |
| *Oceania* | 1,607 | 1,657 | 1,866 | 2,161 | 26,256 | 4,472 | 12.1 |
| Australia | 1,471 | 1,490 | 1,625 | 1,880 | 16,828 | 868 | 9.0 |
| Fiji | 1 | 1 | 1 | 1 | 749 | 298 | 749.0 |
| New Zealand | 135 | 166 | 240 | 280 | 3,337 | 311 | 11.9 |
| *USSR* | 13,657 | 17,000 | 19,485 | 21,064 | 287,322 | 39,102 | 13.6 |

**TABLE E.4**  Irrigated area by region and country *(pages 267–271)*

## DESCRIPTION

The total area of irrigated land changes over time, as populations grow and as the demand for food increases. The total area of irrigated land is given here by continent and country, for 1974, 1979, 1984, and 1989, in thousand hectares. Total population and agricultural population is also given for 1989 for the same regions, in thousands. Agricultural population is defined as all people depending on agriculture for their livelihood, including all agricultural workers and their dependents. Total population numbers come from the United Nations Population Division estimates, adjusted by the Food and Agriculture Organization to incorporate more recently available official population data, particularly for developing countries whose demographic statistics are considered to be reliable. Also shown is the total population of each country per irrigated hectare, for 1989.

## LIMITATIONS

The numbers on irrigated area are a combination of Food and Agricul-ture Organization estimates and other country estimates, with different degrees of accuracy. With respect to the irrigated area data, for Cuba, data refer to the state sector only. For Denmark and Romania, irrigated area refers to land provided with irrigation facilities. For Hungary, these data exclude complementary farm plots and individual farms. For Japan, Korea, and Sri Lanka, the irrigated land data refer to irrigated rice only. Population per irrigated hectare is based on total population for the country, and is not a measure of the actual distribution of people on the surface of the land.

## SOURCE

Food and Agriculture Organization, 1990, *FAO Production Yearbook 1990*, FAO Statistics Series no. 99, vol. 44, prepared by the Statistics Division of the Economic and Social Policy Department, on the basis of information available as of April 1991, Food and Agriculture Organization of the United Nations, Rome. By permission.

**TABLE E.5**  Salinization of irrigated cropland, selected countries

| Country | Percentage of irrigated lands affected by salinization |
|---------|------------------------------------------------------|
| Algeria | 10–15 |
| Australia | 15–20 |
| China | 15 |
| Colombia | 20 |
| Cyprus | 25 |
| Egypt | 30–40 |
| Greece | 7 |
| India | 27 |
| Iran | <30 |
| Iraq | 50 |
| Israel | 13 |
| Jordan | 16 |
| Pakistan | <40 |
| Peru | 12 |
| Portugal | 10–15 |
| Senegal | 10–15 |
| Sri Lanka | 13 |
| Spain | 10–15 |
| Sudan | <20 |
| Syria | 30–35 |
| United States | 20–25 |

**TABLE E.5**  Salinization of irrigated cropland, selected countries

## DESCRIPTION

Salinization occurs when naturally saline lands, or lands with poor drainage, are irrigated. Dissolved salts then accumulate in the upper soil layers, decreasing crop yields and reducing soil fertility. Salinization affects primarily arid and semi-arid lands. Listed here for 21 countries is the proportion of irrigated lands in each country that suffer reduced crop yields from salinization.

A separate figure is presented, which shows the total area irrigated and the area damaged by salinization, for the world's top five irrigators, in hectares $\times 10^6$. These data come from a different source and vary, for some countries, from those in the table.

## LIMITATIONS

No estimate is provided on the severity of salinization, which varies greatly from region to region, or on the date of the estimate. Saline soils are, to some extent, natural to arid and semi-arid regions, even without irrigation. The data do not show the relative importance of the natural state of the soil in the present conditions.

## SOURCES

Data for the Table come from World Resources Institute, 1987, *World Resources 1987*, World Resources Institute and the International Institute for Environment and Development, Basic Books, New York. By permission.

**TABLE E.6** Extent of lands at least moderately desertified, early 1980s

| | Productive dryland types | | | | | | | |
| | Rangelands | | Rainfed croplands | | Irrigated lands | | All productive drylands | |
| | Total area (10⁶ hectares) | Percent desertified | Total area (10⁶ hectares) | Percent desertified | Total area (10⁶ hectares) | Percent desertified | Total area (10⁶ hectares) | Percent desertified |
|---|---|---|---|---|---|---|---|---|
| *Total* | 2,556 | 62 | 570 | 60 | 131 | 30 | 3,257 | 61 |
| Sudano-Sahelian Africa | 380 | 90 | 90 | 80 | 3 | 30 | 473 | 88 |
| Southern Africa | 250 | 80 | 52 | 80 | 2 | 30 | 304 | 80 |
| Mediterranean Africa | 80 | 85 | 20 | 75 | 1 | 40 | 101 | 83 |
| Western Asia | 116 | 85 | 18 | 85 | 8 | 40 | 142 | 82 |
| Southern Asia | 150 | 85 | 150 | 70 | 59 | 35 | 359 | 70 |
| USSR in Asia | 250 | 60 | 40 | 30 | 8 | 25 | 289 | 55 |
| China and Mongolia | 300 | 70 | 5 | 60 | 10 | 30 | 315 | 69 |
| Australia | 450 | 22 | 39 | 30 | 2 | 19 | 491 | 23 |
| Mediterranean Europe | 30 | 30 | 40 | 32 | 6 | 25 | 76 | 39 |
| South America and Mexico | 250 | 72 | 31 | 77 | 12 | 33 | 293 | 71 |
| North America | 300 | 42 | 85 | 39 | 20 | 20 | 405 | 40 |

**TABLE E.6** Extent of lands at least moderately desertified, early 1980s

**DESCRIPTION**

According to the United Nations Conference on Desertification in Nairobi in 1977, desertification is the degradation or destruction of land to desert-like conditions and can include the growth of sand dunes, deterioration of rangelands, degradation of rainfed croplands, waterlogging and salinization of irrigated lands, deforestation of woody vegetation, and declining fresh water availability or quality. Shown here are total areas of productive drylands, in million hectares, and the percentage experiencing moderate to severe desertification, by major regions. "Moderate" desertification is defined as the level at which problems of management and production losses become signficant, with losses of up to 25% of the potential for undegraded land. According to Mabbutt, the rural population directly affected by at least moderate desertification is 280 million people, as of 1983.

**LIMITATIONS**

These data come from a "desertification questionnaire" distributed by the United Nations to all countries. Information provided was patchy and often inconsistent from region to region. This reflects, in part, the lack of simple methodologies for assessing desertification over large areas. Thus, desertification information was presented in the original source on a regional, not a national basis. Areas not generally grazed are excluded from the total area of productive dryland. These areas would increase the total productive drylands area to 3,700 million hectares. No information is given on the extent to which "moderate" desertification can be reversed, given changes in land use.

**SOURCE**

J.A. Mabbutt, 1984, A new global assessment of the status and trends of desertification, *Environmental Conservation*, **11**(2), 103–113, The Foundation for Environmental Conservation, Geneva. By permission.

**TABLE E.7**   Rural populations affected by desertification, by region, early 1980s

| Region | Population affected by desertification ($10^6$ people) | | | | | | | |
|---|---|---|---|---|---|---|---|---|
| | Rangelands | | Rainfed croplands | | Irrigated lands | | Total rural population | |
| | Moderate | Severe | Moderate | Severe | Moderate | Severe | Moderate | Severe |
| Sudano-Sahelian Africa | 13.5 | 7 | 36 | 20 | 1.5 | 0.5 | 51 | 27.5 |
| Africa South of Sudano-Sahelian Region | 8 | 4.5 | 32 | 20 | 1 | 0.5 | 41 | 25 |
| Mediterranean Africa | 4 | 2 | 11 | 6 | 1 | 0.5 | 16 | 8.5 |
| Western Asia | 4 | 2 | 16 | 9.5 | 12 | 4.5 | 32 | 16 |
| South Asia | 9 | 4.5 | 34.5 | 18 | 23 | 6.5 | 66.5 | 29 |
| USSR in Asia | 1 | 0.5 | 1.5 | 0.5 | 4.5 | 1 | 7 | 2 |
| China and Mongolia | 3 | 1 | 4 | 2 | 10.5 | 3.5 | 17.5 | 6.5 |
| Australia | 0.03 | | 0.1 | 0.3 | 0.1 | | 0.23 | 0.03 |
| Mediterranean Europe | 2 | 1 | 13 | 4.5 | 1.5 | 0.5 | 16.5 | 6 |
| South America and Mexico | 4 | 2 | 22.5 | 11.5 | 2.5 | 1 | 29 | 13.5 |
| North America | 1.5 | 0.5 | 2 | 0.5 | 1 | 0.2 | 4.5 | 1.2 |

**TABLE E.7**   Rural populations affected by desertification, by region, early 1980s

## DESCRIPTION

According to the United Nations Conference on Desertification in Nairobi in 1977, desertification is the degradation or destruction of land to desert-like conditions and can include the growth of sand dunes, deterioration of rangelands, degradation of rainfed croplands, waterlogging and salinization of irrigated lands, deforestation of woody vegetation, and declining fresh water availability or quality. Shown here are the rural populations, in millions, affected by either moderate or severe desertification, by major regions. The data are for 1984. "Moderate" desertification is defined as the level at which problems of management and production losses become signficant, with losses of up to 25% of the potential for undegraded land. "Severe" desertification involves production losses of up to 50%. In many regions, populations associated with croplands are most severely affected, reflecting the importance of population pressures as a factor in desertification.

## LIMITATIONS

No information is given on the extent to which "moderate" or "severe" desertification can be reversed, given changes in land use, or on how these populations are actually affected. Only rural populations are given here, though urban areas are also affected.

## SOURCE

J.A. Mabbutt, 1984, A new global assessment of the status and trends of desertification, *Environmental Conservation*, **11**(2), 103–113, The Foundation for Environmental Conservation, Geneva. By permission.

**TABLE E.8**   Characteristics of alternative irrigation systems *(page 275)*

## DESCRIPTION

Growing water shortages in some regions have led to a re-examination of irrigation methods to supplement the many different irrigation methods presently in use. Shown here are seven alternative surface, sprinkler, and drip irrigation systems and their characteristics in a range of areas. Among the different factors evaluated are overall efficiency of water application, applicability by soil, terrain, crop, and water type, labor, energy, and management requirements, durability, and applicability in different weather conditions.

## LIMITATIONS

These qualitative descriptions provide a good overview of the characteristics of several alternative irrigation types. Nevertheless, the advantages and disadvantages of each may depend on some site-specific properties and economic variables not described here.

## SOURCE

J.C. Wade, 1986, Efficiency and optimization in irrigation analysis, in N.K. Whittlesey (ed.) *Energy and Water Management in Western Irrigated Agriculture*, Studies in Water Policy and Management no. 7, Westview Press, Boulder, Colorado, pp. 73–100. By permission.

**TABLE E.8**  Characteristics of alternative irrigation systems

| Site and situation factor | Surface systems | | Sprinkler system | | | Drip systems | |
|---|---|---|---|---|---|---|---|
| | Redesigned surface system | Level basins | Intermittent mechanical move | Continuous mechanical move | Solid set | Emitters and porous tubes | Bubblers and spitters |
| Average efficiency rating | 60-70% | 80% | 70-80% | 80% | 70-80% | 80-90% | 80-90% |
| Soil | Uniform soils with moderate to low infiltration | Uniform soils with moderate to low infiltration | All | Sandy or high infiltration rate soils | All | All | All, basin required for medium and low intake soils |
| Topography | Moderate slopes | Small slopes | Level to rolling | Level to rolling | Level to rolling | All | All |
| Crops | All | All | Generally shorter crops | All but trees and vineyards | All | High value required | High value required |
| Water supply | Large streams | Very large streams | Small streams nearly continuous | Small streams nearly continuous | Small streams | Small streams continuous and clean | Small streams continuous |
| Water quality | All but very high salts | All | Salty water may harm plants | Salty water may harm plants | Salty water may harm plants | All, can potentially use high salt waters | All, can potentially use high salt waters |
| Labor requirement | High, training required | Low, some training | Moderate, some training | Low, some training | Low to high, little training | Low to high, some training | Low, little training |
| Energy requirement | Low | Low | Moderate to high | Moderate to high | Moderate | Low to moderate | Low |
| Management skill | Moderate | Moderate | Moderate | Moderate to high | Moderate | High | High |
| Machinery operations | Medium to long fields | Short fields | Medium field length, small interference | Circular fields, some interference | Some interference | May have considerable interference | Some interference |
| Duration of use | Short to long term | Long term | Short to medium term | Short to medium term | Long term | Long term, durability unknown | Long term |
| Weather | All | All | Poor in windy conditions | Better in windy conditions than other sprinklers | Windy conditions reduce performance; good for cooling | All | All |

**TABLE E.9**  Factors affecting the selection of an appropriate irrigation method

| Irrigation method | Factors affecting selection | | | | | |
|---|---|---|---|---|---|---|
| | Land | Soil | Crop | Climate | Plusses | Minuses |
| Surface | Level or graded to central slope and surface smoothness | Suited for medium to fine textures but not for infiltrability | For most crops, except those sensitive to standing water or poor aeration | For most climates only slightly affected by wind | Low cost, simple, low pressure required | Prone to overirrigate and rising water table |
| Sprinklers | For all lands | For most soils | For most crops, except sensitive to fungus disease and leaf scorch by salts | Affected by wind (drift, evaporation, and poor distribution) | Control of rate and frequency allows irrigation of sloping and sandy soils | Initial costs and water pressure requirements |
| Drip | For all regular and irregular slopes | For all soils and intake rates | For row crops and orchards, but not close-growing crops | Not affected by wind, adapted to all climates | High-frequency and precise irrigation, can use saline water and rough land, reduced evaporation | Initial and annual costs, requires expert management, prone to clogging, requires filtration |
| Microsprayer | For all lands | For all intake rates | For row crops and orchards | May be affected by wind | High-frequency and precise irrigation, less prone to clog | High costs and maintenance |
| Bubbler | Flat lands and gentle slopes | For all intake rates | For tree crops | Not affected by wind | High-frequency irrigation, no clogging, simple | Not a commercial product |

**TABLE E.9**  Factors affecting the selection of an appropriate irrigation method

**DESCRIPTION**

Growing water shortages in some regions have led to a re-examination of irrigation methods to supplement the many different irrigation methods presently in use. Shown here are four factors that will affect the choice of alternative surface, sprinkler, drip, sprayer, and bubbler irrigation systems. The factors include land and soil type, crop type, and climate. Also given is a qualitative summary of the advantages and disadvantages of each. See also Table E.10.

**LIMITATIONS**

These qualitative descriptions provide a good overview of the characteristics of several alternative irrigation types. Nevertheless, the advantages and disadvantages of each may also depend on some site-specific properties and economic variables not described here.

**SOURCE**

D. Hillel, 1987, The efficient use of water in irrigation, World Bank Technical Paper no. 64, The International Bank for Reconstruction and Development (World Bank), Washington, DC. By permission.

**TABLE E.10** Guide for selecting a method of irrigation

| Irrigation method | Topography | Crops | Remarks |
|---|---|---|---|
| Widely spaced borders | Land slopes capable of being graded to less than 1% slope and preferably 0.2% | Alfalfa and other deep-rooted close-growing crops and orchards | The most desirable surface method for irrigating close-growing crops where topographical conditions are favorable. Even grade in the direction of irrigation is required on flat land and is desirable but not essential on slopes of more than 0.5%. Grade changes should be slight and reverse grades must be avoided. Cross slope is permissible when confined to differences in elevation between border strips of 6–9 cm. |
| Closely spaced borders | Land slopes capable of being graded to 4% slope or less and preferably less than 1% | Pastures | Especially adapted to shallow soils underlain by claypan or soils that have a lower water intake rate. Even grade in the direction of irrigation is desirable but not essential. Sharp grade changes and reverse grades should be smoothed out. Cross slope is permissible when confined to differences in elevation between borders of 6–9 cm. Since the border strips may have less width, a greater total cross slope is permissible than for border-irrigated alfalfa. |
| Check back and cross furrows | Land slopes capable of being graded to 0.2% slope or less | Fruit | This method is especially designed to obtain adequate distribution and penetration of moisture in soils with low water intake rates. |
| Corrugations | Land slopes capable of being graded to slopes between 0.5% and 12% | Alfalfa, pasture, and grain | This method is especially adapted to steep land and small irrigation streams. An even grade in the direction of irrigation is desirable but not essential. Sharp grade changes and reverse grades should at least be smoothed out. Due to the tendency of corrugations to clog and overflow and cause serious erosion, cross slopes should be avoided as much as possible. |
| Graded contour furrows | Variable land slopes of 2–25% but preferably less | Row crops and fruit | Especially adapted to row crops on steep land, though hazardous due to possible erosion from heavy rainfall. Unsuitable for rodent-infested fields or soils that crack excessively. Actual grade in the direction of irrigation 0.5–1.5%. No grading required beyond filling gullies and removal of abrupt ridges. |
| Contour ditches | Irregular slopes up to 12% | Hay, pasture, and grain | Especially adapted to foothill condition. Requires little or no surface grading. |
| Rectangular checks (levees) | Land slopes capable of being graded so single or multiple tree basins will be leveled within 6 cm | Orchards | Especially adapted to soils that have either a relatively high or low water intake rate. May require considerable grading. |
| Contour levee | Slightly irregular land slopes of less than 1% | Fruits, rice, grain, and forage crops | Reduces the need to grade land. Frequently employed to avoid altogether the necessity of grading. Adapted best to soils that have either a high or low intake rate. |

*continued*

| Irrigation method | Topography | Crops | Remarks |
|---|---|---|---|
| Portable pipes | Irregular slopes up to 12% | Hay, pasture, and grain | Especially adapted to foothill conditions. Requires little or no surface grading. |
| Subirrigation | Smooth–flat | Shallow-rooted crops such as potatoes or grass | Requires a water table, very permeable subsoil conditions, and precise leveling. Very few areas adapted to this method. |
| Sprinkler | Undulating 1->35% slope | All crops | High operation and maintenance costs. Good for rough or very sandy lands in areas of high production and good markets. Good method where power costs are low. May be the only practical method in areas of steep or rough topography. Good for high rainfall areas where only a small supplemental water supply is needed. |
| Contour bench terraces | Sloping land—best for slopes under 3% but useful to 6% | Any crop but particularly suited to cultivated crops | Considerable loss of productive land due to berms. Requires expensive drop structures for water erosion control. |
| Subirrigation (installed pipes) | Flat to uniform slopes up to 1% surface should be smooth | Any crop; row crops of high-value crops usually used | Requires installation of perforated plastic pipe in root zone at narrow spacings. Some difficulties in roots plugging the perforations. Also a problem as to correct spacing. Field trials on different soils are needed. This is still in the development stage. |
| Localized (drip, trickle, etc.) | Any topographic condition suitable for row crop farming | Row crops or fruit | Perforated pipe on the soil surface drips water at base of individual vegetable plants or around fruit trees. Has been successfully used in Israel with saline irrigation water. Still in development stage. |

**TABLE E.10**  Guide for selecting a method of irrigation *(pages 277–278)*

**DESCRIPTION**
There is a wide choice of irrigation technologies and techniques. The choice for any particular region or crop depends on a range of factors, including topography of the land, the type and value of the crop, and the level of technological sophistication. This table qualitatively describes the applicability of different irrigation methods, given these factors.

**LIMITATIONS**
Little information is given here on cost of water provided by these

different irrigation methods, which will often be a limiting factor in any new development. Similarly, water and soil qualities are not taken into account here.

**SOURCES**
L.D. Doneen and D.W. Westcot, 1984, *Irrigation Practice and Water Management*, FAO Irrigation and Drainage Paper 1, Rev. 1, Food and Agriculture Organization of the United Nations, Rome. By permission.

**TABLE E.11**  Impacts of irrigation development

| Causal activity | Possible impact | Possible remedies |
| --- | --- | --- |
| Surface irrigation | 1. Waterlogging<br>2. Soil salinization<br>3. Increase of diseases<br>4. Degradation of water quality | 1. Increased irrigation efficiency<br>2. Construction of drainage systems<br>3. Disease control measures<br>4. Control of irrigation water quality |
| Sewage irrigation | 1. Contamination of food crops<br>2. Direct contamination of humans<br>3. Dispersion in air<br>4. Contamination of grazing animals | 1. Regulatory control<br>2. Tertiary treatment and sterilization of sewage |
| Use of fertilizers | 1 Pollution of ground water, especially with nitrates<br>2. Pollution of surface flow | 1. Controlled use of fertilizers<br>2. Increased irrigation efficiency |
| Use of pesticides | 1. Pollution of surface flow<br>2. Destruction of fish | 1. Limited use of pesticides<br>2. Coordination with schedule of irrigation |
| Irrigation with high silt load | 1. Clogging of canals<br>2. Raising of level of fields<br>3. Harmful sediment deposits on fields and crops | 1. Avoiding use of flow with high silt load<br>2. Soil conservation measures on upstream watershed |
| High velocity surface flow | 1. Erosion of earth canals<br>2. Furrow erosion<br>3. Surface erosion | 1. Proper design of canals<br>2. Proper design of furrows<br>3. Land leveling<br>4. Correctly built and maintained terraces |
| Intensive sprinkling on sloping land | 1. Soil erosion | 1. Correctly designed and operated system |

**TABLE E.11**  Impacts of irrigation development

**DESCRIPTION**

Intensification of agriculture is often considered a key step in the process of national development, and irrigation is one of the key tools in intensification. Irrigation development, however, can lead to adverse ecological and environmental changes, some of which are qualitatively described in this table. Possible remedies for each impact are suggested.

**LIMITATIONS**

As irrigation impacts are generally cumulative, the listing was drawn from traditional irrigated agricultural systems in the Asia and Pacific region. It is assumed that modern practices, if not adjusted for these problems, will have the same or more severe effects, due to their intensity. The severity and extent of these impacts vary with the specific type of irrigation activity.

**SOURCES**

Economic and Social Commission for Asia and the Pacific (ESCAP), 1987, *Water Resources Development in Asia and the Pacific: Some Issues and Concerns*, United Nations, Water Resources Series no. 62, United Nations Publishers, New York.

Adapted from the United Nations Environment Programme, 1979, *Draft Guidelines on the Environmental Impacts of Irrigation in Arid and Semi-arid Regions*, Nairobi, Kenya.

**TABLE E.12**  Selected environmental effects of agriculture on water quality

| Agricultural practices | Soil | Ground water | Surface water |
| --- | --- | --- | --- |
| Land development, land consolidation programs | Inadequate management leading to soil degradation | Other water management influencing ground water table | Soil degradation, siltation, water pollution with soil particles |
| Irrigation, drainage | Excess salts, waterlogging | Loss of quality (more salts), drinking water supply affected | Soil degradation, siltation, water pollution with soil particles |
| Tillage | Wind erosion, water erosion | | Soil degradation, siltation, water pollution with soil particles |
| Mechanization: large or heavy equipment | Soil compaction, soil erosion | | Soil degradation, siltation, water pollution with soil particles |
| Nitrogen fertilizer use | | Nitrate leaching | |
| Phosphate fertilizer use | Accumulation of heavy metals (such as cadmium) | | Runoff leaching or direct discharge leading to eutrophication |
| Manure, slurry use | Excess accumulation of phosphates, copper (pig slurry) | Nitrate, phosphate combination (by use of excess slurry) | Runoff leaching or direct discharge leading to eutrophication |
| Sewage sludge compost | Accumulation of heavy metals contaminates | | Runoff leaching or direct discharge leading to eutrophication |
| Applying pesticides | Accumulation of pesticides and degradation products | Leaching of mobile pesticide residues and degradation products | |
| Input of feed additives, medicines | Adverse effects depend on input | | |
| Modern building (e.g., silos) and intensive livestock farming | Excess accumulation of phosphates, copper (pig slurry) | Nitrate, phosphate contamination (by use of excess slurry) | Runoff leaching or direct discharge leading to eutrophication |

**TABLE E.12**  Selected environmental effects of agriculture on water quality

**DESCRIPTION**

Agricultural production has a wide range of impacts on soils, and on surface and ground water quality. This table compiles the major environmental impacts of a variety of agricultural practices, including land development, drainage of irrigation water, tillage, mechanization, fertilizer use, the application of pesticides, and several other activities. All impacts are described qualitatively.

**LIMITATIONS**

Only a few environmental impacts are described, and no quantitative information is given. Agricultural practices vary tremendously, depending on region, traditional practices, types of crops planted, irrigation methods used, and many other factors. These variables directly affect the environmental impacts described here.

**SOURCES**

Organization for Economic Co-operation and Development, 1985, *State of the Environment, 1985*, Organization for Economic Co-operation and Development, Paris. By permission.

As quoted in M. Meybeck, D.V. Chapman and R. Helmer, 1989, *Global Freshwater Quality: A First Assessment*, Global Environment Monitoring System (GEMS), World Health Organization and the United Nations Environment Programme, Basil Blackwell, Oxford.

**TABLE E.13**  Land use by different hydrologic water regimes (%)

| | Rainfed land | | | Irrigated land | Naturally flooded land | Problem land |
|---|---|---|---|---|---|---|
| | Low rainfall | Uncertain rainfall | Good rainfall | | | |
| 93 Developing countries (excluding China) | 8 | 13 | 24 | 22 | 11 | 22 |
| Sub-Saharan Africa | 17 | 20 | 31 | 3 | 3 | 26 |
| Near East/North Africa | 14 | 11 | 19 | 29 | 11 | 16 |
| Asia (excluding China) | 6 | 12 | 12 | 32 | 16 | 22 |
| Latin America | 2 | 9 | 50 | 12 | 3 | 24 |

Low rainfall: 1–119 growing days, soil quality marginal to very suitable.
Uncertain rainfall: 120–179 growing days, soil quality marginal to very suitable.
Good rainfall: 180–269 growing days, soil quality suitable to very suitable.
Irrigated land: fully or partly irrigated land.
Naturally Flooded land: land under water for part of the year and lowland nonirrigated paddy fields (gleysols).
Problem land: more than 269 growing days, soil quality marginal to very suitable, excessive rainfall or marginal soil.

**TABLE E.13**  Land use by different hydrologic water regimes

**DESCRIPTION**

Land used for agricultural production varies in its quality and hydrological characteristics. According to the United Nations Food and Agriculture Organization, many developing countries use marginal lands, including lands with too little and too much rainfall, for food production. Presented here for developing countries and regions in the early 1980s is the percentage of land used for agricultural purposes separated by water regime: low, uncertain, and good rainfall, irrigated land, naturally flooded land, and "problem land", defined as having excessive rainfall or marginal soils. For some regions, the dependence on rainfed agriculture is substantial: 68% of the agricultural land in sub-Saharan Africa relies on rainfall and only 3% is irrigated.

**LIMITATIONS**

While the water regime is very important to agricultural productivity,

the actual yields achievable in each of these categories of land vary with the crops planted, the climate, tilling methods, and other agricultural practices.

**SOURCES**

Data from N. Alexandratos (ed.), 1988, *World Agriculture Toward 2000: An FAO Study*, Food and Agriculture Organization of the United Nations, Bellhaven Press, London.

Cited in *An International Action Programme on Water and Sustainable Agricultural Development. A Strategy for the Implementation of the Mar del Plata Action Plan for the 1990s*, 1990, Food and Agriculture Organization of the United Nations. Rome. By permission.

**TABLE E.14**  Crop water requirements and crop evapotranspiration

| Crop | Southwest United States range (mm) | Missouri and Arkansas basin range (mm) | FAO guidelines range (mm) | United States and Canada (Kammerer) range (mm) |
|---|---|---|---|---|
| *Farm crops* | | | | |
| Alfalfa/Forage | 1,060–1,550 | 591–799 | 600–1,600 | 594–1,890 |
| Barley | 378–558 | 405–555 | | 384–643 |
| Broomcorn | 296–351 | | | |
| Buckwheat | | 320–396 | | |
| Cocoa | | | 800–1,200 | |
| Coffee | | | 800–1,200 | |
| Corn (Maize) | 439–607 | 375–558 | 400–750 | 373–617 |
| Cotton | 716–1,070 | | 550–950 | 912–1,050 |
| Emmer | 363–570 | | | |
| Feterita | 296–335 | | | |
| Flax | 375–485 | 448–564 | 450–900 | 381–795 |
| Grains (small) | | | 300–450 | |
| Grass | | | | 579–1,320 |
| Groundnut | | | 500–700 | |
| Kafir | 402–469 | 436–479 | | |
| Millet | 277–332 | 247–287 | | |
| Milo | 293–509 | 332–518 | | |
| Oats | 579–637 | 411–552 | | |
| Oil Seeds | | | 300–600 | |
| Potatoes | 485–622 | 421–518 | 350–625 | 455–617 |
| Rhodes Grass | 1,060–1,350 | | | |
| Rice | | | 500–950 | 920 |
| Safflower | | | | 635–1,150 |
| Sisal | | | 550–800 | |
| Sorghum | 515–634 | 323–448 | 300–650 | 549–645 |
| Soybeans | 506–856 | | 450–825 | 399–564 |
| Sudan Grass | 878–963 | | | |
| Sugar Beets | 539–829 | 488–762 | 450–850 | 546–1,050 |
| Sugar Cane | 1,060–1,390 | | 1,000–1,500 | |
| Sunflowers | | 366–427 | | |
| Tobacco | | | 300–500 | |
| Wheat | 445–683 | 415–549 | 450–650 | 414–719 |
| *Vegetable crops* | | | | |
| Beans | | 396–488 | 250–500 | 396–417 |
| Beans, Snap | 253–439 | | | |
| Beets, Table | 265–418 | | | |

*continued*

| Crop | Southwest United States range (mm) | Missouri and Arkansas basin range (mm) | FAO guidelines range (mm) | United States and Canada (Kammerer) range (mm) |
|---|---|---|---|---|
| Broccoli | | | | 500 |
| Cabbage | 287–454 | | | 437–622 |
| Carrots | 387–488 | | | 422 |
| Cauliflower | 436–539 | | | 472 |
| Corn, Sweet | | | | 386–498 |
| Cucumbers | | 528–1,140 | | |
| Lettuce | 219–411 | | | 216 |
| Melons | 756–1,040 | | | |
| Onions | 223–463 | | 350–600 | 592 |
| Onions, Green | | | | 445 |
| Peas | 369–475 | 415–591 | | 340 |
| Spinach | 244–326 | | | |
| Sweet Potatoes | 539–686 | | 400–675 | |
| Tomatoes | 290–433 | 640–853 | 300–600 | 366–681 |
| *Fruit* | | | | |
| Apples | | 640–792 | | 531–1,060 |
| Avocados | | | 650–1,000 | |
| Bananas | | | 700–1,700 | |
| Cantaloupes | | 457–701 | | 485 |
| Dates | | | 900–1,300 | |
| Deciduous Trees | | | 700–1,050 | |
| Grapefruit | | | 650–1,000 | 1,220 |
| Oranges | | | 600–950 | 933 |
| Plums | | | | 1,070 |
| Vineyards | | | 450–900 | |
| Walnuts | | | 700–1,000 | |

**TABLE E.14** Crop water requirements and crop evapotranspiration *(pages 282–283)*

**DESCRIPTION**

Crop water requirements are typically defined as the depth of water necessary for replacing water lost through evapotranspiration for a disease-free crop, grown in large fields under non-restricting soil conditions, such as soil water and fertility, providing for full production potential. Data are given for two regions in the U.S., for the United States and Canada, and for overall average crop use. All data are presented in millimeters of water applied. Because of the many field factors that affect actual crop evapotranspiration, ranges are given for each crop.

**LIMITATIONS**

Actual water requirements will vary according to climate, crop characteristics, time of planting, length of growing season, and local conditions and agricultural practices. Kammerer's data refer to observed seasonal evapotranspiration for well-watered crops in the United States and Canada. The ranges were prepared from data with varying growing periods and locations. The number of significant figures in the original sources seems excessive, given the great variability in crop water use.

**SOURCES**

Food and Agriculture Organization, 1977, *Guidelines for Predicting Crop Water Requirements*, Irrigation and Drainage Paper no. 24 (revised), Food and Agriculture Organization of the United Nations, Rome. By permission.

J.C. Kammerer, 1982, Estimated demand of water for different purposes, in *Water for Human Consumption*, International Water Resources Association, Tycooly International, Dublin. By permission.

U.S. Department of Agriculture, as cited in F. van der Leeden, F.L. Troise and D.K. Todd, 1990, *The Water Encyclopedia*, Lewis Publishers, Chelsea, Michigan.

**TABLE E.15**  Methods and sources of irrigation in the United States, 1979 and 1984

| Item | 1979 | | | 1984 | | |
|---|---|---|---|---|---|---|
| | Number of farms | Area ($10^3$ hectares) | Meters of water applied | Number of farms | Area ($10^3$ hectares) | Meters of water applied |
| **Methods** | | | | | | |
| All sprinkler systems | 121,749 | 7,442 | 0.43 | 104,641 | 6,833 | 0.40 |
| Center pivot | 24,078 | 3,499 | 0.40 | 32,442 | 3,794 | 0.37 |
| Mechanical move | 29,617 | 2,046 | 0.43 | 25,475 | 1,358 | 0.37 |
| Hand move | 67,344 | 1,500 | 0.43 | 46,885 | 1,181 | 0.49 |
| Solid set and permanent | 16,334 | 397 | 0.61 | 19,694 | 499 | 0.76 |
| All gravity systems | 153,946 | 12,636 | 0.64 | 126,827 | 11,116 | 0.61 |
| Gated pipe | 38,552 | 3,401 | 0.52 | 42,826 | 3,386 | 0.43 |
| Ditch with siphon tube | 47,413 | 3,504 | 0.67 | 59,255 | 4,069 | 0.70 |
| Flooding | 85,351 | 5,731 | 0.76 | 45,045 | 3,661 | 0.67 |
| Drip or trickle | 7,134 | 130 | 0.64 | 11,651 | 339 | 0.58 |
| Subirrigation | 760 | 98 | 0.46 | 2,905 | 252 | 1.16 |
| **Sources** | | | | | | |
| Wells | 120,952 | n/a | 0.46 | 100,703 | 9,833 | 0.43 |
| On-farm surface sources | 44,644 | n/a | 0.49 | 35,982 | 2,383 | 0.55 |
| Off-farm water suppliers | 112,600 | n/a | 0.67 | 98,672 | 6,335 | 0.70 |

n/a not available

**TABLE E.15**  Methods and sources of irrigation in the United States, 1979 and 1984

**DESCRIPTION**

Irrigation is widely used for agriculture in parts of the United States. Shown here are the methods used for irrigation and the sources of water for 1979 and 1984 in U.S. agriculture. Methods include types of sprinkler systems, gravity flow systems, drip irrigation, and subirrigation. Sources of irrigation water include on-farm wells, on-farm surface supplies, and off-farm water supplies. The number of farms and area irrigated using each method is given, together with the average depth of water applied by each method and source. Irrigation area is given in thousand hectares; water applied is given in meters.

**LIMITATIONS**

Wide regional variations in water use are hidden by the averages shown here. More recent data would show a trend toward reducing the use of inefficient irrigation methods, particularly in the western United States, which has experienced drought and agricultural water cutbacks in the late 1980s and early 1990s. These data come from a survey of a 10% sample of farmers reporting irrigation practices in the 1978 United States Census of Agriculture and 5% in the 1983 Census. State data on water distribution are reported for 20 principal irrigation states, representing 94% of the irrigated area in the conterminous United States.

**SOURCES**

R.S. Bajwa, W.M. Crosswhite and J.E. Hostetler, 1987, *Agricultural Irrigation and Water Supply*, U.S. Department of Agriculture, Economic Research Service, Agricultural Information Bulletin no. 532, Washington, DC.

**TABLE E.16**   Area irrigated in the United States with on-farm pumped water, by type of energy, and by region, 1980

| Region | Area irrigated ($10^3$ hectares) | | | | |
|---|---|---|---|---|---|
| | Electricity | Diesel | Gasoline | Natural Gas | Liquid Propane |
| Northeast states | 10.3 | 43.1 | 54.2 | 16.7 | 5.3 |
| Lake states | 240.7 | 171.5 | 43.1 | 0.0 | 11.9 |
| Corn Belt states | 115.8 | 173.8 | 24.7 | 2.6 | 58.9 |
| Appalachian states | 12.1 | 56.4 | 45.5 | 0.8 | 2.6 |
| Southeast states | 389.5 | 796.4 | 84.9 | 0.8 | 136.1 |
| Delta states | 636.3 | 484.7 | 37.6 | 47.3 | 23.7 |
| Northern Plains | 1,325.5 | 1,130.6 | 32.0 | 1,466.0 | 511.5 |
| Southern Plains | 831.6 | 67.1 | 41.5 | 2,511.7 | 199.4 |
| Mountain states | 1,836.4 | 131.5 | 59.6 | 438.8 | 79.5 |
| Pacific states | 2,730.7 | 54.3 | 0.0 | 34.5 | 0.0 |
| | | | | | |
| 48 states | 8,128.9 | 3,109.6 | 423.0 | 4,519.1 | 1,028.8 |
| Alaska | 0.6 | 0.0 | 0.2 | 0.0 | 0.0 |
| Hawaii | 34.5 | 0.0 | 0.0 | 0.0 | 0.0 |
| Total United States | 8,164.0 | 3,109.6 | 423.2 | 4,519.1 | 1,028.8 |

Northern Plains include North Dakota, South Dakota, Nebraska, and Kansas.
Southern Plains include Oklahoma and Texas.
Mountain states include Montana, Idaho, Wyoming, Colorado, New Mexico, Arizona, Utah, and Nevada.
Pacific states include Washington, Oregon, and California.
Lake states include Minnesota, Wisconsin, and Michigan.
Corn Belt states include Iowa, Missouri, Illinois, Indiana, and Ohio.
Delta states include Arkansas, Louisiana, and Mississippi.
Southeast states include Alabama, Georgia, South Carolina, and Florida.
Appalachian states include Kentucky, Tennessee, West Virginia, Virginia, and North Carolina.
Northeast states include Maine, New Hampshire, Vermont, New York, Pennsylvania, Maryland, Delaware, Connecticut, Rhode Island, New Jersey, and Massachusetts.

**TABLE E.16**   Area irrigated in the United States with on-farm pumped water, by type of energy, and by region, 1980

**DESCRIPTION**

Substantial quantities of water for irrigation are pumped from ground water or from surface supplies directly on the farm. Shown here are the total areas irrigated by each of five energy sources, electricity, diesel, gasoline, natural gas, and liquid propane, in 12 regions of the United States in 1980. Areas are measured in thousand hectares. Table E.17 gives the total amounts of each form of energy used for the same regions.

**LIMITATIONS**

Included in this chart are only areas irrigated with pumped water. Sums of regional area exceed total irrigated area because more than one energy source is often used on the same parcel of land. Figures may not add to exact totals due to rounding. Electricity is typically generated elsewhere and consumed on-farm by pumps. No information is given on the form of energy used to generate the electricity.

**SOURCE**

D.M. Lea, 1985, *Irrigation in the United States*, U.S. Department of Agriculture, Economic Research Service, Natural Resources Economics Division, Washington, DC.

**TABLE E.17**  Quantity of energy used in the United States for on-farm pumped irrigation water, by region, 1980

| Region | Electricity (10³ kWh/yr) | Diesel (10³ l/yr) | Gasoline (10³ l/yr) | Natural gas (10³ m³/yr) | Liquid propane (10³ l/yr) |
|---|---|---|---|---|---|
| Northeast states | 6,960 | 12,400 | 14,780 | 6,850 | 1,930 |
| Lake states | 300,030 | 75,885 | 21,210 | 0 | 7,820 |
| Corn Belt states | 91,562 | 40,867 | 8,891 | 650 | 16,210 |
| Appalachian states | 10,554 | 16,610 | 15,840 | 300 | 1,190 |
| Southeast states | 565,825 | 352,130 | 68,792 | 340 | 113,290 |
| Delta states | 662,710 | 151,450 | 14,510 | 20,300 | 11,180 |
| Northern Plains | 1,889,186 | 688,105 | 24,010 | 1,114,300 | 500,846 |
| Southern Plains | 1,481,147 | 46,374 | 35,910 | 1,987,700 | 209,681 |
| Mountain states | 6,371,319 | 147,110 | 26,280 | 951,400 | 95,866 |
| Pacific states | 6,518,027 | 66,780 | 0 | 46,390 | 0 |
|  |  |  |  |  |  |
| 48 states | 17,897,320 | 1,588,930 | 230,230 | 4,128,150 | 958,021 |
| Alaska | 172 | 0 | 19 | 0 | 0 |
| Hawaii | 725,553 | 0 | 0 | 0 | 0 |
| Total United States | 18,623,045 | 1,597,720 | 230,250 | 4,128,150 | 958,021 |

Northern Plains include North Dakota, South Dakota, Nebraska, and Kansas.
Southern Plains include Oklahoma and Texas.
Mountain states include Montana, Idaho, Wyoming, Colorado, New Mexico, Arizona, Utah, and Nevada.
Pacific states include Washington, Oregon, and California.
Lake states include Minnesota, Wisconsin, and Michigan.
Corn Belt states include Iowa, Missouri, Illinois, Indiana, and Ohio.
Delta states include Arkansas, Louisiana, and Mississippi.
Southeast states include Alabama, Georgia, South Carolina, and Florida.
Appalachian states include Kentucky, Tennessee, West Virginia, Virginia, and North Carolina.
Northeast states include Maine, New Hampshire, Vermont, New York, Pennsylvania, Maryland, Delaware, Connecticut, Rhode Island, New Jersey, and Massachusetts.

**TABLE E.17**  Quantity of energy used in the United States for on-farm pumped irrigation water, by region, 1980

**DESCRIPTION**

This table quantifies the total energy used in 12 regions of the United States in 1980 for on-farm irrigation pumping. Data on five energy forms are provided: electricity in kWh × 10³ per year, diesel fuel and gasoline in l × 10³ per year, natural gas in m³ × 10³ per year, and liquid propane, in l × 10³ per year. The total area irrigated by each energy form is given in Table E.16.

**LIMITATIONS**

The figures given here may not add to exact totals because of rounding. Energy used to provide off-farm water to farms is not included. The energy forms used to produce the electricity vary from region to region.

**SOURCE**

D.M. Lea, 1985, *Irrigation in the United States*, U.S. Department of Agriculture, Economic Research Service, Natural Resources Economics Division, Washington, DC.

# F. Water and ecosystems

## TABLE F.1   Wetland terms and types *(page 288)*

### DESCRIPTION

Many different and inconsistent definitions of "wetlands" are in use around the world, complicating the accurate assessment of wetlands extent and loss. The more prevalent terms used to described wetlands are presented below. These definitions have a long history of use and misuse, and are often specific to a particular region or country. For example, the term "swamp" has no direct translation in Russian because of the lack of forested wetlands there, while "bog" is easily translated into Russian.

### LIMITATIONS

The physical and biotic characteristics of wetlands change continu-

ously from one type to another, making any classification somewhat arbitrary. Different regions may use the same term to denote very different ecosystems. For example, in Europe, "swamps" are dominated by reeds, while in the United States, swamps are almost always forested wetlands. Anyone interested in using wetlands data from around the world should take care to understand the distinctions and definitions being use.

### SOURCES

W.J. Mitsch, 1986, *Wetlands*, Van Nostrand Reinhold, New York. By permission.

## TABLE F.2   Wetlands of international importance, 1990 *(page 289)*

### DESCRIPTION

The Convention on Wetlands of International Importance Especially as Waterfowl Habitat (the Ramsar Convention) was signed in 1971. Any party to this Convention who agrees to respect a site's integrity and establish wetlands reserves can designate "Wetlands of International Importance." Those that have done so as of 1990 are listed here by country and wetlands area in hectares.

### LIMITATIONS

The wetlands listed here are a small fraction of total wetlands area. Included here are only those wetlands officially designated under the Ramsar Convention. Other wetlands of ecological importance exist, but are not included in this program. Each country has the responsibility to nominate its own wetlands for this program and participation of

any nation is voluntary. Several errors in the original source were corrected here; for example, Greenland is listed in the original in the totals for North and Central America. Since Greenland is a Danish overseas territory, we have listed it separately under Europe.

### SOURCE

Protected Areas Data Unit of the World Conservation Monitoring Center (WCMC), unpublished data, as cited in World Resources Institute, 1992, *World Resources 1992–93: A Guide to the Global Environment*, in collaboration with the United Nations Environment Programme and the United Nations Development Programme, Oxford University Press, New York. By permission.

1992, Dirk Bryant, World Resources Institute, personal communication, May.

## TABLE F.3   Important wetlands area, for selected countries, 1980s *(page 290)*

### DESCRIPTION

The areas of remaining, high-quality wetlands in two regions of the world are presented here by country. The Western Palearctic Realm includes all of Europe, Central and South Asia bounded by the Yenisei River in the East and the USSR/China and Pakistan/India borders in the south, Turkey, Syria, Lebanon, Israel, Jordan, Iraq, and the five African countries bordering the Mediterranean. The Neotropical Realm includes all of South America, the Caribbean, Central America, and part of Mexico. Many different and inconsistent definitions of "wetlands" are in use around the world, complicating the accurate assessment of wetlands extent and loss. The definition used here is that of the 1971 Convention on Wetlands of International Importance Especially as Waterfowl Habitat: "areas of marsh, fen, peatland or water, whether natural or artificial, permanent or temporary, with water that is static or flowing, fresh, brackish or salt, including areas of marine water the depth of which at low tide does not exceed six meters." Coral reefs are generally excluded. To be considered a wetland of "international importance", a site must (i) support a significant population of waterfowl, threatened species, or peculiar fauna or flora; (ii) be a regionally representative example of a type of wetland or an exemplar of a biological or hydrogeomorphic process; or (iii) be physically and administratively capable of benefiting from protection and management measures.

### LIMITATIONS

While the wetlands in this table are considered to meet the criteria to

be considered "internationally important", the vast majority of them have not received formal protection under the Ramsar (or any other) Convention. Table F.2 lists the much smaller areas of wetlands that have received this formal designation as of 1990. The use of the above criteria in selecting sites was not always possible because of missing data. As a result, some of the wetlands were included in this assessment based on the informed judgement of experts. Information was available for almost all countries of each region, but comprehensiveness varied among the countries. Data for the large countries of South America are considered preliminary. If a site's area was recorded as over or under a given size, the limiting area was used. If a site's area was recorded as a range, the average was used.

### SOURCES

D.A. Scott and M. Carbonell, 1986, *A Directory of Neotropical Wetlands*, International Union for Conservation of Nature and Natural Resources (IUCN) and International Waterfowl Research Bureau (IWRB), United Kingdom.

E. Carp, 1980, *Directory of Wetlands of International Importance in the Western Palearctic*, International Union for Conservation of Nature and Natural Resources (IUCN), and the United Nations Environment Programme, Nairobi.

Data compiled by World Resources Institute, 1987, *World Resources 1987*, Basic Books, New York.

**TABLE F.1**   Wetland terms and types

| | |
|---|---|
| Swamp: | Wetland dominated by trees or shrubs (U.S. definition). In Europe, a forested fen (see below) could easily be called a swamp. In some areas, wetlands dominated by reed grass are also called swamps. |
| Marsh: | A frequently or continually inundated wetland characterized by emergent herbaceous vegetation adapted to saturated soil conditions. In European terminology, a marsh has a mineral soil substrate and does not accumulate peat. |
| Bog: | A peat-accumulating wetland that has no significant inflows or outflows and supports acidophilic mosses, particularly sphagnum. |
| Fen: | A peat-accumulating wetland that receives some drainage from surrounding mineral soil and usually supports marshlike vegetation. |
| Peatland: | A generic term for any wetland that accumulates partially decayed plant matter. |
| Mire: | Synonymous with any peat-accumulating wetland (European definition). |
| Moor: | Synonymous with peatland (European definition). A high moor is a raised bog, while a low moor is a peatland in a basin or depression that is not elevated above its perimeter. |
| Muskeg: | Large expanses of peatlands or bogs; particularly used in Canada and Alaska. |
| Bottomland: | Lowlands along streams and rivers, usually on alluvial floodplains that are periodically flooded. These are often forested and sometimes called bottomland hardwood forests. |
| Wet prairie: | Similar to a marsh. |
| Reed swamp: | Marsh dominated by *Phragmites* (common reed); term used particularly in Eastern Europe. |
| Wet meadow: | Grassland with waterlogged soil near the surface, but without standing water for most of the year. |
| Slough: | A swamp or shallow lake system in northern and midwestern United States. A slowly flowing shallow swamp or marsh in southeastern United States. |
| Pothole: | Shallow marshlike ponds, particularly as found in the Dakotas. |
| Playa: | Term used in the southwest United States for marshlike ponds similar to potholes, but with a different geologic origin. |

**TABLE F.2** Wetlands of international importance, 1990

| | Number of sites | Area (10³ hectares) | | Number of sites | Area (10³ hectares) |
|---|---|---|---|---|---|
| *World*[a] | 506 | 30,379 | *Asia* | 40 | 1,354 |
| | | | India | 6 | 193 |
| *Africa*[a] | 36 | 3,353 | Iran | 18 | 1,088 |
| Algeria | 2 | 5 | Japan | 3 | 10 |
| Burkina Faso | 3 | n/a | Jordan | 1 | 7 |
| Chad | 1 | 195 | Nepal | 1 | 18 |
| Egypt | 2 | 106 | Pakistan | 9 | 21 |
| Gabon | 3 | 1,080 | Sri Lanka | 1 | 6 |
| Ghana | 1 | 7 | Vietnam | 1 | 12 |
| Guinea-Bissau | 1 | 39 | | | |
| Kenya | 1 | 19 | *Europe*[b] | 329 | 3,835 |
| Mali | 3 | 162 | Austria | 5 | 102 |
| Mauritania | 1 | 1,173 | Belgium | 6 | 10 |
| Morocco | 4 | 11 | Bulgaria | 4 | 2 |
| Niger | 1 | 220 | Czechoslovakia | 8 | 17 |
| Senegal | 4 | 100 | Denmark | 27 | 734 |
| South Africa | 7 | 208 | Finland | 11 | 101 |
| Tunisia | 1 | 13 | France | 1 | 85 |
| Uganda | 1 | 15 | German DR | 8 | 46 |
| | | | Germany FR | 21 | 315 |
| *North and Central America* | 40 | 14,101 | Greece | 11 | 107 |
| Canada | 30 | 12,938 | Greenland[b] | 18 | 1,046 |
| Guatemala | 1 | n/a | Hungary | 13 | 110 |
| Mexico | 1 | 47 | Iceland | 2 | 58 |
| United States | 8 | 1,116 | Ireland | 21 | 13 |
| | | | Italy | 45 | 54 |
| *South America* | 5 | 232 | Malta | 1 | 0.011 |
| Bolivia | 1 | 5 | Netherlands | 11 | 306 |
| Chile | 1 | 5 | Norway | 14 | 16 |
| Suriname | 1 | 12 | Poland | 5 | 7 |
| Uruguay | 1 | 200 | Portugal | 2 | 31 |
| Venezuela | 1 | 10 | Spain | 17 | 99 |
| | | | Sweden | 30 | 383 |
| *Oceania* | 44 | 4,516 | Switzerland | 2 | 2 |
| Australia | 39 | 4,478 | United Kingdom | 44 | 173 |
| New Zealand | 5 | 38 | Yugoslavia | 2 | 18 |
| | | | | | |
| | | | *USSR* | 12 | 2,987 |

n/a is not available. Totals may not add to sum of figures due to rounding.
[a]World and Africa totals are corrected from the original, which excluded Mali.
[b]In the original source, Greenland is included in the totals for North and Central America. Greenland is a Danish overseas territory, and is listed here separately under Europe.

**TABLE F.3**  Important wetlands area, for selected countries, 1980s

| Western Palearctic Realm[a] | Number of sites | Total area (hectares) | Neotropical Realm[b] | Number of sites | Total area (hectares) |
|---|---|---|---|---|---|
| Afghanistan | 2 | 40,000 | Anguilla | 10 | 326 |
| Albania | 5 | 32,990 | Antigua and Barbuda | 6 | 4,901 |
| Algeria | 7 | 630,637 | Argentina | 57 | 5,797,930 |
| Austria | 17 | 29,427 | Bahamas | 21 | 428,606 |
| Belgium | 6 | 9,642 | Barbados | 3 | 52 |
| Bulgaria | 5 | 14,500 | Belize | 15 | 58,707 |
| Cyprus | 3 | 10,101 | Bermuda | 10 | 78 |
| Czechoslovakia | 26 | 69,466 | Bolivia | 18 | 4,017,920 |
| Denmark | 39 | 591,730 | Brazil | 38 | 59,789,733 |
| Egypt | 6 | 808,500 | British Virgin Islands | 4 | 614 |
| Finland | 45 | 301,310 | Cayman Islands | 15 | 7,310 |
| France | 25 | 1,321,215 | Chile | 49 | 9,188,713 |
| German DR | 25 | 353,127 | Colombia | 36 | 1,928,389 |
| Germany FR | 49 | 720,525 | Costa Rica | 12 | 82,055 |
| Greece | 10 | 86,580 | Cuba | 17 | 1,746,500 |
| Hungary | 14 | 98,698 | Dominica | 3 | 93 |
| Iceland | 28 | 442,550 | Dominican Republic | 14 | 40,121 |
| Iran | 33 | 1,416,810 | Ecuador | 12 | 722,830 |
| Iraq | 18 | 1,992,500 | El Salvador | 8 | 76,800 |
| Ireland | 28 | 115,104 | Falkland Islands | 9 | 43,500 |
| Israel | 6 | 169,762 | French Antilles | 14 | 11,300 |
| Italy | 32 | 194,551 | French Guiana | 43 | 37,700 |
| Jordan | 1 | 10,000 | Grenada | 7 | 476 |
| Lebanon | 2 | 1,515 | Guatemala | 24 | 220,187 |
| Malta | 1 | 112 | Haiti | 11 | 112,900 |
| Morocco | 10 | 22,985 | Honduras | 6 | 649,000 |
| Netherlands | 22 | 394,019 | Jamaica | 14 | 13,775 |
| Norway | 39 | 151,298 | Mexico | 40 | 3,377,900 |
| Pakistan | 16 | 37,158 | Montserrat | 2 | 20 |
| Poland | 15 | 193,507 | Netherland Antilles | 11 | 5,329 |
| Portugal | 8 | 84,743 | Nicaragua | 17 | 2,111,349 |
| Romania | 11 | 482,527 | Panama | 21 | 646,012 |
| Spain | 10 | 445,000 | Paraguay | 5 | 5,723,528 |
| Sweden | 84 | 1,727,623 | Peru | 43 | 1,318,697 |
| Switzerland | 24 | 175,402 | Puerto Rico | 14 | 12,995 |
| Syria | 1 | 37,500 | St. Kitts-Nevis | 4 | 352 |
| Tunisia | 15 | 868,303 | St. Lucia | 3 | 314 |
| Turkey | 30 | 1,238,150 | St. Vincent | 3 | 1,003 |
| USSR | 12 | 2,837,200 | Suriname | 14 | 1,625,000 |
| United Kingdom | 87 | 449,184 | Trinidad and Tobago | 9 | 21,280 |
| Yugoslavia | 15 | 110,088 | Turks and Caicos Islands | 111 | 64,669 |
| | | | U.S. Virgin Islands | 15 | 979 |
| | | | Uruguay | 12 | 773,500 |
| | | | Venezuela | 29 | 14,447,155 |

[a]No data are given for Libya, Luxembourg, or Liechtenstein.
[b]No data are given for Guyana.

**TABLE F.4**  Important coastal wetlands

| | Wetlands area (km²) | Percent of total regional area |
|---|---|---|
| Mexico | 32,330 | 1.64 |
| Central America | 25,319 | 0.88 |
| Caribbean islands | 24,452 | 9.43 |
| South America Atlantic Ocean coast | 158,260 | 1.13 |
| South America Pacific Ocean coast | 12,413 | 0.53 |
| Atlantic Ocean small islands | 400 | 3.29 |
| North and West Europe | 31,515 | 0.71 |
| Baltic Sea coast | 2,123 | 0.18 |
| Northern Mediterranean | 6,497 | 0.61 |
| Southern Mediterranean | 3,941 | 0.10 |
| Africa Atlantic Ocean coast | 44,369 | 0.56 |
| Africa Indian Ocean coast | 11,755 | 0.16 |
| Gulf States | 1,657 | 0.08 |
| Asia Indian Ocean coast | 59,530 | 1.20 |
| Southeast Asia | 122,595 | 3.42 |
| East Asia | 102,074 | 1.00 |
| Pacific Ocean large islands excluding Australia and New Zealand | 89,500 | 19.39 |
| USSR | 4,191 | 0.02 |
| Total of listed regions | 732,921 | 0.85 |

**TABLE F.4**  Important coastal wetlands

**DESCRIPTION**

Coastal wetlands, including mixed fresh and salt water estuarine systems, are listed here by region, in km². The percentage of total regional area is also given. The table includes all coastal wetlands that could meet the definitions of the official category of international importance, as defined by the Convention on Wetlands of International Importance Especially as Waterfowl Habitat (the Ramsar Convention), signed in 1971. To be considered a wetland of "international importance", a site must (i) support a significant population of waterfowl, unusual fauna or flora, or species that are considered threatened; (ii) be a regionally representative example of a type of wetland or of an unusual biological or hydrogeomorphic process; or (iii) be physically and administratively capable of benefiting from protection and management measures. Only a small fraction of these wetlands have actually received official designation.

**LIMITATIONS**

The vast majority of these sites have no official protection. The listing excludes some major countries, such as the United States, Canada, Australia, and New Zealand. Only the areas of coastal wetlands area given, not all wetlands areas.

**SOURCES**

WMO/UNEP, 1991, *Climate Change: The IPCC Response Strategies*, Island Press, Washington. These data come from a survey paper by the Dutch delegation to the Intergovernmental Panel on Climate Change (IPCC), 1990, A global survey of coastal wetlands, their functions and threats in relation to adaptive responses to sea level rise, Coastal Zone Management Workshop, Perth, Australia; and from the *Directories of Wetlands 1980–1990*, issued by IUCN/UNEP, for 120 countries, excluding, among others, Australia, Canada, New Zealand, and the United States.

**TABLE F.5**   Extent of wetlands, selected OECD countries, 1950 to 1985

| | 1950 (km$^2$) | 1970 (km$^2$) | 1980 (km$^2$) | 1985 (km$^2$) | Losses 1950 to 1985 (%) |
|---|---|---|---|---|---|
| Canada | | | 1,659,240 | 1,271,940 | |
| United States[a] | 432,000 | 396,000 | 380,000 | | 12.0 |
| New Zealand | | | 10,400 | | |
| Austria[b] | | | 93 | 73 | |
| Denmark[c] | | | 3,400 | | |
| Finland | 27,150 | 23,000 | 20,900 | 20,951 | 22.8 |
| France | | | 2,781 | 2,695 | |
| Germany[d] | 2,471 | 1,697 | 1,174 | 1,072 | 56.6 |
| Netherlands[e] | 1,450 | 1,100 | 737 | 659 | 54.6 |
| Norway[f] | | 20,300 | 20,300 | | |
| Spain | | | 10,853 | 10,830 | |
| Sweden[c] | 26,100 | 24,240 | 24,400 | 23,837 | 8.7 |
| Turkey[g] | | 485 | | | |

[a]Data refer to 1954, 1974, the 1980s, and 1954 to 1980s.  Includes estuarine and fresh water wetlands.
[b]Data refer to 1983 and 1986. Includes uncultivated marshes only.
[c]Bogs and fens only.
[d]1985 data refer to 1984. Excludes marshes.
[e]1980 data refer to 1979.
[f]1980 data refer to 1983.
[g]1970 data refer to 1971.

**TABLE F.5**   Extent of wetlands, selected OECD countries, 1950 to 1985

## DESCRIPTION
The areas of wetlands for selected OECD countries are given here for several periods from the 1950s to the mid-1980s, in km$^2$. The notes in the table specify differing assumptions, definitions, and dates of actual measurements. Despite the uncertainties in wetlands area measurements, the trend of significant wetlands loss can be seen. Germany and the Netherlands, for example, have lost over half of certain wetlands between 1950 and the mid-1980s, a trend that is being repeated in many countries around the world.

## LIMITATIONS
This table highlights some of the difficulties of providing good time-series data on wetlands extent. Even among the countries of the OECD, different countries use different definitions of "wetlands" and inconsistent periods of measurement. Trends within countries are more likely to be accurate than comparisons among countries.

## SOURCE
Organization for Economic Co-operation and Development (OECD), 1989, *Environmental Data Compendium – 1989*, OECD, Paris. By permission.

**TABLE F.6** Long-term wetlands loss in selected southern Asian countries, 1880 to 1980

| | | | | | | | Land Area (km$^2$) | | | | |
|---|---|---|---|---|---|---|---|---|---|---|---|
| | 1880 | 1890 | 1900 | 1910 | 1920 | 1930 | 1940 | 1950 | 1960 | 1970 | 1980 |
| Bangladesh | 1,134 | 1,123 | 1,001 | 980 | 988 | 952 | 889 | 823 | 863 | 780 | 722 |
| Brunei | 1,507 | 1,493 | 1,464 | 1,409 | 1,292 | 1,181 | 1,074 | 968 | 875 | 783 | 692 |
| Malaysia | 48,390 | 47,370 | 46,620 | 45,560 | 43,850 | 41,780 | 39,070 | 37,290 | 33,020 | 28,290 | 21,710 |
| Myanmar | 27,830 | 25,850 | 23,140 | 21,040 | 19,440 | 17,680 | 16,500 | 15,620 | 14,690 | 13,600 | 12,050 |
| Northern India[a] | 56,900 | 54,410 | 51,990 | 49,290 | 48,430 | 48,050 | 46,580 | 41,070 | 38,150 | 35,760 | 32,590 |
| Pakistan | 1,980 | 2,660 | 2,850 | 2,680 | 2,660 | 2,610 | 2,640 | 2,590 | 2,500 | 2,320 | 2,130 |

[a]Northern India refers to an area of 1,793,780 km$^2$, 60% of the national area.

**TABLE F.6** Long-term wetlands loss in selected southern Asian countries, 1880 to 1980

**DESCRIPTION**
Long-term data on wetlands extent and loss are difficult to find. Given here are the estimated land area of wetlands for Bangladesh, Brunei, Malaysia, Myanmar, Northern India, and Pakistan, for every ten years from 1880 to 1980. Land area is given in km$^2$.

**LIMITATIONS**
No information is given in the original reference about the definitions of "wetlands" used, the consistency of those definitions over time, and whether the data come from actual measurements throughout the time period shown or from a combination of measurements, model estimates, and guesses. We think it highly unlikely that field measurements, using consistent techniques and definitions, were actually taken precisely every decade for each of these countries. We reproduced the data here, however, because the long-term trends shown are likely to be indicative of actual declines in wetlands areas.

**SOURCE**
United Nations Environment Programme, 1989, *Environmental Data Report*, GEMS Monitoring and Assessment Research Centre, Basil Blackwell, Oxford. (UNEP cites these data as a personal communication from J.F. Richards of Duke University.)

**TABLE F.7**   Descriptions of wetlands regions of the United States

---

*Coastal wetlands*

| | |
|---|---|
| Tidal salt marshes: | Found along protected coastlines in the middle and high latitudes. In the United States, salt marshes are often dominated by the grasses *Spartina* and *Juncus*. Salt marsh vegetation and animals are adapted to the stresses of salinity, temperature extremes, and periodic inundation. |
| Tidal fresh water marshes: | An intermediate in the continuum between tidal salt marshes and fresh water marshes, these near-coast freshwater marshes are subject to the influence of tides. The vegetation in these productive ecosystems includes grasses, annual, and perennial broad-leaved aquatic plants. |
| Mangrove wetlands: | Common to subtropical and tropical regions of the world, these coastal wetlands are named for the salt-tolerant mangrove trees that dominate them. The mangrove ecosystem supports vast quantites of fish and shellfish through crucial developmental stages. Although they can take a wide range of salinity and tides, they require some protection from the open ocean. |

*Inland wetlands*

| | |
|---|---|
| Inland fresh water marshes: | Common throughout the United States, these wetlands are characterized by some combination of emergent soft-stemmed aquatic plants, shallow water, and shallow peat deposits. While they may occur in large regions, they are often smaller and isolated, bordering lakes and slow streams. |
| Northern peatlands: | Deep peat deposits of the once glaciated northeast and north-central United States. Bogs and fens are the two major groups of peatlands. They occur as filled-in lake basins formed in the last glaciation, or as blankets of peat. Noted for nutrient deficiency and waterlogging, peatlands host biological adaptations such as carnivorous plants and nutrient conservation. |
| Southern deep water swamps: | These southern forested wetlands have standing fresh water through most of the growing season, supplied by rainwater or flooding of nearby streams. Their ecosystems are dominated by cypress and gum/tupelo, and have a wide range of nutrient levels and hydrologic conditions. |
| Riparian wetlands: | Streamside ecosystems that are subject to occasional flooding but are usually dry. Referred to as bottomland hardwood forests in the southeastern United States, they also occur in arid and semiarid regions of the country. Their vegetation is diverse, and is controlled in part by the frequency of flooding. Overall productivity is generally higher than that of adjacent uplands because of the periodic inflow of mineral nutrients. |

---

**TABLE F.7**   Descriptions of wetlands regions of the United States

**DESCRIPTION**

Summarized here is one set of definitions of different wetlands regions of the United States. These specific terms often vary from one region to another, complicating the accurate assessment of wetlands extent and loss. This approach to classifying wetlands uses terminology and categories of ecosystems that are generally recognized and about which extensive research literature is available. Regulatory agencies are also familiar with these systems and wetland types.

**LIMITATIONS**

This scheme is simpler and less encompassing than the official National Wetlands Inventory. Anyone interested in using wetlands data should take care to understand the distinctions and definitions being used.

**SOURCE**

W.J. Mitsch, 1986, *Wetlands*, Van Nostrand Reinhold Company, New York. By permission.

**TABLE F.8**  Wetlands area in the United States, 1780s and 1980s
(Ranked by area remaining)

| | State | Total surface area of state ($10^6$ hectares) | Wetlands area ($10^6$ hectares) 1780s | 1980s | Wetlands area lost (%) |
|---|---|---|---|---|---|
| 1 | Alaska | 151.9 | 68.907 | 68.826 | 0.1 |
| 2 | Florida | 15.2 | 8.229 | 4.469 | 46 |
| 3 | Louisiana | 12.6 | 6.557 | 3.556 | 46 |
| 4 | Minnesota | 21.8 | 6.101 | 3.522 | 42 |
| 5 | Texas | 69.3 | 6.478 | 3.082 | 52 |
| 6 | North Carolina | 13.6 | 4.490 | 2.304 | 49 |
| 7 | Michigan | 15.1 | 4.534 | 2.260 | 50 |
| 8 | Wisconsin | 14.6 | 3.968 | 2.158 | 46 |
| 9 | Georgia | 15.3 | 2.770 | 2.145 | 23 |
| 10 | Maine | 8.6 | 2.615 | 2.105 | 20 |
| 11 | South Carolina | 8.0 | 2.597 | 1.886 | 27 |
| 12 | Mississippi | 12.4 | 3.997 | 1.647 | 59 |
| 13 | Alabama | 13.4 | 3.064 | 1.532 | 50 |
| 14 | Arkansas | 13.8 | 3.987 | 1.119 | 72 |
| 15 | North Dakota | 18.3 | 1.995 | 1.008 | 49 |
| 16 | Nebraska | 20.0 | 1.179 | 0.772 | 35 |
| 17 | South Dakota | 20.0 | 1.107 | 0.721 | 35 |
| 18 | Oregon | 25.1 | 0.916 | 0.564 | 38 |
| 19 | Illinois | 14.6 | 3.325 | 0.508 | 85 |
| 20 | Wyoming | 25.4 | 0.810 | 0.506 | 38 |
| 21 | Virginia | 10.6 | 0.749 | 0.435 | 42 |
| 22 | New York | 12.8 | 1.037 | 0.415 | 60 |
| 23 | Colorado | 27.0 | 0.810 | 0.405 | 50 |
| 24 | Oklahoma | 18.1 | 1.151 | 0.385 | 67 |
| 25 | Washington | 18.9 | 0.547 | 0.380 | 31 |
| 26 | New Jersey | 2.0 | 0.607 | 0.371 | 39 |
| 27 | Montana | 38.1 | 0.464 | 0.340 | 27 |
| 28 | Tennessee | 10.9 | 0.784 | 0.319 | 59 |
| 29 | Indiana | 9.4 | 2.267 | 0.304 | 87 |
| 30 | Missouri | 18.1 | 1.961 | 0.260 | 87 |
| 31 | Arizona | 29.5 | 0.377 | 0.243 | 36 |
| 32 | Massachusetts | 2.1 | 0.331 | 0.238 | 28 |
| 33 | Utah | 22.0 | 0.325 | 0.226 | 30 |
| 34 | Pennsylvania | 11.7 | 0.456 | 0.202 | 56 |
| 35 | Ohio | 10.7 | 2.024 | 0.196 | 90 |

*continued*

| State | Total surface area of state ($10^6$ hectares) | Wetlands area ($10^6$ hectares) | | Wetlands area lost (%) |
|---|---|---|---|---|
| | | 1780s | 1980s | |
| 36 New Mexico | 31.5 | 0.291 | 0.195 | 33 |
| 37 California | 41.1 | 2.024 | 0.184 | 91 |
| 38 Maryland | 2.7 | 0.668 | 0.178 | 73 |
| 39 Kansas | 21.3 | 0.340 | 0.176 | 48 |
| 40 Iowa | 14.6 | 1.619 | 0.171 | 89 |
| 41 Idaho | 21.6 | 0.355 | 0.156 | 56 |
| 42 Kentucky | 10.5 | 0.634 | 0.121 | 81 |
| 43 Nevada | 28.6 | 0.197 | 0.096 | 52 |
| 44 Delaware | 0.5 | 0.194 | 0.090 | 54 |
| 45 Vermont | 2.5 | 0.138 | 0.089 | 35 |
| 46 New Hampshire | 2.4 | 0.089 | 0.081 | 9 |
| 47 Connecticut | 1.3 | 0.271 | 0.071 | 74 |
| 48 West Virginia | 6.3 | 0.054 | 0.041 | 24 |
| 49 Rhode Island | 0.3 | 0.042 | 0.026 | 37 |
| 50 Hawaii | 1.7 | 0.023 | 0.021 | 12 |

**TABLE F.8**  Wetlands area in the United States, 1780s and 1980s *(pages 295–296)*

### DESCRIPTION

There have been significant losses of wetlands in the United States over the last two centuries. Listed here are the total estimated wetlands area of each state in the 1780s and in the 1980s. The percentage loss over this time period is indicated, and the states are ranked by their remaining wetlands area.

### LIMITATIONS

The actual area of wetlands in each state in the 1780s is only approximate, because of incomplete surveys at that time. Future changes in official definitions of what constitutes a wetland could alter these numbers, even without further development.

### SOURCES

Cited in The Council on Environmental Quality, 1991, *Environmental Quality: 21st Annual Report (1990)*, US Government Printing Office, Washington, DC.

Data from T.E. Dahl, 1991, *Wetland Losses in the United States: 1780s–1980s*, Report to Congress, U.S. Department of the Interior, Fish and Wildlife Sevice, Washington, DC.

**TABLE F.9** Estimated U. S. federal agency
wetlands holdings, 1985

| Agency | Wetlands ($10^6$ hectares) |
|---|---|
| Fish and Wildlife Service[a] | 13.8 |
| Forest Service[b] | 4.8 |
| National Park Service | 0.77 |
| Bureau of Land Management | 0.57 |
| Army Corps of Engineers | 0.40 |
| Bureau of Reclamation | 0.1 |
| Other Agencies | 0.04 |
| Total | 20.4 |

[a]Includes 11.7 million hectares of Alaskan wetlands.
[b]Includes 3.6 million hectares of Alaskan wetlands.

**TABLE F.9** Estimated U. S. federal agency wetlands
holdings, 1985

**DESCRIPTION**
Approximately one-fifth of the total area of wetlands in the United
States are held by United States federal agencies. Shown here is the
distribution of those holdings in 1985. The vast majority of wetlands
remaining in the United States are in Alaska.

**LIMITATIONS**
Different surveys by different agencies produce significantly different
estimates of wetlands areas. Part of the problem lies in the classification
systems used; part lies in methods of extrapolating limited surveys to
larger areas. Inconsistent definitions of wetlands, and changing regulations regarding the protection of wetlands, has complicated their
management and protection in the United States in recent years.

**SOURCES**
R.E. Heimlich and L.L. Langner, 1986, *Swampbusting: Wetland Conversion and Farm Programs*, U.S. Department of Agriculture Economic Research Service, Agricultural Economics Report no. 551,
Washington, DC, August.

**TABLE F.10**  Fresh water fish catch, by country, mid-1970s and mid-1980s *(pages 299–302)*

## DESCRIPTION

The data in this table include fish, crustaceans, mollusks, and other aquatic animal products (excluding aquatic mammals and plants) taken for all non-recreational purposes, in metric tonnes per year. Country totals include production from aquaculture and quantities caught by vessels flying the national flag but landed in foreign ports. The nominal catch is the live weight equivalent of the landed quantities (i.e., landings of each species adjusted for on-ship processing such as gutting, filleting, drying, and so on). Data are mean annual values for 1974–1976 and 1984–1986.

## LIMITATIONS

Continental aggregates for 1984–1986 may include countries not shown. Continental aggregates for 1974–1976 include listed countries only.

## SOURCES

Cited from United Nations Environmental Programme, 1989, *Environmental Data Report, 1989/90*, GEMS Monitoring and Assessment Research Centre, Basil Blackwell Ltd, Oxford. By permission.

Data from:

Food and Agricultural Organization of the United Nations, 1984, *1984 Yearbook of Fishery Statistics – Catches and Landings 1982*, vol. 54, Food and Agricultural Organization of the United Nations, Rome.

Food and Agricultural Organization of the United Nations, 1988, *1988 Yearbook of Fishery Statistics - Catches and Landings 1986*, vol. 62, Food and Agricultural Organization of the United Nations, Rome.

**TABLE F.10**   Fresh water fish catch, by country, mid-1970s and mid-1980s

| | Catch (tonnes/yr) | | Percent change |
|---|---|---|---|
| | 1974–76 | 1984–86 | |
| *World* | 6,901,800 | 10,464,900 | 52 |
| *Africa* | 1,364,421 | 1,497,500 | 10 |
| Algeria | 17 | | |
| Angola | 8,000 | 8,000 | 0 |
| Benin | 20,370 | 16,300 | −20 |
| Botswana | 1,333 | 1,700 | 28 |
| Burkina Faso | 5,333 | 7,000 | 31 |
| Burundi | 15,287 | 9,847 | −36 |
| Cameroon | 33,167 | 20,000 | −40 |
| Central African Republic | 10,131 | 13,000 | 28 |
| Chad | 110,000 | 111,667 | 2 |
| Congo | 1,000 | 12,513 | 1,151 |
| Côte D'Ivoire | 5,367 | 19,000 | 254 |
| Egypt | 73,880 | 112,336 | 52 |
| Equatorial Guinea | | 417 | |
| Ethiopia | 1,500 | 3,567 | 138 |
| Gabon | 400 | 1,800 | 350 |
| Gambia | 800 | 2,700 | 238 |
| Ghana | 39,983 | 40,000 | 0 |
| Guinea | 1,000 | 1,933 | 93 |
| Kenya | 28,282 | 93,736 | 231 |
| Lesotho | 20 | 14 | −30 |
| Liberia | 4,000 | 4,000 | 0 |
| Madagascar | 41,100 | 44,833 | 9 |
| Malawi | 72,056 | 66,661 | −7 |
| Mali | 95,000 | 58,335 | −39 |
| Mauritania | 13,000 | 6,000 | −54 |
| Mauritius | | 28 | |
| Morocco | 443 | 1,323 | 199 |
| Mozambique | 5,000 | 2,589 | −48 |
| Namibia | 50 | 50 | 0 |
| Niger | 9,636 | 2,450 | −75 |
| Nigeria | 240,067 | 129,946 | −46 |
| Reunion | 0 | 0 | |
| Rwanda | 1,191 | 1,059 | −11 |
| Senegal | | 15,000 | |
| Sierra Leone | 1,067 | 16,333 | 1,431 |

*continued*

| | Catch (tonnes/yr) | | Percent change |
|---|---|---|---|
| | 1974–76 | 1984–86 | |
| South Africa | 100 | 800 | 700 |
| Sudan | 22,500 | 25,695 | 14 |
| Swaziland | | 44 | |
| Tanzania | 166,287 | 253,664 | 53 |
| Togo | 2,637 | 683 | –74 |
| Uganda | 169,300 | 212,200 | 25 |
| Zaire | 108,477 | 100,333 | –8 |
| Zambia | 52,874 | 66,850 | 26 |
| Zimbabwe | 3,767 | 17,090 | 354 |
| | | | |
| *North and Central America* | 144,237 | 246,100 | 71 |
| Belize | 25 | 2 | –92 |
| Canada | 43,051 | 44,143 | 3 |
| Costa Rica | 50 | 271 | 442 |
| Cuba | 1,900 | 16,393 | 763 |
| Dominican Republic | 524 | 1,713 | 227 |
| El Salvador | 1,185 | 1,702 | 44 |
| Guadeloupe | 0 | 0 | |
| Guatemala | 551 | 74 | –87 |
| Haiti | 300 | 300 | 0 |
| Honduras | 171 | 138 | –19 |
| Jamaica | | 945 | |
| Martinique | 0 | 0 | |
| Mexico | 16,671 | 107,636 | 546 |
| Nicaragua | 426 | 86 | –80 |
| Puerto Rico | 0 | 0 | |
| St. Lucia | 0 | 0 | |
| St. Vincent | 0 | 0 | |
| United States | 79,384 | 72,667 | –8 |
| | | | |
| *South America* | 235,726 | 335,833 | 42 |
| Argentina | 11,698 | 9,130 | –22 |
| Bolivia | 1,117 | 4,655 | 317 |
| Brazil | 162,143 | 212,527 | 31 |
| Chile | | 671 | |
| Colombia | 43,582 | 51,866 | 19 |
| Ecuador | | 920 | |
| Guyana | | 800 | |
| Paraguay | 2,700 | 8,500 | 215 |
| Peru | 6,155 | 28,505 | 363 |

*continued*

| | Catch (tonnes/yr) | | Percent change |
|---|---|---|---|
| | 1974–76 | 1984–86 | |
| Suriname | 263 | 191 | –27 |
| Uruguay | 241 | 615 | 155 |
| Venezuela | 7,827 | 17,444 | 123 |
| *Asia* | 3,984,411 | 7,035,533 | 77 |
| Afghanistan | 1,500 | 1,500 | 0 |
| Bangladesh | 606,333 | 587,183 | –3 |
| Bhutan | 1,000 | 1,000 | 0 |
| Brunei Darus | 71 | 134 | 89 |
| Cambodia | 73,900 | 61,667 | –17 |
| China | 1,049,786 | 2,852,306 | 172 |
| Cyprus | 34 | 54 | 59 |
| Hong Kong | 4,402 | 6,014 | 37 |
| India | 788,771 | 1,125,754 | 43 |
| Indonesia | 393,414 | 578,511 | 47 |
| Iran | 6,954 | 24,613 | 254 |
| Iraq | 15,926 | 15,855 | 0 |
| Israel | 14,800 | 14,180 | –4 |
| Japan | 192,561 | 202,172 | 5 |
| Korea DPR | 8,297 | 52,974 | 538 |
| Korea Rep | 52,000 | 103,333 | 99 |
| Laos | 20,000 | 20,000 | 0 |
| Lebanon | 100 | 100 | 0 |
| Malaysia | 2,694 | 7,442 | 176 |
| Mongolia | 300 | 372 | 24 |
| Myanmar | 130,203 | 145,870 | 12 |
| Nepal | 2,500 | 7,793 | 212 |
| Pakistan | 27,299 | 76,283 | 179 |
| Philippines | 233,288 | 557,583 | 139 |
| Singapore | 655 | 305 | –53 |
| Sri Lanka | 11,181 | 32,903 | 194 |
| Syria | 1,317 | 4,160 | 216 |
| Thailand | 155,995 | 165,757 | 6 |
| Turkey | 17,128 | 44,083 | 157 |
| Vietnam | 172,000 | 224,667 | 31 |
| *Europe* | 281,943 | 423,500 | 50 |
| Albania | | 3,600 | |
| Austria | 2,157 | 4,500 | 109 |
| Belgium | | 190 | |

*continued*

| | Catch (tonnes/yr) | | Percent change |
|---|---|---|---|
| | 1974–76 | 1984–86 | |
| Bulgaria | 7,783 | 13,000 | 67 |
| Czechoslovakia | 16,720 | 20,148 | 21 |
| Denmark | 14,949 | 23,637 | 58 |
| Finland | 22,897 | 32,622 | 42 |
| France | | 19,823 | |
| German DR | 13,775 | 20,939 | 52 |
| Germany FR | 15,000 | 24,000 | 60 |
| Greece | 7,693 | 9,745 | 27 |
| Hungary | 30,934 | 37,322 | 21 |
| Iceland | 489 | 538 | 10 |
| Ireland | | 185 | |
| Italy | 19,383 | 43,008 | 122 |
| Netherlands | 3,672 | 3,987 | 9 |
| Norway | 3,260 | 371 | –89 |
| Poland | 22,997 | 30,959 | 35 |
| Portugal | 133 | | |
| Romania | 46,346 | 60,000 | 29 |
| Spain | 15,293 | 25,817 | 69 |
| Sweden | 10,367 | 5,655 | –45 |
| Switzerland | 3,872 | 4,364 | 13 |
| United Kingdom | 101 | 13,463 | 13,230 |
| Yugoslavia | 24,122 | 25,621 | 6 |
| *USSR* | 830,513 | 904,683 | 9 |
| *Oceania* | 1,863 | 5,167 | 177 |
| Australia | 1,499 | 2,463 | 64 |
| Fiji | 156 | 2,628 | 1,585 |
| Guam | | 83 | |
| New Zealand | 93 | | |
| Papua New Guinea | 116 | | |

**TABLE F.11 I**  World fish production, by origin, 1950 to 1985

| Year | Marine (10⁶ tonnes) | Fresh water (10⁶ tonnes) | Total (10⁶ tonnes) |
|------|------|------|------|
| 1950 | 17.6 | 3.2 | 20.8 |
| 1960 | 32.8 | 6.6 | 39.4 |
| 1970 | 59.5 | 6.1 | 65.6 |
| 1975 | 59.2 | 7.2 | 66.4 |
| 1980 | 64.5 | 7.6 | 72.1 |
| 1985 | 74.8 | 10.1 | 84.9 |

**TABLE F.11**  World fish production, by origin, 1950 to 1985

**DESCRIPTION**

The catch of marine and fresh water fish in million tonnes per year are presented for 1950, 1960, 1970, 1975, 1980, and 1985. World fish production by end use (food for human consumption or feed for other uses, including raising livestock) is given every five years from 1950 to 1985.

**LIMITATIONS**

Fish catches fluctuate on an annual basis. The data here are single yearly values for each five-year or ten-year period. No information in the original source is given on level of fishing effort, species caught, or national fish catch.

**SOURCES**

Food and Agriculture Organization, N. Alexandratos (ed.), 1988, *World Agriculture: Towards 2000, An FAO Study*, New York University Press, New York, and Food and Agriculture Organization of the United Nations, Rome. By permission.

**TABLE F.11 II**  World fish production, by end use, 1950 to 1985

| Year | Food (10⁶ tonnes) | Feed (10⁶ tonnes) | Total (10⁶ tonnes) |
|------|------|------|------|
| 1950 | 17.8 | 3.0 | 20.8 |
| 1955 | 23.9 | 4.6 | 28.5 |
| 1960 | 30.8 | 8.6 | 39.4 |
| 1965 | 36.3 | 16.3 | 52.6 |
| 1970 | 39.1 | 26.5 | 65.6 |
| 1975 | 46.0 | 20.4 | 66.4 |
| 1980 | 52.9 | 19.2 | 72.1 |
| 1985 | 59.6 | 25.3 | 84.9 |

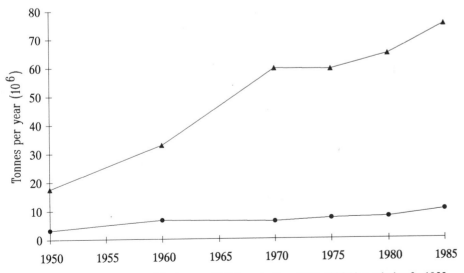

Fig. F–11. World marine (▲) and fresh water (•) fish production, 1950–1985 (data missing for 1955 and 1965). The harvest of fish from the oceans has continued to increase since 1950, though fresh water fish harvests have remained approximately level during that period. See Table F.11 for data.

**TABLE F.12** pH criteria for the protection of European fresh water fish

| pH range | |
|---|---|
| 3.0–3.5 | Unlikely that any fish can survive for more than a few hours in this range, although some plants and invertebrates can be found at pH values lower than this. |
| 3.5–4.0 | This range is lethal to salmonids. There is evidence that roach, tench, perch, and pike can survive in this range, presumably after a period of acclimation to slightly higher, nonlethal levels, but the lower end of this range may still be lethal for roach. |
| 4.0–4.5 | Likely to be harmful to salmonids, tench, bream, roach, goldfish, and common carp which have not been previously acclimated to low pH values, although the resistance to this pH range increases with the size and age of the fish. Fish can become acclimated to these levels, but of perch, bream, roach, and pike, only the last named may be able to breed. |
| 4.5–5.0 | Likely to be harmful to the eggs and fry of salmonids, and adults, particularly in soft water containing low concentrations of calcium, sodium, and chloride. Can be harmful to common carp. |
| 5.0–6.0 | Unlikely to be harmful to any species unless either the concentration of free carbon dioxide is greater than 20 mg/l or the water contains iron salts that are freshly precipitated as ferric hydroxide, the precise toxicity of which is not known. The lower end of this range may be harmful to nonacclimated salmonids if the calcium, sodium, and chloride concentrations or the temperature of the water are low, and may be detrimental to roach reproduction. |
| 6.0–6.5 | Unlikely to be harmful to fish unless free carbon dioxide is present in excess of 100 mg/l. |
| 6.5–9.0 | Harmless to fish, although the toxicity of other poisons may be affected by changes within this range. |
| 9.0–9.5 | Likely to be harmful to salmonids and perch if present for a considerable length of time. |
| 9.5–10.0 | Lethal to salmonids over a prolonged period of time, but can be withstood for short periods. May be harmful to developmental stages of some species. |
| 10.0–10.5 | Can be withstood by roach and salmonids for short periods but lethal over a prolonged period. |
| 10.5–11.0 | Rapidly lethal to salmonids. Prolonged exposure to the upper limit of this range is lethal to carp, tench, goldfish, and pike. |
| 11.0–11.5 | Rapidly lethal to all species of fish. |

**TABLE F.12** pH criteria for the protection of European fresh water fish

**DESCRIPTION**

Critical pH conditions for successful fisheries have been established for European waters by the European Inland Fisheries Advisory Commission (EIFAC). These criteria are listed here by pH range, with details of effects on specific fish and invertebrate families.

here. The information provided here was collected on the basis of data available to researchers. According to the authors, the absence of a reference indicates that insufficient data exist, not that a species is insensitive in that pH range.

**LIMITATIONS**

Different species have different levels of tolerance to pH and chemical contaminants. The stage of the life-cycle is also important in determining effects. Only limited information on these characteristics is given

**SOURCES**

J.S. Alabaster, II and R. Lloyd (eds.), 1980, *Water Quality Criteria for Freshwater Fish*, Food and Agriculture Organization of the United Nations, Butterworth and Co. Ltd., London. By permission.

**TABLE F.13 I** Distribution of lakes and running waters important
for fisheries in the republics of the former Soviet Union

| | Distribution of lakes | | Distribution of running waters | |
|---|---|---|---|---|
| | Total area (10³ hectares) | Proportion (%) | Length (10³ km) | Proportion (%) |
| Russia | 19,998.0 | 81.9 | 416.1 | 79.3 |
| Kazakhstan | 2,883.0 | 11.8 | 10.5 | 2 |
| Kirghizia | 666.3 | 2.8 | — | <0.1 |
| Estonia | 221.6 | 0.9 | 6.2 | 1.2 |
| Armenia | 139.0 | 0.6 | 0.8 | 0.1 |
| Byelorussia | 129.0 | 0.6 | 51.0 | 9.7 |
| Latvia | 98.0 | 0.4 | 6.3 | 1.2 |
| Lithuania | 85.0 | 0.3 | 3.2 | 0.6 |
| Ukraine | 81.0 | 0.3 | 19.0 | 3.6 |
| Uzbekistan | 57.0 | 0.2 | 3.1 | 0.6 |
| Azerbaidzhan | 28.6 | 0.1 | 1.2 | 0.3 |
| Turkmenistan | 14.2 | 0.1 | 3.1 | 0.6 |
| Georgia | 8.8 | — | 1.4 | 0.3 |
| Moldavia | 3.0 | — | 1.0 | 0.1 |
| Tadjikistan | 1.6 | — | 2.0 | 0.4 |
| Total | 24,414.6 | 100 | 524.9 | 100 |

**TABLE F.13 II**   Distribution of small and medium lakes important for fisheries in the republics of the former Soviet Union

| | Total area of small and medium lakes ($10^3$ hectares) | Proportion (%) |
|---|---|---|
| Russia | 13,697.7 | 92.2 |
| Kazakhstan | 541.0 | 3.6 |
| Kirghizia | 46.3 | 0.3 |
| Estonia | 64.6 | 0.4 |
| Armenia | 2.5 | 0.2 |
| Byelorussia | 129.0 | 0.9 |
| Lithuania | 98.0 | 0.7 |
| Latvia | 85.0 | 0.6 |
| Ukraine | 81.0 | 0.5 |
| Uzbekistan | 57.0 | 0.4 |
| Azerbaidzhan | 28.6 | 0.2 |
| Turkmenistan | 14.2 | 0.1 |
| Georgia | 8.8 | 0.06 |
| Moldavia | 3.0 | 0.02 |
| Tadjikistan | 1.6 | 0.01 |
| Total | 14,858.3 | 100 |

**TABLE F.13**   Distribution of small and medium lakes important for fisheries in the republics of the former Soviet Union *(pages 305–306)*

**DESCRIPTION**

The distribution of fresh water lakes and rivers important for fisheries in the former Soviet Union is listed here by republic. Listed are the total area of lakes, in thousand hectares, and the total length of rivers used for fisheries, in km × $10^3$.

Also shown is the distribution of small and medium lakes important for fisheries, by republic. "Small" lakes have an area under 1,000 hectares; "medium" lakes have an area between 1,000 and 10,000 hectares.

**LIMITATIONS**

No data in the original source describe total fish catch, the definition of "important for fisheries," or the species of fish considered "important." According to the authors, the number of lakes and the lengths of rivers important for fisheries are expected to shrink because of the construction of reservoirs and increases in pollution.

**SOURCE**

R. Berka, 1989, *Inland Capture Fisheries of the USSR*, FAO Fisheries Technical Paper no. 311, Food and Agricultural Organization, Rome. By permission.

**TABLE F.14**  The largest lakes in the former Soviet Union supporting capture fisheries

| Lake | Area (10³ hectares) |
|---|---|
| Baikal | 3,390.0 |
| Ladoga | 1,800.0 |
| Balkhash | 1,750.0 |
| Onega | 1,000.0 |
| Issyk-Kul | 620.0 |
| Pskov-Chudskoe | 355.0 |
| Khanka | 350.0 |
| Tchany | 270.0 |
| Sevan | 137.0 |
| Beloe | 112.5 |
| Ilmen | 112.4 |
| Ubinskoe | 50.0 |
| Total | 9,946.9 |

**TABLES F.14**  The largest lakes in the former Soviet Union supporting capture fisheries

**DESCRIPTION**
The twelve largest lakes (greater than 10,000 hectares) supporting capture fisheries in the former Soviet Union are listed here by area, in thousand hectares. While these lakes produce significant quantities of commercial fish, small and medium lakes are considered to be more important overall for fisheries.

**LIMITATIONS**
No data in the original source describe total fish catch in these lakes or the variations in fish catch over time.

**SOURCE**
R. Berka, 1989, *Inland Capture Fisheries of the USSR*, FAO Fisheries Technical Paper no. 311, Food and Agricultural Organization, Rome. By permission.

**TABLE F.15**  Anthropogenic inputs of trace metals into aquatic ecosystems *(page 308)*

**DESCRIPTION**
All industries discharge trace elements into water. Shown here are an inventory of the principal industrial and commercial users of water and producers of solid wastes, annual global discharge of contaminated process water (in billion m³ per year), and emissions of trace metals into aquatic ecosystems (in kg × 10⁶ per year). Ranges are given for all emissions, based on the emissions factors in Table D.7.

**LIMITATIONS**
Where the authors thought the emissions factors estimates in the literature to be too high, they reported the lower end of the concentration range. It is not possible to place an error range on the calculated inventories, according to the authors, though they believe that the exact global values of the metals released into the environment ''are unlikely to differ from those used [in this table] by more than a factor of two or so.''

**SOURCES**
J.O. Nriagu and J.M. Pacyna, 1988, Quantitative assessment of worldwide contamination of air, water and soils by trace metals, *Nature*, **333**, 134–139. Reprinted with permission.

**TABLE F.15** Anthropogenic inputs of trace metals into aquatic ecosystems ($10^6$ kg/yr)

| | Annual global discharge[a] ($10^9$ m$^3$) | As | Cd | Cr | Cu | Hg | Mn | Mo | Ni | Pb | Sb | Se | V | Zn |
|---|---|---|---|---|---|---|---|---|---|---|---|---|---|---|
| **Domestic wastewater[b]** | | | | | | | | | | | | | | |
| Central | 90 | 1.8-8.1 | 0.18-1.8 | 8.1-36 | 4.5-18 | 0-0.18 | 18-81 | 0-2.7 | 9.0-54 | 0.9-7.2 | 0-2.7 | 0-4.5 | 0-2.7 | 9.0-45 |
| Noncentral | 60 | 1.2-7.2 | 0.3-1.2 | 6.0-42 | 4.2-30 | 0-0.42 | 30-90 | 0-1.8 | 12-48 | 0.6-4.8 | 0-1.8 | 0-3.0 | 0-1.8 | 6.0-36 |
| Steam electric | 6 | 2.4-14 | 0.01-0.24 | 3.0-8.4 | 3.6-23 | 0-3.6 | 4.8-18 | 0.1-1.2 | 3.0-18 | 0.24-1.2 | 0-0.36 | 6.0-30 | 0-0.6 | 6.0-30 |
| Base metal mining and dressing | 0.5 | 0-0.75 | 0-0.3 | 0-0.7 | 0.1-9 | 0-0.15 | 0.8-12 | 0-0.6 | 0.01-0.5 | 0.25-2.5 | 0.04-0.35 | 0.25-1.0 | — | 0.02-6 |
| **Smelting and refining** | | | | | | | | | | | | | | |
| Iron and steel | 7 | | | | | | 14-36 | | | 1.4-2.8 | | | | 5.6-24 |
| Nonferrous metals | 2 | 1.0-13 | 0.01-3.6 | 3-20 | 2.4-17 | 0-0.04 | 2.0-15 | 0.01-0.4 | 2.0-24 | 1.0-6.0 | 0.08-7.2 | 3.0-20 | 0-1.2 | 2.0-20 |
| **Manufacturing processes** | | | | | | | | | | | | | | |
| Metals | 25 | 0.25-1.5 | 0.5-1.8 | 15-58 | 10-38 | 0-0.75 | 2.5-20 | 0.5-5.0 | 0.2-7.5 | 2.5-22 | 2.8-15 | 0-5.0 | 0-0.75 | 25-138 |
| Chemicals | 5 | 0.6-7.0 | 0.1-2.5 | 2.5-24 | 1.0-18 | 0.02-1.5 | 2.0-15 | 0-3.0 | 1.0-6.0 | 0.4-3.0 | 0.1-0.4 | 0.02-2.5 | 0-0.35 | 0.2-5.0 |
| Pulp and paper | 3 | 0.36-4.2 | — | 0.01-1.5 | 0.03-0.39 | — | 0.03-1.5 | — | 0-0.12 | 0.01-0.9 | 0-0.27 | 0.01-0.9 | — | 0.09-1.5 |
| Petroleum products | 0.3 | 0-0.06 | — | 0-0.21 | 0-0.06 | 0-0.02 | — | — | 0-0.06 | 0-0.12 | 0-0.03 | 0-0.09 | 0-0.24 | 0-0.24 |
| Atmospheric fallout[c] | | 3.6-7.7 | 0.9-3.6 | 2.2-16 | 6.0-15 | 0.22-1.8 | 3.2-20 | 0.2-1.7 | 4.6-16 | 87-113 | 0.44-1.7 | 0.54-1.1 | 0.4-9.1 | 21-58 |
| Dumping of raw sewage sludge[d] | [6 × $10^9$ kg] | 0.4-6.7 | 0.08-1.3 | 5.8-32 | 2.9-22 | 0.01-0.31 | 32-1.06 | 0.98-4.8 | 1.3-20 | 2.9-16 | 0.18-2.9 | 0.26-3.8 | 0.72-4.3 | 2.6-31 |
| Total input, water | | 12-70 | 2.1-17 | 45-239 | 35-90 | 0.3-8.8 | 109-414 | 1.8-21 | 33-194 | 97-180 | 3.9-33 | 10-72 | 2.1-21 | 77-375 |
| Median value | | 41 | 9.4 | 142 | 112 | 4.6 | 262 | 11 | 113 | 138 | 18 | 41 | 12 | 226 |

[a]The discharges given represent contaminated process waters, and do not include cooling waters.
[b]The wastewater production figure coresponds to about 60 m$^3$ per capita multiplied by the 2.4 × $10^9$ residents in urban and rural areas of the world. The other discharge figures likewise have been derived from the reported water demand per unit tonne of metal smelted or goods manufactured.
[c]We have assumed that 70% of each metal emitted to the atmosphere is deposited on land and the remaining 30% in the aquatic environments.
[d]Worldwide sewage sludge production is estimated to be 30 million tonnes, assuming average sludge production rate of 30 g per capita per day in urban and rural communities. It is believed that 20% of the municipal sludge is directly discharged or dumped into aquatic ecosystems, about 10% is incinerated, and the rest is deposited on land.

**TABLE F.16**  A qualitative trophic characterization of lakes and reservoirs

| Limnological characterization categories | Ultra-oligotrophic | Oligotrophic | Mesotrophic | Eutrophic | Hypertrophic |
|---|---|---|---|---|---|
| Biomass | Very low | Low | Medium | High | Very high |
| Green and/or blue-green algae fraction | Low | Low | Variable | High | Very high |
| Macrophytes | Low or absent | Low | Variable | High or low | Low |
| Production dynamics | Very low | Low | Medium | High | High, unstable |
| Oxygen dynamics | | | | | |
|   Epilimnic | Normally saturated | Normally saturated | Variable saturated | Often over-saturated | Very unstable varying from high, oversaturation to complete lack |
|   Hypolimnic | Normally saturated | Normally saturated | Variable saturated | Undersaturated to complete depletion | Very unstable varying from high, oversaturation to complete lack |
| Impairment of multipurpose uses | Low | Low | Variable | High | Very high |

**TABLE F.16**  A qualitative trophic characterization of lakes and reservoirs

**DESCRIPTION**

The trophic state of lakes and reservoirs can be characterized by a qualitative set of limnological variables. These variables are described for five trophic levels: ultra-oligotrophic, oligotrophic, mesotrophic, eutrophic, and hypertrophic.

**LIMITATIONS**

The variables shown here are qualitative, not quantitative (see Table F.17), but provide a good overview of the nature of the different trophic states.

**SOURCE**

M. Meybeck, D.V. Chapman and R. Helmer (eds.), 1989, *Global Freshwater Quality: A First Assessment*, Global Environment Monitoring System (GEMS), World Health Organization and the United Nations Environment Programme, Basil Blackwell, Oxford. By permission.

**TABLE F.17**   A quantitative trophic characterization of lakes and reservoirs

| Trophic category | Mean total phosphorus (mg/m$^3$) | Mean yearly chlorophyll (mg/m$^3$) | Chlorophyll maxima (mg/m$^3$) | Mean yearly Secchi disc transparency (m) | Secchi disc transparency minima (m) | Oxygen (% saturation)[a] |
|---|---|---|---|---|---|---|
| Ultra-oligotrophic | 4.0 | 1.0 | 2.5 | 12.0 | 6.0 | <90 |
| Oligotrophic | 10.0 | 2.5 | 8.0 | 6.0 | 3.0 | <80 |
| Mesotrophic | 10–35 | 2.5–8 | 8–25 | 6–3 | 3–1.5 | 40–89 |
| Eutrophic | 35–100 | 8–25 | 25–75 | 3–1.5 | 1.5–0.7 | 40–0 |
| Hypertrophic | 100.0 | 25.0 | 75.0 | 1.5 | 0.7 | 10–0 |

There can also be pronounced mesolimnetic oxygen maxima or minima depending on thermal stratification.
[a] Percent saturation in bottom waters depending on mean depth.

**TABLE F.17**   A quantitative trophic characterization of lakes and reservoirs

## DESCRIPTION

The trophic state of lakes and reservoirs can be characterized by different sets of quantitative limnological variables. One set of variables is described here in a system developed for the Organization for Economic Co-operation and Development (OECD) for five trophic levels: ultra-oligotrophic, oligotrophic, mesotrophic, eutrophic, and hypertrophic.

## LIMITATIONS

The variables shown here are quantitative and come from one particular system used by the OECD. Other systems are in use, but each suffers from some shortcomings. The data here apply to lakes and reservoirs in temperate regions. Outside of such climates, these criteria may require modifications. Other limnological characteristics are not accounted for in this system.

## SOURCES

Cited in M. Meybeck, D.V. Chapman and R. Helmer (eds.), 1989, *Global Freshwater Quality: A First Assessment*, Global Environment Monitoring System (GEMS), World Health Organization and the United Nations Environment Programme, Basil Blackwell, Oxford. By permission.

Data from Organization for Economic Co-operation and Development (OECD), 1982, *Eutrophication of Waters, Monitoring, Assessment and Control*, Organization for Economic Co-operation and Development, Paris.

**TABLE F.18**  Problems associated with the eutrophication of lakes, reservoirs, and impoundments

| Problem Areas | Algal blooms and species composition | Excessive macrophyte and littoral algal growth | Altered thermal conditions | Mineral turbidity | Low dissolved solids |
|---|---|---|---|---|---|
| | | | Caused or Indirectly Dependent on | | |
| **Impairment of drinking water quality** | | | | | |
| Taste and odor, color, filtration, flocculation, sedimentation, and other treatment difficulties | Very frequent | At times | At times | At times | |
| Hypolimnetic oxygen depletion, iron, manganese, $CO_2$, $NH_4$, $CH_4$, $H_2S$, etc., formation | Frequent | At times | At times | | At times |
| Corrosive problems in pipes and other man-made structures | Frequent | At times | | At times | At times |
| **Recreational impairment** | | | | | |
| Unsightliness | Frequent | At times | | At times | |
| Hazard to bathers | | Frequent | | | |
| Increased health hazards | At times | At times | At times | | |
| **Fisheries impairment** | | | | | |
| Fish mortality | At times | | At times | | |
| Undesirable fish stocks | Frequent | Frequent | At times | At times | |
| **Aging and reduced holding capacity and flow** | | | | | |
| By silting | At times | At times | | Frequent | |
| Pipe and screen clogging | | At times | | | |

**TABLE F.18**  Problems associated with the eutrophication of lakes, reservoirs, and impoundments

**DESCRIPTION**

The eutrophication of standing bodies of water causes a variety of practical problems. Several of these problems are listed here under four categories: impairment of drinking water quality; impairment of recreational activities; impairment of fisheries; and impairment of water volume or flow. The most common causes and frequency of these problems are shown, including algal blooms, macrophyte and littoral algal growth, altered thermal conditions, turbidity, and low dissolved solids. Health problems can range from minor skin irritations to schistosomiasis, bilharziasis, and diarrhea. In tropical waters and rivers, macrophyte growth can affect navigation and fisheries.

**LIMITATIONS**

These problems are described qualitatively only. Specific problems associated with particular regions or lakes may not be included here.

**SOURCES**

Cited in M. Meybeck, D.V. Chapman and R. Helmer (eds.), 1989, *Global Freshwater Quality: A First Assessment*, Global Environment Monitoring System (GEMS), World Health Organization and the United Nations Environment Programme, Basil Blackwell, Oxford. By permission.

Data from Organization for Economic Co-operation and Development (OECD), 1982, *Eutrophication of Waters, Monitoring, Assessment and Control*, Organization for Economic Co-operation and Development, Paris.

**TABLE F.19**  Selected eutrophic lakes and reservoirs

| Africa | Europe |
|---|---|
| Lake Chad | Mjosa |
| Lake Victoria | Vattern |
| Lake Mariout | Malaren |
| Kariba Reservoir | Lake Paajarvi |
| Lake George | Lake Esrom |
| Hartbeespoort Reservoir | Lake Neagh |
| | Lake Plon |
| Central America | Lac de Nantua |
| Lake Cajititlan | Lake Ladoga |
| Lake Amatitlan | Lake Balaton |
| Lake Valencia | Lake Constance |
| Lake Paranoa | Lake Leman |
| Lake San Roque Reservoir | Lake Zurich |
| Lake Titicaca | Lake Baldegg |
| Poza Honde Reservoir | Lake Sampach |
| | Lake Hallwil |
| North America | Lake Lugano |
| Lake Simcoe | Lake Como |
| Lake Erie | Lake Iseo |
| Lake Ontario | Wahnbach Reservoir |
| Lake Washington | |
| Lake Mendota | Oceania |
| Lake Tahoe | Lake Burley Griffin |
| Lake Memephremagog | |
| Asia | |
| Lake Kasimigaura | |
| Lake Biwa | |
| Lake Suva | |
| Lake Songkhla | |
| Laguna de Bay | |
| Donghu | |
| Xihu | |
| Xuanwuhu | |
| Mochouhu | |
| Moguhu | |

**TABLE F.19**  Selected eutrophic lakes and reservoirs

## DESCRIPTION

Several eutrophic lakes and reservoirs are listed for each continent. Eutrophication is widespread in all continents and is an important water quality issue. If proper action is taken, some lakes that were not originally eutrophic can be returned to a quality close to their original condition.

## LIMITATIONS

The conditions of the lakes listed here vary widely, from severely eutrophic to only slightly eutrophic. For example, Lake Tahoe in the United States is not yet considered eutrophic, but is suffering from significant increases in nutrient concentrations due to development on its margins, and is becoming eutrophic at an increasing rate.

## SOURCE

World Health Organization/United Nations Environment Programme, 1991, *Water Quality*, Global Environment Monitoring System and Assessment Research Centre, London.

**TABLE F.20**  Aral Sea hydrologic data, 1926 to 1990

| Year | River inflow (km³/yr) | Average annual water level[a] (m) | Average annual surface area (10³ km²) | Average annual volume (km³) |
|------|------|------|------|------|
| 1926 | 49.6 | 53.05 | 66.20 | 1,063.20 |
| 1927 | 44.3 | 52.90 | 65.60 | 1,053.36 |
| 1928 | 62.0 | 52.85 | 65.40 | 1,050.09 |
| 1929 | 57.4 | 52.89 | 65.60 | 1,052.72 |
| 1930 | 54.5 | 52.76 | 65.10 | 1,044.25 |
| 1931 | 63.5 | 52.76 | 65.10 | 1,044.25 |
| 1932 | 56.3 | 52.97 | 65.90 | 1,058.09 |
| 1933 | 53.1 | 53.07 | 66.30 | 1,064.72 |
| 1934 | 64.1 | 53.10 | 66.40 | 1,066.71 |
| 1935 | 53.5 | 53.25 | 67.20 | 1,076.79 |
| 1936 | 57.1 | 53.21 | 67.00 | 1,074.11 |
| 1937 | 53.2 | 53.10 | 66.40 | 1,066.81 |
| 1938 | 46.6 | 52.99 | 65.90 | 1,059.56 |
| 1939 | 46.0 | 52.87 | 65.60 | 1,051.69 |
| 1940 | 45.2 | 52.67 | 64.70 | 1,038.75 |
| 1941 | 56.7 | 52.67 | 64.70 | 1,038.75 |
| 1942 | 59.7 | 52.71 | 64.90 | 1,041.35 |
| 1943 | 58.7 | 52.80 | 65.20 | 1,047.21 |
| 1944 | 54.6 | 52.73 | 64.90 | 1,042.67 |
| 1945 | 63.9 | 52.75 | 64.90 | 1,043.97 |
| 1946 | 51.0 | 52.85 | 65.40 | 1,050.51 |
| 1947 | 42.9 | 52.79 | 65.20 | 1,046.60 |
| 1948 | 54.6 | 52.55 | 64.20 | 1,031.19 |
| 1949 | 62.0 | 52.68 | 64.70 | 1,039.60 |
| 1950 | 48.1 | 52.82 | 65.20 | 1,048.73 |
| 1951 | 44.8 | 52.69 | 64.70 | 1,040.32 |
| 1952 | 63.6 | 52.70 | 64.80 | 1,040.96 |
| 1953 | 60.7 | 52.85 | 65.40 | 1,050.77 |
| 1954 | 64.8 | 53.12 | 66.50 | 1,068.73 |
| 1955 | 53.8 | 53.17 | 66.80 | 1,072.07 |
| 1956 | 55.8 | 53.22 | 67.20 | 1,075.43 |
| 1957 | 56.8 | 53.19 | 66.90 | 1,073.42 |
| 1958 | 56.7 | 53.16 | 66.80 | 1,071.42 |
| 1959 | 59.0 | 53.29 | 67.40 | 1,080.18 |
| 1960 | 55.9 | 53.41 | 68.00 | 1,088.34 |
| 1961 | 39.1 | 53.30 | 67.40 | 1,080.93 |
| 1962 | 33.0 | 52.98 | 65.90 | 1,059.84 |
| 1963 | 40.4 | 52.62 | 64.50 | 1,036.62 |

*continued*

| Year | River inflow (km³/yr) | Average annual water level[a] (m) | Average annual surface area (10³ km²) | Average annual volume (km³) |
|---|---|---|---|---|
| 1964 | 49.6 | 52.50 | 64.00 | 1,028.94 |
| 1965 | 27.8 | 52.30 | 63.20 | 1,016.30 |
| 1966 | 42.2 | 51.88 | 61.80 | 990.34 |
| 1967 | 35.6 | 51.57 | 61.20 | 971.37 |
| 1968 | 39.7 | 51.24 | 60.60 | 951.39 |
| 1969 | 70.8 | 51.29 | 60.70 | 954.41 |
| 1970 | 36.5 | 51.42 | 60.90 | 962.32 |
| 1971 | 28.0 | 51.05 | 60.20 | 940.05 |
| 1972 | 30.9 | 50.54 | 59.00 | 909.96 |
| 1973 | 47.7 | 50.23 | 58.50 | 891.83 |
| 1974 | 7.5 | 49.83 | 57.80 | 868.71 |
| 1975 | 12.5 | 49.01 | 56.60 | 822.29 |
| 1976 | 15.0 | 48.28 | 55.70 | 781.63 |
| 1977 | 8.7 | 47.63 | 54.70 | 746.08 |
| 1978 | 22.8 | 47.06 | 53.70 | 715.47 |
| 1979 | 14.1 | 46.45 | 52.60 | 683.38 |
| 1980 | 11.1 | 45.76 | 51.40 | 647.92 |
| 1981 | 9.3 | 45.19 | 50.40 | 619.19 |
| 1982 | 2.0 | 44.39 | 49.00 | 579.99 |
| 1983 | 3.3 | 43.55 | 47.50 | 540.09 |
| 1984 | 9.8 | 42.75 | 46.10 | 503.21 |
| 1985 | 2.0 | 41.95 | 44.40 | 467.69 |
| 1986 | 1.0 | 41.12 | 43.52 | n.d. |
| 1987 | 10.0 | 40.50 | 42.87 | n.d. |
| 1988 | 22.0 | 40.17 | 42.52 | n.d. |
| 1989 | 4.0 | 39.07 | 40.39 | 370.00 |
| 1990 | 6.0 | 38.05 | n.d. | n.d. |
| | | | | |
| Average 1927–1960 | 55.32 | 52.93 | 65.76 | 1,055.71 |
| Average 1961–1974 | 37.77 | 51.63 | 61.84 | 975.93 |
| Average 1975–1985 | 10.05 | 45.64 | 51.10 | 646.09 |

n.d. no data
[a]Above Baltic Sea level.

**TABLE F.20**  Aral Sea hydrologic data, 1926 to 1990 *(pages 313–314)*

## DESCRIPTION

The Aral Sea in Kazakhstan and Uzbekistan in the former Soviet Union has undergone enormous changes in area, volume, and water level because of the consumption of water that used to flow into the Sea. Shown here for the years 1926–1990 are total river inflow and average annual volume of the Aral Sea, in km³, the average annual water level measured in meters above the Baltic Sea, and the average annual surface area in km² × 10³. The average annual data reflect monthly measurements averaged for the year shown.

## LIMITATIONS

Not all of the data come from the same source, so there may be some inconsistencies. A complete water balance should be done using additional information on evaporative losses and precipitation on the lake surface. These data are available from the same source.

## SOURCES

Calculated by P.P. Micklin, 1992 Western Michigan University, from data of A. Ye. Asarin and V.N. Bortnik (1926–1985) and other sources (1986–1990).

**TABLE F.21**  Annual additions to the National Wild and Scenic Rivers System of the United States, 1968 to March 1992 *(page 316)*

## DESCRIPTION

As the number of free-flowing rivers in the United States has been reduced through dam construction and other modifications, public concern led the Congress to create the National Wild and Scenic Rivers Act in 1968. This Act designates selected natural waters in the United States as worthy of protection. By the early 1990s, the Act had protected roughly 10,000 river miles. Rivers are chosen for protection based on their aesthetics or other unusual qualities, and are studied for inclusion by state and federal agencies. The act generally prevents timber cutting, dam construction, and all other major channel development. While boating, fishing, and recreational activities are generally acceptable, they may be controlled during certain times. This table summarizes the number of river miles added to the Act during each year since its inception.

## LIMITATIONS

The level of protection varies by designation: wild, scenic, and recreational, but this table does not distinguish between them. The number of river miles should be taken as an approximation rather than as an absolute, because the designation defines the river sections by their starting and ending points, and length is later estimated through various measurement techniques, which may not be standardized.

## SOURCES

J. Haubert, National Park Service, 1992, personal communication.

E. Foote, American Rivers, Inc., 1992, personal communications.

K. Coyle, 1988, *The American Rivers Guide to the Wild and Scenic River Designation: A Primer on National River Conservation*, American Rivers, Inc., Washington, DC.

**TABLE F.21**  Annual additions to the National Wild and Scenic Rivers System of the United States, 1968 to March 1992

| Year | Miles added | Total miles | Rivers protected |
|---|---|---|---|
| 1968 | 773 | 773 | Middle Fork Clearwater, Eleven Point, Feather, Rio Grande, Rogue, St. Croix, Middle Fork Salmon, Wolf |
| 1969 | 0 | 773 | None |
| 1970 | 95 | 868 | Allagash Wilderness Waterways |
| 1971 | 0 | 868 | None |
| 1972 | 27 | 895 | Lower St. Croix |
| 1973 | 66 | 961 | Little Miami |
| 1974 | 57 | 1,018 | Chattooga |
| 1975 | 127 | 1,145 | Little Beaver, Snake, Rapid |
| 1976 | 465 | 1,610 | Lower St. Croix, New, Missouri, Flathead, Obed |
| 1977 | 0 | 1,610 | None |
| 1978 | 689 | 2,299 | Rio Grande, Missouri, Pere Marquette, Skagit, Upper Delaware, Middle Delaware, North Fork American, Saint Joe |
| 1979 | 0 | 2,299 | None |
| 1980 | 3,363 | 5,662 | Salmon (Idaho), Little Miami, Alagnak, Alatna, Aniakchak, Charley, Chilikadrotna, John, Kobuk, Mulchatna, North Fork Koyukuk, Noatak, Salmon (Alaska), Tinayguk, Tlikakila, Andreafsky, Ivishak, Nowitna, Selawik, Sheenjek, Wind, Beaver Creek, Birch Creek, Delta, Fortymile, Gulkana, Unalakleet |
| 1981 | 1,246 | 6,908 | Lower American, Klamath, Trinity, Eel, Smith |
| 1982 | 0 | 6,908 | None |
| 1983 | 0 | 6,908 | None |
| 1984 | 309 | 7,217 | Verde, Tuolomne, Au Sable, Owyhee, Illinois |
| 1985 | 8 | 7,224 | Loxahatchee |
| 1986 | 139 | 7,363 | Horsepasture, Cache la Poudre, Black Creek, Saline Bayou, Klickitat, White Salmon |
| 1987 | 346 | 7,709 | Merced, Kings, Kern |
| 1988 | 1,555 | 9,264 | Wildcat Creek, Sipsey Fork of the West Fork, Big Marsh Creek, Chetco, Clackamas, Crescent Creek, Crooked, Deschutes, Donner and Blitzen, Eagle Creek, Elk, Grande Ronde, Imnaha, John Day, Joseph Creek, Little Deschutes, Lostine, Malheur, McKenzie, Metolius, Minam, North Fork Crooked, North Fork John Day, North Fork Malheur, North Fork of the Middle Fork of the Willamette, North Fork Owyhee, North Fork Smith, North Fork Sprague, North Fork John Day, Squaw Creek, Sycan, Upper Rogue, Wenaha, West Little Owyhee, White, Bluestone, Rio Chama, North Umpqua, Powder, North Powder, Quartzville Creek, Roaring, Salmon (Oregon), Sandy, South Fork John |
| 1989 | 17 | 9,281 | Middle Fork of the Vermillion |
| 1990 | 52 | 9,318[a] | East Fork of Jemez, Pecos, Clarks Fork of the Yellowstone |
| 1991 | 134 | 9,452 | Niaobrara, Missouri |
| 1992[b] | 520 | 9,972 | Bear Creek, Black, Carp, Indian, Manistee, Ontonagon, Paint, Pine, Presque Isle, Sturgeon (Hiawatha), Sturgeon (Ottawa), East Branch of the Tahquamenon, Whitefish, Yellow Dog |

[a] In 1990 15 miles of the Smith River in California were subtracted from the 1981 designation.
So although 52 new miles were added in 1990, the net change is only 37 miles.
[b] Through March 1992.

**TABLE F.22**  Globally threatened amphibian and fresh water fish species, 1990

| | Amphibians (number) | | Fresh water fish (number) | |
|---|---|---|---|---|
| | Known species | Threatened species | Known species | Threatened species |
| *Africa* | | | | |
| Algeria | | 0 | | 1 |
| Angola | | 0 | 268 | 0 |
| Benin | | 0 | 150 | 0 |
| Botswana | 38 | 0 | 81 | 0 |
| Burkina Faso | | 0 | 120 | 0 |
| Cameroon | | 1 | | 11 |
| Central African Republic | | 0 | 400 | 0 |
| Chad | | 0 | 130 | 0 |
| Comoros | 2 | 0 | 16 | 0 |
| Congo | | 0 | 500 | 0 |
| Côte d'Ivoire | | 1 | 200 | 0 |
| Egypt | | 0 | | 1 |
| Equatorial Guinea | | 1 | | 0 |
| Ethiopia | | 0 | 100 | 0 |
| Gabon | | 0 | 200 | 0 |
| Gambia | | 0 | 80 | 0 |
| Ghana | | 0 | 180 | 0 |
| Guinea | | 1 | 250 | 0 |
| Guinea-Bissau | | 0 | 90 | 0 |
| Kenya | 88 | 0 | 180 | 0 |
| Lesotho | | 0 | 8 | 0 |
| Liberia | | 0 | 130 | 0 |
| Madagascar | 144 | 0 | | 0 |
| Malawi | 69 | 0 | 600 | 0 |
| Mali | | 0 | 160 | 0 |
| Mauritania | | 0 | 15 | 0 |
| Mauritius | 2 | 0 | | 0 |
| Morocco | 32 | 0 | | 1 |
| Mozambique | | 0 | | 1 |
| Namibia | | 0 | 97 | 4 |
| Niger | | 0 | 140 | 0 |
| Nigeria | 19 | 0 | 200 | 0 |
| Senegal | | 0 | 140 | 0 |
| Seychelles | 12 | 3 | | 0 |
| Sierra Leone | | 0 | 130 | 0 |
| South Africa | 95 | 1 | 220 | 28 |

*continued*

| | Amphibians (number) | | Fresh water fish (number) | |
|---|---|---|---|---|
| | Known species | Threatened species | Known species | Threatened species |
| Sudan | | 0 | 120 | 0 |
| Swaziland | | 0 | 45 | 0 |
| Togo | | 0 | 160 | 0 |
| Uganda | | 0 | 300 | 0 |
| Zaire | | 0 | 700 | 1 |
| Zambia | 83 | 0 | 156 | 0 |
| Zimbabwe | 120 | 0 | 132 | 0 |
| *The Americas* | | | | |
| Argentina | 124 | 1 | | 1 |
| Bahamas | 6 | 0 | | 0 |
| Belize | 26 | 0 | | 0 |
| Bolivia | 96 | 0 | | 1 |
| Brazil | 487 | 0 | | 9 |
| Canada | 41 | 0 | | 15 |
| Cayman Islands | | 0 | 1,132[b] | 0 |
| Chile | 38 | 0 | | 1 |
| Colombia | 375 | 0 | | 0 |
| Costa Rica | 151 | 0 | | 0 |
| Cuba | 40 | 0 | | 0 |
| Ecuador[a] | 350 | 0 | | 0 |
| El Salvador | 38 | 0 | | 0 |
| French Guiana | 89 | 0 | | 0 |
| Guatemala | 99 | 0 | | 0 |
| Guyana | 105 | 0 | | 1 |
| Honduras | 57 | 0 | | 0 |
| Jamaica | 20 | 0 | | 98 |
| Mexico | 284 | 4 | | 0 |
| Netherlands Antilles | 2 | 0 | | 0 |
| Nicaragua | 59 | 0 | | 0 |
| Panama | 155 | 0 | | 0 |
| Paraguay | 69 | 0 | | 0 |
| Peru | 235 | 1 | | 1 |
| Puerto Rico | 26 | 1 | | 0 |
| Suriname | 99 | 0 | | 0 |
| Trinidad and Tobago | 15 | 0 | | 0 |
| United States | 222 | 22 | 2,640[b] | 164 |
| Uruguay | 37 | 0 | | 0 |
| Venezuela | 183 | 0 | | 0 |

*continued*

| | Amphibians (number) | | Fresh water fish (number) | |
|---|---|---|---|---|
| | Known species | Threatened species | Known species | Threatened species |
| *Asia* | | | | |
| Afghanistan | | 1 | | 0 |
| Bahrain | | 0 | | 1 |
| Brunei | | 0 | | 2 |
| Cambodia | | 0 | | 5 |
| China | | 1 | | 7 |
| India | 181 | 3 | | 2 |
| Indonesia | | 0 | | 29 |
| Iran | | 0 | | 2 |
| Iraq | | 0 | | 2 |
| Israel | | 1 | | 0 |
| Japan | 95 | 1 | 207 | 3 |
| Laos | | 0 | | 5 |
| Malaysia | | 0 | | 6 |
| Myanmar | | 0 | | 2 |
| Oman | | 0 | | 2 |
| Philippines | 60 | 0 | | 21 |
| Sri Lanka | | 0 | | 12 |
| Thailand | | 0 | | 13 |
| Turkey | 18 | 1 | 555[b] | 5 |
| Vietnam | 80 | 1 | | 4 |
| | | | | |
| *Europe* | | | | |
| Albania | | 0 | | 1 |
| Austria | 19 | 0 | 71 | 2 |
| Belgium | | 0 | | 1 |
| Bulgaria | | 0 | | 3 |
| Czechoslovakia | | 0 | | 2 |
| Denmark | 14 | 0 | 166[b] | 0 |
| Finland | 5 | 0 | 58[b] | 1 |
| France | 29 | 1 | 70 | 3 |
| Germany | 19 | 0 | 70[b] | 3 |
| Greece | | 0 | | 6 |
| Hungary | | 0 | | 2 |
| Iceland | | 0 | | 1 |
| Ireland | 3 | 0 | | 1 |
| Italy | 28 | 7 | 503[b] | 3 |
| Luxembourg | 16 | 0 | 40 | 0 |
| Netherlands | 15 | 0 | 49[b] | 1 |

*continued*

| | Amphibians (number) | | Fresh water fish (number) | |
|---|---|---|---|---|
| | Known species | Threatened species | Known species | Threatened species |
| Norway | 5 | 0 | 172 | 1 |
| Poland | | 0 | | 1 |
| Portugal | 17 | 1 | 39 | 0 |
| Romania | | 0 | | 4 |
| Spain | 24 | 3 | 55 | 2 |
| Sweden | 13 | 0 | 130[b] | 1 |
| Switzerland | 19 | 1 | 52 | 3 |
| United Kingdom | 6 | 0 | 377[b] | 1 |
| Yugoslavia | | 2 | | 5 |
| *USSR* | 34 | 0 | | 5 |
| *Oceania* | | | | |
| Australia | 150 | 3 | 3,200[b] | 16 |
| Fiji | | 1 | | 0 |
| New Zealand | 5 | 3 | 777[b] | 2 |

[a]Includes the Galapagos Islands.
[b]Both fresh and salt water species.

**TABLE F.22**   Globally threatened amphibian and fresh water fish species, 1990 *(pages 317–320)*

## DESCRIPTION

With increased human manipulation of fresh water ecosystems, many aquatic species are being driven toward extinction. Species are susceptible to water quality conditions, such as pH, temperature, clarity, and presence of toxic chemicals. Major engineering projects, such as dams and irrigation systems, physically disrupt breeding and feeding patterns and destroy habitat. Fishing and the introduction of exotic species changes competitive patterns for all inhabitants of the system. This table presents the number of threatened amphibian and fresh water fish species for every country in the world in 1990. The definition of threatened here is based upon the World Conservation Union's classification system, which identifies six distinct categories: endangered, vulnerable, rare, indeterminate, out of danger, and insufficiently known. This listing excludes those species that are either out of danger or extinct. Included in "known species" are species introduced to the region.

## LIMITATIONS

This table should only be considered a starting point for identifying the existence of a problem with world fresh water ecosystems. An accurate list of endangered species has never been compiled because of problems with definitions, lack of adequate biological research, and inadequate reporting. The information provided here can only be based on those species whose population decline has been reported. Thus, note the large difference in species listed as threatened in African countries in comparison with data from the United States – a difference that reflects levels of funding and research rather than actual environmental conditions. Another serious problem with this table, as presented in the original, is the use of a zero (0) for those countries without detailed species information. In fact, a zero (0) here more accurately means that there are no species "known" to be threatened, rather than no threatened species. The assignment of species to particular countries is problematic in that species are found in ecosystems that cross political borders. Thus, no total can be obtained from this information. According to the World Resources Institute, this listing of endangered fish species does not include roughly 250 endemic fish species from Lake Victoria threatened by the introduction of the Nile perch, since details of their ranges and biological characteristics are largely unknown. Other fresh water species than fish and amphibians are not included here, such as threatened fresh water plant and mammal species, and water fowl dependent on fresh water ecosystems. Finally, data for fresh water species are sometimes co-mingled here with data for marine species. These are identified where possible.

## SOURCES

World Resources Institute, 1992, *World Resources 1992–93*, Oxford University Press, New York. By permission.

# G. Water and energy

**TABLE G.1** Hydroelectric generation, capacity, and potential, by country *(page 322–325)*

## DESCRIPTION

Regional and national hydroelectric capacity, generation, and potential are listed for 1989. Two separate estimates are given for installed hydroelectric capacity (in MW) and for hydroelectric generation (in GWh per year). An estimate is also given for hydroelectric capacity under construction and the total "exploitable potential," defined as the energy exploitable under existing technical limitations and economic constraints.

## LIMITATIONS

We provide two different estimates of the same data here because of unreconcilable differences between the data sets. For some countries, the two data sets are the same; for others, there are substantial differences. The estimate of "exploitable potential" far exceeds the total amount of energy that will ever be obtained, because of social and environmental constraints. For the World Resources Institute/United Nations data set, some small countries not listed are included in the regional or continental totals.

## SOURCES

International Water Power and Dam Construction, 1991, *International Water Power and Dam Construction Handbook 1991*, Reed Enterprise, Sutton, UK. By permission.

United Nations, 1989, *Energy Statistics Yearbook 1989*, United Nations Statistical Office, United Nations, New York. As cited in World Resources Institute, 1992, *World Resources 1992–93: A Guide to the Global Environment*, Oxford University Press, New York. By permission.

E. Rodenburg, 1992, World Resources Institute, personal communication.

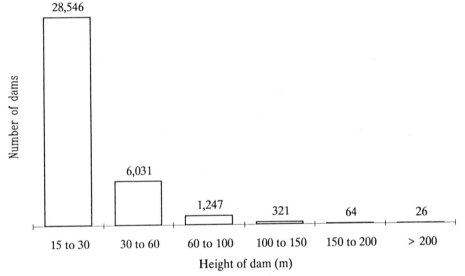

Fig. G–1. Number of dams worldwide, by height. There are over 34,000 dams worldwide greater than 15 m in height. Though the vast majority are between 15 m and 30 m, there are 90 dams higher than 150 m. Data from International Commission on Large Dams, 1988, *World Registry of Dams, 1988*, International Commission on Large Dams, Paris.

**TABLE G.1** Hydroelectric generation, capacity, and potential, by country

| | WRI (1992–93) Installed capacity 1989 (MW) | IWPDC (1991) Installed capacity 1989 (MW) | WRI (1992–93) Hydroelectric generation 1989 (GWh/yr) | IWPDC (1991) Hydroelectric generation 1989 (GWh/yr) | IWPDC (1991) Capacity under construction (MW) | IWPDC (1991) Exploitable potential (GWh/yr) |
|---|---|---|---|---|---|---|
| *Africa* | 18,884 | | 44,156 | | | |
| Algeria | 285 | 285 | 226 | 300 | n/a | n/a |
| Angola | 412 | 400 | 1,355 | 1,335 | 520 | 100,000[a] |
| Burkina Faso | 14 | 14 | 1.7 | 1.7 | 16 | 800[a] |
| Burundi | 32 | 28 | 103 | 93 | 2.8 | 1,445 |
| Cameroon | 528 | 528 | 2,629 | 2,325 | 120 | 115,000[a] |
| Central African Republic | 22 | 22 | 74 | 74 | n/a | n/a |
| Comoros | 1 | | 2 | | | |
| Congo | 120 | 120 | 397 | 233 | n/a | 50,000[a] |
| Côte d'Ivoire | 895 | 885 | 1,250 | 1,290 | n/a | 14,000[a] |
| Egypt | 2,445 | 2,700 | 6,400 | 6,000 | n/a | n/a |
| Equatorial Guinea | 1 | 1 | 2 | 2 | n/a | n/a |
| Ethiopia | 230 | 372 | 655 | 1,004 | 168 | 162,000 |
| Gabon | 125 | 323 | 675 | 699 | 4.4 | 32,500 |
| Ghana | 1,072 | 1,072 | 4,820 | 4,907.9 | n/a | 10,000[a] |
| Guinea | 47 | 47 | 172 | 167 | n/a | 26,000[a] |
| Guinea-Bissau | n/a | n/a | n/a | n/a | n/a | 300 |
| Kenya | 498 | 583 | 2,469 | 2,192 | 140 | 30,000[a] |
| Lesotho | 2.2 | 2.2 | n/a | n/a | 70 | 2,000[a] |
| Liberia | 81 | 81 | 320 | 319 | n/a | 11,000[a] |
| Madagascar | 106 | 106 | 317 | 303.5 | 0.1 | 23,061 |
| Malawi | 146 | 146 | 572 | 564 | n/a | 6,000[a] |
| Mali | 45 | 45 | 170 | 162 | n/a | 10,000[a] |
| Mauritania | n/a | | 25 | | | |
| Mauritius | 59 | 59.3 | 102 | 147.6 | 0 | 110 |
| Morocco | 622 | 615 | 1,157 | 1,167 | 308 | 4,500 |
| Mozambique | 2,078 | 2,360 | 50 | 132.9 | 14.5 | 72,000[a] |
| Nigeria | 1,900 | 1,338.4 | 2,210 | 3,159 | 600 | 40,000[a] |
| Reunion | | 125 | | 530 | n/a | n/a |
| Rwanda | 56 | 56 | 171 | 170 | n/a | 3,000[a] |
| Senegal | 0 | 0 | 0 | 0 | 0 | 2,500[a] |
| Sierra Leone | 2 | 2 | n/a | n/a | n/a | 6,800[a] |
| Somalia | 4.8 | 4.8 | 0.024 | 0.024 | 0 | n/a |
| South Africa | 550 | 600 | 600 | 2,759 | 0 | n/a |
| Sudan | 225 | 225 | 517 | 516 | n/a | 1,900 |
| Swaziland | 42 | 44 | 215 | 220 | 0.3 | 400 |
| Tanzania | 259 | 333.2 | 615 | 1,445 | n/a | 20,000 |
| Togo | 4 | 4 | 5 | 4 | n/a | n/a |
| Tunisia | 64 | 55.4 | 36 | 34 | 0 | 53 |
| Uganda | 155 | 156 | 691 | 644 | n/a | 10,200 |

*continued*

| | WRI (1992–93) Installed capacity 1989 (MW) | IWPDC (1991) Installed capacity 1989 (MW) | WRI (1992–93) Hydroelectric generation 1989 (GWh/yr) | IWPDC (1991) Hydroelectric generation 1989 (GWh/yr) | IWPDC (1991) Capacity under construction (MW) | IWPDC (1991) Exploitable potential (GWh/yr) |
|---|---|---|---|---|---|---|
| Zaire | 2,772 | 2,486 | 5,252 | 5,156 | n/a | 530,000 |
| Zambia | 2,245 | 1,963 | 6,702 | 9,512 | 0 | 30,900 |
| Zimbabwe | 633 | 666 | 2,660 | 3,150 | 0 | 16,000 |
| | | | | | | |
| *North and Central America* | 157,256 | | 598,405 | | | |
| Canada | 57,924 | 58,200 | 291,447 | 287,649 | 4,250 | 592,982 |
| Costa Rica | 735 | 706 | 3,328 | 3,317 | 122 | 37,000 |
| Cuba | 49 | 490 | 82 | 44 | n/a | n/a |
| Dominica | | 2.84 | | 14.2 | 4.84 | 118.3 |
| Dominican Republic | 165 | 165 | 950 | 950 | n/a | 2,517 |
| El Salvador | 405 | 233 | 1,452 | 1,030 | n/a | 3,319 |
| Grenada | | 0 | | 0 | 0 | 20,000 |
| Guatemala | 438 | 438 | 2,089 | 2,088.9 | 0 | 43,370 |
| Haiti | 70 | 54 | 320 | 258.4 | 1.1 | 430 |
| Honduras | 130 | 431 | 880 | 1,914.5 | 0 | 24,000 |
| Jamaica | 20 | 21.7 | 110 | 101.16 | 1.55 | 335 |
| Mexico | 7,825 | 7,734 | 22,950 | 20,800 | 1,590 | 159,624 |
| Nicaragua | 103 | 103 | 268 | 268 | n/a | 17,277[a] |
| Panama | 551 | 551 | 2,181 | 2,848 | n/a | 16,233[a] |
| Puerto Rico | | 85 | | 260 | n/a | n/a |
| United States | 88,746 | 71,820 | 272,021 | 265,061 | 1,138.6 | 376,000 |
| | | | | | | |
| *South America* | 77,175 | | 327,373 | | | |
| Argentina | 6,594 | 6,600 | 15,150 | 13,264 | 4,350 | 390,000 |
| Bolivia | 342 | 310 | 1,270 | 1,142 | 0 | 90,000[a] |
| Brazil[b] | 44,622 | 48,747 | 214,238 | 213,378 | 20,000 | 1,194,900 |
| Chile | 2,290 | 2,300 | 9,603 | 11,458 | 766 | 132,433 |
| Colombia | 6,317 | 6,700 | 29,875 | 24,220 | 1,400 | 418,200 |
| Ecuador | 908 | 915 | 4,918 | 4,934 | 708 | 115,000 |
| Guyana | 2 | 0 | 5 | 0 | 0 | 63,100 |
| Paraguay[c] | 5,440 | 6,148 | 2,784 | 24,303.2 | 2,250 | 78,000 |
| Peru | 2,275 | 3,800 | 10,518 | 11,000 | 210 | 412,000 |
| Suriname | 189 | 189 | 910 | 935 | n/a | 12,840[a] |
| Uruguay | 1,196 | 1,196 | 3,902 | 2,687 | 0 | 4,880 |
| Venezuela | 7,000 | 10,980 | 34,200 | 34,667.4 | 2,540 | 250,000 |
| | | | | | | |
| *Asia* | 118,846 | | 398,440 | | | |
| Afghanistan | 290 | 281 | 758 | 764 | 260 | n/a |
| Bangladesh | 230 | 230 | 735 | 738.9 | 0 | n/a |
| Bhutan | 334 | 336 | 654 | n/a | 1,600 | n/a |
| Cambodia | 10 | 0 | 30 | 0 | 0 | 83,000 |
| China | 30,000 | 34,580 | 109,500 | 118,454 | 24,000 | 1,923,304 |

*continued*

| | WRI (1992–93) Installed capacity 1989 (MW) | IWPDC (1991) Installed capacity 1989 (MW) | WRI (1992–93) Hydroelectric generation 1989 (GWh/yr) | IWPDC (1991) Hydroelectric generation 1989 (GWh/yr) | IWPDC (1991) Capacity under construction (MW) | IWPDC (1991) Exploitable potential (GWh/yr) |
|---|---|---|---|---|---|---|
| Cyprus | 1 | 0.65 | n/a | 0.852 | 0 | n/a |
| India | 18,504 | 17,340 | 63,760 | 57,793 | 7,377 | 600,100 |
| Indonesia | 1,850 | 1,993.8 | 8,600 | 5,300.55 | 1,400 | 709,000 |
| Iran | 1,804 | 1,952 | 6,700 | 7,522 | 61 | 6,784 |
| Iraq | 120 | 100 | 590 | 610 | 1,760 | 70,000 |
| Israel | 0 | 0 | 0 | 0 | 0 | 1,600 |
| Japan | 37,409 | 20,481 | 97,825 | 87,968 | 802 | 130,524 |
| Jordan | 11 | 7 | 29 | 19 | n/a | 87 |
| Korea DPR | 5,000 | 4,600 | 31,750 | 29,100 | n/a | n/a |
| Korea Rep | 2,339 | 1,329 | 4,558 | 2,964 | 165.3 | 3,290 |
| Laos | 200 | 153 | 1,095 | 705 | 45 | 22,638 |
| Lebanon | 246 | 246 | 500 | 610 | n/a | 1,000 |
| Malaysia | 1,437 | 1,580.5 | 6,256 | 5,800.62 | 87.3 | 108,600 |
| Myanmar | 258 | 258 | 1,240 | 1,121 | n/a | 160,000[a] |
| Nepal | 160 | 223.45 | 542 | 706.9 | 17.1 | 144,000 |
| Pakistan | 2,897 | 2,901 | 16,974 | 15,531 | 1,928 | 15,531 |
| Philippines | 2,154 | 2,124 | 6,546 | 6,230.39 | 80 | 11,877 |
| Sri Lanka | 940 | 938 | 2,650 | 2,597 | 169 | 7,255 |
| Syria | 898 | 895 | 4,707 | 1,934.22 | 630 | 4,500 |
| Taiwan | | 1,562 | | 4,840 | 123 | 20,000 |
| Thailand | 2,268 | 2,070.4 | 5,571 | 4,865 | 216 | 5,571 |
| Turkey | 6,598 | 6,565 | 17,939 | 27,450 | 470 | 215,000 |
| Vietnam | 325 | 920 | 2,250 | n/a | 1,440 | n/a |
| *Europe* | 168,949 | | 465,016 | | | |
| Albania | 690 | 1,710 | 3,600 | 5,200 | n/a | 17,000 |
| Austria | 10,838 | 10,200 | 36,137 | 36,136 | 200 | 53,700 |
| Belgium | 1,402 | 92.9 | 362 | 355.7 | 2.57 | n/a |
| Bulgaria | 1,975 | 1,975 | 2,690 | 4,600 | n/a | 15,000 |
| Czechoslovakia | 2,920 | 1,856 | 4,273 | 3,366 | 460 | 10,826 |
| Denmark | 10 | 9.5 | 27 | 4 | n/a | 70 |
| Faeroe Islands | | 27 | | n/a | 0 | 90 |
| Finland | 2,586 | 2,600 | 12,948 | 12,770 | 40 | 20,000 |
| France | 24,815 | 19,800 | 51,160 | 69,600 | 189 | 72,000 |
| German DR | 1,844 | 1,710 | 1,567 | n/a | 0 | 700 |
| Germany FR | 6,861 | 3,560 | 17,000 | 17,550 | 80 | 24,000 |
| Greece | 2,301 | 2,969 | 2,147 | 1,808 | 792.3 | 16,000 |
| Greenland | | 0 | | 0 | 290 | 10,000 |
| Hungary | 48 | 50 | 88 | 158 | 0 | 4,950 |
| Iceland | 756 | 752 | 4,259 | 4,164 | 150 | 31,000 |
| Ireland | 512 | 220 | 991 | 676 | 0.3 | 1,180 |
| Italy | 18,237 | 13,300 | 34,184 | 35,000 | 600 | 65,000 |

*continued*

| | WRI (1992–93) Installed capacity 1989 (MW) | IWPDC (1991) Installed capacity 1989 (MW) | WRI (1992–93) Hydroelectric generation 1989 (GWh/yr) | IWPDC (1991) Hydroelectric generation 1989 (GWh/yr) | IWPDC (1991) Capacity under construction (MW) | IWPDC (1991) Exploitable potential (GWh/yr) |
|---|---|---|---|---|---|---|
| Luxembourg | 1,132 | 28 | n/a | 76 | 0 | 120 |
| Netherlands | 25 | 2.5 | 37 | 10.2 | 12 | 500 |
| Norway | 26,465 | 25,500 | 118,271 | 118,650 | 1,200 | 172,000 |
| Poland | 1,976 | 2,007 | 3,757 | 4,053.1 | 824 | 12,100 |
| Portugal | 3,360 | 3,065 | 5,819 | 5,966 | 918 | 24,000 |
| Romania | 5,583 | 5,853 | 12,628 | 16,470 | 1,000 | 38,000 |
| Spain | 16,223 | 11,678 | 19,530 | 18,936 | 409 | 67,220 |
| Sweden | 15,616 | 16,200 | 72,102 | 70,806 | 200 | 99,000 |
| Switzerland | 11,580 | 11,600 | 29,772 | 30,485 | 270 | 41,000 |
| United Kingdom | 4,163 | 5,218 | 6,970 | 3,912.3 | 0 | 5,200 |
| Yugoslavia | 7,000 | 6,648 | 23,730 | 24,391 | 356 | 71,000 |
| | | | | | | |
| *USSR*[d] | 64,100 | 62,200 | 222,800 | 219,800 | 14,500 | 3,831,000 |
| | | | | | | |
| *Oceania* | 11,891 | | 37,819 | | | |
| Australia | 7,268 | 7,260 | 14,498 | 14,568 | 827 | 30,000 |
| Fiji | 80 | 80 | 330 | 365.6 | 0.5 | 515 |
| New Caledonia | | 75 | | 257 | n/a | n/a |
| New Zealand | 4,287 | 4,587 | 21,900 | 21,891 | 432 | 74,000 |
| Papua New Guinea | 155 | 143 | 455 | 438 | n/a | 98,000 |
| Polynesia | | 10 | | 35 | n/a | n/a |
| Solomon Islands | 0 | 0.03 | 0 | 0.04 | 0.13 | n/a |
| Tahiti | | 20 | | n/a | n/a | n/a |
| Western Samoa | | | 8.25 | 26.933 | 4 | 80 |

n/a not available
[a]Technical capability.
[b]IWPDC figures for Brazil include 11,900 MW capacity for Itaipú and 700 MW under construction.
[c]IWPDC figures for Paraguay include half of Itaipú and half the capacity of Yacyreta under construction (1,500 MW).
[d]IWPDC figures for USSR include note: "Kiev pumped-storage plant in operation; USSR national power programme is under review. Entry for USSR presents calculated data."

**TABLE G.2**   World net hydroelectric generation, 1975 to 1990
($10^9$ kWh/yr)

|                           | 1975  | 1980  | 1985  | 1990  |
|---------------------------|-------|-------|-------|-------|
| *North America*           | 520.7 | 546.9 | 611.0 | 599.6 |
| Canada                    | 202.4 | 251.0 | 300.7 | 292.8 |
| Mexico                    | 15.1  | 16.7  | 26.0  | 23.7  |
| United States             | 303.2 | 279.2 | 284.3 | 283.0 |
| *Central and South America* | 117.2 | 199.1 | 282.1 | 353.4 |
| Argentina                 | 5.1   | 15.0  | 20.4  | 14.3  |
| Brazil                    | 73.8  | 127.4 | 176.5 | 226.8 |
| Chile                     | 6.1   | 7.3   | 10.4  | 9.1   |
| Colombia                  | 9.8   | 14.3  | 21.6  | 31.3  |
| Peru                      | 5.4   | 7.5   | 9.4   | 10.5  |
| Uruguay                   | 1.1   | 2.3   | 6.4   | 3.8   |
| Venezuela                 | 8.8   | 14.5  | 21.1  | 35.0  |
| Other                     | 7.1   | 11.0  | 16.4  | 22.6  |
| *Western Europe*          | 390.5 | 442.4 | 466.2 | 444.7 |
| Austria                   | 23.5  | 28.9  | 31.0  | 33.7  |
| Finland                   | 12.0  | 10.0  | 12.2  | 10.8  |
| France                    | 59.9  | 69.2  | 63.6  | 55.9  |
| Germany, FR               | 16.9  | 20.6  | 17.6  | 18.5  |
| Iceland                   | 2.2   | 3.0   | 3.7   | 4.4   |
| Italy                     | 42.4  | 48.5  | 44.1  | 27.4  |
| Norway                    | 76.6  | 83.0  | 102.9 | 118.5 |
| Portugal                  | 6.4   | 8.0   | 10.7  | 5.5   |
| Spain                     | 26.2  | 30.5  | 32.8  | 16.8  |
| Sweden                    | 57.0  | 58.7  | 69.8  | 71.7  |
| Switzerland               | 34.0  | 33.5  | 32.3  | 25.1  |
| Turkey                    | 5.8   | 11.2  | 11.9  | 22.9  |
| United Kingdom            | 4.9   | 3.6   | 3.3   | 7.0   |
| Yugoslavia                | 19.1  | 27.9  | 24.4  | 22.0  |
| Other                     | 3.7   | 5.7   | 5.9   | 4.6   |
| *Eastern Europe and USSR* | 144.6 | 209.8 | 232.4 | 243.6 |
| Czechoslovakia            | 3.8   | 4.7   | 4.3   | 4.3   |
| Poland                    | 2.4   | 3.2   | 3.9   | 3.3   |
| Romania                   | 8.6   | 12.5  | 11.8  | 10.7  |
| USSR                      | 124.7 | 182.0 | 205.0 | 217.3 |
| Other                     | 5.1   | 7.3   | 7.4   | 8.1   |

*continued*

| | 1975 | 1980 | 1985 | 1990 |
|---|---|---|---|---|
| *Middle East* | 5.5 | 9.6 | 9.5 | 12.6 |
| Iran | 3.4 | 5.6 | 6.3 | 6.8 |
| Other | 2.1 | 4.1 | 3.2 | 5.8 |
| | | | | |
| *Africa* | 37.2 | 62.4 | 45.3 | 43.2 |
| Egypt | 6.7 | 9.7 | 8.0 | 6.0 |
| Zaire | 3.7 | 4.2 | 5.0 | 5.2 |
| Zambia | 5.9 | 9.0 | 9.9 | 5.8 |
| Zimbabwe | 5.3 | 4.0 | 3.1 | 2.6 |
| Other | 15.5 | 35.5 | 19.4 | 23.5 |
| | | | | |
| *Far East and Oceania* | 229.6 | 281.3 | 332.7 | 415.8 |
| Australia | 15.1 | 16.9 | 13.9 | 13.8 |
| China | 44.5 | 57.6 | 92.4 | 113.4 |
| India | 33.0 | 46.1 | 50.5 | 66.8 |
| Japan | 85.0 | 91.2 | 77.5 | 96.4 |
| Korea, DPR | 16.3 | 22.3 | 27.7 | 32.4 |
| Korea, Rep | 1.7 | 2.0 | 3.6 | 6.3 |
| New Zealand | 16.7 | 16.2 | 19.5 | 21.7 |
| Pakistan | 4.6 | 8.6 | 12.1 | 17.9 |
| Philippines | 2.2 | 3.5 | 5.5 | 6.6 |
| Thailand | 3.4 | 3.7 | 3.7 | 5.4 |
| Other | 7.1 | 13.3 | 26.3 | 35.0 |
| | | | | |
| *World total*[a] | 1,445.4 | 1,751.6 | 1,979.2 | 2,112.9 |

[a]Totals may not add to sum of columns due to rounding.

**TABLE G.2** World net hydroelectric generation, 1975 to 1990 *(pages 326–327)*

## DESCRIPTION

The generation of hydroelectricity is given here for major regions and countries of the world in thousand million ($10^9$) kWh per year for 1975, 1980, 1985, and 1990. Data for 1990 are preliminary. Hydroelectric power production includes utility and non-utility production of hydroelectricity. Production is reported on a net basis, that is, on the amount of energy remaining after electricity use by the generating unit.

## LIMITATIONS

The sum of each section here may not equal the total due to independent rounding. The data were gathered by the U.S. Energy Information Agency from national sources. Annual hydroelectricity energy production depends on climatic conditions, installed capacity, and the availability and cost of other forms of generation.

## SOURCES

United States Department of Energy, 1992, *International Energy Annual 1990*, Energy Information Administration, DOE/EIA-0219(90), Washington, DC.

United States Department of Energy, 1986, *International Energy Annual 1985*, Energy Information Administration, DOE/EIA-0219(85), Washington, DC.

**TABLE G.3**   World net electricity generation, by type, 1989
$(10^9 \text{ kWh})$

| | Thermal[a] | Hydro | Nuclear | Geothermal and other[b] | Totalc |
|---|---|---|---|---|---|
| *North America* | 2,292.1 | 578.6 | 604.7 | 16.1 | 3,491.5 |
| Canada | 119.2 | 287.6 | 75.4 | 0.0 | 482.2 |
| Mexico | 85.7 | 22.7 | 0.0 | 4.8 | 113.3 |
| United States | 2,096.6 | 268.2 | 529.4 | 11.3 | 2,895.5 |
| Other | 0.7 | 0.0 | 0.0 | 0.0 | 0.7 |
| | | | | | |
| *Central and South America* | 127.0 | 335.9 | 6.3 | 7.0 | 48.1 |
| Argentina | 28.3 | 15.0 | 4.8 | 0.0 | 227.4 |
| Brazil | 7.4 | 212.1 | 1.5 | 6.3 | 38.0 |
| Colombia | 8.4 | 29.6 | 0.0 | 0.0 | 58.0 |
| Venezuela | 24.2 | 33.8 | 0.0 | 0.0 | 104.8 |
| Other | 58.7 | 45.4 | 0.0 | 0.7 | 476.2 |
| | | | | | |
| *Western Europe* | 1,018.0 | 442.3 | 696.2 | 9.7 | 2,166.3 |
| Austria | 13.1 | 33.0 | 0.0 | 0.0 | 46.1 |
| Belgium | 21.3 | 1.0 | 39.1 | d | 61.4 |
| Finland | 20.2 | 12.8 | 18.0 | 0.0 | 51.1 |
| France | 49.2 | 50.1 | 288.7 | 0.6 | 388.6 |
| Germany, FR | 270.2 | 19.1 | 140.4 | 2.2 | 431.9 |
| Greece | 28.7 | 2.1 | 0.0 | 0.0 | 30.9 |
| Italy | 140.9 | 30.8 | 0.0 | 3.0 | 174.7 |
| Netherlands | 64.6 | d | 3.8 | d | 68.4 |
| Norway | 0.5 | 117.5 | 0.0 | 0.0 | 118.0 |
| Spain | 63.9 | 18.3 | 53.7 | 0.0 | 135.9 |
| Sweden | 5.8 | 71.1 | 62.8 | d | 139.7 |
| Switzerland | 0.6 | 26.9 | 21.5 | 0.0 | 49.1 |
| Turkey | 28.2 | 17.6 | 0.0 | 0.1 | 45.9 |
| United Kingdom | 208.3 | 6.5 | 63.6 | d | 278.4 |
| Yugoslavia | 51.0 | 23.5 | 4.5 | 3.2 | 82.2 |
| Other | 51.3 | 12.0 | 0.0 | 0.7 | 64.1 |
| | | | | | |
| *Eastern Europe and USSR* | 1,597.0 | 249.1 | 274.5 | d | 2,120.5 |
| Bulgaria | 25.4 | 2.7 | 14.6 | 0.0 | 42.6 |
| Czechoslovakia | 56.5 | 4.2 | 23.2 | 0.0 | 83.9 |
| Germany, DR | 105.1 | 1.6 | 11.0 | 0.0 | 117.7 |
| Poland | 131.6 | 3.7 | 0.0 | 0.0 | 135.3 |
| Romania | 59.2 | 12.5 | 0.0 | d | 71.7 |

*continued*

| | Thermal[a] | Hydro | Nuclear | Geothermal and other[b] | Total[c] |
|---|---|---|---|---|---|
| USSR | 1,204.2 | 220.7 | 212.6 | 0.0 | 1,637.5 |
| Other | 15.0 | 3.7 | 13.1 | 0.0 | 31.8 |
| | | | | | |
| *Middle East* | 177.1 | 12.4 | 0.0 | 0.0 | 189.5 |
| Iraq | 21.8 | 0.6 | 0.0 | 0.0 | 22.4 |
| Saudi Arabia | 46.1 | 0.0 | 0.0 | 0.0 | 46.1 |
| Other | 109.2 | 11.8 | 0.0 | 0.0 | 121.0 |
| | | | | | |
| *Africa* | 240.9 | 43.7 | 11.1 | 0.3 | 296.0 |
| Egypt | 28.6 | 6.3 | 0.0 | 0.0 | 34.9 |
| South Africa | 147.8 | 0.6 | 11.1 | 0.0 | 159.5 |
| Other | 64.5 | 36.8 | 0.0 | 0.3 | 101.6 |
| | | | | | |
| *Far East and Oceania* | 1,420.0 | 394.4 | 250.5 | 8.3 | 2,073.2 |
| Australia | 114.6 | 14.1 | 0.0 | 0.0 | 128.7 |
| China | 442.4 | 108.4 | 0.0 | 0.0 | 550.8 |
| India | 177.5 | 63.1 | 3.8 | [d] | 244.4 |
| Indonesia | 30.9 | 8.5 | 0.0 | 0.2 | 39.6 |
| Japan | 405.2 | 90.1 | 174.5 | 1.4 | 671.2 |
| Korea, DPR | 20.4 | 31.4 | 0.0 | 0.0 | 51.8 |
| Korea, Rep | 39.8 | 4.5 | 45.0 | 0.0 | 89.3 |
| Pakistan | 22.7 | 16.8 | 0.1 | 0.0 | 39.6 |
| Taiwan | 43.1 | 6.7 | 27.1 | 0.0 | 76.9 |
| Other | 123.3 | 50.8 | 0.0 | 6.7 | 180.8 |
| | | | | | |
| *World total* | 6,872.1 | 2,056.3 | 1,843.3 | 41.5 | 10,813.3 |

[a]Thermal generation consists of electricity generated from coal, oil, and gas.
[b]Geothermal and other consists of electricity generated from solar, biomass, wind, and other renewable sources.
[c]Sum of components may not equal total due to independent rounding.
[d]Denotes less than 500 million kWh.

**TABLE G.3** World net electricity generation, by type, 1989 *(pages 328–329)*

## DESCRIPTION

The generation of electricity by type is given here for major regions and countries of the world in thousand million ($10^9$) kWh per year for 1989. Energy types are thermal, hydroelectric, nuclear, and geothermal and other. Thermal includes all fossil fuels. All generation data include utility and non-utility sources, according to the original source, but see Limitations, below. Production is reported on a net basis, that is, on the amount of energy remaining after electricity use by the generating unit.

## LIMITATIONS

The sum of each section here may not equal the total due to independent rounding. The data were gathered by the U.S. Energy Information Agency from national sources. Data on "other" generating capacity appear to be less accurate than data for thermal, nuclear, and hydroelectric capacity. For example, the 1989 generation of electricity from "geothermal and other" sources listed for the United States is 11.3 × $10^9$ kWh. But other estimates of the actual 1989 generation of geother-

mal, wind, and solar plants in the U.S. are considerably higher than this – over 32 × $10^9$ kWh. California alone produced over 20 × $10^9$ kWh of electricity from geothermal, wind, and solar (Rader, 1990). The United States also produced substantial biomass and wood waste electricity as well (over 300 × $10^9$ kWh), not included in the "thermal" category above. These differences may come from inaccurate measurements of non-utility or non-government-owned facilities, since a substantial fraction of renewable energy development in the United States is from independent energy producers.

## SOURCES

United States Department of Energy, 1992, *International Energy Annual 1990*, Energy Information Administration, DOE/EIA-0219(90), Washington, DC.

N. Rader, 1990, *The Power of the States: A Fifty-State Survey of Renewable Energy*, Critical Mass Energy Project, Public Citizen, Washington, DC. By permission.

**TABLE G.4**   World electrical installed capacity, by type, January 1, 1990 ($10^6$ kW)

| | Thermal[a] | Hydro | Nuclear | Geothermal and other[b] | Total[d] |
|---|---|---|---|---|---|
| *North America* | 544.6 | 156.8 | 111.4 | 2.3 | 815.1 |
| Canada | 31.1 | 58.5 | 12.6 | 0.0 | 102.2 |
| Mexico | 18.8 | 7.8 | 0.7 | 0.7 | 28.0 |
| United States | 494.5 | 90.5 | 98.2 | 1.6 | 684.7 |
| Other | 0.2 | 0.0 | 0.0 | 0.0 | 0.2 |
| *Central and South America* | 43.7 | 80.3 | 1.7 | 4.8 | 130.5 |
| Argentina | 9.0 | 6.6 | 1.0 | 0.0 | 16.6 |
| Brazil | 2.1 | 44.6 | 0.7 | 4.7 | 52.1 |
| Colombia | 2.2 | 6.7 | 0.0 | 0.0 | 8.9 |
| Venezuela | 10.7 | 7.0 | 0.0 | 0.0 | 17.7 |
| Other | 19.7 | 15.4 | 0.0 | 0.1 | 35.3 |
| *Western Europe* | 272.6 | 155.0 | 116.3 | 2.5 | 546.4 |
| Austria | 4.9 | 10.3 | 0.0 | 0.0 | 15.2 |
| Belgium | 6.4 | 1.4 | 5.5 | c | 13.4 |
| Denmark | 8.4 | c | 0.0 | 0.3 | 8.7 |
| Finland | 6.1 | 2.6 | 2.3 | 0.0 | 11.0 |
| France | 16.2 | 22.9 | 52.5 | 0.2 | 91.8 |
| Germany, FR | 68.6 | 6.9 | 22.7 | 0.8 | 99.0 |
| Greece | 5.9 | 2.3 | 0.0 | c | 8.2 |
| Italy | 31.9 | 16.5 | 1.1 | 0.5 | 50.0 |
| Netherlands | 16.7 | c | 0.5 | c | 17.3 |
| Norway | 0.3 | 26.5 | 0.0 | 0.0 | 26.7 |
| Spain | 19.5 | 15.7 | 7.5 | 0.0 | 42.7 |
| Sweden | 6.6 | 15.1 | 9.8 | c | 31.5 |
| Switzerland | 0.8 | 11.6 | 2.9 | 0.0 | 15.3 |
| Turkey | 8.0 | 6.6 | 0.0 | c | 14.6 |
| United Kingdom | 56.6 | 4.1 | 10.7 | c | 71.4 |
| Yugoslavia | 8.2 | 7.0 | 0.6 | 0.6 | 16.5 |
| Other | 7.3 | 5.8 | 0.0 | c | 13.1 |
| *Eastern Europe and USSR* | 329.0 | 79.5 | 47.2 | c | 455.7 |
| Bulgaria | 6.4 | 2.0 | 2.8 | 0.0 | 11.1 |
| Czechoslovakia | 11.2 | 2.9 | 3.5 | 0.0 | 17.7 |
| Germany, DR | 19.7 | 1.8 | 1.6 | 0.0 | 23.4 |
| Poland | 30.0 | 2.0 | 0.0 | 0.0 | 32.0 |

*continued*

| | Thermal[a] | Hydro | Nuclear | Geothermal and other[b] | Total[d] |
|---|---|---|---|---|---|
| Romania | 17.3 | 5.6 | 0.0 | c | 22.9 |
| USSR | 239.2 | 64.4 | 37.4 | 0.0 | 341.0 |
| Other | 5.2 | 0.7 | 1.7 | 0.0 | 7.6 |
| | | | | | |
| *Middle East* | 58.4 | 3.1 | 0.0 | 0.0 | 61.5 |
| Saudi Arabia | 16.5 | 0.0 | 0.0 | 0.0 | 16.5 |
| Other | 41.9 | 3.1 | 0.0 | 0.0 | 45.0 |
| | | | | | |
| *Africa* | 50.5 | 18.9 | 1.9 | c | 71.4 |
| South Africa | 24.0 | 0.5 | 1.9 | 0.0 | 26.5 |
| Other | 26.5 | 18.3 | 0.0 | c | 44.9 |
| | | | | | |
| *Far East and Oceania* | 336.3 | 121.3 | 43.3 | 1.3 | 502.2 |
| Australia | 27.7 | 7.3 | 0.0 | 0.0 | 35.0 |
| China | 68.0 | 30.0 | 0.0 | 0.0 | 98.0 |
| Hong Kong | 7.5 | 0.0 | 0.0 | 0.0 | 7.5 |
| India | 49.8 | 18.5 | 1.6 | c | 69.9 |
| Indonesia | 9.1 | 1.8 | 0.0 | c | 11.0 |
| Japan | 115.6 | 37.3 | 28.9 | 0.2 | 181.9 |
| Korea, DPR | 4.5 | 5.0 | 0.0 | 0.0 | 9.5 |
| Korea, Rep | 13.6 | 2.3 | 7.6 | 0.0 | 23.5 |
| Pakistan | 5.4 | 2.9 | 0.1 | 0.0 | 8.5 |
| Thailand | 5.6 | 2.3 | 0.0 | 0.0 | 7.9 |
| Other | 29.5 | 13.8 | 5.1 | 1.0 | 49.6 |
| | | | | | |
| *World Total* | 1,635.0 | 614.9 | 321.8 | 11.0 | 2,582.7 |

[a]Thermal capacity consists of coal, oil, and gas.
[b]Geothermal and other capacity consists of solar, biomass, wind, and other renewable sources.
[c]Denotes less than 500 thousand kW.
[d]Sum of components may not equal total due to independent rounding.

**TABLE G.4** World electrical installed capacity, by type, January 1, 1990 *(pages 330–331)*

## DESCRIPTION

The size of electricity generation facilities by type is given here for major regions and countries of the world in kW $\times 10^6$ as of the beginning of 1990. Energy types are thermal, hydroelectric, nuclear, and geothermal and other. Thermal includes all fossil fuels. Capacity data include utility and non-utility sources, according to the original source, but see Limitations, below.

## LIMITATIONS

The sum of each section here may not equal the total due to independent rounding. The data were gathered by the U.S. Energy Information Agency from national sources. Data on "other" generating capacity appear to be less accurate than data for thermal, nuclear, and hydroelectric capacity. For example, the installed capacity of "geothermal and other" listed for the United States is $1.6 \times 10^6$ kW. But the actual installed capacity of geothermal, wind, and solar plants at the beginning of 1990 in the U.S. was considerably higher than this — over 4.3 $\times 10^6$ kW. California alone has $2.79 \times 10^6$ kW of geothermal energy capacity, $1.35 \times 10^6$ kW of wind energy capacity, and $0.3 \times 10^6$ kW of solar electric capacity (Rader, 1990). The United States also has substantial biomass and wood waste electric capacity as well ($7.8 \times 10^6$ kW), not included in the "thermal" category above. These differences may come from inaccurate measurements of non-utility or non-government-owned facilities, since a substantial fraction of renewable energy development in the United States is from independent energy producers.

## SOURCES

United States Department of Energy, 1992, *International Energy Annual 1990*, Energy Information Administration, DOE/EIA-0219(90), Washington, DC.

N. Rader, 1990, *The Power of the States: A Fifty-State Survey of Renewable Energy*, Critical Mass Energy Project, Public Citizen, Washington, DC. By permission.

**TABLE G.5**   The world's major dams and hydroelectric facilities *(pages 333–346)*

**DESCRIPTION**

The following table lists all major dams and hydroelectric plants by country as compiled by the United States Department of the Interior and updated by International Water Power and Dam Construction, United Kingdom. Included in this table are the site name, the year completed and the year initial operation began, the river basin, the dam type, the dam height above the lowest foundation (m), the crest length (m), the dam volume (m$^3$ × 10$^3$), the maximum reservoir capacity (m$^3$ × 10$^6$), the installed capacity (MW), and the planned capacity (MW).

**LIMITATIONS**

Not all dams are listed here. Some valuable data, such as surface area of the dam, hydroelectric head, and annual average energy generation are not provided. When major modifications to a dam are done, the date of initial operation may precede the date of completion by many years.

**SOURCES**

United States Department of the Interior, 1983, *Major Dams, Reservoirs, and Hydroelectric Plants: Worldwide and Bureau of Reclamation*, Bureau of Reclamation, Denver, Colorado.

International Water Power and Dam Construction, 1992, *Water Power and Dam Construction Handbook 1992*, Reed Enterprise, Sutton, UK. Reprinted with permission.

T.W. Mermel, 1992, personal communication.

**TABLE G.5** The world's major dams and hydroelectric facilities

| | Year completed | Year of initial operation | River or basin | Dam type | Height above lowest foundation (m) | Crest length (m) | Dam volume ($10^3$ m³) | Maximum reservoir capacity ($10^6$ m³) | Present rated capacity (MW) | Planned rated capacity (MW) |
|---|---|---|---|---|---|---|---|---|---|---|
| **Albania** | | | | | | | | | | |
| Fierze | 1978 | 1978 | Drin | TE/ER | 167 | 400 | 8,000 | 2,700 | 500 | 500 |
| **Argentina** | | | | | | | | | | |
| Alicura | 1984 | 1982 | Limay | TE | 130 | 880 | 13,000 | 13,215 | 1,000 | 1,000 |
| Casa de Piedra | 1990 | n/a | Colorado | TE | 56 | 10,150 | 16,500 | 4,000 | n/a | 60 |
| Cerros Colorados | 1972 | 1978 | Neuquén | TE/PG | 35 | 7,136 | 21,580 | 43,000 | 450 | 450 |
| El Chocon | 1973 | 1973 | Limay | TE | 74 | 2,270 | 12,200 | 20,200 | 1,200 | 1,200 |
| Loma de la Lata | 1977 | n/a | Neuquén | TE | 16 | 1,500 | 1,500 | 25,100 | n/a | n/a |
| Planicie Banderita | 1979 | n/a | Neuquén | TE | 35 | 350 | 1,194 | 43,000 | n/a | n/a |
| Salto Grande | 1979 | 1979 | Uruguay | TE/PG | 65 | 3,000 | 3,965 | 5,000 | 1,890 | 1,890 |
| **Australia** | | | | | | | | | | |
| Dartmouth | 1979 | | Mitta-Mitta | ER | 180 | 670 | 14,100 | 4,000 | | |
| Talbingo (Tumut 3) | 1971 | 1972 | Tumut | ER | 162 | 701 | 14,490 | 921 | 1,500 | 1,500 |
| Thomson | 1985 | | Thomson | ER | 164 | 1,180 | 13,300 | 1,175 | | |
| **Austria** | | | | | | | | | | |
| Finstertal | 1980 | 1981 | Nederbach | ER | 150 | 652 | 4,500 | 60 | 789 | 789 |
| Gepatsch | 1965 | 1965 | Faggenbach | ER | 153 | 600 | 7,100 | 140 | 392 | 392 |
| Kölnbrein | 1977 | 1978 | Malta | VA | 200 | 620 | 1,525 | 205 | 891 | 891 |
| Zillergründl | 1987 | 1987 | Ziller | VA | 186 | 506 | 1,355 | 90 | 705 | 705 |
| **Belgium** | | | | | | | | | | |
| Coo-Trois Ponts | | | | | | | | | | |
| (Coo I) | 1972 | 1971 | Amblève | TE | 25 | 1,750 | 2,000 | 4 | 450 | 450 |
| (Coo II) | 1980 | 1980 | Amblève | TE | 50 | 1,800 | 2,100 | 4.54 | 621 | 621 |

*continued*

## Brazil

| | Year completed | Year of initial operation | River or basin | Dam type | Height above lowest foundation (m) | Crest length (m) | Dam volume ($10^3$ m$^3$) | Maximum reservoir capacity ($10^6$ m$^3$) | Present rated capacity (MW) | Planned rated capacity (MW) |
|---|---|---|---|---|---|---|---|---|---|---|
| Agua Vermelha | 1979 | 1979 | Grande | TE/PG | 67 | 3,920 | 21,123 | 11,100 | 1,380 | 1,380 |
| Emborcação | 1982 | 1982 | Paranaíba | TE/ER | 156 | 1,507 | 23,644 | 17,500 | 1,000 | 1,000 |
| Estreito | 1969 | 1969 | Grande | ER/PG | 92 | 715 | 4,446 | 1,418 | 1,040 | 1,040 |
| Foz Do Areia | 1980 | 1980 | Iguaçu | TE | 160 | 828 | 14,000 | 6,066 | 1,674 | 2,511 |
| Furnas | 1963 | 1963 | Grande | ER/PG | 127 | 779 | 9,697 | 22,950 | 1,216 | 1,216 |
| Ilha Solteira | 1973 | 1973 | Paraná | TE/PG/ER | 74 | 6,185 | 26,276 | 21,166 | 3,200 | 3,200 |
| Itaipú | 1982 | 1983 | Paraná | TE/PG/ER | 196 | 7,297 | 33,260 | 29,000 | 12,600 | 12,600 |
| Itaparica | 1988 | 1988 | São Francisco | TE/ER | 105 | 4,700 | 13,230 | 10,800 | 1,500 | 2,500 |
| Itumbiara | 1980 | 1980 | Paranaíba | TE/PG | 106 | 6,780 | 37,800 | 17,000 | 2,080 | 2,080 |
| Jupía | 1968 | 1968 | Paraná | TE/PG/ER | 43 | 5,604 | 6,090 | 3,680 | 1,400 | 1,400 |
| Marimbondo | 1975 | 1975 | Grande | TE/PG | 90 | 3,600 | 15,600 | 6,150 | 1,440 | 1,440 |
| Paulo Afonso I | 1955 | 1955 | São Francisco | PG | 19 | 4,125 | 216 | 33 | 1,524 | 3,409 |
| Paulo Afonso IV | 1979 | 1979 | São Francisco | ER/PG | 33 | 7,350 | 6,979 | 128 | 2,460 | 2,460 |
| Salto Osório | 1975 | 1976 | Iguaçu | ER/PG | 67 | 1,200 | 4,280 | 1,240 | 700 | 1,050 |
| Salto Santiago | 1980 | 1980 | Iguaçu | TE/ER | 39 | 1,400 | 10,380 | 6,750 | 2,000 | 2,000 |
| São Simão | 1978 | 1979 | Paranaíba | ER/PG | 127 | 3,600 | 27,479 | 12,500 | 1,613 | 2,680 |
| Sobradinho | 1978 | 1979 | São Francisco | TE/PG/ER | 41 | 8,207 | 18,200 | 34,200 | 1,050 | 1,050 |
| Trés Irmãos | 1990 | 1986 | Tiete | TE/PG | 62 | 3,640 | 15,000 | 13,800 | n/a | 1,292 |
| Tucuruí (Raul G. Lhano) | 1984 | 1983 | Toçantins | TE/ER/PG | 106 | 9,574 | 37,000 | 45,800 | 4,000 | 7,960 |

## Canada

| | Year completed | Year of initial operation | River or basin | Dam type | Height above lowest foundation (m) | Crest length (m) | Dam volume ($10^3$ m$^3$) | Maximum reservoir capacity ($10^6$ m$^3$) | Present rated capacity (MW) | Planned rated capacity (MW) |
|---|---|---|---|---|---|---|---|---|---|---|
| Beauharnois | 1960 | 1966 | St. Lawrence | PG | n/a | n/a | n/a | n/a | 1,574 | 1,574 |
| Bennett, W.A.C. | 1967 | 1968 | Peace | TE | 183 | 2,042 | 43,733 | 70,309 | 2,416 | 2,416 |
| Bersimis No. 1 | 1959 | 1956 | Bersimis | ER | 84 | 643 | 779 | 11,595 | 912 | 1,050 |
| Big Horn | 1972 | | N. Saskatchewan | TE | 150 | 472 | 4,300 | 1,768 | | |
| Caniapiscau Reservoir | 1981 | 1981 | Caniapiscau | | | | | 53,800 | | |
| Churchill Falls | 1971 | 1971 | Churchill | TE | 32 | 5,506 | 1,658 | 32,640 | 5,428.5 | 5,428.5 |

| | Year completed | Year of initial operation | River or basin | Dam type | Height above lowest foundation (m) | Crest length (m) | Dam volume (10³ m³) | Maximum reservoir capacity (10⁶ m³) | Present rated capacity (MW) | Planned rated capacity (MW) |
|---|---|---|---|---|---|---|---|---|---|---|
| Daniel Johnson (Manicouagan 5) | 1968 | 1968 | Manicouagan | MV | 214 | 1,314 | 2,255 | 141,852 | 1,292 | 1,292 |
| Gardiner | 1968 | n/a | S. Saskatchewan | TE | 69 | 5,090 | 65,400 | 9,868 | n/a | n/a |
| Iriquois | 1958 | 1959 | St. Lawrence | VA/PG | 20 | 603 | 134 | 29,960 | 1,880 | 1,880 |
| Jenpeg/Kiskitto | 1975 | | Nelson | | | | | 31,790 | | 1,420 |
| Kemano (Nechako Reservoir) | 1952 | 1952 | Nechako | ER | 104 | 457 | 3,071 | 32,700 | 880 | |
| Kettle Rapids | 1970 | 1970 | Nelson | TE/PG/ER | 61 | 1,704 | 3,448 | 2,529 | 1,272 | 1,272 |
| La Grande 2 | 1978 | 1979 | La Grande | ER | 168 | 2,826 | 23,192 | 61,715 | 5,328 | 5,328 |
| La Grande 3 | 1981 | 1982 | La Grande | ER | 93 | 3,845 | 22,100 | 60,020 | 2,304 | 2,304 |
| La Grande 4 | 1984 | 1984 | La Grande | ER | 128 | 3,750 | 193,000 | 19,400 | 2,650 | 2,650 |
| Long Spruce | 1976 | 1976 | Nelson | TE/PG/ER | 42 | 1,600 | 2,940 | 277 | 1,020 | 1,020 |
| Manicouagan 2 | 1964 | 1965 | Manicouagan | PG | 91 | 692 | 745 | 4,248 | 1,183 | 1,183 |
| Manicouagan 3 | 1975 | 1975 | Manicouagan | TE | 108 | 366 | 9,175 | 10,423 | 1,183 | 1,183 |
| Manicouagan 5PA | 1989 | 1989 | Manicouagan | MV | 214 | 1,314 | 2,255 | 141,852 | 540 | 1,080 |
| Mica | 1973 | 1976 | Columbia | TE/ER | 242 | 792 | 32,111 | 24,700 | 1,736 | 2,610 |
| Missi Falls control | 1976 | | Churchill | n/a | 18 | 2,130 | 190 | 28,370 | | |
| Revelstoke | 1984 | 1984 | Columbia | TE/PG | 175 | 1,620 | 1,600 | 5,180 | 1,843 | 2,700 |
| Saunders/Moses (Niagara) | 1958 | 1958 | St. Lawrence | PG | 47 | 475 | 826 | 808 | 1,824 | 1,824 |
| Sir Adam Beck - Niagara 2 | 1954 | 1954 | Niagara | PG | 17 | 152 | 23 | 50 | 1,328 | 1,328 |
| **Chile** | | | | | | | | | | |
| Los Leones (copper tailings) | 1986 | | Los Leones | TE | 150 | 455 | 7,560 | 106 | 0 | 0 |
| **China** | | | | | | | | | | |
| Baishan | 1984 | 1983 | Songhuajiang | VA/PG | 150 | 664 | 1,630 | 6,510 | 900 | 1,500 |
| Dongjiang | 1989 | 1987 | Laishui | VA | 157 | 438 | 943 | 8,120 | | 977 |
| Gezhouba | 1991 | 1982 | Yangzijiang | PG | 47 | 2,561 | 5,530 | 1,580 | 2,715 | 2,715 |
| Hepu | 1960 | n/a | Xiaojinghe | TE | 40.6 | 832 | 40,090 | 992 | 7.2 | n/a |

continued

| | Year completed | Year of initial operation | River or basin | Dam type | Height above lowest foundation (m) | Crest length (m) | Dam volume (10³ m³) | Maximum reservoir capacity (10⁶ m³) | Present rated capacity (MW) | Planned rated capacity (MW) |
|---|---|---|---|---|---|---|---|---|---|---|
| Liujiaxia | 1968 | 1969 | Huanghe | PG | 147 | 840 | 760 | 6,090 | 1,225 | 1,225 |
| Longyangxia (Longyang Gorge) | 1988 | 1987 | Huanghe | VA | 172 | 342 | 1,750 | 24,700 | 1,280 | 1,280 |
| Sanmenxia | 1979 | 1973 | Huanghe | PG | 106 | 713 | 1,630 | 10,310 | 250 | |
| Tianshengqiao Low (No. 2) | 1990 | 1991 | Nanpanjiang (Hongshui) | PG | 58 | n/a | 1,820 | n/a | n/a | 880/1320 |
| Wangkuai | 1960 | n/a | Haihe | TE | 62 | 1,281 | 17,142 | 1,389 | 43 | n/a |
| Wujiangdu | 1981 | 1981 | Wujiang | VA | 165 | 368 | 1,930 | 2,300 | 630 | 630 |
| Yuecheng | 1970 | n/a | Xhanghe | TE | 53 | 3,570 | 28,000 | 1,090 | 17 | n/a |
| Zhanghe | 1965 | n/a | Juzhanghe | TE | 66.5 | 630 | 46,232 | 2,035 | 52.5 | n/a |
| **China (Taiwan)** | | | | | | | | | | |
| Minghu | 1985 | 1985 | Shuili Creek | PG | 57.5 | 169.5 | | | 1,120 | 1,120 |
| Tehchi (Tachie) | 1974 | | Tachia | VA | 180 | 290 | 456 | 247 | | |
| **Colombia** | | | | | | | | | | |
| Chivor | 1975 | 1976 | Batá | TE/ER | 237 | 310 | 11,174 | 815 | 1,000 | 1,000 |
| Guavio | 1989 | 1992 | Guavio | TE/ER | 246 | 390 | 17,755 | 1,020 | 1,600 | 1,600 |
| Salvajina | 1985 | n/a | Cauca | ER | 160 | 360 | 3,500 | 904 | n/a | n/a |
| San Carlos (Punchina) | 1986 | 1986 | Guapape-San Carlos | TE | 70 | 800 | 6,000 | 72 | 1,280 | 1,600 |
| **Côte d'Ivoire** | | | | | | | | | | |
| Kossou | 1972 | n/a | Bandama | ER | 58 | 1,800 | 5,200 | 27,675 | n/a | n/a |
| **Ecuador** | | | | | | | | | | |
| Amaluza (Daniel Palacios) | 1983 | 1981 | Paute | VA/PG | 170 | 400 | 1,200 | 120 | 1,030 | 1,780 |
| **Egypt** | | | | | | | | | | |
| Aswan High | 1970 | 1967 | Nile | TE/ER | 111 | 3,830 | 44,300 | 168,900 | 1,815 | 2,100 |

*continued*

| | Year completed | Year of initial operation | River or basin | Dam type | Height above lowest foundation (m) | Crest length (m) | Dam volume ($10^3$ m³) | Maximum reservoir capacity ($10^6$ m³) | Present rated capacity (MW) | Planned rated capacity (MW) |
|---|---|---|---|---|---|---|---|---|---|---|
| France | | | | | | | | | | |
| Grand' Maison | 1985 | 1981 | Eau d'Olle | TE/ER | 160 | 550 | 12,500 | 140 | 1,800 | 1,800 |
| Mont-Cenis | 1968 | 1968 | Cenise | TE/ER | 120 | 1,400 | 15,000 | 332 | 360 | 360 |
| Monteynard | 1962 | 1962 | Drac | VA | 155 | 215 | 455 | 240 | 320 | 320 |
| Roselend | 1961 | 1960 | Doron de Beaufort | VA/CB | 150 | 806 | 945 | 187 | 520 | 520 |
| Tignes | 1952 | 1952 | Isère | VA | 180 | 375 | 635 | 230 | 380 | 380 |
| Germany | | | | | | | | | | |
| Markersbach | 1981 | 1979 | Grosse/Kleine Mittweida | TE/ER | 26/55 | 2,600/390 | 3,100/850 | 6.3/7.7 | 1,050 | 1,050 |
| Ghana | | | | | | | | | | |
| Akosombo | 1965 | 1966 | Volta | ER | 134 | 671 | 7,991 | 148,000 | 912 | 1,824 |
| Greece | | | | | | | | | | |
| Kremasta (King Paul) | 1966 | | Achelōos | TE | 160 | 460 | 8,170 | 4,750 | 437 | |
| Mornos | 1976 | | Mornos | TE | 126 | 815 | 17,000 | 780 | | |
| Honduras | | | | | | | | | | |
| El Cajon | 1985 | 1985 | Humuya | VA | 234 | 382 | 1,600 | 6,500 | 300 | 600 |
| India | | | | | | | | | | |
| Balimela | 1977 | | Sileru | TE | 75 | 4,363 | 19,096 | 3,610 | | |
| Bhakra | 1963 | 1963 | Sutlej | PG | 226 | 518 | 4,130 | 9,621 | 450 | 1,050 |
| Dantiwada | 1965 | n/a | Banas | TE/PG | 55 | 3,575 | 41,040 | 464 | n/a | n/a |
| Hirakud | 1957 | 1956 | Mahanadi | TE/PG | 59 | 4,800 | 25,100 | 8,105 | 270 | 308 |
| Idukki | 1974 | n/a | Periyar | VA | 169 | 366 | 460 | 1,996 | n/a | n/a |
| Pong (Beas) | 1974 | | Beas-Indus | TE | 133 | 1,950 | 35,500 | 8,570 | | |
| Ukai | 1972 | | Tapti | TE/PG | 81 | 5,056 | 25,180 | 8,511 | | |

*continued*

| | Year completed | Year of initial operation | River or basin | Dam type | Height above lowest foundation (m) | Crest length (m) | Dam volume ($10^3$ m³) | Maximum reservoir capacity ($10^6$ m³) | Present rated capacity (MW) | Planned rated capacity (MW) |
|---|---|---|---|---|---|---|---|---|---|---|
| **Iran** | | | | | | | | | | |
| Amir Kabir (Karadj) | 1961 | | Karadj | VA | 180 | 390 | 750 | 205 | | |
| Dez | 1962 | | Dez | VA | 207 | 212 | 459 | 3,340 | | |
| Karun 1 (Shahid Abbaspour) | 1972 | 1977 | Karun | VA | 200 | 380 | 1,570 | 3,300 | 1,000 | 2,000 |
| **Iraq** | | | | | | | | | | |
| Mosul | 1983 | n/a | Tigris | TE | 131 | 3,500 | 23,000 | 12,500 | n/a | n/a |
| Razza Dyke | 1970 | | Offstream | TE | 18 | 3,770 | 1,000 | 26,000 | | |
| **Italy** | | | | | | | | | | |
| Alpe-Gera | 1964 | | Cormor | PG | 174 | 528 | 1,700 | 68 | | |
| Ghiotas (Entrácque), Alto Gesso | 1981 | 1982 | Gesso | VA/PG | | | | | 1,183 | 1,183 |
| Edolo | 1983 | 1983 | Oglio | | | | | | 1,021 | 1,021 |
| Lago Delio (Roncovalgrande) (Lago Delio Nord/Sud) | 1971 | 1971 | Casmera | PG | 28.6/36 | 416/158 | 69/26 | 11/11 | 1,040 | 1,040 |
| Place Moulin | 1965 | n/a | Buthier | VA | 155 | 678 | 1,510 | 106 | n/a | n/a |
| Presenzano | 1990 | 1990 | Volturno | | | | | | | 1,000 |
| Santa Giustina | 1950 | | Noce-Adige | VA | 153 | 124 | 112 | 183 | | |
| Spéccheri | 1957 | | Leno di Vallarsa | VA | 157 | 192 | 117 | 10 | | |
| Vaiont[a] | 1961 | | Vaiont | VA | 262 | 190 | 351 | 169 | | |
| **Japan** | | | | | | | | | | |
| Doyo (Matanogawa) | 1986 | 1986 | Doyo | ER | 87 | 480 | 2,650 | 7 | 600 | 1,200 |
| Kassa (Okukiyotsu) | 1978 | 1978 | Kassa | ER | 90 | 487 | 4,450 | 13 | 1,000 | 1,000 |
| Kriyama (Imaichi) | 1978 | 1988 | Nebezawa | ER | 98 | 340 | 2,517 | 7 | 350 | 1,050 |
| Kurobe (No. 4) | 1964 | 1961 | Kurobe | VA | 186 | 489 | 1,582 | 199 | 335 | 335 |
| Kurokawa (Okutataragi) | 1974 | 1975 | Ichi | ER | 98 | 325 | 3,623 | 33 | 1,212 | 1,212 |
| Nagawado | 1969 | 1969 | Sai | VA | 155 | 356 | 660 | 123 | 623 | 623 |

*continued*

| | Year completed | Year of initial operation | River or basin | Dam type | Height above lowest foundation (m) | Crest length (m) | Dam volume ($10^3$ m³) | Maximum reservoir capacity ($10^6$ m³) | Present rated capacity (MW) | Planned rated capacity (MW) |
|---|---|---|---|---|---|---|---|---|---|---|
| Naramata | 1991 | 1991 | Naramata | ER | 158 | 520 | 13,100 | 90 | 12 | 12 |
| Okutadami | 1961 | 1960 | Tadami | PG | 157 | 480 | 1,636 | 601 | 360 | 360 |
| Ouchi (Shimogo) | 1985 | 1981 | Ono | ER | 102 | 340 | 4,459 | 19 | 1,000 | 1,000 |
| Sakuma | 1956 | 1956 | Tenryu | PG | 156 | 294 | 1,120 | 327 | 350 | 350 |
| Seto (Okuyoshino) | 1978 | 1978 | Setodani | ER | 111 | 343 | 3,740 | 17 | 1,206 | 1,206 |
| Shintoyone | 1973 | 1973 | Onyu | VA | 117 | 311 | 348 | 54 | 1,125 | 1,125 |
| Takase (Shin Takasegawa) | 1978 | 1979 | Takase | ER | 176 | 362 | 11,600 | 76 | 1280 | 1280 |
| Tamahara | 1982 | 1982 | Hotchi | ER | 116 | 570 | 5,435 | 15 | 1,200 | 1,200 |
| Tedorigawa | 1979 | 1979 | Tedori | ER | 154 | 420 | 10,050 | 231 | 250 | 250 |
| Kenya | | | | | | | | | | |
| Turkwel | 1991 | 1991 | Suam | VA | 153 | 150 | 170 | 1,641 | 107.4 | 107.4 |
| Luxembourg | | | | | | | | | | |
| Vianden | 1964 | 1964 | Our | n/a | n/a | n/a | n/a | n/a | 1,160 | 1,160 |
| Malaysia | | | | | | | | | | |
| Kenyir | 1985 | 1985 | Trengganu | ER | 155 | 800 | 16,910 | 13,600 | 400 | 400 |
| Mexico | | | | | | | | | | |
| Angostura | 1974 | 1976 | Grijalva | TE/ER | 146 | 323 | 4,030 | 9,200 | 1,100 | 1,100 |
| Chicoasén (Manuel Moreno Torres) | 1980 | 1980 | Grijalva | TE/ER | 261 | 485 | 15,370 | 1,613 | 2,400 | 2,400 |
| Malpaso | 1964 | 1968 | Grijalva | ER | 138 | 478 | 5,100 | 1,300 | 1,080 | 1,080 |
| Mozambique | | | | | | | | | | |
| Cabora Bassa | 1974 | 1975 | Zambezi | VA | 171 | 321 | 510 | 63,000 | 2,425 | 4,150 |
| Netherlands | | | | | | | | | | |
| Afsluitdijk | 1932 | | Zuiderzee | TE | 19 | 32,000 | 63,430 | 6,000 | | |

continued

| | Year completed | Year of initial operation | River or basin | Dam type | Height above lowest foundation (m) | Crest length (m) | Dam volume (10³ m³) | Maximum reservoir capacity (10⁶ m³) | Present rated capacity (MW) | Planned rated capacity (MW) |
|---|---|---|---|---|---|---|---|---|---|---|
| Brouwershavense Gat | 1972 | | Brouwershavense Gat | TE | 36 | 6,200 | 27,000 | 575 | | |
| Haringvliet | 1970 | | Haringvliet | TE | 24 | 5,500 | 20,000 | 650 | | |
| Lauwerszee | 1969 | | Lauwerszee | TE | 23 | 13,000 | 35,575 | 50 | | |
| Oosterschelde | 1986 | | Veerse Gat Oosterschelde | TE/PG | 50 | 9,000 | 50,000 | 2,780 | | |
| **Nigeria** | | | | | | | | | | |
| Kainji | 1968 | 1967 | Niger | TE/PG | 66 | 7,750 | 7,510 | 15,000 | 760 | 1,000 |
| Lower Usuma | 1990 | n/a | Usuma | TE | 49 | 1,350 | 93,000 | 100 | n/a | n/a |
| **Norway** | | | | | | | | | | |
| Kvilldal | 1981 | 1981 | Ulladalsana | | | | | 3,100 | 1,240 | 1,240 |
| Sima | 1980 | 1980 | Bioreira | ER | n/a | n/a | n/a | n/a | 1,120 | 1,120 |
| **Pakistan** | | | | | | | | | | |
| Jari | 1967 | | Jari | TE | 71 | 1,738 | 32,400 | 494 | | |
| Mangla | 1967 | 1967 | Jhelum | TE | 138 | 3,139 | 65,379 | 7,252 | 800 | 1,000 |
| Tarbela | 1976 | 1977 | Indus | TE/ER | 143 | 2,743 | 106,000 | 13,690 | 1,750 | 3,478 |
| **Paraguay** | | | | | | | | | | |
| Itaipú | 1982 | 1983 | Paraná | TE/PG/ER | 196 | 7,297 | 33,260 | 29,000 | 12,600 | 12,600 |
| **Peru** | | | | | | | | | | |
| Gallito Ciego | 1988 | | Jequetepeque | TE | 145 | 750 | 15,000 | 574 | | |
| Mantaro | 1973 | 1975 | Mantaro | PG | 82 | 180 | 157 | 16 | 1,015 | 1,015 |
| Poechos | 1975 | | Chira | TE | 49 | 10,800 | 18,480 | 880 | | |
| **Philippines** | | | | | | | | | | |
| Magat | 1982 | n/a | Magat | TE/ER | 114 | 4,169 | 17,900 | 1,080 | n/a | n/a |

continued

| | Year completed | Year of initial operation | River or basin | Dam type | Height above lowest foundation (m) | Crest length (m) | Dam volume ($10^3$ m$^3$) | Maximum reservoir capacity ($10^6$ m$^3$) | Present rated capacity (MW) | Planned rated capacity (MW) |
|---|---|---|---|---|---|---|---|---|---|---|
| **Romania** | | | | | | | | | | |
| Iron Gates I | 1972 | 1971–72 | Danube | PG | 60 | 1,278 | 3,160 | 2,400 | 1,050 | 1,050 |
| Vidraru | 1965 | 1966 | Arges | VA | 166 | 305 | 480 | 465 | 220 | 220 |
| **South Africa** | | | | | | | | | | |
| Drakensberg | 1982 | 1983 | Mnjaneni (Tugela) | TE/ER | 51/46.6 | 825/500 | 2,900/843 | 36/36.6 | 1,000 | 1,000 |
| Sterkfontein | 1986 | | Nuwejaarspruit (Vaal) | TE | 93 | 3,105 | 18,500 | 2,617 | | |
| **Spain** | | | | | | | | | | |
| Aldeadavila | 1963 | 1963 | Duero | VA/PG | 140 | 250 | 848 | 115 | 1,138 | 1,558 |
| Almendra | 1970 | 1970 | Tormes-Duero | VA | 202 | 567 | 2,186 | 2,649 | 810 | 810 |
| Canales | 1988 | | Genil | ER | 158 | 340 | 7,248 | 71 | | |
| Canelles | 1960 | 1961 | Noguera/Ribagorzana | VA | 150 | 210 | 380 | 678 | 107 | n/a |
| **Switzerland** | | | | | | | | | | |
| Contra | 1965 | 1965 | Verzasca | VA | 220 | 380 | 660 | 105 | 133 | 133 |
| Curnera | 1966 | 1966 | Rein de Curnera | VA | 153 | 350 | 562 | 41 | 330 | 330 |
| Emosson | 1974 | 1975 | Barberine | VA | 180 | 555 | 1,090 | 227 | 620 | 680 |
| Göscheneralp | 1960 | 1960 | Goschenerreuss | TE | 155 | 540 | 9,300 | 76 | 165 | 165 |
| Grande Dixence | 1961 | 1962 | Dixence | PG | 285 | 695 | 6,000 | 401 | 869 | 1,700 |
| Luzzone | 1963 | 1963 | Brenno de Luzzone | VA | 208 | 530 | 1,330 | 88 | 418 | n/a |
| Mauvoisin[b] | 1957–90 | 1957 | Dranse de Bagnes | VA | 250.5 | 520 | 2,030 | 181 | 397 | n/a |
| Zervreila | 1957 | 1957 | Valserrhein | VA | 151 | 504 | 626 | 100 | 237 | 237 |
| Zeuzier | 1957 | 1957 | Lienne | VA | 156 | 256 | 300 | 51 | 92 | 92 |
| **Syria** | | | | | | | | | | |
| Tabqa (Thawra) | 1976 | | Euphrates | TE | 60 | 4,500 | 46,000 | 14,000 | | |

continued

| | Year completed | Year of initial operation | River or basin | Dam type | Height above lowest foundation (m) | Crest length (m) | Dam volume (10³ m³) | Maximum reservoir capacity (10⁶ m³) | Present rated capacity (MW) | Planned rated capacity (MW) |
|---|---|---|---|---|---|---|---|---|---|---|
| **Thailand** | | | | | | | | | | |
| Bhumibol (Yanhee) | 1964 | 1964 | Ping-Chao Phraya | VA/PG | 154 | 486 | 970 | 13,462 | 535 | 710 |
| **Turkey** | | | | | | | | | | |
| Altinkaya | 1988 | 1987 | Kizilirmak | TE/ER | 195 | 634 | 16,000 | 5,763 | 700 | 700 |
| Atatürk | 1990 | 1991 | Euphrates | ER | 184 | 1,820 | 84,500 | 48,700 | | 2,400 |
| Gökçekaya | 1972 | n/a | Sakarya | VA | 158 | 466 | 650 | 910 | 278 | 278 |
| Hasan Ugurlu (Ayvacik) | 1981 | n/a | Yesilirmak | TE/ER | 175 | 405 | 9,223 | 1,080 | 500 | 500 |
| Karakaya | 1987 | 1987 | Euphrates | VA/PG | 173 | 462 | 2,000 | 9,580 | 1,800 | 1,800 |
| Keban | 1974 | 1974 | Euphrates | PG/ER | 207 | 1,126 | 15,585 | 30,600 | 1,330 | 1,330 |
| Menzelet | 1989 | 1989 | Ceyhan | ER | 157 | 420 | 8,700 | 1,950 | 124 | 124 |
| Oymapinar | 1984 | 1984 | Manavgat | VA | 185 | 360 | 676 | 300 | 540 | 540 |
| **Uganda** | | | | | | | | | | |
| Owen Falls | 1954 | 1954 | Lake Victoria/Nile | PG | 31 | 831 | n/a | 2,700,000 | 150 | 180 |
| **United Kingdom** | | | | | | | | | | |
| Dinorwig | 1982 | 1982 | Afon Nant Peris | ER | 70 | 600 | 1,500 | 7,000 | 1,800 | 1,800 |
| **United States** | | | | | | | | | | |
| Bad Creek | 1991 | 1991 | Bad Creek | ER | 108 | 1,314 | n/a | 41 | | 1,000 |
| Bagdad Tailings | 1973 | | Maroney Gulch | TE | 37 | 2,012 | 28,520 | 49 | | |
| Bath County | 1985 | 1985 | Back Creek | TE/ER | 146 | 670 | 17,400 | 44 | 2,100 | 2,100 |
| Blenheim-Gilboa | 1974 | 1974 | Schoharie Creek | TE | 34 | 3,690 | 3,980 | 21 | 1,000 | 1,000 |
| Bonneville | 1981 | 1938 | Columbia | PG | 60 | 820 | 893 | 662 | 1,068 | 1,068 |
| Castaic | 1973 | 1973 | Castaic Creek | TE | 125 | 1,585 | 33,640 | 431 | 1,275 | 1,275 |
| Chief Joseph | 1955 | 1955 | Columbia | PG | 70 | 1,310 | 1,381 | 731 | 2,069 | 2,069 |
| Cochiti | 1975 | | Grande | TE | 77 | 8,785 | 50,230 | 53 | | |

continued

| | Year completed | Year of initial operation | River or basin | Dam type | Height above lowest foundation (m) | Crest length (m) | Dam volume (10³ m³) | Maximum reservoir capacity (10⁶ m³) | Present rated capacity (MW) | Planned rated capacity (MW) |
|---|---|---|---|---|---|---|---|---|---|---|
| Copper Cities Tailings 2 | 1973 | | Tinhorn Wash | TE | 99 | 2,316 | 22,938 | 5 | | |
| Cougar | 1964 | 1964 | S.F. McKenzie | ER | 158 | 488 | 9,939 | 270 | 25 | 65 |
| Davis, Upper | 1950 | 1982 | Colorado | ER | 27 | 2,377 | 2,017 | 37 | 1,200 | 1,800 |
| Don Pedro | 1971 | 1971 | Tuolume | TE/ER | 173 | 579 | 12,806 | 2,504 | 157 | 157 |
| Dworshak | 1973 | 1973 | Clearwater | PG | 219 | 1,002 | 4,931 | 4,259 | 800 | 1,060 |
| Earthquake Lake | 1959 | n/a | Madison | ER | 61 | 366 | 38,228 | 73.3 | n/a | n/a |
| Esperanza Tailings | 1973 | 1973 | Santa Cruz | TE | 37 | 3,231 | 30,355 | 6 | | |
| Flaming Gorge | 1964 | n/a | Green | VA | 153 | 392 | 755 | 4,674 | n/a | n/a |
| Fort Peck | 1937 | 1943 | Missouri | TE | 76 | 6,409 | 96,050 | 23,042 | 185 | 185 |
| Fort Randall | 1952 | 1954 | Missouri | TE | 50 | 3,261 | 38,200 | 6,873 | 320 | 320 |
| Garrison | 1956 | 1956 | Missouri | TE | 62 | 3,658 | 50,845 | 30,097 | 518 | 518 |
| Glen Canyon | 1966 | 1964 | Colorado | VA | 216 | 475 | 3,747 | 33,304 | 1,288 | 1,320 |
| Grand Coulee | 1942 | 1942 | Columbia | PG | 168 | 1,272 | 8,093 | 11,795 | 6,494 | n/a |
| Helms (Courtright) | 1958 | 1983 | Kings | ER | 96 | 263 | 1,193 | 152 | 1,053 | 1,053 |
| Hoover | 1936 | 1936 | Colorado | VA/PG | 221 | 379 | 3,364 | 34,852 | 1,951 | 2,451 |
| Hungry Horse | 1953 | 1953 | S. Fork Flathead | VA/PG | 172 | 645 | 2,359 | 4,278 | 285 | 285 |
| Iroquois | 1958 | n/a | St. Lawrence | PG | 20 | 603 | 134 | 29,960 | n/a | n/a |
| John Day | 1968 | 1969 | Columbia | TE/PG | 70 | 1,798 | 2,164 | 3,256 | 2,160 | 2,700 |
| Kingsley | 1942 | n/a | N. Platte | TE | 52 | 5,420 | 24,450 | 2,467 | n/a | n/a |
| Ludington | 1973 | 1973 | Lake Michigan | TE | 52 | 8,931 | 28,825 | 102 | 1,979 | 1,979 |
| McNary | 1953 | 1954 | Columbia | TE/PG | 67 | 2,244 | 3,668 | 1,100 | 980 | 2,030 |
| Mission Tailings No. 1 | 1973 | | Santa Cruz | TE | 39 | n/a | 40,088 | 57 | | |
| Mossyrock | 1968 | 1968 | Cowlitz | VA | 185 | 502 | 971 | 1,636 | 300 | 300 |
| Navajo | 1963 | | San Juan | TE | 123 | 1,112 | 20,521 | 2,108 | | |
| New Bullards Bar | 1970 | 1970 | N. Yuba | VA | 194 | 677 | 1,988 | 1,196 | 310 | 310 |
| New Cornelia Tailings | 1973 | | Ten Mile Wash | TE | 30 | 10,851 | 209,500 | 25 | | |
| New Melones | 1979 | 1985 | Stanislaus | ER | 191 | 503 | 12,210 | 2,985 | 300 | 300 |
| Northfield | 1973 | 1972 | Briggs | ER | 44 | 3,414 | 306 | 21 | 1,000 | 1,000 |

continued

| | Year completed | Year of initial operation | River or basin | Dam type | Height above lowest foundation (m) | Crest length (m) | Dam volume (10³ m³) | Maximum reservoir capacity (10⁶ m³) | Present rated capacity (MW) | Planned rated capacity (MW) |
|---|---|---|---|---|---|---|---|---|---|---|
| Oahe | 1958 | 1962 | Missouri | TE | 75 | 2,835 | 66,517 | 28,776 | 786 | 786 |
| Oroville | 1968 | 1968 | Feather | TE | 230 | 2,109 | 61,164 | 4,364 | 679 | 679 |
| Priest Rapids | 1959 | 1959 | Columbia | TE | 55 | 2,300 | 2,520 | 180 | 789 | 1,262 |
| Raccoon Mountain | 1979 | 1979 | Tennessee (L. Nickajack) | ER | 37 | 975 | n/a | n/a | 1,530 | 1,530 |
| Ray Roberts (Aubrey) | 1986 | | Trinity (Elm Fork) | TE | 43 | 4,561 | 15,475 | 986 | | |
| Rockhouse Branch Tailings | 1976 | | Rockhouse branch | TE | 152 | 549 | 8,411 | 16 | | |
| Ross | 1949 | n/a | Skagit | VA | 165 | 396 | 703 | 1,770 | n/a | n/a |
| San Luis | 1967 | 1967 | San Luis Creek | TE | 116 | 5,669 | 59,375 | 2,518 | | |
| Saunders/Moses (Niagara) | 1958 | 1958 | St. Lawrence | PG | 47 | 475 | 826 | 808 | 2,028 | 2,418 |
| Shasta | 1945 | 1945 | Sacramento | VA/PG | 183 | 1,055 | 6,445 | 5,615 | 539 | 545 |
| Swift | 1958 | | Lewis | TE | 186 | 640 | 11,774 | 932 | | |
| Trinity | 1962 | | Trinity | TE | 164 | 747 | 22,478 | 3,019 | | |
| Tuttle Creek | 1962 | | Big Blue | TE | 47 | 2,286 | 17,536 | 509 | | |
| Twin Buttes | 1963 | | Concho | TE | 41 | 12,984 | 16,393 | 791 | | |
| Twin Buttes Tailings | 1973 | | Santa Cruz | TE | 73 | 3,444 | 29,514 | 258 | | |
| Upper Mill Branch Tailings | 1963 | | Upper Mill branch | TE | 213 | 338 | n/a | n/a | | |
| Wanapum | 1963 | 1961 | Columbia | TE | 59 | 1,750 | 3,570 | 1,100 | 838 | 1,327 |
| Warm Springs | 1982 | | Dry Creek | TE | 97 | 914 | 22,937 | 470 | | |
| Yellowtail | 1966 | 1966 | Bighorn | VA/PG | 160 | 451 | 1,182 | 1,077 | 250 | 250 |
| **Uruguay** | | | | | | | | | | |
| Salto Grande | 1979 | 1979 | Uruguay | TE/PG | 65 | 3,000 | 3,965 | 5,000 | 1,890 | 1,890 |
| **USSR** | | | | | | | | | | |
| Boguchany | 1989 | n/a | Angara | ER | 79 | 1,816 | 27,360 | 58,200 | n/a | 4,000 |
| Bratsk | 1964 | 1961 | Angara | TE/PG | 125 | 4,417 | 10,962 | 169,270 | 4,500 | 4,500 |

continued

| | Year completed | Year of initial operation | River or basin | Dam type | Height above lowest foundation (m) | Crest length (m) | Dam volume (10³ m³) | Maximum reservoir capacity (10⁶ m³) | Present rated capacity (MW) | Planned rated capacity (MW) |
|---|---|---|---|---|---|---|---|---|---|---|
| Bukhtarma | 1960 | | Irtysh | PG | 90 | 380 | 1,170 | 49,800 | | |
| Charvak | 1977 | n/a | Chirchik | ER | 168 | 764 | 21,600 | 560 | n/a | n/a |
| Cheboksary | 1986 | 1972 | Volga | PG | 42 | 3,497 | 8,350 | 12,800 | 1,404 | 1,404 |
| Chirkey | 1978 | 1974 | Sulak | VA | 233 | 333 | 1,358 | 2,780 | 1,000 | 1,000 |
| Dneprodzerzhinsk | 1964 | n/a | Dnieper | TE/PG | 34 | 35,642 | 22,031 | 2,460 | n/a | n/a |
| Dnieper I/II | 1932 | 1932/74 | Dnieper | PG | 60.5 | 761 | 732 | 3.33 | 648/835 | 648/835 |
| Inguri | 1980 | 1985 | Inguri | VA | 272 | 680 | 3,960 | 1,100 | 1,300 | 1,300 |
| Irkutsk | 1965 | n/a | Angara | TE/PG | 44 | 2,500 | 11,560 | 46,000 | 660 | 660 |
| Ivankovo | 1937 | n/a | Volga | TE/PG | 28 | 9,920 | 15,450 | 1,120 | n/a | n/a |
| Kakhovskaya | 1955 | n/a | Dnieper | TE/PG | 37 | 3,629 | 35,640 | 182,000 | n/a | n/a |
| Kama | 1954 | n/a | Kama | TE/PG | 37 | 2,286 | 27,720 | 12,200 | n/a | n/a |
| Kapchagay | 1970 | n/a | Ili | TE | 52 | 840 | 6,220 | 28,100 | n/a | n/a |
| Khudoni | 1990 | 1990 | Inguri | VA | 201 | 545 | 1,475 | 365 | | 740 |
| Kiev | 1964 | n/a | Dnieper | TE | 68 | 41,185 | 42,841 | 3,730 | n/a | n/a |
| Krasnoyarsk | 1967 | 1968 | Yenisei | PG | 124 | 1,065 | 5,580 | 73,300 | 6,000 | 6,000 |
| Kremenchug | 1960 | n/a | Dnieper | PG | 33 | 12,144 | 31,492 | 13,500 | n/a | n/a |
| Lower Kama (Nizhne Kamskaya) | 1987 | 1994 (est.) | Kama | TE/PG | 36 | 6,200 | 23,000 | 45,000 | 1,248 | 1,248 |
| Mingechaur | 1953 | n/a | Kura | TE | 80 | 1,550 | 15,600 | 16,000 | n/a | n/a |
| Nurek | 1980 | 1976 | Vakhsh | TE | 300 | 704 | 58,000 | 10,500 | 2,700 | 2,700 |
| Rybinsk | 1941 | n/a | Volga | PG | 30 | 628 | 2,545 | 25,400 | n/a | n/a |
| Saratov | 1967 | 1967 | Volga | TE | 40 | 15,260 | 40,400 | 12,900 | 1,360 | 1,360 |
| Sayano-Shushensk | 1989 | 1980 | Yenisei | VA/PG | 245 | 1,066 | 9,075 | 31,300 | 6,400 | 6,400 |
| Shulbinsk | n/a | 1983 | Irtysh | TE | 36 | 570 | 2,700 | 2,390 | 1,200 | 1,200 |
| Talimarjan | 1985 | n/a | Amudaria | TE | 35 | 9,745 | 29,175 | 1,525 | n/a | n/a |
| Toktogul | 1978 | 1974 | Naryn | PG | 215 | 293 | 3,345 | 19,500 | 1,248 | 1,248 |
| Tsimlyansk | 1952 | 1952 | Don | TE/PG | 41 | 13,245 | 33,891 | 24,000 | | |
| Ust-Ilim | 1977 | 1977 | Angara | PG/ER | 102 | 3,725 | 8,866 | 59,300 | 4,320 | 4,320 |

continued

| | Year completed | Year of initial operation | River or basin | Dam type | Height above lowest foundation (m) | Crest length (m) | Dam volume ($10^3$ m$^3$) | Maximum reservoir capacity ($10^6$ m$^3$) | Present rated capacity (MW) | Planned rated capacity (MW) |
|---|---|---|---|---|---|---|---|---|---|---|
| Vilyui | 1967 | n/a | Vilyui | ER | 64 | 600 | 3,740 | 35,900 | n/a | n/a |
| Volga-VI Lenin (Kuibyshev) | 1953 | 1955 | Volga | TE/PG | 45 | 3,781 | 33,869 | 58,000 | 2,300 | 2,300 |
| Volgograd 22nd Congress | 1958 | 1958 | Volga | TE/ER/PG | 47 | 3,974 | 25,932 | 31,500 | 2,563 | 2,563 |
| Votkinsk | 1961 | 1961 | Kama | TE/PG | 44 | 4,982 | 9,330 | 9,400 | 1,000 | 1,000 |
| Zeya | 1978 | 1975 | Zeya | CB | 115 | 758 | 2,160 | 68,400 | 1,260 | 1,260 |
| Venezuela | | | | | | | | | | |
| Guri (Raúl Leoni) | 1986 | 1986 | Caroni | TE/PG/ER | 162 | 11,409 | 77,971 | 138,000 | 10,300 | 10,300 |
| Yugoslavia | | | | | | | | | | |
| Djerdap I | 1972 | 1972 | Danube | PG | 60 | 1,278 | 3,160 | 2,400 | 1,050 | 1,050 |
| Mratinje | 1976 | n/a | Piva | VA | 220 | 268 | 742 | 880 | n/a | n/a |
| Zaire | | | | | | | | | | |
| Inga II | 1979 | 1981 | Zaire | CB | 34 | 176 | n/a | n/a | 1,400 | 1,400 |
| Zambia | | | | | | | | | | |
| Kariba | 1959 | 1959 | Zambezi | VA | 128 | 579 | 1,032 | 180,600 | 1,266 | 1,266 |
| Zimbabwe | | | | | | | | | | |
| Kariba | 1959 | 1959 | Zambezi | VA | 128 | 579 | 1,032 | 180,600 | 1,266 | 1,266 |

n.a. not available

VA, arch; CB, buttress; TE, earthfill; PG, gravity; ER, rockfill
[a]Vaiont (Italy): Reservoir was engulfed by a major landslide in 1963. The dam remains, intact but useless.
[b]Mauvoisin (Switzerland) heightened from 237 to 250.5m, completed in 1990.

**TABLE G.6** The world's major dam and hydroelectric facilities under construction

| | Expected year of completion | Expected year of initial operation | River or basin | Dam type | Height above lowest foundation (m) | Crest length (m) | Dam volume ($10^3$ m$^3$) | Maximum reservoir capacity ($10^6$ m$^3$) | Present rated capacity (MW) | Planned rated capacity (MW) |
|---|---|---|---|---|---|---|---|---|---|---|
| **Albania** | | | | | | | | | | |
| Banje | 1995 | 1995 | Devoll | TE/ER | 100 | 1,350 | 15,000 | 700 | | 60 |
| **Argentina** | | | | | | | | | | |
| Piedra del Aguila | 1993 | 1992 | Limay | PG | 163 | 820 | 2,520 | 11,300 | | 2,100 |
| Yacyretá | 1997 | 1993 | Paraná | TE/PG | 43 | 69,600 | 67,700 | 21,000 | | 2,700 |
| **Brazil** | | | | | | | | | | |
| Campos Novos | 1995 | n/a | n/a | ER | 210 | n/a | n/a | n/a | | n/a |
| Corumba I | 1997 | 1997 | Corumba | TE/ER | 90 | 540 | 4,410 | 1,500 | | 375 |
| Itá | 1997 | 1997 | Uruguay | ER | 125 | 880 | 8,500 | 5,100 | | 1,620 |
| Porto Primavera | 1998 | 1995 | Paraná | TE/PG | 38 | 11,300 | 37,644 | 20,000 | | 1,818 |
| Segredo | 1993 | 1992 | Iguaçu | ER | 145 | 720 | 7,247 | 3,000 | | 1,260/2,520 |
| Serra da Mesa (São Felix) | 1993 | 1996 | Toçantins | TE/ER | 150 | 1,510 | 12,619 | 54,400 | | 1,200 |
| Xingo | | 1994 | São Francisco | ER | 140 | 850 | 13,900 | 3,800 | | 3,000/5,020 |
| **Canada** | | | | | | | | | | |
| La Grande 1 | 1994 | 1994 | La Grande | TE/PG | 60 | 3,175 | 1,500 | 1,200 | | 1,368 |
| La Grande 2A | 1992 | 1991 | La Grande | ER | 168 | 2,826 | 23,192 | 61,715 | | 1,998 |
| Limestone | 1992 | 1990 | Nelson | ER | 40 | 1,890 | 2,900 | n/a | 532 | 1,330 |
| Syncrude Tailings | 1992 | 1992 | | TE | 87 | 18,000 | 540,000 | 424 | | |
| **China** | | | | | | | | | | |
| Dongfeng | | n/a | Wujiang | VA | 168 | 263 | 622 | 9,150 | | 510 |
| Ertan | | n/a | Yalong | VA | 245 | 779 | 4,742 | 5,800 | | 3,300 |
| Geheyan | | 1992 | Qingjiang | VA | 151 | 674 | 3,060 | 3,400 | | 1,200 |

*continued*

| | Expected year of completion | Expected year of initial operation | River or basin | Dam type | Height above lowest foundation (m) | Crest length (m) | Dam volume ($10^3$ m³) | Maximum reservoir capacity ($10^6$ m³) | Present rated capacity (MW) | Planned rated capacity (MW) |
|---|---|---|---|---|---|---|---|---|---|---|
| Guangzhou | 1994 | n/a | Liuxi | ER | 68/43 | 300/127 | 17/17.5 | n/a | | 1,200 |
| Lijiaxia (Lijia Gorge) | | n/a | Huanghe | VA | 165 | 402 | 3,030 | 1,720 | | 2,000 |
| Manwan | 1992 | 1992 | Lancangjiang | PG | 126 | 421 | 2,200 | 920 | | 1,250 |
| Shuikou | 1994 | 1993 | Minjiang | PG | 101 | 786 | 280 | 2,600 | | 1,400 |
| Wuqiangxi | 1995 | 1994 | Yuanshui | PG | 88 | 736 | 5,678 | 2,990 | | 1,200 |
| Xiaolangdi | | n/a | Huanghe | ER | 152 | 1,200 | 12,650 | n/a | | 1,560 |
| Yantan | 1995 | 1992 | Hongshui | PG | 111 | 536 | 3,004 | 2,430 | | 1,210 |
| China (Taiwan) | | | | | | | | | | |
| Mingtan | 1992 | 1992 | Shuili/Sun Moon Lake | | | | | | 0 | 1,600 |
| Congo | | | | | | | | | | |
| Kouilou | 1992 | n/a | Kouilou | VA | 137 | 345 | 390 | 35,000 | n/a | n/a |
| Czechoslovakia | | | | | | | | | | |
| Gabçikovo (Hrusov-Dunakiliti) | | n/a | Danube | TE/PG | 29 | 31,500 | 18,340 | 199 | 720 | 720 |
| Ecuador | | | | | | | | | | |
| Guaillabamba | | n/a | Guaillabamba | VA | 165 | 413 | 704 | 105 | n/a | n/a |
| Greece | | | | | | | | | | |
| Messochora | 1994 | n/a | Acheloos | ER | 150 | 300 | 4,200 | n/a | | 140 |
| Thissavros | 1993 | | Nestos | ER | 172 | 480 | 10,000 | 700 | | |
| Hungary | | | | | | | | | | |
| Bôs (Hrusov-Dunakiliti) | | n/a | Danube | TE/PG | 29 | 31,500 | 18,340 | 199 | 720 | 720 |

continued

| | Expected year of completion | Expected year of initial operation | River or basin | Dam type | Height above lowest foundation (m) | Crest length (m) | Dam volume (10³ m³) | Maximum reservoir capacity (10⁶ m³) | Present rated capacity (MW) | Planned rated capacity (MW) |
|---|---|---|---|---|---|---|---|---|---|---|
| **India** | | | | | | | | | | |
| Kishau | 1995 | 1995 | Tons | PG | 236 | 680 | 9,500 | 1,810 | | 600 |
| Lakhwar | 1991 | 1996 | Yamuna | PG | 204 | 454 | 2,871 | 580 | | 300 |
| Sadar Sarovar | 1994 | 1992 | Narmada | PG | 163 | 1,202 | 7,472 | 9,500 | | 1,450 |
| Tehri | 1997 | 1997 | Bhagirathi | TE/ER | 261 | 575 | 25,645 | 3,540 | | 2,400 |
| Thein (Rajit) | 1992 | 1993 | Ravi | TE/ER | 160 | 565 | 14,213 | 3,280 | | 600 |
| Warna | 1992 | n/a | Warna | TE/PG | 91 | 1,580 | 15,310 | 964 | n/a | n/a |
| **Iran** | | | | | | | | | | |
| Maroun | 1994 | 1994 | Maroun | ER | 165 | 350 | 7,490 | 1,200 | | 145 |
| Siah Bishe | 1996 | 1996 | Chalus | ER | 85/128 | 420/380 | n/a | n/a | | 1,000 |
| **Iraq** | | | | | | | | | | |
| Bekhme | | n/a | Greater Zab | TE/ER | 204 | 600 | 34,000 | 33,000 | | 1,560 |
| **Japan** | | | | | | | | | | |
| Hase (Ohkawachi) | 1992 | n/a | Inumi | PG | 102 | 254 | 538 | 9.6 | | 1,280 |
| Kaore (Okumino) | 1994 | n/a | Itadori | VA | 107 | 341 | 400 | 10 | | 1,000 |
| Miyagase | 1994 | 1994 | Nakatsu | PG | 155 | 400 | 2,000 | 193 | | |
| Nukui | 1997 | n/a | Takiyama | VA | 155 | 382 | 800 | 82 | 0 | 2.3 |
| Urayama | 1996 | 1996 | Takiyama | PG | 155 | 400 | 1,730 | 58 | | |
| **Jordan** | | | | | | | | | | |
| Al-Wehdah | | n/a | Yarmuk | TE/ER | 110/150 | 600 | 8,000/15,000 | 225/520 | | 20 |
| **Korea DPR** | | | | | | | | | | |
| Kumgang | | n/a | North Han | TE | 121.5 | 1,120 | n/a | 9.25 | | 800 |
| Songwon | 1995 | n/a | Chungmangang | ER | 160 | 630 | 1,100 | 3,200 | | n/a |

*continued*

| | Expected year of completion | Expected year of initial operation | River or basin | Dam type | Height above lowest foundation (m) | Crest length (m) | Dam volume (10³ m³) | Maximum reservoir capacity (10⁶ m³) | Present rated capacity (MW) | Planned rated capacity (MW) |
|---|---|---|---|---|---|---|---|---|---|---|
| **Lesotho** | | | | | | | | | | |
| Katse | 1996 | 1996 | Malibamatso | VA | 182 | 700 | 2,200 | 2,000 | | 75/180 |
| **Mexico** | | | | | | | | | | |
| Aguamilpa | 1994 | 1994 | Santiago | ER | 187 | 642 | 14,000 | 6,950 | | 960 |
| Zimapan | 1994 | 1994 | Moctezuma | VA | 200 | 80 | 280,218 | 1,426 | | 280 |
| **Morocco** | | | | | | | | | | |
| Wahada (M'Jara) | | n/a | Ouegha | TE | 87 | 1,600 | 25,000 | 4,000 | | 240 |
| **Paraguay** | | | | | | | | | | |
| Yacyretá | 1997 | 1993 | Paraná | TE/PG | 43 | 69,600 | 67,700 | 21,000 | | 2,700 |
| **Romania** | | | | | | | | | | |
| Gura Apelor (Retezat) | 1992 | 1987 | Riul Mare | ER | 168 | 450 | 10,285 | 225 | 335 | 335 |
| Valea Sadului | | n/a | Jiu | ER | 56 | 7,150 | 18,250 | 306 | | 35 |
| **Syria** | | | | | | | | | | |
| Al-Wehdah | | n/a | Yarmuk | TE/ER | 110/150 | 600 | 8,000/15,000 | 225/520 | | 20 |
| **Turkey** | | | | | | | | | | |
| Berke | | n/a | Ceheyan | VA | 201 | 120 | 729 | 427 | | 510 |
| Çatalan | | n/a | Seyhan | TE | 82 | 894 | 1,700 | 2,126 | | 156 |
| **United Kingdom** | | | | | | | | | | |
| Gale Common Tailings | 2006 | | | TE | 51 | 4,000 | 20,500 | 20.5 | | |
| **United States** | | | | | | | | | | |
| Mayflower Tailings | | | | TE | 170 | 1,450 | n/a | n/a | | |
| **USSR** | | | | | | | | | | |
| Bureya | 1994 | | Bureya | PG | 139 | 810 | 3,561 | 20,900 | | 1,700 |

*continued*

| | Expected year of completion | Expected year of initial operation | River or basin | Dam type | Height above lowest foundation (m) | Crest length (m) | Dam volume ($10^3$ m$^3$) | Maximum reservoir capacity ($10^6$ m$^3$) | Present rated capacity (MW) | Planned rated capacity (MW) |
|---|---|---|---|---|---|---|---|---|---|---|
| Kaishadory | | | Neman | | | | | 41 | | 1,600 |
| Kambaratinsk | n/a | n/a | Naryn | TE/ER | 255 | 560 | 112,200 | 4,650 | | 1,600 |
| Katun | n/a | n/a | Katun | TE/ER | 185 | 760 | 32,700 | 5,800 | | 1,570 |
| Namakhrani I | n/a | n/a | Rioni | VA | 161 | 460 | 1,200 | 560 | | 480 |
| Rogun | n/a | n/a | Vakhsh | TE/ER | 335 | 660 | 71,000 | 13,300 | n/a | 3,600 |
| Tashlyk | n/a | n/a | n/a | | | | | | n/a | 1,800 |
| Zagorsk | n/a | n/a | Kunja | | | | | 28 | 600 | 1,200 |
| Venezuela | | | | | | | | | | |
| Caruachi | 2000 | 1998 | Caroni | TE/ER/PG | 50 | 4,740 | 9,500 | 3.52 | | 2,424 |
| La Vueltosa | 1993 | 1993 | Caparo | TE | 118 | 1,200 | 15,000 | 5,300 | n/a | 1,028 |
| Macagua II | 1993 | 1994 | Caroni | TE/ER/PG | 68.5 | 4,181 | 2,565 | 363 | | 2,560 |
| Tocomo | 2002 | 2000 | Caroni | TE/ER/PG | 58 | 6,240 | n/a | 1,800 | | 2,424 |
| Vietnam | | | | | | | | | | |
| Hoa Binh | 1993 | 1989 | Da | n/a | 128 | 640 | n/a | 9,450 | 960 | 1,920 |

u/c under construction; n/a not available.
VA, arch; TE, earthfill; PG, gravity; ER, rockfill.

**TABLE G.6** The world's major dam and hydroelectric facilities under construction (*pages 347–351*)

**DESCRIPTION**

The following table lists all major dams and hydroelectric plants now under construction by country. These data were compiled by International Water Power and Dam Construction, United Kingdom. Included in this table are the site or dam name, the year the dam is expected to be completed, the year of initial operation, the river basin, the dam type, the dam height above the lowest foundation (m), the crest length (m), the dam volume (m$^3$ × $10^3$), the maximum reservoir capacity (m$^3$ × $10^6$), and the present and planned installed capacity (MW).

**LIMITATIONS**

Only large dams are listed here. Some valuable data, such as surface area of the dam, hydroelectric head, and expected annual energy generation, are not provided. When major modifi-

cations to a dam are done, the date of initial operation may precede the date of completion by many years.

**SOURCES**

International Water Power and Dam Construction, 1992, Water Power and Dam Construction Handbook 1992, Reed Enterprises, Sutton, United Kingdom. Reprinted with permission.

**TABLE G.7**  World's largest dams, by dam volume

| Dam | Year complete | Country | Type | Volume ($10^3$ m$^3$) |
|---|---|---|---|---|
| Syncrude Tailings | u/c (1992) | Canada | TE | 540,000 |
| New Cornelia Tailings | 1973 | United States | TE | 209,500 |
| Pati | u/c (1998) | Argentina | TE | 200,000 |
| Kambaratinsk | u/c | USSR | TE/ER | 112,200 |
| Tarbela | 1976 | Pakistan | TE/ER | 106,000 |
| Fort Peck | 1937 | United States | TE | 96,050 |
| Lower Usuma | 1990 | Nigeria | TE | 93,000 |
| Tucuruí | 1984 | Brazil | TE/ER/PG | 85,200 |
| Ataturk | 1990 | Turkey | TE/ER | 84,500 |
| Guri (Raúl Leoni) | 1986 | Venezuela | TE/PG/ER | 77,971 |
| Rogun | u/c (1991) | USSR | TE/ER | 71,000 |
| Yacyretá | u/c (1994) | Paraguay-Argentina | TE/PG | 67,700 |
| Oahe | 1958 | United States | TE | 66,517 |
| Gardiner | 1968 | Canada | TE | 65,400 |
| Mangla | 1967 | Pakistan | TE | 65,379 |
| Afsluitdijk | 1932 | Netherlands | TE | 63,430 |
| Oroville | 1968 | United States | TE | 61,164 |
| San Luis | 1967 | United States | TE | 59,559 |
| Nurek | 1980 | USSR | TE | 58,000 |
| Garrison | 1956 | United States | TE | 50,845 |
| Cochiti | 1975 | United States | TE | 50,230 |
| Oosterschelde | 1986 | Netherlands | TE/PG | 50,000 |
| Tabqua (Thawra) | 1976 | Syria | TE | 46,000 |
| Aswan High | 1970 | Egypt | TE/ER | 44,300 |
| W.A.C. Bennett | 1967 | Canada | TE | 43,733 |
| Boruca | 1983 | Costa Rica | ER | 43,000 |
| San Rogue | u/c | Philippines | TE | 43,000 |
| Kiev | 1964 | USSR | TE | 42,841 |
| Dantiwada Left Embankment | 1965 | India | TE | 41,040 |
| Saratov | 1967 | USSR | TE | 40,400 |
| Mission Tailings 2 | 1973 | United States | TE | 40,088 |
| Fort Randall | 1956 | United States | TE | 38,380 |
| Kanev | 1976 | USSR | TE | 37,860 |
| Mosul | 1982 | Iraq | TE | 36,000 |
| Kakhovka | 1955 | USSR | TE/PG | 35,640 |
| Itumbiara | 1980 | Brazil | PG/TE | 35,600 |
| Lauwerszee | 1969 | Netherlands | TE | 35,575 |
| Beas | 1974 | India | TE/ER | 35,418 |

*continued*

| Dam | Year complete | Country | Type | Volume ($10^3$ m³) |
|---|---|---|---|---|
| São Felix | 1985 | Brazil | TE/ER | 34,000 |
| Tsimlyansk | 1952 | USSR | TE/PG | 33,891 |
| Volga-VI Lenin | 1955 | USSR | TE/PG | 33,869 |
| Castaic | 1973 | United States | TE | 33,642 |
| Jari | 1967 | Pakistan | TE | 32,800 |
| Mica | 1972 | Canada | TE/ER | 32,111 |
| Kremenchug | 1961 | USSR | TE/PG | 31,492 |
| Esperanza Tailings | 1973 | United States | TE | 30,355 |
| Twin Buttes Tailings | 1973 | United States | TE | 29,514 |

u/c, under construction
ER, rockfill; PG, gravity; TE, earthfill.

**TABLE G.7** World's largest dams, by dam volume *(pages 352–353)*

## DESCRIPTION

The world's largest dams are given here by volume of the dam itself, in m³ × $10^3$. Also shown are the year completed (or whether it is under construction), location, and type of dam. The dam types include: ER, rockfill; PG, gravity; and TE, earthfill.

## LIMITATIONS

Many different types of dams are listed here, including some that are not used for water supply or hydroelectricity generation. For example, the world's two largest dams by volume were built to hold the debris tailings from huge mineral mines. More recent data are needed for dams under 43,000,000 m³ and for dams under construction. Some of those listed as under construction may be cancelled or delayed for significant periods of time.

## SOURCES

United States Department of the Interior, 1983, *Major Dams, Reservoirs, and Hydroelectric Plants: Worldwide and Bureau of Reclamation*, Bureau of Reclamation, Denver, Colorado.

T.W. Mermel, 1991, *Water Power and Dam Construction*, Quadrant House, Sutton, UK, June, p. 67. Reprinted with permission.

**TABLE G.8**  World's highest dams

| Dam | Year complete | Country | Type | Height (m) |
|---|---|---|---|---|
| Rogun | u/c (1991) | USSR | TE/ER | 335 |
| Nurek | 1980 | USSR | TE | 300 |
| Grande Dixence | 1961 | Switzerland | PG | 285 |
| Inguri | 1980 | USSR | VA | 272 |
| Vajont | 1961 | Italy | VA | 262 |
| Chicoasén | 1980 | Mexico | TE/ER | 261 |
| Tehri | u/c (1997) | India | TE/ER | 261 |
| Kambaratinsk | u/c | USSR | TE/ER | 255 |
| Kishau | u/c (1995) | India | TE/ER | 253 |
| Mauvoisin | 1990[a] | Switzerland | VA | 251 |
| Guavio | 1989 | Colombia | TE/ER | 246 |
| Ertan | u/c | China | VA | 245 |
| Sayano-Shushensk | 1989 | USSR | VA/PG | 245 |
| Mica | 1973 | Canada | TE/ER | 242 |
| Chivor | 1975 | Colombia | TE/ER | 237 |
| Kishau | u/c | India | PG | 236 |
| El Cajon | 1985 | Honduras | VA | 234 |
| Chirkey | 1978 | USSR | VA | 233 |
| Bekhme | u/c | Iraq | TE/ER | 230 |
| Oroville | 1968 | United States | TE | 230 |
| Bhakra | 1963 | India | PG | 226 |
| Hoover | 1936 | United States | VA/PG | 221 |
| Contra | 1965 | Switzerland | VA | 220 |
| Mrantinje | 1976 | Yugoslavia | VA | 220 |
| Dworshak | 1973 | United States | PG | 219 |
| Glen Canyon | 1966 | United States | VA | 216 |
| Toktugul | 1978 | USSR | PG | 215 |
| Daniel Johnson | 1968 | Canada | MV | 214 |
| San Roque | u/c | Philippines | ER | 210 |
| Luzzone | 1963 | Switzerland | VA | 208 |
| Dez | 1962 | Iran | VA | 207 |
| Keban | 1974 | Turkey | TE/PG/ER | 207 |
| Lakhwar | 1991 | India | PG | 204 |
| Almendra | 1970 | Spain | VA | 202 |
| Khudoni | u/c (1990) | USSR | VA | 201 |
| Campos Novos | u/c | Brazil | ER | 200 |
| Karun I | 1975 | Iran | VA | 200 |
| Kolnbrein | 1977 | Austria | VA | 200 |

*continued*

| Dam | Year complete | Country | Type | Height (m) |
|---|---|---|---|---|
| Zimapan | u/c | Mexico | VA | 200 |
| Itaipú | 1982 | Brazil-Paraguay | TE/PG/ER | 196 |
| Altinkaya | 1988 | Turkey | TE/ER | 195 |
| New Melones | 1979 | United States | ER | 191 |
| Aguamilpa | u/c | Mexico | ER | 187 |
| Swift | 1958 | United States | TE | 186 |
| Katun | u/c | USSR | TE/ER | 185 |
| Ataturk | 1990 | Turkey | TE/ER | 184 |
| W.A.C. Bennett | 1967 | Canada | TE | 183 |
| Dartmouth | 1979 | Australia | ER | 180 |
| Tiangshengqiao | u/c (1993) | China | ER | 180 |
| Takase | 1978 | Japan | ER | 176 |
| Hasan Ugurlu | 1981 | Turkey | TE/ER | 175 |
| Revelstoke | 1984 | Canada | TE/PG | 175 |
| Don Pedro | 1971 | United States | TE | 173 |
| Thissavros | u/c (1993) | Greece | TE | 172 |

[a]Originally completed in 1957. Raised in 1990.
u/c, under construction
ER, rockfill; PG, gravity; MV, multi-arch; TE, earthfill; VA, arch.

**TABLE G.8** World's highest dams *(pages 354–355)*

**DESCRIPTION**

The world's highest dams are listed here in descending order of height, in meters. Data on dam name, location, type, and year completed (or whether it is under construction) are also given. Only dams greater than 170 m in height are included here. Dam types include: ER, rockfill; PG, gravity; MV, multi-arch; TE, earthfill; and VA, arch.

**LIMITATIONS**

While the height of a dam is often listed as a major characteristic, it is far less important than the "head" created by a dam, where the head is the distance from the top of the water in the reservoir to the top of the water in the tailrace below the powerhouse. The head (together with volume of water flow) is a more precise measure of the amount of energy that can be generated by any facility. Many hydroelectric facilities have a "head" far greater than the height of the dam because they deliver water over large drops through penstocks to powerhouses located at considerable distances from the dam itself. Data on dam "head" are far more difficult to find in the literature.

Some of these facilities are still under construction and may be delayed or even cancelled in the future.

**SOURCES**

United States Department of the Interior, 1983, *Major Dams, Reservoirs, and Hydroelectric Plants: Worldwide and Bureau of Reclamation*, Bureau of Reclamation, Denver, Colorado.

T.W. Mermel, 1991, *Water Power and Dam Construction*, Quadrant House, Sutton, United Kingdom, January, p. 35, and June, pp. 67–77. Reprinted with permission.

**TABLE G.9**  World's largest reservoirs, by reservoir volume

| Name | Country | Capacity ($10^6$ m$^3$) | Year |
|---|---|---|---|
| Owen Falls[a] | Uganda | 204,800 | 1954 |
| Bratsk | USSR | 169,000 | 1964 |
| High Aswan | Egypt | 162,000 | 1970 |
| Kariba | Zimbabwe-Zambia | 160,368 | 1959 |
| Akosombo | Ghana | 147,960 | 1965 |
| Daniel Johnson | Canada | 141,851 | 1968 |
| Guri | Venezuela | 135,000 | 1986 |
| Krasnoyarsk | USSR | 73,300 | 1967 |
| W.A.C. Bennett | Canada | 70,309 | 1967 |
| Zeya | USSR | 68,400 | 1978 |
| Cahora Bassa | Mozambique | 63,000 | 1974 |
| La Grande 2 Barrage | Canada | 61,715 | 1978 |
| La Grande 3 Barrage | Canada | 60,020 | 1981 |
| Ust-Ilim | USSR | 59,300 | 1977 |
| Boguchany | USSR | 58,200 | u/c |
| Kuibyshev | USSR | 58,000 | 1955 |
| Serra da Mesa | Brazil | 54,400 | u/c |
| Caniapiscau Barrage K A 3 | Canada | 53,790 | 1980 |
| Bukhatarma | USSR | 49,800 | 1960 |
| Ataturk | Turkey | 48,700 | 1990 |
| Irkutsk | USSR | 46,000 | 1956 |
| Tucuruí | Brazil | 45,500 | 1984 |
| Vilyui | USSR | 35,900 | 1967 |
| Sanmenxia | China | 35,400 | 1960 |
| Hoover | United States | 34,852 | 1936 |
| Sobradinho | Brazil | 34,100 | 1979 |
| Glen Canyon | United States | 33,304 | 1966 |
| Skins Lake No. 1 | Canada | 32,203 | 1953 |
| Jenpeg | Canada | 31,790 | 1975 |
| Volgograd | USSR | 31,500 | 1958 |
| Sayano-Shushensk | USSR | 31,300 | u/c |
| Keban | Turkey | 30,600 | 1974 |
| Iroquois | Canada | 29,959 | 1958 |
| Itaipú | Brazil | 29,000 | 1982 |
| Loma de la Lata | Argentina | 29,000 | 1977 |
| Churchill Falls (GR-1) | Canada | 28,973 | 1971 |
| Missi Falls Control | Canada | 28,370 | 1976 |
| Kapchagay | USSR | 28,100 | 1970 |

*continued*

| Name | Country | Capacity ($10^6$ m$^3$) | Year |
|---|---|---|---|
| Garrison | United States | 27,920 | 1953 |
| Kossou | Côte d' Ivoire | 27,675 | 1972 |
| Oahe | United States | 27,433 | 1958 |
| Razza Kyke | Iraq | 26,000 | 1970 |
| Rybinsk | USSR | 25,400 | 1941 |
| Longyanxia | China | 24,700 | u/c |
| Mica | Canada | 24,700 | 1972 |
| Tsimlyansk | USSR | 24,000 | 1952 |
| Kenney | Canada | 23,700 | 1952 |
| Ust-Khantaika | USSR | 23,500 | 1970 |
| Shuikou | China | 23,400 | u/c |
| Furnas | Brazil | 22,950 | 1963 |
| Fort Peck | United States | 22,119 | 1937 |
| Xinanjiang | China | 21,626 | 1960 |
| Ilha Solteira | Brazil | 21,166 | 1973 |
| Yacyretá | Argentina | 21,000 | u/c |

u/c, under construction
[a]This capacity is not fully obtained by a dam; the major part of it is the natural capacity of a lake.

**TABLE G.9** World's largest reservoirs, by reservoir volume *(pages 356–357)*

**DESCRIPTION**

The largest artificial reservoirs, by volume of the reservoir, are shown here, together with the year of completion and the country. Reservoir capacity is given in m$^3 \times 10^6$.

**LIMITATIONS**

Reservoir volume varies with season, mode of operation, sedimentation, and other factors. The source does not describe whether the given volumes are maximum capacity, average capacity, or some other measure. The Owen Falls reservoir is not entirely created by a dam; the majority of it is part of a natural lake.

**SOURCES**

International Commission on Large Dams, 1988, *World Registry of Dams*, International Commission on Large Dams, Paris. By permission.

**TABLE G.10**   World's largest hydroelectric plants, by installed capacity

| Name | Year of initial operation | Country | Installed capacity now (MW) |
|---|---|---|---|
| Itaipú | 1983 | Brazil-Paraguay | 11,900 |
| Guri (Raúl Leoni) | 1986 | Venezuela | 10,300 |
| Grand Coulee | 1942 | United States | 9,700 |
| Sayano-Shushensk | 1989 | USSR | 6,400 |
| Krasnoyarsk | 1968 | USSR | 6,000 |
| La Grande 2 | 1979 | Canada | 5,328 |
| Churchill Falls | 1971 | Canada | 5,225 |
| Bratsk | 1961 | USSR | 4,500 |
| Ust-Ilim | 1977 | USSR | 4,320 |
| Tucuruí (Raúl G. Lhano) | 1984 | Brazil | 3,960 |
| Rogun | 1990 | USSR | 3,600 |
| Ilha Solteira | 1973 | Brazil | 3,200 |
| Nurek | 1976 | USSR | 3,000 |
| Mica | 1976 | Canada | 2,660 |
| La Grande 4 | 1984 | Canada | 2,650 |
| Volgograd 22nd Congress | 1958 | USSR | 2,563 |
| Paulo Afonso IV | 1979 | Brazil | 2,460 |
| Cahora Bassa | 1975 | Mozambique | 2,425 |
| Shrum (Portage Mt.) (W.A .C. Bennett) | 1968 | Canada | 2,416 |
| Chicoasén | 1980 | Mexico | 2,400 |
| Gezhouba | 1982 | China | 2,340 |
| La Grande 3 | 1982 | Canada | 2,304 |
| Kuibyshev | 1955 | USSR | 2,300 |
| John Day | 1969 | United States | 2,160 |
| Iron Gates I/Djerdap I | 1970 | Romania-Yugoslavia | 2,136 |

**TABLE G.10**   World's largest hydroelectric plants, by installed capacity

**DESCRIPTION**
The world's largest hydroelectric facilities are listed here, by installed capacity in MW. Also given are the year of initial operation, and the location.

**LIMITATIONS**
Present installed capacity is not necessarily the same as maximum potential capacity. There are plans to increase the capacity at a number of these sites. Total annual energy production depends as much on the flow of water and the operation of the reservoir as on the installed capacity.

**SOURCES**
T.W. Mermel, 1991, The world's major dams and hydro plants, in *Water Power and Dam Construction Handbook 1991*, International Water Power and Dam Construction, UK. By permission.

**TABLE G.11**  Populations displaced as a consequence of dam construction

| Dam | Country | Installed capacity (MW) | Area of reservoir (km²) | Number of people displaced | Date | Source |
|---|---|---|---|---|---|---|
| *Completed* | | | | | | |
| Sanmenxia | China | | | 870,000 | 1960 | Biswas |
| Maduru Oya | Sri Lanka | | 64 | 200,000 | 1983 | WRR 6(1)6, WRD |
| Aswan | Egypt, Sudan | 1,815 | 6,500 | 120,000 | 1970 | WRD, Biswas, IRN |
| Mangla | Pakistan | 600 | | 110,000 | 1967 | TN |
| Kaptai | Bangladesh | | 777 | 100,000 | 1962 | WRR 5(3)10, WRD |
| Damodar (4 projects) | India | | | 93,000 | 1959 | Biswas |
| Nanela | Pakistan | | | 90,000 | 1967 | Biswas |
| Tarbela | Pakistan | 1,750 | 243 | 86,000 | 1976 | WRD, Biswas, IRN |
| Akosombo | Ghana | 882 | 9,000 | 80,000 | 1965 | WRR 6(5)8, IRN, TN |
| Kossou | Côte d' Ivoire | | 1,700 | 75,000 | 1972 | WRD, Biswas |
| TVA (about 20 projects) | United States | | | 60,000 | 1930s to present | Biswas |
| Kariba | Zambia, Zimbabwe | 1,266 | 5,100 | 50,000–57,000 | 1959 | WRD, Biswas, IRN |
| Gandhi Sagar | India | | | 52,000 | | Biswas |
| Itaparica | Brazil | 1,500 | | 50,000 | 1988 | TN |
| Kainji | Nigeria | | | 42,000–50,000 | 1968 | WRD, Biswas |
| Ataturk (Southeast Anatolia Project) | Turkey | | | 40,000 | 1991 | IDN 2(2)1 |
| Bhakra | India | 450 | | 36,000 | 1963 | Biswas, IRN |
| Lam Pao | Thailand | | 400 | 30,000 | 1970 | Biswas, WRD |
| Keban | Turkey | 1,360 | 675 | 30,000 | 1974 | WRD, Biswas |
| Mython (Jharkh) | India | 200 | | 28,030 | 1955 | TN |
| Kedong Ombo | Java, Indonesia | | | 27,000 | 1992 | WRR 6(4)4 |
| Nam Pong | Thailand | | 20 | 25,000–30,000 | 1965 | Biswas, WRD |
| Tucurui | Brazil | 4,000 | 2,430 | 23,871 | 1984 | WRR 6(4)8 |
| Upper Pampanga | Philippines | | | 14,000 | 1973 | Biswas |
| Ruzizi II | Rwanda, Zaire | 40 | | 12,600 | completed | WRR 5(2)8 |
| Manantali | Mali | 200 | | 10,000 | completed | WRR 5(2)5, IRN |

*continued*

| Dam | Country | Installed capacity (MW) | Area of reservoir (km²) | Number of people displaced | Date | Source |
|---|---|---|---|---|---|---|
| Salvajina | Colombia | | 22 | 10,000 | 1985 | IDN 2(1)4, WRD |
| Brokopondo | Surinam | | | 5,000 | 1971 | Biswas |
| Caracol | Mexico | | | 5,000 | 1986 | WRR 6(2)11, WRD |
| Batang Ai | Sarawak, Borneo | 92 | 85 | 3,000 | completed | WRR 5(3)12 |
| Nam Ngum | Laos | | | 3,000 | 1971 | Biswas |
| Netzahualcoyotl | Mexico | | | 3,000 | 1964 | Biswas |
| *Planned or under construction* | | | | | | |
| Three Gorges | China | 30,000 | | 1,200,000 | planned | TN |
| Mahaweli (15 dams) | Sri Lanka | 500 | 360 | 1,000,000 + | 1996 | WRR 5(3)10 |
| Pa Mong | Thailand, Laos | | | 310,000–480,000 | planned | Biswas |
| Kalabagh | Pakistan | 2,400 | | 250,000 | planned | WRR 5(3)4, 4(6)3 |
| Almatti | India | | | 240,000 | uc | TN |
| Srisailam | India | | 247 | 100,000 | uc | TN |
| Chico | Philippines | 1,100 | | 90,000 | postponed | TN |
| Sardar Sarovar | India | 1,450 | 40,000 | 90,000 | uc | WRR 6(4)1, TN |
| Tehri | India | 1,000 | 42.5 | 85,600 | uc | WRR 3(2)8 |
| Narayanpur | India | | | 80,000 | uc | TN |
| Shuikou | China | | | 62,000–67,000 | uc | TN |
| Subarnarekha (Bihar) | India | 10 | | 64,500 | uc | TN |
| Saguling | India | | | 55,000 | uc | TN |
| Karnali (Chisapani) | Nepal | | | 50,000 | planned | TN |
| Soubre | Côte d'Ivoire | | | 40,000 | planned | TN |
| Yacyretá | Argentina, Paraguay | 2,700 | 1,700 | 40,000 | 1996 | WRR 6(1)12, 5(4)5 |
| San Juan Tetelcingo | Mexico | 280 | | 30,000 | planned | WRR 6(1)4 |
| Kotopanjang | Sumatra | | | 22,000 | 1996 | IRN |
| Kayraktepe | Turkey | | | 20,000 | uc | TN |
| Maheshwar | India | 200 | | 14,000 | planned | TN |
| Riau | Sumatra, Indonesia | | | 14,000 | planned | WRR 6(1)4 |

*continued*

| Dam | Country | Installed capacity (MW) | Area of reservoir (km²) | Number of people displaced | Date | Source |
|---|---|---|---|---|---|---|
| Omkareshwar | India | 390 | | 13,000 | planned | TN |
| Bhopalpatnam | India | 1,000 | | 10,000 | postponed | TN |
| Inchampalli | India | 393 | | 10,000 | postponed | TN |
| Wadaslingtan Project (2 dams) | Indonesia | | | 7,500–9,500 | planned | WRR 3(2)5 |
| Bodhghat | India | 107 | | 9,000 | postponed | WRR 3(2)8 |
| Jafuri | Indonesia | 30 | 100 | 8,000 | planned | IDN 1(1)9 |
| Upper Mazaruni | Guyana | | 2,590 | 5,000 | planned | TN |
| Riano | Spain | 680 | | 3,100 | 2010 | WRR 2(3)3 |
| Pak Mun | Thailand | 136 | | 1,000 | uc | WRR 4(6)4 |
| Rio 12 de Outubro | Brazil | 12 | | 347 | planned | WRR 6(2)4 |

u/c, under construction.
Biswas: A.K. Biswas, (1982).
IDN: International Dams Newsletter, Volumes 1(1)–2(5).
IRN: International Rivers Network.
TN: Todd Nachowitz (1992).
WRD: World Register of Dams (1988).
WRR: World Rivers Review, Volumes 2(6)–7(1).

**TABLE G.11** Populations displaced as a consequence of dam construction *(pages 359–361)*

## DESCRIPTION
The construction of dams, the filling of reservoirs, and the subsequent flooding of land is often accompanied by the displacement of populations living in the area. Listed here are major dam projects in which people were removed from their land and communities. The table includes the dam name, country or countries involved, the reservoir area (km²), number of people estimated displaced, installed electric capacity (MW), and the date the dam was completed. These data were compiled from many sources, as described below. Two separate sections are presented: dams completed and dams under construction.

## LIMITATIONS
Many of the figures used were the best estimates available,

based on World Bank data or other estimates. Data on populations displaced are often not accurately kept for political reasons. Data for dams planned or under construction are estimates based on current construction plans and existing local populations.

## SOURCES
These data were compiled from many sources by C. Chiang, Pacific Institute for Studies in Development, Environment, and Security, Oakland, California.
A.K. Biswas, 1982, Environment and sustainable water development, in *Water for Human Consumption*, IRWA Water Resources Series vol. 2.

International Commission on Large Dams, 1988, *World Registry of Dams*, International Commission on Large Dams, Paris. By permission.
International Rivers Network, 1987, *International Dams Newsletter*, vols. 1–2, San Francisco, California.
International Rivers Network, 1987–1991, *World Rivers Review*, vols. 2–6, San Francisco and Berkeley, California.
J. Majot and O. Lammers, 1992 International Rivers Network, Berkeley, California, personal communication.
T. Nachowitz, 1992, Department of Anthropology, Syracuse University, personal communication.

**TABLE G.12** Hydropower and mini-hydropower potential of least developed and selected Pacific island countries

| | Area (km²) | Technical hydrower potential (MW) | Economic hydropower potential (MW) | Existing hydropower generating capacity (MW) | Total installed capacity, 1984[f] (MW) | Cost of completed mini-hydropower per kW installed (U.S. $) | Existing mini-hydro generating capacity (MW) | Identified mini-hydropower project sites — Number | Identified mini-hydropower project sites — MW |
|---|---|---|---|---|---|---|---|---|---|
| *Least developed countries* | | | | | | | | | |
| Afghanistan | 652,090 | 5,000 | | 265 | 433.6[c] | | | | |
| Bangladesh | 143,998 | 1,188 | | 130 | 1,121 | | | 12 | 1.275 |
| Bhutan | 47,000 | 20,000 | | 3.5 | 19.1 | | 3.5 | | |
| Lao People's Democratic Republic | 236,800 | 12,683[e] | | 153.8 | 178.75[g] | | 3.8 | 10 | 45.7 |
| Maldives | 298 | | | | 3.5 | | | | |
| Nepal | 140,797 | | 27,000 | 113 | 138[c] | 1,200–4,000 | 2.72[d] | 16 | 1.8 |
| *Selected Pacific island countries* | | | | | | | | | |
| Fiji | 18,272 | 301 | | 80 | 188.8 | | | 20 | |
| Samoa | 2,935 | | | 4.63 | 18.77 | | 4.63 | 3 | 17.0 |
| Solomon Islands | 28,446 | | | | 8.0[i] | | | 2 | 21.0 |
| Tonga | 699 | | | | 6.55[h] | | | | |
| Vanuatu | 14,763 | | | | 6.37 | | | 2 | 3.94 |
| *Selected developing countries* | | | | | | | | | |
| China | 9,596,961 | 378,000 | | 25,600 | 80,116.9 | 400–1,000 | 8,500[a] | 70,000[a] | |
| India | 3,287,590 | | 89,000 | 13,058 | 38,808.0[c] | 1,250–2,500 | 236.5[b] | | 5,000–10,000 |
| Philippines | 300,000 | 8,000 | | 1,677 | 6,042.9 | 1,300–2,400 | 25–30[c] | 100 | 120 |

[a] As of 1983. Small-scale hydropower with unit capacity below 6,000 kW or plant capacity below 12,000 kW.
[b] As of 1980, 82 mini-hydro stations, up to a unit size of 5,000 kW.
[c] As of 1982.
[d] As of 1984.
[e] Excluding the potential hydropower of the Mekong mainstream projects.
[f] Includes hydro, steam, diesel, gas turbines, and geothermal.
[g] As of 1983. A total of $805 \times 10^6$ kWh was exported to Thailand from July 1984 to June 1985.
[h] As of 1982–83.
[i] As of 1982, excluding capacity of self-generating industries.

**TABLE G.12** Hydropower and mini-hydropower potential of least developed and selected Pacific island countries *(page 362)*

**DESCRIPTION**

A variety of hydroelectric potential and generation data are given here for fifteen countries of the Asia-Pacific region. Included in the table are country area in km$^2$; the technical potential, economic potential, and existing hydroelectric generating capacity, in MW; the total electrical generating installed capacity, in MW for 1984, and a set of data on the cost and size of potential and existing mini-hydroelectric facilities.

**LIMITATIONS**

No standard definition of "mini-hydro" is used in this region, so the mini-hydro numbers here are not strictly comparable. Incomplete information is available on hydroelectric potential and mini-capacity among the different nations of this region.

**SOURCE**

United Nations, 1987, *Water Resources Development in Asia and the Pacific: Some Issues and Concerns*, Economic and Social Commission for Asia and the Pacific (ESCAP), Water Resources Series no. 62, New York.

**TABLE G.13** Exploitable hydropower resources of major river systems in China

| River System | Installable capacity[a] (MW) | Annual energy (GWh) | Percentage of total[b] (percent) |
|---|---|---|---|
| Yangzijiang | 197,243.3 | 1,027,498 | 53.4 |
| Huanghe | 28,003.9 | 116,991 | 6.1 |
| Zhujiang | 24,850.2 | 112,478 | 5.8 |
| Haihe and Luanhe | 2,134.8 | 5,168 | 0.3 |
| Huaihe | 660.1 | 1,894 | 0.1 |
| Rivers in northeast China | 13,707.5 | 43,942 | 2.3 |
| Rivers in the coastal regions of southeast China | 13,896.8 | 54,741 | 2.8 |
| International rivers in southwest China | 37,684.1 | 209,868 | 10.9 |
| Yarlung Zangbo and other rivers in Tibet | 50,382.3 | 296,858 | 15.4 |
| Inland rivers in north China and Xinjiang | 9,969.7 | 53,866 | 2.8 |
| Total | 378,532.4 | 1,923,304 | 100 |

Sum of column may not add to total, due to independent rounding.
[a]The hydroelectric sites with installable capacity of less than 500 kW are not included.
[b]The percentage of the total refers to the annual energy.

**TABLE G.13** Exploitable hydropower resources of major river systems in China

**DESCRIPTION**

An enormous amount of hydroelectricity could, in theory, be generated in China, given its large rivers and mountainous terrain. Listed here are the "exploitable" capacity and the potential annual hydroelectricity generation, in MW and GWh, for different river basins in China. The percentage of total hydroelectric generation that each river basin could provide is also given.

**LIMITATIONS**

No precise definition of "exploitable" is given in the original source, though a figure for total hydropower potential of 676 GW is given, compared to the "exploitable capacity" of 379 GW shown here. This suggests that some measure of economic and technical potential has been applied.

**SOURCE**

Government of China, 1983, *Electric Power Industry in China, 1983*, Scientific and Technical Information Industry, Ministry of Water Resources and Electric Power, People's Republic of China.

**TABLE G.14**  African hydroelectric development, 1920 to 1990

| | Historical development of installed hydro capacity (MW) | | | | | | | | Electricity generation in 1987 | | |
| | 1920–1929 | 1930–1939 | 1940–1949 | 1950–1959 | 1960–1969 | 1970–1979 | 1980–1990 | Total | Thermal (GWh) | Hydro (GWh) | Hydro generation as percentage of total supply |
|---|---|---|---|---|---|---|---|---|---|---|---|
| West Africa | 11 | | | 40 | 1,637 | 407 | 2,040 | 4,135 | 11,163 | 8,853 | 44 |
| East Africa | 17 | | | 172 | 160 | 461 | 589 | 1,399 | 1,855 | 3,721 | 67 |
| Central Africa | 5 | 65 | 34 | 717 | 133 | 566 | 2,111 | 3,631 | 449 | 8,687 | 95 |
| Southern Africa | 31 | 2 | 8 | 293 | 1,195 | 3,309[a] | 1,110 | 5,948 | 4,850 | 15,278 | 76 |
| Total | 64 | | | 1,222 | 3,125 | 4,743[a] | 5,850 | 15,113 | 18,317 | 36,539 | 67 |

[a]Includes Cabora Bassa, Mozambique (2,074 MW).

**TABLE G.14**  African hydroelectric development, 1920 to 1990

## DESCRIPTION

An historical overview of additions to African hydroelectric capacity from 1920 to the present by decades is shown here, including actual hydroelectric and thermal generation in 1987. All four geographical regions produced more than 44% of their electricity from hydroelectric sources; Central Africa produced almost 100% of its electricity from hydroelectric facilities. The countries of each region are:

*West Africa:* Benin, Burkina Faso, Chad, Cote d'Ivoire, Gambia, Ghana, Guinea, Guinea-Bissau, Liberia, Mali, Mauritania, Niger, Nigeria, Senegal, Sierra Leone, Togo

*East Africa:* Djibouti, Ethiopia, Kenya, Somalia, Sudan, Uganda

*Central Africa:* Burundi, Cameroon, Central African Republic, Congo, Gabon, Rwanda, Zaire

*Southern Africa:* Angola, Botswana, Lesotho, Malawi, Mozambique, Swaziland, Tanzania, Zambia, Zimbabwe

## LIMITATIONS

The data for 1980–1990 were based on total installed capacities in 1990 from the IWPDC data base and were not verified in the original source. Note that the 1980–1990 data represent 11 years, rather than the 10 years in each previous column. In the original source, the total for 1980–1990 and the sum of all decadal columns are incorrect due, we believe, to a mathematical error. The totals given here reflect the correct sums. Note the differences between these data and those of Tables G.1–G.4.

## SOURCES

International Water Power and Dam Construction, 1991, *Water Power and Dam Construction*, Quadrant house, Sutton, United Kingdom, March Reprinted with permission

**TABLE G.15**  Reservoir capacity of the United States
(by Water Resources Region)

| Region | Area of region ($10^3$ km$^2$) | Average renewable supply[b] (km$^3$/yr) | Normal reservoir capacity[a] | |
|---|---|---|---|---|
| | | | Total (km$^3$) | As a percent of annual renewable supply |
| New England | 178.7 | 108.5 | 16.0 | 15 |
| Mid-Atlantic | 266.8 | 111.6 | 12.7 | 11 |
| South Atlantic-Gulf | 701.9 | 323.0 | 47.7 | 15 |
| Great Lakes | 347.1 | 102.8 | 8.5 | 8 |
| Ohio (exclusive of Tennessee region) | 414.4 | 193.0 | 24.2 | 13 |
| Tennessee | 111.4 | 57.0 | 13.8 | 24 |
| Upper Mississippi (exclusive of Missouri region) | 469.0 | 106.8 | 15.0 | 14 |
| Mississippi (entire basin) | 3,214.2 | 642.3 | 203.2 | 32 |
| Souris-Red Rainy | 142.5 | 9.0 | 9.9 | 110 |
| Missouri | 1,323.5 | 86.5 | 103.9 | 120 |
| Arkansas-White-Red | 632.0 | 94.9 | 39.2 | 41 |
| Texas Gulf | 461.0 | 45.8 | 30.5 | 67 |
| Rio Grande | 354.8 | 7.1 | 12.8 | 182 |
| Upper Colorado | 266.8 | 20.3 | 46.5 | 229 |
| Colorado (entire basin) | 668.2 | 21.6 | 86.8 | 402 |
| Great Basin | 360.0 | 13.7 | 4.1 | 30 |
| Pacific Northwest | 701.9 | 382.1 | 75.1 | 20 |
| California | 427.4 | 97.1 | 47.8 | 49 |
| Alaska | 1,517.7 | 1,349.5 | 1.8 | 0.1 |
| Hawaii | 15.5 | 10.2 | << 0.1 | 0.0 |
| Caribbean | 10.4 | 7.1 | 0.4 | 5 |

[a]Normal reservoir capacity is about two-thirds maximum capacity.
[b]Average renewable supply is adjusted by adding exports and subtracting imports.

**TABLE G.15**  Reservoir capacity of the United States

**DESCRIPTION**

The total reservoir capacity, in km$^3$, is presented here for each of the 21 water resource regions of the United States. The area of each region in km$^2 \times 10^3$, and the average renewable water supply in km$^3$ per year, are also given. Normal reservoir supply is defined as two-thirds of the maximum reservoir capacity. Average renewable supply includes regional water supply plus water exports less imports. Data are for 1983.

**LIMITATIONS**

The figures for average renewable supply hide substantial year-to-year fluctuations. No data are given on the period of measurement. "Normal" reservoir capacity is not indicative of actual capacity or the monthly variation in water supply. Actual storage can be considerably higher or lower than the normal capacity.

**SOURCE**

United States Department of the Interior, 1984, *National Water Summary 1983 -Hydrologic Events and Issues*, U.S. Geological Survey Water Supply Paper no. 2250, Washington, DC.

**TABLE G.16**   Reservoirs in the United States by primary purpose

| Primary purpose | Number of reservoirs | Maximum storage volume (km³) |
|---|---|---|
| Irrigation | 6,329 | 185 |
| Hydroelectricity | 1,372 | 210 |
| Flood control | 7,776 | 360 |
| Navigation | 187 | 150 |
| Water supply | 7,279 | 325 |
| Recreation | 16,639 | 320 |
| Debris control | 344 | 230 |
| Farm ponds | 4,546 | 240 |
| Others | 4,779 | 130 |

**TABLE G.16**   Reservoirs in the United States by primary purpose

**DESCRIPTION**

Shown here are the primary purpose for which over 40,000 reservoirs in the United States were built. This table does not distinguish between single purpose and multipurpose dams. Maximum reservoir storage is also given, in km³.

**LIMITATIONS**

Reservoirs almost always serve multiple purposes. Separating multi-purpose dams by primary purpose can present a misleading picture, since the declared ''primary purpose'' may not be the major economic purpose. This table was compiled in the late 1970s and should be updated for new facilities and for changes in operating criteria.

**SOURCE**

United States Army Corps of Engineers, 1977, *Estimate of National Hydroelectric Power Potential at Existing Dams*, Institute for Water Resources, Fort Belvoir, Virginia.

**TABLE G.17** Water storage capacity and the capacity lost annually due to sedimentation in the coterminous United States

| Farm production region | Total water storage capacity[a] (km³) | Usable water storage capacity[a] (km³) | Water storage capacity lost (%/yr) | Water storage capacity lost (10⁶ m³/yr) | Sediment originating on cropland (%) | Reservoir sedimentation from cropland (10⁶ m³/yr) |
|---|---|---|---|---|---|---|
| Northeast states | 45.0 | 31.1 | 0.08 | 34.6 | 29 | 10.1 |
| Appalachian states | 73.4 | 37.7 | 0.13 | 93.1 | 29 | 27.0 |
| Southeast states | 90.7 | 58.7 | 0.17 | 157.0 | 33 | 51.8 |
| Lake states | 36.1 | 24.0 | 0.27 | 97.5 | 64 | 62.4 |
| Corn Belt states | 49.0 | 18.7 | 0.26 | 129.2 | 63 | 81.4 |
| Delta states | 52.6 | 24.8 | 0.21 | 107.9 | 41 | 44.3 |
| Northern Plains states | 97.3 | 67.1 | 0.23 | 227.6 | 36 | 82.0 |
| Southern Plains states | 136.0 | 57.5 | 0.19 | 255.7 | 19 | 48.6 |
| Mountain states | 206.0 | 170.3 | 0.18 | 373.0 | 8 | 29.8 |
| Pacific states | 111.8 | 92.1 | 0.49 | 544.5 | 9 | 49.0 |
| 48 states total | 898.0 | 582.1 | 0.22 | 2,020.3 | | 486.3 |

Northern Plains include North Dakota, South Dakota, Nebraska, and Kansas.
Southern Plains include Oklahoma and Texas.
Mountain states include Montana, Idaho, Wyoming, Colorado, New Mexico, Arizona, Utah, and Nevada.
Pacific states include Washington, Oregon, and California.
Lake states include Minnesota, Wisconsin, and Michigan.
Corn Belt states include Iowa, Missouri, Illinois, Indiana, and Ohio.
Delta states include Arkansas, Louisiana, and Mississippi.
Southeast states include Alabama, Georgia, South Carolina, and Florida.
Appalachian states include Kentucky, Tennessee, West Virginia, Virginia, and North Carolina.
Northeast states include Maine, New Hampshire, Vermont, New York, Pennsylvania, Maryland, Delaware, Connecticut, Rhode Island, New Jersey, and Massachusetts.
[a]Reservoirs with 6.2 million cubic meters or more total capacity are included—approximately 98% of U.S. storage.

**TABLE G.17** Water storage capacity and the capacity lost annually due to sedimentation in the coterminous United States

### DESCRIPTION
All reservoirs experience sedimentation through the settling of suspended materials that enter the reservoir in streams and rivers. Reservoirs in regions with high silt loads are losing storage capacity at a rapid rate. This table shows the total water storage capacity (including dead storage) in the United States by farm production region and the useable storage capacity (defined as the amount of water available for release below the maximum controllable level), in km³. The storage capacity lost, in per cent per year and in m³ × 10⁶ per year, is also estimated, together with the fraction of sediment that originates on cropland. Sedimentation rates vary from 0.08% to 0.49% per year and average 0.22% per year for the U.S. Approximately one-fourth of all sediment originates on cropland, though this fraction varies considerably from region to region.

### LIMITATIONS
A great deal of variation in sedimentation rates, causes, and reservoir capacity losses is found among river basins. This variation is hidden in the regional averages shown here. One difficulty in estimating reservoir volume losses is the lack of information on the condition of sediment pools in reservoirs. Sediment pools are portions of reservoirs originally allocated to sediment storage. Information on the amount of remaining capacity in existing sediment pools is not available, so it is not possible to separate the loss of dead storage from the loss of useable storage. No estimates for Alaska and Hawaii are included here.

### SOURCES
B.M. Crowder, 1987, Economic costs of reservoir sedimentation: A regional approach to estimating cropland erosion damages, *Journal of Soil and Water Conservation*, **42**(3), 194–197.

**TABLE G.18**   Installed hydroelectric capacity per unit area inundated, for selected projects

| Project | Country | Rated Capacity (MW)[a] | Normal area of reservoir (hectares)[a] | Area inundated (kW/hectare)[a] |
|---|---|---|---|---|
| Pehuenche | Chile | 500 | 400 | 1,250 |
| Guavio | Colombia | 1,600 | 1,500 | 1,067 |
| Rio Grande II | Colombia | 324 | 1,100 | 295 |
| Paulo Afonso | Brazil | 1,299 | 7,520 | 173 |
| Aguamilpa | Mexico | 960 | 12,000 | 80 |
| Sayanskaya | USSR | 6,400 | 80,000 | 80 |
| Churchill Falls | Canada | 5,225 | 66,500 | 79 |
| Itaipú | Brazil, Paraguay | 10,500 | 135,000 | 77 |
| Grand Coulee | United States | 2,025 | 32,400 | 63 |
| Urra I | Colombia | 340 | 6,200 | 55 |
| Jupía | Brazil | 1,400 | 33,300 | 42 |
| São Simão | Brazil | 2,680 | 66,000 | 41 |
| Tucuruí | Brazil | 6,480 | 216,000 | 30 |
| Ilha Solteira | Brazil | 3,200 | 120,000 | 27 |
| Guri | Venezuela | 6,000 | 328,000 | 18 |
| Paredao | Brazil | 40 | 2,300 | 17 |
| Urra II | Colombia | 860 | 54,000 | 16 |
| Cabora Bassa | Mozambique | 4,000 | 380,000 | 14 |
| Three Gorges | China | 13,000 | 110,000 | 12 |
| Furnas | Brazil | 120 | 135,000 | 9 |
| Samuel | Brazil | 110 | 15,000 | 7 |
| Curua-Una | Brazil | 40 | 8,600 | 5 |
| Aswan High Dam | Egypt | 2,100 | 510,000 | 4 |
| Trés Maria | Brazil | 400 | 105,200 | 4 |
| Kariba | Zimbabwe, Zambia | 1,500 | 510,000 | 3 |
| Sobradinho | Brazil | 900 | 450,000 | 2 |
| Balbina | Brazil | 250 | 124,000 | 2 |
| Akosombo | Ghana | 833 | 848,200 | 1 |
| Brokopondo | Suriname | 30 | 150,000 | 0.2 |

[a]Approximate data only.

**TABLE G.18**   Installed hydroelectric capacity per unit area inundated, for selected projects

**DESCRIPTION**

One indicator of environmental impact of a hydroelectric facility is the amount of land flooded behind a hydroelectric dam. Computed here is the installed capacity of a group of projects per unit area inundated, in kW per hectare. Note the extremely wide range of impacts.

**LIMITATIONS**

The area of a reservoir fluctuates significantly on a seasonal and annual basis, depending on climate, electricity demand and generation, and other water use. These numbers should be considered approximate. The economic or environmental value of the land is not factored into this assessment. Some errors in the original table were corrected here.

There are many possible measures of environmental impacts. Perhaps more revealing is to use the amount of energy a facility generates rather than its installed capacity. This permits comparison with other energy facilities. Such a comparison for a set of United States hydroelectric plants is given in Table G.20.

**SOURCE**

R. Goodland, 1990, The World Bank's new environmental policy on dam and reservoir projects, *International Environmental Affairs*, **2**(2), 109–129.

**TABLE G.19** Annual evaporative water losses from California hydroelectric facilities ($m^3/10^3$ kWh/yr)

| Category | All facilities | Plants over 25 MWe | Plants over 25 MWe |
|---|---|---|---|
| All facilities: range | 0.04–210 | 0.04–160 | 0.2–210 |
| All facilities: median | 5.4 | 2.4 | 14 |
| DH < GSH: range | 0.04–120 | 0.04–120 | 0.2–83 |
| DH < GSH: median | 1.2 | 0.7 | 3 |
| DH > GSH: range | 1.9–210 | 3.6–160 | 1.9–210 |
| DH > GSH: median | 34 | 68 | 34 |

DH, dam height; GSH, gross static head.

**TABLE G.19** Annual evaporative water losses from California hydroelectric facilities

**DESCRIPTION**

There has been considerable debate about the environmental impacts of hydroelectric development as a function of size and type. An important impact is the evaporative water loss from the surface of hydroelectric reservoirs, which represents a consumptive use of water. Presented here for different sets of facilities are the estimated evaporative losses for approximately 100 California hydroelectric facilities, in $m^3$ lost per $10^3$ kWh of electricity generated per year. This measure, rather than one that uses total lifetime energy produced by a plant as the denominator, was chosen because (i) it does not require assumptions about plant lifetime; (ii) it permits normalization and comparison with different types of energy facilities; and (iii) it is easily converted to other indices without confusion. The annual energy values are computed from long-term averages, typically 35 years. The facilities are broken down by size (greater and less than 25 MW installed capacity) and by type (dam height greater and less than gross static head). The gross static head is the distance between the surface of the water in the reservoir and the surface of the water in the tailrace. Ranges and median values are provided.

**LIMITATIONS**

These reservoirs are representative of facilities in the western United States, but may not be characteristic of facilities in other regions. California tends to have greater hydroelectric head per unit land flooded than other regions in the United States due to its mountainous terrain, where many hydroelectric facilities are located. Evaporative losses from open water in this region are approximately 1 m per year. Losses in other regions will range from less than 0.5 m per year to over 3 m per year.

**SOURCE**

P.H. Gleick, 1992, Environmental consequences of hydroelectric development: The role of facility size and type, in *Energy*, Vol. 17, No. 8, pp. 735–747, Pergamon Press, Oxford.

**TABLE G.20**  Land inundated for California
hydroelectric facilities
($m^2/10^3$ kWh/year)

| Category | All facilities | Plants over 25 MWe | Plants under 25 MWe |
|---|---|---|---|
| All facilities: range | 0.04–210 | 0.04–160 | 0.2–210 |
| All facilities: median | 5.4 | 2.4 | 14 |
| DH < GSH: range | 0.04–120 | 0.04–120 | 0.2–83 |
| DH < GSH: median | 1.2 | 0.7 | 3 |
| DH > GSH: range | 1.9–210 | 3.6–160 | 1.9–210 |
| DH > GSH: median | 34 | 68 | 34 |

DH, dam height; GSH, gross static head.

**TABLE G.20**  Land inundated for California
hydroelectric facilities

**DESCRIPTION**

There has been considerable debate about the environmental impacts of hydroelectric development as a function of size and type. An important impact is the amount of land inundated by the development of hydroelectric reservoirs. Presented here for different sets of facilities are estimated land lost for approximately 100 California hydroelectric facilities, in $m^3$ lost per $10^3$ kWh of electricity generated per year. This measure, rather than one that uses total lifetime energy produced by a plant (or the installed capacity as in Table G.18) as the denominator, was chosen because (i) it does not require assumptions about plant lifetime or capacity factors; (ii) it permits normalization and comparison with different types of energy facilities; and (iii) it is easily converted to other indices without confusion. The annual energy values are computed from long-term averages, typically 35 years. The facilities are broken down by size (greater and less than 25 MW installed capacity) and by type (dam height greater and less than gross static head). The gross static head is the distance between the surface of the water in the reservoir and the surface of the water in the tailrace. Ranges and median values are provided.

**LIMITATIONS**

These reservoirs are representative of facilities in the western United States, but may not be characteristic of facilities in other regions. California tends to have greater hydroelectric head per unit land flooded than other regions in the United States due to its mountainous terrain, where many hydroelectric facilities are located. The amount of land flooded was estimated from the surface area of the reservoir. Actual land flooded varies from this number depending on the topography of the site. Additional environmental and ecological impacts extend beyond the limit of the reservoir area, by displacing wildlife, which causes pressures on surrounding habitat.

**SOURCE**

P.H. Gleick, 1992, Environmental consequences of hydroelectric development: The role of facility size and type, *Energy*, Vol 17, No. 8, pp. 735–747, Pergamon Press, Oxford.

**TABLE G.21**  Expected water pollution from selected energy systems *(page 371)*

**DESCRIPTION**

The production and use of conventional energy systems produces a wide range of water pollutants. Those systems that produce significant quantities of pollutants are shown here with the amount of different pollutants emitted per $10^9$ J of energy produced. These figures are average figures for hypothetical energy plants; actual emissions will vary depending on plant characteristics, mode of operation, and the type of pollution controls added. These data were converted from Btu and the number of significant figures in the original were retained here.

**LIMITATIONS**

There are many gaps in this table and there are many portions of energy fuel cycles not included. For example, no data are available for emissions from energy transportation, secondary or onshore oil and gas extraction, nuclear waste disposal, catastrophic nuclear accidents, power plant construction, and so on. There is a wide range in emissions depending on specific plant type and mode of operation. Not all pollutants are included either. Among the more significant omissions are radioactive pollutants, temperature effects, and measures of biological impact such as biological oxygen demand.

**SOURCES**

Compiled in U.S. Department of Energy, 1981, *Energy and Water Resources*, DOE/EV/10154-4, Office of Environmental Programs, Washington, DC.
Data from U.S. Department of Energy, 1980, *Technology Characterizations: Environmental Information Handbook*, DOE/EV/0072, Washington, DC.

**TABLE G.21** Expected water pollution from selected energy systems (tonnes/$10^{15}$ J)

| Pollutant | Uranium hexafluoride conversion | Light water nuclear reactor | High temperature gas reactor | Electric power plant coal | Electric power plant oil | Electric power plant gas | Surface coal mining eastern | Surface coal mining western | Underground coal mining eastern | Coal beneficiation | Oil extraction offshore[a] | Gas extraction offshore[a] | Geothermal vapor dominated[b] |
|---|---|---|---|---|---|---|---|---|---|---|---|---|---|
| Aluminum | | | | 0.26 | 0.98 | 0.97 | 0.03 | 0.005 | 1.5 | 0.03 | | | 73.9 |
| Ammonia | 0.06 | | | 0.05 | 0.06 | | 0.02 | 0.2 | 0.3 | 0.04 | | | |
| Arsenic | | | | | 0.003 | | | | | | | | |
| Bicarbonates | | | 83.8 | | | | | | | | | | |
| Boron | | 12.4 | | | | | | | | | | | 8.60 |
| Cadmium | | | 0.03 | | 0.004 | | | | | | 0.02 | 0.005 | |
| Carbonates | | | 49.2 | | | | | | 119 | | | | 2.7 |
| Chlorides | 0.009 | 1.0 | 17.7 | | 50.5 | 41.2 | | | 10.9 | | 18,583 | 4,852 | 1.7 |
| Chromium | | 0.009 | | 0.009 | 0.03 | 0.03 | | | | | | | |
| Copper | | | 0.07 | | 0.3 | 0.03 | | | | | | | |
| Fluorides | 1.0 | | | | | | | | 0.09 | | | | |
| Iron | 0.002 | | 0.09 | | 0.5 | 0.5 | 0.003 | 0.02 | 0.2 | 0.05 | | | |
| Lead | | | 0.05 | | T | | | | | | | | |
| Magnesium | | | 14.2 | | 60.1 | 59.9 | | | 0.2 | | | | 0.5 |
| Manganese | | | 0.5 | | | | 0.09 | 0.02 | | 0.03 | | | |
| Mercury | | | | | T | T | 0.002 | 0.0009 | 0.02 | 0.03 | T | T | |
| Nickel | 0.009 | | | 3.11 | 3.76 | 3.76 | | | | | | | 0.05 |
| Nitrates | | | 9.8 | 1.61 | 0.5 | 0.51 | | | | | | | |
| Phosphates | | 1.6 | 0.31 | 0.15 | 0.03 | 0.02 | | | | | | | |
| Potassium | | | 6.0 | | | | | | | | | | |
| Selenium | | | | | 0.009 | T | | | | | | | |
| Silicon dioxide | | | 7.3 | | | | | | | | | | 1.9 |
| Sodium | 0.14 | | 11.3 | | 71.4 | 71.4 | | | | | | | |
| Sulfates | 0.18 | | 109 | 35.3 | 230 | 226 | 16.6 | 32.4 | 150 | 15.5 | | | 65.5 |
| Zinc | | | 0.05 | 0.04 | 0.03 | 0.3 | 0.002 | 0.004 | 0.02 | 0.004 | | | 65.5 |

T, trace amounts.

[a]Data values represent annual raw unprocessed water pollutants from average Gulf Coast brine. Concentrations and average brine production per platform, all normalized to $10^9$ J.

[b]No treatment involved, reinjected through wells. Values indicate actual emissions from the Geysers field in California.

**TABLE G.22**   Energy requirements for desalting water

| Technology | Present requirements ($10^6$ J/m$^3$) | Future requirements ($10^6$ J/m$^3$) |
|---|---|---|
| Distillation | 210 | 90 |
| Freezing | 110 | 60 |
| Reverse osmosis (seawater) | 90 | 25[b] |
| Reverse osmosis (brackish water) | 14 | 7 |
| Electrodialysis (seawater) | 150 | 70 |
| Electrodialysis (brackish water) | 20–40 | 10–20 |

[a]The theoretical minimum energy requirement to remove salt from water is 2.8 million J/m3.
[b]With energy recovery.

**TABLE G.22**   Energy requirements for desalting water

**DESCRIPTION**

The greatest barrier to the widespread use of desalination technology is its economic cost, which in large part depends on the amount of energy required to remove salt from water. Shown here are the present energy requirements for desalinating water, using typical equipment commercially available, in J × $10^6$/m$^3$ of water. Data are given for six technologies. Also shown are the potential energy-efficiency improvements that can be achieved given expected changes in technology. In all cases, a reduction in energy requirements of about a factor of two appear to be possible. The theoretical minimum energy required to remove salt from water is $2.8 \times 10^6$ J/m$^3$ (Howe, 1974).

**LIMITATIONS**

These data are for the late 1970s and there have already been improvements in technology. As a result, the most energy efficient versions now available fall somewhere in between the values shown here for present and future requirements. The final cost of desalinated water depends on the type and cost of energy as well on the construction, operating, and maintenance costs of the plant.

**SOURCES**

E.D. Howe, 1974, *Fundamentals of Water Desalination*, Marcel Dekker, New York.

U.S. Department of the Interior, 1978, *Desalting Plans and Progress: An Evaluation of the State-of-the-Art and Future Research and Development Requirements*, Fluor Engineering and Constructors, Inc., Irvine, California.

J.D. Birkett, 1985, Alternative water resources in the Middle East, *U.S.–Arab Commerce*, November/December.

# H. Water and human use

**TABLE H.1** Fresh water withdrawal by region, country, and sector *(pages 374–378)*

## DESCRIPTION

The availability, withdrawal, and use of fresh water resources vary greatly throughout the world. For the continents and countries, this table gives total annual renewable water resources, including water from both internal supplies and external stream flows, and total annual water withdrawals for a specified year, in km³ per year. Also shown for each country are the percentages of renewable fresh water supply that are withdrawn annually, per capita water withdrawals in m³ per person per year, and the percentages of water used in the domestic, industrial, and agricultural sectors.

"Withdrawal" refers to water transferred from its source to its place of use, but not necessarily consumed. The domestic sector includes household and municipal uses, and water used for commercial establishments and public services. Industrial use includes water withdrawn to cool thermoelectric power plants. The agricultural sector includes water for irrigation and livestock. Per capita data were prepared using national population data from the year of the withdrawal data.

## LIMITATIONS

Data on water withdrawals and on how that water is used are vital pieces of information that should be available for every country and region on earth. Yet there are no standards for regular collection and dissemination of these data. As a result, this table is an amalgam of measured, modeled, and derived data for a wide variety of years. Ex-

treme care should be taken when applying these data for other purposes. For example, the data on water withdrawals by country are for different years, and some withdrawal data are over 20 years old. The data on annual renewable water resources by country are for average supply and include water that may flow out of the country. For details on the breakdown of internal and external water flows for each country, see Table A.10. The periods of record and the methods of measurement vary from country to country and are not described in the original source. These averages hide large seasonal and interannual variations in water availability. Data for small, arid countries may be less reliable than those for larger, wetter countries. Some of the water use data are based on national reports and observations; some data were "derived" with computer models using estimates from other data sources (see the section describing problems with derived data in the "About the Data" section). For example, all data provided by the Institute of Geography of the former Soviet Union are "computed," not measured, using other national data, such as area under irrigated agriculture, livestock populations, and precipitation.

## SOURCE

World Resources Institute, 1992, *World Resources 1992–93: A Guide to the Global Environment*, Oxford University Press, New York. By permission.

**TABLE H.1** Fresh water withdrawal by region, country, and sector

| | Annual renewable water resources[a] (km³/yr) | Year | Total fresh water withdrawal (km³/yr) | Fraction of annual renewable water resources withdrawn (%) | Per capita withdrawal (m³/person/yr) | Domestic use (%) | Industrial use (%) | Agricultural use (%) |
|---|---|---|---|---|---|---|---|---|
| *World* | 40,673 | 1987[b] | 3,240 | 8 | 660 | 8 | 23 | 69 |
| *Africa* | | 1987[b] | 144 | 3 | 244 | 7 | 5 | 88 |
| Algeria | 19 | 1980 | 3.00 | 16 | 161 | 22 | 4 | 74 |
| Angola | 158 | 1987[b] | 0.48 | 0 | 43 | 14 | 10 | 76 |
| Benin | 26 | 1987[b] | 0.11 | 0 | 26 | 28 | 14 | 58 |
| Botswana | 18 | 1980 | 0.09 | 1 | 98 | 5 | 10 | 85 |
| Burkina Faso | 28 | 1987[b] | 0.15 | 1 | 20 | 28 | 5 | 67 |
| Burundi | 3.6 | 1987[b] | 0.10 | 3 | 20 | 36 | 0 | 64 |
| Cameroon | 208 | 1987[b] | 0.40 | 0 | 30 | 46 | 19 | 35 |
| Cape Verde | <1 | 1972 | 0.04 | 20 | 148 | 9 | 2 | 89 |
| Central African Republic | 141 | 1987[b] | 0.07 | 0 | 27 | 21 | 5 | 74 |
| Chad | 38 | 1987[b] | 0.18 | 0 | 35 | 16 | 2 | 82 |
| Comoros | 1.02 | 1987[b] | 0.01 | 1 | 15 | 48 | 5 | 47 |
| Congo | 802 | 1987[b] | 0.04 | 0 | 20 | 62 | 27 | 11 |
| Côte d'Ivoire | 74 | 1987[b] | 0.71 | 1 | 68 | 22 | 11 | 67 |
| Djibouti | <1 | 1973 | 0.01 | 2 | 28 | 28 | 21 | 51 |
| Egypt | 58 | 1985 | 56.40 | 97 | 1,202 | 7 | 5 | 88 |
| Equatorial Guinea | 30 | 1987[b] | 0.01 | 0 | 11 | 81 | 13 | 6 |
| Ethiopia | 110 | 1987[b] | 2.21 | 2 | 48 | 11 | 3 | 86 |
| Gabon | 164 | 1987[b] | 0.06 | 0 | 51 | 72 | 22 | 6 |
| Gambia | 22 | 1982 | 0.02 | 0 | 33 | 7 | 2 | 91 |
| Ghana | 53 | 1970 | 0.30 | 1 | 35 | 35 | 13 | 52 |
| Guinea | 226 | 1987[b] | 0.74 | 0 | 115 | 10 | 3 | 87 |
| Guinea-Bissau | 31 | 1987[b] | 0.01 | 0 | 18 | 31 | 6 | 63 |
| Kenya | 15 | 1987[b] | 1.09 | 7 | 48 | 27 | 11 | 62 |
| Lesotho | 4.0 | 1987[b] | 0.05 | 1 | 34 | 22 | 22 | 56 |
| Liberia | 232 | 1987[b] | 0.13 | 0 | 54 | 27 | 13 | 60 |
| Libya | 0.7 | 1985 | 2.83 | 404 | 623 | 15 | 10 | 75 |
| Madagascar | 40 | 1984 | 16.30 | 41 | 1,675 | 1 | 0 | 99 |
| Malawi | 9.0 | 1987[b] | 0.16 | 2 | 22 | 34 | 17 | 49 |
| Mali | 62 | 1987[b] | 1.36 | 2 | 159 | 2 | 1 | 97 |
| Mauritania | 7.4 | 1978 | 0.73 | 10 | 473 | 12 | 4 | 84 |
| Mauritius | 2.2 | 1974 | 0.36 | 16 | 415 | 16 | 7 | 77 |
| Morocco | 30 | 1985 | 11.00 | 37 | 501 | 6 | 3 | 91 |

*continued*

| | Annual renewable water resources[a] (km³/yr) | Year | Total fresh water withdrawal (km³/yr) | Fraction of annual renewable water resources withdrawn (%) | Per capita withdrawal (m³/person/yr) | Domestic use (%) | Industrial use (%) | Agricultural use (%) |
|---|---|---|---|---|---|---|---|---|
| Mozambique | 58 | 1987[b] | 0.76 | 1 | 53 | 24 | 10 | 66 |
| Namibia | 9 | 1987[b] | 0 | 2 | 77 | 6 | 12 | 82 |
| Niger | 44 | 1987[b] | 0.29 | 1 | 44 | 21 | 5 | 74 |
| Nigeria | 308 | 1987[b] | 3.63 | 1 | 44 | 31 | 15 | 54 |
| Rwanda | 6.3 | 1987[b] | 0.15 | 2 | 23 | 24 | 8 | 68 |
| Senegal | 35 | 1987[b] | 1.36 | 4 | 201 | 5 | 3 | 92 |
| Sierra Leone | 160 | 1987[b] | 0.37 | 0 | 99 | 7 | 4 | 89 |
| Somalia | 12. | 1987[b] | 0.81 | 7 | 167 | 3 | 0 | 97 |
| South Africa | 50 | 1970 | 9.20 | 18 | 404 | 16 | 17 | 67 |
| Sudan | 130 | 1977 | 18.60 | 14 | 1,089 | 1 | 0 | 99 |
| Swaziland | 6.96 | 1987[b] | 0.29 | 4 | 414 | 5 | 2 | 93 |
| Tanzania | 76 | 1970 | 0.48 | 1 | 36 | 21 | 5 | 74 |
| Togo | 12 | 1987[b] | 0.09 | 1 | 40 | 62 | 13 | 25 |
| Tunisia | 4.35 | 1985 | 2.30 | 53 | 325 | 13 | 7 | 80 |
| Uganda | 66 | 1970 | 0.20 | 0 | 20 | 32 | 8 | 60 |
| Zaire | 1,019 | 1987[b] | 0.70 | 0 | 22 | 58 | 25 | 17 |
| Zambia | 96 | 1970 | 0.36 | 0 | 86 | 63 | 11 | 26 |
| Zimbabwe | 23 | 1987[b] | 1.22 | 5 | 129 | 14 | 7 | 79 |
| *North and Central America* | | 1987[b] | 697 | 10 | 1,692 | 9 | 42 | 49 |
| Barbados | <1 | 1962 | 0.03 | 51 | 117 | 52 | 41 | 7 |
| Belize | 16 | 1987[b] | 0.02 | 0 | | 10 | 0 | 90 |
| Canada | 2,901 | 1980 | 42.20 | 1 | 1,752 | 11 | 80 | 8 |
| Costa Rica | 95 | 1970 | 1.35 | 1 | 779 | 4 | 7 | 89 |
| Cuba | 35 | 1975 | 8.10 | 23 | 868 | 9 | 2 | 89 |
| Dominican Republic | 20 | 1987[b] | 2.97 | 15 | 453 | 5 | 6 | 89 |
| El Salvador | 19 | 1975 | 1.00 | 5 | 241 | 7 | 4 | 89 |
| Guatemala | 116 | 1970 | 0.73 | 1 | 139 | 9 | 17 | 74 |
| Haiti | 11 | 1987[b] | 0.04 | 0 | 46 | 24 | 8 | 68 |
| Honduras | 102 | 1970 | 1.34 | 1 | 508 | 4 | 5 | 91 |
| Jamaica | 8.3 | 1975 | 0.32 | 4 | 157 | 7 | 7 | 86 |
| Mexico | 357 | 1975 | 54.20 | 15 | 901 | 6 | 8 | 86 |
| Nicaragua | 175 | 1975 | 0.89 | 1 | 370 | 25 | 21 | 54 |
| Panama | 144 | 1975 | 1.30 | 1 | 744 | 12 | 11 | 77 |
| Trinidad and Tobago | 5.1 | 1975 | 0.15 | 3 | 149 | 27 | 38 | 35 |
| United States | 2,478 | 1985 | 467.00 | 19 | 2,162 | 12 | 46 | 42 |

*continued*

| | Annual renewable water resources[a] (km³/yr) | Year | Total fresh water withdrawal (km³/yr) | Fraction of annual renewable water resources withdrawn (%) | Per capita withdrawal (m³/person/yr) | Domestic use (%) | Industrial use (%) | Agricultural use (%) |
|---|---|---|---|---|---|---|---|---|
| *South America* | | 1987[b] | 133 | 1 | 476 | 18 | 23 | 59 |
| Argentina | 994 | 1976 | 27.60 | 3 | 1,059 | 9 | 18 | 73 |
| Bolivia | 300 | 1987[b] | 1.24 | 0 | 184 | 10 | 5 | 85 |
| Brazil | 6,950 | 1987[b] | 35.04 | 1 | 212 | 43 | 17 | 40 |
| Chile | 468 | 1975 | 16.80 | 4 | 1,625 | 6 | 5 | 89 |
| Colombia | 1,070 | 1987[b] | 5.34 | 0 | 179 | 41 | 16 | 43 |
| Ecuador | 314 | 1987[b] | 5.56 | 2 | 561 | 7 | 3 | 90 |
| Guyana | 241 | 1971 | 5.40 | 2 | 7,616 | 1 | 0 | 99 |
| Paraguay | 314 | 1987[b] | 0.43 | 0 | 111 | 15 | 7 | 78 |
| Peru | 40 | 1987[b] | 6.10 | 15 | 294 | 19 | 9 | 72 |
| Suriname | 200 | 1987[b] | 0.46 | 0 | 1,181 | 6 | 5 | 89 |
| Uruguay | 124 | 1965 | 0.65 | 1 | 241 | 6 | 3 | 91 |
| Venezuela | 1,317 | 1970 | 4.10 | 0 | 387 | 43 | 11 | 46 |
| | | | | | | | | |
| *Asia* | | 1987[b] | 1,531 | 15 | 526 | 6 | 8 | 86 |
| Afghanistan | 50 | 1987[b] | 26.11 | 52 | 1,436 | 1 | 0 | 99 |
| Bahrain | <1 | 1975 | 0.31 | >100 | 609 | 60 | 36 | 4 |
| Bangladesh | 2,357 | 1987[b] | 22.50 | 1 | 211 | 3 | 1 | 96 |
| Bhutan | 95 | 1987[b] | 0.02 | 0 | 15 | 36 | 10 | 54 |
| Cambodia | 498 | 1987[b] | 0.52 | 0 | 69 | 5 | 1 | 94 |
| China | 2,800 | 1980 | 460.00 | 16 | 462 | 6 | 7 | 87 |
| Cyprus | 0.9 | 1985 | 0.54 | 60 | 807 | 7 | 2 | 91 |
| India | 2,085 | 1975 | 380.00 | 18 | 612 | 3 | 4 | 93 |
| Indonesia | 2,530 | 1987[b] | 16.59 | 1 | 96 | 13 | 11 | 76 |
| Iran | 118 | 1975 | 45.40 | 39 | 1,362 | 4 | 9 | 87 |
| Iraq | 100 | 1970 | 42.80 | 43 | 4,575 | 3 | 5 | 92 |
| Israel | 2.15 | 1986 | 1.90 | 88 | 447 | 16 | 5 | 79 |
| Japan | 547 | 1980 | 107.80 | 20 | 923 | 17 | 33 | 50 |
| Jordan | 1.1 | 1975 | 0.45 | 41 | 173 | 29 | 6 | 65 |
| Korea DPR | 67 | 1987[b] | 14.16 | 21 | 1,649 | 11 | 16 | 73 |
| Korea Rep | 63 | 1976 | 10.70 | 17 | 298 | 11 | 14 | 75 |
| Kuwait | <1 | 1974 | 0.50 | >100 | 238 | 64 | 32 | 4 |
| Laos | 270 | 1987[b] | 0.99 | 0 | 228 | 8 | 10 | 82 |
| Lebanon | 4.8 | 1975 | 0.75 | 16 | 271 | 11 | 4 | 85 |
| Malaysia | 456 | 1975 | 9.42 | 2 | 765 | 23 | 30 | 47 |
| Mongolia | 25 | 1987[b] | 0.55 | 2 | 272 | 11 | 27 | 62 |
| Myanmar | 1,082 | 1987[b] | 3.96 | 0 | 103 | 7 | 3 | 90 |

*continued*

| | Annual renewable water resources[a] (km³/yr) | Year | Total fresh water withdrawal (km³/yr) | Fraction of annual renewable water resources withdrawn (%) | Per capita withdrawal (m³/person/yr) | Domestic use (%) | Industrial use (%) | Agricultural use (%) |
|---|---|---|---|---|---|---|---|---|
| Nepal | 170 | 1987[b] | 2.68 | 2 | 155 | 4 | 1 | 95 |
| Oman | 2.0 | 1975 | 0.48 | 24 | 325 | 3 | 3 | 94 |
| Pakistan | 468 | 1975 | 153.40 | 33 | 2,053 | 1 | 1 | 98 |
| Philippines | 323 | 1975 | 29.50 | 9 | 693 | 18 | 21 | 61 |
| Qatar | <1 | 1975 | 0.15 | 663 | 415 | 36 | 26 | 38 |
| Saudi Arabia | 2.2 | 1975 | 3.60 | 164 | 255 | 45 | 8 | 47 |
| Singapore | 0.6 | 1975 | 0.19 | 32 | 84 | 45 | 51 | 4 |
| Sri Lanka | 43 | 1970 | 6.30 | 15 | 503 | 2 | 2 | 96 |
| Syria | 36 | 1976 | 3.34 | 9 | 449 | 7 | 10 | 83 |
| Thailand | 179 | 1987[b] | 31.90 | 18 | 599 | 4 | 6 | 90 |
| Turkey | 203 | 1985 | 15.60 | 8 | 317 | 24 | 19 | 57 |
| United Arab Emirates | <1 | 1980 | 0.90 | 299 | 565 | 11 | 9 | 80 |
| Vietnam | 376 | 1987[b] | 5.07 | 1 | 81 | 13 | 9 | 78 |
| Yemen Arab Rep | 1.0 | 1987[b] | 1.47 | 147 | | 4 | 2 | 94 |
| Yemen Dem Rep | 1.5 | 1975 | 1.93 | 129 | 1,167 | 5 | 2 | 93 |
| *Europe* | | 1987[b] | 359 | 15 | 726 | 13 | 54 | 33 |
| Albania | 21 | 1970 | 0.20 | 1 | 94 | 6 | 18 | 76 |
| Austria | 90 | 1980 | 3.13 | 3 | 417 | 19 | 73 | 8 |
| Belgium | 13 | 1980 | 9.03 | 72 | 917 | 11 | 85 | 4 |
| Bulgaria | 205 | 1980 | 14.18 | 7 | 1,600 | 7 | 38 | 55 |
| Czechoslovakia | 91 | 1980 | 5.80 | 6 | 379 | 23 | 68 | 9 |
| Denmark | 13 | 1977 | 1.46 | 11 | 289 | 30 | 27 | 43 |
| Finland | 113 | 1980 | 3.70 | 3 | 774 | 12 | 85 | 3 |
| France | 185 | 1984 | 40.00 | 22 | 728 | 16 | 69 | 15 |
| German DR | 34 | 1980 | 9.13 | 27 | 545 | 14 | 68 | 18 |
| Germany FR | 161 | 1981 | 41.22 | 26 | 688 | 10 | 70 | 20 |
| Greece | 59 | 1980 | 6.95 | 12 | 721 | 8 | 29 | 63 |
| Hungary | 115 | 1980 | 5.38 | 5 | 502 | 9 | 55 | 36 |
| Iceland | 170 | 1987[b] | 0.09 | 0 | 349 | 31 | 63 | 6 |
| Ireland | 50 | 1972 | 0.79 | 2 | 267 | 16 | 74 | 10 |
| Italy | 187 | 1981 | 56.20 | 30 | 983 | 14 | 27 | 59 |
| Luxembourg | 5.0 | 1976 | 0.04 | 1 | 119 | 42 | 45 | 13 |
| Malta | <1 | 1978 | 0.02 | 92 | 68 | 76 | 8 | 16 |
| Netherlands | 90 | 1980 | 14.47 | 16 | 1,023 | 5 | 61 | 34 |
| Norway | 413 | 1980 | 2.00 | <1 | 489 | 20 | 72 | 8 |
| Poland | 56 | 1980 | 16.80 | 30 | 472 | 16 | 60 | 24 |

*continued*

| | Annual renewable water resources[a] (km³/yr) | Year | Total fresh water withdrawal (km³/yr) | Fraction of annual renewable water resources withdrawn (%) | Per capita withdrawal (m³/person/yr) | Domestic use (%) | Industrial use (%) | Agricultural use (%) |
|---|---|---|---|---|---|---|---|---|
| Portugal | 66 | 1980 | 10.50 | 16 | 1,062 | 15 | 37 | 48 |
| Romania | 208 | 1980 | 25.40 | 12 | 1,144 | 8 | 33 | 59 |
| Spain | 111 | 1985 | 45.25 | 41 | 1,174 | 12 | 26 | 62 |
| Sweden | 180 | 1980 | 3.98 | 2 | 479 | 36 | 55 | 9 |
| Switzerland | 50 | 1985 | 3.20 | 6 | 502 | 23 | 73 | 4 |
| United Kingdom | 120 | 1980 | 28.35 | 24 | 507 | 20 | 77 | 3 |
| Yugoslavia | 265 | 1980 | 8.77 | 3 | 393 | 16 | 72 | 12 |
| *USSR* | 4,684 | 1980 | 353 | 8 | 1,330 | 6 | 29 | 65 |
| *Oceania* | | 1987[b] | 23 | 1 | 907 | 64 | 2 | 34 |
| Australia | 343 | 1975 | 17.80 | 5 | 1,306 | 65 | 2 | 33 |
| Fiji | 29 | 1987[b] | 0.03 | 0 | 37 | 20 | 20 | 60 |
| New Zealand | 397 | 1980 | 1.20 | 0 | 379 | 46 | 10 | 44 |
| Papua New Guinea | 801 | 1987[b] | 0.10 | 0 | 25 | 29 | 22 | 49 |
| Solomon Islands | 45 | 1987[b] | 0.00 | 0 | 18 | 40 | 20 | 40 |

Figures may not add to totals due to independent rounding.
[a]Typically includes flows from other countries.
[b]Estimates from Belyaev, Institute of Geography, USSR (1987).

**TABLE H.2** Total water withdrawal by major users, selected OECD countries, late 1980s

| Countries | Total ($10^6$ m³/yr) | Water withdrawal by major uses[a] | | | |
|---|---|---|---|---|---|
| | | Public water supply (%) | Irrigation (%) | Industry (%) | Electrical cooling (%) |
| Austria[e,f] | 2,120 | 24.8 | 2.6 | 23.6 | 47.2 |
| Canada | 43,888 | 11.3 | 7.1 | 9.1 | 55.6 |
| Denmark | 1,462 | 43.1 | — | — | — |
| Finland | 4,000 | 10.6 | 0.5 | 37.5 | 3.5 |
| France[h] | 43,273 | 13.7 | 9.7 | 10.4 | 51.9 |
| Germany[b] | 44,390 | 11.1 | 0.5 | 5.0 | 67.6 |
| Italy[g] | 56,200 | 14.2 | 57.3 | 14.2 | 12.5 |
| Japan[g] | 84,831 | 16.1 | 66.8 | 15.7 | 1.0 |
| Luxembourg | 67 | — | — | — | — |
| Netherlands[c] | 14,471 | 7.7 | — | 1.8 | 63.5 |
| New Zealand[d] | 1,900 | 27.8 | — | — | — |
| Norway | 2,235 | 29.3 | 3.1 | 25.1 | — |
| Spain | 45,845 | 11.6 | 65.5 | 22.9 | — |
| Sweden | 2,996 | 32.4 | 3.1 | 40.2 | 0.3 |
| Switzerland[i] | 709 | — | — | — | — |
| Turkey | 29,600 | 12.8 | 79.1 | 9.8 | — |
| United Kingdom[h,j] | 13,221 | 48.6 | 0.3 | 10.8 | 18.8 |
| United States[h] | 467,000 | 10.8 | 40.5 | 7.4 | 38.8 |

[a]Withdrawals from the four sectors do not necessarily add up to 100%, since "other agricultural uses than irrigation," "industrial cooling," and "other uses" are not covered in this table. Industrial data exclude cooling water withdrawn for industry, except as noted.
[b]Includes West Germany only.
[c]1986 data.
[d]OECD Secretariat estimates.
[e]Industry excludes industrial cooling: ground water withdrawal only; electrical cooling: surface water withdrawal only.
[f]Electrical cooling includes all cooling.
[g]1980 data.
[h]Industry includes industrial cooling.
[i]Withdrawal of lake and spring water only.
[j]Irrigation is total agricultural water withdrawal.

**TABLE H.2** Total water withdrawal by major users, selected OECD countries, late 1980s

## DESCRIPTION

Total annual water withdrawals for some of the countries of the Organization for Economic Co-operation and Development (OECD) are given here, for the late 1980s. Data are provided on total water withdrawals by country in m³ × $10^6$ per year, and on the percentage used for public water supply, irrigation, industry, and power plant cooling.

## LIMITATIONS

Despite being common members of the OECD, the countries in this table use very different definitions for several of the categories described in this table, including "industry" and "irrigation". In addition, total withdrawals are measured for different years, which are not

fully detailed, and using different assumptions, as described more fully in the table notes. This greatly complicates direct country-to-country comparisons.

## SOURCE

Organization for Economic Co-operation and Development (OECD), 1991, *The State of the Environment*, Organization for Economic Co-operation and Development, Paris. By permission.

Organization for Economic Co-operation and Development (OECD), 1989, *OECD Environmental Data: Compendium 1989*, Organization for Economic Co-operation and Development, Paris. By permission.

**TABLE H.3**  Water demand in the Arab region, 1985

| Country and subregion | Water demand[a] ($10^9$ m$^3$/yr) |
|---|---|
| Western | 14.752 |
| Mauritania | 1.662 |
| Morocco | 5.193 |
| Algeria | 3.500 |
| Tunisia | 2.282 |
| Libya | 2.115 |
| Central | 75.040 |
| Egypt | 59.500 |
| Sudan | 13.965 |
| Somalia | 1.500 |
| Dijbouti | 0.075 |
| Eastern | 49.261 |
| Syria | 6.883 |
| Lebanon | 0.859 |
| Jordan | 0.499 |
| Iraq | 41.020 |
| Arab Peninsula | 8.261 |
| Kuwait | 0.804 |
| Saudi Arabia | 3.530 |
| Bahrain | 0.112 |
| Qatar | 0.135 |
| United Arab Emirates | 1.012 |
| Oman | 0.512 |
| Yemen Dem Rep | 0.378 |
| Yemen Arab Rep | 1.778 |
| Total | 147.314 |
| Other Demands[b] | 7.386 |
| Total Demand | 154.700 |

[a]For drinking, domestic, industrial, and agricultural purposes.
[b]Water for other purposes is assumed to be 5% of the total.

**TABLE H.3**  Water demand in the Arab region, 1985

**DESCRIPTION**

Water demands for 21 countries in the Middle East and northern Africa for 1985 are given in thousand million ($10^9$) m$^3$ per year. The countries are divided into four regions: Western, Central, Eastern, and the Arabian Peninsula.

**LIMITATIONS**

The country data on demand given here include drinking water and water for domestic, industrial, and agricultural uses. Shahin adds an additional 5% to the total to cover water requirements in other sectors, which brings total 1985 demand to $155 \times 10^9$m$^3$. Israel was excluded from this table by the original author. ''Palestine'' was included, but no data on water demand were provided. The original source also includes water demand projections for each country for the years 2000 and 2030.

**SOURCE**

M. Shahin, 1989, Review and assessment of water resources in the Arab region, *Water International*, **14**(4), 206–219, International Water Resources Association, Illinois. By permission.

**TABLE H.4** Water resources and use in the Gulf Cooperative Council countries

| Country | Water resources ($10^6$ m³/yr) | | | Water use ($10^6$ m³/yr) | | | | |
| --- | --- | --- | --- | --- | --- | --- | --- | --- |
| | Surface | Ground water recharge | Total renewable | Surface | Ground | Desalination | Treated | Total |
| Saudi Arabia | 3,208 | 2,338 | 5,546 | 450 | 3,000 | 903 | 217 | 4,570 |
| Bahrain | 0.2 | 90 | 90 | | 153 | 16.5 | 0.5 | 170 |
| Qatar | | 55 | 55 | | 90 | 90 | 20 | 200 |
| Kuwait | | 160 | 160 | | 283 | 404 | 80 | 767 |
| Oman | 1,470 | 564 | 2,034 | | 400 | 15 | 8.6 | 424 |
| United Arab Emirates | 365 | 387 | 752 | | 300 | 276 | 0.8 | 577 |

Totals may not add due to rounding.

**TABLE H.4** Water resources and use in the Gulf Cooperative Council countries

**DESCRIPTION**

Renewable fresh water resources and fresh water use are given here for Saudia Arabia, Bahrain, Qatar, Kuwait, Oman, and the United Arab Emirates, which are all members of the Gulf Cooperative Council. The data are for an unspecified period. Water resources available for use come from surface and ground water. Water actually withdrawn for use comes from surface and ground water supplies, and from the use of desalinated water and treated water. All data are in m³ × $10^6$ per year.

**LIMITATIONS**

It is not clear whether these data on water availability and use are for a particular year or averaged over a period of time. No distinction is made here between withdrawals and consumptive use. Compare with the data in Table H.3.

**SOURCES**

A. Akkad, 1990, Conservation in the Arabian Gulf countries, *Journal American Water Works Association*, **82**(5), 40–50.

Cited in U.S. Army Corps of Engineers, 1991, *Water in the Sand: A Survey of Middle East Water Issues*, draft, Office of Strategic Initiatives, Washington, DC.

**Fresh water data**

**TABLE H.5** Water resources used for drinking water supplies in Europe, 1988

| Country | Sources utilized in 1988[a] | | Treated | |
|---|---|---|---|---|
| | Ground water (%) | Surface water (%) | Ground water (%) | Surface water (%) |
| Austria | 96 | 4 | | 100 |
| Belgium | 67 | 33 | 75 | 100 |
| Bulgaria | 40 | 60 | | 100 |
| Czechoslovakia | 60 | 40 | | 100 |
| Denmark | 98 | 2 | 90 | 100 |
| Finland | 49 | 51 | 80 | 99 |
| German DR | 70 | 30 | 70 | 100 |
| Germany FR | 89 | 11 | 62 | 100 |
| Greece | 40 | 60 | | 80 |
| Hungary | 90 | 10 | 30 | 100 |
| Iceland | 87 | 13 | 50 | 100 |
| Israel | 60 | 40 | 80 | 100 |
| Italy | 88 | 12 | 60 | 100 |
| Luxembourg | 66 | 33 | 70 | 100 |
| Malta | 61 | 39 | 90 | 100 |
| Netherlands | 67 | 33 | 90 | 100 |
| Norway | 15 | 85 | 50 | 100 |
| Poland | 60 | 40 | 50 | 100 |
| Portugal | 94 | 6 | 60 | 100 |
| Spain | 20 | 80 | | 90 |
| Sweden | 49 | 51 | 50 | 100 |
| Switzerland | 75 | 25 | | 100 |
| Turkey | 80 | 20 | 40 | 80 |
| United Kingdom | 35 | 65 | 80 | 100 |

[a]Relates to sources used only for potable water supply.

**TABLE H.5** Water resources used for drinking water supplies in Europe, 1988

**DESCRIPTION**

Fresh water used for drinking in Europe comes from ground water and surface water. The relative proportion of each source used in 1988 are given here, together with the fraction treated.

**LIMITATIONS**

No information is given on the standards to which the surface and ground water supplies are treated or the original quality of the water.

The data here apply to sources used only for potable water supply.

**SOURCES**

Unpublished data from the World Health Organization, 1988, Regional Office for Europe, Copenhagen.
Cited in United Nations Environment Programme, 1989, *Environmental Data Report*, GEMS Monitoring and Assessment Research Centre, Basil Blackwell, Oxford. By permission.

**TABLE H.6**   Basic indicators characterizing the protection and rational utilization of water resources in the former Soviet Union, 1980 to 1988 ($km^3/yr$)

|  | 1980 | 1985 | 1986 | 1987 | 1988 |
|---|---|---|---|---|---|
| Total water withdrawals | 323 | 329.8 | 326.3 | 339.5 | 333.7 |
| Of this, from ground water | 30 | 29.1 | 34.0 | 33.9 | 31.4 |
| Losses in transportation[a] | n.d. | n.d | 43.6 | 47.8 | 50.6 |
| Water used for all purposes | 288 | 289.5 | 280.2 | 285.8 | 286.3 |
| Of this, for industrial needs[b] | 104 | 109.5 | 108.7 | 108.7 | 107.2 |
| Treated sewage discharge[c] | 17 | 22.4 | 23.0 | 18.5 | 12.1 |
| Percentage meeting government standards | 46 | 58 | 60 | 47 | 30 |
| Polluted water discharge[d] | 20 | 15.9 | 15.1 | 20.6 | 28.6 |
| Of this, without any cleaning | 13 | 6.9 | 6.5 | 6.7 | 8.1 |

n.d., no data for these years were provided.
[a]Evaporation and seepage losses from water moved from source to site of consumption.
[b]Excluding water for agriculture.
[c]Discharge to government standards.
[d]Without cleaning or with insufficient cleaning.

**TABLE H.6**   Basic indicators characterizing the protection and rational utilization of water resources in the former Soviet Union, 1980 to 1988

**DESCRIPTION**

Total water withdrawals, ground water withdrawals, water losses from evaporation and seepage, and information on water treatment are given here in $km^3$ per year for 1980 and 1985–1988. Total withdrawals and ground water withdrawals decreased slightly toward the end of the 1980s, as has the fraction of water being treated to government standards. Also shown is the volume of water discharged without any cleaning or with insufficient cleaning. This volume has increased substantially in recent years.

**LIMITATIONS**

The data are for water withdrawals, not water consumed (the Russian equivalent is "irretrievable water loss"). The fraction of water being treated to "government standards" has dropped from 60% in 1986 to 30% in 1988. No explanation for this drop is given, nor are details of the applicable government standards described.

**SOURCE**

Goscomstat, USSR, 1989, *Protection of the Environment and Rational Utilization of Natural Resources in the USSR*, Government Committee on Statistics, Statistical Handbook, Moscow (in Russian).

**TABLE H.7** Water withdrawals from natural sources in the former Soviet Union, by republic, 1985 to 1988[a]
(10⁶ m³/yr)

| | 1985 | | 1986 | | 1987 | | 1988 | |
|---|---|---|---|---|---|---|---|---|
| | Total withdrawal[b] | Ground water withdrawal | Total withdrawal[b] | Ground water withdrawal | Total withdrawal[b] | Ground water withdrawal | Total withdrawal[b] | Ground water withdrawal |
| USSR total | 329,797 | 29,073 | 326,347 | 34,001 | 339,490 | 33,889 | 333,687 | 31,397 |
| Russia | 106,547 | 11,366 | 112,962 | 13,828 | 112,107 | 14,314 | 105,826 | 12,551 |
| Ukraine | 30,660 | 4,147 | 34,482 | 5,257 | 32,517 | 5,199 | 30,585 | 4,219 |
| Byelorussia | 2,824 | 1,062 | 2,923 | 1,099 | 2,759 | 1,115 | 2,779 | 1,149 |
| Uzbekistan | 70,621 | 3,060 | 63,284 | 3,758 | 73,338 | 3,267 | 73,878 | 3,244 |
| Kazakhstan | 39,021 | 2,085 | 35,286 | 2,459 | 37,428 | 2,524 | 39,440 | 2,333 |
| Georgia | 4,576 | 713 | 4,629 | 788 | 4,403 | 864 | 3,802 | 1,016 |
| Azerbaidzhan | 15,228 | 1,492 | 15,312 | 1,644 | 15,019 | 1,684 | 14,902 | 1,456 |
| Lithuania | 2,810 | 481 | 3,105 | 499 | 3,233 | 502 | 3,605 | 505 |
| Moldavia | 3,745 | 274 | 4,091 | 305 | 3,855 | 300 | 3,703 | 294 |
| Latvia | 665 | 297 | 690 | 312 | 682 | 313 | 671 | 306 |
| Kirghizia | 9,310 | 947 | 11,762 | 1,022 | 12,297 | 887 | 12,116 | 921 |
| Tadjikistan | 12,926 | 1,378 | 13,438 | 1,304 | 12,943 | 1,136 | 12,761 | 1,126 |
| Armenia | 4,078 | 1,226 | 3,919 | 1,077 | 4,060 | 1,136 | 4,147 | 1,633 |
| Turkmenistan | 24,212 | 393 | 17,190 | 494 | 21,537 | 492 | 22,498 | 479 |
| Estonia | 2,574 | 152 | 3,274 | 155 | 3,312 | 157 | 2,974 | 165 |

[a]Total withdrawals include withdrawals of seawater. In 1988, the withdrawal of seawater for the USSR as a whole was 10,603 million cubic meters. This excludes water lost in water supply systems through seepage and evaporation, which totaled 31,172 million cubic meters in 1988.
[b]Total withdrawals include ground water withdrawals.
Figures may not add to totals due to independent rounding.

**TABLE H.7** Water withdrawals from natural sources in the former Soviet Union, by republic, 1985 to 1988[a]

**DESCRIPTION**

Total water withdrawals and ground water withdrawals are given here in m³ × 10⁶ per year for 1985 through 1988 for all the republics of the former Soviet Union. Total withdrawals and ground water withdrawals have decreased slightly during this period. Total withdrawals include sea water withdrawals, which are about 3% of the total.

**LIMITATIONS**

The data are for water withdrawals, not water consumed (the Russian equivalent is "irretrievable water loss").

**SOURCE**

Goscomstat, USSR, 1989, *Protection of the Environment and Rational Utilization of Natural Resources in the USSR*, Government Committee on Statistics, Statistical Handbook, Moscow (in Russian).

**TABLE H.8** Water withdrawals, use, and losses in the former Soviet Union, by republic, 1986 to 1988 ($10^6$ m$^3$/yr)

| | 1986 | | | 1987 | | | 1988 | | |
|---|---|---|---|---|---|---|---|---|---|
| | Water withdrawal from natural sources | Water used | Water losses in transportation | Water withdrawal from natural sources | Water used | Water losses in transportation | Water withdrawal from natural sources | Water used | Water losses in transportation |
| Russia | 112,962 | 100,971 | 7,910 | 112,107 | 98,654 | 8,449 | 105,826 | 94,848 | 8,766 |
| Ukraine | 34,482 | 30,927 | 2,209 | 32,517 | 29,543 | 1,855 | 30,585 | 28,711 | 1,771 |
| Byelorussia | 2,923 | 2,724 | 49 | 2,759 | 2,721 | 47 | 2,779 | 2,720 | 54 |
| Uzbekistan | 63,284 | 46,058 | 13,338 | 73,338 | 53,125 | 15,716 | 73,878 | 54,777 | 16,881 |
| Kazakhstan | 35,286 | 31,226 | 5,324 | 37,428 | 31,847 | 5,543 | 39,440 | 33,135 | 6,107 |
| Georgia | 4,629 | 3,831 | 830 | 4,403 | 3,559 | 777 | 3,802 | 3,123 | 692 |
| Azerbaidzhan | 15,312 | 12,420 | 4,078 | 15,019 | 12,242 | 3,735 | 14,902 | 12,415 | 4,092 |
| Lithuania | 3,105 | 3,071 | 34 | 3,233 | 3,204 | 30 | 3,605 | 3,578 | 31 |
| Moldavia | 4,091 | 3,980 | 75 | 3,855 | 3,792 | 75 | 3,703 | 3,640 | 61 |
| Latvia | 690 | 649 | 30 | 682 | 646 | 25 | 671 | 651 | 21 |
| Kirghizia | 11,762 | 9,828 | 2,007 | 12,297 | 9,714 | 2,120 | 12,116 | 10,050 | 2,328 |
| Tadjikistan | 13,438 | 12,242 | 1,971 | 12,943 | 11,399 | 1,883 | 12,761 | 11,705 | 1,782 |
| Armenia | 3,919 | 3,136 | 605 | 4,060 | 3,371 | 725 | 4,147 | 3,543 | 674 |
| Turkmenistan | 17,190 | 16,146 | 5,140 | 21,537 | 18,987 | 6,790 | 22,498 | 20,484 | 7,359 |
| Estonia | 3,274 | 3,000 | 18 | 3,312 | 2,955 | 19 | 2,974 | 2,948 | 18 |
| USSR total[a] | 326,347 | 280,207 | 43,617 | 339,490 | 285,760 | 47,790 | 333,687 | 286,328 | 50,637 |

[a]Figures may not add to totals due to independent rounding.

**TABLE H.8** Water withdrawals, use, and losses in the former Soviet Union, by republic, 1986 to 1988

## DESCRIPTION

Total water withdrawn from natural sources, total water use, and water losses in transportation are given here in m$^3$ × $10^6$ per year for 1986 through 1988 for all the republics of the former Soviet Union. Total withdrawals have decreased slightly during this period. Total withdrawals include seawater withdrawals, which are about 3% of the total. Water losses in transportation refer to water lost to evaporation and seepage during the move from the source to the site of consumption. According to the original source, 58% of all seawater used

in Russia, 17% in Kazakhstan, and the remainder in the republics of the Ukraine, Turkmen, Azerbaidzhan, and Estonia.

## LIMITATIONS

The data are for water withdrawal and use, not water consumed. The distinction between water withdrawal and water use is not entirely clear. The most water-intensive sector is agriculture, which accounted for a little under two-thirds of total water withdrawal. The total irreversible withdrawal of water flow for agricultural needs and industry is leading to river depletion. In 1988, such water withdrawal was about 180 km$^3$ for the USSR as a whole.

## SOURCE

Goscomstat, USSR, 1989, *Protection of the Environment and Rational Utilization of Natural Resources in the USSR*, Government Committee on Statistics, Statistical Handbook, Moscow (in Russian).

**TABLE H.9 I**  Offstream water use in the United States, by state, 1990 *(pages 387–388)*

**TABLE H/9 II**  Offstream water use in the United States, by water resources region, 1990 *(page 389)*

## DESCRIPTION

Fresh and saline water withdrawals in the United States for 1990 are given here in two tables: for the 50 states of the United States, the District of Columbia, Puerto Rico, and the Virgin Islands, and for the 21 U.S. water resources regions. Ground water and surface water withdrawals are identified separately, and reclaimed sewage and conveyance losses are also presented. All withdrawal data are in thousand cubic meters per day, 1990 population data are in thousands, and per capita use of water is given in cubic meters per person per day. Total fresh water consumptive use is also shown. Offstream use refers to water withdrawn or diverted from a ground- or surface-water source. Ground water refers to subsurface water from the saturated zone, where the pressure is greater than atmospheric. Surface water includes lake and stream waters. Reclaimed sewage refers to wastewater treatment-plant effluent. Conveyance losses refers to water lost in transit by leakage or evaporation.

## LIMITATIONS

These data are compiled from information from the states and from U.S. water resources subregions and the accuracy of data vary. Most data here are for withdrawals, not consumption; the last column gives consumptive use of fresh water. For the purposes of this table, fresh water is defined as containing less than 1,000 milligrams per liter of dissolved solids, although water containing greater than 500 mg/1 dissolved solids is generally considered undesirable for drinking and most industrial uses. Time series data for some of these categories are shown in Table H.12. These data are considered preliminary by the U.S. Geological Survey.

## SOURCE

U.S. Geological Survey. 1992 (preliminary). *Estimated Use of Water in the United States in 1990.* U.S. Department of the Interior, U.S. Government Printing Office, Washington DC. 1990 numbers provided by W. Solley and H. Perlman, personal communication.

**TABLE H.9 I**  Offstream water use in the United States, by state, 1990

| State | 1990 population (10³) | Per capita fresh water use (m³/person/day) | Withdrawals (includes irrigation conveyance loss) (10³ m³/day) | | | | | | Reclaimed sewage (10³ m³/day) | Conveyance losses (10³ m³/day) | Fresh water consumptive use (10³ m³/day) |
|---|---|---|---|---|---|---|---|---|---|---|---|
| | | | Ground water | | Surface water | | Total | | | | |
| | | | Fresh | Saline | Fresh | Saline | Fresh | Saline | | | |
| Alabama | 4,041 | 7.56 | 1,490 | 34 | 29,100 | 0.0 | 30,600 | 34 | 0.0 | 0.0 | 1,720 |
| Alaska | 550 | 1.96 | 240 | 180 | 836 | 1,170 | 1,080 | 1,350 | 0.0 | 0.4 | 98 |
| Arizona | 3,665 | 6.79 | 10,400 | 2 | 14,500 | 2 | 24,900 | 4.3 | 693 | 3,820 | 16,500 |
| Arkansas | 2,353 | 12.6 | 17,800 | 0.0 | 11,900 | 0.0 | 29,700 | 0.0 | 0.0 | 1,390 | 15,700 |
| California | 29,760 | 4.47 | 55,300 | 1,180 | 77,600 | 43,100 | 133,000 | 44,300 | 503 | 5,910 | 79,100 |
| Colorado | 3,294 | 14.6 | 10,500 | 110 | 37,500 | 0.0 | 48,000 | 110 | 14 | 11,300 | 19,900 |
| Connecticut | 3,287 | 1.23 | 625 | 0.0 | 3,410 | 14,300 | 4,040 | 14,300 | 0.0 | 0.0 | 390 |
| Delaware | 666 | 5.83 | 340 | 0.0 | 3,550 | 1,280 | 3,890 | 1,280 | 0.0 | 0.0 | 220 |
| District of Columbia | 607 | 0.06 | 4 | 0.0 | 30 | 0.0 | 34 | 0.0 | 0.0 | 0.0 | 61 |
| Florida | 12,938 | 2.20 | 17,600 | 0.0 | 10,900 | 39,400 | 28,500 | 39,400 | 643 | 240 | 11,800 |
| Georgia | 6,478 | 3.09 | 3,770 | 0.0 | 16,200 | 250 | 20,000 | 250 | 140 | 0.0 | 3,110 |
| Hawaii | 1,108 | 4.05 | 2,230 | 2 | 2,270 | 5,870 | 4,500 | 5,870 | 23 | 481 | 2,370 |
| Idaho | 1,007 | 74.0 | 28,700 | 0.0 | 45,800 | 0.0 | 74,500 | 0.0 | 0.0 | 27,100 | 23,100 |
| Illinois | 11,431 | 5.94 | 3,480 | 95 | 64,700 | 0.0 | 68,200 | 95 | 0.0 | 0.0 | 2,840 |
| Indiana | 5,544 | 6.43 | 2,350 | 0.0 | 33,300 | 0.0 | 35,700 | 0 | 0.0 | 0.0 | 1,710 |
| Iowa | 2,777 | 3.90 | 1,870 | 0.0 | 8,970 | 0.0 | 10,800 | 0 | 0.0 | 0.0 | 1,030 |
| Kansas | 2,478 | 9.27 | 16,500 | 0.0 | 6,510 | 0.0 | 23,000 | 0 | 20 | 553 | 16,700 |
| Kentucky | 3,685 | 4.43 | 935 | 0.0 | 15,400 | 0.0 | 16,300 | 0 | 0.0 | 2 | 1,170 |
| Louisiana | 4,220 | 8.33 | 5,070 | 2 | 30,100 | 250 | 35,200 | 250 | 0.0 | 340 | 6,020 |
| Maine | 1,228 | 1.64 | 320 | 0.0 | 1,690 | 2,300 | 2,010 | 2,310 | 0.0 | 0.0 | 190 |
| Maryland | 4,781 | 1.16 | 905 | 0.0 | 4,660 | 18,700 | 5,560 | 18,700 | 240 | 0.0 | 481 |
| Massachusetts | 6,016 | 1.28 | 1,280 | 0.0 | 6,400 | 13,200 | 7,680 | 13,200 | 0.0 | 0.0 | 738 |
| Michigan | 9,295 | 4.73 | 2,660 | 17 | 41,300 | 0.0 | 43,900 | 17 | 0.0 | 0.0 | 2,790 |
| Minnesota | 4,375 | 2.83 | 3,020 | 0.0 | 9,390 | 0.0 | 12,400 | 0.0 | 0.0 | 0.0 | 3,300 |
| Mississippi | 2,573 | 4.88 | 10,100 | 0.0 | 2,450 | 1,200 | 12,600 | 1,200 | 4 | 712 | 6,810 |
| Missouri | 5,117 | 4.35 | 2,750 | 0.4 | 19,500 | 4,010 | 22,200 | 4,010 | 0.0 | 0.0 | 2,000 |
| Montana | 799 | 44.1 | 776 | 49 | 34,400 | 0.0 | 35,200 | 49 | 0.0 | 17,500 | 7,910 |

*continued*

Withdrawals (includes irrigation conveyance loss) ($10^3$ m³/day)

| State | 1990 population ($10^3$) | Per capita fresh water use (m³/person/day) | Ground water Fresh | Ground water Saline | Surface water Fresh | Surface water Saline | Total Fresh | Total Saline | Reclaimed sewage ($10^3$ m³/day) | Conveyance losses ($10^3$ m³/day) | Fresh water consumptive use ($10^3$ m³/day) |
|---|---|---|---|---|---|---|---|---|---|---|---|
| Nebraska | 1,578 | 21.4 | 18,100 | 18 | 15,700 | 0.0 | 33,800 | 18 | 0.0 | 8,180 | 16,000 |
| Nevada | 1,202 | 10.5 | 4,010 | 45 | 8,630 | 0.0 | 12,600 | 45 | 42 | 2,330 | 6,400 |
| New Hampshire | 1,109 | 1.43 | 240 | 0.0 | 1,350 | 3,380 | 1,590 | 3,380 | 0.0 | 0.0 | 98 |
| New Jersey | 7,730 | 1.09 | 2,140 | 1 | 6,250 | 40,100 | 8,390 | 40,100 | 0.0 | 0.0 | 799 |
| New Mexico | 1,515 | 8.71 | 6,660 | 0.0 | 6,510 | 0.0 | 13,200 | 0.0 | 0.0 | 2,230 | 7,800 |
| New York | 17,990 | 2.21 | 3,180 | 5.7 | 36,500 | 32,100 | 39,700 | 32,100 | 0.0 | 0.0 | 2,130 |
| North Carolina | 6,629 | 5.11 | 1,650 | 0.0 | 32,200 | 21 | 33,800 | 21 | 64 | 0.0 | 1,480 |
| North Dakota | 639 | 15.8 | 534 | 0.0 | 9,610 | 0.0 | 10,100 | 0.0 | 0.0 | 22 | 863 |
| Ohio | 10,847 | 4.09 | 3,420 | 0.0 | 40,900 | 0.0 | 44,300 | 0.0 | 0.0 | 0.4 | 3,410 |
| Oklahoma | 3,146 | 1.71 | 2,510 | 920 | 2,880 | 0.0 | 5,380 | 920 | 0.0 | 20 | 2,490 |
| Oregon | 2,842 | 11.2 | 2,900 | 0.0 | 29,000 | 0.0 | 31,900 | 0.0 | 45 | 4,810 | 12,000 |
| Pennsylvania | 11,882 | 3.13 | 3,860 | 0.0 | 33,300 | 0.0 | 37,200 | 0.0 | 0.0 | 0.0 | 2,200 |
| Rhode Island | 1,003 | 0.50 | 95 | 0.0 | 409 | 1,490 | 503 | 1,490 | 0.0 | 0.0 | 68 |
| South Carolina | 3,487 | 6.51 | 1,070 | 0.0 | 21,700 | 0.0 | 22,700 | 0.0 | 53 | 0.0 | 1,110.0 |
| South Dakota | 696 | 3.22 | 950 | 0.0 | 1,290 | 0.0 | 2,240 | 0.0 | 0.0 | 240 | 1,310 |
| Tennessee | 4,877 | 7.12 | 1,900 | 0.0 | 32,900 | 0.0 | 34,800 | 0.0 | 3 | 0.0 | 950 |
| Texas | 16,986 | 4.47 | 27,900 | 1,860 | 48,100 | 17,400 | 76,000 | 19,300 | 212 | 2,500 | 34,100.0 |
| Utah | 1,723 | 9.61 | 3,650 | 27 | 12,900 | 350 | 16,600 | 377 | 150 | 2,360 | 8,440 |
| Vermont | 563 | 4.24 | 170 | 0.0 | 2,220 | 0.0 | 2,390 | 0.0 | 0.0 | 0.0 | 110 |
| Virginia | 6,187 | 2.88 | 1,680 | 0.0 | 16,200 | 8,140 | 17,800 | 8,140 | 0.0 | 14 | 848 |
| Washington | 4,867 | 6.17 | 5,490 | 0.0 | 24,500 | 140 | 29,900 | 140 | 0.0 | 3,770 | 10,700 |
| West Virginia | 1,793 | 9.70 | 2,760 | 0.0 | 14,600 | 0.0 | 17,400 | 0.0 | 0.0 | 0.0 | 1,930 |
| Wisconsin | 4,892 | 5.03 | 2,680 | 0.0 | 22,100 | 0.0 | 24,600 | 0.0 | 0.0 | 0.0 | 1,750 |
| Wyoming | 454 | 63.2 | 1,450 | 72 | 27,300 | 0.0 | 28,700 | 72 | 0.0 | 8,140 | 10,300 |
| Puerto Rico | 3,522 | 0.62 | 594 | 0.0 | 1,590 | 9,350 | 2,180 | 9,350 | 0.0 | 64 | 753 |
| Virgin Islands | 102 | 0.34 | 7 | 4.5 | 28 | 579 | 35 | 584 | 0.0 | 0.0 | 5.7 |
| TOTAL | 252,199 | 5.11 | 301,000 | 4,630 | 981,000 | 258,000 | 1,280,000 | 263,000 | 2,900 | 104,000 | 356,000 |

Figures may not add to total due to independent rounding.

**TABLE H.9 II** Offstream water use in the United States, by water resources region, 1990

| Region | 1990 population (10³) | Per capita fresh water use (m³/person/day) | Withdrawals (Includes irrigation conveyance loss) (10³ m³/day) | | | | | | Reclaimed sewage (10³ m³/day) | Conveyance losses (10³ m³/day) | Fresh water consumptive use (10³ m³/day) |
| | | | Ground water | | Surface water | | Total | | | | |
| | | | Fresh | Saline | Fresh | Saline | Fresh | Saline | | | |
|---|---|---|---|---|---|---|---|---|---|---|---|
| New England | 12,797 | 1.40 | 2,630 | 0.0 | 15,300 | 34,700 | 17,900 | 34,700 | 0.0 | 0.0 | 1,550 |
| Mid-Atlantic | 41,541 | 1.93 | 9,990 | 4.5 | 70,000 | 100,000 | 80,000 | 100,000 | 240 | 9.8 | 4,770 |
| South Atlantic-Gulf | 34,732 | 3.64 | 26,900 | 34 | 99,500 | 40,900 | 126,000 | 40,900 | 901 | 260 | 19,500 |
| Great Lakes | 21,406 | 5.73 | 4,580 | 19 | 118,000 | 25 | 123,000 | 43 | 0.0 | 0.0 | 6,210 |
| Ohio | 21,882 | 5.27 | 10,000 | 83 | 105,000 | 2 | 115,000 | 86 | 1 | 2.0 | 7,990 |
| Tennessee | 3,911 | 8.91 | 1,160 | 0.0 | 33,700 | 0.0 | 34,800 | 0.0 | 2 | 0.0 | 1,220 |
| Upper Mississippi | 21,270 | 3.70 | 9,920 | 16 | 68,900 | 0.0 | 78,800 | 16 | 0.0 | 0.4 | 7,420 |
| Lower Mississippi | 7,167 | 9.49 | 31,600 | 2 | 36,500 | 4,240 | 68,000 | 4,240 | 3 | 2,270 | 26,400 |
| Souris-Red-Rainy | 672 | 1.67 | 490 | 0.0 | 628 | 0.0 | 1,120 | 0.0 | 0.0 | 4.2 | 545 |
| Missouri Basin | 10,048 | 14.1 | 32,100 | 140 | 110,000 | 0.0 | 142,000 | 140 | 10 | 34,100 | 45,800 |
| Arkansas-White-Red | 8,250 | 7.07 | 28,100 | 1,100 | 30,200 | 0.0 | 58,300 | 1,100 | 42 | 3,000 | 29,800 |
| Texas-Gulf | 15,239 | 3.35 | 20,700 | 1,510 | 30,400 | 17,400 | 51,100 | 19,000 | 190 | 1,290 | 22,400 |
| Rio Grande | 2,229 | 10.2 | 8,100 | 148 | 14,600 | 0.0 | 22,700 | 148 | 4.2 | 4,050 | 13,100 |
| Upper Colorado | 626 | 42.8 | 481 | 106 | 26,300 | 0.0 | 26,800 | 106 | 2.0 | 6,060 | 9,310 |
| Lower Colorado | 4,747 | 6.18 | 11,700 | 2 | 17,700 | 2 | 29,300 | 5 | 700 | 4,090 | 18,900 |
| Great Basin | 2,182 | 12.5 | 7,460 | 72 | 19,800 | 350 | 27,300 | 424 | 230 | 5,150 | 13,000 |
| Pacific Northwest | 8,912 | 15.4 | 37,000 | 0.0 | 100,000 | 140 | 137,000 | 136 | 45 | 36,600 | 45,800 |
| California | 29,442 | 4.55 | 54,500 | 1,170 | 79,500 | 43,100 | 134,000 | 44,300 | 484 | 6,620 | 78,700 |
| Alaska | 550 | 1.96 | 242 | 182 | 836 | 1,170 | 1,080 | 1,350 | 0.0 | 0.4 | 98 |
| Hawaii | 1,108 | 4.06 | 2,230 | 2.3 | 2,270 | 5,870 | 4,500 | 5,870 | 24 | 481 | 2,370 |
| Caribbean | 3,624 | 0.61 | 602 | 4.5 | 1,610 | 9,920 | 2,210 | 9,920 | 0.0 | 64 | 757 |
| Total | 252,335 | 5.08 | 301,000 | 4,630 | 981,000 | 258,000 | 1,280,000 | 263,000 | 2,900 | 104,000 | 356,000 |

Figures may not add to totals due to independent rounding.

**TABLE H.10 I**  Fresh water withdrawal in the United States, by state and by user, 1990 *(pages 391–393)*

**TABLE H.10 II**  Fresh water withdrawal in the United States, by water resources region and by user, 1990 *(page 393)*

## DESCRIPTION

Total 1990 fresh water withdrawals in $m^3 \times 10^3$ per day are given in two tables for the 50 states, the District of Columbia, Puerto Rico, and the Virgin Islands, and for the 21 U.S. water resources regions. The table also shows fresh water use in eight economic sectors as a percentage of total fresh water use. The sectors include public supply, domestic use, commercial (including military) use, irrigation, livestock, industrial, mining, and thermoelectric withdrawals.

## LIMITATIONS

Public supply includes water withdrawn by public and private water suppliers and delivered to groups of users. See the original source for information on the further distribution of this water. Domestic supply thus refers only to self-supplied fresh water, and not to that obtained from a water agency. No information on saline water withdrawals is presented here, though some states use substantial quantities of saline water, particularly for cooling. Fresh water is defined here as containing less than 1,000 mg/l of dissolved solids, although water containing greater than 500 mg/l dissolved solids is generally considered undesirable for drinking and most industrial uses.

## SOURCE

U.S. Geological Survey, 1992 (preliminary), *Estimated Use of Water in the United States in 1990,* U.S. Department of the Interior, U.S. Government Printing Office, Washington, DC. 1990 numbers provided by W. Solley and H. Perlman, personal communication.

**TABLE H.10 I** Fresh water withdrawal in the United States, by state and by user, 1990

| State | Public supply | Domestic | Commercial | Irrigation | Livestock | Industrial | Mining | Thermo-electric | Total fresh water withdrawals ($10^3$ m$^3$/day) | Fresh water consumptive use ($10^3$ m$^3$/day) |
|---|---|---|---|---|---|---|---|---|---|---|
| | | | | Percent of total fresh water use | | | | | | |
| Alabama | 8.8 | 0.3 | 0.0 | 1.2 | 1.7 | 9.7 | 0.1 | 78.1 | 30,600 | 1,720 |
| Alaska | 32.4 | 2.4 | 6.3 | 0.2 | 0.2 | 39.1 | 8.8 | 10.9 | 1,070 | 98 |
| Arizona | 10.8 | 0.5 | 0.3 | 80.7 | 1.4 | 2.5 | 2.4 | 1.6 | 24,900 | 16,500 |
| Arkansas | 3.9 | 0.7 | 2.8 | 67.0 | 2.4 | 2.3 | 0.0 | 20.9 | 29,700 | 15,700 |
| California | 16.6 | 0.9 | 0.7 | 79.5 | 1.2 | 0.4 | 0.1 | 0.7 | 133,000 | 79,100 |
| Colorado | 5.1 | 0.1 | 0.1 | 91.3 | 1.3 | 0.9 | 0.4 | 0.9 | 48,100 | 19,900 |
| Connecticut | 35.0 | 4.3 | 1.7 | 1.4 | 0.1 | 7.5 | 0.2 | 49.5 | 4,050 | 390 |
| Delaware | 8.3 | 1.1 | 0.2 | 3.1 | 0.2 | 6.3 | 0.0 | 80.7 | 3,900 | 220 |
| District of Columbia | 0.0 | 0.0 | 0.0 | 0.0 | 0.0 | 5.6 | 5.6 | 88.9 | 34 | 61 |
| Florida | 25.6 | 4.0 | 0.7 | 49.5 | 1.0 | 5.4 | 4.2 | 9.7 | 28,500 | 11,800 |
| Georgia | 18.2 | 1.9 | 0.8 | 8.3 | 0.9 | 12.4 | 0.2 | 57.3 | 20,000 | 3,110 |
| Hawaii | 20.0 | 0.8 | 3.4 | 63.4 | 0.6 | 3.6 | 0.1 | 8.0 | 4,500 | 2,370 |
| Idaho | 1.0 | 0.2 | 0.1 | 94.9 | 2.8 | 1.0 | 0.0 | 0.0 | 74,600 | 23,100 |
| Illinois | 10.3 | 0.6 | 1.0 | 0.4 | 0.4 | 2.6 | 0.4 | 84.4 | 68,100 | 2,840 |
| Indiana | 6.4 | 1.3 | 0.7 | 0.5 | 0.5 | 26.3 | 1.0 | 63.2 | 35,700 | 1,710 |
| Iowa | 11.3 | 1.6 | 0.9 | 0.8 | 4.2 | 7.7 | 1.2 | 72.4 | 10,800 | 1,030 |
| Kansas | 6.1 | 0.4 | 0.1 | 68.9 | 1.9 | 0.9 | 0.4 | 21.4 | 23,000 | 16,700 |
| Kentucky | 9.9 | 1.3 | 0.3 | 0.3 | 0.8 | 7.2 | 0.4 | 79.6 | 16,400 | 1,170 |
| Louisiana | 6.7 | 0.5 | 0.1 | 7.6 | 5.9 | 25.4 | 0.4 | 53.3 | 35,200 | 6,020 |
| Maine | 19.9 | 9.2 | 6.4 | 0.3 | 0.3 | 47.7 | 0.7 | 15.4 | 2,010 | 190 |
| Maryland | 54.3 | 4.8 | 1.8 | 2.0 | 1.8 | 4.8 | 1.9 | 28.6 | 5,560 | 481 |
| Massachusetts | 35.2 | 1.8 | 3.6 | 4.9 | 0.1 | 4.3 | 0.2 | 49.8 | 7,680 | 738 |
| Michigan | 12.1 | 1.1 | 0.3 | 2.1 | 0.3 | 14.5 | 0.5 | 69.5 | 43,900 | 2,790 |
| Minnesota | 15.7 | 5.1 | 2.2 | 6.0 | 2.0 | 4.7 | 6.7 | 57.5 | 12,400 | 3,300 |
| Mississippi | 9.6 | 1.0 | 0.5 | 56.6 | 12.4 | 8.1 | 0.1 | 11.6 | 12,600 | 6,810 |
| Missouri | 11.5 | 1.1 | 0.4 | 6.3 | 0.9 | 1.4 | 0.4 | 78.0 | 22,200 | 2,000 |
| Montana | 1.5 | 0.2 | 0.0 | 96.8 | 0.6 | 0.6 | 0.1 | 0.4 | 35,200 | 7,910 |

*continued*

| State | Public supply | Domestic | Commercial | Irrigation | Livestock | Industrial | Mining | Thermo-electric | Total fresh water withdrawals (10³ m³/day) | Fresh water consumptive use (10³ m³/day) |
|---|---|---|---|---|---|---|---|---|---|---|
| Nebraska | 3.4 | 0.5 | 0.0 | 68.2 | 1.6 | 0.5 | 1.5 | 24.4 | 33,800 | 16,000 |
| Nevada | 11.5 | 0.3 | 0.7 | 84.4 | 0.2 | 0.3 | 1.5 | 1.0 | 12,600 | 6,400 |
| New Hampshire | 22.6 | 6.4 | 0.1 | 0.2 | 0.2 | 8.8 | 0.7 | 60.7 | 1,590 | 98 |
| New Jersey | 46.8 | 3.1 | 0.8 | 2.6 | 0.1 | 14.7 | 5.0 | 27.0 | 8,400 | 799 |
| New Mexico | 7.8 | 0.7 | 0.5 | 86.5 | 0.6 | 0.2 | 2.3 | 1.4 | 13,200 | 7,800 |
| New York | 27.7 | 1.1 | 0.6 | 0.5 | 0.2 | 2.6 | 0.4 | 66.6 | 39,700 | 2,130 |
| North Carolina | 9.0 | 1.2 | 0.2 | 1.3 | 2.2 | 4.4 | 1.1 | 80.6 | 33,800 | 1,480 |
| North Dakota | 2.8 | 0.4 | 0.0 | 6.1 | 0.9 | 0.3 | 0.1 | 89.2 | 10,100 | 863 |
| Ohio | 11.1 | 1.1 | 0.3 | 0.1 | 0.3 | 3.0 | 2.1 | 81.6 | 44,300 | 3,410 |
| Oklahoma | 36.3 | 2.9 | 0.4 | 42.3 | 9.2 | 2.5 | 0.2 | 6.3 | 5,380 | 2,490 |
| Oregon | 5.6 | 0.8 | 8.4 | 81.4 | 0.2 | 3.4 | 0.0 | 0.2 | 31,900 | 12,000 |
| Pennsylvania | 17.6 | 1.4 | 0.2 | 0.1 | 0.5 | 19.0 | 2.6 | 58.5 | 37,200 | 2,200 |
| Rhode Island | 76.7 | 3.7 | 4.2 | 1.6 | 0.2 | 9.0 | 5.1 | 0.0 | 503 | 68 |
| South Carolina | 5.9 | 1.7 | 0.0 | 0.9 | 0.4 | 10.5 | 0.2 | 80.3 | 22,700 | 1,110 |
| South Dakota | 12.8 | 1.5 | 2.9 | 66.2 | 7.3 | 2.5 | 6.4 | 0.5 | 2,240 | 1,310 |
| Tennessee | 7.6 | 0.6 | 0.6 | 0.4 | 0.5 | 9.6 | 1.0 | 79.7 | 34,800 | 950 |
| Texas | 15.4 | 0.5 | 0.3 | 42.2 | 1.1 | 4.4 | 0.7 | 35.5 | 76,100 | 34,100 |
| Utah | 11.6 | 0.1 | 0.1 | 82.0 | 0.8 | 2.4 | 0.9 | 2.0 | 16,600 | 8,440 |
| Vermont | 6.2 | 2.7 | 0.6 | 0.1 | 1.0 | 7.0 | 0.6 | 82.1 | 2,390 | 110 |
| Virginia | 15.1 | 2.4 | 0.7 | 0.8 | 0.6 | 10.5 | 1.9 | 68.2 | 17,800 | 848 |
| Washington | 11.1 | 1.3 | 0.3 | 76.2 | 0.4 | 6.3 | 0.0 | 4.2 | 29,900 | 10,700 |
| West Virginia | 3.5 | 1.1 | 0.1 | 0.0 | 0.1 | 2.9 | 11.5 | 81.0 | 17,300 | 1,930 |
| Wisconsin | 9.1 | 1.4 | 0.2 | 2.3 | 1.5 | 7.2 | 0.0 | 78.3 | 24,600 | 1,750 |
| Wyoming | 1.2 | 0.1 | 0.0 | 94.5 | 0.4 | 0.2 | 1.3 | 2.4 | 28,700 | 10,300 |
| Puerto Rico | 70.1 | 1.2 | 0.0 | 24.3 | 1.4 | 1.9 | 0.5 | 0.5 | 2,180 | 753 |
| Virgin Islands | 68.8 | 17.2 | 7.5 | 0.0 | 5.4 | 1.1 | 0.0 | 0.0 | 35 | 5.7 |
| Total | 11.4 | 1.0 | 0.7 | 40.4 | 1.3 | 5.7 | 1.0 | 38.6 | 1,280,000 | 356,000 |

Figures may not add to totals due to independent rounding.

**TABLE H.10 II**  Fresh water withdrawal in the United States, by water resources region and by user, 1990

| Region | Percent of total fresh water use | | | | | | | | Total fresh water withdrawal (10³ m³/day) | Fresh water consumptive use (10³ m³/day) |
|---|---|---|---|---|---|---|---|---|---|---|
| | Public supply | Domestic | Commercial | Irrigation | Livestock | Industrial | Mining | Thermo-electric | | |
| New England | 30 | 3.6 | 2.8 | 2.5 | 0.2 | 10.1 | 0.4 | 51 | 17,900 | 1,550 |
| Mid-Atlantic | 28 | 1.9 | 0.6 | 0.9 | 0.5 | 8 | 1.8 | 58 | 79,900 | 4,770 |
| South Atlantic-Gulf | 15 | 2.0 | 0.4 | 13 | 1.0 | 8 | 1.3 | 59 | 129,000 | 19,500 |
| Great Lakes | 13 | 0.9 | 0.3 | 0.9 | 0.3 | 13 | 0.8 | 70 | 122,000 | 6,210 |
| Ohio | 8.3 | 1.2 | 0.3 | 0.2 | 0.4 | 8 | 3.3 | 79 | 115,000 | 7,990 |
| Tennessee | 5.6 | 0.6 | 0.6 | 0.3 | 2.2 | 13 | 1.0 | 77 | 34,800 | 1,220 |
| Upper Mississippi | 9.1 | 1.8 | 1.3 | 1.9 | 1.3 | 4.6 | 0.7 | 79 | 78,400 | 7,420 |
| Lower Mississippi | 5.8 | 0.5 | 0.5 | 41 | 5.9 | 15 | 0.2 | 31 | 68,100 | 26,400 |
| Souris-Red-Rainy | 24 | 7.5 | 0.1 | 33 | 7.1 | 17 | 2.8 | 9 | 1,080 | 545 |
| Missouri Basin | 4.3 | 0.4 | 0.1 | 66 | 1.1 | 0.5 | 0.7 | 27 | 142,000 | 45,800 |
| Arkansas-White-Red | 9.1 | 0.8 | 1.1 | 54 | 2.3 | 2.4 | 0.5 | 29 | 58,300 | 29,800 |
| Texas-Gulf | 19 | 0.6 | 0.4 | 38 | 1.2 | 5.5 | 1.0 | 35 | 51,100 | 22,400 |
| Rio Grande | 8.9 | 0.4 | 0.3 | 88 | 0.6 | 0.2 | 1.1 | 0.3 | 22,700 | 13,100 |
| Upper Colorado | 1.7 | 0.1 | 0.1 | 93 | 1.7 | 0.1 | 0.7 | 2.5 | 26,800 | 9,310 |
| Lower Colorado | 14 | 0.5 | 0.4 | 78 | 1.3 | 2.2 | 2.2 | 1.4 | 27,700 | 18,900 |
| Great Basin | 8.5 | 0.2 | 0.2 | 88 | 0.5 | 1.5 | 1.2 | 0.4 | 27,300 | 13,000 |
| Pacific Northwest | 4.4 | 0.6 | 2.0 | 88 | 1.7 | 2.8 | 0.0 | 1.0 | 139,000 | 45,800 |
| California | 16 | 0.9 | 0.8 | 80 | 1.1 | 0.4 | 0.1 | 0.7 | 125,000 | 78,700 |
| Alaska | 32 | 2.4 | 6.3 | 0.2 | 0.2 | 39 | 8.8 | 10.9 | 1,080 | 98 |
| Hawaii | 20 | 0.8 | 3.4 | 63 | 0.6 | 3.6 | 0.0 | 8.0 | 4,500 | 2,370 |
| Caribbean | 70 | 1.4 | 0.1 | 24 | 1.5 | 1.9 | 0.4 | 0.4 | 2,210 | 757 |
| Total | 11 | 1.0 | 0.7 | 40 | 1.3 | 5.7 | 1.0 | 39 | 1,280,000 | 356,000 |

Figures may not add to totals due to independent rounding.

**TABLE H.11**   United States water withdrawals, by region, 1960 to 1990 (km³/yr)

| Region | 1960 | 1965 | 1970 | 1975 | 1980 | 1985 | 1990 |
|---|---|---|---|---|---|---|---|
| New England | 8.9 | 10.0 | 13.4 | 19.4 | 18.0 | 22.5 | 19.2 |
| Mid-Atlantic | 37.5 | 46.5 | 62.3 | 71.9 | 71.9 | 60.5 | 65.8 |
| South Atlantic-Gulf | 26.1 | 39.3 | 48.4 | 59.5 | 67.8 | 60.2 | 61.1 |
| Great Lakes | 40.1 | 45.7 | 54.0 | 49.8 | 52.6 | 44.1 | 44.8 |
| Ohio | 33.2 | 41.5 | 49.8 | 49.8 | 52.6 | 43.2 | 42.0 |
| Tennessee | 10.4 | 11.3 | 10.9 | 15.2 | 16.6 | 12.7 | 12.7 |
| Upper Mississippi | 15.2 | 22.1 | 22.1 | 26.3 | 31.8 | 23.4 | 28.7 |
| Lower Mississippi | 7.3 | 7.2 | 18.0 | 22.1 | 29.1 | 24.3 | 26.4 |
| Souris-Red-Rainy | 0.2 | 0.4 | 0.4 | 0.5 | 0.3 | 0.4 | 0.4 |
| Missouri | 29.9 | 28.8 | 33.2 | 48.4 | 54.0 | 47.7 | 51.9 |
| Arkansas-White-Red | 14.4 | 14.4 | 16.6 | 20.8 | 33.2 | 21.2 | 21.7 |
| Texas-Gulf | 30.4 | 22.1 | 29.1 | 30.4 | 23.5 | 26.0 | 25.6 |
| Rio Grande | a | 10.1 | 8.7 | 7.5 | 6.5 | 7.8 | 8.3 |
| Upper Colorado | 19.4 | 9.3 | 11.2 | 5.7 | 11.8 | 10.5 | 9.8 |
| Lower Colorado | b | 9.1 | 10.0 | 11.8 | 12.0 | 10.2 | 10.7 |
| Great Basin | 9.7 | 9.5 | 9.3 | 9.5 | 10.4 | 11.4 | 10.1 |
| Pacific Northwest | 40.1 | 40.1 | 41.5 | 45.7 | 47.0 | 49.1 | 50.2 |
| California | 45.7 | 52.6 | 66.4 | 70.6 | 74.7 | 69.0 | 65.1 |
| Alaska | 0.3 | 0.1 | 0.3 | 0.3 | 0.3 | 0.6 | 0.9 |
| Hawaii | 2.2 | 2.8 | 3.7 | 3.5 | 3.5 | 3.0 | 3.8 |
| Caribbean | 1.7 | 2.4 | 4.2 | 5.7 | 4.6 | 3.8 | 4.4 |
| Total [c] | 374 | 429 | 512 | 581 | 625[d] | 552 | 564 |

[a] Included in Texas-Gulf.
[b] Included in Upper Colorado.
[c] Figures may not add to total due to independent rounding.
[d] This total revised in 1988 to 609 km³/yr.

**TABLE H.11**   United States water withdrawals, by region, 1960 to 1990

**DESCRIPTION**

Total fresh and saline water withdrawals for the 21 United States water resources regions are shown here for 1960, 1965, 1970, 1975, 1980, 1985, and 1990, in km³ per year. Totals include ground and surface water, and exclude reclaimed sewage water.

**LIMITATIONS**

For 1960, the Texas-Gulf region includes the Rio Grande region, and the Upper Colorado includes the Lower Colorado. Data are withdrawals, not consumptive use.

**SOURCES**

All data for 1960 through 1985 come from the U.S. Geological Survey Circulars, *Estimated Use of Water in the United States,* U.S. Department of Interior, U.S. Geological Survey, U.S. Government Printing Office, Washington, DC.

U.S. Geological Survey, 1961, *Estimated Use of Water in the United States in 1960,* U.S. Geological Survey Circular 456.

U.S. Geological Survey, 1972, *Estimated Use of Water in the United States in 1970,* U.S. Geological Survey Circular 672.

U.S. Geological Survey, 1977, *Estimated Use of Water in the United States in 1975,* U.S. Geological Survey Circular 765.

U.S. Geological Survey, 1983, *Estimated Use of Water in the United States in 1980,* U.S. Geological Survey Circular 1001.

U.S. Geological Survey, 1988, *Estimated Use of Water in the United States in 1985,* U.S. Geological Survey Circular 1004.

U.S. Geological Survey, 1992 (preliminary), *Estimated Use of Water in the United States in 1990,* U.S. Department of the Interior, U.S. Government Printing Office, Washington, DC.

Preliminary 1990 numbers provided by W. Solley and H. Perlman, 1992, personal communication.

**TABLE H.12** Summary of estimated water use in the United States, 1900 to 1990 *(page 396)*

**DESCRIPTION**

Water use in the United States since 1900 is given here, in km³ per year. Data on the withdrawals of water by sector are given for 1900 – 1990 for public supply, rural domestic and livestock, irrigation, and industrial uses, including thermoelectric power plant use. The sources of water, including ground and surface water, and saline and fresh water, are given from 1950. Total consumptive use of fresh water is also shown, as are instream volumes of water used for generating hydroelectricity.

**LIMITATIONS**

Older data pertain to a smaller United States than do more recent numbers, so care should be taken in comparing data from before 1955.

Methods for gathering data may also have changed over time. Data from 1900 to 1980 are rounded to two significant figures; data for 1985 and 1990 are rounded to three significant figures.

**SOURCES**

U.S. Geological Survey, 1992 (preliminary), *Estimated Use of Water in the United States in 1990,* U.S. Department of the Interior, U.S. Government Printing Office, Washington, DC.
Preliminary 1990 numbers provided by W. Solley and H. Perlman, 1992, personal communication.
Council on Environmental Quality, 1991, *Environmental Quality: 21st Annual Report,* Washington, DC.

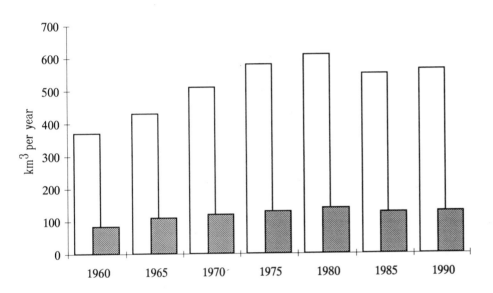

Fig H.12. Withdrawal and consumptive offstream use of fresh water in the United States, 1960–1990 (the entire column represents total withdrawals; the hatched sections represent consumptive use). Consumptive use and total fresh water withdrawals in the United States have increased over time up to 1985, when the first drop in both indicators was reported. Since that time, both have resumed their climb, though at a reduced rate. Values shown here are rounded. See Table H.12 for complete data.

**TABLE H.12**  Summary of estimated water use in the United States, 1900 to 1990 (km³/yr)

| | 1900[a] | 1910[a] | 1920[a] | 1930[a] | 1940[a] | 1945[a] | 1950[b] | 1955[b] | 1960[c] | 1965[c] | 1970[d] | 1975[e] | 1980[e] | 1985[e] | 1990[e] | Percentage change[g] 1985–90 |
|---|---|---|---|---|---|---|---|---|---|---|---|---|---|---|---|---|
| Population (10⁶) | | | | | | | 151 | 164 | 179 | 194 | 206 | 216 | 230 | 242 | 252 | +4 |
| **Offstream use** | | | | | | | | | | | | | | | | |
| Total withdrawals | | | | | | | 250 | 330 | 370 | 430 | 510 | 580 | 610[h] | 551 | 564 | +2 |
| Public supply | 4 | 7 | 8 | 11 | 14 | 17 | 19 | 24 | 29 | 33 | 37 | 40 | 47 | 50.4 | 53.2 | +5 |
| Rural domestic and livestock | 3 | 3 | 3 | 4 | 4 | 5 | 5 | 5 | 5 | 6 | 6 | 7 | 8 | 10.8 | 10.9 | +1 |
| Irrigation | 28 | 54 | 77 | 83 | 98 | 110 | 120 | 150 | 150 | 170 | 180 | 190 | 210 | 189 | 189 | −0.3 |
| Industrial | | | | | | | | | | | | | | | | |
| Thermoelectric power use | 7 | 10 | 12 | 25 | 32 | 44 | 55 | 100 | 140 | 180 | 240 | 280 | 290 | 257 | 269 | +5 |
| Other industrial use | 14 | 19 | 25 | 29 | 40 | 48 | 51 | 54 | 53 | 64 | 65 | 62 | 62 | 42.1 | 41.3 | −2 |
| **Source of water** | | | | | | | | | | | | | | | | |
| Ground | | | | | | | | | | | | | | | | |
| Fresh | | | | | | | 47 | 65 | 69 | 83 | 94 | 110 | 120[h] | 101 | 110 | +8 |
| Saline | | | | | | | | 0.8 | 0.6 | 0.7 | 1.4 | 1.4 | 1.2 | 0.90 | 1.7 | +88 |
| Surface | | | | | | | | | | | | | | | | |
| Fresh | | | | | | | 190 | 250 | 260 | 290 | 340 | 360 | 400 | 366 | 358 | −2 |
| Saline | | | | | | | 14 | 25 | 43 | 59 | 73 | 95 | 98 | 82.3 | 94.2 | +14 |
| Reclaimed sewage | | | | | | | | 0.3 | 0.8 | 1.0 | 0.7 | 0.7 | 0.7 | 0.8 | 1.04 | +30 |
| Consumptive use | | | | | | | | | 84 | 110 | 120[f] | 130[f] | 140[f] | 128[f] | 130[f] | +2 |
| **Instream use** | | | | | | | | | | | | | | | | |
| Hydroelectric power | | | | | | | 1,500 | 2,100 | 2,800 | 3,200 | 3,900 | 4,600 | 4,600 | 4,300 | 4,550 | +8 |

[a] For the years 1900–1945, the area included in the data is not given.
[b] 48 states and District of Columbia.
[c] 50 states and District of Columbia.
[d] 50 states, District of Columbia, and Puerto Rico.
[e] 50 states, District of Columbia, Puerto Rico, and Virgin Islands.
[f] Fresh water only.
[g] Percentages calculated from unrounded numbers.
[h] Revised.

1950–1980 data rounded to two significant figures.
1985 and 1990 data rounded to three significant figures.

**TABLE H.13 I** Ground water withdrawals in the United States, by state and sector, 1990

| State | \multicolumn{9}{c}{Fresh water withdrawals ($10^6$ m$^3$/yr)} | | | | | | | | |
|---|---|---|---|---|---|---|---|---|---|
| | Public supply | Domestic | Commercial | Irrigation | Livestock | Industrial | Mining | Thermo-Electric | Total |
| Alabama | 309 | 39 | 4.3 | 46 | 99 | 43 | 6 | 0.0 | 546 |
| Alaska | 47 | 8.6 | 12 | 0.1 | 0.1 | 7.2 | 5.9 | 6.5 | 87 |
| Arizona | 554 | 44 | 25 | 2,850 | 36 | 54 | 175 | 58 | 3,790 |
| Arkansas | 164 | 70 | 19 | 5,940 | 173 | 137 | 2.5 | 3.3 | 6,510 |
| California | 4,500 | 293 | 79 | 14,800 | 283 | 173 | 18 | 6.4 | 20,100 |
| Colorado | 115 | 26 | 11 | 3,540 | 30 | 46 | 39 | 29 | 3,830 |
| Connecticut | 101 | 64 | 25 | 11 | 1.5 | 26 | 0.6 | 0.3 | 229 |
| Delaware | 46 | 15 | 2.5 | 32 | 2.9 | 25 | 0.0 | 0.7 | 124 |
| District of Columbia | 0.0 | 0.0 | 0.0 | 0.0 | 0.0 | 0.7 | 0.7 | 0.0 | 1.4 |
| Florida | 2,350 | 413 | 69 | 2,680 | 95 | 390 | 413 | 32 | 6,440 |
| Georgia | 323 | 138 | 41 | 363 | 14 | 478 | 12 | 7.2 | 1,380 |
| Hawaii | 305 | 12 | 54 | 276 | 4.7 | 28 | 1.9 | 131 | 813 |
| Idaho | 239 | 66 | 22 | 9,150 | 774 | 235 | 0.8 | 8.4 | 10,500 |
| Illinois | 613 | 159 | 75 | 104 | 84 | 214 | 10 | 12 | 1,270 |
| Indiana | 379 | 163 | 46 | 28 | 36 | 178 | 13 | 17 | 859 |
| Iowa | 323 | 62 | 26 | 29 | 126 | 98 | 1.0 | 17 | 682 |
| Kansas | 240 | 35 | 8.6 | 5,510 | 115 | 69 | 35 | 18 | 6,030 |
| Kentucky | 76 | 69 | 5.8 | 0.7 | 2.2 | 128 | 7.2 | 52 | 342 |
| Louisiana | 380 | 69 | 17 | 623 | 307 | 399 | 0.6 | 55 | 1,850 |
| Maine | 29 | 68 | 5.0 | 0.3 | 0.8 | 14 | 1.0 | 1.9 | 119 |
| Maryland | 105 | 97 | 26 | 26 | 17 | 28 | 29 | 2.5 | 330 |
| Massachusetts | 247 | 51 | 76 | 0.0 | 1.9 | 90 | 0.0 | 0.7 | 467 |
| Michigan | 361 | 170 | 11 | 145 | 26 | 242 | 12 | 3.9 | 971 |
| Minnesota | 401 | 232 | 77 | 217 | 79 | 90 | 3.3 | 3.3 | 1,100 |
| Mississippi | 390 | 46 | 22 | 2,420 | 551 | 199 | 4.0 | 59 | 3,690 |
| Missouri | 256 | 86 | 30 | 463 | 19 | 73 | 35 | 44 | 1,010 |
| Montana | 70 | 21 | 0.0 | 124 | 22 | 41 | 3.6 | 0.0 | 283 |
| Nebraska | 325 | 65 | 0.3 | 6,020 | 149 | 54 | 1.7 | 8.3 | 6,630 |
| Nevada | 144 | 13 | 10 | 1,200 | 1.4 | 13 | 62 | 17 | 1,460 |
| New Hampshire | 47 | 37 | 0.6 | 0.1 | 1.1 | 0.4 | 0.1 | 1.1 | 88 |
| New Jersey | 547 | 94 | 21 | 30 | 2.9 | 73 | 11 | 2.2 | 781 |
| New Mexico | 333 | 33 | 22 | 1,890 | 25 | 6.4 | 106 | 14 | 2,430 |
| New York | 760 | 166 | 40 | 39 | 21 | 117 | 15 | 0.0 | 1,160 |
| North Carolina | 189 | 142 | 23 | 17 | 50 | 87 | 94 | 0.0 | 602 |
| North Dakota | 44 | 17 | 0.1 | 108 | 18 | 2.9 | 4.1 | 0.1 | 194 |
| Ohio | 547 | 181 | 35 | 6.5 | 11 | 170 | 280 | 18 | 1,250 |
| Oklahoma | 111 | 57 | 6.6 | 681 | 48 | 4.6 | 3.0 | 2.5 | 913 |

*continued*

**Fresh water data**

| State | Fresh water withdrawals ($10^6$ m$^3$/yr) | | | | | | | | |
|---|---|---|---|---|---|---|---|---|---|
| | Public supply | Domestic | Commercial | Irrigation | Livestock | Industrial | Mining | Thermo-Electric | Total |
| Oregon | 145 | 79 | 10 | 778 | 4.4 | 43 | 1.4 | 0.0 | 1,060 |
| Pennsylvania | 590 | 195 | 19 | 4.8 | 64 | 249 | 292 | 0.0 | 1,410 |
| Rhode Island | 18 | 6.8 | 5.8 | 0.4 | 0.1 | 3.5 | 0.0 | 0.0 | 35 |
| South Carolina | 109 | 142 | 2.9 | 41 | 17 | 65 | 11 | 1.9 | 390 |
| South Dakota | 72 | 12 | 17 | 195 | 23 | 6.9 | 21 | 0.6 | 347 |
| Tennessee | 372 | 82 | 76 | 21 | 40 | 95 | 11 | 0.0 | 697 |
| Texas | 1,760 | 128 | 70 | 7,720 | 128 | 198 | 128 | 75 | 10,200 |
| Utah | 421 | 6.4 | 5.8 | 702 | 37 | 106 | 51 | 0.0 | 1,330 |
| Vermont | 26 | 23 | 3.9 | 0.0 | 6.4 | 1.4 | 0.4 | 0.6 | 62 |
| Virginia | 95 | 156 | 37 | 11 | 12 | 269 | 32 | 0.6 | 613 |
| Washington | 600 | 144 | 37 | 1,040 | 30 | 144 | 3.3 | 5.5 | 2,010 |
| West Virginia | 59 | 66 | 3.5 | 0.0 | 1.8 | 146 | 728 | 0.6 | 1,010 |
| Wisconsin | 406 | 124 | 15 | 207 | 102 | 80 | 0.3 | 3.3 | 939 |
| Wyoming | 57 | 11 | 1.2 | 327 | 19 | 8.3 | 104 | 1.4 | 529 |
| Puerto Rico | 111 | 4.8 | 0.1 | 75 | 6.6 | 15 | 2.6 | 3.6 | 218 |
| Virgin Islands | 1.4 | 0.3 | 0.6 | 0.0 | 0.4 | 0.1 | 0.0 | 0.0 | 2.8 |
| Total | 20,800 | 4,500 | 1,250 | 70,500 | 3,720 | 5,460 | 2,790 | 725 | 110,000 |

Figures may not add to totals due to independent rounding.

**TABLE H.13 II**  Ground water withdrawals in the United States, by water resources region and sector, 1990

| | Fresh water withdrawals ($10^6$ m$^3$/yr) | | | | | | | | |
|---|---|---|---|---|---|---|---|---|---|
| Region | Public supply | Domestic | Commercial | Irrigation | Livestock | Industrial | Mining | Thermo-Electric | Total |
| New England | 453 | 233 | 113 | 12 | 7.5 | 133 | 1.8 | 4.0 | 958 |
| Mid-Atlantic | 1,934 | 547 | 130 | 141 | 98 | 497 | 290 | 6.4 | 3,640 |
| South Atlantic-Gulf | 3,468 | 910 | 166 | 3,178 | 274 | 1,238 | 531 | 57 | 9,820 |
| Great Lakes | 636 | 390 | 37 | 182 | 70 | 325 | 30 | 4.4 | 1,680 |
| Ohio | 1,069 | 486 | 80 | 39 | 75 | 735 | 1,083 | 90 | 3,660 |
| Tennessee | 151 | 77 | 77 | 5.2 | 44 | 32 | 35 | 0.0 | 421 |
| Upper Mississippi | 1,603 | 513 | 185 | 489 | 300 | 482 | 15 | 29 | 3,620 |
| Lower Mississippi | 968 | 124 | 28 | 8,607 | 981 | 692 | 11 | 104 | 11,500 |
| Souris-Red-Rainy | 47 | 30 | 0.1 | 77 | 22 | 1.8 | 0.3 | 0.0 | 179 |
| Missouri Basin | 844 | 191 | 46 | 9,947 | 334 | 157 | 133 | 69 | 11,700 |
| Arkansas-White-Red | 503 | 163 | 37 | 9,118 | 218 | 93 | 68 | 43 | 10,200 |
| Texas-Gulf | 1,451 | 109 | 65 | 5,485 | 75 | 195 | 116 | 64 | 7,560 |
| Rio Grande | 511 | 32 | 26 | 2,238 | 28 | 15 | 90 | 22 | 2,960 |
| Upper Colorado | 44 | 14 | 7.7 | 44 | 7.3 | 4.0 | 54 | 0.0 | 175 |
| Lower Colorado | 709 | 51 | 32 | 3,095 | 41 | 68 | 192 | 65 | 4,250 |
| Great Basin | 489 | 18 | 10 | 1,948 | 37 | 106 | 106 | 10 | 2,730 |
| Pacific Northwest | 1,004 | 293 | 68 | 10,873 | 816 | 464 | 6.9 | 14 | 13,500 |
| California | 4,504 | 290 | 80 | 14,644 | 278 | 174 | 21 | 6.4 | 20,000 |
| Alaska | 47 | 8.6 | 12 | 0.1 | 0.1 | 7.2 | 5.9 | 6.5 | 87 |
| Hawaii | 305 | 12 | 54 | 276 | 4.7 | 28 | 1.9 | 131 | 813 |
| Caribbean | 112 | 5.1 | 0.6 | 75 | 7.2 | 15 | 2.6 | 3.6 | 221 |
| Total | 20,900 | 4,500 | 1,250 | 70,500 | 3,720 | 5,460 | 2,790 | 728 | 110,000 |

Figures may not add to totals due to independent rounding.

**TABLE H.13 I**  Ground water withdrawals in the United States, by state and sector, 1990 *(pages 397–398)*

**TABLE H.13 II**  Ground water withdrawals in the United States, by water resources region and sector, 1990

## DESCRIPTION

Total fresh ground water withdrawals are given here in two tables. The first lists ground water withdrawals by sector, for all 50 states, the District of Columbia, Puerto Rico, and the Virgin Islands, in m$^3$ × $10^6$ per year. The second gives the same data for the 21 water resources regions of the United States. Sectors include public supply, domestic, commercial, irrigation, livestock, industrial, mining, and thermoelectric cooling uses.

## LIMITATIONS

These data are compiled from information from U.S. water resources subregions and the 50 states and the accuracy of data vary. All data here are for withdrawals, not consumption. Only fresh water withdrawals are listed; a few regions use modest quantities of saline ground water, primarily for power plant cooling. All numbers are rounded to a maximum of three significant figures consistent with the data in the original source. Figures may not add to totals due to independent rounding.

## SOURCES

U.S. Geological Survey, 1992 (preliminary), *Estimated Use of Water in the United States in 1990*, U.S. Department of the Interior, U.S. Government Printing Office, Washington, DC.

Preliminary 1990 numbers provided by W. Solley and H. Perlman, 1992, personal communication.

**TABLE H.14**  Ground water use in the western
United States

| State | Ground water as a percent of total fresh water use[a] | Ground water as a percent of irrigation water use | Agriculture as a percent of consumptive water use |
|-------|----|----|----|
| Arizona | 57 | 58 | 86 |
| California | 38 | 39 | 82 |
| Colorado | 18 | 19 | 92 |
| Kansas | 89 | 92 | 96 |
| Montana | 2 | 1 | 95 |
| Nebraska | 73 | 67 | 97 |
| Nevada | 20 | 17 | 92 |
| New Mexico | 47 | 44 | 86 |
| Oregon | 17 | 14 | 95 |
| Texas | 62 | 70 | 81 |
| Utah | 18 | 10 | 92 |
| Washington | 9 | 4 | 95 |
| Wyoming | 11 | 8 | 96 |

[a]Excludes water use for power plant cooling.

**TABLE H.14**  Ground water use in the western
United States

**DESCRIPTION**

Ground water withdrawals in 13 western United States are shown here as a percentage of total fresh water use and as a percentage of water used for irrigation. There is wide variation among these states: for example, 92% of all water used for irrigation in Kansas comes from ground water, while in Montana only 1% of irrigation water comes from ground water. It also shows the intensive consumptive use of fresh water in the agricultural sector in this part of the U.S. In all states, agriculture accounts for more than 80% of total consumptive use of water, reaching nearly 100 percent in several states. Consumptive water use refers to water that cannot be reused in the same basin due to evaporation, transpiration, deep seepage, or contamination.

**LIMITATIONS**

These data are for the mid-1980s, but not all data are for the same year. Ground water use in some states varies significantly from year to year depending on surface water availability. These data clearly show the intensive use of ground water for agriculture in this semi-arid region, but time-series data might show trends in ground water use and the growing competition for water from urban and industrial users.

**SOURCE**

M. Moench 1991, *Social Issues in Western U.S. Groundwater Management: An Overview.* Pacific Institute for Studies in Development, Environment, and Security, Oakland, California. By permission.

**TABLE H.15**  Ground water resources and use in selected Asian and Pacific countries

| | Ground water utilization ($10^6$ m$^3$/year) by various sectors of the economy | | | | Exploitable potential ($10^6$ m$^3$/yr) | Utilization as percent of exploitable potential | Ground water use as percent of total fresh water use | Ground water extraction cost at wellhead (U.S. $/m$^3$) | Year |
|---|---|---|---|---|---|---|---|---|---|
| | Domestic and municipal | Irrigation | Industry | Total | | | | | |
| Australia | 1,424 | 1,297 | 111 | 2,832 | 37,377 | 8 | | | |
| Bangladesh | 1,710 | 5,997 | 24 | 7,731 | | | | 0.01 | 1979 |
| Guam | | | | 35 | 70 | 50 | 75 | | |
| India | 4,600 | 143,500 | 1,900 | 150,000 | 455,520 | | | 0.02 | 1973 |
| Indonesia | 3,640 | 11 | 33 | 3,684 | | 1 | | | |
| Malaysia | 60% | 5% | 35% | 100% | 23,200 | | | 0.06-0.09 | 1981 |
| Pakistan | | 40,000 | | 40,000 | | | | 0.01 | 1981 |
| Philippines | 2,000 | 1,000 | 1,000 | 3,000 | 33,000 | 9 | 35 | 0.03 | 1983 |
| Thailand | 541 | 202 | 181 | 924 | | | | | |
| Iran | | | | | | | 41 | | |
| Japan | | | | | | | 16 | 0.01-0.04 | 1980 |
| Kiribati | | | | | | | 100 | | |
| Northern Mariana Islands | | | | | | | | 0.03 | 1983 |
| Samoa | | | | | | | | 0.04 | 1983 |
| Mongolia | | | | | 6,000 | | | | |
| Korea Rep | | | | | 15,000 | | | | |
| USSR (Asian) | | | | | 279,000 | | | | |

**TABLE H.15**  Ground water resources and use in selected Asian and Pacific countries

## DESCRIPTION

For 17 Pacific and Asian countries, this table measures exploitable ground water potential and sectoral ground water use in m$^3$ × $10^6$ per year. For Malaysia, these data are listed as percentages. The costs of extracting ground water in these countries is also included in $U.S. per m$^3$. Cost estimates were made for a variety of years, listed in the last column. For several countries, data are also provided on the use of ground water as a percentage of total water use and as a fraction of total exploitable ground water potential.

## LIMITATIONS

No year is given for the ground water utilization data and there are extensive gaps in the data, particularly on sectoral ground water use. This is a problem throughout the world, not just in the region covered by this table. The hydrogeological conditions of this region vary substantially from place to place, complicating assessment of ground water potential. Reliable assessments of ground water availability require a program that involves the study of the ground water flow, the hydrological balance, and the water budget of the aquifers. According to the original source, only four countries of the Asia and Pacific region have adequate ground water programs. The cost figures given are for different years, complicating direct comparison. No definition of the scope of the "Asian USSR" is given.

## SOURCE

Economic and Social Commission for Asia and the Pacific (ESCAP), 1987, *Water Resources Development in Asia and the Pacific: Some Issues and Concerns.* United Nations, Water Resources Series No. 62, New York.

**TABLE H.16 I**  Estimated industrial water use for the world
(in $10^6$ tonnes/yr)

| Industry | Inflow | Outflow | Industry | Inflow | Outflow |
|---|---|---|---|---|---|
| *Iron and steel* | | | *Textile* | | |
| Subtotal | 127,440 | 112,147 | Cotton | 4,129 | 3,856 |
| | | | Wool | 559 | 521 |
| *Other nonferrous metals* | | | Rayon | 4,172 | 3,819 |
| Aluminum | 17,286 | 16,492 | Synthetic | 15,661 | 14,611 |
| Other | 1,895 | 1,213 | Subtotal[a] | 24,621 | 22,908 |
| Subtotal | 19,181 | 17,705 | | | |
| | | | *Rubber* | | |
| *Fertilizer* | | | Natural | 397 | 383 |
| Phosphates | 2,086 | 1,836 | Synthetic | 4,216 | 4,155 |
| Potash | 3,635 | 3,271 | Subtotal | 4,613 | 4,538 |
| Nitrogenous | 4,874 | 4,435 | | | |
| Subtotal[a] | 10,595 | 9,541 | *Petroleum refining* | | |
| | | | Subtotal | 38,395 | 36,475 |
| *Food and agriculture* | | | | | |
| Fruit and vegetables | 307 | 227 | *Miscellaneous* | | |
| Cereal (no wheat) | 340 | 313 | Motor vehicles | 922 | 865 |
| Meat | 1,623 | 1,522 | Coke | 4,731 | 3,642 |
| Butter | 126 | 108 | Cement | 1,920 | 1,799 |
| Cheese | 148 | 129 | Leather | 55 | 44 |
| Fish—canned | 48 | 47 | Subtotal | 7,628 | 6,350 |
| Fish—frozen | 47 | 25 | | | |
| Wheat flour | 250 | 200 | *Total*[a] | 263,489 | 236,997 |
| Sugar, cane | 1,889 | 1,690 | | | |
| Sugar, beet | 760 | 639 | | | |
| Fish meal | 63 | 45 | | | |
| Margarine | 142 | 125 | | | |
| Wine | 8 | 6 | | | |
| Beer | 1,227 | 49 | | | |
| Beverages | 71 | 14 | | | |
| Subtotal | 7,049 | 5,139 | | | |
| | | | | | |
| *Pulp and paper* | | | | | |
| Subtotal | 23,967 | 22,194 | | | |

[a]Total and subtotals as given in original source. May not add to sum of individual industries.

**TABLE H.16 II** Estimated industrial water use by developed and developing countries (in $10^6$ tonnes/yr)

| Industry | World total | | Developed countries | | Developing countries | |
|---|---|---|---|---|---|---|
| | Inflow | Outflow | Inflow | Outflow | Inflow | Outflow |
| Iron and steel | 127,440 | 112,147 | 118,519 | 104,296 | 8,920 | 7,850 |
| Other nonferrous metals | 19,181 | 17,705 | 17,838 | 16,466 | 1,342 | 1,239 |
| Fertilizer | 10,595 | 9,541 | 9,853 | 8,873 | 742 | 668 |
| Food and agriculture | 7,049 | 5,139 | 6,556 | 4,779 | 493 | 360 |
| Pulp and paper | 23,967 | 22,194 | 22,289 | 20,640 | 1,678 | 1,554 |
| Textile | 24,621 | 22,908 | 22,897 | 21,308 | 1,724 | 1,603 |
| Rubber | 4,613 | 4,538 | 4,290 | 4,220 | 323 | 318 |
| Petroleum refining | 38,395 | 36,475 | 35,707 | 33,922 | 2,683 | 2,553 |
| Miscellaneous | 7,628 | 6,350 | 7,094 | 5,905 | 534 | 495 |
| Total[a] | 263,489 | 236,997 | 245,044 | 220,407 | 18,444 | 16,590 |

Error in total outflow from developed countries corrected from original source.
[a]Totals as given in original source. May not add to sum of individual industries.

**TABLE H.16 I** Estimated industrial water use for the world *(pages 402–403)*

**TABLE H.16 II** Estimated industrial water use by developed and developing countries

### DESCRIPTION

Total worldwide water intake and water outflow for a variety of industries are presented in Table H.16I in million tonnes per year as of the mid-1980s. These industries account for about 80% of world industrial water use.

Table H.16II presents the same data broken down by developed and developing countries. Only data for broad industrial categories are given.

### LIMITATIONS

These data are computed from average water use in industries and extrapolated to the world based on industrial production. Ideally, industrial water use data would come from actual reported water use, but only limited mechanisms for such large-scale reporting exist at present. No indication is made here of the quantity of product resulting from this level of water use, and so a comparison across industries would be difficult based on this data alone. Furthermore, the table does not account for technological variation within a given industry, and so these numbers cannot be applied to any one firm. No information is provided about the quality of the outflow and whether it is suitable for reuse elsewhere.

### SOURCES

Cited in M. Meybeck, D. V. Chapman, and R. Helmer, (eds.) 1989, *Global Freshwater Quality: A First Assessment,* Global Environment Monitoring System, (GEMS) World Health Organization and the United Nations Environment Programme, Basil Blackwell, Oxford. By permission.

Data adapted from J.B. Carmichael, and R.M. Strzepek. 1987. *Industrial Water Use and Treatment Practices.* Cassell Tycooly, Philadelphia.

**TABLE H.17** Water use in United States industrial and commercial production

| Industry | Parameters of water use | Units of production | Gross water used | Intake | Consumption | Discharge | Percentage of gross water used for | | |
|---|---|---|---|---|---|---|---|---|---|
| | | | | | | | Noncontact cooling | Process and related | Sanitary and miscellaneous |
| Meat-packing | Liters per tonne carcass weight | l/tonne | 30,000 | 18,100 | 330 | 17,700 | 42 | 46 | 12 |
| Poultry dressing | Liters per tonne ready-to-cook weight | l/tonne | 30,800 | 27,300 | 1,240 | 26,100 | 12 | 77 | 12 |
| Dairy products | Liters per tonne milk processed | l/tonne | 7,060 | 4,320 | 260 | 4,020 | 53 | 27 | 19 |
| Canned fruits and vegetables | Liters per tonne vegetables canned | l/tonne | 82,200 | 39,200 | 3,550 | 35,700 | 19 | 67 | 13 |
| Frozen fruits and vegetables | Liters per tonne vegetables frozen | l/tonne | 93,900 | 58,800 | 1,250 | 57,600 | 19 | 72 | 8 |
| Wet corn milling | Liters per tonne corn ground | l/tonne | 62,100 | 33,300 | 2,680 | 30,700 | 36 | 63 | 1 |
| Cane sugar | Liters per tonne cane sugar | l/tonne | 117,000 | 76,200 | 3,940 | 72,200 | 30 | 69 | 1 |
| Beet sugar | Liters per tonne beet sugar | l/tonne | 138,000 | 46,400 | 1,610 | 44,800 | 31 | 67 | 2 |
| Malt beverages | Liters per liter beer and malt liquor | l/l | 49 | 14 | 3 | 11 | 72 | 13 | 15 |
| Textile mills | Liters per tonne textile fiber input | l/tonne | 291,000 | 125,000 | 12,600 | 113,000 | 57 | 37 | 6 |
| Sawmills | Liters per board foot lumber[a] | l/bd ft | 20 | 12 | 2 | 10 | 58 | 36 | 6 |
| Pulp and paper mills | Liters per tonne paper | l/tonne | 543,000 | 158,000 | 4,920 | 151,000 | 18 | 80 | 1 |
| Paper converting | Liters per tonne paper converted | l/tonne | 27,500 | 16,100 | 1,140 | 15,000 | 18 | 77 | 5 |
| Alkalis and chlorine | Liters per tonne chlorine | l/tonne | 125,000 | 93,100 | 2,820 | 90,200 | 85 | 14 | 1 |
| Industrial gases | Liters per tonne weight of gas | l/tonne | 67,100 | 23,800 | 3,260 | 20,400 | 86 | 13 | 1 |
| Inorganic pigments | Liters per tonne pigments | l/tonne | 408,000 | 206,000 | 6,680 | 199,000 | 41 | 58 | 1 |
| Industrial inorganic chemicals | Liters per tonne chemical products | l/tonne | 60,500 | 19,600 | 1,960 | 17,900 | 83 | 16 | 1 |
| Plastic materials and resins | Liters per tonne plastics | l/tonne | 196,000 | 55,700 | 4,500 | 51,200 | 93 | 7 | |
| Synthetic rubber | Liters per tonne synthetic rubber | l/tonne | 462,000 | 55,100 | 11,700 | 43,300 | 83 | 17 | b |
| Cellulosic man-made fibers | Liters per tonne fibers | l/tonne | 1,930,000 | 564,000 | 38,400 | 525,000 | 69 | 30 | 1 |
| Organic fibers, noncellulosic | Liters per tonne fibers | l/tonne | 843,000 | 319,000 | 8,990 | 310,000 | 94 | 6 | 1 |
| Paints and pigments | Liters per liter paint | l/l | 13.2 | 7.8 | 0.4 | 7.4 | 79 | 17 | 4 |
| Industrial organic chemicals | Liters per tonne chemical building blocks | l/tonne | 520,000 | 227,000 | 11,700 | 216,000 | 91 | 9 | 1 |
| Nitrogenous fertilizers | Liters per tonne fertilizer | l/tonne | 119,000 | 16,700 | 2,930 | 13,800 | 92 | 8 | b |
| Phosphatic fertilizers | Liters per tonne fertilizer | l/tonne | 149,000 | 35,300 | 5,330 | 30,000 | 71 | 28 | 1 |
| Carbon black | Liters per tonne carbon black | l/tonne | 38,400 | 32,900 | 7,390 | 25,500 | 57 | 38 | 6 |
| Petroleum refining | Liters per liter crude petroleum input | l/l | 44 | 6.9 | 0.7 | 6.2 | 95 | 5 | b |

*continued*

**TABLE H.17** Water use in United States industrial and commercial production (pages 404–405)

| Industry | Parameters of water use | Units of production | Gross water used | Intake | Consumption | Discharge | Percentage of gross water used for | | |
| --- | --- | --- | --- | --- | --- | --- | --- | --- | --- |
| | | | | | | | Noncontact cooling | Process and related | Sanitary and miscellaneous |
| Tires and inner tubes | Liters per tire car and truck tires | l/tire | 1,960 | 579 | 53 | 526 | 81 | 16 | 3 |
| Hydraulic cement | Liters per tonne cement | l/tonne | 5,660 | 3,470 | 609 | 2,860 | 82 | 17 | 1 |
| Steel | Liters per tonne steel net tonnes | l/tonne | 261,000 | 159,000 | 5,840 | 154,000 | 56 | 43 | 1 |
| Iron and steel foundries | Liters per tonne ferrous castings | l/tonne | 51,800 | 12,600 | 1,090 | 11,500 | 34 | 58 | 8 |
| Primary copper | Liters per tonne copper | l/tonne | 442,000 | 142,000 | 34,200 | 109,000 | 52 | 46 | 2 |
| Primary aluminum | Liters per tonne aluminum | l/tonne | 410,000 | 99,700 | 1,590 | 98,100 | 72 | 26 | 2 |
| Automobiles | Liters per automobiles | l/automobile | 138,000 | 43,400 | 2,460 | 40,900 | 28 | 69 | 3 |

Percentages may not add due to rounding.

[a] No metric equivalent was found for board foot, which equals 144 cubic inches, or one square foot, one inch deep.

[b] Less than 0.5 percent of gross water use.

**DESCRIPTION**

Water use in United States industrial and commercial production is shown here. Included in the comparison are gross water use, water intake, consumptive use, and discharge, as well as the percentage of gross water used for cooling, processing, and sanitation. Quantities are measured in liters per unit of product, typically a metric tonne, but varying across industries.

**LIMITATIONS**

These data are for the late 1970s and generalize water use for any given industry. Neither the range of water use nor peak water use within an industry are provided, which would give useful information on the potential for improvements in water use efficiency.

**SOURCES**

K.L. Kollar and P. MacAuley, 1980, Water requirements for industrial development, *Journal of the American Water Works Association*, **72**(1), 2–9.

**TABLE H.18**  Water use in United States industry, per employee

| Industry Group | Gross water use per employee (l/day) | Intake per employee (l/day) |
|---|---|---|
| Food and kindred products | 15,540 | 10,360 |
| Tobacco (manufacturers) products | 22,570 | 1,480 |
| Textile mill products | 6,660 | 2,960 |
| Apparel and related products | 370 | 370 |
| Lumber and wood products | 5,920 | 3,700 |
| Furniture and fixtures | 440 | 370 |
| Paper and allied products | 43,930 | 42,920 |
| Printing and publishing | 370 | 370 |
| Chemicals and allied products | 149,110 | 56,240 |
| Petroleum and coal products | 603,100 | 94,350 |
| Rubber and plastic products | 10,730 | 3,700 |
| Leather and leather products | 740 | 703 |
| Stone, clay, and glass products | 11,470 | 5,550 |
| Primary metal industries | 78,440 | 44,030 |
| Fabricated metal products | 2,960 | 1,110 |
| Machinery, except electrical | 3,700 | 1,480 |
| Electrical machinery | 9,250 | 1,110 |
| Transportation equipment | 17,020 | 2,220 |
| Instruments and related products | 4,440 | 1,110 |
| Miscellaneous manufactured goods | 1,110 | 2,960 |

**TABLE H.18**  Water use in United States industry, per employee

**DESCRIPTION**

The average water use per employee is given here for a group of United States industries, as of the late 1970s. Water use is shown as gross water use per employee in liters per day, and intake per employee in liters per day. Intake refers to the volume of water taken into the plant; gross water use is the sum of the intake water plus reused water, and is a measure of the total water required in the manufacturing process.

**LIMITATIONS**

These are overall averages for each industry group. There are wide variations within each industry and within regions. These variations are due to differences in technologies, different prices charged for water, and other factors. The data presented here need to be brought up to date for changes since 1980.

**SOURCES**

K.L. Kollar and P. MacAuley, 1980, Water requirements for industrial development, *Journal of the American Water Works Association*, **72**(1), 2–9.

**TABLE H.19**  Water use by industries in the United States, 1983

| Industry group | Gross water used ($10^6$ m$^3$/yr) | | | Water discharged | | |
| | Total | Intake | Reuse | Total ($10^6$ m$^3$/yr) | Untreated (%) | Treated (%) |
| --- | --- | --- | --- | --- | --- | --- |
| All mineral industries | 12,600 | 4,540 | 8,080 | 3,930 | 31.9 | 68.1 |
| Metal mining | 2,780 | 645 | 2,140 | 504 | 39.7 | 60.3 |
| Anthracite mining | 20 | 8 | 12 | 28 | 12 | 87 |
| Bituminous coal and lignite mining | 449 | 172 | 278 | 440 | 26.2 | 73.8 |
| Oil and gas extraction | 5,500 | 2,280 | 3,220 | 1,800 | 31.0 | 69.0 |
| Nonmetallic minerals, except fuels | 3,860 | 1,430 | 2,430 | 1,150 | 32.6 | 67.4 |
| All manufacturing industries | 128,000 | 38,500 | 90,200 | 33,800 | 54.9 | 45.1 |
| Food and kindred products | 5,330 | 2,560 | 2,880 | 2,090 | 64.5 | 35.5 |
| Tobacco products | 128 | 20 | 108 | 15 | D | D |
| Textile mill products | 1,260 | 503 | 759 | 438 | 52.9 | 47.1 |
| Lumber and wood products | 827 | 326 | 501 | 269 | 63.2 | 36.8 |
| Furniture and fixtures | 26 | 13 | 13 | 13 | 88 | 12 |
| Paper and allied products | 28,200 | 7,200 | 21,000 | 6,700 | 27.1 | 72.9 |
| Chemicals and allied products | 36,500 | 12,900 | 23,600 | 11,300 | 67.0 | 33.0 |
| Petroleum and coal products | 23,400 | 3,100 | 20,300 | 2,650 | 46.2 | 53.8 |
| Rubber and miscellaneous plastics products | 1,240 | 288 | 954 | 237 | 63.6 | 36.4 |
| Leather and leather products | 25 | 23 | 2 | 22 | D | D |
| Stone, clay and glass products | 1,280 | 586 | 689 | 503 | 74.9 | 25.1 |
| Primary metal industries | 22,300 | 8,950 | 13,400 | 8,000 | 58.1 | 41.9 |
| Fabricated metal products | 977 | 248 | 730 | 233 | 48.4 | 51.6 |
| Machinery, except electrical | 1,170 | 455 | 706 | 398 | 67.9 | 32.2 |
| Electric and electronic equipment | 1,270 | 281 | 988 | 266 | 60.5 | 39.5 |
| Transportation equipment | 3,830 | 579 | 3,250 | 528 | 67.5 | 32.5 |
| Instruments and related products | 424 | 113 | 312 | 105 | 49.3 | 50.4 |
| Miscellaneous manufacturing industries | 58.4 | 16 | 42.1 | 15 | D | D |

Percentages may not add to 100.0% due to rounding.
D, Withheld to avoid disclosing data for individual companies.

**TABLE H.19**  Water use by industries in the United States, 1983

## DESCRIPTION

Quantities of gross water used, water reused, and water discharged by United States industries, in m$^3 \times 10^6$, are shown for the year 1983. Total gross water used is the sum of water intake and water recirculated and reused. Also indicated is the fraction of the water discharged that is treated and untreated. The volume of water discharged for some mining practices is greater than intake due to mine water drained and discharged.

## LIMITATIONS

These data are for water withdrawn and reused, which is not the same as water consumed. These estimates need to be brought up to date. No national totals are given because of gaps in the data. Water use is given

in m$^3 \times 10^6$ per year, rather than a quantity of water per unit of production, so these data hide the wide variation in water use efficiency within each industry. See Tables H.21 and H.22. No information is available on the degree of treatment for the discharged water. Data have also been withheld for certain industrial groups.

## SOURCES

Data from United States Department of Commerce, 1985, *1982 Census of Mineral Industries*, Bureau of the Census, U.S. Government Printing Office, Washington, DC.
Cited by F. van der Leeden, F.L. Troise and D.K. Todd, 1990, *The Water Encyclopedia*, Lewis Publishers, Chelsea, Michigan.

**TABLE H.20**   Geographic distribution of water-intensive manufacturing industries in the United States

| Water resources region | Paper ($10^6$ l/day) | Chemicals ($10^6$ l/day) | Petroleum refining ($10^6$ l/day) | Primary metals ($10^6$ l/day) |
|---|---|---|---|---|
| New England | 9,401 | 1,598 | | 643 |
| Mid-Atlantic | 7,104 | 15,736 | 10,704 | 10,341 |
| South Atlantic Gulf | 31,627 | 13,590 | 1,417 | 4,329 |
| Great Lakes | 8,156 | 10,859 | 6,389 | 38,298 |
| Ohio | 2,471 | 21,108 | 3,193 | 33,562 |
| Tennessee | 4,151 | 1,342 | 973 | 395 |
| Upper Mississippi | 2,101 | 2,882 | 2,638 | 2,752 |
| Lower Mississippi | 6,471 | 19,920 | 10,744 | 2,186 |
| Souris-Red-Rainy | 510 | | | |
| Missouri | 99 | 2,527 | 1,986 | 381 |
| Arkansas-White-Red | 1,809 | 3,455 | 7,425 | 1,653 |
| Texas Gulf | 3,814 | 41,425 | 32,996 | 4,495 |
| Rio Grande | | 1,010 | | 103 |
| Upper Colorado | | | 44 | |
| Lower Colorado | 432 | 399 | 40 | 136 |
| Great Basin | | 48 | 614 | 1,949 |
| Columbia-North Pacific | 13,819 | 1,798 | 1,017 | 1,646 |
| California | 4,310 | 2,682 | 8,029 | 795 |
| Alaska | 1,306 | 162 | 51 | |
| Hawaii | | | 310 | |

**TABLE H.20**   Geographic distribution of water-intensive manufacturing industries in the United States

**DESCRIPTION**

The distribution of the four major water-consuming industries in the United States – paper, chemicals, petroleum refining, and primary metals production – is shown here by water resources region. Water withdrawals for each industry are given in million l × $10^6$ per day for 1975.

**LIMITATIONS**

Data are for water withdrawn, not water consumed, and need to be updated. No national totals are given because of gaps in the data. These gaps may mean that the industry is not represented in the region or that no information for that region and industry was reported. Water use is given in l × $10^6$ per day, rather than a quantity of water per unit of production, so these data hide the wide variation in water use efficiency within each industry. See Table H.21.

**SOURCE**

K.L. Kollar and P. MacAuley, 1980, Water requirements for industrial development, *Journal of the American Water Works Association*, **72**(1), 2–9.

**TABLE H.21**   Water use by industry in California, by purpose, 1979

| | | Water intake (%) | | | Total water intake (10$^6$ l/yr) |
|---|---|---|---|---|---|
| Code | Industry group | Cooling | Processing | Sanitary | |
| 20 | Food and kindred products | 34.3 | 63.3 | 2.4 | 218,222 |
| 24 | Lumber and wood products | 28.0 | 71.4 | 0.6 | 51,219 |
| 25 | Furniture and fixtures | 23.4 | 64.7 | 11.9 | 252 |
| 26 | Paper and allied products | 37.5 | 61.9 | 0.6 | 142,269 |
| 27 | Printing and publishing | 68.3 | 13.3 | 17.5 | 120 |
| 28 | Chemicals and allied products | 53.6 | 43.9 | 2.5 | 50,654 |
| 29 | Petroleum and coal products | 84.4 | 15.4 | 0.2 | 271,269 |
| 30 | Rubber and miscellaneous plastics products | 53.8 | 32.5 | 13.6 | 5,690 |
| 31 | Leather and leather products | 8.8 | 84.9 | 6.2 | 581 |
| 32 | Stone, clay and glass products | 11.9 | 86.5 | 1.7 | 44,461 |
| 33 | Primary metal industries | 36.5 | 62.2 | 1.3 | 83,479 |
| 34 | Fabricated metal products | 17.5 | 74.6 | 8.0 | 7,052 |
| 35 | Machinery, except electrical | 14.7 | 57.7 | 27.6 | 8,076 |
| 36 | Electric and electronic equipment | 32.0 | 46.3 | 21.8 | 8,688 |
| 37 | Transportation equipment | 38.7 | 52.9 | 8.4 | 25,397 |
| 38 | Instruments and related products | 41.9 | 40.0 | 18.1 | 2,775 |
| 39 | Miscellaneous manufacturing industries | 46.1 | 20.8 | 33.0 | 631 |

Percentages may not add to 100.0% due to rounding.

**TABLE H.21**   Water use by industry in California, by purpose, 1979

**DESCRIPTION**

Presented here for California are total water intake in m$^3$ × 10$^6$ per year, and the percentages of industrial water withdrawals used for cooling, processing, and sanitary purposes, for 17 industrial groups in 1979. Standard industrial codes (SIC) are shown in the first column.

**LIMITATIONS**

No information was given on levels of industrial production, water use per unit product, or the range of water use within industrial groups. Data are limited to 1979 and need to be brought up to date.

**SOURCES**

State of California, 1982, Water use by manufacturing industries in California 1979, *Department of Water Resources Bulletin*, **124**(3), The Resources Agency, Sacramento, California.

**TABLE H.22**  Water recycling rates in major manufacturing industries in the United States[a]

| Year | Paper and allied products | Chemicals and allied products | Petroleum and coal products | Primary metal industries | All manufacturing |
|------|------|------|------|------|------|
| 1954 | 2.4 | 1.6 | 3.3 | 1.3 | 1.8 |
| 1959 | 3.1 | 1.6 | 4.4 | 1.5 | 2.2 |
| 1964 | 2.7 | 2.0 | 4.4 | 1.5 | 2.1 |
| 1968 | 2.9 | 2.1 | 5.1 | 1.6 | 2.3 |
| 1973 | 3.4 | 2.7 | 6.4 | 1.8 | 2.9 |
| 1978 | 5.3 | 2.9 | 7.0 | 1.9 | 3.4 |
| Projections | | | | | |
| 1985 | 6.6 | 13.2 | 18.3 | 6.0 | 8.6 |
| 2000 | 11.8 | 28.0 | 32.7 | 12.3 | 17.1 |

[a]Number of times a given volume of water is used.

**TABLE H.22**  Water recycling rates in major manufacturing industries in the United States

**DESCRIPTION**

Historical water recycling rates are given for four major American manufacturing industries (paper and allied products, chemicals and allied products, petroleum and coal products, and primary metal industries), and for all United States manufacturing for the years 1954–1978. Projected recycling rates for 1985 and 2000 are given. The recycling rate is the number of times any given volume of water is used.

**LIMITATIONS**

This table was prepared in 1986 with data published in 1981 and 1979.

It therefore needs to be brought up to date. The projections for 1985 and 2000, in particular, need to be updated given new trends and new technologies. The four industries presented in this table do not necessarily reflect recycling rates in other industries. Future improvements in industrial recycling rates will depend on water availability and price, and cost and applicability of new technological developments.

**SOURCE**

S. Postel, 1986, Increasing water efficiency, *State of the World 1986*, Worldwatch Institute, W.W. Norton, New York.

**TABLE H.23** Water requirements for municipal establishments

| Type | Unit | Average use | Peak use |
|------|------|-------------|----------|
| Hotels | Liter/day/square meter | 10.4 | 17.6 |
| Motels | Liter/day/square meter | 9.1 | 63.1 |
| Barber shops | Liter/day/barber chair | 207 | 1,470 |
| Beauty shops | Liter/day/station | 1,020 | 4,050 |
| Restaurants | Liter/day/seat | 91.6 | 632.0 |
| Night clubs | Liter/day/person served | 5 | 5 |
| Hospitals | Liter/day/bed | 1,310 | 3,450 |
| Nursing homes | Liter/day/bed | 503 | 1,600 |
| Medical offices | Liter/day/square meter | 25.2 | 202 |
| Laundry | Liter/day/square meter | 10.3 | 63.9 |
| Laundromats | Liter/day/square meter | 88.4 | 265.0 |
| Retail space | Liter/day/sales square meter | 4.3 | 11 |
| Elementary schools | Liter/day/student | 20.4 | 186 |
| High schools | Liter/day/student | 25.1 | 458 |
| Bus-rail depot | Liter/day/square meter | 136 | 1,020 |
| Car washes | Liter/day/inside square meter | 194.7 | 1,280 |
| Churches | Liter/day/member | 0.5 | 17.8 |
| Golf-swim clubs | Liter/day/member | 117 | 84 |
| Bowling alleys | Liter/day/alley | 503 | 503 |
| Residential colleges | Liter/day/student | 401 | 946 |
| New office buildings | Liter/day/square meter | 3.8 | 21.2 |
| Old office buildings | Liter/day/square meter | 5.8 | 14.4 |
| Theaters | Liter/day/seat | 12.6 | 12.6 |
| Service stations | Liter/day/inside square meter | 10.2 | 1,280 |
| Apartments | Liter/day/occupied unit | 821 | 1,640 |
| Fast food restaurants | Liter/day/establishment | 6,780 | 20,300 |

**TABLE H.23** Water requirements for municipal establishments

**DESCRIPTION**

Average and peak water use is given for 26 different municipal establishments. Units are liters of water per day for a range of variables for each establishment. Peak use is the highest volume measured over an hour, normalized to liters per day. Data are for 1980 for average commercial U.S. establishments.

**LIMITATIONS**

The average values hide wide variations among municipal establishments.

**SOURCE**

J.J. Boland, W.-S. Moy, J.L. Pacey and R.C. Steiner, 1983, *Forecasting Municipal and Industrial Water Use: A Handbook of Methods*, U.S. Army Corps of Engineers, Institute for Water Resources, Fort Belvoir, Virginia.

**TABLE H.24**  Typical household water use in the United States

| Use | Unit | Range |
| --- | --- | --- |
| Washing machine | Liters per load | 130-270 |
| Standard toilet | Liters per flush | 10-30 |
| Ultralow volume toilet | Liters per flush | 6 or less |
| Silent leak | Liters per day | 150 or more |
| Nonstop running toilet | Liters per minute | 20 or less |
| Dishwasher | Liters per load | 50-120 |
| Water-saver dishwasher | Liters per load | 40-100 |
| Washing dishes with tap running | Liters per minute | 20 or less |
| Washing dishes in a filled sink | Liters | 20-40 |
| Running the garbage disposal | Liters per minute | 10-20 |
| Bathroom faucet | Liters per minute | 20 or less |
| Brushing teeth | Liters | 8 |
| Shower head | Liters per minute | 20-30 |
| Low-flow shower head | Liters per minute | 6-11 |
| Filling a bathtub | Liters | 100-130 |
| Watering an 750 square meter lawn | Liters per month[a] | 7,600-16,000 |
| Standard sprinkler | Liters per hour | 110-910 |
| One drip-irrigation emitter | Liters per hour | 1-10 |
| 1/2 inch diameter hose | Liters per hour | 1,100 |
| 5/8 inch diameter hose | Liters per hour | 1,900 |
| 3/4 inch diameter hose | Liters per hour | 2,300 |
| Slowly dripping faucet | Liters per month | 1,300-2,300 |
| Fast-leaking faucet | Liters per month | 7,600 or more |
| Washing a car with running water | Liters in 20 minutes | 400-800 |
| Washing a car with pistol-grip faucet | Liters in 20 minutes | 60 or more |
| Uncovered pool (60 square meters) | Liters lost per month[a] | 3,000-11,000+ |
| Covered pool | Liters lost per month[a] | 300-1,200 |

[a]Depending on climate.

**TABLE H.24**  Typical household water use in the United States

**DESCRIPTION**

In many households in the United States, water has been used freely because costs have been low and supplies have been reliable. In recent years, droughts and growing water demands have led to increased efforts to conserve water by changing patterns of use and by using water-efficient fixtures. This table examines the quantities of water used in a wide range of household fixtures, appliances, and activities. Units vary with item and data have been rounded.

**LIMITATIONS**

Ranges of water use for each fixture or activity are provided. Actual water use will vary considerably, depending on the type of fixture and the style of use.

**SOURCE**

Drought Survival Guide, 1991, *Sunset Magazine*, May, Sunset Publishing Corporation, Menlo Park, California. By permission.
Freshwater Foundation, 1992, Household water uses and abuses, Facets of Freshwater, **21**(9), Wayzata, Minnesota. By permission.

**TABLE H.25**  Water-efficient technologies for the residential and light commercial sectors

| | Ordinary water use | High-efficiency water use | High-Efficiency Models | | | |
|---|---|---|---|---|---|---|
| | | | Median retail price (U.S. $ per item) | Lowest retail price (U.S.$ per item) | Number of models | Number of manufacturers |
| Toilets | 19–26 liters/flush | 3.8–6.1 liters/flush | 191 | 95 | 40 | 24 |
| Showerheads | 15–23 liters/min | 5.7–9.5 liters/min | 15 | 5 | 30 | 16 |
| Residential Faucets | 11–23 liters/min | 1.9–9.5 liters/min | 7 | 2 | 21 | 12 |
| Washing Machines | 150–210 liters/load | 95–110 liters/load | 460 | 460 | 1 | 1 |

**TABLE H.25**  Water-efficient technologies for the residential and light commercial sectors

**DESCRIPTION**

The United States is one of the few industrialized nations not using water-efficient plumbing fixtures widely. Water use in the United States by both typical and highly efficient household water fixtures is shown here. For toilets, showerheads, faucets, and washing machines, this table lists the range of ordinary water use and water use by highly efficient versions. It also shows the number of manufacturers currently making these devices, the median and lowest retail prices for them, and the number of models on the market in late 1991.

"Ordinary water use" describes typical fixtures in place in the United States in 1991. "High-efficiency water use" describes fixtures commercially available in the US. Showerhead and faucet water use reflects nominal flow rate, rated at 80 psi pressure. Actual use is generally lower. Faucets include both lavatory and kitchen faucets. Prices are for 1992 retail.

**LIMITATIONS**

Actual savings from using these fixtures depends on assumptions about the type of use and the types of fixture. No faucets for commercial establishments are listed. Prices are retail; wholesale prices may be significantly lower. Number of models listed included those in the Rocky Mountain Institute catalogue.

**SOURCES**

Compiled by A. Jones, 1992, The Water Program, Rocky Mountain Institute, 1739 Snowmass Creek Road, Snowmass, Colorado, using data from:

C. Laird, 1991, *Water-Efficient Technologies: A Catalog for the Residential/Light Commercial Sector*, 2nd edn., Rocky Mountain Institute, Snowmass, Colorado.

The Water Program, 1991, *Water Efficiency: A Resource for Utility Managers, Community Planners, and Other Decisionmakers*, Rocky Mountain Institute, Snowmass, Colorado.

A. Vickers, 1990, Water-use efficiency standards for plumbing fixtures: benefits of national legislation, *Journal of the American Water Works Association*, **82**(5).

**TABLE H.26**  Characteristics of unconventional water-supply systems *(page 414)*

**DESCRIPTION**

Fresh water is traditionally supplied by pumping ground water and withdrawing water from rivers and lakes. As demands for fresh water supplies increase because of population growth and increasing development, water-short regions are beginning to turn to unconventional methods of supply. This table presents some new water supply techniques, and compares their costs, stage of development, physical requirements, advantages and disadvantages, reliability, and common applications. Cost data are for 1985 in $U.S. per m³.

**LIMITATIONS**

The advantages or disadvantages of each method depend on the alternatives available in a region, the level of technical sophistication of water managers, location, and climate characteristics. For these reasons, the characteristics shown here are only generally applicable.

**SOURCE**

Modified and updated from United Nations, 1985, *Use of Non-Conventional Water Resources in Developing Countries*, Natural Resource/Water Series no. 14, New York.

**TABLE H.26** Characteristics of unconventional water-supply systems

| | Desalination | Transport-tankers | Transport-icebergs | Water reuse | Cloud seeding |
|---|---|---|---|---|---|
| Cost of Water ($1985/m^3) | Brackish $0.25–1.00  Seawater $1.30–8.00 | $1.25–7.50 | Saudi Arabian project $0.02–0.85 | $0.07–1.80 | $0.01 |
| Stage of Development | Moderate to high | Low to moderate | Very low | Moderate to high | Low to moderate |
| Special physical requirements | 1. Source of clean brackish or seawater 2. Method or place to dispose of a brine solution 3. Major equipment and facility construction | 1. Port facilities 2. Storage facilities | 1. Large tugboats 2. A need for water adjacent to a coastal area 3. Deep draught channels 4. Storage facilities 5. Equipment to melt ice and collect water | 1. Source of wastewater 2. Major equipment and facility construction | 1. Suitable clouds 2. Scientific staff 3. Structures and land use to take advantage of increased rainfall |
| Advantages | 1. Proven systems available 2. Wide range of suppliers 3. Provides independence from external sources of supply | 1. Low level of technology required 2. Can be implemented quickly for emergency use | | 1. Proven techniques available 2. Wide range of suppliers 3. Nonpotable applications 4. Can reduce problems associated with present methods of wastewater disposal | 1. Only moderate to small capital investment needed |
| Disadvantages | 1. Requires skilled technicians for operation and repair 2. Generally requires use of foreign exchange to purchase equipment 3. Energy-intensive process | 1. Deep channels and facilities needed for large vessels 2. Must depend on sources outside country. The potential for a supply interruption because of storms, conflicts, boycotts, or strikes is relatively high 3. Can only meet low-volume demand | 1. Essentially untried method 2. Not suitable for small-scale experimentation | 1. Requires some operation and maintenance 2. Improper operation could create the potential for adverse public health effects 3. May not be aesthetically or culturally acceptable | 1. Still in the limited developmental stage with many uncertainties 2. Cannot control the results |
| Certainty of operation | High | High | Low | High | Low to moderate |
| Typical applications for water | Potable Industrial | Potable Agriculture Industrial | Potable Agriculture Industrial | Agriculture Industrial | Potable Agricultural Industrial |

**TABLE H.27** Costs of water supply, $1980

| Technology | ($/acre foot) | ($/$10^3$ m$^3$) |
|---|---|---|
| Diversion projects (interbasin) | 100–200 | 123–246 |
| Storage projects (storage costs only) | 100–200 | 123–246 |
| Distillation | 530–880 | 654–1,085 |
| Freezing | 299–513 | 368–633 |
| Reverse osmosis (for brackish water) | 97–322 | 120–397 |
| Electrodialysis (TDS 2,000–5,000 ppm) | 224–435 | 276–537 |
| Recycling wastewater (AWT) | 162–393 | 200–485 |
| Recycling wastewater (secondary treatment) | 62–104 | 77–128 |
| Ground water development | 72 | 88 |
| Ground water recharge | 95–112 | 118–138 |

TDS, total dissolved solids; AWT, secondary treatment, plus nitrogen, plus phosphate reduction, filtration, and carbon adsorption.

**TABLE H.27** Costs of water supply, $1980

**DESCRIPTION**

A cost comparison is given here of a group of alternative water sources available to end users, in 1980 $U.S. per acre-foot, and per thousand m$^3$. Included are estimates for various desalination processes, wastewater recycling, ground water withdrawal and recharge, and the construction of diversion and storage systems.

**LIMITATIONS**

Valid cost comparisons among different types of technologies are extremely difficult to make, given the differences in capital costs, repayment periods, operation and maintenance requirements, and externalities that are not typically measured in economic terms. The ranges here should not be considered to be the maximum or minimum values, but rather average values for typical facilities built today. For

example, several plans for major water diversions have been proposed with projected costs well above the values of this table; such diversions are unlikely to be built. Also, not all costs in this table are equivalent because some do not include transportation costs to the final user, and operations and maintenance costs are not provided for some of the technologies. Environmental and political costs are also not included. The quality of the water provided by each system varies. Changes in technology over time alter the relative costs of these technologies.

**SOURCE**

Costs compiled in P. Rogers, 1987, Assessment of water resources: Technology for supply, in D.J. McLaren and B.J. Skinner (eds.) *Resources and World Development*, Dahlem Konferenzen, Berlin, John Wiley and Sons, Chichester, UK. By permission.

**TABLE H.28** Production costs and tariffs for drinking water, by country, early 1980s *(pages 416–418)*

**DESCRIPTION**

The cost of building and operating drinking water supply systems varies around the world. Presented here are the costs of constructing and operating drinking water systems for developing countries for the early 1980s. Costs are provided for rural connections, urban household connections, and standpipes in $U.S. per capita. Data are also provided on the cost of water to users in $U.S. per m$^3$, if a tariff is charged, and on whether or not the tariff is progressive, i.e., whether there is an increasing marginal price charged for increasing amounts of water used. Countries listed as having a progressive tariff have one in force in at least one urban area.

**LIMITATIONS**

The data here are based on two questionnaires sent by the World Health Organization in 1980 and 1983. Responses were given in $U.S., and were often a single figure, as opposed to a range of values. Data have never been adjusted for inflation, nor was the exchange rate used in calculations reported. Therefore, this table should only be accepted as a general comparison.

**SOURCE**

World Resources Institute, 1987, *World Resources 1987*, in collaboration with the International Institute for Environment and Development, Basic Books, Inc., New York. By permission.

**TABLE H.28**   Production costs and tariffs for drinking water, by country, early 1980s

| | Facility construction costs ($U.S. per capita) | | | Urban systems | | |
|---|---|---|---|---|---|---|
| | | Urban | | | | |
| | Rural | House connection | Standpipe | Operation cost ($U.S./m³) | Tariff ($U.S./m³) | Progressive tariff |
| *Africa* | | | | | | |
| Angola | 25 | 90 | | | 0.10 | No |
| Benin | 51 | 21[a] | | | | |
| Botswana | | 35[a] | | | | Yes |
| Burkina Faso | 25 | 100 | 40 | 0.23 | 0.40 | Yes |
| Burundi | 17 | 160 | 100 | 0.39 | 0.22 | No |
| Cape Verde | 26 | 36 | 13 | 1.84 | 0.16 | |
| Congo | 200 | 143 | 80 | 0.50 | 0.29 | No |
| Djibouti | | 390 | 150 | 0.40 | 0.55 | Yes |
| Gambia | | | | | | Yes |
| Ghana | 50 | 100 | 80 | 0.30 | 0.20 | Yes |
| Guinea | 21 | 27 | 2 | 0.40 | 0.50 | No |
| Guinea-Bissau | 100 | 160[a] | | 0.50 | 0.50 | No |
| Kenya | 15-70 | 150-300 | 50-150 | | | Yes |
| Lesotho | 100 | 400 | 200 | 1.00 | 0.33 | |
| Liberia | 15 | 92 | | | 0.44 | Yes |
| Libya | | 1,000 | | 0.80 | 0.07 | Yes |
| Madagascar | 38 | 97 | 43 | | | No |
| Malawi | 10 | 75 | 45 | 0.50 | 0.28 | Yes |
| Mali | 38 | 70 | 14 | 0.20 | 0.14 | Yes |
| Malta | 28 | 28 | | 0.51 | 0.30 | Yes |
| Mauritania | | | | 0.62 | 0.68 | Yes |
| Mauritius | | | | 0.25 | 0.20 | |
| Morocco | 115 | 200 | 77 | | | |
| Niger | 47 | 144 | | | | Yes |
| Nigeria | 34 | 81 | 43 | 1.18 | 0.67 | Yes |
| Rwanda | 15 | 120 | 40 | 0.65 | 0.22 | No |
| Senegal | 10 | 13 | 2 | 0.40 | 0.22 | Yes |
| Sierra Leone | 60 | 250 | 200 | 0.80 | 0.20 | No |
| Somalia | 160 | 130 | 90 | 0.50 | 0.75 | Yes |
| Sudan | 17 | 60 | 40 | 0.05 | 0.05 | Yes |
| Tanzania | 56 | 80 | 56 | 0.45 | 0.24 | No |
| Togo | 19 | 126[a] | | 0.66 | 0.31 | Yes |
| Tunisia | 200 | 250 | | | 0.31 | Yes |
| Uganda | 40 | 200[a] | | | | Yes |
| Zaire | 8 | 40 | | | | Yes |
| Zambia | 45-90 | 127 | 82 | 0.34 | 0.22 | Yes |

*continued*

| | Facility construction costs ($U.S. per capita) | | | Urban systems | | |
| | | Urban | | | | |
| | Rural | House connection | Standpipe | Operation cost ($U.S./m$^3$) | Tariff ($U.S./m$^3$) | Progressive tariff |
|---|---|---|---|---|---|---|
| *North and Central America* | | | | | | |
| Bahamas | | 290 | 215 | 1.15 | 2.50 | Yes |
| Barbados | 150 | 170 | 50 | 0.28 | 0.68 | Yes |
| Belize | 125 | 25 | | | | |
| Cayman Islands | | 1,200 | | | 4.40 | Yes |
| Costa Rica | 55 | 80 | | 0.17 | 1.50 | Yes |
| Dominican Republic | 58 | 94 | | 0.05 | 1.70 | Yes |
| El Salvador | 100 | | | | | |
| Guatemala | 48 | 147 | 30 | 0.08 | 0.03 | Yes |
| Haiti | 25 | 120 | 40 | 0.15 | 0.28 | Yes |
| Honduras | 50 | 257 | | 0.19 | 0.24 | No |
| Mexico | 157 | 143 | | 0.09 | 0.06 | Yes |
| Nicaragua | 57 | 116 | 24 | 0.29 | 0.44 | Yes |
| Panama | 60 | 110 | | 0.07 | 0.29 | Yes |
| Trinidad and Tobago | 410 | 350 | 300 | 1.00 | | Yes |
| *South America* | | | | | | |
| Argentina | 170 | 180 | 50 | 0.07 | 0.10 | Yes |
| Bolivia | 88 | 119 | 96 | | | Yes |
| Brazil | 45 | 75 | 25 | 0.10 | 0.13 | Yes |
| Chile | 128 | 170 | | 0.11 | 0.14 | Yes |
| Colombia | 69 | 108 | 30 | 0.04 | 0.05 | Yes |
| Ecuador | 157 | 230 | 85 | | 0.20 | Yes |
| Guyana | 120 | 120 | 100 | 0.08 | 0.03 | Yes |
| Paraguay | 130 | 125 | | 0.24 | 0.19 | Progressive |
| Peru | 52 | 52 | 5 | 0.06 | 0.08 | Yes |
| Suriname | 75 | 180 | 500 | 0.60 | 0.80 | Yes |
| Uruguay | 122 | 122 | 45 | | 0.16 | Yes |
| Venezuela | 104 | | | | | |
| *Asia* | | | | | | |
| Bangladesh | 3 | 40 | 6 | 0.12 | 0.06 | No |
| Brunei | | 240 | | 0.44 | 0.11 | No |
| Cyprus | | | | 0.32 | 0.28 | |
| Hong Kong | | | | | 0.24 | Yes |
| India | 7-70 | 25-70[a] | | 0.16 | 0.10 | Yes |
| Indonesia | 10-15 | 60 | | 0.10 | 0.10 | Yes |
| Jordan | | | | 0.11 | 0.35 | Yes |

*continued*

| | Facility construction costs ($U.S. per capita) | | | Urban systems | | |
| | | Urban | | | | |
| | Rural | House connection | Standpipe | Operation cost ($U.S./m$^3$) | Tariff ($U.S./m$^3$) | Progressive tariff |
|---|---|---|---|---|---|---|
| Korea Rep | 29 | | | | | |
| Laos | 5-50 | 300-500 | 0.3-10 | 0.35 | 0.38 | |
| Macao | 50 | 20 | | 0.18 | 0.28 | No |
| Malaysia | 35-210 | 200 | | 0.18 | 0.25 | Yes |
| Maldives | 12 | | 111 | 1.00 | | No |
| Myanmar | 5-35 | 65-70 | 30-35 | | 0.20 | Yes |
| Nepal | 2-30 | 70[a] | | 0.16 | 0.08 | Yes |
| Pakistan | 19 | 38[a] | | | | Yes |
| Philippines | 37 | 28 | | 0.29 | 0.12 | Yes |
| Qatar | | | | 1.74 | 0.60 | No |
| Saudi Arabia | | 420 | | 1.10 | 0.10 | No |
| Singapore | n.a. | 19 | | 0.18 | 0.25 | Yes |
| Sri Lanka | 12 | 150 | 80 | 0.19 | 0.20 | Yes |
| Syria | 100 | 250 | | 0.25 | 0.13 | Yes |
| Thailand | 4-66 | 61 | | 0.26 | 0.15 | Yes |
| Turkey | 93 | 100 | | 0.20 | 0.25 | Yes |
| United Arab Emirates | 85 | 100 | 30 | 1.32 | 0.90 | No |
| Vietnam | | | | | | No |
| Yemen Arab Rep | 125 | 300 | 250 | 1.10 | 1.40 | Yes |
| Yemen Dem Rep | 240 | 300 | 260 | 0.26 | 0.30 | Yes |
| | | | | | | |
| *Oceania* | | | | | | |
| American Samoa | | 1,000 | | 0.40 | 0.80 | No |
| Cook Islands | 72 | 96 | | 0.30 | 0.00 | No |
| Fiji | 61 | 188 | 233 | 0.35 | 0.18 | Yes |
| Guam | 100 | 205 | | 0.52 | | No |
| Kiribati | 5 | | | | | |
| New Caledonia | | 340 | | 0.10 | 0.18 | Yes |
| Niue | 650[b] | 680[b] | 650[b] | | | No |
| Pacific Island Trust Territory | 120 | 450 | | 0.18 | 0.07 | Yes |
| Papua New Guinea | 15 | 150 | 50 | 0.60 | 0.40 | Yes |
| Solomon Islands | 40 | 130 | | 0.30 | 0.25 | Yes |
| Tokelau | 32 | | | | | |
| Tonga | 48 | 63 | | 0.80 | 0.85 | No |
| Vanuatu | 50 | | | 0.19 | 0.31 | No |
| Western Samoa | 200 | 300 | 150 | 0.75 | 0.30 | Yes |

n.a, not applicable.
[a]Combined figure for house connections and standpipes.
[b]Per house or standpost

**TABLE H.29**   Cost of water production versus Gross National Product for selected countries

| Country | Cost ($/m$^3$) | GNP per capita ($/person) | Percent cost per capita |
|---|---|---|---|
| Cape Verde | 4.65 | 317 | 1.5 |
| Cayman Island | 2.75 | 13,000 | 0.02 |
| Cameroon | 2.00 | 800 | 0.25 |
| Mexico | 1.50 | 2,080 | 0.07 |
| Argentina | 1.50 | 1,929 | 0.08 |
| Netherlands | 1.25 | 9,290 | 0.01 |
| Zambia | 1.05 | 390 | 0.3 |
| Saudi Arabia | 1.00 | 8,850 | 0.01 |
| Sierra Leone | 0.90 | 200 | 0.45 |
| Tonga | 0.80 | 354 | 0.23 |
| Botswana | 0.75 | 840 | 0.09 |
| Togo | 0.66 | 300 | 0.22 |
| Suriname | 0.60 | 3,030 | 0.02 |
| Seychelles | 0.60 | 2,250 | 0.03 |
| Malawi | 0.60 | 170 | 0.35 |
| Papua New Guinea | 0.55 | 649 | 0.09 |
| Tunisia | 0.50 | 1,277 | 0.04 |
| Cook Islands | 0.40 | 7,170 | 0.006 |
| Cyprus | 0.40 | 3,572 | 0.01 |
| Djibouti | 0.40 | 480 | 0.08 |
| Rwanda | 0.40 | 280 | 0.14 |
| Bahamas | 0.37 | 7,556 | 0.005 |
| Laos | 0.35 | 100 | 0.35 |
| Barbados | 0.34 | 4,889 | 0.007 |
| Ghana | 0.35 | 420 | 0.08 |
| Burundi | 0.35 | 230 | 0.15 |
| Mali | 0.33 | 142 | 0.23 |
| Switzerland | 0.33 | 14,764 | 0.002 |
| Afghanistan | 0.30 | 163 | 0.18 |
| Mauritius | 0.29 | 1,020 | 0.03 |
| Myanmar | 0.25 | 188 | 0.13 |
| Singapore | 0.24 | 7,420 | 0.003 |
| Spain | 0.22 | 4,256 | 0.005 |
| Vanuatu | 0.22 | 529 | 0.04 |
| Zaire | 0.22 | 271 | 0.08 |
| Hungary | 0.23 | 1,909 | 0.01 |
| Finland | 0.21 | 10,531 | 0.002 |
| Thailand | 0.21 | 729 | 0.03 |

*continued*

| Country | Cost ($/m$^3$) | GNP per capita ($/person) | Percent cost per capita |
|---|---|---|---|
| Honduras | 0.20 | 720 | 0.03 |
| Korea Rep | 0.19 | 2,032 | 0.009 |
| Haiti | 0.18 | 320 | 0.06 |
| Malaysia | 0.18 | 2,033 | 0.009 |
| Costa Rica | 0.17 | 1,300 | 0.013 |
| Madagascar | 0.17 | 240 | 0.07 |
| Angola | 0.15 | 560 | 0.03 |
| Nicaragua | 0.14 | 770 | 0.02 |
| Morocco | 0.14 | 512 | 0.03 |
| Chile | 0.12 | 1,430 | 0.008 |
| Peru | 0.12 | 1,010 | 0.012 |
| Iraq | 0.12 | 2,964 | 0.004 |
| Bangladesh | 0.09 | 136 | 0.07 |
| Ecuador | 0.09 | 1,160 | 0.008 |
| Western Samoa | 0.09 | 660 | 0.013 |
| Panama | 0.07 | 2,100 | 0.003 |
| Philippines | 0.05 | 585 | 0.008 |

**TABLE H.29** Cost of water production versus Gross National Product for selected countries *(pages 419–420)*

**DESCRIPTION**

The cost of providing water varies widely from country to country, depending on factors such as overall water availability, water quality, the discrepency between location of supply and demand, and the overall efficiency of water use and reuse. Shown here are the cost of providing water, in $U.S. per m$^3$, the per capita gross national product, in $U.S. per person, and the ratio of the two as a measure of the cost of providing water, for selected countries. Some developing countries pay far more to produce water than do industrialized countries, both in real terms and as a fraction of per capita GNP.

**LIMITATIONS**

The countries included here are those responding to a 1985 World Health Organization questionnaire. There may be differences in the base year used, and no detail is given on currency exchange rates prevailing at the time of the questionnaire. It is not clear whether the $1 per m$^3$ quoted for Saudia Arabia is for water obtained by desalination alone or includes ground water withdrawal costs.

**SOURCES**

Cited in T.A. Dabbagh and A. Al-Saqabi, 1989, The increasing demand for desalination, *Desalination*, **73**, 3–26, Elsevier Science Publishers, Amsterdam. By permission.

Data from World Water/WHO, 1987, *The International Drinking Water Supply and Sanitation Decade Directory*, 3rd edn., Thomas Telford Ltd., London.

**TABLE H.30** Countries with desalting plant capacity greater than 20,000 m$^3$/day, as of December 31, 1989 *(page 422)*

**DESCRIPTION**

Desalination is one of the principal alternative sources for fresh water available today. It is used throughout the world, but is most common in the arid Middle East. For example, seven of the ten countries with the greatest installed desalination capacity are in the Middle East. The following table presents the total desalination capacity in m$^3$/day for all 36 countries with more than 20,000 m$^3$/day of capacity, the number of desalination units, and predominant type of plant by both capacity and number of units. All the data are as of 31 December 1989, and include both installed and contracted plants. These 36 nations account for 97% of all desalination capacity and 92% of the total number of units. For more information on desalination techniques and applications, see Chapter 6.

**LIMITATIONS**

The data were compiled from a wide range of sources and depend on the quality of the responses to requests for information. Information for the USSR and China are current through 1988.

**SOURCE**

K. Wangnick, 1990, *1990 IDA Worldwide Desalting Plants Inventory*, Report no. 11, May, Wangnick Consulting, Gnarrenburg, Germany. Reprinted by permission of Wangnick Consulting.

**TABLE H.30** Countries with desalting plant capacity greater than 20,000 m³/day, as of December 31, 1989

| Country | Capacity (m³/day) | Number of Units | Predominant type (by capacity) | Predominant type (by number of units) |
|---|---|---|---|---|
| Saudi Arabia | 3,568,868 | 1,417 | MSF | RO |
| United States | 1,588,972 | 1,354 | RO | RO |
| Kuwait | 1,390,238 | 133 | MSF | MSF |
| United Arab Emirates | 1,332,477 | 290 | MSF | MSF |
| Libya | 619,354 | 386 | MSF | RO |
| Japan | 465,600 | 615 | RO | RO |
| Iraq | 323,925 | 198 | RO | RO |
| Qatar | 308,611 | 59 | MSF | MSF |
| Soviet Union | 299,143 | 53 | ME | RO |
| Bahrain | 275,767 | 126 | MSF | RO |
| Spain | 274,477 | 238 | RO | VC |
| Italy | 266,663 | 141 | MSF | MSF |
| Iran | 260,609 | 218 | MSF | MSF |
| Oman | 186,741 | 79 | MSF | RO |
| Hong Kong | 183,694 | 12 | MSF | MSF |
| Algeria | 176,087 | 123 | RO | RO |
| Netherlands Antilles | 172,626 | 75 | MSF | MSF |
| Korea | 133,220 | 54 | RO | RO |
| Netherlands | 116,618 | 90 | MSF | RO |
| Germany FR | 102,574 | 188 | RO | RO |
| Great Britain | 91,269 | 201 | ME | ME |
| Mexico | 84,095 | 97 | MSF | RO |
| Virgin Islands | 83,821 | 37 | ME | MSF |
| Australia | 81,523 | 107 | RO | RO |
| Israel | 70,062 | 32 | RO | RO |
| Egypt | 67,728 | 110 | RO | ED |
| Malta | 67,099 | 30 | RO | RO |
| German DR | 56,720 | 32 | ME | RO |
| Indonesia | 51,308 | 46 | MSF | MSF |
| South Africa | 48,006 | 30 | VC | VC |
| France | 32,144 | 51 | RO | RO |
| Bahamas | 30,504 | 24 | MSF | RO |
| Greece | 27,499 | 26 | ED | RO |
| India | 25,028 | 34 | RO | RO |
| Tunisia | 22,870 | 39 | RO | RO |
| Gibraltar | 21,140 | 28 | RO | RO |

Data include plants contracted and installed.
ED, electrodialysis; ME, multiple effect; MSF, multi-stage flash distillation; RO, reverse osmosis; VC, vapor compression.

**TABLE H.31**  Desalting plants capable of producing at least 100 m³/day, by country, as of December 31, 1989

| Country | Capacity (m³/day) | Number of units | Predominant type (by capacity) | Predominant type (by number of units) |
|---|---|---|---|---|
| Algeria | 176,087 | 123 | RO | RO |
| Angola | 380 | 1 | ME | ME |
| Antarctica | 300 | 2 | MSF | MSF |
| Antigua | 19,641 | 8 | MSF | MSF |
| Argentina | 2,053 | 7 | ME | ME |
| Ascension Island | 2,128 | 6 | MSF | MSF |
| Australia | 81,523 | 107 | RO | RO |
| Austria | 5,899 | 18 | RO | RO |
| Bahamas | 30,504 | 24 | MSF | RO |
| Bahrain | 275,767 | 126 | MSF | RO |
| Belgium | 5,519 | 9 | RO | RO |
| Bermuda | 10,204 | 27 | RO | RO |
| Brazil | 390 | 2 | ME | ME/RO |
| Bulgaria | 1,320 | 1 | RO | RO |
| Canada | 7,390 | 22 | RO | RO |
| Cape Verde | 6,143 | 10 | MSF | RO |
| Cayman Islands | 9,093 | 11 | RO | VC |
| Chile | 13,210 | 22 | MSF | RO/MSF |
| China | 17,496 | 16 | RO | RO |
| Colombia | 7,465 | 10 | RO | RO |
| Congo | 550 | 2 | VC | VC |
| Cuba | 18,926 | 6 | MSF | MSF |
| Cyprus | 7,710 | 23 | MSF | MSF |
| Czechoslovakia | 11,522 | 11 | RO | RO |
| Denmark | 412 | 2 | ME | ME/VC |
| Djibouti | 404 | 3 | VC | VC |
| Ecuador | 5,904 | 13 | VC | VC |
| Egypt | 67,728 | 110 | RO | ED |
| El Salvador | 378 | 2 | RO | RO |
| Finland | 221 | 2 | ME | ME/VC |
| France | 32,144 | 51 | RO | RO |
| French Antilles | 9,150 | 11 | VC | VC |
| Germany FR | 102,574 | 188 | RO | RO |
| German DR | 56,720 | 32 | ME | RO |
| Gibralter | 21,140 | 28 | RO | RO |
| Great Britain | 91,269 | 201 | ME | ME |
| Greece | 27,499 | 26 | ED | RO |
| Honduras | 651 | 1 | MSF | MSF |

*continued*

Fresh water data

| Country | Capacity (m³/day) | Number of units | Predominant type (by capacity) | Predominant type (by number of units) |
|---|---|---|---|---|
| Hong Kong | 183,694 | 12 | MSF | MSF |
| Hungary | 615 | 2 | MSF | MSF/RO |
| India | 25,028 | 34 | RO | RO |
| Indonesia | 51,308 | 46 | MSF | MSF |
| Iran | 260,609 | 218 | MSF | MSF |
| Iraq | 323,925 | 198 | RO | RO |
| Ireland | 545 | 1 | ME | ME |
| Israel | 70,062 | 32 | RO | RO |
| Italy | 266,663 | 141 | MSF | MSF |
| Jamaica | 1,363 | 2 | RO | RO |
| Japan | 465,600 | 615 | RO | RO |
| Jordan | 8,445 | 13 | RO | RO |
| Korea | 133,220 | 54 | RO | RO |
| Kuwait | 1,390,238 | 133 | MSF | MSF |
| Lebanon | 4,691 | 10 | VC | VC |
| Libya | 619,354 | 386 | MSF | RO |
| Liechtenstein | 151 | 1 | RO | RO |
| Madeira | 800 | 5 | RO | RO |
| Malaysia | 11,873 | 29 | RO | RO |
| Maldives | 400 | 2 | RO | RO |
| Malta | 67,099 | 30 | RO | RO |
| Marshall Islands | 2,050 | 2 | ME | MSF |
| Mauritania | 4,654 | 5 | MSF | VC |
| Mexico | 84,095 | 97 | MSF | RO |
| Morocco | 9,424 | 23 | VC | VC |
| Mozambique | 189 | 1 | RO | RO |
| Namibia | 775 | 2 | MSF | ME/MSF |
| Netherlands | 116,618 | 90 | MSF | RO |
| Netherlands Antilles | 172,626 | 75 | MSF | MSF |
| Nicaragua | 600 | 2 | MSF | MSF |
| Nigeria | 7,070 | 11 | VC | RO |
| Norway | 300 | 3 | ME | ME |
| Oman | 186,741 | 79 | MSF | RO |
| Pakistan | 652 | 4 | Other | Other |
| Peru | 14,852 | 13 | MSF | VC |
| Philippines | 4,071 | 15 | RO | RO |
| Poland | 19,142 | 18 | RO | RO |
| Portugal | 4,420 | 6 | RO | RO |
| Qatar | 308,611 | 59 | MSF | MSF |

*continued*

| Country | Capacity (m³/day) | Number of units | Predominant type (by capacity) | Predominant type (by number of units) |
|---|---|---|---|---|
| Sahara | 7,002 | 2 | MSF | MSF |
| Saudi Arabia | 3,568,868 | 1,417 | MSF | RO |
| Senegal | 132 | 1 | Other | Other |
| Singapore | 19,829 | 28 | RO | RO |
| Solomon Islands | 100 | 1 | RO | RO |
| Somalia | 288 | 1 | RO | RO |
| South Africa | 48,006 | 30 | VC | VC |
| Soviet Union | 299,143 | 53 | ME | RO |
| Spain | 274,477 | 238 | RO | VC |
| Sudan | 1,076 | 4 | ME | MSF |
| Sweden | 542 | 2 | ME | ME |
| Switzerland | 4,103 | 15 | RO | RO |
| Syria | 5,743 | 12 | RO | RO |
| Taiwan | 3,590 | 4 | VC | RO/VC |
| Thailand | 10,102 | 17 | RO | RO |
| Trinidad and Tobago | 189 | 1 | RO | RO |
| Tunisia | 22,870 | 39 | RO | RO |
| Turkey | 1,494 | 8 | RO | RO |
| United Arab Emirates | 1,332,477 | 290 | MSF | MSF |
| United States | 1,588,972 | 1,354 | RO | RO |
| Venezuela | 14,175 | 18 | RO | RO |
| Virgin Islands | 83,821 | 37 | ME | MSF |
| Yemen Dem Rep | 2,053 | 9 | MSF | MSF |
| Yemen Arab Rep | 4,051 | 15 | ME | RO |
| Yugoslavia | 1,102 | 2 | RO | RO |

Data include plants contracted and installed.
ED, electrodialysis; ME, multiple effect; MSF, multi-stage flash distillation; RO, reverse osmosis; VC, vapor compression.

**TABLE H.31** Desalting plants capable of producing at least 100 m³/day, by country, as of December 31, 1989
*(pages 423–425)*

## DESCRIPTION
Desalination is one of the principal alternative sources for fresh water available today and it is used throughout the world. The following table shows total desalination capacity, the number of units, and the predominant type of plant by capacity and number of units, for all countries of the world with desalting plants of 100 m³/day or more. All the data are as of 31 December 1989, and include both installed and contracted plants. For more information on desalination techniques and applications, see Chapter 6.

## LIMITATIONS
Large numbers of ships use on-board desalination technology to supply fresh water. These units are not included in this data set. The data were compiled from a wide range of sources and depend on the quality of the responses to requests for information. Information for the USSR and China are current through 1988. In addition to the plants listed here, more than 200 other units with a capacity exceeding 44,000 m³/day are operating on oil rigs, in military units, or have no country location.

## SOURCE
K. Wangnick, 1990, *1990 IDA Worldwide Desalting Plants Inventory*, Report no. 11, May, Wangnick Consulting, Gnarrenburg, Germany. Reprinted by permission of Wangnick Consulting.

**TABLE H.32** Desalting plants capable of producing at least 100 m³/day, by type of process, as of December 31, 1989

| Process | Number of plants | Percent of total | Capacity (m³/day) | Percent of total |
|---|---|---|---|---|
| Total | 7,536 | 100 | 13,296,597 | 100.1 |
| MSF | 1,063 | 14.1 | 7,442,496 | 56.0 |
| RO | 4,157 | 55.2 | 4,113,015 | 30.9 |
| ED | 1,032 | 13.7 | 677,674 | 5.1 |
| ME | 581 | 7.7 | 617,713 | 4.6 |
| VC | 589 | 7.8 | 368,174 | 2.8 |
| Other | 96 | 1.3 | 46,618 | 0.4 |
| Hybrid | 8 | 0.1 | 22,659 | 0.2 |
| UF | 9 | 0.1 | 8,038 | 0.1 |
| Freeze | 1 | 0.0 | 210 | 0.0 |

ED, Electrodialysis; ME, Multi-effect evaporation; MSF, Multi-stage flash; RO, Reverse osmosis; UF, Ultrafiltration; VC, Vapor compression.

**TABLE H.32** Desalting plants capable of producing at least 100 m³/day, by type of process, as of December 31, 1989

**DESCRIPTION**

The number and capacity of all desalination plants are shown here sorted by the type of desalting process. Total desalting capacity is given in m³ × 10³/day. All land-based plants capable of producing more than 100 m³/day are included. The largest number of plants, 55%, use reverse osmosis technology, but 56% of the total installed capacity is multi-stage flash distillation. All the data are as of 31 December 1989, and include both installed and contracted plants. For more information on desalination techniques and applications, see Chapter 6.

**LIMITATIONS**

Large numbers of ships use on-board desalination technology to supply fresh water. These units are not included in this data set. The data were compiled from a wide range of sources and depend on the quality of the responses to requests for information. Information for the USSR and China are not updated from 1988.

**SOURCE**

K. Wangnick, 1990, *1990 IDA Worldwide Desalting Plants Inventory*, Report no. 11, May, Wangnick Consulting, Gnarrenburg, Germany. Reprinted by permission of Wangnick Consulting.

**TABLE H.33** Desalting plants capable of producing at least 100 m$^3$/day, by source of raw water, as of December 31, 1989

| Source | Capacity (m$^3$/day) | Percent of total | Number of units | Percent of total |
|---|---|---|---|---|
| Total | 13,296,598 | 100 | 7,536 | 100 |
| Seawater | 8,683,779 | 65 | 2,388 | 32 |
| Brackish | 3,549,407 | 27 | 4,026 | 53 |
| Wastewater | 372,161 | 3 | 281 | 4 |
| River water | 352,843 | 3 | 398 | 5 |
| Pure water | 265,861 | 2 | 364 | 5 |
| Unknown | 40,114 | 0.3 | 40 | 0.5 |
| Brine | 32,433 | 0.2 | 39 | 0.5 |

Brackish: includes inland water, 3,000-20,000 mg/l total dissolved solids (TDS).
Brine: includes concentrated seawater, greater than 50,000 mg/l TDS.
Pure water: less than 500 mg/l TDS.
River water: includes other low concentrated saline water, 500-3,000 mg/l TDS.
Seawater: also includes concentrated seawater, 20,000-50,000 mg/l TDS.
Unknown: quality unknown.
Waste: other raw water, i.e., wastewater.

**TABLE H.33** Desalting plants capable of producing at least 100 m$^3$/day, by source of raw water, as of December 31, 1989

**DESCRIPTION**

The number and capacity of all desalination plants are shown here sorted by the source of raw water. Total desalting capacity is given in m$^3$/day. All land-based plants capable of producing more than 100 m$^3$/day are included. Fifty-three per cent of these desalination plants use brackish water, but 65% of all desalination capacity uses sea water. Other sources of water are minor. The definitions of water salinity are given as a concentration of total dissolved solids. All the plant data are as of 31 December 1989, and include both installed and contracted plants. For more information on desalination techniques and applications, see Chapter 6.

**LIMITATIONS**

Large numbers of ships use on-board desalination technology to supply fresh water. These units are not included in this data set. The data were compiled from a wide range of sources and depend on the quality of the responses to requests for information. Information for the USSR and China are not updated from 1988.

**SOURCE**

K. Wangnick, 1990, *1990 IDA Worldwide Desalting Plants Inventory*, Report no. 11, May, Wangnick Consulting, Gnarrenburg, Germany. Reprinted by permission of Wangnick Consulting.

**TABLE H.34** Wind and solar desalination plants with a capacity greater than 10 m$^3$/day *(page 428)*

**DESCRIPTION**

Desalting water is one of the principal alternative sources for fresh water available today, but it remains very expensive because of the large amounts of energy required. Shown here are desalination plants powered by unconventional energy technologies, including wind electric, photovoltaic, and solar thermal, sorted by country. Data are provided on plant capacity in m$^3$/day, the type of desalination process used, the source of water, the date of initial operation, and the source of energy, where available. Three large plants under construction in Libya are also listed. All the data are as of 31 December 1989. For more information on desalination techniques, applications, and energy requirements, see Chapter 6.

**LIMITATIONS**

The data were compiled from a wide range of sources and depend on the quality of the responses to requests for information. Data on solar- and wind-powered have been collected for only a few years and are not likely to be complete.

**SOURCE**

K. Wangnick, 1990, *1990 IDA Worldwide Desalting Plants Inventory*, Report no. 11, May, Wangnick Consulting, Gnarrenburg, Germany. Reprinted by permission of Wangnick Consulting.

**TABLE H.34**   Wind and solar desalination plants with a capacity greater than 10 m$^3$/day

| Country | Capacity (m$^3$/day) | Process | Water supply | Date of operation | Energy source |
|---------|---------------------|---------|--------------|-------------------|---------------|
| *Completed* | | | | | |
| Egypt | 25 | RO | Seawater | 1987 | Wind electric |
| France | 12 | RO | Seawater | 1980 | Wind electric |
| France | 60 | RO | Brackish | 1987 | Collector |
| Germany | 20 | MSF | Seawater | 1986 | |
| Greece | 20 | Other | Seawater | 1967 | Collector |
| Indonesia | 12 | RO | Brackish | 1984 | Photovoltaic |
| Italy | 12 | RO | Seawater | 1984 | Photovoltaic |
| Japan | 20 | RO | Seawater | 1987 | |
| Japan | 15 | RO | Seawater | 1982 | |
| Japan | 16 | ME | Seawater | 1984 | Collector |
| Kuwait | 22 | MSF | Seawater | 1984 | |
| Kuwait | 45 | RO | Brackish | 1988 | Parabolic collector |
| Pakistan | 22 | Other | Seawater | 1972 | Collector |
| Qatar | 24 | RO | Seawater | 1982 | Photovoltaic |
| Qatar | 20 | MSF | Seawater | 1986 | |
| Saudi Arabia | 210 | Freeze | Seawater | 1987 | Point focus |
| Saudi Arabia | 250 | RO | Seawater | 1987 | Line focus |
| Saudi Arabia | 14 | ME | Seawater | 1988 | Heliostat |
| Saudi Arabia | 20 | RO | Seawater | 1988 | Heliostat |
| Spain | 86 | ME | Seawater | 1988 | |
| United Arab Emirates | 500 | ME | Brackish | 1985 | Mirror |
| United Arab Emirates | 80 | ME | Seawater | 1985 | |
| United States | 36 | RO | Seawater | 1987 | Fresnel lens |
| United States | 19 | ME | River | 1987 | |
| United States | 60 | RO | Brackish | 1987 | Heliostat |
| *Planned* | | | | | |
| Libya | 1,000 | RO | Brackish | 1992 | Photovoltaic |
| Libya | 500 | ME | Seawater | 1992 | Parabolic collector |
| Libya | 2,000 | RO | Brackish | 1992 | Wind electric |

ME, multiple effect distillation; MSF, multi-stage flash distillation; RO, reverse osmosis.

**TABLE H.35** Unit cost of desalination alternatives in the Middle East (1984 cost $/m$^3$)

| Alternative | Range | Mean |
|---|---|---|
| MSF (seawater) | 1.07–3.00 | 1.87 |
| MSF (brackish) | 0.53–2.13 | 1.33 |
| RO (seawater) | 1.60–2.67 | 2.13 |
| RO (brackish) | 0.27–1.60 | 0.93 |

MSF, multi-stage flash distillation; RO, reverse osmosis

**TABLE H.35** Unit cost of desalination alternatives in the Middle East

**DESCRIPTION**

The costs of desalinating brackish and sea water using multi-stage flash distillation (MSF) and reverse osmosis (RO) technologies are shown here. A range of costs is given in $U.S. (1984) per m$^3$.

**LIMITATIONS**

The reason for the range of costs is not clearly given in the original text, but has to do with the capacity of the plant, the cost of energy, the salinity of the seawater, and the plant location. Desalination technology is changing very rapidly, and the costs in this table need to be brought up to date. Costs for several other commercial desalination technologies are not given here.

**SOURCE**

United Nations, 1984, *Wastewater Reuse and Its Application in the ECWA Region*, Economic Commission for West Asia (ECWA), Natural Resources Science and Technology Division, New York.

**TABLE H.36**   Desalination applications

| Technique | 0–3,000 ppm | 3,000–10,000 ppm | Seawater 35,000 ppm | Highly saline brines |
|---|---|---|---|---|
| Distillation | T | S | P | P |
| Electrodialysis | P | S | T | P |
| Reverse osmosis | P | P | P | S |
| Ion exchange | P | | | |

P, primary application; S, secondary application; T, technically feasible, not economic.

**TABLE H.36**  Desalination applications

**DESCRIPTION**

Many desalination technologies are commercially available, as described in Chapter 6. For reasons of technology and economics, each method works best on waters of different quality, from highly saline brines to only slightly salty water. Four desalination techniques and the quality of the water they are best suited to treat are described here. A distinction is made between primary and secondary applications. Those approaches that are technically feasible, but not economically attractive, are also described.

**LIMITATIONS**

Several desalination techniques in commercial operation are not included here.

**SOURCE**

Office of Technology Assessment, 1988, *Using Desalination Technologies for Water Treatment*, OTA-BP-O-46, U.S. Congress, Washington, DC.

# I. Water policy and politics

**TABLE I.1**  Agencies collecting water resources assessment data

| Region | Countries collecting data | Surface water quantity | Ground water | Climate | Sediment | Water quality |
|---|---|---|---|---|---|---|
| ECA | 34 | 56 | 30 | 54 | 31 | 30 |
| ECE | 33 | 62 | 54 | 69 | 45 | 62 |
| ECLAC | 27 | 73 | 34 | 64 | 28 | 33 |
| ESCAP | 27 | 59 | 35 | 58 | 37 | 49 |
| ESCWA | 18 | 80 | 54 | 17 | 8 | 13 |
| Total | 139 | 330 | 207 | 264 | 149 | 187 |

ECA, Economic Commission for Africa
ECE, Economic Commission for Europe
ECLAC, Economic Commission for Latin America and the Caribbean
ESCAP, Economic and Social Commission for Asia and the Pacific
ESCWA, Economic and Social Commission for Western Asia

**TABLE I.1**  Agencies collecting water resources assessment data

## DESCRIPTION

Water resources data used in this book and by many others is collected by many national and international organizations. This table shows the number of organizations collecting certain types of basic water information for the various United Nations regions. There are more data sets for a given region than there are countries. This represents situations in which data gathering efforts are duplicated in whole or in part. ECA, Economic Commission for Africa; ECE, Economic Commission for Europe; ECLAC, Economic Commission for Latin America and the Caribbean; ESCAP, Economic and Social Commission for Asia and the Pacific; ESCWA, Economic and Social Commission for Western Asia.

## LIMITATIONS

Many important water resources data are not collected by these organizations and agencies. Detailed information on the kinds of data collected are not given here.

## SOURCE

World Meteorological Organization and the United Nations Educational, Scientific, and Cultural Organization (WMO/UNESCO), 1991, *Report on Water Resources Assessment: Progress of the Mar del Plata Action Plan and a Strategy for the 1990s*, Oxford, UK.

**TABLE I.2**  Development of hydrological networks and data banks *(page 432)*

## DESCRIPTION

Water resources data are collected by a variety of United Nations agencies. For the five major economic regions of the United Nations, this table quantifies efforts to monitor hydrologic conditions of precipitation, discharge, and water quality in 1977, 1987, and 1989. Considerable improvements in monitoring of river discharge and water quality occurred during this period in almost all regions, but monitoring of basic precipitation has dropped since 1987 in four of the five regions. ECA, Economic Commission for Africa; ECE, Economic Commission for Europe; ECLAC, Economic Commission for Latin America and the Caribbean; ESCAP, Economic and Social Commission for Asia and the Pacific; ESCWA, Economic and Social Commission for Western Asia.

## LIMITATIONS

Many important water resources data are not collected by these organizations and agencies. Detailed information on the kinds of data collected are not given here. There is an urgent need for good, consistent, long-term hydrologic data. Unless stations for monitoring these data are funded and maintained for a long period of time, gaps will remain in our understanding of global and regional water balances. Such monitoring is also vital to identifying trends of global change.

## SOURCE

World Meteorological Organization and the United Nations Educational, Scientific, and Cultural Organization (WMO/UNESCO), 1992, Background paper: Water resources assessment and impacts of climate change on water resources, Working Group C, in *International Conference on Water and the Environment, Development Issues for the 21st Century*, 26–31 January, Dublin, Ireland. By permission.

**TABLE I.2** Development of hydrological networks and data banks

|  | 1977 | 1987 | 1989 |
|---|---|---|---|
| **Precipitation Stations** | | | |
| ECA | 4,047 | 4,636 | 3,596 |
| ECE | 49,240 | 50,167 | 48,507 |
| ECLAC | 12,409 | 19,590 | 19,531 |
| ESCAP | 20,980 | 21,027 | 20,422 |
| ESCWA | 4,018 | 4,248 | 4,240 |
| Total | 90,694 | 99,668 | 96,296 |
| **Discharge Stations** | | | |
| ECA | 918 | 1,694 | 1,695 |
| ECE | 9,549 | 24,228 | 23,946 |
| ECLAC | 3,086 | 5,730 | 5,762 |
| ESCAP | 5,923 | 6,282 | 7,023 |
| ESCWA | 1,222 | 1,262 | 1,383 |
| Total | 20,698 | 39,196 | 39,809 |
| **Water Quality Stations** | | | |
| ECA | 123 | 361 | 361 |
| ECE | 15,509 | 40,030 | 42,327 |
| ECLAC | 218 | 1,059 | 1,439 |
| ESCAP | 3,533 | 3,314 | 2,889 |
| ESCWA | 801 | 801 | 821 |
| Total | 20,184 | 45,565 | 47,837 |
| **Hydrological Data Banks Using Computers** | | | |
| ECA | 17 | 23 | 25 |
| ECE | 23 | 28 | 28 |
| ECLAC | 16 | 18 | 21 |
| ESCAP | 12 | 20 | 23 |
| ESCWA | * | * | * |
| Total | 68 | 89 | 97 |

* No comparable data
ECA, Economic Commission for Africa
ECE, Economic Commission for Europe
ECLAC, Economic Commission for Latin America and the Caribbean
ESCAP, Economic and Social Commission for Asia and the Pacific
ESCWA, Economic and Social Commission for Western Asia

**TABLE I.3**  State of water management, by country

| | Water policy statement | Water legislation | National plan | Central coordination | Cost recovery |
|---|---|---|---|---|---|
| *Africa* | | | | | |
| Benin | — | Inadequate | — | – | — |
| Burkina Faso | No | Adequate | No | Yes | — |
| Cameroon | — | Inadequate | — | — | Partial Irrigation |
| Congo | — | Inadequate | — | No | — |
| Côte d'Ivoire | Yes | Adequate | — | Yes | Full |
| Egypt | Yes | Adequate | No | Yes | Partial |
| Ethiopia | Yes | Inadequate | No | Yes | Partial |
| Gambia | — | Inadequate | — | — | — |
| Ghana | No | Inadequate | No | No | Partial |
| Guinea | — | Inadequate | — | — | — |
| Kenya | Yes | Adequate | No | Yes | Partial |
| Liberia | No | Inadequate | No | No | — |
| Malawi | Yes | Adequate | Yes | Yes | Partial |
| Mali | Yes | Adequate | Yes | No | Partial Irrigation |
| Morocco | No | Inadequate | No | No | Partial |
| Niger | Yes | Adequate | Yes | Yes | Partial Irrigation |
| Nigeria | No | Inadequate | No | Yes | — |
| Senegal | No | Inadequate | No | Yes | Partial |
| Sierra Leone | No | Inadequate | No | No | — |
| Sudan | No | Inadequate | No | No | Full |
| Tanzania | Yes | Adequate | — | Yes | — |
| Togo | — | Inadequate | — | — | — |
| Uganda | — | — | No | Yes | — |
| Zaire | — | — | — | No | — |
| Zambia | — | — | — | Yes | — |
| *Latin America and the Caribbean* | | | | | |
| Antigua and Barbuda | No | Adequate | No | Yes | Partial Water Supply |
| Argentina | No | Adequate | No | No | Partial |
| Bahamas | No | Adequate | No | Yes | Partial Water Supply |
| Barbados | No | Adequate | No | Yes | Partial |
| Bolivia | No | Inadequate | No | No | Partial |
| Brazil | No | Adequate | No | No | Full |
| Cayman Islands | Yes | Adequate | Yes | Yes | Full |
| Chile | No | Adequate | No | No | Full |
| Colombia | No | Adequate | No | Yes | Partial |
| Costa Rica | No | Adequate | No | No | Partial |
| Cuba | Yes | Adequate | Yes | Yes | No |

*continued*

|  | Water policy statement | Water legislation | National plan | Central coordination | Cost recovery |
|---|---|---|---|---|---|
| Dominica | No | Adequate | No | Yes | Full |
| Dominican Republic | No | Adequate | No | Yes | Partial |
| Ecuador | No | Adequate | Yes | No | Partial |
| El Salvador | Yes | Inadequate | Yes | Yes | Partial |
| Grenada | No | Adequate | No | Yes | Full |
| Guatemala | No | Inadequate | No | No | Partial |
| Haiti | No | Inadequate | No | No | No |
| Honduras | No | Inadequate | No | No | Partial |
| Jamaica | Yes | Adequate | No | Yes | Partial |
| Mexico | Yes | Adequate | Yes | Yes | Partial |
| Monserrat | No | Adequate | No | Yes | Full |
| Nicaragua | No | Inadequate | No | No | No |
| Panama | No | Adequate | No | Yes | Partial |
| Paraguay | No | Adequate | No | No | Partial |
| Peru | Yes | Inadequate | Yes | Yes | Partial |
| St. Kitts and Nevis | Yes | Adequate | No | Yes | Full |
| St. Lucia | No | Adequate | No | Yes | Full |
| St. Vincent and the Grenadines | No | Adequate | No | No | Partial |
| Trinidad and Tobago | No | Adequate | No | No | Full |
| Uruguay | No | Inadequate | No | No | Partial |
| Venezuela | Yes | Adequate | Yes | Yes | Partial |
| | | | | | |
| *Asia and the Pacific* | | | | | |
| | | | | | |
| Afghanistan | Yes | Adequate | Yes | Yes | Partial Urban |
| Australia | Yes | — | No | Yes | Full |
| Bangladesh | — | Inadequate | — | — | — |
| China | Yes | — | Yes | No | Partial |
| Fiji | Yes | Inadequate | No | No | Partial |
| Guam | Yes | Adequate | Yes | No | Partial |
| Hong Kong | Yes | Adequate | Yes | Yes | Partial |
| India | No | Inadequate | No | Yes | Partial |
| Indonesia | Yes | Adequate | No | No | Full Urban |
| Japan | No | Adequate | No | — | Full |
| Korea Rep | Yes | Adequate | Yes | Yes | Partial |
| Malaysia | No | Inadequate | Yes | No | Partial |
| Myanmar | — | Inadequate | — | — | — |
| New Zealand | Yes | Adequate | No | Yes | Full |
| Niue | Yes | Inadequate | No | Yes | Partial Irrigation |
| Pakistan | Yes | Inadequate | Yes | Yes | Partial Irrigation |
| Papua New Guinea | Yes | Adequate | No | Yes | Full Urban |

*continued*

| | Water policy statement | Water legislation | National plan | Central coordination | Cost recovery |
|---|---|---|---|---|---|
| Philippines | Yes | Adequate | Yes | Yes | Full |
| Samoa | Yes | Inadequate | No | — | — |
| Singapore | Yes | Adequate | Yes | No | Full |
| Sir Lanka | No | Inadequate | Yes | No | Full |
| Thailand | Yes | Inadequate | Yes | — | Partial |
| Tonga | Yes | Inadequate | Yes | Yes | Partial |
| Trust Territory, Pacific Islands | No | Inadequate | No | Yes | Partial Urban |
| Vanuatu | No | Inadequate | No | No | Full |
| Vietnam | Yes | — | Yes | Yes | — |
| *West Asia* | | | | | |
| Bahrain | No | Adequate | No | Yes | — |
| Egypt | Yes | Adequate | Yes | Yes | — |
| Iraq | No | Adequate | No | Yes | — |
| Jordan | Yes | Adequate | Yes | Yes | Partial |
| Kuwait | Yes | Adequate | Yes | No | — |
| Lebanon | No | Inadequate | No | Yes | — |
| Oman | Yes | Adequate | Yes | Yes | — |
| Qatar | No | Adequate | No | No | — |
| Saudi Arabia | Yes | Adequate | Yes | No | — |
| Syria | No | Inadequate | No | No | — |
| United Arab Emirates | No | Adequate | No | Yes | — |
| Yemen | No | Inadequate | No | No | — |

**TABLE I.3**  State of water management, by country  *(pages 433–435)*

## DESCRIPTION

In 1977, the United Nations Water Resources Council adopted the Mar del Plata Action Plan. This plan recommended, among other things, that every country formulate national policies for the use, management, and conservation of fresh water. Included in each program should be research activities, appropriate institutional structures, and laws for the purpose of accelerating development and orderly administration of water resources. The following table surveys action taken within the framework of the Mar del Plata resolutions during the 1980s. For each country, the table looks at success in constructing a water policy statement and a national plan. The adequacy of national water legislation is also rated as adequate or inadequate. Finally, whether or not the country has mechanisms to recover the costs of providing water services and infrastructure is given.

## LIMITATIONS

The details of national water policies, legislation, and coordination are not revealed in a summary table such as this. Far more specific information can be found in the original reference and associated documents related to the Mar del Plata Action Plan. Several countries failed to report adequately on progress; these countries are represented by blanks in the table. Data are only presented here for some developing countries. Similar information on the state of water management programs in other developing nations and in developed nations should also be available.

## SOURCE

United Nations, 1991, Water management since the adoption of the Mar del Plata action plan: Lessons for the 1990s, in *International Conference on Water and the Environment*, United Nations Department of Technical Co-operation for Development, New York.

**TABLE I.4**  Number and area of international river basins, by continent

|  | Number of international river basins | Percent of area in international basins |
|---|---|---|
| Africa | 57 | 60 |
| North and Central America | 34 | 40 |
| South America | 36 | 60 |
| Asia | 40 | 65 |
| Europe | 48 | 50 |
| Total | 215 | 47 |

**TABLE I.4**  Number and area of international river basins, by continent

### DESCRIPTION

There are over 200 international river basins worldwide, encompassing nearly 50% of the total land area of the earth. Summarized here by continent are the number of river basins shared by two or more nations, and the percentage of total continental area in such basins. Australia, which has no international river basins, is excluded.

### LIMITATIONS

These data come from 1978, and should be updated to account for the many changes in international borders since then. For example, the breakup of the Soviet Union has created many new ''international rivers,'' though no compilation of them has yet been published, given the remaining uncertainties over the political status of many of the former republics. Similarly, the disintegration of Yugoslavia has increased the number of international rivers and countries in each basin.

### SOURCES

United Nations, 1978, *Register of International Rivers*, Centre for Natural Resources, Energy and Transport of the Department of Economic and Social Affairs of the United Nations, Pergamon Press, Oxford. By permission.

As compiled and cited by P.H. Gleick, 1989, The implications of global climatic changes for international security, *Climatic Change*, **15**, 309–315, Kluwer Academic Publishers, Dordrecht. By permission.

**TABLE I.5** Number of countries with greater than 75% of total area in international river basins, by continent

| | |
|---|---|
| Africa | 23 |
| North and Central America | 0 |
| South America | 6 |
| Asia | 8 |
| Europe | 13 |

**TABLE I.5** Number of countries with greater than 75% of total area in international river basins, by continent

**DESCRIPTION**

Increasing demands for fresh water over the next several decades will mean a greater dependence on surface water supplies that are shared by two or more nations – so-called "international rivers." Many countries have a large fraction of their total area in international river basins. Shown here by continent are the number of countries with more than three-quarters of their total area in international river basins.

**LIMITATIONS**

These data come from 1978, and should be updated to account for the many changes in international borders since then. For example, the breakup of the Soviet Union has created many new "international rivers," though no compilation of them has yet been published, given the remaining uncertainties over the political status of many of the former republics. Similarly, the disintegration of Yugoslavia has increased the number of international rivers and countries in each basin.

**SOURCES**

United Nations, 1978, *Register of International Rivers*, Centre for Natural Resources, Energy and Transport of the Department of Economic and Social Affairs of the United Nations, Pergamon Press, Oxford. By permission.

As compiled and cited by P.H. Gleick, 1989, The implications of global climatic changes for international security, *Climatic Change*, **15**, 309–315, Kluwer Academic Publishers, Dordrecht. By permission.

**TABLE I.6**  Rivers with five or more nations forming part of the basin

|                      | Number | Area (km$^2$) |
| -------------------- | ------ | ------------- |
| Danube               | 12     | 817,000       |
| Niger                | 10     | 2,200,000     |
| Nile                 | 9      | 3,030,700     |
| Zaire                | 9      | 3,720,000     |
| Rhine                | 8      | 168,757       |
| Zambezi              | 8      | 1,419,960     |
| Amazon               | 7      | 5,870,000     |
| Mekong               | 6      | 786,000       |
| Lake Chad            | 6      | 1,910,000     |
| Volta                | 6      | 379,000       |
| Ganges-Brahmaputra   | 5      | 1,600,400     |
| Elbe                 | 5      | 144,500       |
| La Plata             | 5      | 3,200,000     |

**TABLE I.6**  Rivers with five or more nations forming part of the basin

**DESCRIPTION**

Thirteen international river basins worldwide, including the Lake Chad basin, are shared by five or more nations. These rivers are listed here, together with their total basin area in km$^2$. International river basins encompass nearly 50% of the total land area of the earth. Overall, there are more than 200 rivers shared by two or more nations.

**LIMITATIONS**

These data come from 1978, and should be updated to account for the many changes in international borders since then. For example, the breakup of the Soviet Union has created many new "international rivers," though no compilation of them has yet been published, given the remaining uncertainties over the political status of many of the former republics. Similarly, the disintegration of Yugoslavia has increased the number of international rivers and countries in each basin. Several of the figures in this table would be affected by the recent political changes in eastern Europe.

**SOURCES**

United Nations, 1978, *Register of International Rivers*, Centre for Natural Resources, Energy and Transport of the Department of Economic and Social Affairs of the United Nations, Pergamon Press, Oxford. By permission.

As compiled and cited by P.H. Gleick, 1989, The implications of global climatic changes for international security, *Climatic Change*, **15**, 309–315, Kluwer Academic Publishers, Dordrecht. By permission.

**TABLE I.7** Selected countries dominated by international river basins with low per-capita water availability

| Country | Area in international river basin (%) | Per capita water availability ($10^3$ m$^3$/yr) |
|---|---|---|
| Ethiopia | 80 | 2.39 |
| Gambia | 91 | 4.48 |
| Ghana | 75 | 3.65 |
| Sudan | 81 | 1.31 |
| Togo | 77 | 3.66 |
| Peru | 78 | 1.93 |
| Afghanistan | 91 | 2.76 |
| Iraq | 83 | 2.00 |
| Belgium | 96 | 0.85 |
| Bulgaria | 79 | 1.97 |
| Czechoslovakia | 100 | 1.79 |
| German DR | 93 | 1.01 |
| Hungary | 100 | 0.56 |
| Poland | 95 | 1.31 |
| Romania | 98 | 1.59 |

**TABLE I.7** Selected countries dominated by international river basins with low per-capita water availability

**DESCRIPTION**

Increasing demands for fresh water over the next several decades will mean a greater dependence on surface water supplies that are shared by two or more nations – so-called "international rivers." In regions where water resources are scarce, this increased dependence on shared water is likely to increase the risk of frictions and tensions over water. Many countries with limited fresh water availability also have a large fraction of their total area in international river basins. Shown here are those countries with more than three-quarters of their total area in international river basins and with a per capita water availability of under 5,000 m$^3$ per year.

**LIMITATIONS**

The data on area in international river basins come from 1978, and should be updated to account for the many changes in international borders since then, especially in eastern Europe. No compilation of new international rivers has yet been published, in part because of remaining uncertainties over the political status of many of these states. The data on water availability by country are for 1987 populations and water supply. These data are also of limited reliability because of inconsistencies in the collection of national water information.

**SOURCE**

Compiled by P.H. Gleick, 1989, The implications of global climatic changes for international security, *Climatic Change*, **15**, 309–315, Kluwer Academic Publishers, Dordrecht. By permission.

**TABLE I.8** Largest river basins shared by two or more countries, by continent *(pages 440–448)*

**DESCRIPTION**

For Africa, North and Central America, South America, Asia, and Europe, the following tables list the 15 largest international river basins per continent. The river basin and area in km$^2$ are identified, the countries making up the basin are listed, and each country's share of the basin in km$^2$ and percentage of total basin area are shown.

**LIMITATIONS**

These data come from a 1978 United Nations compendium. Some of the data used to prepare these tables are outdated because of changes in international geopolitical borders and names. For example, the breakup of the Soviet Union has created many new "international rivers," though no compilation of them has yet been published, given the remaining uncertainties over the political status of many of the former republics. Similarly, the disintegration of Yugoslavia has increased the number of international rivers and countries in each basin. Country names have also changed since then, though we have chosen to leave them as listed in the original source, since name changes are sometimes associated with changes in borders and country areas.

**SOURCE**

United Nations, 1978, *Register of International Rivers*, Centre for Natural Resources, Energy and Transport of the Department of Economic and Social Affairs of the United Nations, Pergamon Press, Oxford. By permission.

**TABLE I.8**   Largest river basins shared by two or more countries, by continent

| Continent<br>River basin | Area of basin<br>(km$^2$) | Constituent<br>countries/territories | Area<br>(km$^2$) | (%) |
|---|---|---|---|---|
| *Africa* | | | | |
| Zaire | 3,720,000 | Zaire | 2,310,000 | 62.1 |
| | | Central African Republic | 408,000 | 11.0 |
| | | Angola | 285,000 | 7.7 |
| | | Congo | 255,000 | 6.9 |
| | | Zambia | 175,000 | 4.7 |
| | | Tanzania | 170,000 | 4.6 |
| | | Cameroon | 98,900 | 2.7 |
| | | Burundi | 13,300 | 0.4 |
| | | Rwanda | 4,800 | 0.1 |
| Nile | 3,030,700 | Sudan | 1,900,000 | 62.7 |
| | | Ethiopa | 368,000 | 12.1 |
| | | Egypt | 300,000 | 9.9 |
| | | Uganda | 232,700 | 7.7 |
| | | Tanzania | 116,000 | 3.8 |
| | | Kenya | 55,000 | 1.8 |
| | | Zaire | 23,000 | 0.8 |
| | | Rwanda | 21,500 | 0.7 |
| | | Burundi | 14,500 | 0.5 |
| Niger | 2,200,000 | Mali | 620,000 | 28.2 |
| | | Nigeria | 580,000 | 26.4 |
| | | Niger | 490,000 | 22.3 |
| | | Algeria | 148,300 | 6.7 |
| | | Guinea | 95,000 | 4.3 |
| | | Cameroon | 90,000 | 4.1 |
| | | Upper Volta | 81,700 | 3.6 |
| | | Benin | 50,000 | 2.3 |
| | | Côte d'Ivoire | 25,000 | 1.1 |
| | | Chad | 20,000 | 0.9 |
| Lake Chad | 1,909,820 | Chad | 950,000 | 49.7 |
| | | Niger | 416,000 | 21.8 |
| | | Central African Republic | 214,800 | 11.2 |
| | | Nigeria | 176,000 | 9.2 |
| | | Sudan | 100,000 | 5.2 |
| | | Cameroon | 53,020 | 2.8 |

*continued*

| Continent<br>River basin | Area of basin<br>(km$^2$) | Constituent<br>countries/territories | Share per country | |
|---|---|---|---|---|
| | | | Area<br>(km$^2$) | (%) |
| Zambezi | 1,419,960 | Zambia | 577,600 | 40.7 |
| | | Angola | 260,000 | 18.3 |
| | | Southern Rhodesia | 226,360 | 15.9 |
| | | Mozambique | 161,000 | 11.3 |
| | | Malawi | 110,000 | 7.7 |
| | | Botswana | 40,000 | 2.8 |
| | | Tanzania | 28,000 | 2.0 |
| | | Namibia | 17,000 | 1.2 |
| Orange | 950,000 | South Africa | 570,145 | 60.0 |
| | | Namibia | 250,000 | 26.3 |
| | | Botswana | 99,500 | 10.5 |
| | | Lesotho | 30,355 | 3.2 |
| Juba—Shibeli | 766,500 | Ethiopia | 333,500 | 43.5 |
| | | Somalia | 236,000 | 30.8 |
| | | Kenya | 197,000 | 25.7 |
| Okavango | 529,000 | Botswana | 194,500 | 36.8 |
| | | Angola | 168,000 | 31.8 |
| | | Namibia | 144,500 | 27.3 |
| | | Southern Rhodesia | 22,000 | 4.2 |
| Limpopo | 385,000 | South Africa | 180,000 | 46.8 |
| | | Botswana | 73,000 | 19.0 |
| | | Mozambique | 71,000 | 18.4 |
| | | Southern rhodesia | 61,000 | 15.8 |
| Volta | 379,000 | Upper Volta | 172,500 | 45.5 |
| | | Ghana | 159,000 | 42.0 |
| | | Togo | 19,000 | 5.0 |
| | | Côte d'Ivoire | 13,000 | 3.4 |
| | | Benin | 10,500 | 2.8 |
| | | Mali | 5,000 | 1.3 |
| Senegal | 353,000 | Mali | 163,000 | 46.2 |
| | | Mauritania | 93,000 | 26.3 |
| | | Senegal | 64,000 | 18.1 |
| | | Guinea | 33,000 | 9.3 |

*continued*

| Continent<br>River basin | Area of basin<br>(km$^2$) | Constituent<br>countries/territories | Share per country | |
| --- | --- | --- | --- | --- |
| | | | Area<br>(km$^2$) | (%) |
| Ogooue | 220,700 | Gabon | 195,000 | 88.4 |
| | | Congo | 19,300 | 8.7 |
| | | Cameroon | 4,400 | 2.0 |
| | | Equatorial Guinea | 2,000 | 0.9 |
| Lake Rudolf | 203,300 | Kenya | 106,700 | 52.5 |
| (Lake Turkana) | | Ethiopia | 86,600 | 42.6 |
| | | Sudan | 6,700 | 3.3 |
| | | Uganda | 3,300 | 1.6 |
| Ruvuma | 166,500 | Mozambique | 103,000 | 61.9 |
| | | Tanzania | 60,000 | 36.0 |
| | | Malawi | 3,500 | 2.1 |
| Cuvelai—Etosha | 126,000 | Namibia | 79,000 | 62.7 |
| | | Angola | 47,000 | 37.3 |
| *North and Central America* | | | | |
| Mississippi | 3,250,000 | United States | 3,180,000 | 97.8 |
| | | Canada | 70,000 | 2.2 |
| St. Lawrence | 1,280,000 | Canada | 800,000 | 62.5 |
| | | United States | 480,000 | 37.5 |
| Nelson—Saskatchewan | 990,000 | Canada | 871,200 | 88.0 |
| | | United States | 118,800 | 12.0 |
| Yukon | 765,000 | United States | 481,950 | 63.0 |
| | | Canada | 283,050 | 37.0 |
| Colorado | 615,000 | United States | 608,850 | 99.0 |
| | | Mexico | 6,150 | 1.0 |
| Columbia | 610,000 | United States | 506,300 | 83.0 |
| | | Canada | 103,700 | 17.0 |
| Rio Grande (Bravo del Norte) | 550,000 | United States | 302,500 | 55.0 |
| | | Mexico | 247,500 | 45.0 |
| Fraser | 260,000 | Canada | 249,600 | 96.0 |
| — | | United States | 10,400 | 4.0 |
| Grijalva-Usumacinta | 120,000 | Mexico | 81,000 | 67.5 |
| | | Guatemala | 39,000 | 32.5 |

*continued*

| Continent<br>River basin | Area of basin<br>(km²) | Constituent<br>countries/territories | Share per country | |
|---|---|---|---|---|
| | | | Area<br>(km²) | (%) |
| Yaqui | 70,000 | Mexico | 67,000 | 95.7 |
| | | United States | 3,000 | 4.3 |
| Stikine | 56,700 | Canada | 51,600 | 91.0 |
| | | United States | 5,100 | 9.0 |
| St. John | 51,800 | Canada | 34,200 | 66.0 |
| | | United States | 17,600 | 34.0 |
| San Juan | 39,350 | Nicaragua | 26,780 | 68.1 |
| | | Costa Rica | 12,570 | 31.9 |
| Coco (Segovia) | 24,800 | Nicaragua | 20,590 | 83.0 |
| | | Honduras | 4,210 | 17.0 |
| Motagua | 12,570 | Guatemala | 10,660 | 84.8 |
| | | Honduras | 1,910 | 15.2 |
| *South America* | | | | |
| Amazon | 5,870,000 | Brazil | 3,715,000 | 63.3 |
| | | Peru | 935,000 | 15.9 |
| | | Bolivia | 700,000 | 11.9 |
| | | Colombia | 340,000 | 5.8 |
| | | Ecuador | 125,000 | 2.1 |
| | | Venezuela | 50,000 | 0.9 |
| | | Guyana | 5,000 | 0.1 |
| La Plata | 3,200,000 | Brazil | 1,425,000 | 44.5 |
| | | Argentina | 990,000 | 30.9 |
| | | Paraguay | 407,000 | 12.7 |
| | | Bolivia | 238,000 | 7.4 |
| | | Uruguay | 140,000 | 4.4 |
| Orinoco | 966,000 | Venezuela | 626,000 | 64.8 |
| | | Colombia | 340,000 | 35.2 |
| Essequibo | 147,000 | Guyana | 113,190 | 77.0 |
| | | Venezuela | 33,810 | 23.0 |

*continued*

| Continent<br>River basin | Area of basin<br>(km$^2$) | Constituent<br>countries/territories | Share per country | |
|---|---|---|---|---|
| | | | Area<br>(km$^2$) | (%) |
| Lake Titicaca—Poopo system | 114,000 | Bolivia | 59,300 | 52.0 |
| | | Peru | 49,900 | 43.8 |
| | | Chile | 4,800 | 4.2 |
| Courantyne | 72,100 | Suriname | 36,900 | 51.2 |
| | | Guyana | 35,200 | 48.8 |
| Maroni | 66,000 | Suriname | 37,000 | 56.1 |
| | | French Guiana | 29,000 | 43.9 |
| Lagoon Mirim | 55,700 | Uruguay | 32,500 | 58.3 |
| | | Brazil | 23,200 | 41.7 |
| Catatumbo | 34,840 | Colombia | 19,300 | 55.4 |
| | | Venezuela | 15,540 | 44.6 |
| Oyapock | 30,270 | French Guiana | 16,194 | 53.5 |
| | | Brazil | 14,076 | 46.5 |
| Baker | 25,700 | Chile | 20,350 | 79.2 |
| | | Argentina | 5,350 | 20.8 |
| Lauca | 23,500 | Bolivia | 20,700 | 88.1 |
| | | Chile | 2,800 | 11.9 |
| Patia | 22,540 | Colombia | 22,400 | 99.4 |
| | | Ecuador | 140 | 0.6 |
| Chira | 16,220 | Peru | 8,740 | 53.9 |
| | | Ecuador | 7,480 | 46.1 |
| Aysen | 15,300 | Chile | 14,400 | 94.1 |
| | | Argentina | 900 | 5.9 |
| *Asia* | | | | |
| Ob | 3,010,000 | USSR | 2,955,000 | 98.2 |
| | | China | 55,000 | 1.8 |
| Yenisei | 2,530,000 | USSR | 2,200,000 | 87.0 |
| | | Mongolia | 330,000 | 13.0 |

*continued*

| Continent<br>River basin | Area of basin<br>(km²) | Constituent<br>countries/territories | Share per country | |
|---|---|---|---|---|
| | | | Area<br>(km²) | (%) |
| Amur | 1,900,000 | USSR | 995,000 | 52.4 |
| | | China | 845,000 | 44.5 |
| | | Mongolia | 60,000 | 3.2 |
| Ganges—Brahmaputra | 1,600,400 | India | — | — |
| | | China | — | — |
| | | Nepal | 140,800 | 8.8 |
| | | Bangladesh | 113,000 | 7.1 |
| | | Bhutan | 47,000 | 2.9 |
| Indus | 980,000 | Pakistan | — | — |
| | | India | — | — |
| | | Afghanistan | 70,000 | 7.1 |
| | | China | — | — |
| Tarim | 980,000 | China | 945,000 | 96.4 |
| | | USSR | 35,000 | 3.6 |
| Euphrates—Tigris[a] | 884,000 | Iraq | 362,500 | 41.0 |
| | | Iran | 238,500 | 27.0 |
| | | Turkey | 163,000 | 18.4 |
| | | Syria | 120,000 | 13.6 |
| Mekong | 785,500 | Laos | 199,500 | 25.4 |
| | | Thailand | 180,000 | 22.9 |
| | | China | 174,000 | 22.2 |
| | | Cambodia | 149,000 | 19.0 |
| | | Vietnam[b] | 60,500 | 7.7 |
| | | Myanmar | 22,500 | 2.9 |
| Amu-Darya | 653,000 | USSR | 503,000 | 77.0 |
| | | Afghanistan | 150,000 | 23.0 |
| Hsi | 436,000 | China | 419,000 | 96.1 |
| | | Vietnam[b] | 17,000 | 3.9 |
| Irrawaddy | 396,000 | Myanmar | 345,000 | 87.1 |
| | | India | 33,000 | 8.3 |
| | | China | 18,000 | 4.5 |

*continued*

| Continent<br>River basin | Area of basin<br>(km²) | Constituent<br>countries/territories | Share per country | |
| --- | --- | --- | --- | --- |
| | | | Area<br>(km²) | (%) |
| Helmand | 386,000 | Afghanistan | 300,000 | 77.7 |
| | | Iran | 78,000 | 20.2 |
| | | Pakistan | 8,000 | 2.1 |
| Salween | 270,000 | China | 143,000 | 53.0 |
| | | Myanmar | 110,000 | 40.7 |
| | | Thailand | 17,000 | 6.3 |
| Kura–Araks | 225,000 | USSR | 140,000 | 62.2 |
| | | Turkey | 57,000 | 25.3 |
| | | Iran | 28,000 | 12.4 |
| Ili | 176,000 | USSR | 111,000 | 63.1 |
| | | China | 65,000 | 36.9 |
| *Europe* | | | | |
| Danube[c] | 796,250 | Romania | 233,000 | 29.3 |
| | | Yugoslavia | 179,000 | 22.5 |
| | | Hungary | 93,030 | 11.7 |
| | | Austria | 79,549 | 10.0 |
| | | Czechoslovakia | 66,369 | 8.3 |
| | | Germany FR | 56,000 | 7.0 |
| | | Bulgaria | 42,500 | 5.3 |
| | | USSR | 41,500 | 5.2 |
| | | Switzerland | 2,788 | 0.4 |
| | | Italy | 2,000 | 0.3 |
| | | Poland | 364 | 0.0 |
| | | Albania | 150 | 0.0 |
| Wisla | 193,000 | Poland | 175,500 | 90.9 |
| | | USSR | 14,800 | 7.7 |
| | | Czechoslovakia | 2,700 | 1.4 |
| Rhine | 168,757 | Germany FR | 100,500 | 59.6 |
| | | Switzerland | 27,500 | 16.3 |
| | | France | 23,700 | 14.0 |
| | | Netherlands | 10,000 | 5.9 |
| | | Austria | 2,900 | 1.7 |

*continued*

| Continent<br>River basin | Area of basin<br>(km$^2$) | Constituent<br>countries/territories | Share per country | |
|---|---|---|---|---|
| | | | Area<br>(km$^2$) | (%) |
| | | Luxembourg | 2,600 | 1.5 |
| | | Belgium | 1,400 | 0.8 |
| | | Liechtenstein | 157 | 0.1 |
| Elbe | 144,500 | German DR Republic | 80,000 | 55.4 |
| | | Czechoslovakia | 49,700 | 34.4 |
| | | Germany FR | 12,600 | 8.7 |
| | | Austria | 1,400 | 1.0 |
| | | Poland | 800 | 0.6 |
| Oder | 125,900 | Poland | 103,800 | 82.4 |
| | | German DR | 13,000 | 10.3 |
| | | Czechoslovakia | 9,100 | 7.2 |
| Rhône | 95,300 | France | 88,000 | 92.3 |
| | | Switzerland | 7,300 | 7.7 |
| Douro | 94,500 | Spain | 78,600 | 83.2 |
| | | Portugal | 15,900 | 16.8 |
| Neman | 86,300 | USSR | 78,500 | 91.0 |
| | | Poland | 7,800 | 9.0 |
| Ebro | 84,440 | Spain | 83,600 | 99.0 |
| | | Andorra | 440 | 0.5 |
| | | France | 400 | 0.5 |
| Tejo | 82,000 | Spain | 56,500 | 68.9 |
| | | Portugal | 25,500 | 31.1 |
| Vuoksa | 76,000 | USSR | 41,400 | 54.5 |
| | | Finland | 34,600 | 45.5 |
| Po | 74,300 | Italy | 70,600 | 95.0 |
| | | Switzerland | 3,700 | 5.0 |
| Guadiana | 61,400 | Spain | 53,800 | 87.6 |
| | | Portugal | 7,600 | 12.4 |

*continued*

| Continent<br>River basin | Area of basin<br>(km$^2$) | Constituent<br>countries/territories | Share per country | |
| | | | Area<br>(km$^2$) | (%) |
|---|---|---|---|---|
| Maritsa | 56,000 | Bulgaria | 32,700 | 58.4 |
| | | Turkey | 14,600 | 26.1 |
| | | Greece | 8700 | 15.5 |
| | | | | |
| Garonne | 53,425 | France | 52,900 | 99.0 |
| | | Spain | 500 | 0.9 |
| | | Andorra | 25 | 0.0 |

[a]The government of Iraq proposed that the Euphrates and Tigris be listed as separate basins, and provided the Secretary-General with the following data:

| | | | | |
|---|---|---|---|---|
| Tigris | 3,788,834 | Iraq | 220,000 | 58.0 |
| | | Iran | 110,000 | 28.8 |
| | | Turkey | 48,000 | 13.0 |
| | | Syria | 834 | 0.2 |
| | | | | |
| Euphrates | 400,000 | Iraq | 240,000 | 60.0 |
| | | Turkey | 105,000 | 26.3 |
| | | Syria | 55,000 | 13.7 |

[b]This was formerly the Republic of Vietnam.
[c]Data from the Danube Commission gives an area of 817,000 km$^2$ for the whole basin; presumably it is based upon more detailed maps.

# J. Units, data conversions, and constants

## METRIC PREFIXES

| PREFIX | ABBREVIATION | MULTIPLE |
|---|---|---|
| deka- | da | 10 |
| hecto- | h | 100 |
| kilo- | k | 1000 |
| mega- | M | $10^6$ |
| giga- | G | $10^9$ |
| tera- | T | $10^{12}$ |
| peta- | P | $10^{15}$ |
| exa- | E | $10^{18}$ |
| deci- | d | 0.1 |
| centi- | c | 0.01 |
| milli- | m | 0.001 |
| micro- | μ | $10^{-6}$ |
| nano- | n | $10^{-9}$ |
| pico- | p | $10^{-12}$ |
| femto- | f | $10^{-15}$ |
| atto- | a | $10^{-18}$ |

## LENGTH (L)

**1 micron ( )**
= $1 \times 10^{-3}$ mm
= $1 \times 10^{-6}$ m
= $3.3937 \times 10^{-5}$ in

**1 millimeter** (mm)
= 0.1 cm
= $1 \times 10^{-3}$ m
= 0.03937 in

**1 centimeter** (cm)
= 10 mm
= 0.01 m
= $1 \times 10^{-5}$ km
= 0.3937 in
= 0.03281 ft
= 0.01094 yd

**1 meter** (m)
= 1000 mm
= 100 cm
= $1 \times 10^{-3}$ km
= 39.37 in
= 3.281 ft
= 1.094 yd
= $6.21 \times 10^{-4}$ mi

**1 kilometer** (km)
= $1 \times 10^5$ cm
= 1000 m
= 3280.8 ft
= 1093.6 yd
= 0.621 mi

10 millimeters = 1 centimeter
10 centimeters = 1 decimeter
10 **decimeters** (dm) = 1 meter
10 meters = 1 dekameter
10 **dekameters** (dam) = 1 hectometer
10 **hectometers** = 1 kilometer

**1 mil**
= 0.0254 mm
= $1 \times 10^{-3}$ in

**1 inch** (in)
= 25.4 mm
= 2.54 cm
= 0.08333 ft
= 0.0278 yd

**1 foot** (ft)
= 30.48 cm
= 0.3048 m
= $3.048 \times 10^{-4}$ km
= 12 in
= 0.3333 ft
= $1.89 \times 10^{-4}$ mi

**1 yard** (yd)
= 91.44 cm
= 0.9144 m
= $9.144 \times 10^{-4}$ km
= 36 in
= 3 ft
= $5.68 \times 10^{-4}$ mi

**1 mile** (mi)
= 1609.3 m
= 1.609 km
= 5280 ft
= 1760 yd

**1 fathom** (nautical)
= 6 ft

**1 league** (nautical)
= 5.556 km
= 3 nautical miles

**1 league** (land)
= 4.828 km
= 5280 yd
= 3 mi

**1 international nautical mile**
= 1.852 km
= 6076.1 ft
= 1.151 mi

## AREA ($L^2$)

**1 square centimeter** ($cm^2$)
= $1 \times 10^{-4}$ $m^2$
= 0.1550 $in^2$
= $1.076 \times 10^{-3}$ $ft^2$
= $1.196 \times 10^{-4}$ $yd^2$

**1 square meter** ($m^2$)
= $1 \times 10^{-4}$ hectare
= $1 \times 10^{-6}$ $km^2$
= 1 **centare** (French)
= 0.01 are
= 1550.0 $in^2$
= 10.76 $ft^2$
= 1.196 $yd^2$
= $2.471 \times 10^{-4}$ acre

**1 are**
= 10 $m^2$

**1 hectare** (ha)
= $1 \times 10^4 m^2$
= 100 are
= 0.01 $km^2$
= $1.076 \times 10^5$ $ft^2$
= $1.196 \times 10^4$ $yd^2$
= 2.471 acres
= $3.861 \times 10^{-3}$ $mi^{-2}$

**1 square kilometer** ($km^2$)
= $1 \times 10^6$ $m^2$
= 100 hectares
= $1.076 \times 10^7$ $ft^2$
= $1.196 \times 10^6$ $yd^2$
= 247.1 acres
= 0.3861 $mi^2$

**1 square inch** (in$^2$)

= 6.452 cm$^2$
= 6.452 x 10$^{-4}$ m$^2$
= 6.944 x 10$^{-3}$ ft$^2$
= 7.716 x 10$^{-4}$ yd$^2$

**1 square foot** (ft$^2$)

= 929.0 cm$^2$
= 0.0929 m$^2$
= 144 in$^2$
= 0.1111 yd$^2$
= 2.296 x 10$^{-5}$ acre
= 3.587 x 10$^{-8}$ mi$^2$

**1 square yard** (yd$^2$)

= 0.8361 m$^2$
= 8.361 x 10$^{-5}$ hectare
= 1296 in$^2$
= 9 ft$^2$
= 2.066 x 10$^{-4}$ acres
= 3.228 x 10$^{-7}$ mi$^2$

**1 acre**

= 4046.9 m$^2$
= 0.40469 ha
= 4.0469 x 10$^{-3}$ km$^2$
= 43,560 ft$^2$
= 4840 yd$^2$
= 1.5625 x 10$^{-3}$ mi$^2$

**1 square mile** (mi$^2$)

= 2.590 x 10$^6$ m$^2$
= 259.0 hectares
= 2.590 km$^2$
= 2.788 x 10$^7$ ft$^2$
= 3.098 x 10$^6$ yd$^2$
= 640 acres
= 1 section (of land)

**1 feddan** (used in Egypt)

= 4200 m$^2$
= 0.42 ha
= 1.038 acres

---

## VOLUME (L$^3$)

**1 cubic centimeter** (cm$^3$)

= 1 x 10$^{-3}$ liter
= 1 x 10$^{-6}$ m$^3$
= 0.06102 in$^3$
= 2.642 x 10$^{-4}$ gal
= 3.531 x 10$^{-3}$ ft$^3$

**1 liter** (1)

= 1000 cm$^3$
= 1 x 10$^{-3}$ m$^3$
= 61.02 in$^3$
= 0.2642 gal
= 0.03531 ft$^3$

**1 cubic meter** (m$^3$)

= 1 x 10$^6$ cm$^3$
= 1000 liter
= 1 x 10$^{-9}$ km$^3$
= 264.2 gal
= 35.31 ft$^3$
= 6.29 bbl
= 1.3078 yd$^3$
= 8.107 x 10$^{-4}$ acre-ft

**1 cubic dekameter** (dam$^3$)

= 1000 m$^3$
= 1 x 10$^6$ liter
= 1 x 10$^{-6}$ km$^3$
= 2.642 x 10$^5$ gal
= 3.531 x 10$^4$ ft$^3$
= 1.3078 x 10$^3$ yd$^3$
= 0.8107 acre-ft

**1 cubic hectometer** (ha$^3$)

= 1 x 10$^6$ m$^3$
= 1 x 10$^3$ dam$^3$
= 1 x 10$^9$ liter
= 2.642 x 10$^8$ gal
= 3.531 x 10$^7$ ft$^3$
= 1.3078 x 10$^6$ yd$^3$
= 810.7 acre-ft

**1 cubic kilometer** (km$^3$)

= 1 x 10$^{12}$ liter
= 1 x 10$^9$ m$^3$
= 1 x 10$^6$ dam$^3$
= 1000 ha$^3$
= 8.107 x 10$^5$ acre-ft
= 0.24 mi$^3$

**1 cubic inch** (in$^3$)

= 16.39 cm$^3$
= 0.01639 liter
= 4.329 x 10$^{-3}$ gal
= 5.787 x 10$^{-4}$ ft$^2$

**1 gallon** (gal)

= 3.785 liters
= 3.785 x 10$^{-3}$ m$^3$
= 231 in$^3$
= 0.1337 ft$^3$
= 4.951 x 10$^{-3}$ yd$^3$

**1 cubic foot** (ft$^3$)

= 2.832 x 10$^4$ cm$^3$
= 28.32 liters
= 0.02832 m$^3$
= 1728 in$^3$
= 7.481 gal
= 0.03704 yd$^3$

**1 cubic yard** (yd$^3$)

= 0.7646 m$^3$
= 6.198 x 10$^{-4}$ acre-ft
= 46656 in$^3$
= 27 ft$^3$

**1 acre-foot** (acre-ft or AF)

= 1233.48 m$^3$
= 3.259 x 10$^5$ gal
= 43560 ft$^3$

**1 Imperial gallon**

= 4.546 liters
= 277.4 in$^3$
= 1.201 gal
= 0.16055 ft$^3$

**1 cfs-day**

= 1.98 acre-feet
= 0.0372 in-mi$^2$

**1 inch-mi$^2$**

= 1.738 x 10$^7$ gal
= 2.323 x 10$^6$ ft$^3$
= 53.3 acre-ft
= 26.9 cfs-days

**1 barrel (of oil)** (bbl)

= 159 liter
= 0.159 m$^3$
= 42 gal
= 5.6 ft$^3$

**1 million gallons**

= 3.069 acre-ft

**1 pint** (pt)

= 0.473 liter
= 28.875 in$^3$
= 0.5 qt
= 16 fluid ounces
= 32 tablespoons
= 96 teaspoons

| | |
|---|---|
| **1 quart** (qt) | = 0.946 liter |
| | = 57.75 in$^3$ |
| | = 2 pt |
| | = 0.25 gal |
| **1 morgen-foot** (S. Africa) | = 2610.7 m$^3$ |
| **1 broad-foot** | = 2359.9 cm$^3$ |
| | = 144 in$^3$ |
| | = 0.0833 ft$^3$ |
| **1 cord** | = 128 ft$^3$ |
| | = 0.453 m$^3$ |

## VOLUME/AREA (L$^3$/L$^2$)

| | |
|---|---|
| **1 inch of rain** | = 5.610 gal/yd$^2$ |
| | = 2.715 x 10$^4$ gal/acre |
| **1 box of rain** | = 3,154.0 lesh |

## MASS (M)

| | |
|---|---|
| **1 gram** (g or gm) | = 0.001 kg |
| | = 15.43 gr |
| | = 0.03527 oz |
| | = 2.205 x 10$^{-3}$ lb |
| **1 kilogram** (kg) | = 1000 g |
| | = 0.001 tonne |
| | = 35.27 oz |
| | = 2.205 lb |
| **1 metric ton** (tonne or te or MT) | = 1000 kg |
| | = 2204.6 lb |
| | = 1.102 ton |
| | = 0.9842 long ton |
| **1 dalton (atomic mass unit)** | = 1.6604 x 10$^{-24}$ g |
| **1 grain** (gr) | = 2.286 x 10$^{-3}$ oz |
| | = 1.429 x 10$^{-4}$ lb |
| **1 ounce** (oz) | = 28.35 g |
| | = 437.5 gr |
| | = 0.0625 lb |
| **1 pound** (lb) | = 453.6 g |
| | = 0.45359237 kg |
| | = 7000 gr |
| | = 16 oz |
| **1 short ton** (ton) | = 907.2 kg |
| | = 0.9072 tonne |
| | = 2000 lb |
| **1 long ton** | = 1016.0 kg |
| | = 1.016 tonne |
| | = 2240 lb |
| | = 1.12 ton |
| **1 stone** (British) | = 6.35 kg |
| | = 14 lb |

## TIME (T)

| | |
|---|---|
| **1 second** (s or sec) | = 0.01667 min |
| | = 2.7778 x 10$^{-4}$ hr |
| **1 minute** (min) | = 60 s |
| | = 0.01667 hr |
| **1 hour** (hr or h) | = 60 min |
| | = 3600 s |
| **1 day** (d) | = 24 hr |
| | = 86400 s |
| **1 year** (yr or y) | = 365 d |
| | = 8760 hr |
| | = 3.15 x 10$^7$ s |

## DENSITY (M/L$^3$)

| | |
|---|---|
| **1 kilogram per cubic meter** (kg/m$^3$) | = 10$^{-3}$ g/cm$^3$ |
| | = 0.062 lb/ft$^3$ |
| **1 gram per cubic centimeter** (g/cm$^3$) | = 1000 kg/m$^3$ |
| | = 62.43 lb/ft$^3$ |
| **1 metric ton per cubic meter** (te/m$^3$) | = 1.0 **specific gravity** |
| | = density of H$_2$0 at 4°C |
| | = 8.35 lb/gal |
| **1 pound per cubic foot** (lb/ft$^3$) | = 16.02 kg/m$^3$ |

## VELOCITY (L/T)

| | |
|---|---|
| **1 meter per second** (m/s) | = 3.6 km/hr |
| | = 2.237 mph |
| | = 3.28 ft/s |
| **1 kilometer per hour** (km/h or kph) | = 0.62 mph |
| | = 0.278 m/s |
| **1 mile per hour** (mph or mi/h) | = 1.609 km/h |
| | = 0.45 m/s |
| | = 1.47 ft/s |
| **1 foot per second** (ft/s) | = 0.68 mph |
| | = 0.3048 m/s |
| **velocity of light in vacuum** (c) | = 2.9979 x 10$^8$ m/s |
| | = 186,000 mi/s |
| **1 knot** | = 1.852 km/h |
| | = 1 nautical mile/hour |
| | = 1.151 mph |
| | = 1.688 ft/s |

## VELOCITY OF SOUND IN WATER AND SEAWATER

(assuming atmospheric pressure and seawater salinity of 35,000 ppm)

| Temp. °C | Pure water, m/s | Seawater m/s |
|---|---|---|
| 0 | 1,400 | 1,445 |
| 10 | 1,445 | 1,485 |
| 20 | 1,480 | 1,520 |
| 30 | 1,505 | 1,545 |

## FLOW RATE (L$^3$/T)

| | |
|---|---|
| **1 liter per second** (l/sec) | = 0.001 m$^3$/sec |
| | = 86.4 m$^3$/day |
| | = 15.9 gpm |
| | = 0.0228 mgd |
| | = 0.0353 cfs |
| | = 0.0700 AF/day |
| **1 cubic meter per second** (m$^3$/sec) | = 1000 l/sec |
| | = 8.64 x 10$^4$ m$^3$/day |
| | = 1.59 x 10$^4$ gpm |
| | = 22.8 mgd |
| | = 35.3 cfs |
| | = 70.0 AF/day |
| **1 cubic meter per day** (m$^3$/day) | = 0.01157 l/sec |
| | = 1.157 x 10$^{-5}$ m$^3$/sec |
| | = 0.183 gpm |
| | = 2.64 x 10$^{-4}$ mgd |
| | = 4.09 x 10$^{-4}$ cfs |
| | = 8.11 x 10$^{-4}$ AF/day |
| **1 cubic decameters per day** (dam$^3$/day) | = 11.57 l/sec |
| | = 1.157 x 10$^{-2}$ m$^3$/sec |
| | = 1000 m$^3$/day |
| | = 1.83 x 10$^6$ gpm |

= 0.264 mgd
= 0.409 cfs
= 0.811 AF/day

**1 gallon per minute** (gpm)
= 0.0631 l/sec
= 6.31 x $10^{-5}$ $m^3$/sec
= 1.44 x $10^{-3}$ mgd
= 2.23 x $10^{-3}$ cfs
= 4.42 x $10^{-3}$ AF/day

**1 million gallons per day** (mgd)
= 43.8 l/sec
= 0.0438 $m^3$/sec
= 3785 $m^3$/day
= 694 gpm
= 1.55 cfs
= 3.07 AF/day

**1 cubic foot per second** (cfs)
= 28.3 l/sec
= 0.0283 $m^3$/sec
= 2447 $m^3$/day
= 449 gpm
= 0.646 mgd
= 1.98 AF/day

**1 acre-foot per day** (AF/day)
= 14.3 l/sec
= 0.0143 $m^3$/sec
= 1233.48 $m^3$/day
= 226 gpm
= 0.326 mgd
= 0.504 cfs

**1 miner's inch**
= 0.025 cfs (in Arizona, California, Montana, and Oregon: flow of water through 1 $in^2$ aperature under 6-inch head)
= 0.02 cfs (in Idaho, Kansas, Nebraska, New Mexico, North Dakota, South Dakota, and Utah)
= 0.026 cfs (in Colorado)
= 0.028 cfs (in British Columbia)

**1 quinaria** (ancient Rome)
= 0.47 - 0.48 l/sec

## ACCELERATION (L/$T^2$)

**standard acceleration of gravity**
= 9.8 m/$s^2$
= 32 ft/$s^2$

## FORCE (ML/$T^2$ = Mass x Acceleration)

**1 newton** (N)
= kg·m/$s^2$
= $10^5$ dynes
= 0.1020 kg force
= 0.2248 lb force

**1 dyne**
= g·cm/$s^2$
= $10^{-5}$ N

**1 pound force**
= lb mass x acceleration of gravity
= 4.448 N

## PRESSURE (M/$L^2$ = Force/Area)

**1 pascal** (Pa)
= N/$m^2$

**1 bar**
= 1 x $10^5$ Pa
= 1 x $10^6$ dyne/$cm^2$
= 1019.7 g/$cm^2$
= 10.197 te/$m^2$
= 0.9869 atmosphere

= 14.50 lb/$in^2$
= 1000 millibars

**1 atmosphere** (atm)
= standard pressure
= 760 mm of mercury at 0°C
= 1013.25 millibars
= 1033 g/$cm^2$
= 1.033 kg/$cm^2$
= 14.7 lb/$in^2$
= 2116 lb/$ft^2$
= 33.95 feet of water at 62°F
= 29.92 inches of mercury at 32°F

**1 kilogram per sq. centimeter** (kg/$cm^2$) = 14.22 lb/$in^2$

**1 inch of water at** 62°F
= 0.0361 lb/$in^2$
= 5.196 lb/$ft^2$
= 0.0735 inch of mercury at 62°F

**1 foot of water at 62°F**
= 0.433 lb/$in^2$
= 62.36 lb/$ft^2$
= 0.833 inch of mercury at 62°F
= 2.950 x $10^{-2}$ atmosphere

**1 pound per sq. inch** (psi or lb/$in^2$)
= 2.309 feet of water at 62°F
= 2.036 inches of mercury at 32°F
= 0.06804 atmosphere
= 0.07031 kg/$cm^2$

**1 inch of mercury at 32°F**
= 0.14192 lb/$in^2$
= 1.133 feet of water at 32°F

## TEMPERATURE

**degrees Celsius or Centigrade** (°C)
= (°F-32) x 5/9
= K - 273.16

**Kelvins** (K)
= 273.16 + °C
= 273.16 + ((°F-32) x 5/9)

**degrees Farenheit** (°F)
= 32 + (°C x 1.8)
= 32 + ((K-273.16) x 1.8)

## ENERGY (M$L^2$/$T^2$ = Force x Distance)

**1 joule** (J)
= $10^7$ ergs
= N·m
= W·s
= kg·$m^2$/$s^2$
= 0.239 calories
= 9.48 x $10^{-4}$ Btu

**1 calorie** (cal)
= 4.184 J
= 3.97 x $10^{-3}$ Btu (raises 1 g $H_2O$ 1°C)

**1 British thermal unit** (Btu)
= 1055 J
= 252 cal (raises 1 lb $H_2O$ 1°F)
= 2.93 x $10^4$/kWh

**1 erg**
= $10^{-7}$ J
= g·$cm^2$/$s^2$
= dyne·cm

| | |
|---|---|
| **1 kilocalorie** (kcal) | = 1000 cal<br>= 1 Calorie (food) |
| **1 kilowatt-hour** (kWh) | = 3.6 x $10^6$ J<br>= 3412 Btu<br>= 859.1 kcal |
| **1 quad** | = $10^{15}$ Btu<br>= 1.055 x $10^{18}$ J<br>= 293 x $10^9$ kWh<br>= 0.001 Q<br>= 33.45 GWy |
| **1 Q** | = 1000 quads<br>≈ $10^{21}$ J |
| **1 foot-pound** (ft-lb) | = 1.356 J<br>= 0.324 cal |
| **1 therm** | = $10^5$ Btu |
| **1 electron-volt** (eV) | = 1.602 x $10^{-19}$ J |
| **1 kiloton of TNT** | = 4.2 x $10^{12}$ J |
| **1 $10^6$ te oil equiv.** (Mtoe) | = 7.33 x $10^6$ bbl oil<br>= 45 x $10^{15}$ J<br>= 0.0425 quad |

**POWER ($ML^2/T^3$ = rate of flow of energy)**

| | |
|---|---|
| **1 watt** (W) | = J/s<br>= 3600 J/hr<br>= 3.412 Btu/hr |
| **1 TW** | = $10^{12}$ W<br>= 31.5 x $10^{18}$ J<br>= 30 quad/yr |
| **1 kilowatt** (kW) | = 1000 W<br>= 1.341 horsepower<br>= 0.239 kcal/s<br>= 3412 Btu/hr |
| **$10^6$bbl (oil)/day** (Mb/d) | ≈ 2 quads/yr<br>≈ 70 GW |
| **1 quad/yr** | = 33.45 GW<br>≈ 0.5 Mb/d |
| **1 horsepower** (H.P.or hp) | = 745.7 W<br>= 0.7457 kW<br>= 0.178 kcal/s<br>= 6535 kWh/yr<br>= 33,000 ft-lb/min<br>= 550 ft-lb/sec<br>= 8760 H.P.-hr/yr |
| **H.P. input** | = 1.34 x kW input to motor<br>= horsepower input to motor |
| **Water H.P.** | = H.P. required to lift water at a definite rate to a given distance assuming 100% efficiency<br>= $\frac{\text{gpm x total head (in feet)}}{3960}$ |

**EXPRESSIONS OF HARDNESS**

| | |
|---|---|
| **1 grain per gallon** | = 1 grain $CaCO_3$ per US gallon |
| **1 part per million** | = 1 part $CaCO_3$ per 1,000,000 parts water |
| **1 English, or Clark, degree** | = 1 grain $CaCO_3$ per Imperial gallon |
| **1 French degree** | = 1 part $CaCO_3$ per 100,000 parts water |
| **1 German degree** | = 1 part CaO per 100,000 parts water |

CONVERSIONS

| | |
|---|---|
| **1 grain per US gallon** | = 17.1 ppm, as $CaCO_3$ |
| **1 English degree** | = 14.3 ppm, as $CaCO_3$ |
| **1 French degree** | = 10 ppm, as $CaCO_3$ |
| **1 German degree** | = 17.9 ppm, as $CaCO_3$ |

Source: F. van der Leeden, Fred L. Troise and David Keith Todd, 1990, *The Water Encyclopedia,* 2nd edn., Lewis Publishers, Inc., Chelsea, Michigan.

**WEIGHT OF WATER**

| | |
|---|---|
| 1 cubic inch | = 0.0361 lb. |
| 1 cubic foot | = 62.4 lb. |
| 1 gallon | = 8.34 lb. |
| 1 imperial gallon | = 10.0 lb. |
| 1 cubic meter | = 1 tonne. |

**DENSITY OF WATER**

| **Temperature** °C | °F | **Density** gm/cm³ |
|---|---|---|
| 0 | 32 | 0.99987 |
| 1.667 | 35 | 0.99996 |
| 4.000 | 39.2 | 1.00000 |
| 4.444 | 40 | 0.99999 |
| 10.000 | 50 | 0.99975 |
| 15.556 | 60 | 0.99907 |
| 21.111 | 70 | 0.99802 |
| 26.667 | 80 | 0.99669 |
| 32.222 | 90 | 0.99510 |
| 37.778 | 100 | 0.99318 |
| 48.889 | 120 | 0.98870 |
| 60.000 | 140 | 0.98338 |
| 71.111 | 160 | 0.97729 |
| 82.222 | 180 | 0.97056 |
| 93.333 | 200 | 0.96333 |
| 100.000 | 212 | 0.95865 |

Density of seawater: approximately 1.025 gm/cm$^3$ at 15°C.

Source: F. van der Leeden, Fred L. Troise and David Keith Todd, 1990, *The Water Encyclopedia,* 2nd edn., Lewis Publishers, Inc., Chelsea, Michigan.

# Index

# Index

**463**

*See also* economic activities; industrial
effluents

infections. *See* water-related diseases

inland fresh water marshes (United States),
described, 294

inland wetlands, regions defined, 294
*See also* wetlands; wetlands areas

inorganic pollutants, drinking water standards
for (table), 226–227

Institut de Droit International (Institute of
International Law), 98
on the principle of equitable utilization, 98
resolutions on shared water resources, 98

Institute of International Law. *See* Institut de
Droit International

internal runoff, 15
areas of (table), 123

International Boundary Commission (United
States, Mexico), 96

international controversies over shared water
resources. *See* water-related disputes

International Drinking Water Supply and
Sanitation Decade, 3, 33, 89, 90

International Fund for Agricultural
Development, on small farmer-managed
systems, 63

International Hydrological Decade, 3

international law. *See* international water law

International Law Association (ILA), 98
on equitable apportionment, 98
on existing reasonable use, 98
resolutions on rules and recommendations
concerning international law, 98

International Law Commission (ILC), 98
articles on the Law of the
Non-Navigational Uses of International
Watercourses, 98–99
on equitable utilization, 99

international politics
and hydrologic data, 119
and water-related disputes, 100
*See also* water-related disputes

international river basins
with five or more nations forming part of
(table), 438
international disputes over, 9, 92–97
largest shared by two or more countries
(table), 439–448
number of and percent of area in, by
continent (table), 436
number of countries with greater than 75%
of total area in (table), 437
percent of world land area in, 9, 92

international rivers, 437
*See also* international river basins; *and
individual rivers by name*

international water law, 9–10, 97–99, 110
need for clarity, 97
role in resolving water-related disputes, 10,
93–94, 99, 108
*See also* customary international law;
international water treaties

international water treaties, 10, 92, 97–98
examples, 94–97

invertebrates (fresh water)
and acidic waters, 45
diversity, 41

ion-exchange processes
for desalinization, 70

for treating drinking water, 36

ions in rainwater, sources of (table), 231

ions in river water, average concentration of
(table), 233

Iraq, and the Greater Anatolia Project, 93

irrigated area
by continent and individual countries
(table), 19
by continent (table), 266
global (table), 265–266
gross, defined, 57
irrigated with on-farm pumped water, by
region, in the United States (table), 285
net, defined, 57
per capita in the world, 6, 57
graph, 57
by region and country (table), 267–271
in the top 20 countries and the world
(table), 56
under microirrigation in leading countries
and the world (table), 61
*See also* cropland

irrigation
20th century expansion and slowdown, 6,
56, 57
in ancient civilizations, 6, 68
characteristics of alternative irrigation
systems (table), 274–275
costs
of adding new capacity, 57
environmental, 6, 58–59
crop water requirements and crop
evapotranspiration (table), 282–283
cropland use by region and country (table),
259–264
current trends, 6, 56–58
drainage contamination problems, 58, 87
efficiency, 56, 57
improving, 6, 60–62
energy used for on-farm pumped irrigation
water, in the United States, by type
(table), 286
future contribution, 57
ground water overpumping, 6–7, 58–59
ground water withdrawals
in selected Asian and Pacific countries
(table), 401
in the United States (tables), 397–398,
399, 400
inequities in distribution, 57–58
irrigation development impacts (table), 279
lands at least moderately desertified (table),
273
lending decline, 57
problems, typical, 61–62
reforms, need for pricing, 62
and the rural poor, 57, 58, 59
rural populations affected by desertification
(table), 274
salinization of irrigated cropland (table),
272
scheduling, 62
shift in thinking required, 57, 58
sources in the United States (table), 284
support of cash crop over food staple
production, 57
unsustainability, 58–60
urban water demands versus, 7
use of wastewater in, 63, 87

water withdrawals
by continent and individual countries
(table), 19
in selected OECD countries (table), 379
in the United States (table), 391–393
in the world, 58
and world crop production, 56, 57, 58
*See also* agriculture; dams; irrigated area;
irrigation systems

irrigation associations, 62

irrigation systems, 60–61
alternative
characteristics of (table), 274–275
need for, 62, 81–82
factors affecting the selection of (table), 276
guide for selecting (table), 277–278
lack of maintenance, 57
as military targets, 109
small-scale alternatives, 62–63
in the United States (table), 284

Islamic Water Resources Management
Network, on water demands and water
availability in the Arab region, 84

Israel
disputes with Jordan over the Jordan River,
9, 92–93
use of wastewater in irrigation, 63
water management policy, 82, 92–93
water predicament, 60, 82, 92–93, 108n,
109

**J**

James Bay hydroelectric project, mercury
pollution from, 5, 34–35

Japan, mercury pollution in, 28

Johnston Plan, 92, 93

joint mechanisms, in water-related disputes,
10

Joint Technical Committee (Egypt and
Sudan), 94

Joint Technical Regional Rivers Committee
(Iraq and Turkey), 93

Jordan
disputes with Israel over the Jordan River,
9, 92–93
plan for a unity dam with Syria, 93
water predicament, 60, 93

Jordan River, 92
diversion works on, 92
international disputes over, 9, 92–93

juvenile water, 15

**K**

karst habitats, 45

Kesterson National Wildlife Refuge
(California), selenium contamination of,
58, 87

Kumgansan Dam, as a military threat, 109

Kurdish Workers' Party (PKK), 93

**L**

lake basins. *See* drainage basins

lakes
characteristics of the major lakes of the
world (table), 161–165
distribution of global water in (table), 41
as a global water reserve (table), 13
salinity, principal sources and causes of
(table), 232